1 MONTH OF
FREE
READING

at
www.ForgottenBooks.com

By purchasing this book you are eligible for one month membership to ForgottenBooks.com, giving you unlimited access to our entire collection of over 1,000,000 titles via our web site and mobile apps.

To claim your free month visit: www.forgottenbooks.com/free750031

ISBN 978-0-365-20433-6
PIBN 10750031

This book is a reproduction of an important historical work. Forgotten Books uses
state-of-the-art technology to digitally reconstruct the work, preserving the original format
whilst repairing imperfections present in the aged copy. In rare cases, an imperfection in
the original, such as a blemish or missing page, may be replicated in our edition. We do,
however, repair the vast majority of imperfections successfully; any imperfections that
remain are intentionally left to preserve the state of such historical works.

THE SCOTTISH GEOGRAPHICAL MAGAZINE

THE SCOTTISH

GEOGRÁPHICAL

MAGAZINE

PUBLISHED BY THE ROYAL SCOTTISH GEOGRAPHICAL SOCIETY

PROFESSOR JAMES GEIKIE, LL.D., D.C.L., F.R.S., HON. EDITOR

W. A. TAYLOR, M.A., F.R.S.E., ACTING EDITOR

VOLUME XII : 1896

EDINBURGH

PRINTED BY T. AND A. CONSTABLE, PRINTERS TO HER MAJESTY

AT THE UNIVERSITY PRESS

1896

ROYAL SCOTTISH GEOGRAPHICAL SOCIETY.

SOCIETY'S HALL:
QUEEN STREET, EDINBURGH.

ROYAL SCOTTISH GEOGRAPHICAL SOCIETY.

CONDITIONS AND PRIVILEGES OF MEMBERSHIP.

IT is provided by Chapter I. § IV. of the *Constitution and Laws* of the Royal Scottish Geographical Society, that—

"*The Ordinary Members shall be those who are approved by the Council, and who pay the ordinary annual subscription, or a composition for life-membership.*"

The Annual Subscription is One Guinea, and is payable *in advance* at the commencement of each Session. A single Contribution of Ten Guineas constitutes Life-Membership. Ladies are admitted as Members.

The Official Year, or Session, of the Society is from November 1st to October 31st. New Members are required to pay the Subscription for the Session in which they join the Society, at whatever period, and they are entitled to receive the ordinary publications of that Session. Resignations, to take effect, must be lodged with the Secretary *before* the commencement of a new Session.

The privileges of Membership include admission (with one Guest) to the Ordinary monthly Meetings of the Society, and the use of the Library and Map-Room. Non-resident Members may borrow books from the Library, but they must defray the cost of transit both ways. Each Member is entitled to receive, free by post, the *Scottish Geographical Magazine*, which is published monthly by the Society.

Branches of the Society have been established in Glasgow, Dundee, and Aberdeen, where periodical Meetings are held.

CONTENTS.

VOL. XII : 1896.

CONTENTS.

No. III.—MARCH.

No. IV.—APRIL.

No. V.—MAY.

No. VI.—JUNE.

No. VII.—JULY.

No. VIII.—AUGUST.

CONTENTS.

No. IX.—SEPTEMBER.

No. X.—OCTOBER.

No. XI.—NOVEMBER.

No. XII.—DECEMBER.

ERRATA.

Page 28, line 11 from bottom, *for* Pore *read* Pory
,, 46, ,, 23 from bottom, *for* Orthography *read* Orography.
,, 238, ,, 19 from top, *for* in *read* is.
,, 372, ,, 14 from bottom, *for* Riesengebirge *read* Isergebirge.
,, 479, ,, 4 from bottom and line 24, *for* Condreau, *read* Coudreau.
,, 481, ,, 6 from bottom, *omit* the words in brackets.
,, 495, ,, 5 from top, *insert* Geographical *before* Congress.
,, 526, ,, 4 from bottom, *for* Goodson *read* Goodison.
,, 529, ,, 22 from bottom, *for* Rev. J. C. Robinson, *read* Rev. C. H. Robinson.
,, 552, last line, *for* Society *read* Council.
,, 464, line 9 from top, *for* Loango *read* Loanda.
,, 581, the last sentence should read :—The only systematic series of pictures advertised has been arranged by the energetic professor of Colonial Geography, whose Chair is, not in this country, which has more colonies than any other, but in the Sorbonne— M. Marcel Dubois.
,, 659, top line, *for* 1883 *read* 1893.

atted
THE SCOTTISH
GEOGRAPHICAL
MAGAZINE.

TARHUNA AND GHARIAN, IN TRIPOLI, AND THEIR ANCIENT SITES.

By H. S. COWPER, F.S.A.

(With Illustrations.)

THE vilayet of Tripoli, in Barbary, lying between longitude 10° and longitude 25° E., is perhaps less known than any other part of North Africa—I would venture to say even than Morocco; and this is rendered more remarkable by the fact that on either side of it we find countries held in the strong control of European Powers, and quickly donning the somewhat incongruous garb which Mohammedan States must always assume when they come under Western influence.

Of these two neighbours, Tunis and Egypt, Tripoli has much more in common with the latter. Tunis has the rugged hills, the fertile plains, and, as far as its northern half is concerned, the uncertain climate of Algiers and Morocco. Tripoli, on the other hand, is, like Egypt, in the Libyan Desert; and also its climate is more like that of Egypt than of any other country. There is, however, one great difference between the two—a difference, moreover, of vast importance. Egypt is the child of the Nile, born of her, supported by her, and dependent on her. Tripoli is riverless; so that while in Egypt a vast population has lived and worked for untold ages, for thousands of miles along her river banks, in Tripoli the settled population has been confined to the coast, and to such large fertile tracts as happen to exist or to have existed in modern or ancient times.

Now, the whole province of Tripoli, as shown upon our maps, covers a very large area. There is, first of all, what may now be called Tripoli proper, extending along the coast from Tunisia to the southernmost point of the Gulf of Sidra.   West of this comes Barka, the Hellenic Cyrenaica, or Pentapolis, of much less area, but full of historical interest; while to the south of Tripoli itself lies Fezzan, a wild and inhospitable desert, intersected by wadis, and crossed by caravan tracks which converge on Tripoli from various points of the Sudan.   Fezzan has, however, certain oases and towns.

It might be expected that the first of these divisions, being the one in which the capital is situated, would be the most explored, but this hardly appears to be the case.   For although in the old days, before Tripoli was a forbidden country to Europeans, there were numerous exploring parties which started from Tripoli for the interior, they seem, with but one or two exceptions, to have attempted but little examination of the country near the capital, and not to have paused or swerved from the direct road until either Ghadames, Ghat, or Murzuk was reached.

So that while Barka has drawn students of antiquity by its well-known ancient sites, and Murzuk and the surrounding part of Fezzan find a place and description in the pages of our travellers, because it was the first town from Tripoli, we fail to discover much of interest about that range of hill-country lying inland from the coast, between Nalut on the west and Lebda on the east, at a distance varying from 50 to 100 miles.

I hope to show in the following notes that these hills, through which I made a brief journey in March last, are full of interest, and well deserve notice at the hands of both geographers and archæologists.   Certain parts of them are rich in evidences of a hitherto uninvestigated civilisation, and the varied physical features of the tract itself put it before the monotonous and inhospitable desert which lies beyond it.

The features of the coast here are somewhat peculiar.   In the first place, we have next to the sea a fertile strip containing the capital and a line of villages built in palm groves.   Wells are abundant, and the population is settled, commercial and agricultural.   Behind the palm groves come in some parts corn lands and pastures, but elsewhere, as behind Tripoli itself, the palms are immediately succeeded by a sandy waste.   This strip narrows towards the east where the hills approach the coast, but behind the capital it is about 15 miles in width; it consists of wave-like ridges of blown sand, yielding to the foot, and consequently difficult to travel on.   There are small oases here and there, and a few wells, perhaps of Roman origin, for the use of travellers.   Yet, in spite of its present desolation, there is evidence that this dreary region once supported a considerable population.   The traveller, as he crosses it, continually notices patches of brown hard ground where the sand is blown away.   These spaces are the sites of former human occupation, and worked flints and Roman potsherds and *tesseræ* are to be found in them.

South of the desert strip we come to a more fertile region, consisting of a grassy plain, or in some places gentle slopes, which extend up to the base of the hill range itself.   South of Tripoli, on the direct road to Gharian, there is a gently ascending plain some 20 miles in width and rising from about 150 to 700 feet above the sea.   South-east of Tripoli, outside the Wadi Doga in Tarhuna, where I entered the district, we find

the desert succeeded by gentle slopes and undulating country which
is about 400 to 600 feet above the sea, and has a width of 10 miles to
the base of the true hills.

Although Tripoli has no perennial streams, it has plenty of wadis,
which are dry in summer, but after winter storms have running water in
them.   Whether the rainfall is ever sufficient nowadays to carry a stream
out to sea, as it once did, or whether the feeble streams are sucked up
and evaporated before they ever reach the coast, I am not prepared to
say.   But some of them, like Wadi Raml, can be traced from the hills to
the coast.   The point where they leave the hills and enter the plain is
designated the "Fum" (i.e. the mouth), as Fum Doga and Fum Raml,
the points where the wadis of those names disengage themselves from
Jebel Tarhuna.

The total length of the hill chain from Nalut on the west (where it
is about 90 miles from the coast) to where it approaches the sea at Homs
appears to be about 200 miles, and the principal divisions, enumerating
them from the west, are Nefusa, Yefrin, Gharian, and Tarhuna.   Of the
first two I know nothing from personal experience, and therefore I
propose to confine myself to the districts of Tarhuna and Gharian, through
which my route lay last spring.

As far as I am able to ascertain, the only map published as yet, which
gives in any detail or with any accuracy a representation of this district,
is one in the first volume of Barth's *Travels in North Africa*.   Barth,
indeed, travelled from one end to the other of the Tripolitan chain, and
the information recorded by him in his chapters on the district, as well
as in the map, is of the most valuable description.   He was not an
archæologist, and did not specially search for ancient sites, but those he
passed upon his route he made notes of and conscientiously recorded.
The only other traveller who can be said to have made investigations of
any sort was Von Bary, who died in 1877 at Ghat.   I shall have occasion
to refer again to the work of these two explorers in treating of the
ancient sites of Tarhuna.

The total width of that part of Tarhuna through which I travelled
is about 40 miles, measured from Jebel Msid on the east to Wadi Wif on
the west.   South of Msid the boundary between Tarhuna and Mesalata
appears to pass south-west, so that the measurement to the south may
be much greater, but there appears to be no trustworthy source of
information available as to its limits beyond here.   The district between
these two points may be divided into two parts—the Tarhuna hills themselves, broken up by numerous wadis passing through them towards the
north, and the Tarhuna plateau lying beyond them, which from an
archæological point of view is the most interesting part of the whole
district.

The Tarhuna hills are of no great elevation in any part, the highest
points being probably those near the boundary of Gharian, where they
rise to about 1900 feet.   Close to where Wadi Raml emerges from them
on to the plain, they are pushed forward into a sort of headland or elbow,
the principal point of which is Kashi Maruf, which is apparently the
nearest point in the hill range to the capital.

The principal valleys which intersect the hills are the Wadis Raml and Saghia (which join outside on the plain), Wadi Doga and Wadi Msid. Of these, the Wadi Doga, about eight miles in length from its " Fum " to Kasr Doga, where it debouches on to the plateau, is, I venture to think, the most important, although hitherto unmentioned by any traveller, or marked in any map. My route from Tripoli to Tarhuna lay south-east across the desert strip, passing a small oasis called Neshieh, and crossing the Wadi Raml exactly at its junction with Wadi Saghia. On the morning of the third day I found myself at the Fum Doga, and the ancient sites in this valley took a long day to make even a hasty examination of. On the east and west of the Fum lie respectively Jebel Ahmar and Jebel Aref, which appear to be about 250 feet above the mouth of the valley, and therefore close on 1000 feet above sea-level.

The Wadi Doga is not a water-course like most of the Tripoli wadis, but a wide green valley bordered on either side by low hills, and through its centre winds a ravine with a rocky bed some 150 feet below the valley level. The valley rises towards the south, and the hills are also higher, so that Kasr Doga, at the valley head, is 1350 feet, and Jebel Arva, nearly opposite it, about 1750 feet above sea-level.

At the head of the valley stands the magnificent Roman mausoleum called Kasr Doga, which Barth has already described and figured, though without doing it justice, in his book. Immediately south of this lies the Tarhuna plateau, where I camped with the family of my guide, and spent five days examining and photographing the ancient sites.

The plateau, or rather that part of it which I examined, stretches about 26 miles from the east, near the south end of the Wadi Kseia, to the base of the Tarhuna hills, where they rise again from the plain towards Gharian. It is an undulating grassy country, averaging about 1250 feet above sea-level, but falling gently to the west. The western half appears to be nearly waterless, but that towards the east is covered with ancient sites and has numerous encampments of Arabs. As throughout all Tarhuna, the country is now practically treeless; so that a solitary *batum* tree of fine proportions called El Khadra, "The Green," forms a landmark for miles. The easternmost part of the plain here is called Ferjana, and on the north side of it are several wadis draining towards the south. Of these, Wadi el Menshi is the most important, —a wide, shallow depression almost in the same line as Wadi Doga. North-east of Ferjana lies Wadi Daun, running east and west, and connected by Kurmet el Hatheia with the Wadi Kseia, which is rather a plain among the Tarhuna hills than a water-course.

After examining the sites on the plateau and in Ferjana, I rode north-east by a small valley, called Shaahbet el Kheil, through Wadi Daun, where there is a running stream, and by the above-named Kurmet el Hatheia into the Kseia, at the north-east end of which I camped at the base of Jebel Msid. The summit of this hill is 1450 feet above the sea and about 600 feet above the valley. On the top is a curious little ruined building, apparently of Arab origin, and seemingly half castle, half mosque. This, and a tiny mosque on the plateau, are about the only buildings of post-Roman date I saw in Tarhuna.

SENAM EL NEJM, FERJANA (TRILITHON)

SENAM AREF (GENERAL VIEW)

From the summit of Msid could be seen the town of Kusabat, apparently about seven miles distant to the north-east, and embosomed in palms. Elsewhere all that met the eye was barren hill ranges or treeless plains, except to the south-west, where at our feet was the sheltered Kseia through which we had just passed.

The Tarhuna district is inhabited throughout by tribal Arabs, who appeared to me to be of pure stock. Negroes are hardly ever seen. The Arabs are in all cases tent-dwellers or troglodytes, and are not nomadic. The system is a patriarchal one, in which each family, the father and his wives with the sons and their wives, all occupy a long row of tents. These tent villages are apparently quite stationary, as that of my guide appears to be exactly in the same place as when Von Bary visited it nearly twenty years ago. In summer, however, the villagers leave the goat's-hair tents and occupy wattle huts scattered about among the crops. These are the summer shealings, " buyut el saif."

The underground dwellings are a characteristic feature both in Gharian and Tarhuna. In no part that I visited were they so common as the tents, and many were empty, being occupied only in the hot weather. It does not appear, indeed, that they belong to any special tribe. In all cases they are simply chambers hollowed out below the level of the ground, sometimes single rooms, sometimes two or three opening into each other. The system recalls the Ethiopian troglodytes of Herodotus.

In person the Tarhunis are handsome, dignified, and courteous. The men, without exception, wear the long white barracan of Tripoli, and always travel armed. They are extremely particular about their women, who are so carefully hidden that throughout my stay I never once saw the face of a young female. At the same time I saw no trace of anti-Christian feeling, and in Tarhuna was always treated with the greatest courtesy—due no doubt to some extent to the fact that I was travelling with a guide who was one of a highly respected family of merábuts.

The chief trouble to the stranger at one of these camps is the pack of white fierce dogs which lie round the tents. These animals render it impossible to move out of one's tent by night without an escort ; and, curiously, they attack indiscriminately stranger or inhabitant, so that all the Arabs carry long sticks with which to beat them off without mercy when they enter the camp. Once within the tent doors the traveller is left unmolested.

In my guide's encampment I found the rites of religion were not neglected, which might be due to some extent to the semi-sacred character of his family. A mueddin chanted regularly the call to prayer, and frequently in an evening a little line of worshippers would perform their devotions under the direction of my guide.

The occupations of the Tarhunis, and also of the Arab hillmen farther west, are agriculture and stock-raising. The wheat is generally unenclosed, but here and there are pretty little enclosed fields or gardens (henshir). A few figs are cultivated; but in Western Tarhuna and Gharian the esparto harvest is an important crop. The grass is gathered.

by hand, the harvester winding each handful round a piece of wood, and then tearing it away. The domestic animals are chiefly camels and sheep; cattle and horses are less common.

Although the Arabs are physically a fine race, and do not seem to lack independence of character, it does not appear that the Turkish authorities have now much difficulty with them. Although the Turks are no more beloved by these people than by other Arabs, it does not seem to be found necessary to have many forts or military depôts to control them. Of this I am not, however, well qualified to speak, for I was not able to visit any of the Turkish stations. Of these, the principal is Kasr Gharian, but there are also soldiers at Kasr Tarhuna, and farther west. The Tripolines generally appear to have little anti-Christian intolerance, and, from what I have seen of them, I am inclined to believe that they would not highly resent European control.

The real object of my journey, and in every way the most remarkable feature of the district, was the series of ancient sites of megalithic temples. The information hitherto at hand on this subject was of the most meagre description, and consisted of the notices by Barth and Von Bary. The former, in his *Travels and Discoveries in North and Central Africa*,[1] gives notes or brief descriptions of some eight or nine sites, and also engravings of two trilithons. Von Bary, on the other hand, made an excursion into the very district, and visited about the same number, and his notes on the subject, unaccompanied by either plans or drawings, are to be found in the *Revue d'Ethnographie*.[2] In neither case was the examination of the sites made by these travellers exhaustive, and the sites visited were by no means among the most interesting or important. Without their writings, however, it might never have been known that these remarkable remains existed.

In all I was able to visit and examine nearly sixty sites, and numerous others I heard of or sighted at a distance, but found no opportunity of visiting. The districts where they are most numerous are the Wadi Doga, the Tarhuna plateau, and Ferjana, and also the Kseia valley. The western and more arid part of the plateau seems to contain but few examples, and although I found one or two on my route through that part of Tarhuna which adjoins Gharian, I lost them altogether after entering Gharian. It should, however, be remarked that in Gharian my people were strangers, and did not know the country, and a variety of reasons, which need not be entered into, made it impossible for me to make many inquiries in this district.

The sites are mostly called "Senam" (the Arabic for idols), this word being first applied to certain megalithic structures resembling the trilithons of Stonehenge, and more loosely to the groups of ruins generally. Often where there is no Senam proper, the site is termed "buni gedim" (old buildings), signifying a ruinous site, whether Roman or earlier. The following is a list of the principal sites visited :—

---

[1] In five volumes, 1857.

[2] *Senams et Tumuli de la Chaîne de Montagnes de la Côte Tripolitaine.* Par le Dr. Edwin von Bary, *Revue d'Ethnographie*, vol. ii. p. 426. Paris, 1883. Translated from the German, *Zeitschrift für Ethnologie*, Leipzig, viii. 378-385.

WADI DOGA (UPPER PART, LOOKING NORTH)

WADI EL HAMMAM

Wadi Doga.[1]—S. Bu Saiedeh.   S. Aref.   Kasr Fasgha.   Kasr Semana.
S. Argub el Mukhalif.   S. Shaahbet el Shuaud.   S. Henshir Aulad Ali.
Kasr Borimzeh.   Kasr Gharaedamish.   Kasr Doga (two sites near)
and Roman Mausoleum.

The Tarhuna Plateau.—S. M'aesara.   S. el Ragud (two sites).   Sajit el
Haj Ibrahim.   Ras el Id.   Ras el M'shaaf.   S. Um el Yuluthenat.
Kom es Las.   Henshir M'zuga.   Henshir el Buglah.   Henshir Bu
Ajeneh.   Kom Nasr.   Henshir Maagel.   S. el Jereh (two sites).
Kom el Lebet.   Kom el Saud.   El Khadra.   Kom el Khadajieh.
S. el Thubah.   S. el Bir (two sites).

Ferjana in Tarhuna.—Kasr Zuguseh.   S. Bu Hameda.   S. el Nejm, and
five others.

Wadi Kseia.—S. el Gharabah.   Henshir el Mohammed (and three others
at Jebel Msid).   S. Bu Mateereh.

Wadi Daun.—Kasr Daun.   Roman dams, and several Senam sites.

Western Tarhuna.—S. el Megagerah.   Wadi Wif, and two others.

Of the sites visited, those in the Wadi Doga are, upon the whole, the best
preserved.   Yet in spite of the great number available for examination,
it is impossible to say exactly what their original form was.   What we
find, more or less ruined, at every Senam, are the following :—firstly, a
great rectangular enclosure of magnificent masonry, but seldom preserved
to any height.   Generally the enclosed space is divided at intervals by
lines of short square columns, which, in a few instances, carry rudely
designed but well-worked capitals.   Secondly, the Senams proper.   These
are tall megalithic structures, trilithonic in shape, with jambs and cap-
stone, but the jambs are frequently constructed of more than one stone, and
they are always placed at intervals close to, or in line with, the enclosure
walls.   As a rule, the side facing the enclosure is carefully dressed, while
the other side is left rough, and in the jambs are always square perfora-
tions, apparently formed with a view to support some sort of wooden
structure.   The Senams vary from 6 to 15 feet in height, and are erected
on carefully prepared footing stones.   They were not doorways, for the
passage between the jambs averages but $16\frac{1}{2}$ inches ; and they were in
some way connected with ritual, for often right before them we find a
massive altar flush with the ground.

The Romans adopted and used these sites, and apparently preserved
the Senams.   Of this there is abundant evidence, and a few phallic
sculptures which were found, all showing Roman influence, may possibly
point to the form of rites the Romans found in use here.   I have, in a
paper read before the Anthropological Section of the British Association,
called attention to the design upon many Babylonian seals, representing a priest worshipping at an altar, at the back of which is a figure
which is simply a reproduction of a Senam, and have ventured to suggest
that in both the Tripoli structures and the form on the seals, we have
a symbol akin to the "Asherah" of the Old Testament, which was the

---

[1] S=Senam.

emblem of the goddess of fertility, and was worshipped in connection with Baal and Moloch. It is not here the place to discuss the *pros* and *cons* of this theory; but it may be accepted, without discussion, that these structures were symbolical, and very intimately connected with the rites of these remarkable sites.

I cannot venture here any more into the question of the probable builders of the Senams, than into that of the ritual that prompted them; for, in the first place, a great difficulty presents itself in disentangling the original work from that of Roman, and possibly also of Phœnician, date. The trilithons in themselves look older than the masonry of the enclosures, but I venture to say that it is only the idea—the traditional idea of the worship of great stones—that necessitated the erection of these strange monuments, by builders who were masters of the art of masonry.

In our present stage of knowledge of these sites, lacking as we do any such evidence as even superficial excavations will often furnish, it would then be rash to theorise far as to their origin. We have the accounts of Herodotus as to the tribes which occupied this part of Libya, and it is known that Phœnician settlements of a pre-Carthaginian date existed in North Africa, so that, taking into consideration the character of the masonry found at the Senams, coupled with the strange megalithic forms, it would be a matter of no surprise to me, if, when these ruins can be efficiently explored, they are proved to be the work of Libyan races largely influenced by contact with the arts and crafts of Phœnician civilisation. The trilithonic monuments may turn out to owe their origin to the same feelings which caused the erection of the upright dual columns which existed before the Cyprian temple of Paphos, and which had again their parallel in Jachin and Boaz, which stood before the temple of Solomon.

It is, however, most remarkable that in Mediterranean countries no distinct analogy can be traced between any other groups of megaliths and those on the "high places" of Tarhuna. Indeed, neither in the Algerian dolmens, in the Maltese temples, nor among the "taulas" and "talayots" of Minorca can we find much, if anything, which seems to elucidate the mystery. Strange as it may seem, it is none the less the fact, that the only monuments now standing which parallel at all the Senams of Tarhuna, are the great trilithons of Salisbury Plain. The key to Stonehenge may perhaps be found in the Senams of Tripoli,—but who is to find the key to the Senams?

There is, however, one thing which we must not lose sight of. Here, in these hills of Tarhuna, in a country so feebly supplied at the present day with water that it but ill supports a sparse population of Arab peasantry, we find abundant traces of a race possessing all knowledge requisite for building elaborate structures of dressed masonry, and professing some form of religion the observance of which necessitated the construction of elaborate places of worship. In the present dried-up condition of the land, it is sufficiently evident that, since the days when these were built, a great climatic change must have taken place. Treeless and riverless as it now is, the country could not support such a population, or foster such a civilisation, as we find traces of in these ruins. Evidence tells us, too, that in Roman times the population was nearly, if not quite, as dense in Tarhuna as when the Senam builders were alone. Most of the Senam sites are Romanised, the wells are probably in many cases Roman, and Barth has shown how the Romans can be traced far into the desert by their monuments. A monument like the mausoleum at Kasr Doga would never have been erected by them if the place had then been a wilderness. We may, indeed, be sure that in these days, here as in other parts of North Africa, the now dry wadis ran with perennial streams, the wells were full of water, and the hills covered with forests of subtropical timber.

The question is, When did this disafforesting and consequent drying up of the land take place? The answer no doubt is, that it began with the Arab invasion, and went on until it was complete. It was the same in Tunisia; it was the same in Morocco. The Arabs, with all their fine qualities and with their lofty religious aims, have been to North Africa like a cloud of locusts. They have taken no thought for the future, and they have turned Barbary from a garden into a wilderness.

From Jebel Msid, in the Kseia, I retraced my steps to the Tarhuna plateau by way of Wadi Daun, which I followed up till it came out on to the plain. In the widest part of this valley is an ancient castle, Kasr Daun; and both above and below this, and also in the lateral valleys of Shaahbet el Kheil and Kurmet el Hatheia, are to be seen strong rubble walls built across the valleys at frequent intervals. These appear to be the remains of an elaborate system of damming of the Roman period. Wadi Daun has, indeed, numerous fragmentary remains of this period.

From the Tarhuna plain I rode south-west to the Western Tarhuna hills, which I entered by a pass between two points called Jebel Jumma and Ras el Aswad. From here to Wadi el Ghan, at which point I commenced my return journey to Tripoli, the country is of a totally different character. My road wound at an elevation of about 1500 feet through a rough broken country, of which the hill points were some 500 feet higher than my track. Through these hills run numerous water-courses towards the coast. First, behind Jebel Jumma, we come to Wadi Hammam, then a lesser, Shaahbet el Zeraghwanieh, and next, a beautiful little valley embosomed in the hills, called Wadi Wif. At this point the boundary between Tarhuna and Gharian is reached, and here I found the last example of a Senam. Marching in a north-westerly direction from here, I crossed several more important wadis, and camped for two days in the beautiful Wadi el Ghan, where there was the luxury of running water.

After leaving the plain, I had the unpleasant experience of travelling for four days in a *gibleh* wind. Its exhausting qualities can hardly be imagined by those who have not experienced it. The force of the south wind is almost a gale, and, besides its intolerable heat, it is filled with imperceptible dust from the desert. Its peculiarity is that it is also laden with electricity, so that whenever I moved I found that my white woollen burnous emitted a distinct crackling sound. In Tarhuna and Gharian there is no shade, and no water that a European can safely drink without boiling, and I thought regretfully of the fresh, almost frosty, air I had lived in on the Tarhuna plateau. A camel man fell out with sunstroke, and my photographic films and stationery warped so with the heat that they could hardly be manipulated.

The Wadi el Ghan, where I camped, is the southern prolongation of the important Wadi el Haera, which leads direct to the capital. No greater contrast can, however, be imagined than that between this and the Wadi Doga, by which I entered the hills. Here, instead of a wide green plain surrounded by hills, I found a wild and romantic ravine. As we picked our way among crags and boulders, I could not but admire the varied contours of the scenes before me. Here, by the stagnant pools left from the winter torrents, great sections of pudding-stone could be seen overlapped by limestone. There, from the torrent bed, now dry and stony, rose grand cliffs of limestone and sandstone pierced in one or two places by caves, which I could see were used as dwellings ; and at one place rose a cliff of ochreous clay, coloured a deep red with iron ore.

At about ten miles from my camp in Wadi el Ghan, the Wadi el Haera widens out and the traveller finds himself on the plain. Here lie three curious isolated hills, which, rising from a level plain, look exactly like islands. Indeed, the whole of the Tarhuna and Gharian ranges, rising as they do in a long line of bluffs, instantly reminds the observer of sea cliffs, and the sensation on arriving at them from Tripoli is exactly that of approaching the coast from the sea. I venture indeed to suggest to those who study the geology of North Africa that this indeed is really the history of this region, and that once the waves of the Mediterranean covered both desert strip and green plain, and washed the Tarhuna and Gharian hills.

The three isolated hills, my guide said, were called Battus, M'dawar, and Mamureh, in which I am by no means sure he was correct. Barth's map marks them as Bates, Smaera, and Gedæa ; and Captain Lyon describes Bates and Smeeran as close together.

The distance from the mouth of Wadi el Haera to Tripoli is about seventeen hours' travelling at camel speed, or about forty-two miles. About half-way a track falls into the fertile Wadi Mejenin, a wide depression in the plain, more highly cultivated than any part I had yet seen. About ten miles beyond Bir el Sbeia, in this wadi, the sandy desert patch is encountered, and this continues up to the palm groves of the capital.

ASHANTI AND THE GOLD COAST.

By W. Scott Dalgleish, M.A., LL.D.

(*With Map.*)

There was good reason to believe that the Ashanti war of 1873-74, which ended in the defeat of the king's troops, in the burning of Ku-massi, his capital, and in the infliction of a heavy fine, had given the Ashanti a sharp lesson which they would not soon forget. Unfortunately that belief has been belied. The old trouble has broken out again, and last month a British force was despatched to the Gold Coast to repeat the treatment which, twenty-one years ago, was thought to have been final—so difficult is it to apply to savage races the ordinary calculations of diplomacy.

It is not the purpose of this article to give a history of our relations with Ashanti, to narrate the incidents of the last war, or even to go with any detail into the circumstances which have made the present war necessary. These matters may be left to the historian, the diplomatist, and their necessary concomitant the newspaper correspondent. It may, however, enable the readers of a geographical magazine to follow with greater interest and intelligence the progress of events, if we put before them a brief description of the physical aspect of the country which our troops have to traverse, of its climatic conditions, of its commercial resources, and of the character and customs of its people.

This information is the more pertinent because the causes, both of the last war and of the present one, are to a large extent geographical. Ashanti forms the hinterland of the British Gold Coast Protectorate, and the chief grievance of the Ashanti for a quarter of a century has been that the British Protectorate has formed a strict barrier between them and the sea, and has thus interfered with their trade, both legitimate and illegitimate, especially the latter. And here it is necessary to refer briefly to one or two historical facts which have a bearing on this point. Down to 1850, the Danes and the Dutch, as well as the British, had settlements on the Gold Coast, and it was obviously of advantage to the Ashanti that there were three rival European powers on the Coast, whom they could play off against one another in making terms for the conduct of trade, and for other privileges. But, in 1850, the British Government purchased the Danish forts, and in 1871 that Government concluded a convention with the Netherlands by which the Dutch settlements were transferred to Great Britain. Thus the entire Gold Coast, from Aflao, beyond the river Volta on the east, to Newtown beyond Apollonia on the west, became practically British territory, and the Ashanti were hemmed in seawards by a single and formidable European power. Their position was very similar to that of the Transvaal and the Orange Free State in South Africa; but it was aggravated by complications. There were tribal jealousies between the Ashanti and the Fanti and the Elmina, and these jealousies were intensified when the Fanti and the Elmina and other tribes near the Guinea Coast were brought within the British Protectorate. These were the proximate, if not the direct, causes of the war of 1873.

The defeat and failure of King Kwoffi Kari-kari (popularly known as Koffu Kalkali) led to his dethronement in 1874, and he was succeeded by his brother Mensah in the following year. Mensah's cruelties led to his dethronement in 1883, and he was succeeded by his nephew Kwaku Dua II. in accordance with a law common to the Ashanti and the Fanti, that the son of the sister of the king inherits in preference to his direct descendants. Kwaku Dua II. died after reigning only a few months, and he was succeeded by his brother Kwaku Dua III., or Prempeh, the present sovereign. The disturbed state of the country during his reign has called for British intervention in the interest of peaceful trade. Sir Francis Scott's mission in 1893 appeared to be

successful, but the troubles broke out again. King Prempeh was at no
pains to conceal his anti-British feeling; and he was perhaps emboldened
by the idea that he might make better terms with the rulers of French
Guinea. At this point the British Government required him to receive
a British Resident at his court. That demand he rejected in April 1895;
and, as subsequent negotiations failed to bring him to terms, the present
expedition has been undertaken in order to enforce the claim.

Now a few geographical facts may lead to a better understanding of
what is to follow. The first fact to be noted is that the Gold Coast lies
wholly within the Torrid Zone, its most southerly point—Cape Three
Points—being rather less than 5° north of the Equator. It has therefore
a tropical climate, and as the heat is accompanied with a great amount of
moisture, both on the coast and in the inland forest region, the climate is
very unhealthy, and is especially trying and dangerous to Europeans.
The wet season—or rather seasons, for there are two, the one from April
to July, the other from October to the end of November—is the more
severe, though the temperature is then lower. The dry season, extending
from December to March, though more trying in respect of the greater
heat, is less unhealthy because of the dryness, and of the prevalence of the
Harmattan wind.[1] Small-pox and dysentery are the scourges from which
the country suffers most, and the latter often attacks Europeans.[2]

The present expedition has been timed, it will be seen, so as to give
it the benefit of the dry season. It may also effect its purpose in time
to allow the produce of the present year to be sent down to the coast
before the recurrence of the wet season in April—a point to which West
African traders attach a good deal of importance.

The British Crown Colony and Protectorate of the Gold Coast extend
along the coast of Upper Guinea for 350 miles, while the average breadth
inland is 50 miles, the maximum being 150 miles. The area of the
Crown Colony, consisting of the lands adjacent to the settlements on the
coast, is 15,000 square miles. The total area, including the Protectorate,
is estimated at 46,600 square miles, or about two-thirds of the area of
England and Wales, with a population of a million and a half, of which
only 200 (some say 150) are Europeans.[3]

The British Protectorate forms a compact wedge between French
Guinea on the west and the German Protectorate east of the river Volta.
It includes a number of dependent native States. Those on the coast,
from west to east, are Apollonia, Ahanta, Denkera, Fanti (Tshi group),
Akra (Ga), and Awuna (Ewe); while inland there are Wassau, Tufel,
Assin, Akim, Akwamu (Tshi), and Krobo (Ga). All these are south of the
river Prah. North of the Prah are Upper Denkera and Ashanti, which
includes the Adansi, Dadiasi, and other tribes.[4]

---

[1] C. P. Lucas, *Historical Geography of the British Colonies*, vol. iii. p. 207.

[2] *Fanti and Ashanti*, by Brackenbury and Huyshe, pp. 84-85.

[3] Lucas, and *Statesman's Year-Book*. Other authorities estimate the sea-board at 250
miles, and the area at 20,000. See Huyshe, p. 94.

The Tshi-speaking tribes, among which is the Ashanti, and the Ewe-speaking natives
are described by Major Ellis in *The Tshi-speaking Peoples of the Gold Coast of West Africa*
and *The Ewe-speaking Peoples of the Gold Coast of West Africa*.

The chief British settlements on the coast—from west to east—are Newtown, Axim, Dixcove, Elmina, Cape Coast Castle, the former, and Akra, the present, headquarters of the British power;[1] between which lie Anamaboe and Saltpond, the latter being one of the chief commercial centres of the colony.[2]

The coast is absolutely destitute of harbours. It is sandy and surf-beaten, and ships can find safe anchorage only by lying a mile or two from the shore. This, however, involves very little risk, as the coast is free from stormy winds. Mr. H. M. Stanley draws the following picture of the coast and its surf-storms:

" By hugging the shore closely we were enabled to detect beauties that are never seen by passengers travelling on the steamers. Tiny nut-brown villages modestly hiding under a depth of green plantain fronds, and stately silk cotton-trees, which upheld their glorious crowns of vivid green foliage more than fifty feet above the tallest palm-tree ; depths of shrubbery wherein every plant struggled for life and breathing-space with its neighbour, through which the eyes attempted to penetrate in vain beyond a few feet; tracts of tall wavy grasses, tiger, spear, and cane, fit lurking-places for any wild beast of prey, varied by bosky dells, lengthy winding ravines literally choked with vegetation, and hills on the slope of which perhaps rested a village of a timid suspicious sub-tribe. And if the eye, ever in search of the picturesque and novel, ever roving for strange scenes, chanced to fall on the long line of surf, and the ear listened to the sonorous deep-sounding thunder, the mind recurred to the Psalmist's expressive words: ' They that go down to the sea in ships, that do business in great waters, these see the works of the Lord, and his wonders in the deep.'

" The surf on the African coast is ever a wonder and a danger. There is no other coast in any part of the world a thousand miles in length without possessing a cove or harbour where a ship could anchor secure from being rocked by the surf waves. Try along the whole of the Grain, Ivory, the Gold, and Slave Coasts, and there is not one port. But fortunately for ships trading to these places, there is seldom a hurricane or a gale blowing, so that they are able to anchor about a mile from shore. There is never any dead calm. The sea is ruffled in the morning, by the breeze from oceanward : during the night it is moved by the land breeze, so that ships anchoring in the roadsteads are ever to be seen rolling uneasily ; they are never at·rest. Unceasingly the long line of waves is to be traced, rolling onwards towards the shore, gathering strength as they advance nearer, until, receiving the ebbing waters flowing from the beach from preceding seas, there is a simultaneous coiling and rolling, and at once the long line of water is precipitated with a furious roar on the land. Where the water meets a rock a tall tower of spray and foam is suddenly reared, the wave line is broken and is in mad confusion. Where the beach is smooth sand, you may trace a straight unbroken line of foam nearly a mile long.

" One may easily understand then the trouble and annoyance trade

---

[1] A. B. Ellis, *A History of the Gold Coast of West Africa.* Map.   [2] Lucas, p. 205.

undergoes on such an inhospitable coast. An ordinary ship's boat is useless ; it would be a drowning matter for a crew of sailors unaccustomed to the surf to attempt to land anywhere along the coast between Sherbro and Lagos, a distance of nearly 1200 miles. The mouths of the insignificant rivers which feed the sea along this distance offer just as dangerous impediments to an ordinary ship's boat as the beach does. A long line of mighty breakers run across the mouth of each river and form a bar, which is almost certain death to cross, except in boats especially constructed for the peculiar work, and native canoes. These surf-boats have no straight stems or sterns, the keel is in the shape of a bow, which allows the advancing wave to be well under the boat before its crest lifts it aloft to precipitate it in the deep trough left in the track. A straight-stemmed boat cuts the wave, and its divided crest as it falls unites and swamps it. Many and many a poor sailor, ignorant of these things, has lost the number of his mess on this coast, and the history of trade on it is fraught with many a doleful tale." [1]

Though gold gave its name to this part of the Guinea Coast, and though it certainly abounds in the alluvial form in the region, especially in Wassau and Ashanti, it is no longer, if it ever was, the leading product. " As an article of export it ranks far below palm-oil and rubber." [2] " No Johannesburg has yet sprung up upon the Gold Coast, if only because the mining industry is sadly hampered by want of roads and of facilities of transport." [3]   " Timber, mainly mahogany, is exported in increasing amount from Axim, Dixcove, Chama, and Secondee." [4]   The trade of the Gold Coast has latterly made great strides. The average yearly value of the exports in the years from 1886 to 1891 was £477,000. In 1892 the value rose to £665,000, of which nearly three-fourths represented trade with the United Kingdom, which also contributes two-thirds of the imports, amounting in 1892 to £597,000. Cotton goods head the imports, the second place being taken, unfortunately, by rum, which comes mainly from the United States. Of the exports, 42 per cent. consist of palm-oil and palm kernels, 25 per cent. of indiarubber, and only 15 per cent. of gold. [5]

The following are the returns of the leading exports for 1893 : [6]—

Palm-oil,	.	.	.	£178,954
Palm kernels, .	.	.	.	183,910
	Total palm,	.	.	£362,864
Indiarubber, .	.		.	218,162
Other kernels,			.	80,721
Gold-dust,	.	.	.	79,099
	Total, .	.	.	£740,846

These figures show that the trade of the Gold Coast is steadily increasing, and that it is deserving of attention. According to *The Statesman's Year-Book*, the gold workings are now being systematically developed.

---

[1] *Coomassie and Magdala*, pp. 73, 74.   [2] Lucas, p. 207.   [3] *Ibid.* p. 208.
[4] Lucas, p. 209.   [5] *Ibid.* p. 210.   [6] *Statesman's Year-Book*.

Something now falls to be said of the nature of the country. There are only two routes (for they can scarcely be called roads) from the coast to Kumassi—the one, from Cape Coast Castle, is that followed by Sir Garnet Wolseley in 1873 ; the other, from Akra, coincides with that taken by Sir John Glover in his flank march from the south-east in the same year. These lines of march were simply war-paths hewn through dense bush and forest, impenetrable to man in their natural state. Two other roads are used by the Ashanti for trade purposes, the one passing through Wassau, the other, the most westerly of all, leading to Apollonia and Axim.

In the neighbourhood of Cape Coast Castle, the coast shows stretches of reddish land covered with vegetation, broken by dwarf cliffs and scaurs, and lined on the shore with clean sand. Behind this there is a background of undulating hills, paps, and hummocks, all covered with dense bush. Farther north, the rising ground is covered with forests of great trees—cottonwood, ironwood, dum palm, bamboo, and papaw— cottonwood-trees forming the most prominent feature, as they reach a height of 150 feet, with 50 feet of circumference. The tropical plants include cotton, indigo, coffee, rice, sugar-cane, maize, yams, plantains, bananas, cassavas, guavas, and pine-apples. Flowering shrubs, convolvuli, and exquisite parasitical plants abound in tropical profusion.[1]

Here is Mr. Stanley's description of the forest region of Mansu, about thirty miles north of Cape Coast Castle :—

"The forest grew denser, loftier ; there were cotton-woods of extraordinary height, and girth, and lengthy prolongations of buttresses to support the colossal stems ; there were sturdy teaks with lower crowns of foliage, and tamarinds spreading out in broad terraces of lighter green, gum-trees upreared their dark foliage, and slender palm-trees shot their graceful heads to view, all of which become more frequent, assume taller proportions and greater girth, as we move on. The road is all a glade ; an impervious depth of leafage, formed by the commingling branches overhead, shelters us from the sun. The engineers have done their work so well that it becomes a pleasure to walk on the smooth, broad road, though there are parts where the corduroys make it unpleasant for a short time. Steeps and descents, up and down hills, vary the uniformity of the level forest land and jungle ; a few rippling streams which we have to cross during the day are not unpicturesque reliefs to eyes jaded with the flanking walls of vegetation. . . . Perhaps a truer idea of this country may be gained if I state that from Cape Coast to Mansu the land, if stripped of its thick garniture of forest and jungle, would appear to be undulating, some land-waves higher than others, some troughs, or hollows, or dells, deeper than others ; but there is a general uniformity of undulation, all of which is covered with forest, undergrowth of all tropical trees, and plants. This intense block-like mass of vegetation spreads out on each side of the road, and the road, shaded by the commingling and embracing branches, is more like a tunnel than a highway. . . . The march to Sutah, and thence to Yan Coomassie (Assin), was through a

---

[1] *Fanti and Ashanti*, p. 112.

forest, where powers of description fail.   In density and wild luxuriance
it eclipsed everything I have ever seen in Africa.   I do not suppose that
anything in Brazil even can give one a more thorough realisation of
tropical luxuriance than a march through the district that intervenes
between Yan Coomassie (Assin) and Mansu." [1]

The following is Mr. Stanley's description of the hills, and of the
view from the summit :—

" From the summit of the Adansi hills we obtained a better idea of the
nature of the country which we had left behind than we had while
traversing it, and the nature of that portion which we are destined to
travel through in a few days if we proceed to Coomassie was revealed
to us.

" The Adansi hills appear to be a line of truncated hills, rising to an
equal altitude, and spreading out crescent-shaped eastward and westward,
inclining northerly, as if to hem in that land in which we are informed
Coomassie lies.   Each hill is wooded, each dip and depression of land is
wooded, high land and low land ; far and near, north, south, east, and
west, everywhere the forest land of Ashantee north of the Adansi hills
heaves and rolls, wave after wave, varicoloured and uneven, now a ridge,
then a hollow.   South it bears the same aspect ; west it is a line of peaks ;
east it is alike.

" At our feet, as we gaze down from a height of 1500 feet above sea
level, that contiguous to us presents us with a beautiful play and mixture
of colours, the mixture of the sere with the green, and the rich autumn
hue with the spring.   There are terraces of tamarinds, great wide-
spreading branches like parachutes, globes of silk cotton with pale green
leaves, round uprising towers of teak leafage, flat extents of ordinary
vegetation, deep hollows, which plantain and palm fronds combine to fill,
until the eye, tracing the variety and form of the foliage, insensibly is
carried away to where the colour of the vegetation is lost amidst purple
haze and blue ether.

" This scene is similar to that which might be obtained by a sail in
a balloon from Cape Coast to Coomassie.   The eye could not pierce
through the heavy drifts of foliage heaped up in ridges, globes, pyramids,
or the thick carpet which covers the hollows.   Indeed, looking down from
this height on such a scene, the power or force of the eyesight is compelled to give way to that of the mind ; for, after an ineffectual effort of
the eyesight, 'tis the mind which observes here.   It is the unbroken
expanse of foliage that baffles the eye, subdues its power and vitality,
and compels the mind to an indolent and careless observation of the
whole." [2]

The forest region is continued without variation, apart from the
billowy nature of the land surface, southward to the river Prah, which
forms the southern boundary of the kingdom of Ashanti.   It is thus
described by Captain Huyshe :—

" The Prah river rises in a mountain between Akeya and Quahoo
(two provinces lying N. and N.E. of Coomassie), near a little *croom*

---

[1] *Coomassie and Magdala*, pp. 116-120.   [2] *Ibid.* pp. 161-162.

called Sumtassoo. It is looked upon as a tutelary god, or fetich, and is generally called 'Bossom Prah,' or the Sacred Prah. Human sacrifices are made to it at its spring-head, said to be a large gaping rock half-way up the mountain. Its course is first southerly, then south-west, until a few miles beyond the town of Prahsue, where it bends again to the south, and falls into the sea at Chamah, 25 miles west of Cape Coast Castle.

"At the mouth there is a somewhat dangerous bar, with a depth of water varying according to the season from three to five feet. In the rainy season the current is constantly running out; in the dry season there is a three hours' flood and nine hours' ebb. The river is navigable for small craft for some distance—about two days' journey—when navigation is stopped by rapids and waterfalls, occasioned by a tract of hilly country. For about 60 miles of its course, counting from its bend at Prahsue up-stream in an N.E. direction, it forms the boundary between Ashanti and the British Protectorate.

"Opposite Prahsue the river Prah flows between steep banks 40 to 60 feet high, with a strong current of three or four miles per hour; and is estimated to be from 4 to 7 feet deep, and 40, 80, and 100 yards broad—a discrepancy doubtless owing to the season of the year, and the amount of the previous rainfall. The ford is below some shelving rocks in the centre of the river, and crosses in a slanting direction up-stream: the bottom is a rich clay, free from pebble, and covered with a bed of light sand. At this point it is described as being a 'noble' river, flowing through a richly-wooded country, with beautiful scenery resembling the Thames at Richmond!"[1]

The next physical feature of the country that claims attention is the ridge of the Adansi hills, 40 miles north of the Prah. These hills are thus described by Captain Huyshe:—

"The Adansi or Moinsey range of hills constitute by far the most serious physical obstruction between Cape Coast Castle and Coomassie; indeed they are the only hills worthy of the name. Very little is known as to their extent. . . . The hills cross the Coomassie road at about thirty-nine miles N. of the Prah, the ascent on the south side being abrupt, rugged, and broken into deep gullies. On the summit, 1600 feet above the sea-level, is a small undulating table-land of only a few hundred yards in extent, and there the path leads down the north slope at a much more favourable incline, the country on that side being at a higher elevation than that on the south side.

"There are said to be two other paths across this range besides the main route: one, called the King's Road (because it is usually followed by the king of Ashanti), leaves the main path near the village and stream of Ansah, and passing round to the eastward, rejoins the path at Fomanah; the other leaves the main path at the foot of the ascent, and passes round to the westward. It therefore appears probable that the point at which the main path crosses the hills is the steepest and worst, though perhaps the most direct; and that a more feasible road might be found on one side or the other—illustrating the old proverb, that, 'the longest way round is the shortest way home.'"[2]

---

[1] *Fanti and Ashanti*, pp. 113-115.   [2] *Ibid.* pp. 116-117.

The forest land extends northward to some 40 or 50 miles beyond Kumassi, where it gives place to vast plains reaching to the Kong country.

It may be inferred from this that Ashanti, as regards its physical character, differs but little from the countries lying south of it. It is said to have an area of 70,000 square miles, and its chief characteristic is that it is covered with dense and all but impenetrable bush and forest. Clearings have of course been made where villages are built, and these are numerous north of the Adansi hills. The largest of these clearings is on the site of Kumassi, the capital, which is described as an oblong, measuring four miles in circuit. Though built on the side of a hill of ironstone,[1] its surroundings are said to be swampy and unhealthy, owing to the overflowing of a small river which flows beside it. Hence there arise dense fogs which cover the city in the morning and the evening. The dysentery thus caused is most apt to attack natives coming from the coast.

Before the burning of the town in 1874, its streets were described as broad and clean, and ornamented with banyan-trees that afforded welcome shade. The principal buildings were the king's palace, which was solidly built of quarried stone, and covered an area of five acres; the fortress, called Bantammah, containing the royal treasures and the tombs of the kings, as well as the munitions of war; and the king's summer palace, situate about two miles north-east of the city. The ordinary houses consist of wattle-work walls plastered with clay and white-washed, and with roofs of palm-leaves.

Of the taste displayed by the Ashanti in their workmanship and their homes, Mr. Stanley speaks strongly. He says :—

" Many little things which we see about us evince the taste and industry of the Ashantees. Take one of their stools, for instance, and examine it. Formerly it was one square block of white wood, very like sycamore. An artisan has chiselled and shaped a beautiful stool which any drawing-room might possess for its unique shape, design, and perfection of workmanship. The seat is crescent-shaped, the ends of which when we are seated in it come up half-way to the hips ; a central column, beautifully carved, resting on a flat board twelve or fifteen inches in length by about eight in breadth, supports it ; on each side of the column is a side support chipped until it resembles a lace-work pattern. There is art in this stool, and whether it is original with the Ashantees, or borrowed from strangers, it is certainly a most interesting specimen of woodwork, the whole of which is cut from a solid block of wood. I think I have seen the shape of a stool similar to it painted on one of the tombs of Thebes.

" Take again this soup or water ladle, and regard the designs which ornament the handle. A European turner would be proud of the work, yet all that finical carving and cutting was done with an iron knife of native manufacture. Yonder are some specimens of earthenware, the figures on which are very praiseworthy, and here at my

---

[1] *Fanti and Ashanti,* p. 120.

feet are sandals excellently done; nay, you might almost declare upon oath that they are as good as anything worn by a middle-class Turk or Egyptian. Sandals! At the mere repetition of the word one's thoughts revert to the inhabitants of Egypt, Syria, and Asia Minor. Why not to Tunis, Tripoli, the Touaregs of the desert, to the Moors of Timbuctoo? Then, if these sandals of such shape are one of the necessities which intimacy with Moorish visitors taught the Ashantees, why may not this elaborate scroll and flower design on the walls have been an æsthetic taste imparted to them by the Moors? Is it only an ingenious deduction, or may it not be likely that this semi-civilised taste has been imparted to them by long contact with Moors?"[1]

And here is an inventory of Ashanti furniture drawn by the same practised hand :—

"An examination of the interior of the houses revealed to us much of the domestic life of the Ashantees. We saw the sleeping apartments, those set apart for reception, and those used for store rooms. These last were filled with a strange medley of articles—hide chairs, carved stools, hide bedsteads, wooden basins, pipes, drums, snuff-mills, ladles, platters, spoons, large earthenware crockery, bolsters stuffed with silk cotton, antelope skins, bags of leopard skins, rolls upon rolls of hard-pressed tobacco, a heap of Indian corn on the cob, spiked walking-staffs, iron knives and cleavers, curious teak boxes studded over with brass tacks, old locks, keys, gun flints, leathern pouches and belts, forsaken accoutrements and disused rusted guns, and many more things too numerous for detail.

"Some of the houses contained long suspicious rolls covered with long grass and tightly bound, which, on being opened, were discovered to be dried corpses of chiefs or heads of influential families. There were six of these bodies found altogether, but as soon as they were discovered they were taken out of the houses and deposited in the jungle."[2]

One of the most curious customs of the Ashanti, and indeed of all the native tribes at the Gold Coast, is that of intramural sepulture. They bury their corpses in the basement of their dwelling-houses, in the same way as the tombs of the kings are in the vaults of the Bantammah. The arrangement may suit the notion of keeping the quick and the dead near each other; but it is open to the gravest objection on sanitary grounds.

The Ashanti follow the custom of the Jews and other Oriental peoples in adhering to the system of polygamy, a system which is destructive of domestic life and purity. The king, it is said, has always 3333 wives—a mystical number which appears to have some subtle significance.[3]

The most shocking feature of Ashanti life is the prevalent custom of human sacrifice. Governor Hill says that "it is revolting to humanity to know that, as a common average, at least 3000 persons are sacrificed annually to their superstitious rites, and, on some occasions, 3000 have been slaughtered on one royal grave."[4] On this subject, Colonel Ellis has the following passage :—

---

[1] *Coomassie and Magdala*, pp. 167, 168.
[3] *Fanti and Ashanti*, p. 105.
[2] *Ibid.* p. 170.
[4] *Ibid.* p. 25.

" In September of the same year (1881) a report reached Cape Coast that King Mensa had put two hundred young girls to death, in order to use their blood for kneading the clay required for the repairs of his palace, and this aroused some attention in England, as by Article 8 of the Treaty of Fomana, the king had promised to do all in his power to check human sacrifices. The truth of the report was denied ; but though the particulars were inexact, there is no doubt that human sacrifices on a large scale had taken place, the occasion being the death of Yah Affileh, sister of the queen-mother. She was buried at Bantama, and for four weeks, girls and women, in parties of ten or twelve, were put to death ; while at the termination of the funeral ceremonies, her tomb was built up of clay kneaded with human blood. Up to this time the 8th Article of the Treaty of Fomana had really been a dead letter, and Captain Lonsdale reported that human sacrifices frequently took place in Kumassi, though generally in secret. The necessity of putting a stop to these practices had been impressed upon Buakji Tintin, who was intrusted with a message to that effect to the king; but it afterwards transpired that he falsely rendered it, and told Mensa that persons who had spoken disrespectfully of the king and the royal family, or had broken the great oath, or committed adultery, might be reserved for sacrifice." [1]

As to the theory of this custom, Mr. Stanley makes the following suggestion :—

" The road was marked with the dead fallen in the affair of the morning ; and each village had its human sacrifice lying in the middle of the path, for the purpose of affrighting the conquerors. The sacrifice was of either sex, sometimes a young man, sometimes a woman. The head, severed from the body, was turned to meet the advancing army, the body was evenly laid out with the feet towards Coomassie. This laying out in this manner meant no doubt, ' *Regard this face, white men, ye whose feet are hurrying on to our capital, and learn the fate awaiting you.*' " [2]

And to this may be added his gruesome picture of the Golgotha of Coomassie :—

" After some thirty paces we arrived before the dreadful scene, but it was almost impossible to stop longer than to take a general view of the great Golgotha. We saw some thirty or forty decapitated bodies in the last stages of corruption, and countless skulls which lay piled in heaps and scattered over a wide extent. The stoutest heart and the most stoical mind might have been appalled.

" Bowdich, in 1817, speaks of this death-grove, and we heard along the road from Cape Coast hither strange reports of it. M. Bonat says he has seen some two or three hundred slaves slain at one time, as customary after the death of the king's sister, and during his long residence as a prisoner in Coomassie he has frequently seen as many as a dozen slaves executed in the most barbarous manner and dragged to the grove, now filled with the relics of the butcheries which have been

---

[1] A. B. Ellis, *A History of the West Coast of Africa*, pp. 371-2.
[2] *Coomassie and Magdala*, p. 213.

MAP OF THE
GOLD COAST COLONY
AND
ADJACENT TERRITORIES.

Scale 1 = 4,000,000
English Miles

EXPLANATION OF COLOURING

Forests

Grass & Cultivated Lands

The Edinburgh Geographical Institute

J.G.Bar

going on year after year. If it is true that about a thousand slaves, offenders, rebels, and others are executed annually, we may form an approximate idea of the number of victims which have contributed to swell the terrible death-roll of the Coomassie Golgotha since the time of Sy Tutu, the founder of the present dynasty, who established his capital here in the middle of the eighteenth century. At the rate of a thousand victims a year, it would be no exaggeration to say that over 120,000 people must have been slain for 'custom' since Ashanti became a kingdom!"[1]

It is thus evident that, though the practice of offering human sacrifices may have been discontinued for a time, in accordance with the treaty of 1874, it was resumed; and there is no doubt that the civil troubles and dynastic feuds of recent years have been attended with much cruelty and bloodshed. There are, therefore, humane as well as political reasons for putting a check on the sanguinary rule of the tyrants of Kumassi; and a British Protectorate over the country would probably be effective for that purpose.

HAUSALAND

By the Rev. CH. H. ROBINSON, M.A.

(With a Map.)

AFRICA has been crossed and recrossed so many times within the last twenty-five years, and so much light has as a result been shed even into its darkest recesses, that it seems scarcely credible that there should still remain a people so numerous as to form one per cent. of the whole population of the globe, whose country and language are still almost completely unknown, at least as far as the general public are concerned. Nor can the neglect with which this people has so long been treated be justified by the fact that they are so low in the scale of civilisation, or their country so devoid of interest, that the study of them and of their surroundings might not unreasonably be postponed till the rest of Africa has been opened up. So far, indeed, is this from being the case, that, in the opinion of any who have had an opportunity of instituting a comparison, the Hausas are superior both intellectually and physically to all other natives of Equatorial Africa. The reason why they have been so long neglected is that their country has been, and to a certain extent still is, cut off from intercourse with Europeans by two physical obstacles of more than ordinary magnitude. Hausaland proper extends, roughly speaking, from latitude 8° N. to 14° N., and from longitude 4° E. to 11° E. Of the two possible ways by which a traveller from the coast can approach this territory, the shortest and most obvious is to ascend the river Niger for 300 miles and thence proceed overland; the distance to

---

[1] *Coomassie and Magdala*, pp. 231-2.

Kano, the most important town in the Hausa States, being about 400 miles from the river.   The reason why this route has so seldom been attempted is chiefly because of the great loss of life which has been experienced since the opening up of the lower portion of the river Niger, alike by missionaries, traders, and explorers, in their efforts to penetrate the interior.   The other possible way is by crossing the Great Sahara from the Mediterranean coast.   The distance to Kano by this route is nearly 2000 miles, a large part of which is across an almost waterless desert.   After spending a year in North Africa in the studying of the Hausa language, and in making inquiries as to the possi- bility of crossing the Sahara, I came to the conclusion that this route is for the present impracticable, and accordingly determined to attempt that by way of the Niger.

Accompanied by two other Englishmen, I reached the mouth of the Niger in July 1894, and having ascended the Niger and Binué rivers for 400 miles, set out to walk to Kano.   Owing to delays of various kinds, it took us four months to reach Kano, where we remained for upwards of three months studying both the language and the people. Kano has well been called the Manchester of Central Africa.   Its market is the most important in the whole of Tropical Africa, and its manufactures are to be met with from the Gulf of Guinea on the south to the shores of the Mediterranean on the north, and from the Atlantic on the west to the Nile or even the Red Sea on the east.   The city is fifteen miles in circumference, and on any ordinary day from 25,000 to 30,000 people are to be found at a time in its market-place.   Monteil, who visited it two years before us, gives it as his opinion that 2,000,000 people pass through it in the course of the year.   The importance of the place is due first to the native industries which it contains, and secondly to its trade.   Its industries consist chiefly of the weaving of cloth from native-grown cotton, and in the making and dyeing of various articles of clothing therefrom.   It would be well within the mark to say that Kano clothes more than half the population of the Central Sudan, and any European who will take the trouble to ask for it, will find no difficulty in purchasing Kano-made cloth at towns on the coast as widely separated from one another as Alexandria, Tripoli, Tunis, and Lagos.   Shortly before reaching Kano we obtained a practical illustration of the splendid physical strength which the average Hausa possesses.   When we first started on our march towards Kano we had with us none but Yoruba and Nupé porters.   Our loads were at that time made up into packs weighing from 50 to 60 lbs.   Never a day passed without these men coming to complain that their loads were too heavy for them to carry. On reaching Zaria we got rid of these porters, and before engaging Hausas in their place, I had all our luggage repacked and made up into loads weighing 90 lbs. each.   When the Hausas came in to feel the weight of the loads prior to bargaining for wages, half a dozen of them, on feeling the weight of a 90 lb. pack, put it down disdainfully, and asked to be allowed to carry two, on the understanding, of course, that they would thus earn double wage.   These six men carried a load of about 180 lbs. each a distance of 95 miles, going from 10 to 12 miles a day on an average, and neither they, nor any one of the rest, ever complained that their loads were too heavy. The man who carried the heaviest of all would at times perform a dance with the load on his head. The reader will the more easily realise the weight of 180 lbs. when he is reminded that this is exactly three times the weight of luggage allowed to a third-class passenger on an English railway. It is in part because of their splendid physique that the Hausas are in such great request on the coast as soldiers, and that they have been selected to form part of the forthcoming Ashanti expedition. In appearance the Hausas are as black as any African natives, but their lips are not as thick nor their hair as curly as those of the ordinary negro. Their tradition is that in very early times they came from the far east, away beyond Mecca, a tradition to which a study of their language would appear to lend some support. Hausa is the *lingua franca* of the Central Soudan, extending from the Sahara to the pagan tribes near the Gulf of Guinea, and from the Egyptian Sudan to the French colony of Senegal. It is estimated, as was above stated, that upwards of fifteen millions speak this language, many of whom can read and write it in a modified form of Arabic character. During our stay in Kano, we succeeded in obtaining a considerable number of Hausa manuscripts, consisting chiefly of religious poems. These will be published in facsimile, with translations and notes attached, in a few months' time by the Cambridge University Press. The only rivals which this language may be said to possess on the continent of Africa, whether in point of numbers speaking it or in the richness of the vocabulary, are Arabic and Swahili; and looking forward into the still distant future of the continent, it seems not improbable that if we add English to the three already mentioned, Hausa will form one, and that by no means the least important, of the four languages which are destined to supersede all the multitudinous dialects now spoken throughout the continent, and which at present form so great an obstacle to its development and civilisation.

The height of Kano above sea-level is 1425 feet, the greatest altitude of any town in Hausaland being about 2500 feet. The comparatively low elevation of the country generally, tends to suggest that it can never be colonised by Europeans in the ordinary sense of the word, though it may be governed and its resources developed by British subjects, to the mutual advantage of the governors and the governed. The one obstacle to the development of the country is the slave-trade. There is no portion of Africa, or indeed of the world, where slavery and slave-raiding are being carried on on so large a scale, and so entirely unchecked by any European influence. If Hansaland contains one out of every hundred people in the world, it is no exaggeration to say that one out of every three hundred people in the world is a Hausa-speaking slave. Every town of any size possesses its slave-market; in the market of Kano there are usually about 500 at a time on sale. The tribute paid by the smaller to the larger towns, and by the larger to Sokoto, consists almost entirely of slaves. The great mass of these are obtained, not from without the Hausa territories, but from villages and towns the inhabitants of which are of the same tribe and speak the same language as their captors. It is scarcely possible to exaggerate the harmful influence of slavery as thus carried on ; its result, indeed, is to bring upon the country nearly all the evils of a perpetual civil war. During the course of our march from Loko on the Binué to Egga on the Niger, *via* Kano, a distance of 800 miles, we had frequent opportunities of proving the accuracy of this statement. Soon after leaving Loko we entered the town of Nassarawa, where we were compelled to wait till the chief had returned from a slave-raiding expedition on which he was then absent; on reaching Jimbambororo, a village a few miles further on, we were told that its chief was not " feeling sweet," owing to the fact that that very morning twenty of his people had been seized and carried off as slaves by the people of an adjacent town. On leaving this village, we passed a spot where two days before fifteen native merchants had been seized and carried off as slaves; and again, shortly before reaching Katchia, we were shown another point on our path where, within the previous two days, a similar fate had befallen five other travellers. On reaching the town of Zaria, in the market of which we saw about 200 slaves exposed for sale, we were once again informed that the chief was absent on a slave raid. During our stay in Kano, as many as a thousand slaves were brought into the town as a result of another such expedition. In the course of our march from Kano to Bida, we passed so many towns and villages that we ceased to keep count of them, all of which had been recently destroyed, their inhabitants having been sold as slaves, and this not by any foreign invader, but by the chief in whose territory the places themselves were situated. By claiming, as Great Britain has done, the whole of these territories as being within her special "sphere of influence," she has claimed for herself one of the most valuable portions of Tropical Africa, but she has at the same time assumed a moral responsibility for the wellbeing of the millions of oppressed slaves who are to be found in this country.

In my book on Hausaland, to be published this month, which contains an account of my recent journey, with illustrations of the people and country, I have discussed the possibility of putting an end to this diabolical trade, and of opening up the country to legitimate trade and enterprise.[1] At the time of the Norman Conquest of England, this people, if their history be correct, were engaged in the manufacture of cloth similar to that which they at present produce, and were in many respects more civilised then than our own countrymen. Whether we regard them simply as an African people of unique interest owing to their ancient and independent civilisation, or whether we regard them as our recently acquired subjects for whose welfare we have made ourselves responsible; or lastly, whether we regard them from the point of view of the missionary or of the trader, they will be equally seen to deserve from us the help which, albeit unconsciously, they so greatly need.

---

[1] *Hausaland ; or Fifteen Hundred Miles through the Central Soudan,* by C. H. Robinson M.A. Published by Sampson Low, Marston, & Co.

OBITUARY: 1895.

By J. W. M'CRINDLE, M.A., M.R.A.S.

Sir HENRY CRESWICK RAWLINSON, Bart., G.C.B., etc. etc.—On the 5th of March last, at the advanced age of 85, the bearer of the above honoured name passed away from life—a man of a noble and commanding presence; a man of rare endowments and great achievements, illustrious as a soldier, a diplomatist, a scholar, a geographer, and an archæologist; a man whose like will not soon be seen again. He was the son of Mr. Abram Rawlinson of Chadlington Park, Oxfordshire, and was born there on April 5th, 1810. He was educated at Ealing, in Middlesex, and on being nominated to an Indian cadetship, went out to Bombay, in 1827, in the same ship with Sir John Malcolm, the new Governor of that Presidency. On reaching India he at once entered on the study of the native languages, which he learned so rapidly that in less than a year after joining his regiment he was appointed to the post of its interpreter. After serving for five years, he was selected as one of the instructors whom the Indian Government sent to Persia to organise the Shah's army. In the course of this military mission, on which he was engaged from 1833 to 1839, he rendered great services to geography, for which he received the Founder's Medal of the Royal Geographical Society. The terms in which his services were acknowledged bore that he had "explored with great zeal, perseverance, and industry the provinces of Luristan, Khusistan, and Azerbaijan, and the mountain ranges which divide the basin of the Tigris from the elevated plains of Central Asia." In the Afghan campaign he held at Kandahar the responsible and arduous post of Political Agent, and when he and Nott, who held the military command, were both ordered to withdraw from that station, these gallant officers, instead of marching southward to the Indus, boldly marched northward to Kabul, where they joined their forces to those under Pollock, in September 1842. Rawlinson's next appointment was that of Political Agent at Baghdad, with the rank of lieutenant-colonel. In 1851 he was promoted to Consul-General, and four years afterwards returned to England. In 1856 he was made a K.C.B., and nominated a Crown Director of the East India Company. This was also the year of his retirement from the service. In 1858 he was made a member of the new India Council, and a year later went again to Persia, now as Envoy and Minister, and with the rank of major-general. He soon found it necessary, however, to resign this post, and after a few months returned to England. "Henceforth," says his friend Sir F. J. Goldsmid, "Sir Henry Rawlinson became, as it were, an institution in his own country; a leader of Central Asian politicians; a mentor to Oriental students; a referee on Persian questions to successive administrations, whether Conservative or Liberal. His home work was mainly to turn to account the experiences of his Eastern career of nearly thirty years, for the benefit of his fellows, in no small degree across the Channel as well as at home."

Of his contributions to literature and learning we have scarcely left ourselves room to speak. His most memorable achievement, as every one knows, was his decipherment of the cuneiform inscriptions on the Bisitun Rock, the *Oros Bagistanou* of Diodorus. He published a complete account of his discoveries in this connection in the tenth volume of the *Journal* of the Royal Asiatic Society. For his paper on Ekbatana he was rewarded with the Gold Medal of the Geographical Society. He was elected a member of the former society in 1847, filled the office of its President from 1878 to 1881, and was its Director for the last 33 years of his life. It was into the pages of the journals of these societies that he chiefly poured out the stores of his vast and varied learning. The number of his contributions between 1864 and 1880 has been estimated at upwards of 130. His *England and Russia in the East,* which consists of a series of papers on the political and geographical condition of Central Asia, has been pronounced to be a "marvel of wisdom and profundity in respect of theory and statement of fact."

The honours and distinctions which he received, besides those already specified, are too numerous to be here recorded. We can only mention that our Edinburgh University, at its tercentenary celebration, conferred on him the degree of LL.D.

Sir EDWARD H. BUNBURY, Bart.—This accomplished scholar, the author of the *History of Ancient Geography*—a work which will long preserve his memory—died at Brighton on the 4th of March last, in his eighty-fifth year. He was the second son of General Sir H. Bunbury of Mildenhall and Barton in Suffolk, and his mother was a niece of Fox the celebrated statesman. He received a home education along with his elder brother, whom at a late period of his life he succeeded in the baronetcy. At the age of sixteen he was taken along with this brother to Italy that their minds might be enlarged by travel, and their taste for art cultivated by the study of its masterpieces. On returning to England, Edward went to Cambridge, where he took high honours, becoming in 1833 Senior Classic and Chancellor's Medallist. In 1841 he was called to the bar, and from 1847 to 1852 sat in Parliament on the Liberal side, as member for Bury St. Edmunds. He delighted in social intercourse, and was popular in various circles, both on this and on the other side of the Channel; yet so strongly was he prepossessed with the love of classical learning, that neither politics nor society with all their attractions could ever divert him from its pursuit. He was also a collector of coins, and is said to have been an accomplished numismatist.

. The results of his studies he gave to the world, partly in the noble work to which we have already referred, and partly in numerous articles which he contributed to the pages of Smith's well-known classical dictionaries of geography, mythology, and biography. These articles have been well characterised by the President of the Royal Geographical Society as models of accuracy and exhaustive erudition, and he tells us at the same time that Professor Freeman used to say of them that when he saw the initials "E. H. B." he knew that the information would be correct, and that further research was unnecessary. In 1839, Mr. Bunbury became a Fellow of the R.G.S., and served on its council in 1846-47. He died unmarried, and was succeeded in the baronetcy by his nephew, the present Sir Henry Bunbury.

JAMES DWIGHT DANA.—By the death of Professor Dana, which took place in New York on the 14th of April last, America has lost her most eminent naturalist, who in the course of his long life, which extended to 82 years, made multifarious additions to our knowledge of the sciences of zoology, mineralogy, and geology. He was a native of Utica, in the State of New York, and entered Yale College, where he specially directed his attention to the natural sciences and mathematics, and where he graduated in 1833. He had soon afterwards an opportunity of visiting the Mediterranean, having been appointed instructor in mathematics to midshipmen in the United States navy. On returning to America he became assistant to Professor Silliman, whom he afterwards succeeded in the Chair of Mineralogy at Yale. In 1836 he was appointed to be one of the scientific members of the United States Exploring Expedition, which, under Commodore Wilkes, sailed from New York in August 1838, and, after a cruise round the world, returned thither in June 1842. Throughout this cruise he acted in the capacity of mineralogist and geologist to the expedition, but ultimately it devolved upon him to observe and report on all the natural history. He wrote three official reports—one on Zoophytes, a second on Geology, and a third on Crustacea—while he supplied besides a great number of papers based upon facts which had fallen under his observation, or on conclusions to which his investigations had led him. The publication of these reports and papers made his reputation, and to his experiences during the cruise may be ascribed the fact that for the rest of his life his attention was more or less occupied with such subjects as corals, coral islands, volcanoes, mountain-making, and cephalisation. With regard to coral islands, he decided in favour of the views held by Darwin, which he supported by some original arguments of great force. His interest, again, in cephalisation may be traced to the studies on which he entered for the preparation of his report on Crustacea.

In 1850 he was elected to be Professor Silliman's successor, but it was not till 1855 that he entered on the duties of his office. In that year he addressed, in the capacity of its President, the American Association for the Advancement of Science. His famous work, entitled *A System of Mineralogy*, appeared so early as 1837. It has run through numerous editions, and is the standard authority on that science. In 1863 the first edition of his *Manual of Geology* was given to the world. The author in this work seeks to account for the configuration of the Earth, as we find it, by tracing to the action of the existing laws of nature the successive changes through which it has passed from its primordial to its present condition. It has run through several editions, the last of which was finished only a few months before the author's death. In this revision we have a complete account of the science as it is at present known. Geologists on this side of the Atlantic find in the *Manual* their chief source of information regarding the geology of America.

In 1846 Dana became one of the editors of the *American Journal of Science*, which had been founded by Professor Silliman, his father-in-law, and in 1878 this serial, which is the best known of the scientific periodicals of the United States, passed entirely into his hands. His contributions to this journal were numerous, and frequently had reference to the glacial question—one in which he was very deeply interested.

Professor Dana was remarkable for the breadth and richness of his knowledge, for his thoroughness in research, the soundness of his judgment, and his mental grasp of principles. He was kind and considerate, and treated with the utmost fairness all with whose opinions he found himself unable to agree.

His services to science received recognition abroad as well as in his own country. He was a member of the Academy of Sciences of Paris, a Fellow of the Royal Society, and of the Academies of Berlin and Munich. He had the Copley Medal awarded him in 1877, and the Wollaston Medal of the Geological Society in 1872.

ROBERT BROWN, PH.D., F.L.S.—The premature and sudden death of Dr. Robert Brown, so well known as a botanist, a traveller, and author or editor of numerous popular works on geography and other sciences, has caused deep regret in every circle in which his name was known. He died at Streatham on the 25th October last, in his fifty-fourth year. He was the only son of Mr. Thomas Brown of Campster, in Caithness, and studied at the University of Edinburgh, where he gained prizes and medals and took his M.A. degree. He studied afterwards at Leyden, Copenhagen, and Rostock, from the last of which universities he subsequently obtained the degree of Ph.D. In 1861 he visited Spitzbergen, Greenland, and Baffin's Bay, and from 1863 to 1866, as botanist to the British Columbia Expedition, his investigations extended from the West Indies and Venezuela to Vancouver Island, Alaska, and the Bering Sea coast. Afterwards, in 1867, he joined Mr. E. Whymper in his expedition to the west coast of Greenland, where he collected botanical specimens and formed a theory, which still holds its ground, regarding the erosive power of ice. He then returned to Scotland, where he was successively lecturer on Natural Science in the High School and in the School of Arts of Edinburgh, and in the Mechanics Institute of Glasgow. Finally, in 1876, he went to London. Here he led a life of great literary activity, being a member of the editorial staff of the *Standard* daily newspaper, contributing articles to the *Academy*, throwing off scientific memoirs and reviews without number, penning original works, and editing those of others. His chief productions are *Countries of the World*, in 6 vols.; *Peoples of the World*, in 6; *Africa*, in 4; and a *Manual of Botany*. Among the works which he edited are *Our Earth and its Story*, in 3 vols., and *Science for All*, in 5. One of his last literary labours was that of editing for the Hakluyt Society a new edition of Pore's *Leo Africanus* with an introduction and elaborate notes.

In his intervals of leisure Dr. Brown frequently visited the Barbary States, where he found pleasant occupation in studying the history and bibliography of the southern shores of the Mediterranean. His researches resulted in a work which was published in the last volume of the *Supplementary Papers* of the Royal Geographical Society. It is entitled *Bibliography of Morocco*, and was produced in conjunction with Sir L. R. Playfair.

Dr. Brown was a member of many learned societies in Great Britain, Europe, and America, and had been President of the Royal Physical

Society, and also of the Edinburgh Field Naturalists' Society, the founding of which he had been the first to suggest. We must not omit to mention in conclusion that his name has been given to several new botanical species, also to a mountain, a mountain range, a river, a cape, and an island.

HENRY SEEBOHM.—This distinguished naturalist and traveller died in London on the 26th of November last, from the effects of an attack of influenza by which he had been prostrated some six months previously. A few weeks before he passed away he attended a meeting of the British Ornithologists' Club, where eggs of the grey plover and the little stint, brought from Kolguev island, were exhibited by their discoverer, Mr. Pearson. He had himself, twenty years previously, in company with Mr. Harvie-Brown, gone on an expedition to the valley of the lower Pechora in a quest for the eggs of these very birds—a quest which had proved successful.

Mr. Seebohm was of Quaker parentage, and was a native of Bradford, in Yorkshire. That his mind had a strong inherent bent towards natural history was shown even in his earliest years by the close attention with which he studied every kind of animal found about his home. He received his education at the Friends' School in York, and for many years after he left it conducted successfully the business of a steel manufacturer in Sheffield. The calls and care of business had, however, no effect in detaching him from his favourite pursuit—that of ornithology, and having in view to write a work on the "History of British Birds," he made occasional visits to various parts of Europe for the purpose of observing the habits and migrations of the feathered tribes. One of these excursions was made to the valley of the Yenisei, and his researches there resulted in discoveries of great scientific value.

Mr. Seebohm was a prolific author. He wrote many papers on the classification of birds—a subject which engaged much of his attention. In 1888 he produced a splendid work on the geographical distribution of plovers and snipes, having qualified himself for writing this by visiting South Africa in order to study these birds in their winter retreats. The work by which he will probably be best known is his *History of British Birds and their Eggs*, a work of which a writer in the *Times* observes, that none such has been written since the days of Macgillivray. We must not omit to mention that Mr. Seebohm was also the author of two very pleasant volumes of travel, entitled respectively *Siberia in Europe*, and *Siberia in Asia*. In May 1894 Mr. Seebohm lectured to the Society in Edinburgh, describing a journey with Captain Wiggins to the Yenisei.

HUGH FRANCIS CLARKE CLEGHORN, M.D., LL.D., F.R.S.E.—This distinguished botanist, to whom belongs the honour of having organised the Indian Forest Department, died on the 16th of May last at his estate, Stravithie, Fifeshire. His father was the Administrator-General in the Supreme Court of Madras, and in that city he was born in August 1820. He was brought up at Stravithie, and was educated at the High School of Edinburgh, and at the University of St. Andrews. On leaving that seat of learning he returned to Edinburgh to study medicine, and graduated there in 1841. The following year he went out to Madras, where he was attached to the General Hospital, and had good opportunities for studying Indian diseases. Coming home on sick leave, after six years' service, he was shipwrecked on the way, and landed at Cape Town. He made only a short stay in this country, but in the course of it he assisted in preparing the catalogue of raw products for the great Exhibition of 1851. On returning to India he was appointed Professor of Botany at Madras, and a few years afterwards entered on the task of organising for India a Forest Department. With this in view he spent three years in examining the forests of the Himalayas, including those of Kashmir and the Trans-Indus territory. When Dr. Brandis, who was the Inspector-General of Forests, left for home, Dr. Cleghorn received the acting appointment. This was the last office he held in India, from which he took his departure in 1869, five years after his father's death, who had left him the family estate. His services to the Indian Government were duly acknowledged in an official minute at the time of his retirement. When settled at Stravithie he spent his time in ways useful to the public, taking part in every philanthropic movement in his neighbourhood. He was a member of the Edinburgh Botanical Society, and a president for several years of the Royal Scottish Arboricultural Society, and the founding of the Forestry Lectureship in Edinburgh University was largely due to his exertions. He was a member of our Society from its formation, and served for some time on its council. To its library he occasionally presented works, some of very considerable value.

JAMES JACKSON.—Mr. Jackson, who was noted for the great services which he rendered gratuitously to the Paris Geographical Society as its *Archiviste Bibliothécaire*, died on the 17th of July last in his fifty-second year. He was born in France, but, as his name shows, was of English parentage. During his tenure of the office referred to, he did much to improve the library of the Society, even adding to its contents at his own expense. To the Society's collections of photographic views he added a series of no fewer than 17,000, besides 2000 portraits of travellers and geographers. To our own Society, also, of which he was an Hon. Fellow, he presented several hundred excellent photographs. He was the author of a *Liste Provisoire de Bibliographies Géographiques Spéciales*, of *Socotra*, *Notes Bibliographiques*, and of a useful *Tableau de diverses Vitesses*. In his last will he bequeathed a sum of 100,000 francs to be divided equally among nine geographical societies, of which one was the London Royal Geographical Society. To the pages of that Society's *Geographical Journal* we are indebted for the foregoing particulars.

Dr. ROBERT ANSTRUTHER GOODSIR.—To the regret of a large circle of friends both in Edinburgh and elsewhere, Dr. R. Goodsir passed away in the middle of January last. He was trained as a banker, but afterwards adopted medicine—a profession in which his brother John rose to great eminence. Another brother, Harry, having accompanied Sir John Franklin on his last and ill-fated expedition, perished along with his gallant leader. Robert accordingly joined one of the search-expeditions, and in the capacity of surgeon sailed on board Penny's ship *Advice*. Of this voyage he published an account, which had a large circulation. In 1850 he accompanied another search-expedition for Franklin, and thereafter spent much of his time in travelling, till at last, in 1885, he settled down to a quiet life in Edinburgh. He died at the age of 71. Some of his letters and other documents are now in the possession of the Society.

H. MORITZ WILLKOMM.—This distinguished botanist and geographical explorer died on the 26th of August last, near Niemes, in Bohemia, at the age of 74. He was born near Zittau, in Saxony, and at Leipzig pursued the study of medicine and of natural science. After visiting the Spanish peninsula, to examine its botany and the structure of its mountains, he came back, in 1852, to Leipzig, where for some time he taught botany. He was called thence to Tharandt, and subsequently to Dorpat, where, in recognition of his services to science, the title of Imperial Russian Councillor of State was bestowed upon him. From Dorpat he went to Prague, in 1873, to occupy the Chair of Botany in its ancient university, and act at the same time as director of its famous botanical garden. Here he was employed till 1893, when he was pensioned off. During his tenure of office he frequently revisited the Peninsula, from which, and from the Balearic isles, he brought home with him rich botanical collections. He did also good work with his pen—writing treatises on botany, in which he dealt especially with the descriptive side of the science ; writing also geographical works, one of which was a comprehensive description of Spain and Portugal, and another a description of the Böhmerwald, which was at the time covered with primeval forest. The Austrian Government, in acknowledgment of his merits, appointed him a corresponding member of the Imperial Academy of Sciences in Vienna.

Dr. VALENTINE BALL, C.B., F.R.S.—By the untimely death, on the 15th of June last, of Dr. V. Ball, Director of the Science and Art Museum of Dublin, that city has to deplore the loss of one of her most eminent men of science. He was born in Dublin, the second son of Dr. Robert Ball, and younger brother of Sir Robert Ball, the well-known astronomer. Having completed his university education at Trinity College, in his native city, he proceeded, in 1864, to the East, where he found the kind of work most congenial to his nature in serving on the staff of the Indian Geological Survey. The memoirs of that survey are enriched by many valuable papers from his pen on the mineral products of the country, as viewed mainly in their economic aspects. His work on *Jungle Life in India* shows, however, that his sphere of observation was by no means restricted to inanimate nature, but that in the course of his journeys he studied the plant- and bird-life of the country, and marked the ways and characteristics of Indian races of which very little had been previously known.

On his return to Dublin in 1881, he was appointed Professor of Geology and Mineralogy in the University, and, after a two years' tenure of this office, succeeded Dr. Steele in the Directorship of the Science and Art Museum. This institution was variously improved, developed, and extended under his rule. Having in 1884 visited the principal museums of Canada and the United States, he drew up a valuable report on what he had observed. Last year again he visited various Continental museums, and on these was preparing a report when his career was cut short.

In 1883, when it devolved upon him to deliver the Presidential Address to the Royal Geological Society of Ireland, he chose for his subject " A Geologist's Contribution to the History of Ancient India "; and in the year following he read a paper before the Royal Dublin Society, in which he took up a kindred topic, "The Identification of the Animals and Plants of India which were known to early Greek Authors." The writer of this notice may be pardoned for stating that in both papers Dr. Ball makes frequent reference to his *Annotated Translations of the Greek and Roman Authors who have written upon India.* As Dr. Ball naturally took a great interest in the history of famous diamonds and other precious stones, he was led to edit a new edition of the *Travels* of Tavernier, who in his several journeys to the East invariably travelled as a dealer in gems and other valuable articles of small bulk.

Dr. Ball devoted himself so unsparingly to the duties of his responsible post that it is to be feared his resolute application to work had the effect of materially shortening his life. We may conclude this notice with quoting the words of one of his colleagues : " Though mourning his removal while still in his prime, his colleagues and friends feel that he leaves them the example and inspiration of unflagging devotion to duty."

Dr. KARL VOGT.—By the death of Karl Vogt, which occurred on the 6th of May last, a high-minded politician, a powerful orator, and one of the great masters of science has been lost to the world. Vogt was born at Giessen in 1817, and was educated first in his native town, and afterwards at Berne, where he graduated in medicine and took his doctor's degree. In 1839 he was employed at Neufchâtel in assisting L. Agassiz in preparing his *Natural History of Fresh-water Fishes* and his *Embryology of Salmon.* From Neufchâtel he proceeded to Paris, where he spent two years. He paid next a brief visit to Italy, and on returning thence to Giessen obtained a Chair in its University. Having been returned as a Deputy to the German National Assembly, he so energetically supported the National Party that he was deprived of his professorship, and retiring to Berne and to Nice, devoted himself to the study of zoology. He was appointed in 1849 to a professorship in Geneva, and was still holding it at the time of his death. He has made contributions of the highest value to the sciences of anthropology, zoology, and geology. Two of his works made a great impression on men's minds, his *Science and Superstition,* and his *Animal Communities.*

WILLIAM PARKER SNOW, who was purser on board the *Prince Albert,* the vessel which Lady Franklin despatched in quest of her husband, died on March 12th last in his seventy-seventh year. In 1851 he published an account of the search voyage, and six years afterwards he published another volume entitled *Two Years' Cruise off Tierra del Fuego and Patagonia*. In this work he reported what he had done in connection with the Fuegian Mission. The rescue of the Franklin Expedition was a subject which engrossed nearly all his thoughts, and he long clung to the hope that there might still be survivors. The sad story of his life is very touchingly told in a short obituary memoir by the President of the R.G.S., Mr. Clements Markham, who, when a midshipman, had met him in the Arctic seas. The memoir will be found in the May number of the *Journal* of that Society.

PROCEEDINGS OF THE ROYAL SCOTTISH GEOGRAPHICAL SOCIETY.

MEETINGS IN DECEMBER.

A meeting was held in Edinburgh on Dec. 5th, when Mr. Lionel W. Lyde, M.A., gave a lecture on "The Teaching of Geography," describing the method he has adopted in the Glasgow Academy. Dr. Scott Dalgleish presided.

Mr. C. E. Borchgrevink lectured in Edinburgh on Dec. 19th, his subject being the recent voyage of the *Antarctic* to the South Polar regions, of which accounts have been published on pp. 372 and 648 of vol. xi.

After the lecture, the Marquis of Lothian proposed a vote of thanks to the lecturer, at the same time expressing a hope that more interest would be taken in scientific research in the Antarctic regions, and that more pressure would be brought to bear on those who had the power of promoting a scientific expedition to the south.

Dr. John Murray, who occupied the Chair, spoke of the interest connected with a scientific expedition to the South Polar seas, and regretted that Mr. Goschen felt unable to recommend his colleagues to spend money, or spare a vessel for such an undertaking, while the East was in its present disturbed state. The Royal Geographical Society of London were considering the practicability of sending out an expedition independently of the Government, at an early date. If the money necessary could not be collected, a party might, perhaps, be sent out with an expedition that was proposed for killing blue whales. With £5000 the Chairman believed that arrangements might be made to send out twelve men with a commercial expedition, who would be landed on the Antarctic continent and taken off in the following year.

Mr. Borchgrevink addressed the Glasgow Branch on Dec. 20th. Mr. Paul Rottenburg presided, and Mr. R. Gourlay, Hon. Treasurer, proposed a vote of thanks to the lecturer.

A Christmas lecture to young people was given at Edinburgh on Dec. 27th. Mr. J. C. Oliphant was the lecturer, and his subject was Iceland.

LECTURES IN JANUARY.

On Jan. 9th, Dr. A. Markoff will lecture in Edinburgh on "Towns of Northern Mongolia."

Dr. Markoff will lecture at Dundee on Jan. 6th, and at Aberdeen on Jan. 7th, on "Russian Armenia and Prospects of British Trade there."

The Rev. Walter Weston will deliver a lecture in Glasgow on "Travel and Exploration in the Japanese Alps," on Jan. 21st.

This address will also be given in Edinburgh on Jan. 23rd.

GEOGRAPHICAL NOTES.

By The Acting Editor.

EUROPE.

The Origin of the Baltic Sea.—This formed the subject of an address delivered by Professor R. Credner before the Congress of German Naturalists and Physicians at Lübeck, on September 20th, and the address is published in the *Geogr. Zeitschrift*, Jahrg. i. Nos. 10 and 11. He remarks, in the first place, on the configuration of the sea bottom, consisting of a succession of deep hollows, as described by Professor Pettersson, and exhibited on the chart facing p. 620 in vol. x. Around it also are many depressions filled with water, such as the lakes of Finland and the Wener, Wetter, Mälar, Onega and Ladoga lakes, the last of which reaches down to 1200 feet, a depth surpassed in the Baltic sea itself in only a few places. In geological formation the basin of the Baltic belongs to two very different regions of the continent. From the Kattegat to the shores of the Arctic Ocean, it is composed almost entirely of archæan crystalline rocks—granite, gneiss, etc.,—and is a part of Europe which dates from the earliest period of the Earth's history. Rising above the sea as dry land as early as the Palæozoic age, it has since then been considerably reduced. The series of strata of Silurian and Devonian rocks which once covered it has, except in a few sheltered spots, been abraded and carried off. It is only at the edges, in the Blekinge district, on Öland, Gotland, and in the Russian Baltic provinces, that the Palæozoic covering still remains. The southern part of the Baltic basin is of quite a different character; younger deposits of Mesozoic and Tertiary age—limestone, marl, sandstone, slate, and clay—form the basis. The Quaternary deposits are of great service in giving evidence of the changes that have taken place in this region. They are of two distinct formations; the one group, consisting chiefly of loamy and sandy accumulations without any definite structure, but full of fragments of rocks of northern origin, are the ground moraines of vanished glaciers. In the second group are comprised sand, gravel, and clays, which exhibit stratification and a sorting-out of the material according to its weight and the size of its component fragments, and have, therefore, been laid down in water. Of special importance are those deposits of this group which indicate, by the remains of animal and vegetable life they contain, the extension of the water in the Baltic basin in inter-glacial or post-glacial times.

Having described the present condition of the Baltic basin, Professor Credner considers (1) by what processes the depression in which the Baltic sea lies was first formed; and (2) by what means and under what circumstances the Baltic sea attained its present form. With regard to the first question, the author states that the Baltic region lies in highlands carved out by faults. This is proved by the various elevations at which the same formation is found. For instance, the white chalk, which in Rügen and Möen rises about 500 feet above sea-level, is found in the adjacent sea 100 feet or more below the surface. Clefts, subsidences, and displacements of all kinds are found in every part of the basin. In many places the base-rock is seen to have undergone large disturbances to considerable depths, and the hollows running into the land as inlets from the sea can be traced far into the country.

By this process the formation of the Baltic basin was commenced, but its further development was the work of different agents, the ice-masses of the Glacial Periods. It was this powerful agent that stripped the Scandinavian peninsula of its upper strata, leaving the archaic rocks rounded, polished, and scored, and carried the detritus into the heart of Germany and as far as Kieff in Russia (see Professor Geikie's paper, vol. viii. p. 375). In the Baltic basin its action had the effect of rounding off and grinding down rugged forms due to dislocations, reducing the elevations that obstructed its movement, and deepening and widening the hollows already existing. Naturally it is in the southern portion, in the domain of the softer Silurian, Devonian, and Mesozoic rocks, that the basin attains its greatest breadth, whereas where the granite and porphyry of the Åland islands offered greater resistance to the passage of the ice, the bed is narrowed and at the same time deepened.

Again, the ice-sheet exercised another influence in forming the basin, by accumulating the detritus along its edges. One of the most striking effects of this accumulation was the shutting off of the Baltic bed from the continent, from Lithuania to Jutland. Though the core of the ridge that hems it in on this side is formed in many parts by an elevation of the base-rock, its greatest portion is built up of detritus carried down from the north by the glaciers and their streams. As in other glacial regions, so also in the Baltic, the central area is the domain of erosion and the periphery of accumulation. Hence the different character of the surface in the north, where the archæan rocks are exposed with all the characteristic marks of glacial erosion, while towards the south they are more and more hidden under a layer of detritus.

Lastly, how and under what circumstances has the Baltic in its present form come into existence? In its present extent and confirmation it is of very recent origin, dating not further back than the conclusion of the Glacial Period, and therefore from a time when man was already an inhabitant of Central Europe. Tectonic disturbances have, indeed, frequently recurred, and in the southern part of the basin water and land have repeatedly changed places, but these changes have been only temporary, and in particular a long period intervened, lasting through the latter half of the Tertiary to the beginning of the Ice Age, when dry land occupied the site of the waters that subsequently filled the Baltic bed. The seas also which during the intervals between the epochs of glaciation filled the Baltic basin were only temporary, for they were swept away by the succeeding glaciers. The bed these seas covered had a very different relief from that of the existing basin. The proof of later disturbances is evident in the island of Rügen, where the lower morainic accumulations lie evenly on the chalk, showing that the faults were not in existence when the ice-sheet lay above the island. These disturbances must have taken place at the retreat of the last ice-sheet, for its ground moraines alone, represented by an upper boulder-clay, cling around the upstanding blocks of the chalk.

The Baltic, then, as a permanent sea, was the work of the last glaciation, but it was not the Baltic of to-day. It was first an icy sea with an Arctic fauna, and then a lake with a pronounced freshwater fauna, before it became, as now, a land-locked sea of brackish water. These changes were brought about by sinking and elevation of the land of more or less extent. Two great periods of subsidence may be distinguished, each followed by one of elevation, the latter of which has not yet ceased to make itself felt on the Swedish and Finnish coasts. While the first subsidence, being of considerable depth, let in freely the salt water of the ocean, the second caused the formation only of a brackish-water sea. So also the first great elevation isolated the Baltic into a freshwater lake, while the second failed to cut off communication entirely with the ocean.

The formation of the sea followed closely on the retreat of the ice, but the water then extended no farther southwards than a line from Skåne to Bornholm. In consequence of the succeeding elevation the marine animal forms died out, and new brackish-water, and finally freshwater, forms took their place. A few survived the change of circumstances and the diminution of the salinity, and formed the parent stock of the interesting residuary fauna discovered by Lovén, consisting chiefly of true marine Crustacea and fishes, which inhabit the central and northern parts of the Baltic and the Wener and Wetter lakes, the remains of an old strait. In consequence of a greater elevation in the northern parts of the basin, the sea, or rather lake, spread itself out over portions of its basin that had been dry, and, swelled by its rivers, sought a fresh outlet to the sea. At this time, according to Munthe, were formed those winding hollows which are characteristic of the relief of the sea-bottom in the neighbourhood of the Belts, and a communication was opened with the Kattegat.

Again the bed of the sea sank, and, as in the former case, to a greater depth in the north than in the south. But whereas the former subsidence amounted in Angermanland to 885 feet, the latter did not exceed 330 feet. Consequently the water of the north sea flowed in less freely, and covered only a small strip on the Finnish and Swedish coast, beyond the limits of the present Baltic. The Belts and Sound, which during the existence of the freshwater lake served only as drainage channels, were deepened and allowed the salt water to flow in under the surface current which flowed outwards. But the flow of the salt water was checked by another slight rise of the ground, and confined to the Belts, the Sound being rendered so shallow as to allow only the upper current to pass through it. Consequently the marine forms migrated from the north and east to their present haunts in the south and south-west, and their place was taken by organisms that live in brackish and sweet water; since then the configuration and extension of the Baltic sea have been practically unchanged.

In Professor Credner's paper, of which the foregoing is only a meagre sketch, the numerous proofs of the processes described are fully detailed, and the smaller modifications that have occurred in later times are also discussed. Every paragraph is full of interesting information.

Rumania.—M. Michel Holban, Consul of Rumania, has communicated a sketch of the country to the Geographical Society of Geneva (*Le Globe*, Tome vi., No. 2). The land rises in steps from the banks of the Danube and Pruth to the Carpathians, and may be divided into three zones—the southern zone formed of large plains, the central zone occupied by vine-covered hills, and the northern zone covered by mountains, clothed with forests, and rich in mines. The climate is one of extremes, the thermometer falling in winter to −20 or −30° F., for the country is exposed to the winds from Russia, and rising in summer to 95° or 104°. There is scarcely any spring, but the autumn is magnificent; and at that season most of the rain falls. Snow commences in November or December and is very abundant.

The territory of Rumania extends over an area of 50,600 square miles, and supports a population of probably six millions at the present day. Bukarest had 230,000 inhabitants in 1894; it covers a large area, for many of the houses are surrounded by gardens. The appearance of the town is partly modern, partly oriental, but it is changing rapidly under the influence of modern progress. The Rumanians have retained the old Roman customs in their festivals, and the national type is the same as in the Roman Campagna. Public instruction is well provided for. The railways, which belong to the State, have an aggregate length of 1550 miles.

Rumania is essentially an agricultural country, and is very fertile. Two-thirds of the people work in the fields, and more than 14,000,000 acres are under cultivation. Large properties are the rule ; the State owns immense estates, acquired by the secularisation of church property. The peasants were freed in 1856, and a law of 1864 granted to each peasant a croft of from 7½ to 17 acres, taken from the landowners, who were compensated by the State. Rumania possesses several mines of coal, anthracite, and salt, the latter of which is exported in large quantities. Petroleum wells are worked to a certain extent, but are capable of supplying all Europe with oil. Industries are yet in their infancy, but are making good progress. Paper, cloth, soap, candles, and brandy are made, and wood is sawn. The foreign trade is of the value of nearly £15,000,000 in exports (chiefly cereals) and £17,000,000 in imports.

Pygmies in Europe.—Professor J. Kollmann has examined a settlement near Schaffhausen, which was inhabited in the Palæolithic, the Neolithic, and the Metallic periods. The second, or Neolithic stratum, is about 16 inches deep and its contents show that the reindeer no longer existed at that latitude, but the remains of the stag, roe, black bear, moor-ox, and other animals are found. Of human remains, besides skeletons of normal-sized persons, bones have been unearthed which must have belonged to dwarfs. Calculating from the lengths of three femora, Professor Kollmann has obtained the following heights :—1416 mm. (about 4 feet 7¼ inches) ; 1355 mm. (4 feet 5¼ inches) ; 1500 mm. (4 feet 11 inches). The average stature, therefore, is 1424 mn. (about 4 feet 8 inches), or less than that of the Veddahs of Ceylon, which, according to Messrs. Sarasin, is 1575 mm. (5 feet 2 inches). The average height of even the Andaman islanders is a little more.

While Professor Kollmann was engaged at Schweizersbild, Signor Sergi, assisted by Dr. Mantia, discovered some living pygmies in Sicily and Sardinia, of stature generally under 1500 mm., and with a cranial capacity 400 to 500 cc. less than that of the taller European races. In several districts of these islands they form 14 per cent. of the population.—*Journ. of the Anthrop. Institute*, November 1895.

ASIA.

Lake Gokcha.—M. E. S. Markoff was commissioned in 1894 by Lieutenant General Zhilinski to examine this lake with a view to its utilisation in the irrigation of parts of the Erivan Government. In area about 530 square miles, Gokcha has the form of an irregular triangle with sinuous shores, and stretches from northwest to south-east. Its length is 44½ miles and its maximum breadth 24½. Two promontories, the Narodus and Adatapin, divide the lake into two unequal parts, of which the larger, at the south-eastern end, has an area of about 380 square miles, with a length of 26 miles, and an average breadth of 17 ; the smaller basin is about 17 miles long and 8 to 9 miles broad, and covers an area of 150 square miles. Gokcha lies at a height of 6340 feet above sea-level, and is larger than any of the mountain lakes of Europe. It is surrounded by ramifications of the Anti-Caucasus, rising to heights of between 10,000 and 11,000 feet. On the north-east the banks are high and fall steeply to the water ; farther south the mountains recede and give place to a broad, deep, and flat valley, in which lies the small Ghilli lake, separated from Gokcha by a series of parallel beaches. A short and narrow channel connects the two lakes. On the south side a narrow strip of land intervenes between the mountains and the lake, increasing somewhat in breadth here and there, especially on the western side of the south-eastern basin. To the north the mountains again close it, and only in one place, at the north-western extremity of the lake, leave a broad passage by which the water of the lake escapes through the river Zanga to the valley of the Araxes.

Two hundred and fifty soundings were taken which showed that the centre of the larger south-eastern basin forms a plain lying at a depth of 164 feet below the surface. This depth was found close up to the steep north-eastern bank, but on the south and south-west the contour recedes from the shore, especially at the southern corner, where a river deposits its alluvium on the bottom. The smaller basin consists of a deep narrow hollow with a steep side to the north-east, and a more gentle slope to the south-west. The line of greatest depth lies close to the north-eastern shore, the maximum sounding (277 feet) being found, however, to the east of the island Sevang.

The surface temperature ranged during the three months of observations—July to September—ranged between 59° and 70° F., and the differences of the temperatures at the same hour of the day were comparatively slight. As to the temperatures at various depths below the surface, M. Markoff ascertained the existence of a layer with a constant temperature of 4° C. (39·2° F.)—the temperature of maximum density—and also found that as in other deep lakes there was a *Sprungschicht*, or layer where the temperature suddenly sinks. The transparency of the water was tested by discs of seven colours and white, and samples were obtained of the deposits of the bottom. Limnological stations were established at the north-western corner of the lake, and another five miles from the former on the road to Novo-bayazet.—*Izvestiya* of the Russ. Geog. Soc., Tom. xxxi. No. 1.

The Mineral Wealth of Transcaspia.—M. Sokoloff has discovered that petroleum is one of the principal minerals in Transcaspia. It exists in several places, particularly on the islands Cheleken and Nefte-dagh. The area over which it is found on the latter island has an extent of 162 acres; the borings yield between 9 and 10 tons of oil daily. Ozokerit is extracted on both islands, and a thick layer of sand occurs which contains 20 per cent. of asphalt. In the mountains Kara-tau, Tuarkyr, and at other places, are large deposits of brown coal. One hundred and thirty miles from Geok-tepe are hills where nests of "sulphur stone," a kind of quartz which yields 60 per cent. of sulphur, are imbedded in strata of schist. But the difficulties of transport are a hindrance to the exploitation of these deposits. Rock salt is of frequent occurrence, but is chiefly worked on Cheleken, and carried as ballast by vessels sailing from Azun-ada. Ironstone and green copper ore are found in the Kara-tau and in the Great Balkans, gypsum occurs near Krasnovodsk, and lithographic stone near Askhabad.—*Globus*, Bd. lxviii. No. 21.

The Periyar Irrigation Works.—The south-west monsoon, leaving Africa a hot dry wind, absorbs large quantities of moisture on its passage across the Indian Ocean, which is driven in the form of clouds against the Western Ghâts, and precipitated on the coastal zone, very little moisture passing into the interior. The north-east monsoon also loses most of its moisture before it reaches Southern India. Consequently, while the coastal region receives 100 to 190 inches of rain annually, the inner districts do not receive more than 25 to 40. As long ago as 1808 Captain Caldwell, engineer of the Madras Government, proposed to equalise the distribution of water by diverting towards the east one of the rivers which run westwards. The Periyar, one of the largest of those streams, rises not far from the Vaigai, which flows south-eastwards into the Bay of Bengal. The former frequently causes disastrous floods in the coastal regions on the west, whereas the Vaigai is nearly dried up, and in some years hardly reaches the sea. In 1808 the Madras Government was not in a position to carry out Captain Caldwell's scheme; but in 1887 the work was commenced after plans drawn out by Colonel Pennycuick, and is now completed. A reservoir 8000 acres in extent has been formed by a dam 155 feet high. This reservoir is capable of containing thirteen million cubic feet of water, half of which is available for irrigation. The water is conducted by a tunnel 2216 yards long through the hills, and distributed by a network of canals over 220 square miles of fields in the Madura plain. The cost of the work has been about half a million sterling.

The Surgut District in Western Siberia.—The small town of Surgut, lying on the banks of the Ob in lat. 61° 14′ N. and long. 90° 59′ E., and about 2500 miles from St. Petersburg, is the chief town of a huge District of the Government of Tobolsk. This District covers an area of some 85,000 square miles, is drained by many mighty streams, contains thousands of lakes and swamps, and is almost deserted, for its inhabitants number barely 7500. The southern part is clothed with primeval forest, and the northern consists of boundless *tundra*. The climate is very severe, the thermometer frequently standing at − 58° F. for weeks together, while a temperature of − 13° is considered mild. The terrible north-east wind, to which the District is much exposed, adds to the horror of the climate. A winter of eight to ten months is followed by a short, warm, and damp summer. The humidity is increased to no small degree by the inundations of the Ob and its great tributaries, which frequently convert the greater part of the District into a swamp during the summer, and call into existence innumerable swarms of gnats—a torment to man and beast. In spite of a summer heat of 80° to 100°, the ground never thaws to a greater depth than one to three feet. The forests are for the most part composed of conifers, and birches grow only to the south of the Ob. Fur-bearing animals of many kinds, elk, stags, etc., blackcock, ducks, geese, and other birds haunt the forests, while the waters abound in fish.

The population consists of not more than two hundred Russians, established in the town of Surgut and in five villages, and of Ostiaks, a few Samoyeds, a Jewish family, and a couple of banished Poles. The town of Surgut consists of a couple of dozen wooden houses blackened with smoke, and an old, miserable-looking church standing on the high bank of the Ob. On three sides it is encompassed by dark forests; on the fourth it enjoys a wide prospect over the river. The only route in this forest wild is the stream, traversed in summer by boat, and in winter in sledges. In spring and autumn all communication with the outer world is cut off.

The Russian of Surgut has irregular, sharp features, an ugly mouth and expressionless eyes, and, covered with the usual crust of dirt, does not present an ideal of beauty. His figure is short and squat, but he has a certain air of courage and defiance that he has gained in his struggle with the hostile forces of Nature. The Ostiak, on the other hand, gives at first sight the impression that he is of a dying race. His long arms hang helplessly down, his face has a dirty grey tinge, his broad mouth with thin white lips, his oblique eyes and coarse unkempt hair form no attractive whole.

The climate renders agriculture impossible. Some twenty years ago an attempt was made at the village of Yuganskoie to grow corn, naturally without success.

The forest is common property, and any man may fell a tree in it or hunt the wild animals. But the Siberian stone pine is protected for the sake of its edible nuts. Every man has his traps in the forest, and each village possesses fishing grounds. Cattle and horse grazing are important occupations, and the hay mown on the meadows of the District is usually sufficient to feed the animals, but occasionally the cows have to be fed on salt fish.

For gathering the nuts of the stone pine the men form parties. Each man brings his contribution of provisions, etc., and the party proceeds by boat along the Ob to some favourable spot in the latter half of July. One man ascends the trees, often with the help of climbing irons; while the others collect the cones into bags, which are then dried in the sun that the nuts may be easily extracted. At the latter end of September the men return home, and each member of the association receives £8, or in very good years as much as £10, as his share of the profits.

Besides the hunting of fur-bearing animals, which are killed with guns and caught in various traps, and, by the Ostiaks, with bow and arrows also, the catching of birds, especially ducks, is a very profitable occupation. The ducks are caught by a net, which is spread out over trees or a framework of poles, and is pulled down on the approach of the birds, entangling some of them in its meshes. In this way a man catches eighty, or even as many as two hundred or three hundred, ducks in a day.

The shipping on the Ob has called a new employment into existence, that of supplying wood for the engines; but the business is in the hands of a few well-to-do "sweaters," and the labourer receives a very trifling remuneration.

Handicrafts do not flourish in Surgut, and most of the artisans are immigrants or convicts. Barter with the natives is a more favourite occupation. A fair is held at Surgut in winter, and in May and June there are markets in Yuganskoie and Lariatskoie. The Ostiaks come into Surgut in the evening, and quickly drive into the courtyards of the merchants. There they are provided with spirits, and the bargaining is frequently prolonged till daybreak, when the natives leave the town as noiselessly as they entered, and return to their camp ten or twelve miles distant. The goods offered them in exchange for their furs, fish, feathers, pine nuts, etc., are usually of very inferior quality and excessively dear; and false weights and measures are employed to cheat them still further.

The houses of the Russians in Surgut are clean and orderly in outward appearance. The boards are covered with mats, the walls are papered, and the ceiling whitewashed. But the appearance is deceptive, and the traveller who ventures to lie on the neat down bed finds that it swarms with vermin. Bread and fish are the chief ingredients of Surgut dishes, the latter being served up in a variety of ways. Fish oil is used in the preparation of all the dishes. But still more disgusting is the salt fish which, salted in autumn, begins to decay in spring, and in this state is eaten with great relish. Brick tea supplies the chief drink, but spirits also are consumed in large quantities.

Intellectually the Russian of this country is in a very low condition; as a rule he cannot read or write, and is prejudiced against education. He is full of superstition, believes in innumerable spirits, in witches, and in the return of departed spirits to their earthly abodes. The Russians observe all the ceremonial regulations of the Greek Church, and keep the fasts strictly, but are quite incapable of understanding its dogmas, not distinguishing between saints and God, for instance.
—Peter von Stenin in *Deutsche Rundschau*, Jahrg. xviii. Heft 3.

AFRICA.

African Travellers.—During the past twelve months, Miss Kingsley has been travelling in West Africa, chiefly for the purpose of forming a collection of fishes and of studying fetichism. From Old Calabar, Miss Kingsley went to the French settlement on the Gaboon. She ascended the Ogowe, and crossed the Sierra de Cristal, mountains ranging from 7000 to 8000 feet in height. Subsequently she ascended Cameroons Peak, 13,700 feet high.

Dr. Donaldson Smith, who left England in the summer of 1893, has returned from East Africa. Being prevented by the Abyssinians from marching directly to Lake Rudolf, he turned southwards across the Webi Shebeli and the Jub, and then marched eastwards through the rich and well-watered territory of the Bor'ni, who on more than one occasion attacked the expedition. On leaving their country he marched northwards, crossed the Konso mountains, and explored the small Abala lake. Passing round the northern end of Lake Stefanie, and discovering a large river that discharges itself into the lake, Dr. Smith came to Lake Rudolf, and explored the country for a considerable distance to the north of the lakes, coming among people who had never seen a white man. The journey back to the coast was made by way of the Tana river. Dr. Smith is to read a paper this month before the Royal Geographical Society, and no doubt Miss Kingsley will shortly give a full account of her travels.

The People of Madagascar.—A lecture of M. E. T. Hamy on the natives of Madagascar is published in the *Revue Scientifique* of September 21. The island measures 940 miles from north to south, and has an average breadth of 290, which dimensions give an area of about 273,000 square miles. The estimates of the population vary considerably; the latest, by M. Catat, is about seven and a half millions. The density is small, for there are vast solitudes in the island, particularly between the coastal zone and the high plateaus. On the latter, which occupy a comparatively small proportion of the surface of the island, the population is densest, more than two million souls having their homes here. The Antisaika of the east coast are also closely packed, the density being according to M. Catat nearly ninety-seven per square mile.

Among these people, the observer finds himself at first in great perplexity as to their ethnological affinities. The Malagasy is not a Negro; his features are very different from those of the Africans, but have a resemblance to the Malay type. He is not yellow, and his hair is a mass of fluffy tresses. He is a singular mixture of very different elements, some of which call to mind the natives of the great archipelago of Asia. All the Indonesian peoples present a number of physical, intellectual, and moral traits which are found more or less intact among the Malagasy. All the Malagasy tribes, without exception, speak a dialect closely allied, according to Müller, to that of the Battaks of Sumatra, and differing locally only by some slight changes of pronunciation. The clothing is also the same in Madagascar and in the Malay Archipelago, and the most essential household utensils—the pestle and mortar, the bamboo water-vessels, etc., and the *valiha*, a sort of guitar—are identical in Madagascar, Timor, and Laos. The sarcophagus of the Sakalava may be found again in the Philippine Islands, and the method of tattooing and other rites and ceremonies have more affinity with those prevailing in the East than in Africa.

While among the people long established on the island, as the Sakalava, these characteristics are more or less modified, they are well preserved among the Antimerina, incorrectly termed Hovas, who, in number some 800,000 to 1,200,000,

inhabit the elevated country of Imerina or Emyrne. According to M. Grandidier the term Hova denotes the citizen class, as opposed to the Andriana or nobility. It is about the end of the eighteenth century that the Hovas first appear in the history of Madagascar. They were a small people, superior in intelligence to their neigh-bours, skilled in the working of iron and the weaving of fabrics, but a prey to intestine quarrels which hindered their development. Their original home was in the south-east. First they occupied Menabe, and thence gained the plateaus of the interior. The Sakalava, indeed, say that the Hovas were driven from more fertile lands on to the plateaus, which were certainly healthy but desolate. Many of the higher nobility of the Hova State have all the characteristics of the pure Malay type, but lower in the social scale the Negro element becomes more and more marked. The immigrant Hovas or Merinas were probably at first few in number, and brought few of their women with them. Hence, by intermixture with the native races, the original type became much obscured, and now all shades of colour may be seen from yellow to black. The Merinas are of weaker build than the other Malagasy, and in other respects they exhibit the same differences as the Malays and Indonesians.

The Betsileo are regarded by most travellers as Hovas slightly modified, the Negro element being very conspicuous. The Sakalava, who occupy all the north of the island and the east coast as far as Onehali, are composed of numerous tribes, the Antankara to the north of Antongil Bay, and the Antsianaka to the north of Imerina, being the most important. Some of these have unmistakable Indonesian features, while others—the Bazas of the interior, for instance—have the thick lips and woolly hair of the Negro. A third element also, the Semitic, has at various epochs mingled with the Indonesian and Negro. Jews formerly inhabited Sainte-Marie, and Arabs have settled among the Antanosy and the Antaimoro.

AUSTRALASIA.

Meteorological Observations at Alice Springs.—Alice Springs is a station of the transcontinental telegraph line from Port Darwin to Adelaide, and stands on the south side of the M'Donnell range close to the Tropic. During the years 1881-1890 the atmospheric pressure, temperature, and humidity have been recorded every three hours during both day and night. Professor Hann reports on these records in the *Meteorologische Zeitschrift*. The barometrical observations are useless, for they are reduced to sea-level, whereby the daily variations are completely changed. In summer Alice Springs is one of the warmest places in the world. The mean temperature in January, reduced to sea-level, is 90·9° F. ; the mean for July is 57° ; and for the year 75·6°. The absolute maximum and minimum are 117° and 23°, and the mean annual range is 88·2°. The yearly rainfall is barely 9 inches. Most of it falls in summer, the quantity measured from December to February being 4·76 inches, or nearly 53 per cent. of the total. The winter is dry, whereas in Adelaide there are the usual winter rains in that season.—*Geogr. Zeitschrift*, Jahrg. i. Nos. 10 and 11.

Norfolk Island.—This island, of which little is heard since it has been occupied by some of the descendants of the mutineers of the *Bounty*, is described by M. Paul Wenz in the *Comptes Rendus*, No. 13, 1895, of the Paris Geogr. Soc. It lies in 29° 3′ 45″ S. and 167° 58′ 6″ E. long., 980 miles east-north-east of Sydney, and about midway between New Zealand and New Caledonia. It is 5 miles long by 3 broad, and has an area of 8600 acres. Captain Cook, who discovered it in 1776, speaks of it as wooded to the water's edge ; but now hills covered with pasturage alternate with thickets of indigenous trees. The only forests of any extent grow on Mount Pitt (1027 feet high) and on the chain to which it belongs, and also on the low ground at the north-west.

The soil, of a dark brown chocolate colour, is very fertile, and will support almost all tropical plants as well as those of temperate climes. Coffee, bananas, sugarcane, all kinds of vegetables, oranges, lemons, vines, mulberries, mangos, and other fruits thrive luxuriantly. Of the flora peculiar to the island may be mentioned the famous Norfolk pine (*Araucaria excelsa*), a palm (*Areca Baveri*), and a tree-fern (*Alsophila excelsa*). The great defect of the island is its inaccessibility ; the rocks rise precipitously from the sea, and only two points, one at the north and the other at the south, can be approached by vessels, even in calm weather.

Norfolk Island was colonised by New South Wales in 1788, when Lieut. P. G. King landed there a party of convicts and free men. The little colony made rapid progress, and was prosperous in 1800, when it was abandoned, the inhabitants being gradually removed to lands granted them in New South Wales and Tasmania. Between 1806 and 1826 it was entirely deserted, but in the latter year it was made a convict station, and was thus utilised until 1855, when it was again abandoned, as transportation to the colonies was abolished. About this time the Pitcairn islanders found that their isle was no longer able to support its increasing population, and petitioned that Norfolk Island should be ceded to them. Accordingly in 1856 a Government vessel carried 199 of them to their new abode, where, with the exception of six families who returned to Pitcairn Island, they and their children have lived ever since.

The government of the island is in the hands of three functionaries elected annually by the people, who receive authority to act from the Governor of New South Wales. Taxes are unknown, and the only revenue is derived from a few insignificant fines. The men between eighteen and sixty years of age are expected to do four days' work each between January and June, chiefly on the roads. The chief magistrate receives a salary of £25 a year, the postmaster £8, and the registrar of lands £5. The doctor and chaplain are paid in part from the interest of accumulated funds invested at Sydney. The laws could be printed on a couple of sheets of foolscap paper, but they answer every purpose ; crime is unknown, and there is no prison.

The chief occupations of the people are whale-fishing and agriculture. The exports consist of whale oil, wool, horses, onions, potatoes, and fruit, and the imports of clothing, spices, and fishing tackle. Intoxicating liquors are prohibited. The annual value of the trade is at least £6000. Kingston, named after Lieut. King, in the south of the island, is still the central point. Most of the old houses are inhabited, and one or two public buildings, but the rest is a heap of ruins. Most of the people live in small wooden houses scattered over the island. Each couple, on their marriage, receive a grant of twenty-five acres of land. Of the 4000 acres thus alienated, 600 only are cultivated, the remainder consisting of pasturage on which cattle, sheep, and horses are fed. The climate is very healthy and disease is rare ; the mortality is 9 per 1000.

The Melanesian Mission holds 4000 acres, ceded to it thirty years ago with the consent of the people, and the island is the residence of the Bishop of Melanesia. In December 1894 the total population was 899, of whom 661 were Norfolk Islanders and 238 belonged to the Mission.

GENERAL.

The Use of the Plane-Table in Exploration.—We are indebted to Mr. Henry Gannett, Chief Topographer U.S. Geological Survey, for the following note :—

" The above subject received some attention at the recent Geographical Congress in London, and as the United States Geological Survey has had an extensive and varied experience in the use of this instrument, the results may be of value to the discussion. This Department has in daily use fully 100 plane-tables, some of them adapted for the most accurate work, others for the cruder work of filling in details. All of the topographic work of the organisation, with the exception of the primary triangulation, is done with these instruments. With them half a million square miles, comprising every variety of country, have been mapped. They have been used upon scales ranging from four miles to an inch up to 800 feet to an inch.

" The instrument is not only applicable to all scales, all kinds of country, and all methods of work, but it is the best whatever the conditions may be. It is fully as accurate as any; it is the most rapid, and it produces the best results.

" The plane-table in use upon the Geological Survey is fitted with what is known as the Johnson movement, which is a modification of the ball and socket. Two cups play one within the other for levelling, and the instrument is clamped in level by a screw which binds them together. A third cup permits the azimuth movement, and this is clamped in a similar manner. The tripod, made of split legs, combines rigidity and lightness in the highest degree. The tripod and movement weigh but eight pounds. The board is made of pine or cherry, one-half inch in thickness, with cleats across the ends to prevent warping. These cleats are not glued on, but simply dovetailed, as otherwise they would cause the board to warp. The body of the board must be allowed to shrink and expand freely. Its weight depends, of course, on its size. The boards most in use on the Geological Survey are 24 by 30 inches, but these are too large to be conveniently carried on horseback. For traversing, a compass is set in a narrow box in the edge of the board.

"The alidade, in its crude form, consists of a ruler with a pair of open sights, which may be hinged, so as to turn down when not in use, and is not more than 8 to 12 inches long. In the higher forms a telescope is attached to the ruler by a column, and this telescope bears a vertical arc or circle for the measurement of heights, and turns in a sleeve, for the adjustment of vertical collimation. In this form the ruler is 18 to 24 inches long. It is graduated to suit the scale of the map to be made, so that distances in miles and fractions can be read on it. The paper used is heavy drawing paper, mounted on cloth. For the best work, where it is necessary to avoid all *irregular* shrinking and expansion, it is mounted in two thicknesses, with the grain of the paper in contrary directions. The paper is fastened to the board by thumb screws or tacks, and, if the needle is used for orienting, the paper should not be removed until the work on it is completed.

" The plane-table and alidade are described more extensively and figured in *Monograph* 22 (*Manual of Surveying*), United States Geological Survey, to which those interested in the matter are referred for further details regarding its construction.

" The manipulation of the instrument, which is very simple, is well shown in the illustration on p. 643 of vol. vii., where also the methods of orientation are explained.

" In surveying linear features, such as roads, trails, etc., it is more convenient to use the direction and distance method, or traverse method, instead of the intersection method, or triangulation. In the traverse method, the table must be oriented by compass, because in orienting by sighting from station to station there is rapid accumulation of error in direction, since the traverse stations are commonly close together. Distances may be measured by any available means, preferably by counting wheel revolutions, if on a road, otherwise by stadia, if that is sufficiently rapid, while in reconnaissances distances may be roughly measured by pacing or by the rate of travel of one's riding animal. The direction is simply plotted immediately on the table and the distance measured off on the course, whether it be a fore or a back sight. Points off the line are intersected by sights from the line, using it as a base line, and sketching is carried on coincidently with the location of points. Many thousand miles of road are annually traversed by the topographers of the United States Geological Survey, and the work is carried on at an average rate of 10 linear miles per day, per man. All surveying is by one or the other of these two methods, by triangulation or by traverse, and the plane-table is equally applicable to either of them.

"Exploratory work generally takes the form of continuous journeys from point to point, and the results are commonly route maps, made by traverse of the route. with location and sketching of as much topography on either side as can be seen. For such work the plane-table is admirably adapted. It is light and portable, easily and quickly handled, its manipulation is of the simplest and can be taught to any topographer of fair intelligence in a day. With its use, the map is made on the spot, from the country as copy. Errors, gaps, and weak spots in the work are discovered while on the ground, when they can readily be corrected. There is practically no office work, no tedious plotting of notes after they have become cold. Thus half the labour of making the map is saved.

"Having written thus in favour of this instrument, it remains to state its drawbacks. There is but one. It cannot be used in the rain.

"As to the cost of the outfit, it may be said that the best form of plane-table, including a telescopic alidade, need not cost more than £30, and from that down to £5, for the simplest. With the first, angles can be measured and laid down with greater accuracy than with a minute-reading theodolite and the best protractor."

MISCELLANEOUS.

The preparatory works for the long-talked-of railway from Vladikavkaz across the Caucasus to the Transcaucasian system have now been commenced.

On Saturday, November 16th, British Bechuanaland was incorporated with Cape Colony. A strip of Khama's country 16 miles wide, running from the Mpakwi to the Maklutsi river, will be granted to the Chartered Company.

The proprietors of the *Dundee Courier* are sending out Mr. Wm. Blyth to inquire into the conditions of agriculture in Argentina, and especially to ascertain the circumstances which enable that country to send wheat to the British markets at so low a price. It is hoped that the information obtained may prove useful to British farmers.

Mr. and Mrs. Littledale have made an attempt to reach Lhasa. They travelled from Cherchen to Tengri-nor past the Garing-cho near Gagalinchen, discovered by Captain Bower. When within two days' journey from the capital of Tibet they were stopped by the Lamas, and were obliged to go back to Gagalinchen, whence they made their way by Rudok and Leh to Srinagar.

A new monthly publication has been started by the U.S. Weather Bureau, entitled *Climate and Health*, under the editorship of Dr. W. F. R. Phillips. It shows in tables and charts the barometric pressure, temperature, etc., with their departures from the normal ranges, etc., for each week, and the numbers of deaths from various diseases during the same periods, as well as the morbidity.

A road is to be made from the Caspian Sea to Tabriz and Persia, and the engineers appointed to survey the routes have decided in favour of one from Astara through Ardebil to Tabriz, as entailing less expense than if Lenkoran were chosen for the point of departure. It is, however, intended to construct a road along the coast from Astara to the island Sara, near Lenkoran, so that the roadstead of Sara may be used in stormy weather by vessels laden with goods for Persia.—*Geogr. Zeitschrift*, Jahrg. i. Nos. 10 and 11.

The German Commission for the exploration in the Antarctic met in Berlin on November 5th. It discussed a plan of action, and selected Kerguelen Island as the point from which an expedition should sail in two vessels towards the Pole. A meeting for the same object was also held by the Berlin Gesellschaft für Erd- kunde, and Herr von Payer, who was present at both meetings, was asked to abandon his Greenland voyage and take the command of the South Polar Expedi- tion. As the Viennese seem disposed to follow the lead of Berlin, a combined expedition is likely to be organised.—*Deutsche Rundschau*, Jahrg. xviii. Heft3.

A fully equipped meteorological observatory is now completed on the Brocken, 3742 feet above sea-level. It occupies a three-storied building at the north wing of the hotel. The construction and fitting up of the observatory has been super- vised by Dr. Assmann. The importance of the site lies in its position on the track of barometric depressions. Most of the cyclonic systems which cross the North Sea have the Brocken on their southern edge and Ben Nevis on their northern, and therefore it is highly probable that fresh progress in weather prognostications will result from the combined results of these stations.—*Deutsche Rundschau*, Jahrg. xviii. Heft 3.

A preliminary prospectus has been issued of The New Physical Atlas, prepared under the direction of Mr. J. G. Bartholomew, Hon. Sec. R.S.G.S., parts of which are now approaching completion. It is based on Berghaus' *Physikalischer Atlas*, and, as at present arranged, will consist of 120 plates, grouped into five volumes dealing with Geology, Orthography and Hydrography, Meteorology, Magnetism, Botany, Zoology, Ethnography and Demography respectively. Each subject will be under the direction of a special editor, those selected up to the present being Prof. James Geikie, Dr. John Murray, Sir Archibald Geikie, Dr. Buchan, Prof. C. G. Knott, and Prof. Bayley Balfour. Members of this Society will have the advantage of obtaining the volumes at a reduced price.

Mr. Theodore Bevan writes to protest against the discovery of the Purari river being ascribed to the Rev. James Chalmers. He was styled the discoverer of this river in an address delivered by Sir W. MacGregor in Edinburgh, and pub- lished by the London Missionary Society. Also in his *Report* for 1894 the Administrator spoke of Mr. Chalmers as "the original discoverer of the Purari outlets at the sea." Mr. Bevan does not deny that he was the first explorer to see these outlets, but quotes from the *Sunday at Home* for 1887 to show that Mr. Chalmers thought they belonged to the Fly river. In the same year, 1887, Mr. Bevan, as stated in the *Report* above cited, "first of Europeans, saw the Purari above the delta," and named it Queen's Jubilee river.

The U.S. Weather Bureau has issued a *Bulletin*, No. ii. Part 2, containing the *Report of the International Meteorological Congress held at Chicago, Ill.*, August 21-24, 1893. It is edited by Oliver L. Fassig, and contains the second instalment of the papers read at the Congress held under the auspices of the Congress Auxiliary of the World's Columbian Exposition. An interesting account of the rise and pro- gress of meteorology in the United States is given in a series of papers by such authorities as Professors Abbe, Langley, Waldo, and others. Mr. J. G. Symons,

F.R.S., contributes an important paper on English meteorological literature, 1337-1499, from which it appears that the earliest journal of the weather was one kept at Driby, Lincolnshire, by the Rev. Walter Merle, from 1337 to 1344. Most of the early works on meteorology described were of an astro-meteorological nature. Section v. contains six papers on Agricultural Meteorology, and Section vi. eleven memoirs on Atmospheric Electricity and Terrestrial Magnetism.

NEW BOOKS.

Ice-Bound on Kolguev: a Chapter in the Exploration of Arctic Europe to which is added a Record of the Natural History of the Island. By AUBYN TREVOR-BATTYE, F.L.S., F.Z.S., etc. With numerous Illustrations and 3 Maps. Westminster : Arch. Constable and Co., 1895. *Price 21s. net.*

The island of Kolguev, in the author's opinion, has never been connected with the mainland, which is separated from it by a channel about 50 geographical miles wide and less than 30 fathoms deep, but is a recent upheaval. In the northern portion, covering two-thirds of the whole area, the ground is higher, presenting peat-covered or bare ridges intersected by gullies, and enclosing small lakes and swamps. The remaining third consists of a dead flat covered with grass, bog, or peat, and reaching to the sea. The two highest hills have only an elevation of about 250 feet. They are essentially heaps of sand, terminating in peaks and ridges, and the winter snow was still on them on August 26. There are no traces of moraines, and nothing to suggest that the island ever supported a sheet of ice. Owing to shallow water round the island and bars at the mouths of the rivers, it is difficult of access, save to vessels of small draught, and then perhaps only at Sharok on the south-east coast.

During the three months spent at Kolguev, which the natives seem to know only as "the island," the author had some opportunity of studying Samoyede life. And he was fortunate from the very first in lighting upon a highly amiable family, far cleaner and more civilised than the average. He describes their method of netting wild geese on a large scale, and when he himself was present they captured no less than 3325 at one time. They still use bows and arrows occasionally, and make good practice at 40 paces. But they do not seem to use the forked arrow-heads and the bone guard for the wrist that their brethren on the mainland employ. Once he saw a Samoyede make an excellent finger-ring by pouring some white soft metal, melted in a hollow piece of wood, into a mould, which consisted of a groove cut round a stick ; but their bullet moulds are made of stone set in a pair of wooden scissors. At draughts Mr. Battye and his English companion were quite unable to hold their own, being invariably beaten by the natives.

The narrative gives much incidental information about the habits of the birds in those regions, for it was zeal in the service of ornithology that tempted the author so far afield.

An Account of Palmyra and Zenobia, with Travels and Adventures in Bashan and the Desert. By Dr. WILLIAM WRIGHT, Author of the *Empire of the Hittites,* etc. With 80 Illustrations and 32 full-page Illustrations. London, Edinburgh, and New York : Thomas Nelson and Sons, 1895. Pp. xviii + 394. *Price 7s. 6d.*

Dr. Wright has produced an extremely vivid and humorous picture of Oriental travel. The greater part of the book is occupied by a description of the visit which the author, along with Hon. Mr. Berkeley and others, paid to Palmyra in

1874, and this is supplemented by less interesting notes on a ride through Bashan and the country of the Druses, in which Dr. Wright was accompanied by Dr. Harper. The route chosen from Damascus to Palmyra was well to the north, past Saidenaya, where the miraculous picture, painted by St. Luke, continues "to distil a fluid very efficacious for eye complaints"; Maloula, where you can still drink the "wine of Helbon" mentioned by Ezekiel; Yabrond, where the cultivation of madder has been depressed by the discovery of aniline dyes; and Karyetein, which is apparently the Hazar-enan of Scripture (Num. xxxiv. 9-10). One-fifth of the inhabitants are Christians, being Syrian Jacobites. Dr. Wright picks up a number of interesting local words on the road. Thus: *wa'al* = ibex, *bint* = girl, *seil* = dry bed of torrent, *dairagi* = curlew, *dabbous* = shillelagh, *khowieh* = blackmail (ironically called "brotherhood") paid by the poor Syrian peasants to the Bedouin robbers. Turkish misrule is flagrant—excessive and arbitrary taxation, and no protection given in return. "What the Turkish caterpillar leaves the Bedawi locust devours." But the feature of Dr. Wright's book is his artistic sketch of the ruins and the noble architecture of Palmyra, the Grand Colonnade, the Temple of the Sun, the Tomb Towers, the Temple of the King's Mother (belonging to the time when Tadmor became Hadrianopolis), and many beautiful arches, doorways, ceilings, capitals, etc. The illustrations of these are finely executed. Dr. Wright also describes the Ephca, or great warm underground stream, and claims to have discovered a scarab of the time of Tirhakah (B.C. 688). It is still possible to distinguish the early stone-work of Solomon's time from later work. A few chapters are devoted to the profoundly interesting history of Palmyra, with the great dramatic figures of Valerian, Longinus, Odainathus, Sapor, and the brilliant Zenobia (otherwise called Sitt Zeinab and Bath-Zabbai). The political and commercial position of the city, between Persia and Parthia on the one hand and Rome upon the other, gave it peculiar strength. The policy of the buffer State was always important to both sides, and what Pliny calls "an almost impassable desert" protected it from hostile aggression, and enhanced its value as a shelter on a great trade route.

A Visit to Bashan and Argob. By Major ALGERNON HEBER-PERCY. With Prefatory Note by Canon TRISTRAM. London : Religious Tract Society, 1895. Pp. 175. *Price 7s. 6d.*

This is a pleasant account of a brief visit made by Major Percy, and his wife and sons, from Damascus to Bosra and Salcah and back. Except so far as colonised by the Druses, the land of King Og seems still to be in a disturbed and dangerous condition. The route chosen was along a line between Trachonitis, the Argob of the Old Testament, now known as the Lejah (which appears to consist of rocky ravines at the edge of a volcanic field), and the Ardal Bathanieh, one of the richest districts in Syria. Burckhardt and Dr. Porter have been the chief explorers. Some years after the Lebanon massacre of 1860, the Turks suppressed all the Christian schools in Bashan. Major Percy's photographs of the ancient Cyclopean, and the more recent classical and Christian, architecture are excellent and interesting, especially those of the towers and walls at Bosra, Salcah, and 'Ayun, and of a fine doorway at Deir Nileh.

Rambles in Japan, the Land of the Rising Sun. By H. B. TRISTRAM, D.D., LL.D., F.R.S., Canon of Durham. London : The Religious Tract Society, 1895. Pp. 304.

There is a great family likeness among books written on Japan by the passing tourist. There are the inevitable first impressions, beyond which some never get.

There are the curio shops, the temples, the glimpses of open-air life, the earth-quakes, the Forty-Seven Ronin, and the "wonderful progress" made by the Japanese in Western civilisation—all affording too ready a text for the *soi-disant* observant stranger. No doubt it is impossible to write a book on Japan without touching on these everlasting themes ; and possibly every book so written has established its *raison d'être* by being the eye-opener of at least one eager intelligent mind. In Canon Tristram's modest book, which is largely a transcript of his private journal, we meet of course with the oft-told tales ; but we also find a great deal of fresh and original matter which makes the book a welcome addition to Orient literature. In the accomplishment of his primary object of studying the work of the Church Missionary Society, he visited parts of the country lying quite outside the usual tourist beat ; and to the elucidation of all he sees he brings the eye of the naturalist and the training of the investigator. The numerous botanical and zoological notes, for which Canon Tristram need hardly have apo-logised, give a special interest to a book so well written in other respects. Of the ten chapters the most interesting geographically, because the freshest, are probably the last three, which treat of *Shikoku*, the *Island of Kyushu*, and *Aso San and the Geysers of Yunotan*. To the reader interested in the Christianisation of Japan, and especially in the work of the C.M.S., Canon Tristram's *Rambles* will of course appeal most strongly. A good feature of the book is its accuracy and freedom from foolish generalisation or wild panegyric in regard to things Japanese. We have noticed only two mistakes—curiously enough, within two pages of one another. The Central Tabernacle, in which Canon Tristram delivered before a large audience of educated Japanese a lecture on Egyptian Discoveries, belongs not to the American Episcopal Methodists (as he puts it), but to the Canadian Methodists, and was built and organised almost entirely through the exertions of one man. Then, after giving a list of the various mission bodies in Japan, Canon Tristram concludes : "At this time I do not think there are any British Nonconformists." As a matter of fact, there were and are both English Baptists and Scottish United Presbyterians. The book is enriched with forty-five illustrations of subjects bearing on Japanese scenery, architecture, modern customs, antiquities, and religions.

Reise nach Innerarabien, Kurdistan und Armenien, 1892. Von Baron EDUARD NOLDE. Braunschweig : F. Vieweg und Sohn, 1895. Pp. xv + 272.

The author of this volume died, after an adventurous life, in May last in London by his own hand. He fought with the Carlists in 1872, and with the Chilians against Peru in 1877. After 1888 he spent most of his time in travel, principally in Africa and the East.

The first part of this narrative was printed in *Globus*, and a summary was given in the June number of the *Magazine* (p. 308). After penetrating farther into the interior of Arabia than any other European, except Palgrave, Baron Nolde travelled northwards through Bagdad, Mosul, and Bitlis to Trebizond. He has much that is interesting to say about camels and the native inhabitants—the Kurds, Armenians, and Hamavands—and is able to point out some errors in Kiepert's map of the country north of Jezireh-ibn-Omar.

A pleasing portrait of the traveller and a sketch map of the country he traversed are annexed.

Charles Lyell and Modern Geology. By Professor T. G. BONNEY. London : Cassell and Co., 1895. 8vo. Pp. 224.

In this well-written volume Professor Bonney gives an interesting sketch of the life and labours of Lyell, and of the great influence which his published work has had upon geology. In the compass of a few pages the author has succeeded in outlining not only the life-work of his hero, but in presenting his readers with an excellent *résumé* of the rise and progress of geographical science. There is nothing in the book with which geologists are not already familiar, but it is well to have the facts of Lyell's career, and the methods he pursued as an observer, so clearly set forth. The lesson these teach is one which no student of science can neglect. Lyell was not only an excellent and laborious observer and an admirable exponent, but to the end of his long life he kept his mind open to new impressions. At an age when most men have retired from work and become more or less averse to, or incapable of, following the later developments of their particular line of study, Lyell continued to show all the zeal of his earlier years—keeping pace with the advance of his science and modifying and expanding his views to the end. He never became fossilised.

The Story of the Sea. Edited by Q., assisted by Prof. J. K. LAUGHTON, H. O. ARNOLD-FORSTER, M.P., W. LAIRD CLOWES, HERBERT W. WILSON, etc., etc. London, etc. : Cassell and Co., 1895. Pp. viii + 760.

Many stories might be written of the sea, and therefore the title does not give much information about the contents. We find that it is the story of man on the sea, containing narratives of great naval fights, shipwrecks, and other disasters, and descriptions of ships, guns, lighthouses, lifeboats, signals, etc. The information is useful and interesting, and the illustrations are, as a rule, good and instructive. The book is a storehouse of facts, which will be of some use for reference, and it will be an excellent Christmas present for young people. A more systematic arrangement would, we think, have been better. The development of navigation might have been traced up from its infancy, instead of which chapters on Great Naval Disasters, Duels on the Sea, Perils of Ice and Flame follow one another without any rational sequence, Early Methods of Naval Warfare coming, not first, as one would expect, but in the tenth chapter.

De Garebegs te Ngajogyàkartà. Door J. GRONEMAN. Met Photogrammen van Cephas. 's-Gravenhage : Martinus Nijhoff, 1895. Pp. 87.

In this volume the author describes fully the ceremonies observed at Ngajogyàkartà in Java, beginning with the festivals which precede the birthday of Mohammed, and then taking the others as they follow one another in the Javanese year. As he remarks, these will one day have passed away, and then this work will have even greater ethnological interest than at present. The musical arrangements, processions, weapons of the life guard, etc., are described in great detail ; and the numerous photographs by the court photographer assist the text in producing a clear picture of the scenes. The constant use of Javanese words is a decided drawback, but one, perhaps, not easy to avoid, as they describe things or persons to which there is nothing corresponding in European countries. In publishing this volume the K. Institut voor de Taal-Land-en Volkenkunde van Nederlandsch-Indië, as by other publications, deserves the thanks of ethnologists for preserving records of native customs which are destined soon to be modified at least, if not to pass away altogether.

Glimpses of Peebles ; or, Forgotten Chapters in its History. By Rev. ALEX. WILLIAMSON, F.S.A. Scot. With Illustrations. Selkirk : George Lewis and Co., 1895. Pp. xvi + 322. *Price 4s. 6d.*

There is not much that is of interest or value to the geographer in this volume, unless it be regarded as a record of the social and material changes which a typical provincial town of Scotland has undergone during the last one hundred and thirty years. The aim of the author,—himself a Peeblean—has been to put the present generation of natives in possession of scenes, facts, and characters which are fast passing out of recollection ; and his tasteful volume will no doubt serve this purpose very well. The narrative is perhaps apt to run too much in the ecclesiastical groove.

Rifle and Spear with the Rajpoots: being the Narrative of a Winter's Travel and Sport in Northern India. By Mrs. ALAN GARDNER. Illustrated. London : Chatto and Windus, 1895. Pp. xvi+336.

The authoress travelled with her husband on a sporting tour in India during the cold weather of 1892-93. They first went to Kashmir, whence they descended through Chamba to the Salt Range of the Punjab ; and then, having been the guests of Kapurthalla and Patiala, made their way through Delhi, Dholpur, and Muttra to Rajputana. The writer of this brief notice is well acquainted with much of the ground, and can testify to the accuracy of the descriptions of the country, the people and the sport. The authoress had to endure many discomforts and even hardships, but her pages sparkle with fun and good humour, and her book is altogether delightful reading.

Voyage en France. Par ARDOUIN-DUMAZET. *3ème Série, Les Îles de l'Atlantique— d'Arcachon à Belle Isle.* Pp. 314. *4ème Série, Les Îles de l'Atlantique— d'Hoédic à Ouessant.* Pp. 318. Paris et Nancy : Berger-Levrault et Cie, 1895. *Price 3 fr. 50 c. each vol.*

These are the third and fourth volumes of a series already favourably noticed for their valuable features as topographical handbooks. The author deals with the country in its natural divisions, and, in a series of letters, written on the spot, sketches all the leading features of interest to the traveller. They are cleverly written, and being the direct impressions of a keen and skilled observer, have a force and originality seldom met with in such works. The present volumes are illustrated with numerous cuttings from the *Carte de l'État-major*, which are a valuable addition.

North-Western France (Normandy and Brittany). By AUGUSTUS J. C. HARE, Author of *Paris, Walks in Rome,* etc. London : George Allen, 1895. Pp. 410. *Price* 10s. 6d.

We have on former occasions commended Mr. Hare's descriptive guide-books to districts of France for their thoroughness, their excellent method, their fine literary flavour, and their delicately beautiful illustrations. We can confidently repeat concerning his *North-Western France* all that was said in praise of previous volumes. It is as far above an ordinary guide-book as a history by Froude or by Gardiner transcends a commonplace compilation. That is to say, it bears the stamp of the author's individuality, and has all the value and all the charm of an original creation. This volume is specially interesting to Englishmen, as it treats of Normandy, the home of a long line of English kings, and of Brittany, which was the Lesser Britain of an earlier age. How valuable would be a series of detailed topographical descriptions of the British Islands treated on similar lines, and on the same scale !

The Statistical Year-Book of Canada for 1894. Tenth year of issue. Issued by the Department of Agriculture. Ottawa : Government Printing Office, 1895. Pp. 1134. Many tables and sketch-map.

A statistical year-book so well known as this does not lend itself to detailed review. A new undertaking may be criticised, but after the lapse of years the form of the publication becomes stereotyped and only changes of minor importance occur. So it is with the Year-Book of Canada ; still we must commend the care with which it is compiled. It is well up to date in all respects. The new features are a chapter on Railways, an Account of the United States of America, being the second of a series on "Countries with which Canada does business," and a chapter devoted to some of the outstanding events of the year. The chapter on the Early History and that on the Constitution and Government have been expanded. The map shows the British Possessions and the principal commercial routes.

A digest of the mining laws is given in connection with the chapter on Minerals, and should prove of use.

Canada, we are proud to notice, seems to flourish ; long may it do so !

Stanford's Compendium of Geography and Travel (new issue). *Africa: Vol. II., South Africa.* By A. H. KEANE, F.R.G.S. London : Edward Stanford, 1895. 11 Maps and 92 Illustrations. Pp. 655 and Index. *Price* 15s.

It was only in our June number (p. 324, 1895) that we noticed, more briefly than we could have wished, the first volume of this work. We gave it great praise, and this, the second and concluding volume is, we are glad to remark, as good, if not better, than the first. Both author and publisher may well be content with their endeavours. The numerous maps and nearly all the illustrations are excellent.

We need not dwell on the points we previously mentioned, viz., that this is a completely new work, and no mere reprint or revisal of the old well-known Stanford's *Africa.*

The author has had very great difficulty in keeping pace with the kaleidoscopic political changes which are taking place in Africa, and yet he has been remarkably successful, though his task was not to be envied, for pages often had to be either modified or rewritten whilst the book was passing through the press.

Geography in its widest sense has been dealt with ; prominence is given to history, political questions, and ethnology—and one great feature of the book is that the real student can from its pages refer to the original sources of information, which indeed will be often needful, as the necessary compression is very great.

The term "South Africa" is a little elastic, for it includes Cameroons, all the African islands, French Equatorial Africa, etc., as well as the "South Africa" of popular conception.

The author has been very happy in his treatment of political questions, and has held the balance even and shows no bias.

The description of the various races is terse and vivid, and gives at a glance, so to speak, a word-picture of varied tribes, a task of no little difficulty, and one which shows by its success the grasp which Mr. Keane has of his vast subject. The "prospects" of the various areas dealt with are perhaps hardly so well conceived ; a little more space was wanted, for the condensation is too great for the average reader to form a just conclusion, we think. For instance, dealing with the prospects of British East Africa (p. 592), we find a page of generalities, which hardly give the anxious inquirer what he wants to know ; but this may, perhaps, be hypercritical.

In conclusion, no book could be better adapted for giving a general and comprehensive knowledge of Africa, and, as has been indicated, whosoever requires further light has his path made plain for him.

Deutschlands Kolonien, ihre Gestaltung, Entwickelung und Hilfsquellen. Von ROCHUS SCHMIDT. 2nd vol. With over 100 Illustrations and 6 Maps. Berlin : Verlag des Vereins der Bücherfreunde (Schall und Grund). N.D. Pp. 430 and Index.

We had occasion to favourably review Mr. Schmidt's first volume (see p. 381 of vol. xi.). Although he is not personally acquainted with the districts dealt with in this volume, he has been careful to select his authorities with care and discrimination. He deals with the German colonies of Cameroons, Togoland, and the South Seas. An accurate and readable description is given of the districts in question and their capabilities. The illustrations are good and' the maps fair. We gather from a perusal of the pages that the development of these colonies is progressing rapidly, and it is expected that ere long they will prove of considerable value to the German Empire.

Handbook for Asia Minor, Transcaucasia, Persia, etc. Edited by Major-General Sir CHAS. W. WILSON, R.E., K.C.B. London : John Murray, 1895. Pp. 416. With Maps.

The general arrangement of this work is the same as in Mr. Murray's other well-known Handbooks, *viz.*, in Routes. By far the greater part of the book is devoted to Asia Minor, a region with which the editor became personally familiar during his Consul-Generalship in that country. The general results of his experience and observations are condensed into an excellent introduction which occupies the first 88 pages. The routes are for the most part the work of the military officers who served under him as Consuls and Vice-Consuls in different parts of Asia Minor. The remainder of the Handbook, about a fourth of the whole, relates to portions of Syria, Chaldea, Transcaucasia, the Turkish Islands, and Persia.

The name of the editor is a sufficient guarantee that the whole work of compilation has been done in a careful and scholarly manner. In such a work, however, absolute accuracy in every detail is practically unattainable, even when, as in the present instance, the greatest pains have evidently been taken to come as near perfection as possible. Mistakes like *Pat-i-tak*, p. 326, for *Pah-i-tak*, and *Dasht-i-Arzen*, p. 334, for *Dasht-i-Arjen*, are probably due to indistinct handwriting. Errors of this kind would either be neutralised or eliminated, and the value of the work as a book of reference greatly enhanced, were the Eastern proper names printed in Arabic as well as in Roman letters, and the proof-sheets revised by competent natives of the countries concerned. The trouble and expense would of course be considerable, but not, we believe, disproportionate to the advantage. The mis-statement at p. 341 that the important town of Kashan was "recently (1895) almost destroyed by earthquake," is no doubt due to an error in the transmission of a telegram that appeared in the London newspapers, in which *Kashan* was erroneously substituted for *Kuchan*, a small town on the Turcoman frontier. The book is furnished with good maps of Asia Minor, but, strangely enough, with none of Persia, beyond what appears in the small scale general map of Turkey in Asia. The use, in the Persian part of the Handbook, of the Turkish word *Khan*, instead of the Persian word *Caravanserai*, is somewhat inappropriate, as it would not be understood by Persian-speaking people. From a praiseworthy desire to make the Handbook as portable as possible the inner margin has unfortunately been unduly narrowed, thereby making it difficult to read to the end of the lines.

All these remarks, however, refer to matters of comparatively slight importance. In all essential particulars the Handbook is an excellent one, and cannot fail to be of great practical use to travellers in the extremely interesting countries to which it is a guide.

Thorough Guide Series : Scotland, Part I.—Edinburgh, Glasgow, and the High-
lands, etc. By M. J. B. BADDELEY, B.A. 45 Maps and Plans by Bartholomew. Eighth Edition, thoroughly revised. London : Dulau and Co., 1895. Pp. xxxii + 334.

The new edition of Mr. Baddeley's *Thorough Guide* to Scotland (Part I., including the Highlands) contains many important additions and corrections. When compared with the third edition—the latest in the Society's library—it shows an addition of upwards of 50 pages to the letterpress, and of 15 maps and plans. The latter include maps or plans of Stirling, of Bridge of Allan, etc., of Perth, of Crieff and Lochearnhead, of Dunkeld, Pitlochry, and Aberfeldy, of Kingussie and neighbourhood, of Grantown and neighbourhood, of Inverness, of the West Highland Railway, of Loch Awe, and of Staffa and Iona. There are also view-maps from Ben Nevis and Ben Lomond, and hotel plans of Edinburgh and Glasgow. At the same time the text has undergone a very thorough revision. Every one who has had occasion to consult Mr. Baddeley's guide-books knows how largely their value depends on the "thoroughness" of the author's personal examination of the routes traversed and described. Indeed his acquaintance with the by-ways of travel is amazing, and the present edition of "Scotland" surpasses all its predecessors in accuracy and exhaustiveness of detail. The whole work is excellently planned, and successfully and suggestively worked out. Two pre-liminary sections, treating respectively of cycling routes and golf courses,—the latter numbering nearly one hundred, of which details of fees, membership, number of holes, and other particulars are added—are of great interest and practical value. Yet the golf record is not quite complete, as it makes no mention of the Barnton course, the Duddingston course, the Gullane course, the North Berwick course, the North Queensferry course, and several others that might have been mentioned. As far as it goes, however, the list is useful and interesting. The general and index map of Scotland is contained in a pocket at the end of the volume, and that is a distinct convenience. All the maps and plans are carefully engraved and clearly printed.

Hazell's Annual for 1896 : *a Cyclopædic Record of Men and Topics of the Day.*
Edited by W. PALMER, B.A. (Lond.). London : Hazell, Watson, and Viney, 1896. Pp. 678.

The changes in this year's issue, the eleventh, are of the usual kind. There are articles on the prominent questions of the day, and biographies of eminent men. Among the latter one would expect to find M. Pasteur, and there should, we think, be a reference to the sketch published in last year's volume. In the article on Geographical Progress, Captains Decoeur and Toutée are referred to, but Captain Lugard is not mentioned. The explorer of the Sahara is called Fourneau, instead of Foureau ; M. Fourneau explored the Sangha. On the whole the volume is well edited, and contains a mass of valuable information. The only wonder is that in such a vast collection of the most varied details so few errors should have crept in.

Scotland: Picturesque and Traditional. By GEORGE EYRE-TODD. London :
Cassell and Co., 1895. Pp. 320. *Price 6s.*

In this beautifully illustrated volume Mr. Eyre-Todd describes an unconventional journey, undertaken in the autumn of 1894, between Melrose and Inverness, the author progressing with map and walking-stick, and exploring the main interests of central Scotland. In his narrative the author displays literary grace as well as historical and archæological erudition, whilst the lavish manner in which the work is illustrated should make it one of the most attractive yet published in Scotland. Of course so wide a subject excuses a few slips, as when the author observes that "Sandyknowe farmhouse remains much as it must have stood in Scott's time" : for, had he read the last published *Transactions of the Berwickshire Naturalists' Club*, he would have found evidence that not a single stone of Scott's grandfather's farmhouse has been left standing. He is also somewhat bold in making Burns the literary progenitor of Goethe, seeing that the latter had formed his style and influence before Burns had published anything. We are afraid that, in this instance, Mr. Eyre-Todd is *plus Écossais que les Écossais.*

Pictorial New Zealand. London : Cassell and Co., 1895. Pp. 301. *Price 6s.*

In his Preface to this beautifully illustrated and most attractive volume, Sir W. B. Perceval, Agent-general for New Zealand, remarks :—"The material progress of the colony, many evidences of which are afforded in the following work, cannot fail to favourably impress the visitor (to New Zealand). He must never forget that all that he sees has been accomplished during the reign of Queen Victoria, and that before then New Zealand was only known to a few whalers and missionaries. The twenty million sheep supplying fresh mutton for the English dinner-table, the two hundred butter factories, built on the most modern and scientific plan, from which cheese and butter of finest quality are distributed in Europe ; the ten million acres of land under cultivation, the large area of magnificent forests, immense coalfields, minerals of every kind, a prosperous, energetic, and happy people supplying their own wants, and exporting surplus products to the value of £10,000,000 per annum—all this, and a great deal more, will convince him that New Zealand affords a most striking example of successful colonisation, and that her immense and varied resources, which are only now beginning to be developed, must infallibly make her—and this in the lives of many of the readers of this work—the most prosperous and powerful colony of our empire." We may add that this splendid prophecy will only be realised provided the financial institutions of New Zealand acquire a solidity which they have not yet attained.

Travel and Adventure in Northern Queensland. By ARTHUR C. BICKNELL.
London : Longmans, Green, and Co., 1895. Pp. 219. *Price 15s.*

The author informs us that in the year 189— he undertook an expedition to Queensland, and in this work he gives an account of his travels, camp life in the bush, and the manners and customs of the natives whom he observed. The result is a most interesting volume, illustrated with numerous engravings from sketches by the author, whose portrait is given in an admirable photogravure. In his chapter on cannibalism he recounts a conversation he had with a Queensland Black, whom he asked what he liked best to eat ? He replied, " Human flesh." " You mean," said Mr. Bicknell, " the flesh of the white man ?" " No," was the answer, " I mean black man. White man too salt—terrible nausea." The author concluded he had eaten both, and subsequently learnt that whilst the Blacks never eat the head of a foe, they consider the kidneys the greatest delicacy.

NEW MAPS.

ASIA.

CHINA, Vertheilung der Niederschläge in ——, in den einzelnen Jahren 1885-1894, zur Darstellung der Wetterverlegung.

Deutsche Rundschau, Jahrg. xviii. Heft 3.

Compiled by H. W. Krebs to show the northward movement of drought, etc., in successive years.

AFRICA.

CENTRAL AND SOUTH AFRICA, Bartholomew's New Map of ——. *Price 2s.;*
mounted on cloth 3s.   *Edinburgh Geographical Institute*, 1896.

This excellent map has appeared annually for several years past. No very striking changes have taken place in the past twelve months, but the map shows the latest alterations, notably in South Africa. In this issue the gold-fields of Mashonaland and Matabeleland are marked.

ASHANTI EXPEDITION, Map to illustrate the ——, in West Africa. *Price,*
paper 6d.; cloth 2s.   ·   *W. and A. K. Johnston, Edinburgh.*

A useful map at the present time, containing most of the principal places in the Gold Coast Colony and Protectorate. It might have been extended a little, up to the eastern and western boundaries.

ATLAS.

SCHWEIZ, Topographischer Atlas der ——, im Massstab der Original Aufnahme, nach dem Bundesgesetze vom 18 Dezember 1868 durch das eidg. topogr., Büreau gemäss den Direktionen von Oberst Siegfried veröffentlicht.

Lief. XLV :—

No. 382	Isenthal.	No. 437 *bis*	Bouches de la Dranse.
„ 420	Ardez.	„ 433	Begnins.
„ 430	Les Plats.	„ 474	Vouvry.
„ 434	Bière.	„ 474 *bis*	Pas de Morgins.
„ 436	Aubonne.	„ 514	Locarno.
„ 436 *bis*	Rolle.	„ 537	Brissago.

CHARTS.

DE GUINEA EN EQUATORIAL STROOMEN, voor iedere Maand afzonderlijk bewerkt volgens de gegevens von 2900 Journalen gehouden aanj boord van Nederlandsche Schepen.

H. G. Bom, Zeevaartkundige Boekhandel, Amsterdam, 1895.

The above title indicates in general the contents of this collection of charts, and the number of logbooks from which the data are taken shows that the compilation has a sure foundation. Not only the currents, but the temperature, winds, etc., are included, and various marks are used to distinguish strength, direction, etc. The explanatory letterpress is, fortunately, printed in French as well as Dutch. The work is published under the auspices of the K. Nederlandsch Meteorologisch Instituut, and does credit both to the compiler and printer.

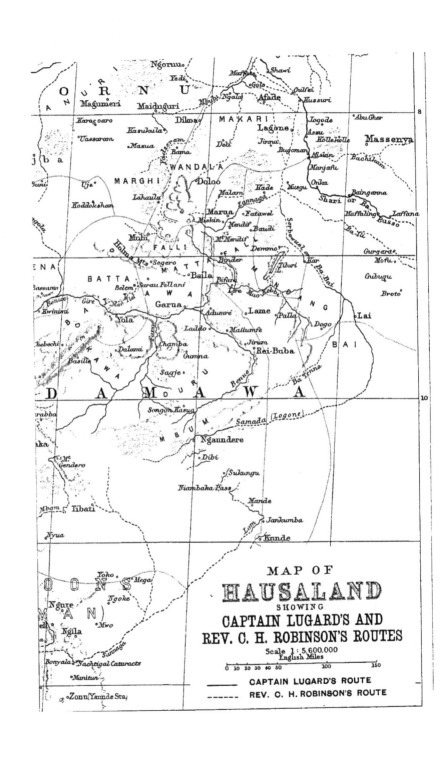

MAP OF
HAUSALAND
SHOWING
CAPTAIN LUGARD'S AND
REV. C. H. ROBINSON'S ROUTES

Scale 1 : 5,600,000
English Miles

0 10 20 30 40 50   100   150

———— CAPTAIN LUGARD'S ROUTE
------ REV. C. H. ROBINSON'S ROUTE

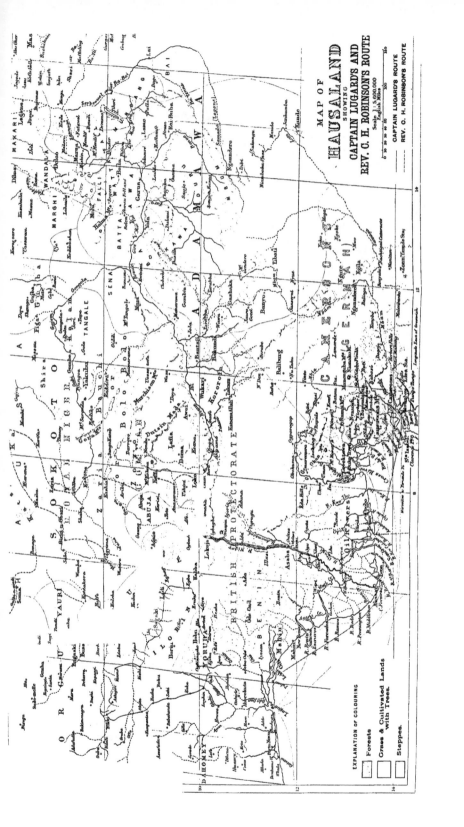

MAP OF
HAUSALAND
SHOWING
CAPTAIN LUGARD'S AND
REV. C. H. ROBINSON'S ROUTE

Scale 1:5,600,000
English Miles

CAPTAIN LUGARD'S ROUTE
REV. C. H. ROBINSON'S ROUTE

EXPLANATION OF COLOURING

Forests

Grass & Cultivated Lands
with Trees.

Steppes.

THE SCOTTISH

GEOGRAPHICAL

MAGAZINE.

THE TOWNS OF NORTHERN MONGOLIA.

By DR. ANATOLIUS MARKOFF.

(Read at a Meeting of the Society in Edinburgh, on January 9th.)

MONGOLIA seems a very long way off to the ordinary Englishman, and a proposal to make a trip of discovery thither would probably strike him with more dismay than an invitation to go lion-hunting in Africa. Now, in Russia at the present time, the country is regarded in a different light. Mongolia is being opened up, and Russians of to-day look upon it as a region worth visiting, and its antiquities as worthy of study. We shall therefore soon know more about this neglected part of the world, and as a student of Mongolia myself, and of its language and literature, I expect very important results from the scientific investigations now, and about to be, carried on.

The great fault of most of our existing sources of information on Mongolia—books, pamphlets, etc.—is, that every traveller follows unquestioningly in the footsteps of his predecessors, and generally repeats what has been said before, so that all descriptions of Mongolia read like so many compilations—the later from the earlier. However, now some progress is being made, and as savants are taking an interest in the country, we shall not, it is to be hoped, have to lament much longer this unsatisfactory state of affairs.

Another curious fact which may be noticed is, that when an author does enter into a description of a Mongolian town—a rare occurrence— it is generally a collection of personal impressions, loosely strung together, and in the ordinary globe-trotter's style, with its "I saw this" and "I saw that," "I went to such a place," and so on *ad infinitum.* What comes under his nose he writes down, and fills in the gaps from other authors. The result is, naturally, that we find much diffuse writing in such descriptions, but little information. The inner life of Mongolia, the history of its towns and people—in fact, all the essential characteristics of the country—are waiting till some skilful diver plunges beneath all racial prejudices, and studies and sees the whole social system through Mongol literature, and with the eyes of a Mongol. When that time comes we shall know something really worth knowing of this country. Most of us have not yet reached that point of philosophical perfection when we can, as Burns says—

"·See oursel's as others see us";

and when a European sees, say, the ordinary portable house of the Mongol, always ready for moving bodily, if wished, he is apt to think of the European way of moving and its accompaniments—Pantechnicon vans, railway trucks, broken crockery, beer-money, and all the traditional rites and customs thereof. The Mongol method, therefore, is carefully noted down by him as being "peculiar," "strange," "funny," and such like, whereas it might have occurred to him, if he had viewed matters from an impartial standpoint, that perhaps the Mongol method was as reasonable as, or more reasonable than, the European. It is true that I am going to give you a few personal impressions myself; but as I have lived with Mongols, and am therefore, if I may say so, somewhat tinged with the Mongol spirit, my view of matters will not be that of a counsel for the prosecution, but impartial—at least, to the best of my ability. In my paper to-day I shall confine myself to the towns of Northern Mongolia, or Khalkha, in their order from east to west—Urga, Uliasutai, and Kobdo—which I visited in 1889. The expedition to which I was on that occasion attached had no scientific object, but was one of those commercial expeditions which are often despatched by rich Russian merchants to inquire into the markets of Asia. Much money is expended on them, but the result fully compensates for the outlay. By this means a network of trade routes is extended over Eastern Asia and Persia. These parties are generally accompanied by a linguist, and, as I had just passed my examination before the Russian Board of Examinations, I was engaged in that capacity on the occasion referred to. My travels extended over a period of five months. And now to my narrative.

URGA.

I will not trouble you with a detailed description of the route taken, but will pass on at once to Urga, the chief town of Mongolia (47° 54' 8" N. lat. and 106° 57' 22" E. long.), which is situated at the confluence of the rivers Tola and Selba, 180 miles south of the Russian frontier town Troitskosavsk, generally included under the name of its close neighbour, Kiakhta.

The Mongol name "Örgö" (pronounced Urga by Europeans), which means literally a palace, or a tent of some high personage, is unknown to the natives. They call the place Da Khurie or Yekhe Khurie, that is, "The Large Monastery." Its proper name is Rebun Ghejilin.

Urga has a twofold importance; firstly, as a commercial centre established as far back as the seventeenth century, and secondly, since a more recent date, as the religious capital of Mongolia.

At the present time Urga consists of three distinct towns : (1) the monastery where the Khutukhta, or Primate of Mongolia, resides; (2) Gandan, in which are the temples of the Tsanit, and whither students resort to study Buddhistic theology with a view to admittance into the ranks of the Lamas; and (3) the Maimachen, or business quarter. All three are situated at a distance of about 1½ miles from the right bank of the Tola. The Selba flows through the monastery from north to south, but does not approach the other parts of the town. Urga lies in the midst of a great valley, 18 to 20 miles long from east to west, and 4 to 8 miles broad from north to south. Mountains are visible on all sides. To the south rises the majestic Bogdo Ula; to the east Bain Kutul; to the north Chingiltu Ula and Dalan Dabkhur, with the Kundui pass between them ; and to the west the Sanwin Ula, separated from the Bogdo Ula by the river Tola.

Of these the Bogdo Ula is the most important. It is held in great veneration by the Mongols, as might be inferred from the meaning of its name—"Holy Mountain." It is clothed from top to bottom with splendid pine forests, and is noteworthy as marking the boundary of forest vegetation ; no trees grow to the south of it. Bogdo Ula has been held sacred from ancient times, and now it is forbidden to fell trees on it, to hunt in its forests, or plough its slopes. Consequently, it abounds with animals of all kinds, birds and big game. Capital punishment may not be inflicted within sight of the mountain, and therefore all condemned criminals are sent to Dolon-nor or Kalgan. All mountains, indeed, are objects of veneration to the Mongols, especially the high ones ; but Bogdo Ula has a further claim to distinction as the reputed birthplace of Temuchin, or, as he was afterwards called, Chinghis Khan, to whom a sacrifice is offered up every year at its foot.

There are no sufficient data respecting the population, nor is it easy to estimate its number, as it fluctuates considerably. As regards the Lamas, however, it is certain that there are about 12,000 in the town. They are divided into 28 *aimaks*, each constituting a distinct parish with its own temple, round which the houses are grouped. In the centre of all these parishes stand the palace of the Khutukhta or Gheghen; the cathedrals *Tsokchin* and *Duchin Galabyin Sumê*, in which all the inmates of the monastery, irrespective of their parochial domiciles, pray together ; the temples belonging to the various colleges of (*a*) astronomers, (*b*) students of the *Tarni*, and (*c*) of medicine; and lastly the temple *Maidari* and the temple of Abatai Khan.

All these temples and colleges are immensely rich, for the Mongols have been, and still are, exceedingly pious, and freely give large donations to their religious institutions in the belief that they will thereby ensure a happier existence after death than they enjoy in this life. Moreover, these offerings confer a right to certain privileges. When a Mongol comes to Urga he generally puts up at the college to which he has given a donation, and one of his sons is received as a Lama into the same college, for as a rule one member of a Mongol family joins the ecclesiastical order.

The temples, to which all who are acquainted with Mongolian life and customs are admitted, are divided into two parts—the holy place and the holy of holies, or, as they are called here, the "Temple of Services" and "the Residence of the Sacred," in the latter of which stands an altar.

In the cathedral all the Lamas meet together four times a year: on New Year's Day, on the anniversary of Shighemuni's first announcement of the Four Truths, when a commemoration sermon is preached, and the Buddhist bible, Ganjur, is read ; on the Maidari festival ; and on the day when a sacrifice is offered to the Khutukhta.   The Lamas are summoned together by a shell and horn blown from a kind of minaret called *Burein-shata*.   The Tsokchin has a very rich treasury, which is only surpassed by that of the Khutukhta.   The various ecclesiastical vestments, books, and church vessels in the temples and colleges may be numbered by tens of thousands.

We will now pass on to the individual temples.   The Duchin Galabyin Sume, one of the finest, is built in a rich Tibetan style of architecture, and has a gilded cupola hung about with innumerable silver bells, which are set in motion by the slightest breath of air.   As this cathedral is quite inaccessible to non-Buddhists, nothing is known of its interior.   The services are conducted by Tibetan monks.   There is a general assembly of the Lamas in this temple only on special occasions, such as the death of the Khutukhta or on Maidari-day.

Of the other temples two are worthy of special mention—the Barun-örgö, or chapel of Abatai Khan, who introduced Buddhism into Mongolia, and the Maidari temple.   The former is scarcely a temple properly speaking, for it is not furnished as such.   There are no seats or mats for the Lamas, and only three *burkhans* (idols).   It is, in fact, simply a very old *yurta*, or Mongol tent, which, according to tradition, belonged to Abatai Khan.   In it are preserved ancient relics, and it serves as a museum of Mongolian archæology.   The idols are clumsily carved of wood, and have hideous faces—whether to frighten the worshipper into giving further donations, I cannot say.   There is also an ancient throne surrounded by six figures representing Mongol heroes contemporary with Abatai ; and every Mongol who enters the temple lays his head on the shoulders of the figures, believing that he thus imbibes some of their strength.   On the walls around are hung ancient weapons, pieces of armour, arrows, bows, sabres, etc.

The Maidari temple, the largest and loftiest of all, was built between 1820 and 1836.   The *burkhan*, or idol, is a typical example of Chinese art.   It is a figure seated on a throne ornamented with lions, and is thirty feet high.   The material is brass heavily overlaid with gold.   The figure is hollow, and is filled with leaves on which the prayers of the faithful are inscribed.   This image was made at Dolon-nor, and was paid for by money collected throughout Mongolia.   The building of the temple was intrusted to Chinese workmen, but the plan was drawn in Tibet, and the style is pure Tibetan.   The walls are constructed of logs with an outer covering of sheet iron—the use of the latter being suggested to the Mongols by seeing it employed in the building of the Russian consulate, then in course of erection. The Maidari *burkhan* stands in the centre of the temple, surrounded by five other large images. On large shelves and in cupboards round the walls are deposited 10,000 images of Buddha made before 1799. Above these Buddhas are chairs on which the sacred books of the Ganjur and Danjur are placed.

Only Lamas dwell in the monastery, and no women are allowed within its walls, except the old women who cook for the colleges, one to each.

Outwardly, the monastery, intersected by two broad streets crossing each other and innumerable alleys, is not unattractive to the European eye. As usual in Asiatic countries, there are no windows looking into the street, but only walls, sometimes as much as twenty or thirty feet high, are seen. Silk ribbons inscribed with prayers decorate the houses of high ecclesiastical dignitaries.

Life in the monastery is very monotonous, for the Lamas spend most of their time in the temples or in their tents. The market-place outside the walls presents a marked contrast. It is a square surrounded by Russian and Chinese shops, where nearly all the trade of the place—mostly wholesale—is transacted. Besides the shops, there are also some 80 to 120 felt huts, in which chibouks and other pipes, tobacco-pouches, toys, beads, earthenware, mirrors, boots, and belts are sold. The Mongols have monopolised the trade in second-hand goods, and sell all kinds of curiosities. A large number of women are employed in the manufacture of Lama caps, which requires considerable skill. The inhabitants of the steppes come to Urga to sell their cattle, horses, and sheep, as well as fresh hay, which fetched a high price when I was there (in February 36 English pounds cost 3 shillings). In this market, also, Chinese smiths work at their forges, and Chinese cooks dispense roast-mutton, steak-pies, fruit-pies, ginger-bread, etc., in their refreshment tents. Near these tents musicians play on their violins, banjoes, flutes, and other instruments, while pilgrim Lamas read their prayers aloud, shaking their staffs all the time to make the bells on them tinkle. In the small Chinese restaurants opium-smoking is indulged in—in fact, wherever Chinese are, there is also opium. Beggars are never seen, for the Mongols have quite recovered from the disasters they suffered in 1870 at the hands of the Dungans. Police officers, called *khia*, or night-watchmen (*mandchi*), are always posted in the market-place, and are responsible for the maintenance of order.

Notwithstanding that the streets are used as receptacles for refuse, dirt is not as noticeable here as in Kalgan, or, to judge from descriptions, in Pekin. Innumerable packs of dogs prowl about and act as public scavengers, devouring anything and everything, even to the corpses of their own unfortunate fellows.

The second quarter of Urga, also with a Lama population, stands on a small hill a mile to the west of the monastery, the interval being occupied by the shops and houses of Chinese and Mongolian residents. In this, the recognised centre of Buddhistic learning in Mongolia, students

attend the *Tsanit*, the High School of Buddhistic philosophy. Externally
this school or college closely resembles the monastery. There is the same
broad square with two large temples ; the same yellow enclosure for the
palace—only smaller ; the same narrow streets ; the same crowded dwellings built in square blocks, where the Lamas live ; and the only difference is that the palace of the Gheghan stands out more conspicuously,
not being so much hidden by other buildings.

The third quarter, the Maimachen, is situated to the east, 3 miles from
the monastery, and is to all intents a distinct town. It has 5000 inhabitants. At a distance it looks like one huge square. It has no
enclosing wall, but all the houses are surrounded by hedges. From the
side towards the monastery start three large parallel roads, which
traverse the whole town, and were formerly terminated by gates. The
interior consists of a Chinese town in the centre, called Toro (inner),
and a Mongol town around it, called Gada (outer). The two are
separated by a high stockade with seven gates, which are all closed at
sunset, after the well-known Chinese custom. The water for daily use
it supplied by the river Tola, 2½ miles distant. In this connection we
find an instance of Mongolian conservatism, for nothing will induce the
Mongols to carry water into the town, as the Chinese do, by means of
oxen, horses, or donkeys, but they prefer to toil back with it on their
shoulders in the old laborious fashion. There are no trees about the
town ; the forest begins at a little distance and abounds with roebuck,
boars, deer, and wolves. The Mongols hunt them with their old flint-
locks and pistols.

The streets of the Maimachen are broad, but crooked and irregular.
As they are not paved, they are channels of liquid mud in the rainy
season, and simply unbearable owing to the dust during the summer.
Canals are, indeed, constructed from the river Uliatai for the purpose of
watering the streets ; but in summer, when the water is wanted, the river
is nearly dry, and the water is plentiful in the canals only when the
streets are thoroughly drenched with rain. A causeway runs on either
side of the canal, and there is a bridge over the canal opposite each
house. In some streets there are a few white-willow trees. The plan
of the houses is as follows :—first, there is an outer enclosure of upright
larch poles daubed over with clay so as to make a good solid wall. The
court is entered by a broad and high gateway, usually surmounted by a
canopy. On the gate itself are inscribed in Chinese characters such
words as " happiness," " long life," " fortune," etc., which are painted
afresh every New Year's Day. Having passed through the gate we
enter the court, where goods of all kinds are stored, and the shop stands
in front of us. Behind the shop is a larger court containing the kitchen,
pantry, and other domestic offices. The first court is surrounded by
galleries, and often has a quaint and pleasing appearance, especially in
summer, when the galleries are covered with flowers. Fir-trees are
planted in winter, and then the Chinese shop looks like a winter-garden.
The large windows are covered with paper, for, if the owner comes into
possession of a piece of glass, it is promptly used for decorative purposes,
a flower or mythological scene being painted on it. In one corner of the

shop stands a large bench, generally covered with carpets, and, beside it, a stove of brick or iron. In the Chinese quarter are two inns, where Chinese merchants coming from a distance put up. The tariff is a brick of tea (worth about 1s. 3d.) for a day's board and lodging. No vehicles are regularly kept for hire, but a horse and saddle or a Chinese cart can always be procured.

The Mongol quarter is more lively than the Chinese. Here we see goods exposed for sale in booths or tents, or sometimes simply laid out on a board. Meat of various kinds, poultry, and occasionally hares, are sold. The Mongol women also take part in trade, selling berries, butter, milk, cheese, toys, Lama caps, etc.

The Mongol houses are very different from the Chinese. In the first place, they are not so large, and have no canopy over the gate, which is smaller and is always kept closed. As the Mongol does not use carts, the broad Chinese gate is unnecessary. In the centre of the court-yard is generally a felt tent, where the family takes up its quarters in winter, the houses being too slightly built for that season. All the artisans are Chinese; they build the houses, ovens, and stoves, and make the bricks and fell the timber required for the work.

The Mongol is very frugal and abstemious; a little flour, meal, and tea are all the sustenance he wants, and these can be obtained at a very small cost. The Chinaman's needs are greater; he consumes vegetables and fruits—melons, cabbage, carrots, parsley, potatoes, radishes, onions, garlic, etc. He also burns wood, whereas the Mongol prefers dried cow or camel dung, called *argal*, for fuel.

The Russian consulate is undoubtedly the finest building in Urga, and it stands on a hill on a commanding site. The consul has an important post. He is on intimate terms with the clergy, and knows the country and its people thoroughly.

With respect to religion the inhabitants are divided into two sects:— (1) Taoists, who are all Chinese; and (2) Buddhists—mostly Mongols, with a sprinkling of Chinese. There are three Chinese temples, and one Mongolian Buddhist. In the centre of the Chinese town there is a theatre, which is very popular.

As to the morality of the people, not a good word can be said for it. Diseases are rife; as there are no hospitals, the wretched victims of vice often drop down dead in the street. If not claimed by the relations, the bodies are torn to pieces where they lie by the ravenous pariah dogs which infest the place. For this state of affairs the whole blame rests, of course, on the Lamas; they are not allowed to marry, and the natural consequences ensue. The governor of the town is always a Manchu, and concerns himself only with thefts, murders, etc. Offences against morality are not regarded as crimes; public opinion sanctions them, and the law follows suit.

The Mongol burial-ground lies outside the town. I have called it a burial-ground, but the term is hardly accurate. Deceased Mongols are simply laid in the appointed plot and covered with a few rags; the dogs and wolves complete the work.

ULIASUTAI.

The second largest town of North Mongolia is Uliasutai, in 47° 44′ 4″ N. lat. and 96° 53′ 46″ E. long. It is the administrative centre of the country, being the seat of the Governor-General, or Tziantziun. There is a great contrast between Urga and Uliasutai. While the former is the religious centre and is like a huge monastery, the latter is the political capital and a fortress. It was founded when the Mongols and Manchus were at variance with their southern brethren, the Ölöts. We find Uliasutai mentioned as the seat of the Governor-General as far back as 1734. The town always took precedence of Kobdo, the place we shall describe next, especially during the wars with the Dungans, but as it was the seat of the Chinese government, the offices of which were mostly in the possession of Manchus, the Mongols had never much affection for Uliasutai. Besides, all lawsuits were tried in the Chinese courts, and Chinese justice is a thing to wonder at and avoid. The chief temple here is that of Ghurandi, which was built as late as 1787, and is called by the Mongols the temple of Ghesser, their national hero. At the Dungan invasion this was the only building which was not destroyed.

The population of the town is chiefly composed of the soldiers, who number some 3000 to 4000. Most of them are Chakhars and Solons from Southern Mongolia, for only very recently have Northern Mongolians enlisted in their ranks. Their term of service is for three years, but they frequently remain for nine years or even longer. Besides the soldiers, there are a few traders who live in their own private houses. The town, therefore, consists of two parts—the fortress, wherein the houses of the Governor-General and the *ambans*, the government buildings and the barracks are situated, and the commercial town, which stands opposite the fortress on the other side of the river Chinghistu-gol. The wall of the fortress, 14 feet high, is built of piles and earth covered on both sides with clay. All business is carried on in the Maimachen, which is a group of 150 houses ranged along two intersecting streets. The largest buildings are two inns which also serve as bazaars. The goods sold are chiefly brick-tea, Chinese, Russian, and American cotton goods, flour, rice, pork, fish, spirits, wine, oil, and tobacco. The merchant engages a room at the inn and hands over his wares to the innkeeper, who receives a commission for finding customers. Whether the merchant or the innkeeper sells the goods, the latter receives his commission, but he generally sells them, as he knows all the buyers personally. There is no money currency, but all exchanges are effected by barter, which makes the trade very profitable to the merchant.

The innkeeper often gives credit, but in such cases he is responsible to the merchant, to whom he guarantees payment. One inn is called *Yukhundi*, the other *Baturdian*. The latter was established in 1869 with Russian capital, when the Russians with their consul Pavlinof first came to Uliasutai. The Chinese merchants did not settle here until after the Dungan rising, and have at present only four shops, while the Russians have nine. The Chinese sell silk and cotton goods of Chinese,

Russian, and American manufacture, ready-made clothes, boots, tobacco-pouches, wood and earthenware (spoons, plates, and dishes). The rich merchants of Kukukhoto have twelve shops here, but sell their wares through travelling agents, who dispose of them chiefly in the country or in the neighbourhood of a monastery. The goods received in exchange are sold elsewhere to Russians and Chinamen for cash, and large profits are thus realised. The Mongols sell nothing but meat and milk. The craftsmen, all Chinese, comprise smiths, carpenters, and wood-carvers, who make wooden spoons, bowls, plates, etc. The material is transported by Mongols from forests twenty-five miles distant. There are two public buildings in the Maimachen—the police-office and a Buddhist temple. The temple is nearly in ruins, as it was plundered by the Dungans. The police officers, besides their proper functions of preventing and detecting crime, act also as custom officers. They inspect all goods entering Uliasutai, and assess the duty according to the number of camels needed to carry them. They also collect a house-tax from every Chinese resident.

Scattered around the town are farms worked by Chinese companies. The labourers are generally Chinamen, but there are a few Mongols among them. Their wages are £7 a year and board. Corn is sown in April and harvested in August. Wheat is the principal crop, and yields 100 to 120 bags for every 10 bags of seed; the worst land produces 50 bags. Vegetables are also raised, and thrive so luxuriantly that, after the local demand has been satisfied, there is a surplus which is exported to Guchen and Barkul. The Chinese banking firms make large profits near this town, lending cash and receiving payment in kind, and charging 36 per cent. interest on loans. If the principal be not returned in the first year, the interest is added to it, and the original loan is doubled by some marvellous process in the second year. As the Mongol considers it a sacred duty to pay his debts, not having attained to the Western freedom of conscience in this respect, it may easily be imagined that he lives under a most burdensome yoke.

The country round Uliasutai is very picturesque, especially the river valleys. The forests supply wood and charcoal, the latter being burned by Mongols. All the tar and turpentine are wasted, as the trunks are simply burned in large pits dug out in the mountains.

The Mongols generally keep to certain manual employments, working as labourers, water-carriers, dung-gatherers, wood-cutters, etc. They very seldom enter into trade. Most of them are shepherds and herdsmen attending to their cattle and other animals in the country round the town. The herds are numerous, and, in particular, there are large herds of horses and camels, which are used for military purposes.

From Uliasutai starts the great post road to Kalgan and Urga.

KOBDO.

The geographical position of this town is given by the Russian geographer, Colonel Matusovski, as 48° 0′ 5″ N. lat., and 91° 36′ E. long. Chinese historians mention it for the first time in 1731. It was founded after the defeat of Galdan. Kobdo is situated in the broad valley of the Buiantu, and consists of a fortress with a small garrison and a commercial town. In the fortress, which is surrounded by a dry moat, live all the Government officials, including the *amban*. In its northern part are large magazines for the storage of provisions, the temple of Ghesser (the only one in the place), and a Mohammedan mosque. An inscription on a stone in front of it informs the stranger that the temple was rebuilt in 1862. The magazines are in a more or less dilapidated condition. The prison, a typical building, is surrounded by a stockade of larch poles covered with clay in the usual fashion. On the top of this wall is a square coping made of thin rods. The gates are left open during the day and closed at night. In the court is a house, called *baishin*, occupied by the warders. This is connected with the prison, which is entered through the guard-room. The prison consists of a single room underground, divided by wooden gratings into compartments, in which the prisoners sit. A corridor down the middle allows warders and visitors to inspect the prisoners. Against the walls, which are of earth, stand wooden posts with iron chains, by which prisoners of importance are secured. The chains are attached to the neck and feet only, or to the neck, feet, and waist, and sometimes to the wrists as well. At night the prisoners' feet are fastened in wooden stocks, so that the poor wretches cannot even move, much less lie down. Smoking is allowed in the day-time. Prisoners convicted of trivial offences enjoy greater liberty, and are even allowed to take a walk during the day in the town, carrying, however, a kind of prison badge in the form of a large board (*cangue*) round the neck. At night they must return, and then they are put in the stocks, but are allowed to rest in a reclining position. In winter all the prisoners are covered up with one large blanket, which is suspended from the ceiling during the day. There are no stoves.

The town is a hundred yards to the south of the fortress, and a mile from the river. Most of the water used is drawn from wells. Kobdo is remarkable for its cleanliness—that is, according to the Mongolian standard of cleanliness, which is very different from the European. Nine merchants live in the town, but trade is not very flourishing, for the country around is becoming poorer every year. Chinese merchants come every season from Pekin, bringing the newest articles in the market. There are also many retail shops, the occupants of which gain a more or less precarious living by cheating, which they have evidently studied with much diligence and success. Kobdo has two antiquarian curiosities—two stones sent to the monastery in 1768 by the Emperor Tsian-lun. Tradition says that the inscription on one of them, in three languages, was incised by the Emperor himself. It is practically a narrative of the Emperor's own deeds, and concludes with the statement that he had rebuilt the monastery. A few months ago a Russian consulate was established in Kobdo.

With the construction of the Siberian railway a new era will commence in Mongolia, and Russia will play a prominent *rôle* in these parts of the world ; and, understanding the East, as she does, better than other nations, making no attempts to proselytise, but studying the languages, customs, and manner of life of the people, and learning their tastes with a view to supplying their demands, she is bound to succeed in her enterprises.

And now a few words about Mongolia's neighbours—the Chinese and Japanese. I think that this question, which is now called the "Question of the Far East," is properly understood only in Russia; we Russians have been studying the Mongols since 1224, and know the aspirations of the Mongol race. If Great Britain knew, as we do, that the Japanese, after being forced to open their ports to European commerce, convoked a secret council of the nobility, in which they resolved to fight the Europeans in every way, English people would take a very different view. At this secret council it was decided to send the ablest men in Japan to Europe and America to study everything, and on their return to teach their fellow-countrymen to turn out every commercial product that their teachers themselves could supply. And Japan is able to compete, because her people live more simply and work longer than Europeans. It is Japan's aim to destroy the foreign trade. When the Japanese came to Europe, Europeans (except Russians) were only too ready to show them what clever men they were. It was forgotten that these Japanese learners kept a pencil and note-book. How well they learned the lessons that Europe had to teach they showed in the last war. German and English merchants see already the rising danger on the side of the Japanese. I saw, quite recently in London, merchants who told me that within the present year the Staffordshire earthenware trade and the carpet industry would be completely ruined, because the Japanese have sent large quantities of these goods to London at a ridiculously low price. But this is only the beginning. The Japanese will extend their trade to Europe, for they actually think of beating Europe in manufactures. Russia understood these aims, and therefore stepped in between the Japanese and China, and wished England to join her, as the danger was one common to European trade and to Christian principles. Russia, true to her mission, which was to defend civilisation and Christianity in deeds, not words, took up the struggle without Britain's aid, though she regretted that it was withheld. Then there were hopes raised by the Government of Lord Rosebery and by the visit of the Prince and Princess of Wales to Russia, that the two nations would come to a better understanding. Russia showed her goodwill by settling the Pamir question—which she would have had no inducement to settle, had her intentions been at all hostile. And what was your first answer to our goodwill? You annexed Chitral in the face of your promises.

I believe that the so-called isolation of England, at the present time, is only due to her misunderstanding with Russia. I regret that so little is known among English people of Russia's history and aims. Should that history be fully written in the future, two points will be clear. The first is that Russia has been always the bulwark of Europe against Asiatic hordes trying to destroy European civilisation, and the second, that she has always fought for Christianity in deeds, not words, against heathens and Moslems. The latter question is, at the present time, nearly at an end. If England had not prevented Russia, by the Crimean war, and in the Treaty of Berlin, from driving out the Turk from Europe, there would not have been such a problem to-day. The Bulgarian and Armenian massacres would never have happened, as Russia was going to grant self-government to all the subject Christian peoples of Turkey— Servians, Bulgarians, Macedonians, Armenians, and the rest of them. In the present Armenian question, England has only reaped what she has sown. Russia alone is able to solve that question, and Russia is prevented from doing it, as she was prevented seventeen years ago from accomplishing what England cannot solve now.

The Treaty of San Stefano would have given the Christian peoples of Turkey all that they are striving for ; but Russia could step in only at the request of the whole of Europe.

I wish it were better understood in this country that, if you would cease to meddle in our internal affairs, which you do not understand at all, having never attempted to gain a knowledge of our country, or to study our history, you would be, in Russia, the most popular of nations. We, in Russia, study your language, your history, and your customs, and admire them. You listen with eagerness only to the stories brought over by Poles and Nihilist conspirators, for whom Russia has been made too hot. It is these one-sided views of Russia which you adopt. We have in reality the same views as you. We have sufficient territory, and do not want any more. We desire only to develop our resources, and for that development peace is necessary. Together, Britain and Russia might secure the blessings of peace and civilisation for the whole world. And now is the time for such an alliance. Circumstances may occur at any moment to make it too late. Think what an alliance it would be—an alliance between the greatest naval power in Europe, and the greatest military power. The strength of such an alliance would be the surest guarantee of European peace.

Both countries have the same aim—the spread of civilisation. And what a splendid sight it would be to see England and Russia advancing hand in hand in the march of progress, by striving to give peace to, and bestow the blessings of civilisation on, Asia !

THE STORY OF THE NORTH-EAST PASSAGE.

IN the May and July numbers of the *Magazine* the progress of Arctic discovery in search of the North-West Passage was narrated. With this the enterprise of which we propose to treat is intimately connected, for in both cases the object was to open a route to the East from the north of Europe, which would be shorter than that round the Cape of Good Hope. The project of sailing to Cathay by the north-east seems to have been attempted first ; for though late in life Sebastian Cabot asserted that the object of the voyage of his father and himself in 1497 was to search for a route to the East, nothing of such a plan is mentioned in the patent granted by Henry VII., and, as it was not then known that a continent lay in the way, such a route could not be called a North-West Passage, as the term was applied later. But though the search was first commenced to the eastward, it was not till our own day that the problem was actually solved by Baron Nordenskiöld, and then almost the whole of the northern coast of Asia had been discovered, principally by Russian explorers.

The belief prevailing at the beginning of the sixteenth century that the coast of Asia trended rapidly southward from the frontier of Russia to Cathay, probably dates from very ancient times. When it was believed that the ocean encircled the earth like a ring, the northern shore of Asia, of which nothing was actually known, was naturally supposed to curve round to the south, as may be seen in some of the maps accompanying Dr. Murray's paper in vol. ix. pp. 561 to 586. The only writer of antiquity who gives any precise details concerning the northern coast is Pliny, who died in A.D. 79. Of the northern coast of Europe he knows nothing, simply speaking of that region as a part of the world cursed by Nature, and plunged in thick darkness, and alluding to the legend of the Hyperboreans. The Caspian sea he describes as an inlet of the Scythian ocean, and states that from the Caspian and the Scythian promontory the coast bends to the east, skirting lands partly inhabited by Scythians and partly deserts haunted by wild beasts, up to the promontory called Tabis. These promontories figure on maps down to the early part of last century, and correspond, apparently, the Scythian to the peninsula of Taimyr, and Tabis to the Chukchi. Pliny also mentions an island named Zazata, which is placed by cartographers between those peninsulas [*Nat. Hist.* vi. 17 (19)]. Later authorities, such as Ptolemy (about A.D. 150), do not pretend to know anything of the northern regions of the Old World, though they had discovered that the Caspian was an enclosed lake unconnected with the Arctic Ocean.

As is well known, the cosmographers of the Middle Ages, far from progressing in knowledge of the world, rather retrograded to the most primitive notions of its outlines. Marvellous *Imagines Mundi* were constructed, in which Jerusalem and the Terrestrial Paradise played a prominent part. Exploration, however, made some advance in the northern regions. The Färöe Islands, Iceland, and Greenland were discovered, and respecting the north-western extremity of Europe some curious information was given by Octher to King Alfred. About the year 890, this Octher, who stated that he dwelt farthest north of any Normans, in a country called Helgoland, made a voyage to find out how far the land stretched northwards, and whether there were any inhabitants beyond him. In three days' sail he came as far north as the whalers were wont to venture; after three days more he found the land trending to the east, and had to wait for a westerly wind before he could make further progress, and sailing onwards four days more, having the land on his starboard the whole way, he then found that the coast turned southwards, in which direction he came in five days to the mouth of a large river, perhaps the Dwina, the limit of his exploration. On the northern coast he saw Finns who lived by fishing and hunting, while on the river banks dwelt the Biarmias, who tilled their land.   He did not dare to go ashore, and therefore had little to report about them, but believed that they spoke the same language as the Finns.

Little more additional knowledge had been acquired of the northern coast of Europe even on the eve of the great maritime discoveries in the fifteenth and sixteenth centuries.   Edrisi (1154), one of the great Arabian geographers, who restored the art of cosmography by reviving the method of longitudes and latitudes, shows considerable knowledge of the Baltic and the towns of Norway, mentioning Finland and even Tavastland, but he conceives Norway to be an island.

A map of Dacia, Norwegia, and Gota, first published in 1482 to supplement Ptolemy's, shows the island Margerester, which is evidently Mageröe, but represents Greenland as continuous with Europe, separated from Norway by a long gulf.

On Martin Behaim's globe (1492), the White sea is represented, but it has no more definite name than *Mare Congelatum.*

And during all these centuries still less new information was obtained about the eastern extremity of the Old World.   In the thirteenth century the friars Plano de Carpini and William de Rubricis, and afterwards Marco Polo, travelled to the court of the Great Khan, and the great distances they travelled and their statements that the kingdom of Tartary was bounded by the Northern Ocean, probably misled cartographers into placing Karakorum at no great distance from the northern coast.   To the last mentioned is due, Dr. Chr. Sandler believes (*Zeitschrift für Erdkunde,* 1894), the Anian strait, which appears first in the year 1566 on the map of Bologninus Zalterius and afterwards on those of Ortelius, Mercator, and others, as bounding Asia on the north-east.   He says that a gulf, called Cheinan, crossed in sailing west, and a little south-west from Zaitum, stretches northward for two months' journey, washing the shores of Ania and Toloman, and contains many islands and rivers with golden sands. "This gulf is so large, and so many people live in it, that it almost seems like another world."   From these words Dr. Sandler thinks that some misty report of another continent to the east had reached the ears of Marco Polo.

Leaving now these speculations and hazy glimpses of the Northern ocean, we come to actual exploration, commencing with the voyage of Sir Hugh Willoughby and Richard Chancellor in the year 1553.   Sailing in May from the Thames, the fleet of three vessels sailed up the coast of Norway, and Sir Hugh Willoughby, after losing sight of his consorts during stormy weather, advanced into the Arctic Ocean as far as 72° N. lat., where he discovered an uninhabited land, which, according to Admiral Beecher, must have been that part of Novaia Zemlia named by the Russian, Admiral Lütke, the Goose Coast,[1] and then, returning to the coast of Lapland, perished with all his crew at Arzina in the year 1554.   Richard Chancellor, more fortunate, reached Vardö, and thence journeyed into Russia, gathering much information concerning the people and their commerce, and obtaining the privileges which led to the establishment of the Muscovy Company.

---

[1] *The Polar Regions.* By Sir John Richardson, LL.D., p. 55, Edinburgh, 1861.

In 1556 Stephen Burrough sailed in the *Serchthrift*, and, after giving its name to the North Cape, skirted the Kola peninsula, and, passing the Kanin Noss, discovered Kolguief, and entered the mouth of the Pechora. Thence he turned northwards, visiting Novaia Zemlia, where Russian traders spoke to him of the route to the Ob, and sailing across to the island of Vaigatz. He was therefore the first to discover an entrance into the Kara sea.

In the year 1580, Arthur Pet and Charles Jackman were sent in two barks to search for a passage to the Ob, discovered the Yugor strait, and entered the Kara sea, but were prevented by the ice from reaching the goal of their enterprise. In connection with this project Mercator wrote in a letter to Richard Hakluyt that beyond the island of Vaigatz and Novaia Zemlia lay a great bay enclosed on the farther side by the mighty promontory of Tabin, in the midst of which were the mouths of great rivers, which passed through the country of Serica, and were, he believed, navigable into the heart of the continent, so that by this route goods could be transported into England out of Cathay, Mangi (Southern China), and other kingdoms. Of the existence of the Tabin promontory he is persuaded, not only out of Pliny, but also other writers and certain maps.

The next discoverers to appear on the scene were the Dutch, who, jealous of the success of English merchants in the trade with Russia, fitted out three ships in 1594, and sent them, under the command of Cornelison Nai, to explore in northern waters. Two of these vessels sailed into the Kara sea through the Yugor strait, and returned with the belief that they had sailed to the longitude of the Ob, and had not been far from the Taimyr peninsula. The captain of the third vessel was the famous Barentz or Barentszoon, who on a subsequent voyage discovered Bear Island and sighted Spitzbergen, and then rounding the northern end of Novaia Zemlia, and being imprisoned in the ice on the east side of that island, was the first explorer to pass a winter in the Arctic seas. He contributed, however, nothing essential to the opening up of the North-East Passage. Indeed, after this time, the exploration of the northern coast of Asia was undertaken by the Russians, the English and Dutch turning their attention to the north and west.

In 1580, the year that Pet and Jackman sailed to the north, the Cossack Yermak Timofeieff, despatched by Ivan Vasilievich, subdued the Tartar khanate of Sibir, on the banks of the Irtish and Ob, and, penetrating still farther into the interior, defeated Kuchum, the Khan of the Tartars. He thus commenced the conquest of the country for the Russians, who, before the end of the following century, had extended their dominion to the Pacific, annexing Kamchatka in 1696.

Russian traders had visited the Ob from a much earlier date, dragging their boats overland from the Karskaia Gulf, and a trader named Anika had sent out parties thither overland. Purchas gives an account of this man, whom he names Oneeko, and states that he had obtained his information from Hakluyt. Perhaps this is the Unekius mentioned in a letter written in 1581 by a certain John Balak to Mercator, which is published in Hakluyt's *Collection of Voyages*. A Netherlander, named

Alferius, or Oliver, travelled with Yacovius and Unekius overland to the river Ob, and he speaks of the lake of Kittay beyond the sources of the Ob. This lake receives a river, the Ardoh, and is said to border the country of the Carrah Colmak, which, it is added, is Cathay. From the narrative of Jenkinson's journey to Boghar (Bokhara) in 1558, it appears that this lake was the Sea of Aral, for he speaks of the Oxus as a tributary of the river Ardock, which, after a course of 1500 miles, falls into the lake of Kithay. He has, however, a better notion of the distance to Cathay than Alferius, for he gives it as 60 days' journey from Boghar.

As soon as the Russians had established themselves on the Ob, the exploration of the interior proceeded at a rapid pace; Kopilof reached the shores of the Eastern sea in 1639, and in 1644 Staduchin obtained the first information about the Chukchi people. Much was also done in this century towards the exploration of the coast, the Cossack, Yelissei Busa, sailing in 1636 down the Lena to its mouth, and skirting the shore to the Olenek. The same navigator also in 1638 sailed for five days eastwards, and passed a winter on the banks of the Yana. The following year he prosecuted his voyage to the Chendoma. In the same year the Indigirka was discovered by Ivanoio, who surveyed the coast to the Alaseia, 163° E. long. In 1648 Dishnef, Ankudinof, and Alexeief sailed from the Kolyma and rounded the Sviatoi Noss, called afterwards Shelagskii Noss by Wrangell (not the point between the Lena and Indigirka). Deshnef's boat was at length driven ashore in the Gulf of Anadyr, having passed through Bering strait eighty years before Bering discovered this passage between the Old and New World. Alexeieff also passed through the strait, and his vessel being wrecked on the Chukchi shore, some of the crew were killed by the natives, while the remainder made their way to Kamchatka, where they also were at length murdered.

The first voyage of Vitus Bering, a Dane in Russian service, took place in 1728. By order of the Empress Catherine, he built a vessel at Nizhnii Kamchatkoi, and, accompanied by Lieutenants Chirikof and Spanberg, passed the island of St. Lawrence and sailed to East Cape in lat. 67° 18', from which point he perceived that the coast trended westwards. The other side of the strait was discovered in 1730 by Krupishef and Gvosdef, who were driven from Bering's farthest point past St. Lawrence Island westward, and skirted the American coast southwards for two days. Bering sailed again in 1740, and also in 1741, when he made the American coast in lat. 58° 28' N., and died on the return voyage on Bering's Isle. On this voyage Steller found his sea-cow, *Rhytina Stelleri*.

Meanwhile a systematic exploration of the northern coast of Siberia had been planned. Three expeditions were equipped: one to sail from Archangel to the Ob, another to follow the coast from the Ob to the Yenisei, and the third to start in two boats from the mouth of the Lena, the one party having the task of exploring westwards towards the Yenisei, while the other was to make for Bering strait. The first expedition, under the command of Lieutenants Muravief and Pavlof, was unsuccessful in 1734 and in 1735, but in 1736, under the command of Malgyn and Skuratoff, reached the Kara river, and in the following year the goal of its enterprise. The second expedition was led by Lieutenant Ovzin, and also started in 1734, but did not reach the Yenisei until 1737. The pilot Minin took the command in 1738 and surveyed the eastern shore of the Gulf of Yenisei, and in the following year ascended the river to Yeniseisk, besides making a sledge expedition northwards to Cape Stergelof in 1740. In the third expedition the Lieutenants Pronchish-cheff, Lasinius, Khariton and Dmitri Laptef extended their labours from 1735 to 1743. During the first few years the ice impeded progress towards the west, though lat. 76° 48' was reached. Pronchishcheff and his wife died on their return to their winter quarters near the mouth of the Olenek. In 1739 Khariton Laptef sailed up to Cape St. Faddei in lat. 76° 47' N., according to his reckoning, and returning south entered the Khatanga river. He wintered in the Bludnaia, one of its tributaries. Laptef reached the Taimyr lake overland in 1741, and followed the Taimyr river to the sea which it entered, according to his observations, in lat. 75° 36' N., and advanced to Cape Taimyr. He then surveyed the western coast of Taimyr. The eastward division of this expedition was conducted by Dmitri Laptef, who surveyed the Bear Islands and extended his voyage to the Baranoff rock. In 1742 the northern extremity of the Taimyr peninsula was reached on sledges by Lieutenant Cheliuskin, an officer of Pronchishcheff's party, by whose name it is now generally known, though it is sometimes called the North-East Cape, a name given to it by Wrangell.

Accordingly, by the middle of last century the whole of the coast of Siberia had been, to a certain extent, explored by the Russians, and the only part which had not been skirted by sea was the extremity of the great peninsula of Taimyr, extending much farther to the north than any other part of Asia, and consequently more closely beset by ice. So great were the difficulties of navigation in this region, that Laptef felt convinced that Cape Cheliuskin would never be doubled by a vessel, and more than a century and a quarter elapsed before he was proved to be a false prophet.

Meanwhile the coast was examined more closely, and the islands in this part of the Arctic Ocean were, one after the other, added to the map. Liakhof Island, to the north of Sviatoi Noss, was discovered as early as 1710 by Yakof Permiakof, but owes its name to a merchant who, in 1770, happening to visit Sviatoi Noss on business, saw a herd of reindeer coming over the ice from the north. Travelling northwards for 47 miles, he came to an island, and 13 miles further, to a second. Here he was stopped by the roughness of the ice. The large quantity of · mammoth tusks he found on the islands induced him to make a second journey in 1773, but it was not till 1823 that Lieutenant Anjou travelled round the whole group of New Siberia and determined their positions. This officer, seeing vapour rising to the north-west, supposed that open water existed in that direction. Vaporisation, however, takes place whenever ice cracks, even where it is thick and solid. Anjou did, indeed, see open water on more than one occasion, but his attempts to travel northwards and eastwards from the Liakhof group were frustrated by the thinness of the ice.

About the same time Baron von Wrangell made repeated expeditions across the Polar sea and along the coast from the Kolyma eastwards. On arriving at Shelagski Noss he heard from a Chukchi that on a clear day snow-covered mountains could be seen towards the north, and that herds of reindeer sometimes came across the ice from that direction. This land Wrangell attempted to reach in sledges, but was checked by broken ice. Sighted in 1849 from Herald Island by Captain Rollet of H.M.S. *Herald*, and again in 1867 by Captain Thomas Long, an American whaler, it received the name of Wrangell Land, and was supposed to form part of a great Arctic continent, until it was discovered by Commander Delong of the *Jeanette*, in the winter of 1879-80, to be an island of limited extent.

Thus in 1878, when Baron von Nordenskiöld sailed from Sweden in the *Vega*, fitted out by Baron Oscar Dickson, to accomplish the North-East Passage, the coast was known in general outline; whereas the navigators who sought for the North-West Passage in the first half of this century had to ascertain whether a route by water to the north of America actually existed, and to find its position. There was no doubt that sea extended along the whole of the coast of Asia, and the only question was whether ships would be able to make their way through the ice, especially round the huge, far-stretching peninsula of Taimyr. Here, of course, the coast, as laid down in Russian maps, was found to be very inaccurate, and several bays and some new islands were discovered. Cape Stergelof proved to lie in latitude 75° 26′ N., and considerably to the west of the position hitherto assigned to it. Cheliuskin's determination of the cape named after him, 77° 34′ N. lat., proved to be very near the truth; but Laptef's outline of the peninsula was drawn too far to the east on both sides, especially on the eastern, the breadth of the peninsula being twenty to twenty-five miles broader than was supposed. Cape Cheliuskin is formed by a low tongue of land divided into two promontories by a bay. The western point lies, according to Professor Nordenskiöld's observations, in 77° 36′ 36″ N. lat. and 103° 25′ 30″ E. long., and the eastern a little farther north, in latitude 77° 41′ N. and longitude 104° 1′ E. The voyage of the *Vega* occurred so recently that it is not necessary to give a full account of this successful undertaking and its valuable results, which may be collected from the books of the explorers and from several periodicals. Suffice it to say that, having sailed from Karlskrona on June 2nd, 1878, the expedition made such progress that it reached the mouth of the Lena on August 27th, when hopes were entertained that Bering strait would be passed before the winter set in. These hopes were, however, disappointed, the *Vega* being frozen in at the end of September near Koliuchin Bay, on the Chukchi coast, not much more than 100 miles from the strait. Fortunately, no damage was done to the vessel, and in the summer of 1879 it was able to continue its voyage, calling at Yokohama in September, and it returned to Sweden on April 24th, 1880, having completed the circumnavigation of Eurasia.

The North-East Passage had, then, been found, but as a route to the East was as impracticable as that to the north-west. As a trading route along the shores of Siberia, however, it may prove of no little importance. Large rivers pour their waters in summer into a wide extent of ocean, instead of into narrow channels between large islands such as confine the ice on the northern coast of America, and clear an open channel during the summer months. The voyage to the Yenisei has been accomplished by Norwegian vessels, and with great success by Captain Wiggins, who again sailed last summer in the *Lorna Doone* with rails for the great Siberian railway. Of the course between the Yenisei and the Lena, Professor Nordenskiöld expressed his opinion that one voyage along it was possible every year. At the northern extremity of the Taimyr peninsula only bay ice was encountered in August, and that in small quantities. The *Vega* performed the voyage from Dickson's Harbour, near the mouth of the Yenisei, to the Lena, about 1000 miles, in eighteen days, including four days spent at Taimyr Island and one at Cape Cheliuskin; and the *Lena*, fitted out by M. Sibiriakof, which accompanied the *Vega* to the mouth of the Lena, made the voyage from Tromsö to Yakutsk in fifty-five days. The route from the Lena to Bering strait Professor Nordenskiöld regarded as less favourable than either of the others, but he thought it might be used for the transport of heavy and cheap goods from the Pacific. Wood is obtainable everywhere, and the engines of the steamers used in the traffic should be adapted for burning it.

THE OROGRAPHY OF CENTRAL ASIA AND ITS SOUTH-EASTERN BORDERLANDS.[1]

M. V. A. OBRUCHEF's late explorations in Central Asia have extended over an enormous tract of country, and have enabled him to give a very complete account of a large part of Mongolia and the adjacent provinces of China. Leaving Kiakhta on September 27th, 1892, he returned westwards in 1894, reaching Kulja on October 10th, after a journey lasting two years and two weeks, during which he traversed about 9300 miles in Eastern and Central Mongolia, Ordos, the southern parts of the Ala-shan and Dzungaria, the Chinese provinces of Chi-li, Shan-si, Shen-si, and Kan-su, the eastern Kuen-lun and Nan-shan with the northern part of Zaidam, the upland region of Pei-shan and the southern slope of the Tian-shan with the Liukchiun depression.

Richthofen has divided Asia into three regions—the centre, the periphery, and a transitional zone. Central Asia he defines as the continental region of ancient hydrographic basins having no outlet to the ocean. It may be said to extend roughly from the slopes of the Tibetan plateau on the south to the Altai on the north, and from the Pamir on the west to

---

[1] Abstract from article by M. Obruchef in the *Izvestiya* of the Russian Geographical Society, No. iii., 1895.

the huge rivers of China and the Khinghan range on the east. The peripheral region includes all the countries where the drainage flows to the ocean, or to the Aral and Caspian seas and other lakes left behind by the retreat of the waters of the ocean. This region encircles Central Asia ; but between them lies a third region, in which during the latest periods some tracts have been transferred to the periphery through their drainage forcing its way to the ocean, while others, the connection having been cut off, have been added to the central basin. In the former case the country has to a large extent lost the peculiarities of the central region, while in the second it has not yet acquired the characteristics of the inland zone, and therefore these tracts do not strictly belong to either.

While maintaining that the greater part of the Aral and Caspian basin, or Russian Turkistan and the Transcaspian province—concerning which Richthofen was in doubt whether to include them in the peripheral or transitional region—differs geologically and geographically so little from Central Asia that there are no valid grounds for separating them from it, M. Obruchef accepts for the present the above definition of Central Asia in its general outlines. It includes, then, almost all Mongolia, except the northern hilly and wooded country where spring the great rivers of Siberia, Ordos, Ala-shan and Pei-shan—that is, the desert of Khami and the Gashiun Gobi, lying between the Eastern Tian-shan on the north, and the Altyn-tagh on the south ; and, lastly, Dzungaria and the Tarim basin, or Chinese Turkistan. With the last and most of Dzungaria, M. Obruchef does not deal, as his travels did not take him through them.

Central Asia is a mountainous rather than a level region, and the explorations of the last twenty-five years have shown the fallacy of the view formerly prevailing, that the interior of the Asiatic continent—the desert or steppe of Gobi—was in general an extensive plain of a desert or steppe-like character with a tract of dry sand in its central and lowest parts. Besides the two great mountain systems of the Eastern Tian-shan and Southern Altai, containing summits rising beyond the limit of eternal snow, and the Ala-shan, also distinguished by its great altitude though exhibiting no perpetually snow-covered peak, the greater part of Central Asia is traversed by a complete network of ranges, hills, and mounds, not surpassing in general an absolute height of 10,000 feet, and a relative height of 2000 to 3000.

Naturally, in an immense area like that of Central Asia, stretching over 40 degrees of longitude and 12 of latitude, these ranges have not everywhere the same extension. They are the residues of ranges formed at various geological periods, and of different kinds of rock ; and, besides their original difference of development, the various degrees in which they have been subjected to the erosive action of sea-waves, precipitation, and atmospheric agencies, or in which, owing to their component rocks and their relative hardness, they have resisted disintegration, have produced great variety in their configuration. Consequently, Central Asia is covered with elevations which, with the exception of a few principal ranges such as the Eastern Tian-shan, Southern Altai, etc., run in different directions, and, as a rule, are of no great length, but frequently have a cross section relatively considerable. There are also tracts of low mountains or hills where it is impossible to distinguish the general line of elevation; and also isolated groups of pointed, flat, or dome-shaped summits are met with, composed of granite, porphyry, trachyte, basalt, and other volcanic rocks of various age and structure. Lastly, there are tablelands composed of the horizontally stratified deposits of the Han-hai, the vanished Central Asian sea, either isolated or adjoining on one or more sides higher ranges and groups of older formation.

The usual directions of the ranges of this region run from east to west, west-north-west to east-south-east, and east-north-east to west-south-west, sometimes approaching to north-east and south-west. Few run from the north-north-west or north-north-east; some chains with a general direction east to west bend round to the south, so that their extremities run east-north-east and west-north-west respectively, as for example the Tostu in Central Mongolia and some of the chains of Pei-shan. A marked peculiarity of many chains of Central Asia is that they stand on high broad pedestals insensibly sloping down to the low central parts of the depressions. Frequently the elevation of the pedestal above the lowest part of the adjoining hollow is twice or even thrice that of the ridge above the upper boundary of the pedestal. The mountain proper has undergone extensive degradation, and is full of crags and ravines, while the pedestal has gentler contours, and the broken jagged outline of the upper part of the range here passes suddenly into a flattish curve that runs with its convexity downwards to the bottom of the adjoining hollow. Such is, at least, the general configuration.

Between the elevations of this region are scattered hollows of very varied dimensions. Sometimes they take the form of valleys six miles broad and more, stretching some tens of miles between two ranges; but more frequently they are basins or troughs of various shapes, enclosed on all sides, or communicating with one another by broad or narrow passes so as to form a succession or network.

EASTERN MONGOLIA.

Eastern Mongolia exhibits a very diversified surface. Its northern portion must be included in the transitional zone of Richthofen; from an absolute height of 2600 feet near Kiakhta it gradually rises to 3770 feet near Urga, and the summits rise above the principal valleys to a relative height of 1000 to 1500 feet. It is distinguished in general by gentle slopes and the flat-topped, conical, more rarely dome-shaped, summits of its generally continuous ridges. Mangatai between the rivers Bain-gol and Kara-gol, Tologoitu to the north, and Bogdo-ula to the south, of Urga, are the highest ranges. The beds of the streams are usually enclosed between steep banks 6 or 8 feet high, where löss is exposed to view overlaid with alluvial sand, mud, and clay. Löss of still greater thickness, 15 to 20 feet, is seen in the steep sides of dry gullies and in the waterless valleys so common between Kiakhta and Urga. Primi-tive rocks, somewhat wildly distributed in this locality, are exposed only on the ridges and upper slopes in the form of crags and cliffs.

Accumulations of rocky fragments are rarely met with, except at the foot of very steep slopes, the products of degradation being covered with a thick layer of löss.

Immediately to the south of Urga begins the Central Asian region, cut off from hydrographic connection with the ocean, the first fifty miles from the ford of the Tola, however, forming a threshold to the present Gobi or Inner Mongolia. As far as the slopes of the Narin-ula the road passes through a country considerably eroded, similar to that on the north of Urga, with broad, flat troughs and hollows separated from one another by ranges of flat mountains, which in most cases deserve only the name of hills. From the Narin-ula the path descends to a flat-bottomed basin, thirteen miles in diameter, containing the salt lake Narin-kul, beyond which a slightly undulating plain, named Sakhir-ukhe, stretches south-westwards to the Sansyr-ula. Two groups of rocky mountains, the Sansyr-ula and the Bogdo-ula, rising to an altitude 500 to 600 feet above the plain on the north, are the prominent elevations of the hilly country skirting them from north-east to south-west, where outcrops of archaic rocks (granite) are frequently seen in crags and heaps on the principal summits, and fragments and splinters on the innumerable hills and mounds, among which the temple Choirin-sume is picturesquely situated to the south-west of Bogdo-ula. This part of the abraded surface of the country has been, to some extent, moulded by the subsequent action of wind and water into groups and ranges of flat hills alternating with hollows of various forms and dimensions. On the top of nearly every hill runs a slightly projecting crest of archaic rocks, generally weathered into angular blocks and splinters. Some of the hollows are of considerable breadth, and six to eight miles long, such as those which contain the salt lakes Dutkhygen-nor and Altkhyn-nor. The country between the southern foot of the Bogdo-ula and the plain Daitkhyn-tala gradually sinks 750 to 800 feet. Across this plain, which lies to the south of Altkhyn-nor, the route runs for a distance of twenty-three miles to the well of Sain-usu, beyond which, as far as the well of Ar-Kheiles, the country is, on the whole, more mountainous. The road crosses several chains of flat-topped mountains between valleys usually considerably broader, to which they descend with very gentle slopes. The absolute height of the valleys varies within small limits, being approximately 2900 to 3200 feet, and an absolute height of 2000 feet, of which Richthofen speaks, was never met with by M. Obruchef.

The next section of the journey, from Ar-Kheiles to the large salt lake Iren-dabasu-nor, was again over flat hills rising more or less prominently above broad, flat lowlands, with in one place a small tableland, Shire-Khairkhan. In this tract are met with almost all the forms of surface seen in the Gobi; the absolute height is 3000 to 3300 feet.

From the Iren-dabasu-nor begins a region of broad tablelands alternating with flat lowlands which stretches to the mountains Tabun-tukhum. The absolute height is much the same, attaining to 3300 to 3600 feet, while the culminating points of the plateau rise 330 to 500 feet above the hollows. The tablelands present the appearance of a nearly level steppe, while the intervening lowlands sometimes assume the form of a slightly undulating steppe, sometimes rise into flat mounds. The plateaus have slopes more or less steep, and their edges are sometimes broken by ravines ascending from the lowlands to the plateau, so that the latter projects here and there in promontories, or totally separated mounds in the form of complete or truncated cones, like those found in the deserts of Africa and Arabia.

With the mountains of Tabun-tukhum commences a more diversified region extending to the hills of Tabun-boglo, and containing mountains of greater elevation, between which lie flat basins sometimes extending to a length of seven to nine miles. In the southern part these basins generally contain small lakes of fresh or brackish water. On the whole, the country ascends considerably, from an altitude of 3770 feet at Tabun-tukhum to 4760 at the hills of Tabun-boglo.

Lastly, the southern border of Eastern Mongolia, on the line of M. Obruchef's route, from Tabun-boglo to the pass before Kalgan, is a gently undulating steppe, with numerous fresh and salt water lakes in broad, flat basins. The absolute height ranges from 4560 to 4760 feet. To the north of the road rise isolated groups of mountains; to the south are seen flat elevations skirted by hills. In these the road attains its greatest altitude of 5330 feet, and then rapidly descends to the narrow valley of Kalgan, at the mouth of which it is only 2710 feet above sea-level.

As regards the directions of the ranges between Urga and Kalgan, M. Obruchef affirms that they exhibit no regularity, and he cannot agree either with Richthofen and others that the crests run from the west to east, or with Fritzsche that they follow for the most part a direction from west-south-west to east-north-east. The latter he observed in very few places, and then mostly in the case of hills not arranged in distinct chains bordered by valleys, but forming extensive groups, among which ran a line of more prominent summits composed of harder rocks—a dyke of quartz or diorite, or a mass of quartzite. In the southern part, too, where the mountains between Tabun-tukhum and Tabun-boglo attain a greater elevation, no regularity in the disposition of the ranges was apparent; nor have reports on the geological structure yielded any evidence to prove the existence of chains running from east to west. The above view is confirmed by the latest and best of existing maps, that published by the Russian General Staff, on which, along the post-roads between Urga and Kalgan, and between Kerulen and Dolon-nor, short and long ranges are represented as running, not only from east to west, but in other directions also, and the tablelands and other elevations are drawn with very irregular outlines.

CENTRAL MONGOLIA AND THE ALA-SHAN.

Central is distinguished from Eastern Mongolia by a larger number of ranges with strongly pronounced strikes—east to west, west-north-west, and north-east. The absolute heights of the more important lowlands varies from 3400 feet in the basin of Gashiun-nor to 5580 feet in the low tract adjacent to the Southern Altai. The altitude of the ranges does not exceed 8200 feet, so that the relative height is not more than 2600 to 3200 feet (mountains of Gurban-Saikhan), and is, as a rule, 650 to 1600 feet. On his journey from the lower course of the Edzin-gol to the Mongolian *yamen* in the Dzolin mountains, M. Obruchef passed two systems of ranges, ridges, and groups, forming apparently an eastern prolongation of the Eastern Tian-shan. Sometimes the elevations are united into a continuous belt, sometimes they are separated by intervening valleys and basins, or, again, fall into distinct groups lying on the same line of dislocations and on the same pedestal. The southern system near the traveller's route comprises the mountains Tostu, Dushe, Noin-boglo, Dzun-adun-kherikte, Gurban-tumyl, Khorimtsuk, and Khokhshuka, while the northern consists of the Ghelbin-tu, Siavri, Dzolin, and Argalin-tu, distinct groups of lofty heights rising from a common pedestal. Between these systems stretches a long valley more than twenty miles broad, in the midst of which the irregular groups Dengh and Tzurum-tai run from the east-north-east.

Between the Dzolin mountains and the mission at San-to-ho on the Whang-ho, M. Obruchef crossed the Tsokho and Shara-tologoi, running from the west-north-west, and the Butyghyn, Kara-obo, Main-tu, Tabun-tukhum, Bain-shanti, and Kara-Naryn-ula, with directions between north-east and east-north-east. These elevations do not for the most part form continuous chains with clearly marked crests carved into peaks and saddles, but a collection of short ridges and groups mounted on a common pedestal and intersected by a labyrinth of ravines, gullies, dales, and basins, through which the path winds, rarely crossing the higher passes. Between the ranges of more clearly marked orientation are scattered elevations of irregular contour, either forming very hilly tracts, or in groups and ridges between valleys and basins. The tablelands built up of the horizontal deposits of the Han-hai, characteristic of certain parts of Eastern Mongolia, are absent here, or, more correctly speaking, are confined to a small area between the mountains Noin-ula and Tabun-takil, where, however, they are much broken by innumerable gullies descending to the Shara-muren, the only real river in Central Mongolia. Consequently, this region is, more markedly even than Eastern Mongolia, a land of hills and mountains dominating hollows.

A similar form of relief characterises the Northern Ala-shan region, which bounds Central Mongolia on the south; but the direction of its ranges is from north-east to south-west. To the south of the Ala-shan, bordering the belt of the oases of Kan-su along the foot of the Nan-shan, are broad expanses of loose sands (Syrkhe, Tyngeri, etc.), bare deserts of clayey sand mingled with detritus, studded with small groups of elevations, which become more numerous to the west, near the valley of the Edzin-gol.

The Ala-shan chain extends from north-north-east to south-south-west for a distance of 160 or 170 miles. From the parallel of Ta-pa on the Whang-ho to the north, it rises on both sides rapidly from the adjacent plains to a height of 4200 to 4800 feet on the west, and 3400 to 3700 on the east. Notwithstanding its small breadth—seventeen miles on the average—the chain is of striking grandeur, for its main crest rises 5000 to 6000 feet above the surrounding plains, while the highest peaks, Baian-tsumbur and Bugu-tui, attain an absolute height of 10,600 to 11,600 feet. The pass on the road from Fu-ma-fu to Nin-sa lies at an altitude of 10,000 feet, and is nearer to the western foot of the mountins than the eastern. The other pass, accessible to beasts of burden, lies on the road from Fu-ma-fu to Phin-lo-sa. It is only by these two passes that the range can be crossed, owing to the steepness of its slopes and its general wild and alpine character. There is no perpetual snow on the crest, and even in winter the snow lies only on the northern slopes, melting on the southern side soon after it has fallen. On the latitude of Ta-pa, about 38° N., the chain ends in a low ridge over which the bridle-road from Lian-chau to Nin-sa passes, at an altitude of probably not more than 5000 to 6000 feet above sea-level. Farther south mountains again appear, but considerably lower than the Ala-shan (not more than 7000 to 8000 feet in absolute height), less steep and rocky, but much more desolate. They occupy an area of some seventy miles from north to south, by fourteen to twenty from east to west, and probably mark the intersection of the dislocation systems of the Ala-shan and Nan-shan. To the north the Ala-shan gradually approaches the Whang-ho, and north of the town Shi-tzi-tze (in Mongolian Kara-Khoto) it actually skirts the bank, whereas on the latitude of Nin-sa it is twenty to twenty-five miles distant.

To the Ala-shan, westward beyond the valley of the Edzin-gol, succeeds the desert of Khami and the Gashiun Gobi, which at the time of the travels of the brothers Grum-Grzhimailo received the name of Pei-shan. This mountainous region was not indeed discovered by these travellers, for some information about it was given by Piasetski, who accompanied the expedition of Sosnovski in 1874-75, and it is fairly indicated on the map of the second and third journeys of Przhevalski, who actually speaks of the mountains "Bei-sian." Nor is the astonishment of G. E. Grum-Grzhimailo at entering, beyond Kufi, into a country of considerable elevation, rising in some parts into ranges of no small relative altitude, at all comprehensible, seeing that the absolute height of the pass by which he crossed the mountain Taben-moo is only a little over 705 feet more than that given by Przhevalski for the well of Ma-lian-chuan, and that the other elevations of the highlands between Kufi and An-si range from 4600 to 6000 feet, or only a few hundred feet above those on Przhevalski's more westerly route.

Other notices also, unaccompanied, indeed, by barometrical observations, portray this part of the Pei-shan as an extensive undulation studded with mountains. The brothers Grum-Grzhimailo, though not the original discoverers, certainly explored it fully, especially its eastern part, for which there were no hypsometrical records. At the same time, M. Obruchef cannot agree with the view, repeatedly expressed in their writings, that the Pei-shan to the east of An-si joins on to the mountain system of Nan-shan, and on the north-east is connected with the mountains of Khami, thus bridging over the interval between the Altai and Nan-shan. With the Tian-shan system the Pei-shan is actually connected, not, however, at the place these explorers point to, but to the west of the meridian of Toksun. With the Nan-shan east of An-si is connected, not the broad undulations of the Khami desert included by the Grum-Grzhimailo under the name of Pei-shan, but a low distinct ridge lying to the south of this region on the boundaries of the Sa-chau and Mo-min plains.

On the east this undulation extends to the Edzin-gol, and in all probability is prolonged beyond it, and still farther, becoming less conspicuous, possibly merges into the Ala-shan. Westwards it runs right to the valley of the Tarim, and, contracting somewhat, is continued from the Bagrach-kul and the river Khaidu-gol, in two ridges which lose themselves in the offshoots or longitudinal ranges of the Tian-shan. The breadth of this undulation from north to south is nowhere considerable, comparatively, ranging from 130 miles on the meridians of Liukchiun and Yu-myn to 80 on that of Sa-chau and at the Edzin-gol, while its absolute height is considerable, ranging in some parts from 5000 to 7000 feet. The ranges placed on this great pedestal do not rise in general more than a few hundred feet above the valleys, though in a few exceptional cases they have a relative height of as much even as 3000 feet (Ta-ma-tun-shan, In-va-shan, Lugan-shan). Still lower are the passes; in most cases with hardly noticeable slope.

As regards configuration, what has been said of Central Asia may be repeated with reference to the Pei-shan. Instead of clearly defined ranges we find a broad belt of low mountains and hills, a labyrinth of valleys and hollows with very irregular outlines and with promontories and bays like those of a coastline; but among these irregularities can also be detected ranges running west-north-west, east to west, or east-north-east, those running from east to west being also combined with others of lesser height lying in a direction from west-north-west to east-north-east. The plains and hollows exhibit the same irregularity, and, as usual in Central Asia, are sometimes arranged in a series or network.

The Pei-shan is skirted on the north and south by broad, deep lowland tracts, running also along parallels, which separate it on one side from the Nan-shan, and, on the other, from the eastern Tian-shan. The southern hollow, 3500 to 4700 feet in absolute height, extending from west-north-west to east-south-east, runs from the Khala-chi (Kara-nor), situated to the north-west of Sa-chau, up the course of the river Buluntsir, and then, passing the dried-up beds of the Khua-Kheitza and Alak-chi lakes, crosses the Edzin-gol between the towns of Tin-ta-sy and Mo-min, and apparently stretches still farther eastwards, embracing the dry lake-beds of Chan-nin-hu and Yui-hai (placed on maps 100 to 110 miles too much to the north), and terminating in the sands of the southern Ala-shan, in the vicinity of the town of Chen-fan, where also there are small lakes. Between the meridians of the towns of Yu-min and Sa-chau the breadth is considerable, and the low country borders immediately on the outer rampart of the Nan-shan, enclosing the succession of oases at its foot. It is here forty to seventy miles across. In its lowest part, traversed by the ramifications of the Su-lei-ho, are cultivated fields and broad patches of reeds and tamarisks, together with salt-marshes and sand-dunes. West of the meridian of Yu-min, the low country is shut off from the Nan-shan by a line of rocky, desolate ranges also called by the inhabitants of the oases Pei-shan, but on Chinese maps Khe-li-shan. The most important of these is known locally as the Lun-tu-shan, Lun-shan, and Lun-kou-shan, and is the most northern of the chains of the Nan-shan system, which, to the east of the meridian of Yu-min, changes its direction from west-north-west to east and west, and approaches the range following on the south, the Richthofen. With the same character as its western portion, this end of the southern lowland, bordering the Pei-shan, continues across the Edzin-gol and its affluent, the Ta-pei-ho, and thence to the meridian of the town of Kan-chau traverses a country unexplored by Europeans. East of the town of Chen-fan it becomes for the most part a desert of clay and sandy *débris*. Thus these lowlands extend, with some interruptions, from the Kara-nor to the town of Chen-fan, or across nine degrees of longitude.

On the northern side the Pei-shan is bounded by a similar lowland tract, which separates it from the eastern Tian-shan. Its breadth is from twelve to more than thirty miles, and from a depression 200 to 300 feet below sea-level on the west it gradually rises to an absolute height of 2000 to 3000 feet on the meridian of Khami; in a still more easterly prolongation apparently are situated the lakes Sogo-nor and Gashiun-nor, which receive the waters of the Edzin-gol, at a height of 3200 to 3300 feet. Though containing a few ridges, groups, and series of hills, it completely shuts off the Pei-shan from the eastern Tian-shan eastwards of the meridian of Urumchi.

Ordos, cut off from Central Asia proper by the bend of the Yellow river, is also distinguished in some measure from the regions just described by its geological structure. The western parts crossed by M. Obruchef may be divided into two belts; from the Yellow river to the hill Burkhan-dabasun-obo it presents flat hills and mounds, rarely attaining an altitude of 600 to 1300 feet, composed of friable sandstone, and in places, especially near the river, covered with heaps and dunes of sand. The southern belt, extending to the boundary of Ordos, chiefly consists of sandy accumulations, among which are found basins containing dried-up or vanishing lakes of salt or fresh water. Similar sands skirt the left bank of the river and stretch northwards to the foot of the Kara-Naryn-ula. Some variation in the relief of Western Ordos is afforded by a high range, called Arbiso and Khantageri, skirting the right bank of the river between Kara-Khoto and Dyn-kou. Apparently it belongs to the Ala-shan system, though not forming a prolongation passing over to the other side of the river, as represented on maps, but a distinct eastern range. Its culminating summits rise 3000 feet above the rest of Ordos, but it diminishes southwards, breaking off between Kara-Khoto and the latitude of Nin-sa, beyond which springs up the chain of Arshanula, only a few hundred feet in height. These elevations are composed of rocks considerably older than the sandstones of Ordos, and are rocky and desolate in appearance.

In concluding his sketch of Central Asia, M. Obruchef gives a brief account of the surface. The base-rock is exposed to view in all parts, but only where it is washed by running water, though but seldom, or covered with sand, does it retain its normal colour and consistency. Owing to the scarcity of vegetation and the dryness of the air in a severe continental climate, the range of temperature is very considerable in the uppermost strata, even within the space of twenty-four hours, especially in spring and autumn, and the rapid alternations of heat and cold continued through many centuries has shattered the rocks into small fragments which, though remaining *in situ*, are really separated from one another by minute cracks and fall to pieces at a light blow of the hammer. The rocks are composed of coarse grains of different colour and composition, being granite, gneiss, conglomerate, or sandstone, disintegrated for some inches below the surface. Consequently, on the more gentle slopes and on the flat hills and mounds the ground is covered with a thin layer of gravel and angular fragments of the rocks beneath, mixed with small quantities of sand and clay, and therefore, while it is easy to ascertain of what rocks the hills are composed, it is hard to examine their bedding and their interconnection.

This gravel, clay, and minute fragments of rock is the most widespread soil of Central Asia. The central portions of the low grounds have a rough clayey soil, more or less bare, and covered with a white saline efflorescence, and sometimes contain bitter pools surrounded by piles of sand. On the outskirts of Central Asia, also, the surface is somewhat altered; for instance, in Eastern Mongolia, to the north and south of the Gobi proper, are steppes of löss.

The distribution of atmospheric precipitation, more abundant in the eastern parts, which are comparatively nearer to the ocean than the western, is indicated by the vegetation. In Eastern Mongolia there are no true deserts, but the ground is covered with grass, sometimes, indeed, so scanty that the soil peeps out between the stalks, but still sufficient to feed horses and cattle, and on the splendid steppes of the southern border to yield a crop of hay for winter use. In Central Mongolia shrubs are the prevailing forms of vegetation—*Ephedra, Caragana*, saltwort, wormwood, etc.; grass is much rarer. On the eastern borders of Pei-shan and Ala-shan, and in Eastern Dzungaria are found patches of true desert. In Central Asia trees are seldom seen; near the running water at its outskirts grow thickets of willow, osier, and tamarisk, and to the north, on the northern slopes of the mountains, appear larches, aspens, and birches.

One of the peculiarities of the country is the scarcity of flowing water. Occasionally are found streams bordered by a belt of more luxuriant vegetation than grows elsewhere, but these soon lose themselves in the sand, break up into numerous arms, and irrigate beds of reeds and shrubs, or empty themselves into salt lakes. Water for the use of men and animals is generally obtained from springs and wells springing up in gorges and in the middle of hollow basins.

The Eastern Tian-shan.

This immense range, the Karluk-tagh of the Turks, forms a climatic barrier between the cold Dzungaria and the warm Kashgaria. To the east of Urumchi the Tian-shan forms a massive, rocky, and on the southern flank very barren, range, consisting of several steep-sided chains. Its principal crest, springing from the insignificant hills of Dun-shan, that lie across the great road from Turfan to Urumchi, quickly reaches its culminating point in the gigantic group of Bogdo-ola (the Turpanat-tagh of the Turks), a twin peak rising 17,000 to 18,000 feet above sea-level, and covered with eternal snow. Beyond this peak it sinks considerably, being probably, to judge from the small amount of snow that lies on its southern flank, not more than 13,000 feet in height. To the east of Bogdo-ola the main chain approaches, and at length completely coalesces with, a second, more southern, chain, the Jargoz, forming a connection between the parts of the Tian-shan east and west of Urumchi. This chain runs unbroken from west to east, skirting on the south the basin containing the lakes Seiopu and Aidin-kul, whereas the main chain on the meridian of Urumchi dwindles down to low hills. The Jargoz is marked by very steep sterile flanks and a rocky, in some parts very jagged, crest, rising to an elevation of 9000 to 10,000 feet.

East of the meridian of 60° E. of Pulkova (nearly 90° 20′ from Greenwich), the whole range begins to sink, and on the meridian of 62° no longer presents any peaks above the snow-line. On the meridian of the station I-van-chuetze it begins to rise again. West of this meridian, to the low and deeply eroded section bounding on the north the basin of Tzi-ghe-tingze is attached the high and steep Chogluchai chain, which at its eastern end shuts in the basin on the south, and runs from west-north-west to south-south-east. Here also the above-mentioned southern range is reduced to low rocky hills, and between the two is left a broad barren space enclosing the basin Tzi-ghe-tingze. These mountains G. E. Grum-Grzhimailo regards as the advanced ridges of the Pei-shan, and hence locates here a close connection between the Pei-shan and Tian-shan; but there are no grounds, orographical or geological, for grouping these ranges with the Pei-shan, from which they are separated by broad tracts of lowlands.

Between the meridians of I-van-chuetze and Khami the range increases in altitude, the southern range now sending up peaks above the snow-line, and maintaining the same steep and rocky character. Some deep indentations cut it into short segments, giving passage to the bridle-paths from Jigdy, Tu-pu, and Khami to Barkul. On the meridian of Khami is a very broad gap, to the east of which the range rises for the last time into the region of eternal snow, forming a high but not very extensive group, called by G. E. Grum-Grzhimailo the mountains of Khami or Karlyk-tagh. Its crest, broken by a few narrow necks, rises above the snow-line, and its rocky peaks probably attain an altitude of 14,000 to 15,000 feet. Lastly, past the meridian of 65° the Tian-shan falls rapidly, and is changed into ridges of typical Central Asian character, sometimes rocky, sometimes smooth, of which the tectonic and orographical continuation may, perhaps, be found in the Tostu and Noin-boglo mountains in Central Mongolia. West of the meridian of Khami the foot of the range is displaced a little to the south, and in the Karlyk-tagh group the two parallel ranges of the western portion can no longer be traced.

The dreary southern slope of the Eastern Tian-shan, poor in vegetation and running water, is skirted by a zone of most desolate wastes. The rivers and streams, formed of water from melted snow and springs, are generally lost in the alluvium of their beds at no great distance from the foot of the mountains, and are continued only by dry courses where a few bushes grow. On the larger streams and at the mouths of the ravines are small oases, which drain the water from the streams and leave their beds dry. Between these beds the ground over many square miles of area is destitute of vegetation, and consists alternately of sand, clay, pebbles, and fragments of rock. This belt of desert is succeeded on the south by a line of oases, inhabited by Turks, Dungans, and Chinese. Here, side by side with perfectly sterile desert, are seen thickly peopled tracts, rich in vegetation and highly cultivated. The oases are well supplied with water that wells up out of the soil, for the deep layer of pebbles and detritus, engulphing almost all the water that streams down from the mountains, is thinned out in some places, on the slope to the northern flank of the Tus-tagh and Chiktym-tagh, while in others it is broken up into troughs and gullies; in the former case we have the Turfan and Mor-gol oases, in the latter those of Khami. Between the last Turfan oasis at Chiktym and the nearest Khami oasis at Liao-dun, there is an interruption in the succession for a distance of some eighty-five miles, for the water that descends from the Tian-shan south of Tzi-gha-tingze, where the range is low, is insufficient to do more than feed a few wells and springs.

The third belt has been already mentioned as the broad expanse separating the Pei-shan from the Tian-shan. Beginning to the west of Toksun, near the meridian of 88° 20′ E., it runs eastwards, now expanding to a breadth of more than thirty miles, now contracting to twelve, and sinks in the neighbourhood of Toksun and Liukchiun to a depth of 200 to 300 feet below sea-level. Between these places it is covered with vegetation kept alive by the water of the Tian-shan which has not been already appropriated to the irrigation of the Turfan oases. From the foot of the Yamshi-tagh, the Tuiuk-tagh and the Syrkyp-tagh on the north to where the land slopes up to the northern fiank of the Chul-tagh, extends a stony desert over which are scattered oases with pasture lands and gardens beside the brooks and springs. To the east of Liukchiun the character of this hollow changes. Bare hills of loose sand occupy nearly the whole breadth of the valley, and, attaining a height 650 feet above the adjoining country, look at a distance like a mountain range, and hence are known as the Kum-tagh or Sha-shan (sandy mountains), and on the map of the brothers Grum-Grzhimailo are even represented as an offshoot of the Chul-tagh, from which, however, they are really separated by a passage more than three miles broad, through which runs the road to Khami. At the eastern end of Kum-tagh the two branches re-unite, and the valley, regaining its normal proportions, runs on like a huge trough to the Shona-nor, studded over with flat hills, ridges, and tablelands of the red deposits of the Han-hai, which are typical examples of the work of deflation. Beyond the lake the bottom rises above sea-level, and gradually ascends towards the east, finally joining the heights of the Pei-shan.

M. Obruchef also describes the country east of Khami, on the northern slopes of Pei-shan, besides giving a long account of North-Western China which must, for the present at any rate, be passed over. One interesting point, the rainfall at Liukchiun, may be briefly noticed. At this meteorological station from the middle of November 1893 to September 8th, 1894, a sprinkling of rain fell five or six times, and three times there was a fall of barely one inch of snow, which disappeared on the following day; and in the Khami oasis the precipitation appears, from inquiry, to be no greater. Yet the dry beds and troughs evidently excavated by running water prove that heavy showers must have occurred, if only once in several years, and the natives speak of such deluges of rain.

THE BRITISH GUIANA FRONTIER.

VENEZUELA claims all the territory of British Guiana west of the Essequibo, alleging that this river formed the boundary of the Captaincy-General of Venezuela up to 1810, when the Spanish-American provinces revolted against the mother-country.

On the other hand, Great Britain has always held that the territories of Demerara, Berbice, and Essequibo, formally ceded to her by the Dutch in the Treaty of Paris, 1814, embraced the whole basin of the Essequibo and its tributaries, as well as those smaller rivers on which the Dutch had had posts or trading stations. These claims are further supported by subsequent occupation and exploration. Thus the British territory would extend westwards to the thin dark line on the map.

Between 1840 and 1844 Robert Schomburgk was commissioned by the British Government to explore the outskirts of British Guiana, and drew the Schomburgk line enclosing those points where he found remains of Dutch posts or which were formerly, according to the evidence of natives, under Dutch influence. This line, as shown in the map, is drawn from a map in Richard Schomburgk's *Reisen in Britisch-Guiana in den Jahren* 1840-1844 (Leipzig, 1847), compiled from surveys executed by Sir R. Schomburgk and deposited with his map in the Colonial Office. As indicated by the broken line in the map, it starts from the mouth of the Amacuru river (marked Victoria Point on Schomburgk's map), and, crossing the Cuyuni, proceeds southwards and eastwards to the source of the Essequibo and the Corentyne river.

Lastly, between the extreme boundary claimed by Great Britain and the Schomburgk line is another, marked on the map with dots and dashes, which has also been associated with Schomburgk's name. From Mr. G. G. Dixon's map, published in the *Geographical Journal*, Jan.-June, 1895, we learn that it is " shown on Sir Robert Schomburgk's Map (1875) "; but on what authority it was drawn on this map—for Schomburgk died ten years before its publication, and does not seem to have visited Guiana after 1844—we are unable to state.

In its southern part, however, from the source of the Acarabisi creek, it coincides with the boundary lines proposed by British Foreign Secre-

Scottish Geographical Magazine 1896

taries during the last fifty years. In 1844 Lord Aberdeen proposed a line beginning at the mouth of the river Moroco and running straight to the confluence of the Barama and Waini; thence it ascended the Barama and Aunama to the nearest point to the Acarabisi creek to which it crossed, and then followed the same course as the "extended Schomburgk line," and, from Roraima, the Schomburgk line proper, to the Essequibo and Corentyne.

Again in 1881 Lord Granville laid down a frontier starting on the coast at a point twenty-nine miles east of the Barima river and running south to the Arikita hill, on the parallel of lat. 8° N., whence it ran due west to the Schomburgk line, leaving it again farther south for the Acarabisi; but Lord Rosebery in 1886, beginning at a point on the west of the Waini river, drew his frontier, as far as Roraima, almost entirely within the Schomburgk line (*Correspondence between the Venezuelan Government and H.B.M.'s Government*, Caracas, 1887).

Lord Salisbury maintains the unimpeachable right of Great Britain to the possession of "the territory within the Schomburgk line"; the question of the ownership of the rest of the Cuyuni basin he is willing to submit to arbitration. (See Lord Salisbury's despatch of November 26, published in *The Times* of December 18.)

Robert Hermann Schomburgk was born in 1804 at Freiburg in Prussian Saxony. In 1835 he was appointed by the London Geographical Society to the leadership of an exploring expedition in British Guiana, during which he discovered the magnificent water-lily, the *Victoria regia*. In 1840 he returned to Guiana to survey the colony for the British Government. This work he accomplished most successfully, and at its conclusion received the honour of knighthood. He was subsequently British Consul at San Domingo and British representative at the Siamese Court. The latter post he resigned in 1864, and in 1865 he died at Schöneberg near Berlin.

PROCEEDINGS OF THE ROYAL SCOTTISH GEOGRAPHICAL SOCIETY.

Lectures delivered in January.

Dr. Markoff lectured, on January 6th, at Dundee, on "Russian Armenia and Prospects of British Trade there." Mr. John Robertson presided, and the Rev. D. M. Ross proposed a vote of thanks to the lecturer.

The following evening Dr. Markoff addressed the Aberdeen Branch on the same subject. Mr. James Spence took the Chair, and Bishop Douglas proposed a vote of thanks.

At Edinburgh, on January 9th, Dr. Markoff delivered the lecture on "The Towns of Northern Mongolia" published in this number. Lord Dalrymple presided.

On January 21st the Rev. Walter Weston delivered a lecture in Glasgow on "Travel and Exploration in the Japanese Alps." Professor G. A. Smith presided.

This lecture was repeated in Edinburgh on January 24th, when Professor Calderwood presided.

On January 30th, in Edinburgh, Sir David P. Chalmers, late Chief Justice of British Guiana, delivered a lecture on British Guiana. Professor Butcher was in the Chair.

LECTURES IN FEBRUARY.

Miss Kingsley will address the Society in Edinburgh on February 6th at 4.30 P.M. Her subject will be the results of her recent travels on the West Coast of Africa.

Miss Kingsley will also lecture to the Glasgow Branch on February 7th.

At Aberdeen a lecture on Japan will be delivered by Mr. Jas. Troup, H.B.M. Consul at Yokohama, on February 6th.

A meeting will be held in Edinburgh on February 20th at 8 P.M., when Professor Clifford Allbutt, F.R.S., will lecture on "Ancient Trade."

PRESENTATIONS TO THE LIBRARY.

At a meeting held on January 17th the following list of books, maps, etc., presented to the Library during the year 1895, was read before the Council, which desires to record its thanks to the private donors who have added these volumes, and also pamphlets, etc., to the Society's collection :—

	Presented by
Navigationi et Viaggi, raccolti da M. G. B. Ramusio. 3 vols.	
Raccolta di Documenti e Studi publicati dalla R. Commissione Colombiana. Part v. vol. ii. . .	Clements R. Markham, Esq.
Philosophical Transactions of the Royal Society of London. 53 vols.	H. M. Cadell, Esq.
On the Geographical Distribution of Diseases in Africa, by R. W. Felkin, M.D.	The Author.
Report on Oceanic Circulation, by Alex. Buchan, M.A., LL.D. ("*Challenger*" Reports) . . .	The Author.
Lieut. Irving of H.M.S. "Terror," edited by Benjamin Bell, F.R.C.S.E.	R. C. Bell, Esq.
Palæolithic Man in N.-W. Middlesex, by J. Allen Brown, F.G.S.	J. Allen Brown, Esq.
The North-West Passage and the Plan for the Search for Sir John Franklin, by John Brown . .	
Voyages of Discovery in the Arctic and Antarctic Seas, by R. M'Cormick, R.N. 2 vols. . . .	
Manuel de Conversation en Trente Langues, par Dr. E. Poussié	The Author.
Travels and Researches in Chaldæa and Susiana, by W. Kennett Loftus; *A History of the Kingdom of Guatemala,* by Don. D. Juarros; and several other narratives of Travel—21 volumes in all .	J. E. Mathieson, Esq.
Lakes of North America, by Professor I. C. Russell .	The Author.
Guide to Scotland. Part I., by M. J. B. Baddeley .	The Author.
Old Edinburgh Geologists, by R. Richardson, . .	The Author.
Journal of the R. Asiatic Society. 22 Parts . .	Coutts Trotter, Esq.
Carte del' Empire Ottomane, 1822 . . .	"
Chart showing the Route of the "Experiment Snow," 1785-6	

Chart of the N.-W. Coast of America and N.-E. Coast of Asia, explored in the years 1778 and 1779 .	*Presented by* Coutts Trotter, Esq.
Map of the River Quorra from Rabba to the Sea, and a portion of the River Tchadda, 1833 . .	
Forma Urbis Romæ, delineavit R. Lanciani. Fasc. 1-3	Hugh Rose, Esq.
Map of the Arctic Seas, 1854	G. H. Johnston, Esq.
Atlas des Lacs Français, par A. Delebecque . .	The Author.
Mapa Geogr. y Corografico de Bolivia . . .	Alex. Taylor, Esq.
Four Japanese Maps of the Islands of Japan . .	A. B. Young, Esq.
Twenty Photographs of Early Navigators, Old Maps, etc., with slides of the same . . .	E. Delmar Morgan, Esq.
Thirteen Photographs of Views near the Musart Pass, Tian-shan	
Nineteen Photos of Views, etc., in New Guinea, Fiji, and the New Hebrides	Baron F. von Mueller.
Photograph of the Ruins of the Convent of San Francisco, Argentina	Andrew Mackintosh, Esq.
Photographs of Cape Adare and N. Possession Island .	H. J. Bull, Esq.

GEOGRAPHICAL NOTES.

By The Acting Editor.

EUROPE.

The Climate of the British Empire in 1894.—There is a note and table on this subject in *Symons's Monthly Meteor. Magazine* for December, which concludes with the following summary :—

Highest temp. in shade, .	107°·0 at Adelaide on Nov. 26.
Lowest ,, ,, .	− 46°·1 at Winnipeg on Jan. 24.
Greatest range in year, .	141°·9 at Winnipeg.
Least ,, ,, .	25°·8 at Colombo, Ceylon.
Greatest mean daily range, .	23°·1 at Winnipeg.
Least ,, ,, .	10°·6 at Bombay.
Highest mean temp., .	80°·8 at Colombo.
Lowest ,, ,, .	36°·5 at Winnipeg.
Driest station, . .	Adelaide, mean humidity 63.
Dampest ,, . .	Esquimalt, ,, 88.
Highest temp. in sun, .	177°·0 at Trinidad.
Lowest temp. on grass,[1] .	− 11°·2 at Toronto.
Greatest rainfall, .	77·46 in. at Colombo.
Least ,, .	18·12 in. at Winnipeg.
Most cloudy stations, .	London and Esquimalt ; average amount 6·3.
Least ,, ,, .	Bombay and Grenada ; average amount 4·0.

The above, of course, refers only to the records regularly kept and published.

Relation of Humidity to Atmospheric Pressure.—At the Royal Society of Edinburgh, on January 6th, Dr. Buchan spoke of the high temperature of September

---

[1] Not recorded at other Canadian stations.

last at the Ben Nevis Observatories. After indicating briefly the weather of the month, which was strongly anticyclonic—the pressure being above the average every day except from the 9th to the 12th and on the 18th—he discussed at length the weather of the last three days, when the meteorological peculiarities were most strongly accentuated. On these days the atmosphere at Fort William was all but saturated, except in the middle of the day, when the increased heat of the sun's rays raised the temperature without adding to the vapour. The same condition prevailed over the whole of Scotland at low levels. Professor Tait and other physicists have been of the opinion that when the moisture in the atmosphere is in the condition of pure vapour it is practically diathermous—that is, offers no resistance to the passage of the sun's rays. Now, in the last days of September the atmosphere at the summit of Ben Nevis was excessively dry, and the barometer reading at the top, when reduced, was half-a-tenth of an inch higher than at Fort William, while the difference of temperature was 4° F. With this temperature the difference of pressure should have been one-hundredth instead of five times that amount, while, if the dryness had continued to the foot of the mountain, the readings would have been about the same. Dr. Buchan pointed out the importance of the result as regards the detection of coming storms, and Professor Tait remarked that a great principle had been for the first time established, namely, that the specific gravity of the air increased with its dryness.

Finnish Lapland.—Herr Nath, a landscape painter, and Lieut. Schnackenberg travelled through this country last summer, and made several valuable observations. Some of these tend to prove that the chief weathering agent in the north is the acids of the vegetable soil ; they produce their effect by removing the ferruginous binding matter from the soil. In warmer climates these acids have no apparent effect, unless in the mountains and in very poor soil, but in the north they are more powerful than carbonic acid. The formation of moors from swampy woods was observed over a very large tract. The displacement of the tree limit is due to a large extent to the action of man, and especially to forest fires. The birch has disappeared from the neighbourhood of Vadsö, where it occurs in large quantities in the peat, and the pine has vanished from the country north of Raututunturi, having been destroyed chiefly by fire.—*Verh. der Gesell. für Erdkunde zu Berlin*, Bd. xxii. Nos. 8 and 9.

The Deepest Boring in the World.—In vol. ix. p. 484 the boring at Paruschowitz, near Rybnik, in Upper Silesia, was mentioned as the deepest in the world. On May 17th, 1893, it had reached the depth of 6569¼ feet. When work was recommenced after a few weeks about another yard was bored, and then the rod broke short off, and could not be pulled up. The work, which had cost £3611, was therefore discontinued. The bore-hole at Schladebach, near Merseburg, is not so deep by 836 feet, but as Paruschowitz stands 833 feet above sea-level, and Merseburg 335, the boring at the former place has penetrated only 338 feet nearer to the earth's centre. The boring at Paruschowitz is a Government undertaking, and its object was to inquire into the coal seams. Eighty-three seams have been found, with an aggregate depth of 293½ feet. The temperature records, 384 in number, show a very irregular increase of temperature ; the average was 1° F. for every 62·23 feet.—*Geogr. Zeitschrift*, Jahrg. ii. Heft 1.

The Lakes of Tver, Pskof, and Smolensk.—The lakes of the districts of Ostashkof in the Government of Tver, and of Toropets in the Government of Pskof, lie in a region of moraines. The country exhibits unmistakable signs that it was once covered with the ice of a great Scandinavian glacier. Morainic *débris* in layers of various depths, consisting of loam and sand mixed with fragments of crystalline rocks, or heaps of stones intermingled with sand, are found all over the country. These deposits are not always laid down evenly, but in general rise in a series of oval hills following the supposed direction of the glaciers, or sometimes at right angles to their course, being in the latter case terminal moraines, and showing the limits of the ice at different epochs. This configuration of the country and the impermeability of the clay and loam present conditions favourable for the formation of lakes, which are common here, as in all other formerly glaciated regions.

In the districts of Toropets and Ostashkof lakes are never found in caldron-shaped hollows. But a characteristic example of this type occurs in the Government of Smolensk. Near the town of Bielsk (Bieloi ?) lies the Bezdonnoie lake, with a length of 185 yards, and a breadth of 130. It looks like a small caldron or broad funnel, and is 40 feet deep.

In this part of the country a broad, shovel shape, or an irregular, circular form, is the rule, such as that of the largest of the lakes that have been examined—the Dwinie with the Velikoie (33 square miles) and the Shiskizkoie (36 square miles). They are shallow, none being over 26 feet deep, and most of them only 13 to 20 feet. Some of the smaller ones are rapidly turning into swamps, and many have no doubt already disappeared. Long narrow lakes are less frequent. In general they run from north, north-west, or north-east to the opposite points of the compass, though there are a few, the Selskoie for instance, which extend east and west. They have, in general, a greater depth than those of the former type, though their area is much smaller. The Vin is 121 feet deep. But their average depth is much less, and the uneven bottom consists of a succession of elevations and channels, which show that these lakes have been created by the union of small converging streams. The most characteristic lakes extend in the direction of the vanished glaciers, and the hollows in their beds have probably been eroded by the water melted from the glaciers. Unlike the former class, these lakes have seldom any outlet.

Temperature observations have been made in many of the lakes at various depths. In the shallower lakes there is little variation, the whole mass of water having a summer temperature of 59° to 64° F. In the Svaditsa, which may be considered the type of a warm lake, the temperature to a depth of 30 feet varies only $\frac{1}{2}$° to $5\frac{1}{2}$°, but there is rapid fall from 64° to 59° between the depths of 20 and 30 feet. These observations were taken in the afternoons in the latter half of June. There was a very perceptible change in the evening, the surface being cooled in a higher ratio than the water below, while the deeper strata were comparatively warm, the temperature at the bottom being 61·3°. The Besdonnoie lake, which is far smaller but somewhat deeper, exhibited a complete contrast in its temperature conditions. Here at eight o'clock in the evening, with an air temperature of 56·3°, the temperature of the surface was 64°. Between 6$\frac{1}{2}$ and 19$\frac{1}{2}$ feet it fell from 62° to 41·4°, or nearly 1·6° per foot, and in the lower half of the lake there was the latter low temperature. The difference of temperature in the two lakes is very striking, for at the same level it was 16° to 22° when the temperature of the air was the same. One of the causes of this discrepancy is no doubt the greater depth of the Besdonnoie lake compared to its area. Another is that while the Swaditsa lake is fed exclusively by small streams, the Besdonnoie receives most of its water from springs. Some of the other lakes also exhibit peculiarities of temperature.

The fauna of the deeper, again, is different from that of the others. Bleak are found in the deeper parts of the Vin lake, and also, it is said, in the Otolovskoie, and the catching of them is pursued as an industry.—*Globus*, Bd. lxviii. No. 21.

Bosnia and Herzegovina.—The *Vorläufige Ergebnisse der Volkszählung vom 22 April* 1895 show that the population of these provinces, on the date named, amounted to 1,565,357, as against 1,336,091 on May 1st, 1885. The increase, therefore, has been 229,266, or 17·16 per cent. The towns in particular have received a large addition to their population, especially the capital, Sarajevo, where the numbers for 1885 and 1895 are ¡26,268 and 41,173 respectively, giving an increase of 11,445, or 43·57 per cent. Dónja-Tuzla has received an addition of 42·13 per cent., and in the other chief towns of departments the increase has been fairly large, except in Travnik. As regards the individual districts, the results of the two censuses cannot readily be compared, for several of the departments have received an addition of territory, while others have been considerably reduced. Again, in 1885 all the residents were counted, whether present or absent on the census day, while in the census of 1895 only those in the country on the particular day (natives and strangers) were included.

ASIA.

The Little Ararat.—In the autumn of 1894 Professor A. Arzruni visited the southern part of Transcaucasia for the purpose of examining its singular volcanic region. Being joined at Erivan by Lieutenant Pastukhof, whose observations of temperature on Mount Ararat have been noticed in these pages, he rode to the saddle between the two summits of Ararat and encamped at the Sardar-Balagh or Governor's Spring, so named because during the Persian rule the Sardar of Erivan used to resort thither in summer. Next morning M. Pastukhof set off for the glacier on the Great Ararat, and advised Professor Arzruni to scale the lower peak. This summit, 13,210 feet high, is even more troublesome to climb than the other, owing to its steepness, but the way is much shorter and the ascent can be made in six hours. In some places the slope is as much as 38°, at others huge blocks of stone have to be surmounted, and again the climber stumbles about in loose rubbish. This rubbish is not volcanic ash, but is produced by the disintegration of the lava. In some places it consists of sand, in others of stone fragments and large blocks which threaten every moment to roll down.

Though several scientists, among them some geologists, have ascended the mountain, none has described its crater. It has a regular semi-circular edge which slopes rather steeply both on the outside and inside. The inner side is formed of fine ash, which also covers part of the bottom. In the middle of the crater rises a cone of later origin, now weathered into a mass of fissured blocks. At its foot was found a clear spring of water with a temperature of 30° F. The apex has been shattered by the frequent thunderstorms that burst over the summit. The eastern half of the crater edge is gone, having been carried away, no doubt, by the later lava streams, which, though now fissured and eroded, still raise their rugged and imposing outlines above the plain. The rock of both the Great and Little Ararat is andesite, of every variety of colour and grain, successive eruptions have evidently poured forth materials essentially the same. There is very little vegetation on the mountain. At its north-western foot stands a small birch wood, but higher up the slopes are bare, save for a few herbs. A kind of forget-me-not, with an odour of heliotrope, grows in the shelter of the blocks where there is a little moisture, and a peppermint is fairly common. The temperature of the air on the summit at 4.38 p.m. was 51° F.

On the descent the party came across a huge furrow on the north-western side. In it lie masses of ice of various dimensions, protected from the sun by the rubbish overlying them. In time, however, the heat gains the day, and the loosened mass falls as an avalanche down the gully.

The fulgurites of the Little Ararat, often described already, suggest the conclusion that the maximum discharges of electricity take place at a certain elevation depending on the latitude. This level is at 13,000 feet on the Caucasus, which happens to be that of the snow-line. Probably the form of the summit has something to do with the frequency and violence of storms, for fulgurites are not common on Aragatz, the four summits of which are not much higher than the Little Ararat. Another singularity is a burial-ground on a flat space just below the uppermost cliffs of the summit. One of the gravestones is in the form of a St. Andrew's cross. It has been asserted that some of them bear inscriptions in the Tartar language and in Arabic characters, but Professor Arzruni did not see any.—*Verh. der Gesell. für Erdkunde zu Berlin*, Bd. xxii. Nos. 8 and 9.

The Kafirs of the Hindu Kush.—Colonel Holdich writes on the origin of this people in the *Geographical Journal* for January. They claim for themselves a Greek origin, or at least some of the tribes known under the name of Kafirs claim this distinction. Probably they are a mixed race, chiefly of Tajak blood, who formerly dwelt in the lowlands of Badakshan, but have been forced into the mountains by the encroachments of landgrabbers. Many of the Kafir tribes are hardly known at all, most of our information relating to the Kamdesh Kafirs of the lower Bash-gol, who were visited by Sir George Robertson. He described the people as of Aryan type (see vol. x. p. 539).

In Arrian's history of the invasion of India by Alexander the Great, the monarch is said to have marched from Dyrta (Dir) towards the Indus, and to have entered that part of the country which lies between the two rivers Kophenes (Kabul) and Indus, where Nysa is said to be situated. This city was, according to its inhabitants, built by Dionysos and named after his nurse, while the adjacent mountain Merus commemorated his miraculous birth. Further, in his *Indika*, Arrian distinguishes between the Assakenoi (who inhabited the upper Swat valley east of Nysa) and the Nysaioi. The former he describes as "not men of great stature like the Indians . . . not so brave nor yet so swarthy as most Indians." "The Nysaioi, however, are not of an Indian race, but descendants of those who came to India with Dionysos." This Dionysos, according to Megasthenes, was a beneficent conqueror who taught the Indians to make wine and cultivate their fields, and was regarded after death as a deity. Colonel Holdich quotes other passages from ancient authors, and mentions that among the inscriptions in the Kunar valley are some that recall a Greek alphabet of archaic type. There is also a sculpture which may, perhaps, represent a Bacchic procession.

Not long ago, when surveying in the Kunar valley, Colonel Holdich heard the natives sing a verse of a hymn to their god, Gish, in which Gir-Nysa, or the mountain of Nysa, was mentioned. Among other names that occurred in it was Katan Chirak, an ancient town in the Minjan valley of Badakshan, now in ruins. It was the first large place taken by the Kafirs, and is apparently held as symbolical of victory. Thus the Kamdesh Kafirs are connected with Badakshan. From these and minor indications Colonel Holdich concludes that the Kafirs are a very ancient western race, and descended from the Nysæans. The notes on the Nysæans referred to above may be found in Mr. M'Crindle's translations of Arrian and Megasthenes, and in his *Ancient India : Its Invasion by Alexander the Great.*

The Survey of India.—During the year 1893-94 the principal triangulation in Upper Burma on the meridian of 96° 30′ E. long. was extended northwards over a distance of ninety miles, reaching the parallel of 24° 40′ N., and embracing an area of 1480 square miles. Along the mouth of the Indus beacons were set up at suitable intervals to furnish a basis for the operations of the Marine Survey, and seventeen of them were fixed by means of a secondary series of triangles carried along the coast. Topographical surveys were executed in the Bombay Presidency, Baluchistan, and the Himalayas. The survey of the Indus in Sind was continued, and a detachment was formed to complete that of the Yafi'i country in the Aden peninsula. The topographical survey of Burma was commenced. The areas sketched on various scales amounted in the aggregate to 16,072 square miles. Traverse surveys were carried out over 3572 square miles in the Central Provinces and Bengal. Reconnaissance parties were engaged in Persia, Arabia, and geographical surveys were executed in Upper Burma over 10,870 square miles, part in the southern Shan States, part on the Anglo-Siamese boundary, besides 2550 square miles in the Chin-Manipur hills and 1370 in the Kouke country.

The appendix to the *Report* contains several papers of considerable interest. Colonel Holdich describes the antiquities, ethnography, and history of Las Bela and Makran in Baluchistan, and Captain S. G. Burrard deals with latitude operations and local attraction, particularly to the south of the Himalayas and along the coast. The locus of points where the influence of the Himalayas first disappears runs through Lucknow, Aligarh, Hansi, and Sialkot. Nothing is known of the deflection between Lucknow and the Brahmaputra. Lucknow, Aligarh, and Hansi are all about 125 miles from the nearest point of the Himalayas, which in each case lies to the north-west, whereas Sialkot is only fifty-five miles distant. At Amritsar, only fifty-five miles off, there is a southerly attraction of 4″, and this is the more remarkable because flat unbroken plains lie to the south and west. A subterranean source of attraction apparently exists in the desert around Montgomery, affecting the plumb-line at both Amritsar and Mooltan. There are also good reasons for believing that along the coast of India, from Bombay to Coconada, the plumb-line is attracted towards the ocean, though in the Bay of Bengal there is no sign of such attraction.

The Pamir Boundary Commission.—Before the present commission was appointed, the Russian triangulation had been extended to Osh, about 200 miles from Lake Victoria, and, on the Indian side, several peaks south of the Hindu Kush and Mustagh ranges had been fixed by Colonel Tanner. On the journey to the boundary, Colonel Wahab who, under Colonel Holdich, was in command of the survey party, was able to fix a series of Himalayan peaks from the Benderski pass. At the first meeting of the British and Russian surveyors, it was decided to select two points for a comparison of the latitudes, longitudes, and altitudes obtained, and a pillar set up at the east end of Lake Victoria and the Gumbaz, or dome of Kizil Robat, were chosen. At the pillar the Russian value for latitude was 37° 26′ 10″, and the English 37° 26′ 30″, or a difference of 20″. At Kizil Robat Gumbaz the difference was only 10″. As regards longitude the results were much closer, the difference being 4″ at the pillar and 6″ at Kizil Robat. Forty well-marked points were fixed by triangulation, 4800 square miles of topography were sketched on the quarter-inch scale, and 250 miles of special survey to illustrate doubtful points. The altitudes ascertained by the Russian and English surveyors were singularly close, the altitudes of Kizil Robat (12,780 feet) being identical, and in the case of Lake Victoria differing by only 40 feet.—*Geographical Journal*, January.

Prince Henry of Orleans.—This explorer has arrived at Sadiya in Assam after a journey of 2000 miles from Tonkin, 1500 of which were through unexplored country. Ascending the Red river to Man-hao, he visited Mong-tse, and then, turning westwards, travelled through country inhabited by tribes recognising the nominal suzerainty of China but virtually independent. Of these may be mentioned the Honins, who dwell exclusively in the high mountains, and the Païs, of pure Laotian race, living mostly in the warm valleys. Eight days of arduous travel from Semao brought the party to the Mekong at Tian-pi, here flowing between precipitous cliffs 4000 feet high and more. The country through which the river flows in this part of its course is uninhabited, for there are no meadows on its banks where cultivation can be pursued. The current is moderate, but frequent rapids and rocks render navigation impossible. With a breadth of 100 to 160 yards, it has enormous depths; a sounding line of 130 feet did not reach the bottom. This character the river maintains from the 22nd to the 23rd degree of latitude. Marching northward, Prince Henry came to Ta-li-fu, whither he had been preceded a month before by Captain Davis.

Of the journey up to this town some notes are given in a letter from one of the party, M. Émile Roux, published in the *Tour du Monde* of October 12th. Further details have appeared recently in the *Times* of Jan. 11th. From these we learn that the source of the Irrawadi was discovered about thirty miles north of lat. 28°, between 98° and 99° E. long. The main stream, known locally as the Tourong, becomes lower down the N'maika of British geographers (see vol. viii. p. 260). Eleven streams in all drain the mountain range south of Tibet, three being longer than the others. The Tourong, the largest of all, lies farthest east. The Mekong, Salwin, and Tourong are separated by high ranges preventing any possible connection between them. The Prince and his companions crossed seventeen ranges between Tsikon on the upper Mekong and Sadiya. Some of these were 13,000 feet high, while adjacent ranges showed still higher peaks.

Siam.—On Jan. 15th an agreement between Great Britain and France, regulating the affairs of Siam, was signed in London. By this it is agreed that neither of the Powers will, without the consent of the other, send troops into the basins of the rivers Pecha, Buri, Mei-klaung, Menam, and Bang-Pa-Kong (Petriou), and their respective tributaries, together with the extent of coast from Muong-Bang-Tapan to Muong-Pase, including the basins of the rivers on which these places are situated, and of those entering the sea along the intermediate coast, and also including the territory lying to the north of the basin of the Menam, and situated between the Anglo-Siamese frontier, the Mekong and the eastern watershed of the Me-Ing (Nam-Ing). These stipulations, however, are not intended to interfere with the rights France acquired by treaty with Siam in 1893 over a zone of sixteen miles along the right bank of the Mekong, and over the navigation of the river.

From the mouth of the Nam-Huok (or Mehok) northwards to the Chinese frontier, the *Thalweg* of the Mekong is to form the boundary between the British and French possessions.

AFRICA.

Tribal Names in German South-West Africa.—Herr D. P. H. Brincker, a missionary, makes some interesting remarks, in *Globus*, Bd. lxviii. No. 24 a, on the origin of certain forms now generally accepted in Europe. The form Dámara is due to Captain Alexander, who was struck with the peculiar dress and appearance of two women in the camp of the chief Yonker. In answer to his question who they were, he received as answer, Damara. Now, the syllable *ra* is simply a suffix form-ing the femine dual, and Dama*ra* signifies two female Damas. So also Náma*ra* would signify two Nama women, while more than two would be called Nama-*ti*. The Namas call all the neighbouring Bantu tribes Dama*n*, the termination *n* including both genders. The Herero (Ovaherero) they call Koma-Daman (*Koma*=beast), from their wealth in cattle; the Bechuana, Buri-Daman, because they prefer goats; and the Ovambo, Brood-Daman, because they sometimes eat bread. The name

Namaqua arose from an error similar to that which introduced Damara. Nama*ka* is the masculine dual ; the plural is Nama*ku*, and the form common to both genders Nama*n*. The spelling of Nama*qua* shows that it had a Portuguese origin.

At one time little attention was devoted to the correct forms of native names, and therefore they were frequently mutilated. Thus we have Zulu for Ama-zulu, Fingo for Ama-fingo, Pondo for Ama-pondo, Swazi for Ama-swazi, Herero for Ova-herero, Swaheli for Wa-swaheli. Yet the prefixes in Ov-ambo, Be-chuana, Ba-suto, etc., have been retained.

Again, Swakob is a corruption of the Nama name Tsoa-chaub or Tsoa-chaob, (*Tsoa*=winding, and *chaub* meaning a certain kind of dirt), and Bulawayo should, according to Herr Brincker, be written *Ubuluayo*. Even the name Congo is not quite right, being pronounced by the natives *Kuango*, with a very short *a* (nearly like Kwango).

But the greatest liberty is that which has been taken in applying the collective name Bantu to all the dark-brown peoples of South and Central Africa. At first they were called the *Bunda* peoples ; but when it was discovered that the Ova-mbu-nda dwelling in Angola were a single tribe only of this great family, the name was dropped and Bantu adopted instead. Only the Kaffir tribes, however, have the form *Aba-ntu* to denote "men," while the others have the forms *o-va-ndu*, *'va-ntu*, *va-tu*, and *ba-tu*. The name Bantu, a mutilation of Ova-mbunda, may then pass when applied to Kaffirs, but it should not be given to other tribes.

AMERICA.

A Newly-discovered River in Canada.—Dr. Robert Bell of the Geological Survey of Canada, to whom we are indebted for the excellent description and map of Labrador lately published in the *Magazine*, travelled last year direct from Ottawa to James Bay, discovering a river more than 500 miles long. Dr. Bell heard of this river from natives as far back as 1887, and sent his assistant, Mr. A. S. Cochrane, to survey its upper course, hoping to complete the survey in the following season ; but more urgent work interfered with his plans. In the *Summary Report* of the Geological Survey for that year Mr. Cochrane's route is sketched. Having made a track survey of the northern arm of Grand lake, he crossed the watershed, and explored the chain of lakes which leads down to Shabogamog lake, which proved to be over thirty miles long, and then followed the river which drains this lake for ten miles northwards, where a series of rapids begin. It was a large stream, fully equal to the Ottawa just below Grand lake, and was at the time supposed to be the upper course of a river which enters Hannah Bay.

Dr. Bell travelled up the Gatineau river and its branch the Desert, and then, crossing to the Upper Ottawa, descended to Grand lake. Beyond the height of land, running very close to the lake, he came to the stream which becomes a large river before it reaches the sea. Within the first 100 miles it is joined by the Méjiskun, which has its source near the head of the St. Maurice river. It is now as large as the Ottawa above Lake Temiscaming, and continues to receive important affluents, especially from the west. In this section it is wide and sluggish, but deep—30 to 40 feet on an average. Presently it flows into Lake Mattakami, which also receives at its opposite (eastern) end the Waswanipi, a very large river. The outlet is on the northern side, and thence the water flows in a tolerably straight course to the head of Rupert Bay, which may be called its estuary, and which also receives the Broadback, a large branch from the east.

The height of land between Grand lake and the headwaters of the new river has an elevation of about 1000 feet, and the descent is very gradual as far as about

100 miles from the sea, where the current becomes very rapid and dangerous to canoes. The watershed is not a mountain range or even a ridge, and the removal of a few feet of sand would allow the waters of the Ottawa above Lake Temiscaming to flow northwards as they did in remote geological times. The drainage area of the new river is considerably larger than the whole of England, being about 300 miles long by 260 broad, and having an extent of 60,000 miles or more. The country is generally level or gently undulating, but an isolated hill was occasionally seen from the river. As a whole it may be described as a plateau, with an elevation of 1000 feet along the watershed, diminishing to 400 feet at 100 miles from the mouth of the river, and then rapidly falling to James Bay. It is covered with forest of black and white spruce, pitch-pine, tamarack, balsam, cedar, white birch, etc. The climate is fairly good, and the soil is capable of producing grain, hay, and root-crops in abundance. The precipitation is greater than in the regions to the south and west, and, when the discharge of the river is gauged, it will probably be found to exceed that of the Ottawa.

The mouth of the river is named the Nottaway on maps, but its supposed course is wrongly indicated, for it has been unknown except by a few Indians, and even by them only in sections. Mr. Archibald Blue, Director of Mines for Ontario, has, therefore, appropriately suggested that it should be called the "Bell" river.

Observations of the Temperature of the Earth's Crust in America.—Professor Agassiz has communicated some of the results of experiments in the copper mines of Calumet and Hecla, on the Keweenaw peninsula, Mich., to the *American Journal of Science.* A depth of 4711 feet was reached, and the temperature was observed at various levels. Only those at depths of 105 feet (59° F.) and 4580 feet (79° F.) are reported. Thus, the interval corresponding to a difference of one degree was 223·7 feet, while the most accurate of the Schladebach observations gave only 72·2 feet. But the conditions in the Calumet boring were of a very exceptional nature, and, therefore, Professor Agassiz is perhaps hardly justified in basing on these observations a calculation of the thickness of the crust. In 1886 Wheeler published six temperature observations from this neighbourhood—at lesser depths, indeed—and found that the thermal interval diminished from 122 feet to 76·4 in a direction from the Lake Superior. It seems, therefore, probable that the lake water penetrates into the rock and lowers the temperature.—*Pettermanns Mitt.,* Bd. xli. No. 12.

The Inambari.—The Madre de Dios, Aquiry, and other streams lying to the east of the Andes properly belong to Bolivia; but of late the Peruvians have been active in this country, and appear to almost consider it their own. The Bolivians affirm that the boundary runs from the source of the Yavari in lat. 7° 30′ S. to the confluence of the Madre de Dios with the Inambari, and then follows the latter to its source. In order to lay claim to this territory, and also to obtain some information about the country, an expedition was despatched at the end of March 1894, under the leadership of Dr. Ramon Paz, a copy of whose report has been kindly sent to the Society by Señor Ballivian.

The expedition left Riberalta (or Ribeira Alta) on March 24th, and ascended the Madre de Dios in a steam-launch, calling at various indiarubber establishments on the way, the chief of them being El Carmen, which has a population of 1000 souls, and collects about 170 tons of rubber yearly. A stay was also made at Monteverde; a few miles further are the Vasquez rapids, which, however, as the water was high, revealed their presence only by agitation of the surface and eddies.

On April 13th the launch reached the mouth of the Inambari. The party ascended the river for not more than four or five miles, and therefore did not bring back much information about it. Indeed, the expedition seems to have been very badly equipped for scientific exploration, not having even a barometer for the measurement of altitudes. Dr. Ramon Paz has, however, in his report collected some useful notes on this river and the Madre de Dios from other sources. The Rio Inambari, says the first of Peruvian geographers (Raimondi, no doubt), rises on the northern slope of the cordillera near the snow-fields of Poto, which are the source of the Rio Crucero on the other side of the range. Flowing northwards, it passes the settlement of Sina, the name of which it assumes. After receiving the tributaries Quiaca and Sandia, it is known as the Huarihuari, and a few leagues farther turns westwards. It receives in this part of its course numerous tributaries, in the valleys of which large quantities of gold have been obtained. Below the mouth of the Pullani the Huarihuari becomes the Inambari, by which name it is known in the province of Carabaya.

According to Colonel Pando, who had been in this region the previous year, the mouth of the Inambari lies in 12° 42′ S. lat. and 69° 43′ W. long. Below, the Madre de Dios runs in a direction on the whole north-eastward as far as Independencia, where it turns to the east; from the mouth of the Sena or Manupari to the confluence with the Beni its general direction is east-north-east. The rapids of Camacho (lat. about 11° 30′) and Vasquez (lat. 12° 30′) will always render navigation impossible when the water is low, unless a channel can be made through them. No river of the country contains so many islands as the Madre de Dios; there are sixty-three between its confluence with the Beni and the highest known point, above the mouth of the Inambari. A few miles above the Heath river is a lake, named by Colonel Pando after Padre Armentia, and another larger one, known as Montaño or Viego from its discoverer, extends from the Buzeta rocks to the vicinity of Maravillas, and communicates with the river by two channels, one of which is navigable. Both these lakes lie on the left side of the river.

The Inambari is the most important known affluent. Dr. Ramon Paz measured it roughly at the narrowest point near the confluence, and found its breadth to be about 200 yards; at the mouth itself it is probably 330. The Heath, which enters the main stream some thirty miles lower down, has a breadth of not more than seventy-five yards, but it appears to be very deep. Whereas these two tributaries have reddish waters, the Manupari or Sena is black. It has a smaller volume of water than the Heath, but is navigated at all seasons by canoes for eight days' journey, up to the gum-collecting centres. The breadth of the Madre de Dios is in some places, usually just above a group of islands, as much as 1000 yards broad, but the average is 300 to 500. It discharges nearly double the quantity of water brought down by the Beni.

Señor C. F. Fiscarrald, in a letter published in the report, speaks of a march of fifty-five minutes he made from the upper waters of the Camisea, a tributary of the Urubamba, to Mano, one of the headwaters of the Madre de Dios, which is capable of navigation by river steamers.

Chilian Railways.—The southern terminus of the main trunk line of Chile is now at Temuco, but surveys and plans for its continuation to Puerto Montt, 270 miles farther, have been prepared, and it is expected that by the end of 1897 direct railway communication will be established between Valparaiso and Puerto Montt. A branch line, of 20 miles in length, will run from Antilhue to Valdivia. The section from Richi-Ropulli to Osorno, a distance of 61 miles, was to be finished by the end of last year, and the rails of the branch line have been laid.

The Government lines now open to traffic have an aggregate length of 693 miles, and the private lines an extension of 807 miles. The Government main trunk line is divided into three sections, namely :—(1) Valparaiso to Santiago, 141 miles ; (2) Santiago to Talci, 184 miles ; (3) Talci to Victoria, 368 miles. These lengths include the branch lines, and make a total of 693 miles. It is expected that the railway to Puerto Montt will open up a large trade with Argentina, for the latter, not having any coal deposits, must sooner or later resort to Chile for her supply of fuel. Wood is going up in price and will soon become too dear to be used as fuel. A railway could easily be built over the Andes from Puerto Montt.—*South American Journal*, November 16th.

MISCELLANEOUS.

The section of the Anatolian railway from Afium through Karahissar to Ak-Sheher was opened in November last.

A geological survey of Egypt is to be commenced this year under the direction of Captain H. G. Lyons, R.E., who is now superintending the excavations at Philæ.

An error occurs in the numbering of the parallels of latitude on the map of Hausaland, published in the January number. They should be numbered upwards —4, 6, 8, 10, 12, and 14.

The German explorer, Otto Ehlers, has been drowned in the Heath river. He started from Bayern Bay with the object of crossing into British New Guinea. —*Deutsche Rundschau*, Jahrg. xviii. Heft 4.

The body of Lieut. Eivind Astrup, Peary's companion in his Greenland expeditions, has been found in the Lille Elvedal. Lieut. Astrup left Jerkin, at the foot of Snehættan in the Dovrefjeld, shortly before Christmas, and nothing was heard of him after.

The *Annals of Scottish Natural History* for January contain several articles of interest, chiefly ornithological. Of geological importance is a short paper by Mr. James Beunie on Arctic Plant-Beds in Scotland. The notes contain many details respecting particular species.

P. A. Colombaroli has compiled an elementary grammar of the A-Sandeh language, which has been published in the *Bull. de la Soc. Khédiviale de Géographie*, ive Série, No. 6. A reprint presented by the author contains also French-A-Sandeh and A-Sandeh-French vocabularies.

Dr. Cook, whose project was noticed in these pages, started for the south in December. His expedition consists of sixteen men, six of whom are scientists. He has two vessels of 100 tons each. His goal is the bay of Erebus and Terror, 700 miles south of Cape Horn.—*Revue Française et Exploration*, January.

Spain and Japan, having had some disagreements respecting the possession of certain islands lying between the Japanese Archipelago and the Philippines, have agreed that the Bashi Channel shall form the boundary between the possessions of the two States.—*Boll. della Soc. Geogr. Italiana*, Fasc. 12, 1895.

The Royal Geographical Society of Antwerp has issued a report of the Congrès de l'Atmosphère held under its auspices in 1894. The papers, by MM. Plumandon, Durand-Gréville, V. Ventosa, Hovent, Lancaster, Le Clément de St. Marcq, and Wouvermans, treat for the most part of winds, storms, and atmospheric vortices.

Signor Guido Boggiani, whose work on the Caduvei we noticed not very long ago, has forwarded a vocabulary of the Guana language common to several tribes in Matto Grosso and the Gran Chaco. Signor Boggiani has also prepared vocabularies of the Chamacoco and Zamuco dialects, all of which are published by the Academia dei Lincei.

The islands Huahine and Bolabola, or Borabora, of the Society group in the Pacific, were annexed, in January 1895, by the French, and joined to the protectorate of Tahiti. Bolabola was discovered in 1722 by Roggeveen, and is distinguished by an imposing mountain mass more than 3000 feet high and beautiful vegetation. Huahine was discovered by Captain Cook in 1769. Both the islands are of insignificant extent.—*Globus*, Bd. lxxiii. No. 22.

The U.S. Commission for the examination of the Nicaragua canal project has reported that the canal cannot be commenced without further investigation. The cost of the work is roughly estimated at £27,000,000 sterling, or nearly double the estimate of the company, and the verdict on the work already executed is unfavourable. Under these circumstances there is no room to hope that a connection between the Atlantic and Pacific will be established during the present century.

We have received the fifth volume of the *Anales del Instituto Fisico-geografico Nacional de Costa Rica*, edited by the Director of the Institute, Señor H. Pittier. It contains his meteorological record for 1892. We have before alluded to these records, the only observations made regularly in Central America. The remainder of the volume consists of translations of Professor Gabb's report on the exploration of Talamanca in 1894, and A. von Frantzius' description of the right bank of the Rio San Juan published in *Petermanns Mitteilungen* in 1862.

Since 1889 the Geographical Society of Lisbon has been advocating the proposal to celebrate in 1897 the fourth centenary of the departure of Vasco da Gama on the voyage during which he rounded the Cape of Good Hope and discovered the ocean route to India. The idea has been adopted by the Portuguese nation, and a commission has been appointed, with the King as its honorary president, to make arrangements for celebrating the occasion with due solemnity, and the co-operation of men of all nations is invited to render the commemoration worthy of an event of so much importance in the history of the world.

To a periodical printed by the General Staff of St. Petersburg for private distribution, General Tselenoi has contributed a paper on the distribution of the Armenians. Armenians dwell in nine vilayets of Asiatic Turkey, but form the majority of the population only in five cantons around Lake Van, and are everywhere on the decrease. A larger number live in Russian territory. In the whole of the Caucasian Governments they make up 13 per cent. of the population, namely 20 per cent. in Transcaucasia, and 1 per cent. in Ciscaucasia. In five districts there are more than 50 per. cent. of Armenians.—*Globus*, Bd. lxix. No. 7.

In the *Imperial Institute Journal* for December is described a curious tree, which, far from being damaged by fire, seems rather to flourish where forest fires are frequent. This tree is common in Columbia and other parts of South America, and is called the Chaparro (*Rhopala obovata*). It grows to a height of 15 or 20 feet, and its distorted trunk measures 9 to 12 inches in diameter. The tree is found in most parts of Columbia, and in Tolima, in particular, it abounds on the hill slopes at an altitude of 1000 to 3500 feet. Its bark consists of a series of integuments, the outermost performing no organic function, but serving as a protection from fire.

The *Geographical Journal* for January contains particulars of the voyage of a bottle which was picked up on the shore of Flores, and was forwarded to the Royal Geographical Society through the United States Hydrographer. It was thrown overboard from the steamship *Huntcliffe* of West Hartlepool at 10.30 A.M. on March 3rd, 1895, in lat. 24° 8′ N. and long. 82° 11′ W., and was found at Flores on Dec. 4th. It could not, therefore, have been more than 214 days in the water. Probably it did not follow the most direct course, or it would have gone ashore at Bermuda. If it travelled along the southern edge of the Gulf Stream, the distance would be about 3000 miles, and the average speed 15 nautical miles in 24 hours. This confirms the opinion, already entertained, of the high velocity of the west-running current in the western half of the North Atlantic.

The enormous territorial area of Canada is made strikingly manifest by the recent action of the Dominion Government in dividing the unorganised and unnamed portions of the Dominion into provisional districts. The territory east of Hudson's Bay, having the province of Quebec on the south, and the Atlantic on the east, is to be hereafter known as "Ungava." The territory embraced in the islands of the Arctic Sea is to be known as "Franklin." The Mackenzie river region is named "Mackenzie," and the Pacific Coast territory lying north of British Columbia, and west of Mackenzie, is called "Yukon." The extent of Ungava and Franklin is undefined. Mackenzie covers 538,600 square miles, and Yukon 225,000 square miles, in addition to 143,500 square miles added to Athabasca and 470,000 to Keewatin. The total area of the Dominion is estimated at 3,456,383 square miles.—*The Times*, October 22nd.

The Geographical Association.—The formation of this Association and its aims were noticed briefly in vol. ix. p. 485. At that time its object was defined in general terms to be "The Improvement of the Status and Teaching of Geography in Schools." In a leaflet lately received, the means adopted to this end are set forth: they are (1) to organise and supply to schools joining the Association improved aids and materials for instruction; (2) to act as a medium for the exchange of ideas and opinions between members, and for the dissemination of successful methods of teaching Geography; (3) to draw up a systematic List of Standard and Recent Geographical Works, other than school text-books, for the guidance of teachers; and (4) to co-operate as far as possible with the Geographical Section of the Teachers' Guild, so as to prevent overlapping, and to enable the members of each body to benefit by the work done by the other.

As a result of correspondence with the Educational Committees of the Royal Geographical Society and of the Royal Colonial Institute, the advisability has been recognised of "addressing a joint request to the various examining bodies, asking them to assign a larger proportion of marks to Geography, and to arrange that the papers may lend themselves less to unintelligent 'cramming'"; and with a view to obtaining the support of a large body of teachers, without which no such representations can be effective, the opinion on certain questions of masters in Secondary Schools was elicited by a circular.

In this circular it was suggested (1) that papers in Geography should be set and looked over by geographical experts; (2) that the principles of Physical Geography should form part of every examination (in Geography); (3) that the subject "Geography" as set is too wide and vague, and that a subdivision of it would be a great advantage, so as to include, besides the principles of Physical Geography, the geography of some continent; (4) that in competitive examinations Geography ought to be compulsory, or to receive a sufficient number of marks to make it "pay." In general, these suggestions received the approval of teachers of geography ; an exact definition of Physical Geography was asked for, and, while geography was considered essential in army examinations, it was doubted whether it was wise to add to the long list of subjects in certain other examinations.

The Committee has already taken in hand the matter of lantern-slides, and more than 250, illustrating diagrammatically the physical and commercial geography of various parts of the world, have been prepared by the Hon. Sec., Mr Dickinson ; and some 300, illustrating scenery, have been prepared from photographs lent to the Association, and are arranged in sets corresponding to the above. These will be lent to the representatives of schools joining the Association.

Members of the Association have also the advantages of receiving the *Geographical Journal* at a reduced price, and of obtaining a reduction on all goods bought or hired from the agents of the Association, Messrs. G. Philip & Son.

The methods of teaching specially advocated by the Association are those which tend to instil into the pupil a thorough comprehension of geographical principles, as opposed to a mere knowledge of isolated facts. In a pamphlet entitled *Geography as a School Subject*, published at the request of the Association, Mr Dickinson shows how Geography can thus be applied to the training of the mind. The study of climate he considers to be of fundamental importance ; excellent lessons in deduction may be found in working out the climate of a particular country from the general laws of climate modified by surface features and other local peculiarities ; and from the climate, the productions of the country and its suitability for human habitation and industry may in turn be deduced in their general outlines. Accordingly, Mr Dickinson deals first with climate in general, and sketches out a syllabus for teaching its factors—the sun's heat, atmospheric circulation, etc.—accompanied by a list of diagrams. In the latter pages the principles thus learnt are applied locally to British Australasia.

NEW BOOKS.

The Climates of the Geological Past, and their Relation to the Evolution of the Sun. By Eug. Dubois. London : Swan Sonnenschein and Co., 1895. 8vo. Pp. viii + 167. *Price 3s. 6d.*

In this essay the author attempts to explain the great climatic changes of geological time, which he thinks have been brought about by variations in the solar radiation. In the first part of his work he discusses the evidence of geology as to the character of the climate of the globe in earlier ages, and has no difficulty in showing that up to Tertiary times a more uniform distribution of heat obtained than is now the case. In the relatively short Tertiary period a general cooling gradually supervened, which eventually culminated in the glacial conditions of the Pleistocene. These cold conditions, however, were not persistent, but were interrupted at least twice by inter-glacial epochs characterised by a genial, temperate climate, similar to that which prevailed in pre-Glacial, and again in post-Glacial, times. Dr. Dubois discusses very briefly the several hypotheses which have hitherto been adduced in explanation of the evidence, and dismisses them as being all more or less inadequate. He thinks that the various telluric agencies which have been appealed to could never have sensibly modified the general heating of the earth during the time it has been inhabited, and that the more uniform distribution of heat up to the Tertiary period can only be accounted for by the total amount of heat received by the earth from the sun having been greater than at present. In short, the author believes that the great climatic vicissitudes of the past find a satisfactory explanation in the history of the sun's energy. Referring to the classification of stars into *white* (bluish-white), *yellow*, and *red* stars, and to the inference of many astronomers that these indicate successive stages in stellar evolution, Dr. Dubois concludes that our sun (a yellow star) is already far advanced in life. More than half (58·5%) of all known stars belong to the first or bluish-white class. The second class (yellow) contains 33·5%, while the red stars form only about 8% of all visible stars. Comparatively few stars form the transition from the first to the second class; while those that must be ranged between the second and the third class are relatively more numerous. The nebulæ from which the white stars have been evolved are not nearly so numerous as the stars— occurring in fewer thousands than the stars in millions. On the assumption that the evolution of celestial bodies takes place in a perfectly uniform manner, we are able, according to Dr. Dubois, to estimate the duration of the typical stages from the observed proportion which exists between the number of stars of the different classes. Thus the white stage may be fixed at 58·5%, that of the yellow stage at 33·5%, and that of the red stage at 8% of the mean total luminous existence of a star. The nebular stage, which could not have lasted 0·1% of the total time of evolution, need not be taken into account ; and similarly the stars that are transitional between different stages, and are relatively few in number, may be neglected. According to this view, therefore, our sun must have passed about three-fifths of its life as a white star. How long it has existed as a yellow star cannot be determined astronomically or physically. But the geological evidence shows that the present thermal condition of the earth was developed during the Tertiary period, and was fully attained at the beginning of the Pleistocene age. It would appear, therefore, that the sun entered its yellow stage at a comparatively recent date, and that it has two-fifths of its life as a luminous and heating star still to run. The passage from the white to the yellow stage was comparatively rapid, as is shown by the comparatively rapid cooling of the climates of the Tertiary period : and this stage in the evolution of the sun is represented, according to our author, by the relatively small number of stars, which, like Procyon, form the transition from the first to the second stage. The subsequent epochs of considerable temporary cooling, leading to glaciation wherever favourable conditions existed, must have been induced by very considerable reductions of the solar radiation. Such changes would necessarily be marked by great disturbances in the condition of the sun, which would be rendered apparent by its colour and its spectrum. Are there any stars, then, at present in this "glacial intermediate period" of the second stage? According to Dr. Dubois the condition of the sun during our glacial epochs is represented by the stars which are transitional between the second and the third class, or perhaps by stars belonging to the latter class. "Everything," says our author, "seems to indicate that during the yellow stage in long oscillations, always during a relatively short time, chemical combinations occur by which the colour of the star becomes reddish (or red), and broader and darker lines, indicative of a denser atmosphere, and dark bands or columns, indicative of chemical combinations, appear in the spectrum. By the otherwise unimportant and regular decrease of the radiation during the yellow stage, these oscillations will repeat themselves probably for a long time without any considerable lengthening or shortening of the intermediate periods, and only shortly before the end of the sun's life the intermittent cool period will grow rapidly, and the body of the sun grow permanently red, and at last dark." Dr. Dubois having adopted the view that the white stage of the sun probably did not last more than ten millions of years, endeavours in conclusion to meet the objections which have been urged by geologists and biologists to the results reached by Lord Kelvin and others as to the relatively limited duration of time within which all geological changes must have taken place. In this discussion he displays much ingenuity, and if his arguments do not quite carry conviction, they at least serve to show that the demands of a certain school of geologists have been excessive. His hypothesis of past climatic changes is most attractive, but speculative to a degree. The geological evidence seems from some points of view to support it, but many doubts and difficulties suggest themselves. If it could be shown that wide-spread glacial conditions obtained for some time in what are now temperate and sub-tropical latitudes during the Palæozoic era—that is to say, during the white stage of the sun—M. Dubois' hypothesis would be shattered. And that such a cold stage did in all probability obtain in Permo-Carboniferous times is apparently becoming more and more clearly made out. But the author may be congratulated on this able and suggestive essay, which is sure to be widely read by every one interested in the history of earth and sun.

Adamaua; Bericht über die Expedition des Deutschen Kamerun-Comitees in den Jahren 1893-4. Von Dr. SIEGFRIED PASSARGE. 7 Maps, 21 Plates, 294 Illustrations. Pp. 549. Notes and Index. Berlin : Geographische Verlagshandlung von Dietrich Reimer, 1895.

Dr. Passarge was the surgeon and scientific observer of the expedition sent out by a Berlin Committee, under the leadership of Herr von Uechtritz, to prevent the "Hinterland" of Cameroon being affected by Lieutenant Mizon's French Expedition. The Royal Niger Company approved of the German Expedition, and, subject to certain stipulations, conveyed it to Yola free of charge. If one may judge from the splendid book now before us, the expedition was most fortunate in its doctor, for the scientific and geographical information contained in it will compare more than favourably with the accounts of other African exploring parties. The publisher has left nothing undone to present the matter in the most favourable manner, for maps, illustrations, and text are produced in a lavish and most commendable style.

We may pass over the description of the marches, for although very well conceived, they of necessity differ but little from the usual accounts of African travel. Turning, however, to the scientific parts of the volume, we find a wealth of detail and of well-expressed information, a considerable part of which is either quite new or else presented with uncommon insight and vigour. Our author writes, too, with commendable brevity, and profuse illustrations are brought into requisition. The following rough translation will illustrate his style. It refers to an interview with the Sultan Siberu's minister : " When we entered the house, the minister lay upon the divan, the upper part of his body being uncovered. Uechtritz and Mr. Bradshaw found seats on wooden boxes, but as there was no seat for me I placed myself without more ado on the ministerial divan at his Excellency's feet. This familiarity did not seem to be approved of, for at a sign an old *madugu* brought me also a box. Then greetings commenced with innumerable *lafias*, and *ssanus*, accompanied by the clapping of hands. The minister had then a longish palaver with the *madugu*, which naturally we did not understand. It was interesting to observe the man. His stalwart body lay negligently at full length upon the divan. His high forehead was wrinkled, the head hoary, the long face framed by a grey beard, the nose fleshy and arched, the lips broad and thick. Tired, yet on the alert, from time to time he closed his red, bleared eyes. As much cunning as vulgarity were indicated by his features. Phlegmatic and drawling are the words which he carelessly addresses to the old *madugu*, whilst leaning upon his right elbow. The *madugu*, squatted before him upon the earth, occasionally agrees with him by nodding the head and exclaiming, 'Hakka ni ! hakka ni !' His conversation becomes more lively, more expressive, his face lights up, hands and arms rapidly gesticulating. Now the voice is low and deep, then higher and higher, louder and louder ; rapidly and still more rapidly the words pour forth like the sparks from a burning squib. Suddenly the acme of his speech is reached ; sharply and with emphasis the last word is shot out : the squib has burst. As a sequel, a quiet murmur follows, then deep silence. The great orator is spent, and sinks back upon his couch, grasping lazily the short pipe which a slave has in the meantime freshly filled. This Herr Akal is a great orator, and a great dramatist ; that was clear to us at once. But what, after all, was the meaning of the long speech ? . . .

"We could now examine the house at leisure. It is a large round building of about five meters in diameter, composed of a mud wall about a man's height, and covered with a roof of grass. The two doors are not exactly opposite one another. To one side of the entrance stands the bed. The bedroom is separated from the rest of the room by a mud wall the height of a hand, and is sprinkled with gravel. Opposite to it stands the horse, tied to a peg ; between the bed and the horse sit the company. The Akal's house is typical of the dwellings of the rich. The custom of keeping the horses in the house is no doubt of Arab origin. It was just the hottest time of the day, and a drink of sour milk, which the Akal's chief wife brought us, was therefore extremely welcome. This lady was no longer young, but was still pretty, and wore her hair in the usual fashion of the Yola beauties, a high twist on the top of the head, and two rolls of hair on the temples."

Throughout the volume great attention is paid to ethnological and anthropological matters ; the illustrations of the various types of people are of exceptional merit, and the care with which the observations have been made is self-evident in almost every page. Dr. Passarge does not believe that Cameroons is a colonisable country, but he is of opinion that it will pay to develop by means of trading stations. We are sorry to notice that he mentions spirits as an article of barter, as well as arms and ammunition. All European nations should combine to prevent the import of these articles into Africa. He holds that at present it would be most inadvisable to interfere with the existence of slavery, and indeed says that the domestic slaves are far better off than European labourers, who, he thinks, need Christian charity far more than they. His description of the condition of matters at Lagos is amusing, and he evidently does not approve of the quasi-civilisation which there obtains, and opines that it must take many generations before the people will have developed sufficiently to do without the oversight and support of Europeans. He thinks that the sharp differentiation of class, which exists in India and which obtains in America, ought to exist on the west coast of Africa, and believes that from the point of view of governing it would be far easier to deal with Negro Mohammedans than Negro Christians.

We cannot enter into more details. Suffice it to say that the book is well worthy of attention, and that we can sincerely congratulate the author upon his successful work.

Lakes of North America. By ISRAEL C. RUSSELL, Professor of Geology, University of Michigan. Boston, U.S.A., and London : Ginn and Co., 1895. 8vo. Pp. viii + 125.

This is a welcome addition to our knowledge of lakes. The author, from his long connection with the U.S. Geological Survey, has enjoyed exceptional facilities for this particular study. Much of the material he lays before us, therefore, is the fruit of his own research ; but he judiciously combines with these results the observations made by other students of physiography in North America. The work is excellently illustrated, and is written in a clear and popular style, the subject being treated under the following heads :—*Origin of Lake Basins: Movements of Lake Waters and the Geological Functions of Lakes: Topography of Lake Shores: Relation of Lakes to Climatic Conditions: The Life-Histories of Lakes : Studies of Special Lacustral Histories.* In discussing the question of the origin of lakes, the author does not advance any new views, but recognises, as others have done, that lakes have originated in many different ways. It may be mentioned that he strongly supports the late Sir A. C. Ramsay's view that many rock-basins have been ground out by glacial action.

Columbretes. Prague : Heinrich Mercy, 1895. Pp. 177.

The princely volume before us is by the same illustrious author as described the Lipari islands reviewed in this Magazine. The Columbretes islands (lat. 39° 56′ N., long. 0° 40′ E.) are in the Mediterranean, about 40 miles off the east coast of Spain, and some 70 miles to the north-east of Valencia. They consist of three main islets, viz., Columbrete Grande (the largest), La Ferrara, and La Horadada, with some smaller islets (El Bergantia, etc.), farther south. The author subjects the Columbretes to the same minute analysis as he applied to the Liparis, and details their meteorology, geology, flora and fauna, fishery and sport, with an account of the lighthouse on Columbrete Grande. " All the Columbretes," he remarks, "are composed of the same geological elements, viz., volcanic tuff, similar to peperino, in the form of slightly inclined layers or beds, crowned by a great mass of black cellular basalt, which represents apparently an ancient lava-stream that overflowed from the crater of the main island." We can only express the hope that so erudite an author will visit some of the volcanic islands of the west coast of Scotland and devote to them volumes as thorough and splendid as those he has dedicated to the Liparis and Columbretes.

Aegypten, 1894. Von A. FREIHERRN VON FIRCKS. Berlin : Geographische Verlagshandlung von Dietrich Reimer, 1895. Pp. 300 and Map. *Price* 3 M.

The author of this book spent a long holiday in Egypt, and before doing so made himself acquainted with a considerable amount of literature upon the country. The volume before us is the first part of the work which he is publishing to give expression to the information he has gleaned both by reading and by personal observation. In it he deals with the country and the people, the army and the fleet. He does not refer to ancient Egypt, dealing only with Egypt of the nineteenth century. The book may be best described as a monograph. The information is well compressed and systematically arranged. It gives an accurate description of the country, its plants and animals, and its population, and shows that the author has been at considerable pains to balance conflicting opinions and to reach as far as possible the truth.

Les Industries Nationales. Thèmes de Conférences par PAUL VIBERT. Paris et Nancy : Berger-Levrault et Cⁱᵉ., 1895. Pp. 386.

The industries of France have, within recent years, undergone a great change. Some of those for which she was once famous have disappeared, owing to a change of taste, or to new processes of manufacture having been discovered by science ; others are being killed by foreign competition. Among the men who have called attention to this state of things stands foremost M. Paul Vibert, an ardent patriot, who for the last ten years has endeavoured to rouse his countrymen to a sense of the danger that threatens their national industries through foreign competition. In *Les Industries Nationales* he has gathered together a series of graphic sketches on the rise or decay of a great many French industries, dwelling with special emphasis on the injuries done to them by foreign competition. The chief offenders, in his eyes, are the Germans, who flood the market, even within France itself, with their cheap and nasty imitations of French goods. His language becomes passionately eloquent on this subject, and the epithets he bestows on the Germans and their products are by no means flattering. French goods, of course, are incomparably superior ; and even the beer, which is imported from Germany in ever increasing quantities, could, he thinks, be made by the French quite as well, if not better. It goes without saying that the English are still most formidable rivals. Here is an instance : the French grow the Halfa grass in great quantities in their African possessions, then sell it to the English, and buy it back again in the shape of paper. M. Vibert's book is of great interest to the student of political and social science because of the light it throws on French character and methods. A foreigner cannot help being struck by a singular lack of enterprise on the part of the French. M. Vibert has to admit that his countrymen are not always as enterprising as their English and German competitors ; but, according to him, other causes are mainly responsible for the present condition of things, viz., dearness of labour in Paris, Government interference, the policy of protection and insufficient utilisation of the colonies. He thinks that if the French, instead of gambling in gold shares and other speculative paper, were to employ their national capital in the construction of a ship canal between Bordeaux and Marseilles, in making Paris a seaport, in building a railway accross the Sahara, and in enlarging their commercial navy, France would again march at the head of the artistic and industrial movement of the end of this century.

The Making of the Nation, 1783-1817. By FRANCIS A. WALKER, Ph.D., LL.D. With Maps and Appendices. London : Sampson Low, Marston and Co., 1896. Pp. xv + 314. *Price 7s. 6d.*

This work possesses both a general and a special value : it is a readable and reliable record of a critical period of American history, and it is marked off from the ordinary narrative by containing incidentally much sound teaching concerning industrial development and general State economy, even as the reputation of the writer would lead us to expect.

The constitutional significance of the epoch is great, since it witnessed the birth of the nation, although the Convention of 1787 had itself " dodged the vital question." The administrative interest of the period centres round the terms of office and the political capacity of Washington, John Adams, Jefferson and Madison. The first is written of with a well-toned enthusiasm, and also with pardonable excessive praise : "to-day all nations revere our first president as the finest and noblest character of political history." Hardly less enthusiastically, but more critically, reference is made to Washington's Farewell Message as one of the nation's most precious legacies, dissuading from entangling foreign alliances, and counselling "respect for law, the sacredness of national credit, moderation in party feeling, public and private virtue." Mr. Walker has brought out much more clearly than is usual in historical works the connection between national development in general—industrial, economic, social—and administrative and constitutional in particular. Land tenure on a popular basis, an agricultural class directly interested in the soil (when "the great manufacture of the United States was the manufacture of farms"), and the mechanical and inventive genius of the American people laid the ground-work, during these years of constructive activity, of the rise and greatness of the United States, and provide a means of illustrating the American political system by reference to the inherent and applied capacities of the race.

Neumayr's Erdgeschichte. New Edition. By Professor VICTOR ULIG. Vol. I., pp. 693 with 396 Illustrations ; Vol. II., pp. 700 with 513 Plates and Figures. Bibliographisches Institut : Leipzig and Vienna, 1895.

In bringing out the new edition of the well-known and popular work on Geology by Neumayr, Professor Ulig has steadily kept in view the aims of the original author, and has evidently devoted his best endeavours to bringing the matter of the former edition as nearly up to date as the eminently progressive character of the science permitted.

The first volume is devoted to General Geology, under which the editor includes Physical Geology, Dynamic Geology, and Petrography. The second volume deals with Descriptive Geology, under which are classified Stratigraphical Geology, the outlines of Palæontology, and Topographical Geology. In the concluding part of the second volume, Professor Ulig deals with Minerals and Rocks from an economic point of view.

The book is very fully illustrated by figures printed with the text, many of which are excellent, being wood-cuts faithfully copied from photographic or other reliable representations of the facts to be illustrated. There are also several maps, and, in addition, a series of chromo-lithographic plates, many of them brilliantly coloured, and well calculated to interest a large number of the class of readers for whom the book is intended. The type is very clear and readable ; and each volume is accompanied by full indexes, which render it easy to find any passage to which the reader may wish to refer.

Supplement to the Year-Book of the Imperial Institute: A Statistical Record of the Resources and Trade of Colonial and Indian Possessions of the British Empire. London : Offices of the Imperial Institute, 1895.

Instead of issuing a new edition of the Imperial Institute Year-Book for 1895 the Executive Council have published a supplement containing the most recent statistics and particulars about the various colonies and possessions. The same arrangement is adopted in this supplement as in the Year-Book, and side references to the pages of the latter tell where the amended information should be inserted. Two maps are given—one of the railways, telegraphs, and steamship routes to India, on rather too small a scale ; and one of Tasmania on a scale of fifteen miles to the inch. The Somali Coast Protectorate is still to be found under Asia.

Aarbog for Dansk Kulturhistorie. Udgiven af POUL BJERGE. Aarhus : Jydsk Forlags-Forretning, 1895.

We are glad to see that this excellent little "Year-book" shows no falling off in interest or thoroughness of workmanship. The papers, although based almost exclusively upon documentary materials, are neither erudite nor dry-as-dust productions, but eminently readable accounts of private life and popular manners, beliefs, and customs in Denmark during the centuries that are past. It is only incidentally that it treats of geographical knowledge.

Catalogue of the Library of the Royal Geographical Society, containing the Titles of all Works up to December 1893. Compiled by Hugh Robert Mill, D.Sc. Pp. viii + 833.

The last complete catalogue was published in 1865, and was supplemented by lists of accessions. The present catalogue contains 521 pages of names of books and authors in alphabetical order, and three appendices devoted to collections of voyages and travels, Government and other miscellaneous publications, and transactions of societies and other periodicals. The work has been carefully compiled and revised, and is printed in clear type and on good paper.

Oliver and Boyd's Edinburgh Almanac and National Repository for the Year 1896. Edinburgh : Oliver and Boyd. Pp. 1182.

The plan of this indispensable almanac is the same as in former years, and it has only been necessary to bring the work up to date. This has been carefully done ; and those changes which occurred too late to be recorded in their proper places have been collected in one page immediately after the title. Of the new matter an abstract of the new Fishery Act for Scotland may be mentioned as most nearly connected with geography, on its commercial side.

An Almanack for the Year of our Lord 1896. By JOSEPH WHITAKER, F.S.A. London : Whitaker and Co. Pp. 744.

Many changes have been rendered necessary in this year's issue in consequence of the accession of a Conservative ministry to power. Several other enlargements have also been made in the lists, etc. The articles of most interest to geographers are those on Geographical Progress and on Boundary Treaties and Territorial Changes, by Mr. Ravenstein. The former is a useful summary, but contains one or two errors in the spelling of names. The latter, which is illustrated by several sketch-maps, is very complete. We think, however, that the boundary between British and Dutch New Guinea is not very clearly defined, and that the question concerning the "Portland Canal" in British Columbia should have been explained rather more fully.

Bishop Heber, Poet and Chief Missionary to the East, Second Lord Bishop of Calcutta, 1783-1826. By GEORGE SMITH, C.I.E., LL.D., author of *William Carey, D.D.,* etc. London : John Murray, 1895.

Bishop Heber, though he has now been dead some seventy years, is still remembered—for breadth of view and amiability of disposition even more than for his intellectual capacity. His life was written shortly after his death by his widow, and published in two quarto volumes. Few readers can be found to struggle through such a work, especially so many years after the subject of the memoir has been laid in his grave, and therefore a concise biography containing matter which is of general and lasting interest ought to be sure of a welcome. The author of the *Conversion of India* has most fittingly undertaken the task and produced the present interesting volume. Heber's life at Oxford, his parish work, and his travels in India brought him into contact with men of very diverse character and into widely different surroundings, and therefore the materials contained in his letters and *Journal* are unusually varied. From these a careful selection has been made in order to produce a complete, though concise, picture of the man, his life and work.

Many of the illustrations are from the original wood-engravings cut from Heber's sketches.

NEW MAPS.

ASIA.

NAN-SHAN, Schematic Map of the Mountain System of the ——, compiled from the surveys of V. Obruchef and V. Roborovski.
Izvestiya of the Imp. Russian Geogr. Soc., No. 3, 1895.

THE SIAMESE MALAY STATES. Scale 1 : 5,000,000 (78·9 miles = 1 inch).
The Geographical Journal, Nov. 1895.

WEST BORNEO. Stromkarte von ——. Nach den Aufnahmen der topographischen Brigade der Niederländischen Armee 1886-1895, mit Angabe der Stationen und Routen der Niederländischen Expedition in den Jahren 1893 und 1894, von Professor Dr. G. A. F. Molengraaff. Massstab 1 : 2,000,000.
Petermanns Mitteilungen, Tafel 14, 1895.

AFRICA.

CONGO FRANÇAIS. Carte dressée à l'échelle de 1 : 1,500,000. Par J. Hansen, Cartographe du Congo Français. 2 Feuilles.
Presented by the Ministère des Colonies, Paris.

This map is compiled from the surveys of a large number of French, German, Belgian, and one or two British, explorers. It embraces all the country from the mouth of the Congo to Old Calabar, and extends eastwards to Djabbir on the Welle. In an inset, on the scale 1 : 5,250,000, is shown the country between Lake Tsad and the Nile. The routes of the traveller are marked in red—a very useful addition to a map of a country not yet thoroughly opened up.

SOUTH AFRICA, The Castle Line Map of ——. About 4 ft. by 2 ft. 9 in.
Issued by Donald Currie and Co., London.

A large amount of information of various kinds is contained on this wall-map. Railways constructed and projected, navigable waterways, elevation, rainfall, mines, both of gold and other minerals, the haunts of wild game, and the localities where grazing of cattle, sheep, or goats is chiefly pursued, are indicated. The only things we miss are telegraph lines, especially that from the Cape through Fort Salisbury and across the Zambesi.

AMERICA.

VENEZUELAN QUESTION, Map to illustrate the ——, 1896. *Price, coloured,* 1s. ; *cloth and case,* 2s. *W. and A. K. Johnston, Edinburgh.*

Shows the boundary claimed by Venezuela, the Schomburgk line, and the extreme boundary claimed by Great Britain.

AUSTRALASIA.

HATZFELDTHAFEN (Kaiser Wilhelm Land), Das Hinterland von ——. Auf Grund der Aufnahmen und Vermessungen der ¡Landmesser V. Brixen und Linnemann entworfen und gezeichnet von P. Langhans. Massstab 1 : 100,000. Nebenkarte : Hatzfeldthafen. Mit Benutzung der Vermessungen S. M. Kreuzer *Adler.* Massstab 1 : 40,000.
Petermanns Mitteilungen, Tafel 13, 1895.

THE SCOTTISH
GEOGRAPHICAL
MAGAZINE.

TRAVELS ON THE WESTERN COAST OF
EQUATORIAL AFRICA.

By Miss M. W. KINGSLEY.

(Read at a Meeting of the Society in Edinburgh, January 1896.)

MY aim in visiting West Africa this time was to get together a general collection of fishes from a West African river north of the Congo, for the terrific current of this river makes a great impression on distribution. There is an equally interesting difference in the forms of the native religion in districts north, and districts south, of Calabar; and my own chief interests in West Africa being fetich and native law, I was anxious to go on with local observations on these subjects—observations which I had commenced in 1893. I received the greatest assistance in this from Miss Mary Slessor, that most wonderful and brave lady who lives at Oköyön, and whose guest I was for some part of my visit of five months in Calabar, and whose humble admirer I shall always be. I left Calabar in April and went down to the Congo Français, where I knew, from previous experience in Kacongo, that the influence and advice of my friend Mr. C. G. Hudson would be of great service to me. It was, and I started fishing in the Ogowé, pottering up that lovely river, receiving every hospitality from the Mission Evangélique, to above the Alemba rapid. I then pottered down, and made my way down the Rembo Ongo into the Karkola river. Thence to Lake N'Covi, and then overland through Efuma, Egaja, and Esun to Ndorko, on the river Rembwe. From Ndorko I went up the river to Acondgo, and from there down to Glass, where, well knowing my wanderings through the Fan country would not meet with approval, I said I had done it from scientific motives. "No, Miss Kingsley," said my English friends, "you fell into the hands of those Fans, and they took you touring about their country like a circus."

I admit there is some truth in the statement. After arriving at Glass, I
got a small schooner and a native crew through the kindness of the
Rev. Dr. Nassau, and sailed to Corisco island, in order to see the
annual fishing in the lakes in its interior. This done, we sailed back
to Glass, and on our way called—that is to say, were driven ashore
—at divers places. I will not dwell on this voyage, for it is fuller
of sensational incident than scientific interest, because I was commanding the vessel and the South Atlantic was rough. At the end of
September I reached Victoria, per s.s. *Niger*, and owing to the kindness
and help given me by Herr Von Lücke, the Vice-Governor, and Herr
Liebert, commanding the garrison post at Buëa, I was enabled to ascend
Mungo Mah Lobeh, the Great Cameroon. The usual route up this
exceedingly beautiful peak of 13,760 feet is from its sea face, at Bibundi.
I went up the south-east face, through Buëa, because I was anxious to see
the trend of the Rumbi and Umon ranges of mountains. These ranges
are practically one and the same, and I believe will be found continuous
with, and of the same formation as, the Sierra de Cristal. They are
entirely distinct from the Cameroon mountain, whose relations are the
volcanic islands of Fernando Po, 10,190 feet, Principe, 3000 feet, San
Thome, 6913 feet, and possibly also Ascencion, St. Helena, and the
Tristan d'Acunha group—mushrooms as regards age compared with the
Rumbi and Sierra de Cristal ranges. From Victoria, on my return from
the mountain, I was given a passage on His Imperial Majesty's gunboat
Nachtigal, which was then going round with the German officer represent-
ing the Cameroon colony on the delimitation commission between it and
the Niger Coast Protectorate, and thus luxuriously I landed back at Old
Calabar, whence I came home in one of my old friends the British
African steamers, landing in Liverpool on the 31st of November.

The Ogowé river is certainly the part of my journey that alone
deserves your attention, this magnificent river not having been much
visited by Englishmen since Du Chaillu made his journeys in its regions.
Certainly Du Chaillu is nearly as good an authority on it as he was in
1865, for no one who has any personal acquaintance with the country
will dispute his assertions. The only important difference that has been
made is, that the Fans are now nearer the coast all along the line from
Panavia Bay to Cape Sta. Clara.

This is the largest river that debouches on the west coast between
the Congo and the Niger, and it is, strictly speaking, the largest
equatorial river in the world, for its course lies mainly on the Line.
Its delta terminates at Fernan Vaz, some thirty miles or so below the
Equator. That long stretch of it between N'djoli and Boué is about ten
miles below the Line, and the region of its source is the inland plateau of
the Congo Français (visited in 1878 by M. de Brazza). In some places
the sources are not more than 120 miles from the Alima, which is an
affluent of the Congo. The basin of the Ogowé is roughly estimated at
130,000 square miles, and during the long wet season its average discharge is about 1,750,000 cubic feet per second. Monsieur Krüger says
its main stream is 1300 kilomètres long. Its affluents have been but
little explored; the most important among them are the N'guni, falling

in on the south bank just above Lambarene, the Okana on the north bank, near Kondo Kondo, and the Ivindo on the north bank at Boué.

The river is more similar to the Congo than to the Niger, save that, unlike the Congo, it has an immense delta. This delta commences at Lambarene, 130 miles from the sea. The delta region is exceedingly interesting, both in its flora, fauna, and fetich; it is called Kama country, and is better known in England as Gorilla land. Its main population consists of malarial microbes and mosquitoes, and it is supremely damp. Indeed, the whole of it, save the strange bubble-shaped hills you find in it, is under water from Lambarene to the sea, where the Ogowé comes down during the long wet season. There are a quantity of lakes in connection with the entire course of the river; the most extensive and important of these are those in the Lambarene district, and the largest, as far as is at present known, is the Lake of Islands, Eleziva Zonangué: the next largest is Lake Azingo. The rise of the Ogowé in the wet seasons is great, and remarkable for commencing a month before the rains fall on the river itself.

The rise in the long wet season is from 18 to 20 feet in the Talagouga narrows; in the short wet season from 12 to 15 feet. The river is navigable for small steamers for 206 miles, up to N'djoli; above this place the rapids commence, and you have to take a canoe for the next 500 miles. I will not enter into the subject of those rapids, but confine myself to the new route I followed to the Rembwe, disclaiming any intention either to recommend it, or appropriate any credit to myself for taking it. When I was told no one had been that way before, I said, "Wise men!" I started on it from Kangwe in Lambarene, but I will not go into details about the route as far as the Adjuma town on the Rembo Ongo, called Arewuma. From this town, with divers diversions and disasters on sand-banks, we paddled west to the entrance of the Karkola river, a swift, narrow, but deep stream that leaves the Rembo Ongo on the north side. The Rembo goes on, still to the west, just below the Karkola, dividing up into three streams, one falling into the main Ogowé at Ashuka, one into the Nunghi, and one into the Ngumbi, down in Kama country, the whole series of rivers being made of the divided waters of the Ogowé. Kama country, I should observe, commences at Ashuka, and terminates at the sea.

The Karkola river seems to be formed of the overflow from the Rembo Ongo and from Lake N'Covi, and the stream running west from the district of Lake Azingo, and it most probably falls into the river Nazareth, but it has not been explored. We turned into it from the Rembo Ongo, and, after being carried down at a rapid pace by its current for some distance, turned up into what we should call in Norfolk a straith, which took us into a broad expanse of water, out of which rose sand-banks in every stage of exposure, from the one which had only just come to the surface on the falling of the waters in the dry season, to that which had held its head above water for years, and which, by carefully collecting *débris* from the waters, was pretending not to be a sand-bank at all, but a real island, and so had persuaded trees to grow on it. The majority of the islands and the bank of the river all through this district were now covered with the beautiful tender green hippo grass, which springs up annually in the dry season, and which is the favourite food of both the hippopotamus and the manatee, and amongst which the great black hornbills can be seen busily hunting for their favourite food—small snakes and frogs. From my point of view the richness of animal life in this broad stretch of water and sand-bank was almost cloying. We were skirting a long stretch of high hippo grass, my crew cheerily singing their boat song, when an immense hippo rose up in the grass alongside, about six feet from us, stared calmly, and then yawned a yawn a yard wide, and grunted the news of our arrival to his companions, who also rose up, and strolled through the grass with the flowing grace of Pantechnicon vans. Passing by rapidly we came to a big fellow on a bed of crushed-down grass, right on the edge of the bank, with several little black-and-white birds running over him. He lazily deigned to raise his head, and seemed when he had done so to be lazily speculating whether he had not better get the rest of himself up, and put a summary stop to respectable hippopotami being disturbed in their afternoon nap. We left that bank hastily, and crossed to the other just in time to see another monster come dripping and shining up out of the water where our canoe would have been if we had held our course. I have once been upset by a hippo—in fun, I believe—but I was glad not to repeat the experience. We thought it wise to put an island between us and them, and hoped that to the hippo out of sight meant out of mind.

We shot into a narrow channel between a low island and a large sand-bank, and that sand-bank had on it as fine specimens of the West African crocodile as you could wish to see. They also were having their siesta, stretched sprawling on the sand with their mouths wide open ; one old lady had a lot of young crocodiles running over her, evidently playing like a lot of kittens, and the heavy musky smell from them was most offensive. We did not, however, complain aloud about this, because we felt hopelessly in the wrong in intruding on these family scenes, and apologetically hurried past. When we were out of earshot, I asked one of my Adjuma crew if there were many gorillas, elephants, leopards, and bush-cow round here ? "Plenty too much," said he, and I wished myself in England, at the same time regretfully remembering that the last word a scientific friend had said to me before I left home was, "Always take measurements, Miss Kingsley, and always from the adult male." I had neglected magnificent opportunities of getting record measurements of hippos and crocodiles, and I can only excuse myself by saying I had mislaid my yard measure, and that I felt the crew would not have liked it. The banks all round this expanse of water seemed of light-coloured sandy clay, and we made our way across it into a channel that came into its eastern extremity, without any further misadventure, save that I nearly had the back of my head blown away by one of the crew's guns going off accidentally.

The river we ran up into, zig-zags about, and then lays a course south-east and by east. It is studded with islands, thinly clad with forest. It was a lonely-looking land. In front of us was a low range of mountains, and they were seen in the distance to bend to the north-west. We passed out of this channel into a melancholy, but exceedingly beautiful, lake, which my crew told me was Lake N'Covi. I went over the name several times with them before putting it down, because I did not know it; but they were all of a tale, and quite sure it was "Lake N'Covi." I have not, I regret to say, subsequently been able to find the name on a chart or map; but it had been visited, I heard from the natives, by a French officer, a long time ago. The only name I can find down on this map, prepared by the French authorities, that could fit with a traveller passing this way is that of F. Tenaille D'Estais, 1882. The lake he visited is called on the map Ebouko, but that name was not used by either the Adjuma or the resident Fan, nor is it on the map placed sufficiently west of Azingo. So I will confine myself to saying that it is an exceedingly beautiful place to go into, as I did, late in the afternoon. The rich golden sunlight, followed by the short-lived but glorious flushes of colour, played over the scene as we paddled north-north-east, the canoe leaving a long trail of frosted silver behind her on the mirror-like water, and each stroke of the paddle sending down the air with it to come up again in luminous bubbles of silver white, not in swirls of mud as is usual in West African waters. The rim of the lake is wreathed in all directions with nobly forested mountainous hills, indigo and purple in the dying daylight. On the north-north-east and north-east these hills come down directly into the lake; on the north, south-west, and south-east there is a band of level forested ground, behind which the hills rise. There are many beautifully wooded islands and dwarf cliffs, and we made our way to one of these towards the north-east part of the lake.

There was a large village on the level top, and a steep dwarf cliff overgrown with verdure came down to the small beach, which was covered with grey, waterwashed rocks, and two or three canoes lay on them. There seemed to be some discussion going on in the village that required a deal of shouting. My men, after they had driven the canoe near the beach, stowed their paddles and picked up their guns, slinging on their ammunition bags, and slipping the beautiful covers made of gorilla and antelope skin off the locks of their guns. One of my men, who thought he had a trade friend among the Fans at this town of M'fetta, got on to the beach. By this time the inhabitants had become cognisant of our arrival, and, abandoning what certainly ought to have been a mass meeting to remonstrate with the local authorities on the unsanitary condition of the town, they came swarming down, a seminaked, brown mass of humanity, to attend to us, evidently regarding us as an imperial question. Things did not look restful; every man among them was armed with a gun with its sheath off the lock, and there were no ladies. They drew up about twenty paces from us in silence. My man, who had first gone ashore, was joined by myself and our head boatman, and the men shouted out the name of the Fan friend they hoped to find, and for some terribly long minutes they shouted in vain, and the rest of the crew became nervous. I did not, for I had been nervous since those hippos. What made us most uncomfortable was that the Fans had not given us, as they should have, a greeting, nor would they answer my ingratiating salutations save with a grunt. At last the crowd opened, and a fine-looking middle-aged man came to the front, clad in a twist of calico, and a bunch of leopard and wildcat tails hanging from his shoulder by a strip of leopard skin; my man went for him with a rush, as if he were going to clasp him to his ample bosom, but held his hands almost touching the Fan's shoulders, and the Fan, who grunted feelingly, closed his hands and held them so as to all but clasp the Adjuma. Then another of my men made a rush for the crowd and went through great demonstrations of affection with another gentleman, whom he recognised as a trade friend, and whom he had not expected to meet here—and we all breathed again.   Peace having been proclaimed, conversation became general, and then we disappeared into a mob of men and a fog of noise, and went up the hill. You would have thought, from the vehemence of the shouting and gesticulation, that we were going to be torn forthwith to shreds; but not a single hand touched me, the crowd opening out in front and closing in behind as I passed up the steep path to the village. The noise redoubled in violence when we reached it, for we fell in with the ladies and children, and the dogs. Each child, as soon as it saw my white face, gave a howl, and fled headlong into the nearest hut, and, I fear from the continuance of the screams, had fits. It was an exceedingly filthy village, built of bark, like all Fan villages. The remains of a crocodile that had been killed the week before last, piles of fish offal, and portions of an elephant, hippo, or manatee—I cannot tell you which, because it was so high—united to make a most impressive stench. Taken all together, I cannot recommend the accommodation at M'fetta, and I will go into no more details about it, which were many, especially as I had to engage some of its inhabitants to act as porters and guides, none of my men ever having crossed from the Ogowé to the Rembwe by this route.

I took three Fans, elephant hunters by profession, with me next morning. They honestly said they could only guide me as far as the big town of Efuma; they had never been beyond, but they would come with me all the way if I would guarantee their safety. This I agreed to, and made the agreement to pay off at a sub-factory of Messrs. Hatton & Cookson, which I knew was on that river in charge of a native trader, and we started off in the canoe again, going across the lake to the upper end, and then into a channel set with manatee traps, and out into a broader bit of river, walled by the lank dark forest apparently in all directions, but really winding through it. This river, the Fans said, came from Lake Azingo, in a valley among the mountains; we ran our canoe into the bank on the south side, among the line of dark foliaged herbs that grew along the water edge, and got out on to, or more strictly, into, a most fitting introduction to the sort of country we were to spend our time on before we saw the Rembwe, namely, knee-deep black slime.

The whole of our path across this piece of country lay through the great gloom of Africa's equatorial forest belt. This forest is a region of great fascination and charm for me. Our first day's march in it was among ebony and giant redwood, no palm-trees showing, save, now and again, my old enemy, the climbing palm (*Calamus*), on its long excursions up one tree, down another, and up again ; here and there, when it reached the sunlight, bursting into a plume of long fronds, out of whose centre rose the long thin spike of a young frond, and covered all over every part of its cable-like stem with a perfect fur of thorns.   After the first day we struck the western spur-hills of the Sierra de Cristal, and the forest became even more interesting.   For hours we would pass among an apparently infinite series of columns of uniform height, about 100 feet. At the top of these the great boughs sprang out and interlaced, forming a canopy or ceiling, which dimmed the light, even of the equatorial sun, to such an extent that no undergrowth could thrive in the gloom.   In other places, we were among vast buttressed tree stems, and from their far-away summits hung great lines of bush ropes, some as straight as a plumb-line, and others intertwined among each other, like so many struggling serpents which had become, at the height of their combat, fixed by some magic spell, almost all of them as bare of foliage as a ship's wire rigging.   These bush ropes, I noticed, were usually carried up with the growing tree, so can hardly be called climbers.   The india-rubber vine, however, is a true climber, and it abounds in this region.   These stretches of forest were made up of spindle-stemmed trees, among which I often noticed the remains of some forest giant, whose death by lightning, or by his superior height having given the demoniac tornado wind an extra grip on him, had allowed sufficient light to come into the bottom of the forest, so that the young saplings, which had been living a half-starved life for years, when the light came, shot up.   They seemed to know that their one chance was getting to the level of the top of the forest without a moment's delay.   No time to grow fat in the stem, or send out side branches, or any such vanity ; up, up to the light level, and he that reached it first won in this game of life or death—for when he reached it, he spread his crown of upper branches, and shut off again the life-giving light from his competitors, and they paled off and died, or dragged on an attenuated existence, waiting for another chance.   Now and again we in the under gloom knew that, far away above us, there was another world—a world of blossom, scent, and beauty—which we saw as little of as the earthworm in a flower-bed ; around us the ground would be strewn with cast blooms, sometimes thick, wax-like, glorious cups of orange and crimson, each one of which told us that some of the vast trees were showing a glory of colour to heaven alone ; sometimes pure white stephanotis-like flowers telling us the bare, twisted, festooned cables were rubber vines, which had burst into blossom when they had found the sun.

I feel justified in stating that the track was bad, for I consulted the natives, who freely stated it was "bad too much," and "no man fit to pass this way in the rains."   This was self-evident, for even in the dry season the swamps were all that could be managed.   There were representatives of the three chief classes of West African bog.   The broad deep one was the best to tackle, because it made a break in the forest, so that the sun could get down to it and bake a crust over it, on which you could go, if you went quickly—a minute's pause in one place meant going through. The next best was the shallow knee or waist-deep affair, through which you could wade. The worst, the most frequent in this region, was the deep narrow one, so shaded that the sun could not form a crust over it; these required great care, and took up a good deal of time. Whichever of us happened to be at the head of his party, when we struck one of these, used to go down into the black, batter-like ooze, and try and find a ford, going on into it carefully until the slime was up to the chin; sometimes we made three or four attempts in vain, and then had to come back to our own bank and go higher up and try again, the other members of the party sitting quietly on the bank until the ford was found. We several times came across stretches of shallow swamp where elephants had been rolling and bathing. I, on one occasion, when going on ahead, came across a party of eight thoroughly enjoying themselves, and lay down and watched them. Passing over these places was difficult, for their great footmarks, in which you could have placed a bamboo arm-chair, were filled with water, and the rest of the ground was rolled hard and slippery; above all, those elephants left their ticks behind them. I will not enlarge on this subject, nor on the leeches, but I shall never forget either, particularly an experience in the great tidal swamp we struck south-west of Ndorko, which connected with the Rembwe. We waded two hours through it, up to our chins all the time, and came out with a sort of astrachan collar of leeches, which we removed with trade salt; indeed, our appearance on entering Ndorko was more striking than beautiful, each of us being encased in mud, which was streaked with blood and bespangled with flies; fortunately there is no white society at Ndorko or anywhere on the Rembwe.

Bad as the swamps were, the hillsides were worse. They were at abrupt angles, and wherever they were exposed to the full force of tor-nado winds there were terrific falls of timber, ancient and modern, over which we had to climb—terribly scratchy, dangerous work, for when a man missed his hold down he went, sometimes for six or seven feet, sometimes for fifteen or twenty, before he reached the rotten stuff under-neath; and, when one got there, there were more snakes and scorpions, etc., than one had any use for—I speak from experience—and then one had to be hauled with bush rope up through the sticks, which had been turned the wrong way by the down journey. Added to this, the sky being open above, the sun came down on us who were hauling or being hauled, and as we were hot enough with our exertions and the steam-laden atmosphere, and heavy with the stench of the swamps between the surrounding hills, this sun heat was an unwelcome addition.

The inhabitants of the region were entirely Fan, with a few scattered communities of pygmies. Of these latter I have little knowledge. Both the Fans and the Adjumas detest them, although they do not molest them much, partly, I fancy, from fear of the pygmies' poisoned arrows. I inquired of my good friend Wiki, the renowned elephant hunter, as to the reason of his dislike of pygmies, and his observations, as usual with Wiki more powerful than select, amounted to, "Oh, they are so low—degraded, you know. Upon my word, one can hardly tell them from monkeys; and then look at the way they live!" I had no opportunity of hearing a pygmy's opinion on the Fans. I expect they would have said disagreeable things about cannibalism. The main haunt of these pygmies is, however, not on this northern side of the Ogòwé region, but on the southern, round the upper waters of the N'guni above the falls of Samba, where they live on good terms with the Fans' first cousins, the Bakili tribe. Regarding the Fans themselves I could say much, for I made them my chief study while in the Congo Français. From my point of view, they were lamentably common-sense people, and although surrounded on all sides with intensely superstitious tribes, had very few superstitions of their own. One amusing incident I came across : the chief of a Fan town had died, and his fellow-chief—there are usually two or three in a village—in order to do his deceased *confrère* honour, called in the neighbouring tribe of N'comi, who are renowned for their death-dances. He addressed the people of the village, saying, "I have sent for these N'comi, so that our late chief may have due honour paid him. We Fans do not pretend to go in for this sort of thing, or understand it, but our good neighbours thoroughly understand the affair, and how it ought to be managed." So the people formed a ring and watched the weird, complicated N'comi death-dance for an hour or so, and then began one by one to retire to bed, feeling, I fear, bored. When the N'comi noticed the audience getting thin, they remonstrated ; they said it was the Fans' chief they were doing it for in a neighbourly way, and the least the Fans could do was to watch them appreciatively ; so the chief had the people who had left early turned out again, but it ultimately ended in the N'comi retiring highly disgusted, as usual, with "those Fans."

The Fans are an immense tribe who have quite recently appeared on the border of the known regions of the Congo Français. Where they have come from no one exactly knows, save that it is from some region to the north-east and by east of the Ogowé. Du Chaillu was, I believe, the first European to come into contact with them, and he perceived that they were in a state of migration. In his day, about 1863-1865, they were still far up in the interior, but now they are in the coast regions all the way along from the Batanga regions to the northern shore of the Gaboon estuary—in some places, as at Cape Esterias, right down on the seashore itself. They are, on the whole, a fine race, and one occasionally sees magnificent specimens of humanity, both male and female, among them. Their colour is light bronze ; many of the men have beards, and albinoes are rare among them. The average height of both sexes is 5 ft. 6 in.—5 ft. 8 in. Their countenances are very expressive, and once you have been with them you can never mistake a Fan, and I often amused my white friends by picking out a Fan from a crowd of Benga, Mpongwe, or Galwa. But it is in their mental characteristics that their difference from these lethargic, dying-out tribes is most marked. The Fan is full of fire, temper, intelligence and go, very teachable, rather difficult to manage, quick to take offence ; but, I ought to confess, people who know him better than I do say he is a treacherous, thievish, murderous cannibal. I never found him treacherous or thievish, and I like him better than any African, as a tribe, I have yet met. He is a cannibal, not from superstitious motives like the negro tribes: he just does it in his common-sense way. Man's flesh, he assures me, is very good, and he wishes I would try it; and I must say this for him, he does not buy slaves to kill and eat, as some of the Middle Congo tribes that I know of do. Indeed, with slaves he has nothing to do; he neither keeps them nor sells them, like the Galwa and Mpongwe. He is always very much abused for eating his relations, but he does not really do this : he will eat his next-door neighbour's relation and allows his next-door neighbour to eat his, but even then he always keeps a little something as a memento—a foot or hand, some hair, or eyeballs; unfortunately he keeps them hung up in a pine-apple fibre bag in his living and sleeping rooms, and the odour is not pleasant, particularly if the little something is the record of a recent, or comparatively recent, loss in his family. At one of the Fan towns I passed through, for instance, between Lake N'comi and the Rembwe, I noticed a smell, which I decided had an organic origin, in the room I was occupying for the night. As no one was about, I investigated, and knocking off the hot end of the rush-light on the floor, saw three bags hanging from the roof pole. Carefully noticing how they were tied up, I unfortunately shook out the contents of the first one into my only hat; it was a hand, three big toes, four eyes, two ears, and some other fragments ; the hand was quite fresh, the others were in various stages of decay. Bad as no doubt cannibalism is, what I did object to among the Fans was the idea that it was a grand thing for a village to possess a white man's eyeball. I hate superstition in this horrid form; besides, it might have extended, I feared, to white women's eyeballs too. It is almost needless to say that cannibalism is not allowed, nor practised, in districts close to the Government stations. I myself doubt whether there is half as much cannibalism near a station of the Congo Français as there is near stations in the Oil Rivers. Of course it is more difficult to suppress in the Rivers, because it is connected with the secret societies, etc.

I have not space here to tell you of the way I went hunting elephants, and fishing, and rubber collecting, etc. with these Fans— experiences I hope to repeat ; but I do hope I may have better ground to go over than that between N'covi and Ndorko, for that piece of country seems to me to have been made as an obstacle race-track for giants of the olden time, and to have fallen into bad repair, and I sincerely hope not to fall down any more game traps ; they are most unpleasant, and act not only as traps for leopards, antelopes, and boars, but as fortifications for the villages. Nearly every village in the Fan country is on none but fighting terms with its next-door village, and all the villages we passed through in this bit of country were elaborately guarded with pits, felled trees, and stakes driven into the path, besides being situated close to a river and its attributive swamp. We always knew when we were approaching a village by the person who was leading disappearing with a crash through the path. The rest of the party would then hurry up, slip the skin sheaths from their gun locks, see their powder was all right, loosen the trowel or long leaf-shaped knives in their snake-skin scabbards, and then haul out the victim from the pit and tie him up, where necessary, with cool green leaves, for he was pretty sure to be hurt somewhere, because of the ebony spikes fixed in the bottom of the pit.   I went down all sorts of pits, and personally prefer elephant pits to fall into, because they are V-shape, and you can get out unaided, and regard the pits constructed entirely for human enemies as the worst, because they have a bed of spiky thorns at the bottom of them.   All pits, save the elephant ones, are formed like a bag, small at the top, and larger at the bottom, so getting out of them unaided is almost impossible. My chief method of getting into these and other of the severer forms of affliction and terror was by going on alone in front, while the Fans were sitting down having one of their frequent snacks. I was compelled to do this, otherwise I could never have kept pace with them, for they were infinitely the most rapid walkers of any Africans I have ever come across.

Personally I got on very well with my elephant hunters, and indeed with the Fans in general, but I had no end of trouble with them regarding my canoe men and my three Fans.   I fear that M'fetta, the town in which I engaged them, was the worst in the world, and I fear I engaged the three worst characters in it, for there was not a single crime that my three men were not taxed with having committed, and not only they themselves, but their maternal ancestors (paternal do not count) before them.   Fan punishment is killing and subsequent eating, and I therefore used to have to stand hour after hour, dead tired with the day's march, wet through with its swamps and rivers, surrounded by sand flies and mosquitoes, pleading and arguing for their lives.

Nothing but my interest in native law enabled me to live through these word-swamps of palaver, and become the renowned Criminal Court lawyer I am among the Fans ; unfortunately, this was an empty honour, for so low in the scale of civilisation were these savages, that they had no idea of paying lawyer's fees.   But they had one that, as they had got an able lawyer with them, they had better make the most of the opportunity, and they deliberately led me into a village where one of the three had got an outstanding charge of murder, another a case of fraud over an ivory transaction, and the other a terrible wife palaver, depending on me to pull them through, while they bought indiarubber a leaf a ball cheaper than it could be bought elsewhere.   Really, when I found this out, I regretted that I had not allowed the village Fans just to nibble them slightly.   But, now I am calmer, I only feel glad that I was able to do something for those men, for they did much for me, teaching me woodcraft as only a Fan can : how to make a fire in the drenching rain, how to do without lucifer matches—this was an awful lesson, by the by, because Keva kept his flint and steel and tinder in one and the same bag as his snuff and gunpowder, and we had a nice little explosion, and came near sneezing our heads off into the bargain.   Then they taught me how to track game, and in the course of his instructions Wiki showed me a gorilla family.

It was just on the borders of the Egaja plantations.   The party were sitting down having a snack as usual, and Wiki, also as usual, was looking after bush rope.   He came to me, and signed to me to be quiet and follow him ; we crept through the bush for twenty

yards or so, and then lay down and wormed our way still more
cautiously forward — Wiki first, I following in his trail under the
koko leaves. After fifty yards of this we stopped, and I saw, about
thirty yards off, five gorillas busily employed in pulling down plantains and general depredation. There was one old male, one young,
and three females ; one female had clinging to her back a young fellow
with beautiful wavy black hair with just a kink in it. The big male
was crouching on his heels, with his long strong arms hanging down, resting the back of his hands on the ground. An old lady was tearing to
pieces and eating a pineapple, while the rest of the party were pulling
down plantains, destroying more than they ate. They kept up a
whinnying, chattering noise, and I noticed their reach of arm was
immense. When they passed from one plantain-tree to another across
the clear ground they waddled along in a most inelegant style, dragging
their long arms knuckle downwards on the ground. I should think
the old male and female were both well over six feet, but this was
another case in which I failed to take measurements. I watched them
intently, until I was disturbed by a quaint noise from Wiki ; looking
round at him, I saw to my horror his face convulsed and his hand
clutching his throat. Forming the opinion that he intended having a
fit, I became anxious ; he rolled his head to and fro for a second, and
then buried his face in a heap of dried rubbish at the foot of a plantain clump, clasped both his hands over it, and gave an explosive sneeze.
The gorillas let go all, and gave a queer sound between a squeal and a
roar, and the ladies and young folk "went for bush one time" ; the old
male rose right up and looked full at where we were. Wiki went off
into a paroxysm of falsetto sneezes the like of which I have never heard ;
nor I fancy had the gorilla, who, doubtless thinking, as one of his black
co-relatives would have thought, that the phenomenon "favoured Duppy,"
went off after his family with a celerity that was amazing as soon as he
reached the forest. I have seen various wild animals one time and
another in their native wilds, but I have never seen anything so grand
as a gorilla going through the forest. It is a powerful, graceful, superbly
perfect trapeze performance.

Regarding the river Rembwe, I will only say that when I reached it I
found that the mangrove line was much farther inland on it than on the
Ogowé ; this is caused by the great body of salt water brought far inland
by the estuary of Gaboon, and the fact that the currents of the Rembwe
and its neighbour the Como are sluggish, not extremely swift and powerful
like the Ogowé's, and that the body of fresh water which these two rivers,
considerable though they are, bring down to the estuary is far inferior to
that brought down by the most lovely river I have ever seen, the Ogowé.

My ascent of Mungo Mah Lobeh, the Throne of Thunder as the
natives call it, the Great Cameroon as the white people call it, was made
from the south-east face. I was the second "party" to ascend this face ;
the first was composed of the first officer and the doctor of H.I.M. ship
Hyæna. I may remark that I did not "discover," as I have seen
stated, seventy volcanic craters on it, but I have heard that they are
there. Instead of being in any way hindered by the German authorities,
I received from them every help, good advice, and kindness.

BRITISH GUIANA:

A Sketch.

By Sir David P. Chalmers, recently Chief Justice of British Guiana.

(Read before the Society on Jan. 30th, 1896.)

I FEEL considerable diffidence in appearing before a Geographical Society, as I cannot claim the qualifications of a discoverer or explorer of unknown regions. I stated these difficulties to your very courteous Secretary when he did me the honour of conveying to me the wish of your Council that I should read a paper, and his reply, after referring to the Council, was to the effect that my objection had been overruled. I shall therefore proceed to put before you some particulars about our colony of British Guiana, which I hope may at the present time have some interest for you.

Long before Great Britain possessed any colony in South America, the whole country lying along the Atlantic seaboard on the north-eastern shoulder of that continent, between the two great rivers the Orinoco and Amazon, and bounded inland by what were then Spanish provinces—Grenada and New Andalusia—was known by the general name of Guiana, derived, it is believed, from the Indian name of a small river, the Wa-ini or Gwa-ini, which enters the Atlantic near the mouth of the Orinoco. Towards the middle of the fifteenth century, small Dutch trading ships began to visit the West Indies. Some of them sailed along the coasts of the mainland, which lay immediately to the south of the islands, and opened trade with the aboriginal inhabitants, bartering knives, axes, beads, and the like, for cotton and other products of the country. As early as 1580 these voyagers had established a trading post at the mouth of the Pomeroon, and another on the Abary river, both within what is now our colony of British Guiana. In 1595, that chivalrous and adventurous Englishman, Sir Walter Raleigh, set sail in his own ship, accompanied by two smaller ones, in search of "that mighty rich and beautiful empire of Guiana, and of that great and golden city which the Spaniards call El Dorado and the Indians Manoa." He explored part of the Orinoco, and, after various adventures, returned to England, bringing back a most glowing account of the wealth and resources of the country, its mines of gold and precious stones, and the "Imperial City of El Dorado." During the evil times that fell upon Raleigh after this, and in his long imprisonment, he still cherished dreams of Guiana and the Golden City, and when at length he was released, he again led an expedition to the land of hope. But this expedition resulted in disastrous failure, and the English Crown obtained no dominion in Guiana at that time. In the early part of the last century, the Portuguese, French, and Dutch claimed sovereignty over different portions of this country. The Dutch held the northern part, extending from the mouth of the Orinoco south-eastwards to the river Maroni, a distance of about 530 miles. The French possessions extended from the Maroni south to Cape Orange, a distance of somewhat over 200 miles. The Portuguese held the remainder of the country as far as the mouth of the river Amazon. But the representatives of the several nationalities had only made actual settlements in some places on the coast, from whence they exercised a more or less widely extended influence over the Indians, who were the real inhabitants of the country, and at the time of which I am speaking far outnumbered the strangers.

The colony which we know as British Guiana was settled by the Dutch as three colonies, named from the three rivers upon which the first beginnings of occupancy took place, the colony of Essequebo, the colony of Demerara, and the colony of Berbice. In order the better to understand the position, political as well as economical, of our colony, it is necessary to carry our view backwards a little to the times of the Dutch settlers. It was they who rendered it possible for us to utilise those amazingly fertile lowlands which form the sugar-producing portions of the colony. "Every acre," it has been well said, "has been the scene of a struggle with the sea in front and the floods behind"; and the whole system of embankments, canals, and sluices, to which we owe the safety from inundation, and the possibility of cultivating our coast and river-side lands, is due to Dutch skill, industry, and perseverance. The first beginnings of trade were indeed very small. Early in the seventeenth century the principal post in the colony of Essequebo was on an island called in Dutch, Kyk-over-al—literally, see-over all—situated about fifty miles from the sea, in a tributary of the Essequebo river. At this time one ship in a year was sufficient to carry on the whole trade between the colony of Essequebo and the mother country. Gradually the cultivation of the land made progress, and contemporaneously the trade over sea increased, so that when we come down as late as 1791, we find that the whole of the coast lands, from the mouth of the Essequebo river as far north as the mouth of the Pomeroon, had been granted out, and were being cultivated as plantations. The settlement of the colonies of Demerara and Berbice, somewhat later in date, was on very similar lines with that of Essequebo. The first settlements were made by traders, and these settlements were followed by appropriation for cultivation of the lands lying along the sea-coasts and river banks.

The colonies met with many vicissitudes in their earlier growth, and they suffered in the wars. In 1781 they capitulated to the British. In 1782 the British admiral in the West Indies, having to cope with the Americans, French, Dutch, and Spaniards, had his hands too full to garrison the colonies in Guiana sufficiently, and they surrendered to a greatly stronger French force. In 1783 they were given back at the peace to the Dutch, but surrendered to the British in 1796. In 1802, at the Peace of Amiens, they were again given back to the Dutch, but again surrendered to Great Britain in 1803, and have been British possessions continuously since then, and the rights of Great Britain were fully recognised in the treaty arrangements made at the Peace of 1814.

The coastline of the present colony of British Guiana—or rather perhaps I should say, that which we claim as British Guiana—extends in a south-easterly direction from the mouth of the Amakura river in the estuary of the Orinoco in N. latitude about 8° 75′ to the mouth of the river Corentyn, which separates British Guiana from that part of the Dutch colonies still retained by them under the name of Dutch Guiana. Leaving the coast at this point, the boundary follows the course of the Corentyn in a southerly direction until it reaches a point about 1° 50′ of N. latitude. Then it proceeds south-west along the borders of Brazil, following the line of the Acarai mountains until it reaches to about 75′ of N. latitude, which is its extreme southern point, and from thence the boundary follows a north-westerly direction, bordering at first with Brazil and then with Venezuela. It coincides for part of the way with the Takutu and Cotinga rivers, until the mountain Roraima is reached, and then it is said to proceed north-east to the Imataca range and onward in a northerly direction to the Amakura river, and to follow this river to the Orinoco. The Brazilian, as well as the Venezuelan, boundary has not yet been definitely settled. Our coastline is somewhat over 300 miles in length, not following the indentations and prominences closely; the greatest depth of inland measurement in a straight line from north to south is about 550 miles, and the greatest breadth from east to west somewhat over 300 miles, but the average is considerably less, the outline on the Venezuelan and Brazilian sides being irregular. These measurements are taken from a map recently published by Stanford, and probably vary a little from the measurements which may be made on some other maps. The superficial area is computed as being about 109,000 square miles, which is more than the area of England, Wales, and Scotland taken together. So far, therefore, as territorial extent is concerned, this, the only British possession on the continent of South America, is no insignificant one.

The journey to Guiana is not difficult. Starting from Southampton, a run of eleven and a half days across the Atlantic in a direction south-west by south, or nearly so, would carry us to Barbadoes, which lies somewhat to the north and west of Guiana, and after a short delay there for transhipment to an inter-colonial steamboat, a course of rather less than two days' further steaming would bring us to Georgetown, the principal town of the colony. The impression of Guiana which a voyager receives on approaching it is not very favourable. Long before coming in sight of land we observe a change in the colour of the water on which we are sailing. The deep transparent blue of the Atlantic, which is so beautiful a feature for the last five or six days previous to reaching Barbadoes, is no longer seen; the water seems shallower and much less bright. When we reach within forty or fifty miles of Guiana, there is a still greater change. We come to a rippling line in the water, as when two currents meet, but are slow to mingle, and on the further side of the line the water appears nearly opaque and of a brownish yellow colour. The rippling line is not always seen, and then the change from clear to opaque takes place more gradually. This discoloration is produced by immense quantities of fine sand and *débris* of decayed vegetation which is brought down by the rivers of Guiana, and also by the greater rivers the Amazon and Orinoco, and is continually kept in motion by the currents which are always flowing over the shelving coast

of this part of South America. The position of Georgetown is made out
by means of a floating beacon or light-ship, which is also a pilot station,
moored ten or twelve miles from the shore. At this distance the land
seems perfectly level, and would be barely discernible except for tall
chimney-stalks appearing on the horizon, which show the locality of some
sugar manufactories. A little nearer, the feathery tops of palm-trees
catch the eye, and then the features of the land we are approaching
become more distinct. Wide expanses of vivid green mark the position
of sugar-cane fields; white painted walls and roofs of houses show amongst
groves of palms and other trees, and in a very short time after leaving
the light-ship we run into the Demerara river, which forms an excellent
natural harbour a mile and a half in width, in which a multitude of
ships can find safe anchorage. Landing at one of the numerous
jetties in the river—or stellings, as they are still called, according to
the old Dutch nomenclature—we find ourselves in a city laid out
in wide streets running north and south, and east and west, which
intersect each other at right angles, forming rectangular blocks of
moderate dimensions. These streets give free access to the breezes,
which blow with much regularity. In many of them shady trees afford
a shelter from the too predominant sunshine. In some streets are broad
canals, reminding us even now of the Dutch founders. These canals
serve a useful purpose in rapidly carrying off the superfluous water during
rainy weather. Pink lotus flowers, and a hyacinth-coloured water-lily
floating in these water-courses give variety to the colouring, and the
giant leaves and flowers of the *Victoria regia* are seen in many places.
Private dwelling-houses are of all sorts and sizes, without the slightest
architectural pretensions, but following regular lines along the streets.
Each stands in a plot of garden-ground, most of them bright with
flowers and shaded with trees, and larger or smaller according to the
means and inclinations of the owners. Among the conspicuous buildings
is Queen's College, a Government foundation which does exceedingly good
educational work. At the end of the same street is the Cathedral of the
Diocese of Guiana. The original building has been removed and replaced
by a much larger one, built of timber and concrete, one of the largest
—I am not sure that it is not quite the largest—structures of the
kind which has ever been built. St. Andrew's Church is the principal
Scotch church in Georgetown. It is a somewhat curious feature in
the ecclesiastical settlement in Guiana that, although the colony as
a whole is a bishop's see, under the jurisdiction of the Archbishop of
Canterbury, the Scotch Church is not only concurrently endowed with
the Anglican, but has also a territorial status; Anglican and Presbyterian
parishes alternate throughout the colony, each parish lying contiguous
on either side to parishes of a different denomination. I believe this
arrangement has not led to any friction. Roman Catholic and Wesleyan
churches are also assisted by grants of public money. The offices of the
various Government departments are located in a brick and concrete structure—one of the few of the kind in Georgetown. The building material
is mainly wood, both on account of the scarcity and expense of other
material, and of the difficulty of making sufficient foundations in the

soft and yielding subsoil for brick or stone buildings. We may also notice Water Street, the principal business street in Georgetown, and the Sea Wall, important as a bulwark against possible incursions of the sea and also as a promenade. Time does not permit us to linger in Georgetown; but I may tell you before leaving it, that it is a town of somewhat over 50,000 inhabitants, that it is the seat of government, the principal port for imports and exports, that it is governed by a Mayor and Corporation, that it contains several churches besides those I have mentioned, a town hall, an hospital capable of accommodating 700 patients, courts of justice, schools, alms-houses, a Royal Agricultural and Commercial Society, possessing an excellent library and museum, a Chamber of Commerce, an Institute of Mines, two Banks having the privilege of issuing their own notes, and was lighted by electric light four years before Edinburgh began to resort to this method of lighting. There is a large and well-managed Botanic Garden, which not only serves purposes of amenity and beauty, but also important economic uses in the assistance it gives to agriculture, both by supplying plants and seeds of the most approved descriptions, and by devising by experiments new methods of culture.

It would be a very incorrect idea of British Guiana if we were to think of it, as a dweller in Georgetown might perhaps be apt to do, as merely a level tract of seaboard and riverside land, mostly occupied with highly cultivated cane-fields and sugar manufactories. The flat belt in which the cane cultivation occurs occupies only a small portion of the colony. It varies in width from a few miles in some places to about forty or fifty miles in others. It is an alluvial deposit of great depth, resting upon granite. It is so thoroughly alluvial, consisting of clay and vegetable *débris* and perhaps a little sand, that it yields no stones large enough for road making; such roads as exist are made of burned clay and shells mixed together, and in some places of granite brought by water carriage from higher ground on the rivers. This flat land slopes very gradually upward as we go inland; it is generally bounded by a row of low sandhills, and beyond these we enter upon a wide uncultivated region. The general elevation here is above that of the alluvial belt, but parts are low and swampy. Within this region a profitable trade has been carried on in timber, and it has been partially denuded in those parts which lie within easy distance of ports of shipment, and adjacent to the banks of rivers and creeks, where the indispensable aid of water carriage is available. In places remote from this means of transport immense breadths of forests remain untouched. The soil and climate are so fertile that wherever the large trees are cut down a thicket speedily springs up, almost impenetrable in its density, composed of smaller trees, shrubs, and creepers. In process of time the trees of stronger growth may overcome the weaker undergrowth and become valuable timber trees, but as yet there is no systematic forestry. Beyond, and at a higher level than this first forest region, there are huge tracts of primeval forest. These tracts are above the cataracts, which are met with on all the great rivers at varying distances from the sea, these cataracts being in fact the sudden descent of the rivers from the

higher inland plateau. The last region, and the farthest from the sea, is
that of the savannahs. It is continuous with the grass-covered plains
so frequent in the central parts of South America.   The general
appearance of the savannahs is not unlike those wider parts of the
English downs, where the rolling surface is broken by a few stunted
hawthorns or clumps of tall furze. But in the hollows between the
ridges of the savannahs, instead of the fir, beech, and hazel coppices of
our own country, there are groves or clumps of Eta palms, or other
tropical trees, or shrubs. The principal mountains of the colony are
found in the savannah region. The term savannah is also applied to a
variety of grassy plains, more or less marshy, met with near the creeks
and rivers in the forest region, and even descending to the seaboard.
The aspect of some of these lower savannahs I was very familiar with in
travelling on the road between Georgetown and New Amsterdam for the
purpose of holding the circuit court at the latter place. The road ran
through a perfectly level plain, stretching for many miles, bordered on
the one side by a belt of mangrove and courida trees fringing the sea-
shore, and on the other by the trees of the forest region. The plain was
covered in rainy weather with tall rank grass and other plants, much
frequented by stork-like birds, the most conspicuous of which was the
Negrokop, a bird of dark grey plumage, standing almost four feet high ;
he would stalk along with majestic strides, ever and anon bending his
long neck to pounce upon the smaller animals on which he fed ; sometimes these birds were seen also in flocks. Alligators used to be seen in
the watercourses bordering the road, but as they are timid animals,
always hiding themselves when they can, those seen gave no adequate
representation of the real numbers.   Although in droughts these
savannahs became very parched, and the herbage almost disappeared,
they were much used for the pasturing of cattle, and great herds grazed
on them.   In the rainy weather so much water lay on the surface that
it was not uncommon to see cattle up to their necks in water, suffering
no inconvenience apparently, but busily cropping the succulent grasses
and other herbage which rose above the surface. The most eastern
portion of the colony does not show so distinctly the sort of terrace-like
structure found in the central part. It is an inclined plain descending
gradually from a height of about 800 feet to the sea-level. The plain
extends westwards and southwards and connects with the savannahs of
Brazil.

The only one of these regions in which any considerable number of
inhabitants are found is the alluvial belt next the coast. Small parties of
woodcutters visit the forests, and live there for weeks or months, but their
residence only lasts until their work is accomplished. Since about 1884 it
has been found that gold could be profitably worked in various places
in the interior, and considerable numbers have betaken themselves to this
attractive form of industry ; but these gold-seekers and gold-diggers are
companies of adventurers having no settled homes where they work, and,
like the woodcutters, they return from time to time to their permanent
dwelling-places in the coast region. The regions of primeval forest and
savannah, so far as they can be said to be inhabited, are inhabited only

by the Indians, and possibly a very small number of settlers. The Indians live widely apart, and moreover change their places of residence from time to time, even going over into Brazil or Venezuela, so that it has been difficult to form any close estimate of their numbers; but authorities most capable of judging have put the total number usually in Guiana as from 12,000 to 20,000, the latter estimate being probably nearer the truth. These Indians are of different tribes, known by different names, and distinguished by considerable differences in their physical appearance and in their habits, but there seems no doubt that they all belong to the great race forming the aboriginal inhabitants of America. It would take too long to dwell upon the interesting questions connected with the Indians. The population of the colony, apart from Indians, was estimated in 1892 to consist of 141,531 Negroes and half-castes, 105,463 East Indians, 3714 Chinese, and 12,166 Portuguese and natives of Madeira and the Azores. The Europeans are only a small number, probably not more than 2000 in all. Their real number could not be ascertained by a census of the usual kind, as they are never all in the colony at any one time. The nationalities represented are English, French, German, Dutch, Irish, and Scotch.

Viewing our colony thus generally, as a land of wide forests and savannahs, the rivers are an important feature. They are not only the drainage channels which carry the superfluous rain-water to the sea, but also the highways leading from the inhabited parts into the interior. There is a good road along a portion of the sea-coast, about 120 miles in all, and also extending a few miles inland from the mouths of the Demerara, Essequebo, and Berbice rivers; but with these exceptions there are no roads. The traveller who would visit the interior must take his way on the rivers and their tributary creeks and then through the forests. It is not always that he can find a path where he needs it. The axe and the felling knife are essential. He must be prepared often to cut his own path through the thick bush or jungle. Perhaps I should mention here—lest I would be representing the colony as less advanced than it really is—that there is a railway, about twenty-one miles in length, running parallel to the coast road out of Georgetown, serving an important sugar district, and paying, I believe, good dividends, and that when I left Guiana there were other railway schemes under consideration. Including the river Corentyn, a splendid river, navigable for 150 miles, which, as I have already mentioned, is the boundary between British and Dutch Guiana, our colony is drained by five great rivers and six or seven smaller ones, not including tributaries or small streams. Taking these rivers in their order from north to south, the next to the Corentyn is the Berbice, navigable for ships of light draft for 150 miles, and having a total course of about 300 miles. The Demerara is navigable for eighty-five miles, and its entire course is about 200 miles. The Essequebo, although the largest of the rivers, has the shortest navigable course, being interrupted by cataracts at about sixty miles from the sea; it rises in the most southern part of the colony, and reaches the sea after a course of about 600 miles. Its estuary is twenty miles wide at its widest part. In this estuary are situated three large islands; one of them is twelve miles long by three miles across. Numerous other islands, some large, some small, said to equal in number the days of the year, are scattered along the course of this river. It has three very large tributaries : the Mazaruni, which rises in the near neighbourhood of Mount Roraima, and has a course of 200 miles before joining the Essequebo ; another, the Cuyuni, a familiar name in the boundary dispute, rises on the borders of Venezuela, and forms, for about 140 miles, the boundary between territory which is claimed as indisput-ably belonging to Great Britain, and other territory the title to which is considered doubtful. Its course, until it joins the Essequebo, is not much short of 300 miles. The third large tributary, the Rupununi, has a course of 220 miles. These great rivers all run nearly parallel to each other from south to north, and all enter the ocean in a direction oblique to the coast. Smaller rivers are the Canje, an affluent of the Berbice, having great depth of water, and running almost a level course for 150 miles ; the Abary, Mahaicoay, and Mahaica creeks, running into the Atlantic between the Berbice and Demerara rivers, and the Pomeroon, Wa-ini and Barama falling into the Atlantic to the northward of the Essequebo. The Barima, navigable for at least eighty miles, and the Amakura, which is one of the boundary rivers, join the estuary of the Orinoco.

The rainfall in Guiana is very large, being on an average nearly 100 inches in the year at those stations on the coast where registers have been kept, and probably a good deal more in the interior, and it is a natural result that the flow of fresh water in the rivers during the rainy season is enormous. I have myself observed it in the Demerara and in the Berbice rivers (with which I happen to be most familiar) running with the rapidity of a mill race. In the Demerara river it is estimated to run at ebb-tide seven or eight knots an hour in the rainy season. This current makes it a work of some difficulty and risk to go alongside ships in the harbour, and has sometimes led to serious accidents. I have already alluded to the quantity of detritus brought down to the sea by the river floods. A result of this and of the very shallow and shelving slope of the coast is the formation of banks not only at the river mouths, where they are a serious hindrance to navigation, but more or less all along the shores. These deposits frequently shift with the shifting of the currents. At one time there will be a secure rampart protecting the low-lying land from the inroads of the sea, and again, after some months, perhaps, the rampart will be washed away and breaches even made in the land behind. The necessity of protecting by artificial ramparts the coasts of the plantations forms one of the anxieties and a fertile source of expense in carrying on a sugar plantation. And not only must the plantation be protected from the sea in front, but also from the flood-water coming from the direction of the forest behind. Ramparts or dams all round the plantation, and trenches to carry away the masses of rain-water, are essential.

The character of the scenery of the rivers is, of course, very different when above the influence of the tides from what it is in the lower reaches. Mangroves, with their strange aerial roots, fringe the banks where the salt water is a predominating influence. But even in the tidal waters the scenery is often of great beauty. The effect in sailing over these waters is as if passing through a series of lakes. Where the wood-cutter has not too recently disturbed the landscape, the waters are bordered by a dense thicket of huge trees of varied forms and leafage, the brilliant blue sky is reflected in the almost still waters, the intense sunlight brings out the keenest contrasts of light and shade. Going inland, the rivers are narrower and more picturesque. Being nearer, the forms of the vegetation can be better seen and are more appreciated; rocks and islands diversify the scenery. Cataracts occur on all, or nearly all, the rivers at a certain distance from the coast; the exceptions are those which have their course in the delta of the Orinoco. Lovers of Nature would find endless enjoyment in studying the varying picturesque aspects of the cataracts. Persons engaged in such occupations as wood-cutting or gold-seeking look upon them as so many obstructions to navigation, which no doubt they are. Unfortunately many of the cataracts are associated with disastrous boat or canoe accidents.

The fall which is perhaps best known by name is the Kaieteur. The name is derived from an Indian tradition, which narrates how an old man, having become troublesome to his family beyond what they considered could reasonably be borne, was placed by them in a small canoe set adrift in the river a little above the fall, over which it was, of course, speedily carried. The canoe itself, the tradition says, was changed into a pointed rock, which is still seen at the fall, and the basket-work portmanteau, or peggall, in which the family had considerately packed the old man's belongings, became a square rock, also still seen. It illustrates the unexplored character of Guiana that this fall, which occurs on the Potaro, a large tributary of the well-known Essequebo, was unknown except to the Indians until 24th April 1870, when it was first seen by Mr. Barrington Brown; and when Mr. Brown made his discovery known in Georgetown, great surprise was manifested, and almost doubts were raised. I quote a few sentences written by Mr. Brown on the first freshness of the discovery:—"We came within the roar of a large fall. I observed that heavy masses of vapour were drifting before the north-east wind, making the trees, grass, and shrubs on our right dripping wet. This came from the great fall to which we were in close proximity, but which was hidden by a grove of trees. Making a detour to our right through the grove, we came on the flat rocks at the head of the fall, and walking to the edge of the precipice, down which the water was precipated, I gazed with wonder and delight at the singular and magnificent sight that lay before me. Not being prepared for anything so grand and startling, I could not at first believe my eyes, but felt that it was all a dream. There, however, was the dark silent water, passing slowly but surely to the brink of a great precipice, and breaking into ripples as it approached its doom; then, curving over the edge in a smooth mass of a brownish tinge, changing into a snow-white fleecy foam, it was precipitated downwards into a black seething caldron hundreds of feet below. Of how many hundreds I could not with certainty make out; but, judging from the appearance of the river below the basin, I thought it must be near upon 900 feet. . . . The river was swollen by the late rains, and was over 100 yards in width, containing a vast volume of water, which, as it reached the basin below, sent up great creamy spirts of foam with a steady thundering roar." Subsequent accurate measurements showed the first estimate to be not far from correct. The river makes one perpendicular leap of 741 feet, and then falls by a sloping cataract 88 feet further. The perpendicular leap is thus almost exactly seven times the height of the Dean Bridge, at Edinburgh, which, I think, is 106 feet above the bed of the river.

The forest trees resemble our own Spanish chestnuts, oaks, and laurels —I mean in general aspect, not botanically. The trees are on 'a much larger scale than our own. Palms bulk considerably, sometimes in groves, sometimes in smaller groups. There is great variety in the colouring of the foliage, and the brilliant colours of flowers also occur at some seasons. For instance, there is a magnificent avenue of trees in the Botanic Garden which I formerly mentioned. These trees, about April and May, are covered to their summits with blossoms of a bright salmon colour. Creepers carrying bright-tinted flowers add to the array of colour; but flower-covered trees and creepers occur only at intervals. Many of the trees are deciduous, but, so far as I have observed, there is no winter sleep of the plant, as in temperate climates. As soon as a tree has shed its mature leaves, the fresh young buds are ready to burst forth, so that in about a week the tree is clothed again with fresh verdure. As this sort of fall of the leaf and renovation does not take place at one time, but is going on throughout the year, it adds to the variety of tints in the forest. The varieties of timber hitherto exported have been chiefly the mora and greenheart; but india-rubber trees, and many others producing curious and beautiful timber, and also medicinal plants, grow in great abundance. It is a most impressive sight to look on the outside of a great reach of forest. This, of course, is only possible when there is some sufficient height from which a view of the lower-lying forest can be obtained. I remember one occasion when I had the good fortune to obtain a view of this sort. Standing on the top of a hill which rises in complete isolation upon a level plain, traversed by the Barima river, I looked over an expanse of forest, bounded in one direction by some hilly ridges, hazily, yet sufficiently visible at a distance, as I calculated, of about fifteen to twenty miles, in another direction stretching away, apparently without limit, until it merged in the horizon. I can use no simile so appropriate as an ocean of forest. Even the ocean swell was simulated by the effect of slight irregularities of the ground, whilst the tossing of the tree-tops by the wind in the nearer distances, their bright foliage rising in glistening points of light, produced a semblance of the superficial wavelets.

The mountains of Guiana are so far from the coast, and so difficult of access, that only a few enterprising travellers have visited them. There are two parallel chains running in the direction east to west, the Pacaraima and the Merume mountains. There are other mountains with very unpronounceable names, not equalling in magnitude second-class Scotch mountains. The Pacaraima range culminates in that extraordinary mountain, Roraima, which until recently was believed to be inaccessible.

This mountain rises 5000 feet above the elevated tableland, and 8609 feet above the sea. The upper part consists of a table-top surmounting a mural precipice, estimated at about 2000 feet high. Schomburgk says: "These stupendous walls are as perpendicular as if erected with a plumb. line." Considerable masses of water fall at different points. The Indian songs and traditions are full of allusions to this mountain. They had a tradition that the spirits of the mountain would for ever prohibit its ascent by man. Amongst scientific men there was an idea that on this tableland, cut off as it was from all communication with the world, animal and vegetable forms of primitive type might exist, different from anything known on the ordinary surface. The mystery was dispelled by a successful ascent, made on 18th December 1884, by Mr. im Thurn, Commissioner of the North-West district of Guiana. Resolved on making a thorough exploration and examining the precipice from different points of vantage, he at length discovered a sort of ledge on the rock by which it was practicable to ascend, and in this way reached the summit, where he was able to remain long enough to make important observations. The primeval forms, however, which it was thought possible might exist, were not found, but various plants, hitherto undescribed, were collected. A most interesting account of this ascent was published in a Guiana Journal (*Timehri*; June 1885). At least three other ascents have since been made.

Time will not permit me to say much of the fauna of Guiana. Amongst the animals useful for food are various rodents, such as the labba, accourie, water-haas. Bush-hogs, tapirs, and deer of several varieties are also plentiful. Amongst beasts of prey the most important are the jaguars and ant-bears. Monkeys are numerous and of several varieties, and snakes are characteristic and abundant, although from their timid habits less seen than might be expected. They include the camoodie, a species of boa, which has been found thirty feet long, and more than three feet round, the rattlesnake and the deadly labarria, besides many other varieties, the greater proportion of which are not venomous. Insect life abounds, and the country is particularly rich in birds, as many as 1200 varieties having been named, and the list not being complete. Some of the birds are of remarkably beautiful plumage, their bright colours even being a feature sometimes in the landscape. I have seen flocks of flamingoes, of brilliant vermilion colour, circling and poising themselves like rooks amongst the tree-tops on a river bank, and at other times in ordered battalions on their aerial journeys—the older birds in their bright uniforms on the outside of the formation, the young ones in grey undress in the centre. And I remember also seeing in the North-West district flights of parrots, some bright green, others—the large macaws—showing most brilliant hues of blue and yellow.

The climate of Guiana is equable and moderate for a tropical one. There is no such burning heat as is known in the plains of India. The highest marking of a perfectly shaded thermometer is said to be 90°. I do not myself remember to have seen it over 87° in the course of sixteen years, and very seldom as high as that. In the colder months— that is, from December to April or May—it seldom rises above 81°

in the shade, and for all purposes of indoor life, as far as sensation is concerned, the temperature is perfect. The lowest range is about 72° or 71° between five and six in the morning, and then a feeling of chilliness is produced. But the average minimum is not less than 76°. These are temperatures which occur at the sea coast. In the forests and on the higher grounds the minimum is considerably lower, and the maximum is higher. The comparatively moderate temperature is no doubt due in part to the trade winds, which blow from the north-east with great regularity. They diminish the sensation of heat as much as, or perhaps more than, the actual temperature. The length of the nights is also a mitigating factor. The sun is below the horizon for about twelve hours at the longest day and twelve and a half hours at the shortest. I believe the hothouse at Kew where the *Victoria regia* is kept —which most of you know very well—is meant to be of the temperature which prevails in a Guiana forest in the lower levels; but the high temperature of the Kew house is much more felt, because we go suddenly into it from the cool air outside, whereas one is gradually introduced to the temperatures of the Tropics.

It is, perhaps, as good a practical test of a climate to observe what people do, as to study meteorological instruments. It may illustrate the nature of the Guiana climate when I mention that there is a flourishing cricket club at Georgetown. They play matches amongst themselves, and, when opportunity offers, with teams from the West India Islands, and recently have played with an English team. I have never heard of any mischief resulting. Lawn tennis is also a favourite amusement, but it is generally played after four in the afternoon, when the sun's heat is much diminished. It was, and I believe is, a practice for life assurance companies to charge considerably higher rates for Guiana than for temperate climates, and there is a widespread impression, probably not without foundation in most cases, that the climate exercises some strain upon European constitutions, to counterbalance which it is advisable to make occasional visits to cooler countries.

Thunderstorms are frequent at the commencement of the rainy weather, and also, although in a less degree, at its termination. Lightning is very vivid, and is often seen streaming from cloud to cloud without coming to the ground. It has been remarked that few instances have been noticed of injury done either to buildings or to animals. It has been surmised that certain species of palm-trees, which tower to great heights, and have very moist and succulent leaves and stems, act as conductors, but there has been no scientific verification of this. It should also be mentioned, when on the subject of meteorology, that hurricanes, such as those which have from time to time done great mischief in the neighbouring West India Islands, are unknown in Guiana, so far as discoverable from the memory of any inhabitant or from any written record. It would detain you too long were I to enter upon the conjectures by which it has been endeavoured to explain the prevailing immunity.

It is time to say a word as to the industries of the colony. These are named in three words—sugar, gold, and—a long way after—timber. Although the soil is probably capable of growing every product of the

Tropics, and although other products than sugar are cultivated to a slight extent, sugar has from a variety of causes come to be far and away the most important—indeed, as things are at present, to be the only cultivation of vital importance. In the palmy days of the industry, a sugar estate was as valuable as a gold mine. Not only did the proprietors receive very large revenues from their estates, but their managers and attorneys in the colony—for owners were much addicted to absenteeism—lived in wealth and luxury. The sugar estates were laid out and cultivated by the Dutch settlers by means of slave labour, and their cultivation was carried on by their British successors by the same instrumentality, until the public conscience of Great Britain rebelled against the system. It is strange nowadays to come face to face, as one does in old newspapers and records of the colony, with the working of a system of slavery. The prædial slaves were, as to their legal status, the appurtenances of the plantation to which they belonged. If the plantation changed owners, the slaves changed their masters. Curious advertisements may be read of lots of slaves for sale, in which their sound health is guaranteed, and their qualifications are described as trained carpenters, blacksmiths, field hands, or otherwise. Even at the present day there are persons living in the colony who were slaves, and some of them regretfully contrast their present circumstances with the older times, when they lived without cares, and were comfortably provided for in old age by their masters as well as when in full strength and vigour. The system was suddenly brought to an end, and there ensued a not altogether unnatural revulsion on the part of many of the new freedmen against steady or hard work in any form. The consequence was that it was difficult to induce them to engage as hired labourers, except at excessive rates of wages. Even on these terms they were not always reliable, and would strike work and hold out for impossible wages at critical periods, as when the crops had reached a stage at which it was of vital importance they should be rapidly reaped. To remedy this state of things a system of bringing East Indians to Guiana as estate labourers was set on foot, which still continues. These immigrants are engaged in India on a system of indenture, under the supervision of the Indian Government, and are taken charge of by a Government department in Guiana upon their arrival there. This department regulates, under local laws, the allocation of the labourers to the plantations, their housing, wages, work, medical care, and other incidents of their position. On completing the term of five years' indenture the immigrant is entitled to a free return passage to India if he chooses to claim it. Many of the immigrants then take their return passages and go back to India, generally carrying with them considerable savings. A good many, after spending some time in India, return to the colony under fresh engagements. Others reindenture in the colony, and others quit the regulation labour and live in the colony on their own responsibility, many finding employment on the plantations. The expense of importing labourers is borne partly by the sugar proprietors, partly by the general community.

Notwithstanding this artificial labour system, the Negro retains his position. He is of stronger physique than the East Indian, and preferred for the hardest work. Negroes also, for the most part, fill the more responsible positions in the manufacturing processes, and in managing machinery. Negroes and coloured men have, if not quite a monopoly, yet a very large share of all the handicraft work carried on in the colony, as carpenters, blacksmiths, and the like. They become excellent workmen when properly trained. This is not the time to enter into a discussion of the many-sided Negro question, which is not, nor is likely to become, a pressing one in British Guiana. The Emancipation Acts of half a century ago found the Negroes of Guiana, as well as of the West India Islands, in a condition in which it was scarcely possible for them to derive full benefit from their new-found freedom. The state of subjection which they had been under for centuries, whilst it had obliterated the tribal ideas and organisations of their own country, in which there was much to foster a regard for others and many of the virtues, had supplied very little instead, beyond a certain amount of secured material well-being. There was very little to develop the conscience or intelligence of individuals, much less to raise anything like a high general standard of feeling or opinion. But many agencies of improvement are now at work; it is unquestionable that improvement has set in, and there is good hope for the future. The new industry in gold seeking and collecting in the interior of the country has powerful attractions for a considerable class of Negroes and half-castes, who, although possessed of much desultory energy, are repelled by the monotony of fixed employments. Many of these have benefited in a rational way by the increase of well-being which gold has brought to them, whilst the older sugar industry has been partially disorganised by the competition of the younger one. That difficulty, of course, admits of remedy.

But the sugar industry has a much more serious grievance in the payment by Foreign Governments of bounties to their own people on their exports of beetroot sugar, whilst the importation of Guiana and West India sugar into the beet-producing countries is barred by heavy duties. These bounties operate adversely to our colonial sugar-growers in two ways: directly, they enable the grower of beetroot sugar to under-sell the colonial grower in the market; indirectly, by stimulating the over-production of beet sugar in foreign countries, they drag down the ruling average of price to a level at which it ceases to be remunerative. Much of this bounty-fed sugar finds its way to Great Britain, and while it reduces in the meantime the market price of sugar, it has inflicted great injury on all who are interested in the production of sugar in the colonies. Instead of the plantations producing large profits, many have been found unworkable except at a loss, and have been abandoned to become wilderness; whilst even as regards the plantations most favourably placed as to soil and situation, it is a most anxious question to those whose fortunes are at stake whether any margin is left at all after paying the expenses, which, under any circumstances, are very heavy. The proprietors and their managers seem to have done all that manly endeavour and skill can do to maintain their position, by improved methods of cultivation, by improved methods and machinery of manufacture, and by retrenching expenses in all possible directions. There are some indications that the bounty-paying countries may grow tired of a system in which they tax their own citizens in order that sugar may be sold cheaply to foreigners. Time will show.

As to gold I shall not say much. In 1884 the export notified to the Customs was 250 ounces; in 1888 it was 14,570 ounces; in 1895 it was 122,935 ounces. There is no doubt that gold exists over a large part of the colony. Hitherto it has chiefly been found on the Essequebo river and its tributaries, and to the northward and westward. But the practical problem is to find gold in any given locality in sufficient richness of deposit to give remunerative results. It must be remembered that great expense is incurred in conveying the labourers to the spot, as also their food, and all tools and machinery. Every requisite must be conveyed by boat carriage on the rivers, and then by manual labour, through paths that often must be cut through the forest. In some places—notably on the Essequebo, Potaro, Cuyuni, and Barima rivers and their tributaries—gold has been found in paying quantities, but the deposits successfully worked have been alluvial; and such deposits in the course of continued working become exhausted. Those who have been successful in the industry have had the good fortune to find fresh deposits when the earlier ones were no longer profitable. How long this may continue is a question on which even experts hesitate to give an opinion. There are, it is said, veins of quartz rich in gold in some places, and at least one company is or will be engaged in quartz crushing. I have, however, no recent information as to the result.

In concluding this sketch I may inform you that the exports of Guiana in the year 1893-94 were of the total value of £2,358,917, and the imports were of the value of £1,920,715. The public revenue from all sources was £602,762 and the expenditure £566,833, and the public debt was of the respectable amount of £867,875. The colony, you will have perceived, has great possibilities, and its value may well be deemed to lie rather in its resources than in their actual present development.[1]

ARABIA AND ABYSSINIA IN ANCIENT TIMES.

By J. W. M'CRINDLE, M.A., M.R.A.S.

THE author of the work noted below[2] has made the early history of Arabia his special study. He has frequently travelled in that country, and made himself familiar with its language, and with its old records and monuments, which throw light on its past before Islam had as yet arisen. The first fruits of his researches he published in 1890, in a two-volume work, entitled *Sketches of the History and Geography of Arabia from*

---

[1] The remainder of this paper related to the dispute as to the boundary between British Guiana and Venezuela, in which the United States has interfered, and is not deemed suitable for this journal.—ED.

[2] *Die Abessinier in Arabien und Afrika*, von Dr. Eduard Glaser. München, 1895.

the Earliest Times to the Prophet Mohammed, with an Appendix on Abyssinian History. For his materials he was but little indebted to the native historians, for he found their works, so far as they refer to the times before Mohammed, to be but a tangled mass of legends, fables, and fancies, from which it would be as hopeless to extract solid facts as to gather grapes from thorns.   The Abyssinian writers likewise detail legends useless for the purposes of history, while but little reliance can be placed on the lists of kings which are here and there to be found in their writings.   The theological writers, again, of the Abyssinian Church, disdained to employ their pens on the profane history of bygone centuries. Dr. Glaser, therefore, turned for his materials to references in Scripture and the classical writers, and above all to inscriptions on monuments found both in Arabia and Abyssinia, many of which he has himself copied, interpreted, and elucidated by comparing them with each other or with intimations gathered from other sources.   The work now before us is, like its predecessor, full of new facts and most surprising discoveries.   Its main object is to show that the original home of the Ethiopians or *Habashat* was not Africa, but was that part of Arabia now called Mahra, which extends eastward from Hadhramaut along the coast in the direction of Oman.   Now, as Abyssinia is called *Habesh* even by the Arabs themselves, the question at once suggests itself, How is it proved that a kingdom of the Habashat ever did exist in Arabia? Glaser argues that this is a fact beyond dispute, as this kingdom is mentioned in several Sabæan inscriptions, and in particular in what is called the " Treaty Inscription," which he shows cannot have been composed later than 100-50 B.C., but probably earlier.   Should it be objected that this kingdom may have been that of Axum, which, as we know from the *Periplûs*, was in existence at that time, the answer is, that Axum was then too insignificant to have figured in that treaty as an independent power.   The Habashat, moreover, are no doubt identical with the Abaseni of Uranius, a Greek writer on Arabia, who lived in a city called Abissa, which Ptolemy's geography assigns to Sachalitis, a district of South Arabia famous for its frankincense.

The classical writers to whom Dr. Glaser has turned for light on his subject are chiefly Herodotus, Eratosthenes, Agatharchides, Diodorus, Strabo, Pliny, and Basil, the author of that most valuable work, the *Periplûs of the Erythræan Sea.*

These writers unfortunately have left us no systematic details of the countries lying to the south beyond Egypt, but present us merely with stray or isolated facts, to which it is sometimes difficult if not impossible to assign a date.   This silence cannot be attributed to ignorance, for the vast consumption of aromatics by the Egyptians, Syrians, and Greeks compels us to suppose that they must have had some degree of knowledge of the countries from which these highly esteemed products came. We may not indeed be warranted in asserting that these nations sent their ships some twenty centuries B.C. to bring them home their supplies of frankincense, but yet we know from inscriptions that in the most primitive times the merchants of Southern Arabia traded to Gaza and to Egypt, and that the incense producers on the opposite African coast followed their example. We know, in fact, from the accounts given by the Egyptians of their voyages to that coast, which they called Pwent (Punt), that they were accompanied by some of its natives on their homeward voyage to the Delta.   The knowledge acquired by the Egyptians in their visits to the Straits regarding the regions which lay still farther south·could not fail to penetrate to the Greeks.   When Herodotus tells us of the Ethiopian Macrobii, whom he locates on the shores of the Southern Sea, he probably means thereby the inhabitants of Somaliland and the adjoining Abyssinian tract.   It would further appear that in old times there were, in Ethiopia proper, powerful kings who, in the eighth century B.C., subjected Egypt to their sceptre.   Occasionally, also, glimpses· were obtained of the Upper Nile region to which Axum belongs, even as early as the time of Homer.   Herodotus, again, knew about the 240,000 Egyptian warriors who, in the second half of the seventh century B.C., withdrew from Egypt to Ethiopia in disgust at their unjust treatment by King Psammetichus.   They were called *automoloi*, this being the Greek word for "deserters."   Our author points out that, at this day, there is close to Massâwa a place called *Hotumlu*, and although he will not assert that the king of the Ethiopians settled the fugitives in that place, which would have been too small for their numbers, yet the close similarity of the names, which cannot possibly be accidental, serves as a finger-post to indicate where we must look for the territory which was assigned to the immigrants.   From their story, as well as from the accounts which Herodotus has given us concerning the Ethiopians inhabiting the shores of the Southern Sea, we learn the important fact that the powerful kingdom of Napata, which the Greeks called Ethiopia, and which, for some time, held all Egypt under its sway, extended itself in the second half of the seventh century as far very probably as the coast of the Red Sea— nay, even the inhabitants of Somaliland were in the time of Herodotus (about 450 B.C.) regarded as Ethiopians—which name, moreover, had long before appeared in the Homeric poems as equivalent to the *Kash* of the Egyptians and the *Cush* of Scripture.   How far this kingdom of Kash or Kesh stretched to the south and the south-east can, in the present state of our knowledge, or rather ignorance, of the interior of Africa, not be determined.   It maintained its existence until the invasion of Egypt by the Persian king, Cambyses, who penetrated southward to the capital Napata, which he captured and destroyed and stripped of its importance. Thereafter the name of Meroe cropped up, and this became the residence of the Ethiopian kings.   There is some uncertainty as to the site of their former residence Napata, but it is generally placed near Mount Birkel, at the eastern extremity of that great bend of the Nile which skirts the desert of Bayuda.   It was the farthest point south beyond Egypt to which the arms of Rome penetrated.   Meroe continued to hold some importance till the middle of the first century, when it disappears from history.   The Blemmyes then step into notice, and later the Christian kingdoms of Nubia [1] and of Aloa, while in another direction, in the same

---

[1] The Nubae of the Classics were predatory nomads, who neither occupied the whole of what is now called Nubia nor were confined to its limits.

century, information regarding the Axumite kingdom begins to appear, this being derived from inscriptions and also from notices found in the *Periplûs*, of which the date, according to Glaser, is between 56 and 67 A.D., although hitherto a date of some fifteen years later has generally been assigned to it.

While the Greeks, as has been shown, designated the countries which lay to the south of Egypt, and included Napata, Meroe, Axum, and Somaliland, by the vague appellation of Ethiopia, three names are met with in the indigenous nomenclature which were applied with like vagueness to the same vast regions. These are Pwent, Habesh, and Kash or Kesh. They appear not to have been known to the Greeks, or were at least not current among them. Kash, under the form of Kasu, and Habesh are met with in the Axumite inscription of the fourth century,[1] which is written in the Geez character. Habesh is a term which claims a very high antiquity, for it has been found by Dr. W. Max Müller in one of the hieroglyphic inscriptions of Egypt, where the "Habsti of the Divine Land" are identified with the people of Pwent— that is, of east Somaliland—from which Egyptian ships, so early as the second pre-Christian millenium, brought cargoes of frankincense and other aromatics necessary in especial for embalming the dead. In course of time the name of *Hbst* or *Habashat* was applied to the greater part of the Somali peninsula. In some of the Egyptian inscriptions, indeed, Pwent is made to lie on both sides of the sea, and would thus seem to designate the Arabian as well as the African incense country. That the two were politically connected is known from the *Periplûs*, but, notwithstanding, Pwent can never have designated southern Arabia, for the name never occurs in any of its numerous inscriptions. Habashat, it has been seen, does so occur, and that too at a comparatively early period, and it may be assumed that the oldest name of the Somali peninsula, and of several other parts of Africa, especially Abyssinia, was *Habashat*, while a part of south Arabia was also so called. In fact, the people of Kash, of Pwent, and of Habashat were, from very ancient times, conceived of as forming an ethnic unity, and in this collective capacity were designated by the Greeks *Ethiopians*. Dr. Glaser rejects the etymology generally assigned to this name (aithô, *I burn*, and ôps, *the visage*), and suggests another which we cannot but think will eventually supersede it. *Habashat*, he explains, when taken strictly, signifies nothing else than "collector"— collector, that is, of aromatics, spices, and gums. It is natural, then, to suppose that *Ethiopia* must have had a somewhat similar signification, and Dr. Glaser suggests that it is derived from *atyûb* (pronounced atyôb), the plural of *taib*, one of the Arabic names for frankincense. That the name of Ethiopia was not confined to Africa, but was also applied to the frankincense districts of Arabia, is proved by the already mentioned bi-lingual inscription of Axum of the fourth century, a copy of the Greek text of which was long ago taken by Mr. Salt, while the other text in the Geez or old Ethiopian character became known only quite recently from

---

[1] Dr. Glaser, after a careful investigation, has fixed the date of this inscription at between 352 and 355 A.D.

the copy of it taken by Mr. Bent. The Greek text begins thus :—
"Aizanas, King of the Axômites and the Homêrites, and of Raidân, and
of the Ethiopians and Sabæans, and of Sileê and of Tiamô, and of the
Bugaïtes and of Kâsu, King of Kings, son of the invincible Arês," etc.
Now, in the other text, the word which represents the *Ethiopians* is
Habashat, and since their name has, both before and after it, the names
of Arabian races, it is clear that they also belonged to Arabia.

This inscription further shows us that the king of Axum, who, in
the first century of our era, was only a petty African chief, had, in the
fourth century, become the sovereign of a great empire which embraced
dominions both in Africa and Arabia. Fortunately we know how this
empire was first established from an earlier inscription which has been
preserved for us by Cosmas Indicopleustes, an Egyptian monk of the
sixth century, who, in his earlier years, had seen much of the world as
a travelling merchant, and in his later wrote a work called *Christian
Topography*, in which he took occasion to record many valuable facts
which had fallen under his observation during his secular career. In
one of his voyages he came to Adule (now Thulla or Zula, near Annesley
Bay), which was then the seaport of Axum, and was much frequented
by traders from Alexandria and the Elanitic Gulf. Near this place, on
the road to the capital, was to be seen a chair made out of a single
block of a costly kind of white marble, and covered all over with a
Greek inscription. Behind this chair lay a tablet of basanite stone,
which was similarly inscribed, recording an expedition undertaken by the
third Ptolemy, surnamed Euergetes, in which he conquered the greater
part of Asia, while the chair inscription enumerated a series of conquests
by an Axumite king on both sides of the Red Sea, made, as Glaser has
been able to show, towards the end of the third century after Christ.
Cosmas was requested by the governor of Adule to make copies of both
inscriptions for the king, and while doing so he made copies for himself,
which he afterwards transferred to the work we have already mentioned.
The chair inscription throws so much light on the early geography of
the regions concerned that we must quote it in full. It will be seen
that its beginning had either disappeared, or escaped the notice of
Cosmas, who erroneously supposed that it was a continuation of the
narrative of Ptolemy's warlike expeditions :—

"Having after this, with a strong hand, compelled the nations bordering on my kingdom to live in peace, I made war upon the following nations,
and by force of arms reduced them to subjection. I warred first with
the nation of Gazê, then with Agamê and Sigyê, and having conquered
them I exacted the half of all they possessed. I next reduced Aua and
Tiamô, called Tziamô, and the Gambêla, and the tribes near them, and
Zingabênê and Angabe and Tiama, and Athagaûs, and Kalao and the
Semênai, who live beyond the Nile, on mountains difficult of access,
where winter reigns all the year round, with frosts, hailstorms, and snows
in which you sink knee-deep. To attack these nations I passed the
river and effected their reduction. I next subdued Lasine and Zaa, and
the Gabala tribes who inhabit mountains with steep declivities, abounding with hot springs ; then Atalmô and Bega, and along with them all

the contiguous tribes. I marched next against the Tangaïtae, who adjoin the Egyptian border, and having reduced them I made a footpath to open a way by land from that part of my dominions into Egypt. I next reduced Annine and Metine, tribes living among steep mountains. My arms were next directed against the Sesea nation, who had retired to a high mountain, to which access was difficult; but I blockaded the mountain all round, and forced them to descend and make surrender. I then selected for myself the best of their young men and their women, with their sons and daughters, and all besides that they possessed. The Rhausi I next brought to submission—a barbarous race spread over wide, waterless plains in the interior parts of the frankincense country. Thereafter I encountered and vanquished the Solate, and left them with instructions that they should guard the coast. All these nations, protected though they were by mountains, which seemed to be impregnable, I conquered in engagements in which I was myself present. When they submitted I restored them their lands, subject to the payment of tribute. Besides these, many other tribes voluntarily submitted, and became likewise tributary. And I sent a fleet and land forces against the Arabitae and Cinoedocolpitae who dwelt on the other side of the Red Sea, and having reduced the sovereigns of both I imposed upon them a land tribute, and charged them to take measures to make travelling safe both by sea and by land. In this way I subdued the whole coast from Leukê Kômê to the country of the Sabæans. I, first and alone of the kings before me, subdued all these nations. For this success I now offer my thanks to my mighty god Arês who begat me, and by whose aid I reduced all the nations bordering on my own country, on the east to the country of frankincense, and on the west to Ethiopia and Sasu. Of these expeditions some were conducted by myself personally, and ended in victory ; and as for the others I intrusted them to my officers. Having thus established peace throughout my empire, I came down to Aduli and offered sacrifice to Zeus, Arês, and Poseidôn, whom I entreated to befriend all who go down to the sea in ships. Here also I reunited all my forces, and setting down this chair in this place, I consecrated it to Arês in the twenty-seventh year of my reign."

When this inscription is compared with the bi-lingual of King Aizanas, already quoted, it will be seen that in the interval of about sixty years, by which their dates are separated, the power of Axum in Arabia had been very largely extended. The original conquest had been limited to the maritime province of Hedjaz, but to this had been added by the time of Aizanas the most rich and flourishing divisions of all Arabia, Yemen, and Hadhramaut. In the Adulitic inscription, the number of the places and tribes we find entered amounts to twenty-three, and of these our author has shown that one-half are Sabæan names which originally designated places or tribes in Arabia. This confirms the fact, otherwise known, that long ere yet the Queen of Sheba had visited King Solomon, the people of her country had crossed to the other side of the sea and established colonies, which were the means of introducing civilisation into the regions lying to the southward of Egypt. The date, of near the close of the third century after Christ, which Glaser assigns to the inscription, is that which we think will be generally adopted. Professor D. H. Müller of Vienna, however, ascribes its authorship to Zoskales, who was the king of Axum when the writer of the *Periplûs,* in the course of his trading voyage along the African coast, visited the port of Adule, somewhere about the year 60 A.D. That writer, however, describes such a lawless state of things existing in his time in Hedjaz as bars the supposition that that province could have been then subject to such a capable strong-handed ruler as the author of the inscription must have been. Zoskales, or Za-hakale, as his name appears in the native lists of the Axumite kings, is described in the *Periplûs* as "a man of penurious habits and grasping at more, but otherwise a gentleman and a good Greek scholar," like the dog in Burns. His scholarship might account for the Greek on the chair, but can it be supposed that a petty chief like this, who lived penuriously and was only bent on increasing his hoards, was the great conqueror who dictated the inscription? He, like Zoskales, came down from Axum to Adule, but he came not, like him, to drive hard bargains with Egyptian traders, when swopping with them the raw products of his dominions [1] for their manufactured goods.

Most of the place-names in the inscription have been identified, for which thanks are especially due to the late M. Vivien de Saint-Martin, whose conclusions have met with general acceptance.

Gazê.—This represents Agazi, which was at one time a designation of that portion of the Abyssinian plateau of which the slopes command the Red Sea above Massâwa. The learned language of Abyssinia is called *Geez,* Agazi having been at one time used as a synonym for Abyssinia—at least in works of learning.

Agamê.—This still designates a province of Tigré lying to the eastward of Axum. It is rich and fertile.

Sigyê.—The name may be connected with Tzigam, an Agaou tribe seated to the west of Lake Tana. The traditions of the tribe, however, connect them with the Agaous of the Takazzé. The Agaou people are the aborigines of the Abyssinian plateau.

Aua.—This is called Auê by Nonnosus, an envoy sent by the Emperor Justinian to the king of Axum in 531 A.D. We learn from the itinerary of his journey that Auê lay half-way between Adule and Axum. Adoua still preserves the name: Ad'oua = city of Oua, the capital of Tigré.

Tiamô appears as Tiama in the Vatican codex of the *Christian Topography.* In the form Tiamô it is found several times in Abyssinian geographical inscriptions. It designates a prefecture in Tigré, adjacent to Agamê.

Gambêla.—This is the valley of Jambêla in the province of Enderta.

Zingabênê.—If this has been erroneously written for Zingarêne, as Dillmann conjectures, then it would be identical with Zangarên in Hamâsên, a district not far from Massâwa.

Angabe and Tiama cannot be identified.

Athagaûs is probably Addago, a district which lies to the left of the Takazze, below the mountains of Semen.

---

[1] Elephant tusks, tortoise-shells, and rhinoceros hides.

Kalaa seems to correspond to Kalawe, a district adjacent to that just mentioned. Glaser, however, would identify it with Koloe, a town distant a three days' journey inland from Adule, according to the *Periplûs*.

Semênai.—The inscription gives this name in exact accordance with its present orthography. The mountains of this remarkable district rise to a height of 15,000 feet. When the Semênai are said to live beyond the Nile, the river meant is the Takazze, called by the ancients the Astaboras.

Lasinê, Zaa, and Gabala cannot now be identified. Cosmas notes that though they were ancient names they were still current in his day. Lasinê, Saint-Martin conjectures, may be the district of Baséna, on the northern frontier of Tigré.

Atalmô and Bega.—The former cannot be identified. Bega refers to the ancient race of the Bedjas or Bodjas. They are the nomads who, under the name of Bicharieh, are found in many parts of the great Nubian desert between the Nile and the Red Sea.

Tangaïtes.—This is a Greek form of the name of the people who inhabited the country of *Taka*, which is watered by the united streams of the Takazze and Atbara. The Tangaïtes were the most powerful of the Beja tribes.

Annine and Metine.—The fact that these two tribes inhabited a mountainous region shows that their position was eastward towards the Red Sea. The countries next conquered lay quite in a different quarter.

Sesea.—This is in Somaliland, of which the coast region, extending to the Aromatic Cape (now Guardafui), and called anciently Barbaria, was the cinnamon-bearing country of the Greeks and Romans. One of the principal Somali tribes, called Issa, seems to preserve the name of Sesea.

The Rhausi—are very probably the Rhapsii of Ptolemy (B. IV. viii.). Their name exists, little altered, in that of the Arussi, a large tribe in the interior, to the south of Abyssinia, and one of those which carry on a regular traffic with the coast.

Sasu—lay in the south-east part of the Somali peninsula, not far from the Italian colony of Hobia (Oppia, Obbia), situated on the coast a little above the fifth parallel of north latitude. This is beyond doubt the actual position of Sasu, but this is inconsistent altogether with the state-ment in the inscription that Ethiopia and Sasu formed the *western* boundary of the conqueror's dominions. Dr. Glaser has solved this difficulty by taking *Sasu* to be an error in transcription for *Kasu*. Now *Kasu* was shown by Dillmann to be a far *westward* territory, since in the bi-lingual Axumite inscription, where it occurs, it admits of being located only in or near Meroê, probably in the Khartum district.

Arabitae.—This name denoted originally the wandering tribes of the northern deserts of Arabia and the commercial people along the northern part of the eastern coast of the Red Sea. Glaser thinks that the Arabites and Arabians are perhaps different, the former denoting the inhabitants of Arhab.

Cinoedocolpitae.—The name of this people, written exactly as here is found in Ptolemy. They occupied Hedjaz, now the Holy Land of Arabia, containing the cities of Mecca and Medina. Towards their northern frontier was situated the port of Leukê Kômê, from which at one time the costly commodities received from India and Southern Arabia were transmitted in caravans the size of armies to Petra of the Nabathæans. It has been identified with the port of Hauara in about 25° N. latitude. The Sabæans occupied Northern Yemen, above the Himyari or Homerites, in about 15° N. latitude.

There are many other subjects of great interest and importance discussed in Dr. Glaser's work to which the want of space prevents us from even adverting. There is, however, one point which we must notice before concluding—the time, namely, at which the Habashat left their homes in Mahra and passed over into Africa. It is a question difficult to answer, and Glaser can only give as an approximate date, either the middle or some time in the latter half of the first century before Christ. The Parthian kings had by that time made themselves complete masters of the frankincense country, thus compelling such of the Habashat as could not brook a foreign yoke to seek homes for themselves elsewhere.

We may add that an inscription (copied by Bent) of later date than the famous bi-lingual, but written by the same king, Aizanas, shows that toward the latter part of his reign the Axumite power in Southern Arabia had begun to wane. Before the end of the fourth century it had entirely collapsed. In the annals of the Christian Church, the reign of Aizanas is memorable as that in which Frumentius, who had been ordained Bishop of Axum by Athanasius, prosecuted with great vigour and success the work of converting the natives to Christianity.

THE NEW BOUNDARIES IN FURTHER INDIA.

In 1893 the French attacked Siam, and by a treaty, signed on October 1st, extended the frontier of Cochin China to the Mekong, obtaining also other concessions, which were fully set forth in an article by Mr. Coutts Trotter in vol. ix. p. 449.

Again, by a treaty with China, signed at Pekin in June last year, the French added to Tonking (1) the principality of Deo-van-tri, of which Laï-chau is the capital; (2) the district of Pu-fang; (3) the region of Muong-U, which commands the waterway of the Nam-U; and (4) the country of the Pa-fat-sai, whence all the surrounding peoples obtain their supply of salt. By the same treaty certain commercial advantages were granted by the Chinese to French traders.

But the most important changes have been effected by the Anglo-French treaty, signed on January 15th, the chief points of which were given on p. 97. By this the independence is guaranteed of that part of Siam comprising the valleys of the Menam and certain smaller rivers. This area is indicated by the dark-shaded portion of the map. The remainder, marked by a lighter shade, is in the meantime Siamese territory, but it is not secured from aggression by any stipulations in the treaty. On the west it includes the Siamese states of the Malay penin-

NEW BOUNDARIES IN FARTHER INDIA

The Edinburgh Geographical Institute

J. G. Bartholomew

sula only partially shown in the map, as well as the strip on the Burmese frontier.

In the north the project of forming a buffer state has been abandoned. The British Government has resigned to France the portion of Kiang Kheng lying to the east of the Mekong, and has evacuated Muong Sing. But China retains the part of Kiang Hung on the east of the same river, which was ceded to it by Great Britain.

The right of control over a belt twenty-five kilomètres broad (about sixteen miles) on the right bank of the Mekong, granted to France by the treaty with Siam in 1893, is expressly recognised by the new treaty.

PROCEEDINGS OF THE ROYAL SCOTTISH GEOGRAPHICAL SOCIETY.

LECTURES DELIVERED IN FEBRUARY.

On February 6th, Miss Kingsley was present at a meeting of the Society in Edinburgh, when the account of her travels, forming the first article in this number of the *Magazine*, was read. Dr. John Murray took the Chair.

Miss Kingsley's paper was read in Glasgow on the following evening. Mr. Robert Gourlay presided.

At Aberdeen, on February 6th, Mr. James Troup, H.B.M. Consul at Yokohama, lectured on "The Industrial and Commercial Development of Japan." Lord Provost Mearns presided, and Mr. A. P. Hogarth proposed a vote of thanks to the lecturer. This paper will be published in the *Magazine*.

On February 19th Sir David Chalmers delivered in Dundee the address on "British Guiana" he gave at Edinburgh in January. Mr. J. S. Ross presided.

The following evening he addressed the Aberdeen Branch, Dr. C. B. Davidson taking the Chair.

In Edinburgh, on February 20th, a meeting was held, at which Professor Clifford Allbutt, F.R.S., lectured on "Ancient Trade." The Chair was occupied by Sir Thomas Clark, Bart.

LECTURES IN MARCH.

The lectures at present arranged are as follows :—

Sir David Chalmers will deliver his lecture in Glasgow on March 2nd, at 8 P.M.

Mr. G. Seymour Fort will lecture in Edinburgh on South Africa on March 5th, at 4.30 P.M.

He will also address the Aberdeen Branch on March 6th, the Dundee Branch on March 9th, and the Glasgow Branch on March 10th.

On March 19th Mr. A. Montefiore will lecture in Edinburgh, at 8 P.M., on the Jackson-Harmsworth Arctic Expedition.

On March 16th Mr. R. M. Routledge, Stipendiary Justice of Trinidad, will lecture on the Falkland Islands.

GEOGRAPHICAL NOTES.

By The Acting Editor.

EUROPE.

The Climate of Portugal.—The adjoining diagrams of temperature have been drawn by Senhor J. C. de Brito Capello, Director of the Lisbon Meteorological

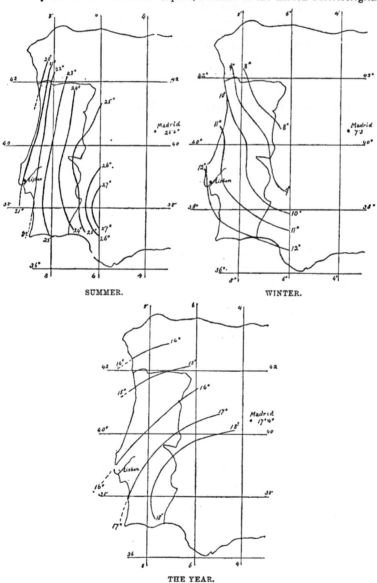

SUMMER.   WINTER.

THE YEAR.

Observatory, and have been kindly placed at the disposal of the Society by Prof. Carlos de Mello. The observations have been reduced to sea-level by deducting 1° C. for every 200 mètres of altitude in winter, 1° C. for every 160 m. in summer, and 1° for every 180 m. from the annual means. Prof. de Mello also furnishes the following table of meteorological data in Portugal, the Azores, and Madeira :—

Meteorological Stations.	Positions.		Altitudes in m.	Temperature C.		Daily Range	Mean Temp.	Temperature Extremes.		Humidity $\frac{9h+3h}{2}$	Rain in mm.	Years of Observation.
	Lat. N.	Long. W.		Max.	Min.			Max.	Min.			
Montalegre,	41° 49′	7° 40′	1027	14·04	4·11	9·93	9·08	34·0	−11·2	69·7	1080·5	1880-87
Moncorvo,	41° 10′	7° 1′	415	17·53	9·23	8·30	13·38	35·0	−7·2	79·7	605·2	1878-87 1866-85
Oporto,	41° 9′	8° 35′	100	19·55	10·80	8·75	15·17	38·2	−0·8	75·0	1171·2	{ Humidity for 14 years only
Viseu,	40° 39′	7° 57′	494	18·07	8·88	9·19	13·47	40·0	−5·0	61·0	1427·9	1879-86, except '82 1866-85
Guarda,	40° 32′	7° 14′	1039	14·03	6·31	7·72	10·17	34·6	−8·6	80·3	913·8	{ Humidity for 14 years only
S. da Estrella,	40° 25′	7° 35′	1450	10·21	3·93	6·28	7·07	29·8	−11·6	71·86	3081·9	1883-86
Coimbra,	40° 12′	8° 23′	141	19·46	10·70	8·76	15·08	40·4	−2·6	72·1	913·5	1869-88
Campo Major,	39° 2′	6° 59′	288	22·95	10·29	12·66	16·62	45·0	−3·6	56·4	545·3	1866-84
Lisbon,	38° 43′	9° 9′	95·4	19·33	12·38	6·95	15·85	38·8	−1·5	71·24	746·5	1856-85
Evora,	38° 35′	7° 52′	326	21·22	10·91	10·32	16·06	42·9	−3·5	67·4	642·2	1874-86
Lagos,	37° 6′	8° 38′	13	21·73	13·08	8·66	17·40	38·3	+0·5	67·9	519·6	1866-80
Angra do Heroismo,	38° 39′	27° 14′	44	19·09	14·42	4·67	16·74	28·5	+4·5	80·9	1041·1	1871-85
Ponta Delgada,	37° 45′	25° 41′	20	20·35	14·10	6·25	17·23	30·0	+3·4	72·4	921·7	{ 1866-85, except 1880 and 1881
Funchal,	32° 38′	16° 55′	25	21·19	15·56	5·63	18·37	32·7	+6·5	66·4	688·6	1866-88

The Highest Point of Northern Europe.—Which is the highest mountain in Norway has hitherto been a doubtful question. Colonel Herzberg has now settled it by determining the heights of the Galdhöpig or Galdhötind (8399 feet), of the Glittretind (8379 feet) and the Knutholstind (7818 feet). The differences of altitude have been determined with accuracy, but the absolute heights obtained are not quite reliable, as the observer had to base his operations on points fixed by deficient barometer observations and road levellings. Colonel Herzberg has lately expressed his opinion that in time the Glittretind may surpass the Galdhöpig in altitude, for the latter has no room on its summit for accumulation of snow, while the Glittretind is covered with a somewhat extensive snow-field more than 23 feet deep. The depth of snow increases or diminishes with the meteorological conditions, and it is quite possible that in the course of a few years it may increase by more than the 20 feet which is now the difference in the heights of the two mountains.—*Geogr. Zeitschrift*, Jahrg. ii. Heft 1.

ASIA

Assam and Burma.—An expedition has set out from Assam to examine the passes over the Patkoi range between Assam and the Hukong valley, with a view to the construction of a railway to Burma. Mr. Way, the engineer-in-chief, has already surveyed other likely routes, and expects that the Patkoi range will present fewer obstacles than the hills of Manipur and Chittagong. The country is very unproductive, and is inhabited by tribes who are not very submissive to the British Government. On Nov. 4th a camp was pitched on the top of the Patkoi, over 4000 feet high, whence, when the jungle had been cleared away, the country could be easily surveyed. About the same time Mr. M'Intyre started to reconnoitre the passes east of the Non-yong lake. A column from Mogaung was working its way eastwards to meet Mr. Way's party.—*Geogr. Journ.*, Feb.

From the Mekong to the Brahmaputra.—In the *Tour du Monde* of February 15th are published later letters of M. Roux, continuing the narrative of the Prince of Orleans' party (see p. 96). They left Ta-li-fu on June 16th, and struck the Mekong at Pei-long-kiao. The country to the west was extremely mountainous, and travelling across it exceedingly difficult; and, therefore, after reaching the Salwin at Luku, the party recrossed the range which separates the two rivers by a pass 11,800 feet high, and followed the Mekong northwards for a distance of 180 miles, to the town of Attentse. At Tse-ku, a little lower down the river, they found two French missionaries, who assisted them in procuring Tibetan porters, and they left this town on September 10th. The western range was crossed this time by a pass 12,500 feet high, dominated by a peak to which the name of Francis Garnier was given. It took the party ten days to reach the Salwin, but still more difficult were the subsequent marches. The routes are nothing but trails, quite impassable by baggage animals, and dangerous even to travellers on foot. The beds of torrents are frequently the only possible line of march. On November 19th the Prince and M. Roux reached the Khamti country, having crossed eight branches of the Irrawadi. They were then in country that had been visited by Europeans—by Mr. Gray as recently as 1893 (see vol. x. p. 205). Their troubles, however, were by no means over, for, having chosen a route somewhat to the north of that followed by Mr. Gray, and being attacked by fever, caught in the Khamti plain, they were delayed, and their provisions ran short. Along more than ninety miles of this route there are no villages, and therefore provisions had to be taken with them.

In 1882 the Pundit Krishna (A-k) travelled into the country of the Mishmis, crossing successively the Mekong and Salwin, and arrived at the sources of the Lobit or Brahmaputra. On his return he reported that the Zayul Khanung range was the watershed of the Irrawadi, which could not therefore rise, as had been supposed, in Tibet (see vol. i. p. 368). This view is confirmed by the explorations of the French travellers who skirted the same range on the south, and saw distinctly the sources of the Telo, one of the chief headwaters of the Irrawadi. The Turang, east of the Telo, and about equal in volume, rises a little farther to the north, the range running from east-north-east to west-south-west. The peaks of the range cannot be less than 16,000 to 19,000 feet.

AFRICA.

Wallega.—The Swiss engineer, Alfred Ilg, has lately visited this auriferous region at the instance of King Menelik. It lies to the south-west of Abyssinia on both sides of the river Yabus, an affluent of the Nile, and its mineral riches were exploited in ancient times. King Menelik has come into the possession of the country within the last few years, partly by conquest, partly through the voluntary submission of the inhabitants. No European had previously penetrated into this district, and therefore Herr Ilg is the only one who knows it from personal observation. The distance from Addis Abeba, the present capital of Abyssinia, to Wallega is about forty-five miles, and Herr Ilg took forty days to accomplish the journey there and back. The road runs through the country of the Galla, traversed by mountains similar to the Jura. It has an excellent climate which promotes a luxuriant vegetation. The fauna also is rich. Among the chief rivers is the Hauash, which rises near to the watershed between the Red Sea and the Mediterranean. A few miles to the west of its source lie, according to Herr Ilg, the headwaters of the river Jub. From the same neighbourhood the Guder runs northwards to the Blue Nile in an enormous cañon, 2600 feet deep, while the Diddesa flows northwestwards through a valley one hundred miles long and twenty to thirty broad. Beyond this extraordinary fertile tract begins the escarpment, about 4300 feet high, that leads to the plateau of Wallega ; dense thickets of bamboo render the ascent very difficult. The trachyte, which underlies the lower country almost to the exclusion of other rocks, here gives place to granite, syenite, felspathic and mica-schists. The tableland is broken up into a number of flats of various dimensions, and small elevations worn by continual degradation into low rounded hills, with a diameter of about half a mile, among which run innumerable water-courses, usually hidden by rank vegetation, so that it is best to follow the tracks worn by elephants. There is no elevated point from which a general notion of the country can be obtained ; it seems to slope from south-east to north-west, for the streams run in that direction. There is no cultivation, the inhabitants occupying themselves solely in gold-washing in the beds of the small brooks, particularly in the valley of the Daber. But this valley is so enclosed that only Blacks can defy the fevers that prevail in it. The Galla do not stay there for more than two weeks at a time, returning then to more elevated regions. Herr Ilg found about one hundred men at work ; they gain twelve to sixteen shillings in the fortnight, risking attacks of fever and dysentery. The implements being of the rudest description, and the extraction of the gold being conducted in no systematic manner, and without the aid of chemical processes, a large part of the precious metal is wasted. Fifteen to twenty cwt. of pure gold is obtained annually, which is formed into small cylinders about four inches long, or rings, and is sent by the Nile into Egypt, or through Abyssinia to the sea, where it is exchanged for salt. The inhabitants of Wallega barter this salt again for grain with the peoples to the west of them. Menelik intends to take measures for a more rational development of the mineral resources of this country.—*Boll. della Soc. Geogr. Italiana*, vol. ix. Fasc. 1.

The Uluguru Mountains.—Dr. Stuhlmann has fully described these mountains in the *Mitt. aus den Deutschen Schutzgebieten*, Bd. viii. Heft 3. In the northern part of German East Africa the traveller comes, a few days after leaving the coast, across a mighty mountain system, formed of crystalline gneiss. It constitutes the much eroded edge of the plateau that occupies the greater part of the country. At the extreme north, separated by a narrow plain from the coast, are the highlands of Usambara, and farther south the Nguru mountains, and the highlands of Usagara rise more to the west. Still farther south the margin of the elevated country approaches again more closely to the coast. A massive elevation lies in front of the southern part of Usagara, separated from it by the broad plain of the Mkatta river. This is the mountainous country of Uluguru, which only in the south is in some measure connected with Usagara by a tract of low hills. To the south the plateau again moves westwards. Except where the above-mentioned hills are situated, the Uluguru mountains are surrounded on all sides by plains which appear to the observer almost completely flat. On the north-west the mountains rise rapidly and suddenly from the Mkatta plain, 1300 feet above sea-level, to a height of 8200 feet ; while on the east and south a broad hilly country 1600 to 2000 feet in altitude intervenes between the plain and the mountains. Here the range slopes steeply to the west, while the eastern declivity is more gentle, and cut up by numerous transverse valleys.

The limit between the plains and mountains does not in any way correspond to the geological formation. The line between the gneiss and the Mesozoic rocks runs nearly straight from north-north-east to south-south-west. It approaches the coast in the north, and to the south of Uluguru appears to bend sharply to the west. In the eastern part of the mountains the boundary follows almost exactly the course of the Ruon river, as it turns southward after emerging from the mountains. Here occur red clay-schists, with a strike from north to south, and a slight dip towards the east. Farther north the geological boundary probably skirts the eastern slope of the spurs, for on the road between Bagamoyo and Morogoro *Septaria* are found, and old coralline limestone occurs in a river bed. East of the clay-schists is a wide plain covered with a greyish-brown clayey soil, where no rocks are to be found, from which rises an elevated ridge (Gongarogwa) of oolite. West of Gongarogwa is seen a violet-grey or reddish quartzite, while to the east lie the broad plains of late Jurassic, or perhaps even more recent, formation.

The outer ridges of Uluguru, rising to a height of 2000 feet, consist partly of quartz which exhibits a crystalline structure. Often a rock rises out of the ground with a rounded form as though a glacier had once covered this tract. But these rounded surfaces are doubtless the result of atmospheric agencies, which have such a remarkable effect on this kind of rock. Gneiss is also of common occurrence. The central mass is composed of gneiss, having a strike to north-north-east and a dip towards the east-south-east. Mica is often found in the north and north-east, and small particles of graphite are numerous. On the east, west, and south the central elevation is begirt by a broad zone of lower heights. In the north-east they form a kind of plateau, intersected by deep erosion valleys and supporting isolated hills, while in the south they are arranged in three parallel ranges. To the west they are connected with the highlands of Mgunda. The central mass forms in the northern part a high chain with steep summits, falling steeply to the plain on the west, and full of steep-sided valleys, where rise the headwaters of the Ngerengere brook, while the eastern slope, drained by the feeders of the Ruon, is much more gentle. The southern half of the mountains consists of a large plateau-like mass, attaining a height of 7800 feet, and with steep declivities on all sides.

A large part of the outer range is covered with deep black soil, which seems to be the product of the disintegration of the underlying rocks. This soil also occurs in the central mass, but here laterite covered, where the forest is still standing, by a thick layer of humus, is more frequent. This steppe forest is the most common form of vegetation, while in the valleys round the central chain are rich tropical woods. Primeval forest clothes the slopes from a height of 5900 to 6200 feet upwards, much of it covered with mosses, owing to the abundant mist. The natives cultivate maize in the valleys, where the humidity allows sowing and reaping to be carried on all the year round. *Cajanus Indicus*, gourds, and *sorghum* are also grown, and *Colocasia* and beans thrive up to the limit of human habitation (about 6200 feet). Rice grows up to 3000 feet or more, and papaws are common. Coffee and wheat may be cultivated in the lower hills, and cocoa, vanilla, and tea would find suitable situations in the damp hollows. For grazing the vegetation is not well adapted.

Lake Chiuta.—Mr. Robert Codrington, Assistant Collector in the Zomba district, left Chikala Fort on August 6th to visit this lake. He followed the Mikoko river, and encamped at Chenapini, on the edge of the Ntoradenga swamp, and the next night at the Panakatope swamp, having marched across the north end of the dried-up bed of Lake Chilwa. The next day he came to the southern end of Lake Chiuta, and he spent August 9th on the lake, visiting Chiuta island, which is rocky and unfit for cultivation. The south-western extremity of the lake is a swamp of considerable extent. The water is called by the natives a river, and the name Chiuta is applied to the island, and in a general way to the whole district. Six miles to the north is Chechikweyo's town, where the river is broadest and almost entirely free from grass. Its depth, both here and at other points visited by Mr. Codrington, varied from three to twelve feet, the most frequent measurements being five to six feet. There is a succession of villages along the bank for thirty-three miles farther, to Chenapulu's, near which the Litande and Lujende leave the Chiuta. The water is here quite open and free from grass, and is about 700 yards across.—*Geographical Journal*, February.

AMERICA.

The Drainage of the Valley of Mexico.—This undertaking, which was alluded to in vol. x. p. 658, was to be completed at the end of last year. A full account of it is given by Signor C. Poma, Italian Consul in Mexico, in the *Boll. del Ministero degli Affari Esteri* for October. The valley of Mexico is an immense basin lying at an altitude of fully 7800 feet above sea-level, and with an area of 2050 square miles. In outline it is more or less elliptical, but narrows towards the north. To the east and south rise the highest mountains—Popocatepetl (smoke mountain), Ixtaccihuatl (white woman), and the Ajusco—while to the west the basin is bounded by the Sierra de las Cruces and Monte Alto. To the north the country gradually rises till it meets with low hills having ravines of easy access, such as the Nochistongo and Tlila, through which, at the end of the sixteenth century, it was sought to conduct the water, by one method or another, out of the valley. The latter was formerly covered with lakes, of which those of Xochimilco and Chalco to the south-east of the city, the extensive lake of Texcoco on the east, the San Cristobal and Xaltocan on the north-east, and the Zumpango in the north, still remain. The waters of these lakes have been a perpetual source of danger to the capital, which has frequently been visited by terrible inundations, and the streams springing up in the mountains, especially the rivers de las Avenidas de Pachuca and Cuautitlan, which flow to the north where the plateau has a gentle rise, pour their waters back to the lake of Texcoco in time of flood, and have frequently threatened with disaster the city of Mexico, distant only three miles.

The Aztecs, then, who, coming from the north in the fourteenth century, founded their capital of Tenochtitlan on a large island, and extended it further on piles, soon found themselves compelled to protect their city from inundation. The lake on which it stood, and which has since disappeared, was a part of the lake of Texcoco up to about 1450, when Netzahualcoyotl, king of Texcoco, constructed a dyke ten miles long from north to south, dividing the lake into that of Texcoco on the east and that of Mexico on the west, and protecting the Aztec capital from the waters of Texcoco, rendered the more dangerous by the great streams from the north-east. This dyke was destroyed by Cortes during the siege of Tenochtitlan.

In the same period Montezuma Ilhuicamina, the fifth Aztec king, constructed a dyke to divide the great lake to the south-east of Tenochtitlan into two parts, which received the names of Xochimilco and Chalco, and it was by this dyke that Cortes advanced to the centre of the valley. As the waters of the western lake of Xochimilco rise above those of Chalco in the rainy season, while in the dry season the current turns in the other direction, sluices were made in the dyke by which the Aztecs were able to regulate the flow. Still, though the ground to the south of the capital was at a lower level, the waters, being always rather high, could menace the capital through the strait of Mexicalzingo.

From 1450 to 1850 many schemes were initiated or completed with a view to the prevention of floods, the viceroys turning their attention to the problem of providing an outlet for the water out of the valley, whereas the Aztecs thought only of protecting their capital. During the Spanish dominion the most favoured project was to conduct the waters through the ravine of Nochistongo, between the hills of Sinocque and Rinconzapotl on the left and San Sebastian and Tetzontli on the right, to the river Tula, and thence to the ocean. This grand scheme was never carried out in full, though the waters of the Cuautitlan were drained off in this direction and thus prevented from flowing southwards to the lake of Texcoco.

The plan now adopted is, with some modifications, that of the engineer Francisco de Garay, and consists of a great canal, which, leaving the city of Mexico towards the north, skirting the western shore of Lake Texcoco, crossing those of San Cristobal and Xaltocan, and cutting through the extreme western point of Lake Zumpango, pours the water into a tunnel in the gorge of Tlila, whence it flows through the river Tequizquiac into the Tula, which enters the Gulf of Mexico near Tampico. The canal starts from the Porta di San Lazzaro, the extreme western quarter of Mexico, and has a total length of 29 miles and 100 yards. For about the first 12½ miles it has a breadth of 6 yards, having to receive only the water of the canal of San Lazzaro, which has for centuries drained the city of Mexico, while the lower section, which is fully 7 yards wide, has also to contain the overflow of the Texcoco and the other lakes in the northern part of the valley. The discharge is calculated at over 176 cubic feet per second in the first section, and at nearly 616 in the second and in the tunnel. The depth at the beginning is 18 feet, and increases towards the end to 67 feet.

The tunnel is situated on the side of the hills of Acatlan, and terminates near the village of Tequizquiac. Its length is 6 miles and 120 yards, and its height 14 feet. It is calculated to admit of a current of 636 cubic feet per second. On emerging from the tunnel the water will flow through a cutting nearly two miles long to the river Tequizquiac.

It is not intended to drain the lakes of the valley entirely, which might lead to excessive dryness of the air, but only to prevent them from exceeding their usual limits in the rainy season. The effects of the undertaking on the climate have not been entirely foreseen, even by the most competent judges, but it is certain that the sanitary conditions of the valley will be ameliorated, and it is hoped that at some time the canal may serve for the purpose of irrigation, as suggested by Humboldt. The capital will feel the advantageous effects immediately, for the drainage into the Lake Texcoco has been imperfect and uncertain, and when the lake rose it overflowed into the city, carrying with it the germs of disease. The mortality has been 60 per thousand, this high rate being chiefly due to the unsanitary condition of the lower parts of the town, though the want of proper care of the innumerable children helps to swell the number.

OCEANS.

The Prince of Monaco's Investigations.—The Prince of Monaco started on a cruise in the *Princesse Alice* on May 23rd, and arrived at the Havre on Aug. 16th. A report of his work has now been published in a pamphlet entitled *Sur la deuxième campagne de la 'Princesse Alice,'* being a reprint from the proceedings of the Académie des Sciences. Scientific work was carried on from June 17 to August 12th, along a course across the area included between 37° and 49° N. lat., and 11° and 31° W. long.

Thirty-five soundings were taken, down to a depth of 2865 fathoms, the bottom temperature was observed on twenty occasions, down to a depth of 1202 fathoms, and fourteen samples of water were collected from the bottom, down to the lowest depths sounded. A new apparatus (Buchet's), designed to catch fish while the vessel is moving with a speed of not more than seven knots an hour, was tried, but neither this nor the trawl could be used, for the singular reason that during almost the whole voyage, from long. 21° W. to the Azores, and thence to the Gulf of Gascony, the sea was covered every night by a layer of medusæ (*Pelagia noctiluca*), which soon clogged the apparatus and threatened to break the net with their weight. Altogether, this cruise, the Prince of Monaco states, has been the most productive of any he has conducted. He is now planning more extensive researches which will demand the employment of several vessels working in concert.

MISCELLANEOUS.

It is proposed to extend the Buenos Ayres Great Southern Railway from Bahia Blanca to Ñeuquen on the eastern slope of the Andes. The new section will have a length of 350 miles, and will bring up the total length of the railway to 1751 miles.—*The South American Journal*, Feb. 1.

A syndicate has been formed in Canada to promote a short route to British Columbia through Hudson's Bay. Surveys have already been made for a railway from Fort Churchill in the direction of Winnipeg. It is claimed that by this route 1300 miles will be saved between Liverpool or Glasgow and Vancouver.— *The Nautical Magazine*, Dec.

A calculation has been made of the quantity of energy generated by the avalanche on the Gemmi in September last. The result obtained is 14,200 foot-tons, and this energy is sufficient to maintain 90,000 incandescent lamps, each of fifteen candle power, for five hours daily during a whole year.—From the *Electric World* in the *Revue Scientifique* for Jan. 11th.

In August last an official of the Bosnia-Herzegovina National Museum found extensive ruins of a Roman town on the left bank of the Skelani. The position of the forum was indicated by inscriptions and the bases of the statues of emperors and state officials. ¡Two miles to the north of the town was a temple of the Capitoline deities.—*Mitth. der k.k. Geogr. Gesell. in Wien*, Bd. xxxviii. Nos. 9 and 10.

The canal which since 1820 has connected Cincinnati on the Ohio with the town of Toledo on the Erie lake is to be enlarged to admit vessels of deeper draught. It is expected that the traffic on the deepened canal will amount to fourteen million tons, while on the Suez Canal it is only eight million tons. When the overwhelming importance of a waterway for large vessels between the Mississippi and the chain of lakes is considered, this estimate may not seem too large.—*Geogr. Zeitschrift*, Jahrg ii. Heft 2.

A Russian expedition under MM. Slutin and Bogdanovich is to explore the Sea of Okhotsk and Kamchatka during the years 1896 and 1897. Bogdanovich will devote his attention to the gold deposits along the shores of the sea, which, when worked by the latest and most approved methods, are expected to become very profitable; while Dr. Slutin will study the whale, cod, and herring fisheries. The main object of these investigations is to open up sources of natural wealth by which the present poverty-stricken condition of the inhabitants may be ameliorated. —*Geogr. Zeitschrift*, Jahrg. ii. Hept 1.

The smallest republic in the world is the island Tavolara, known to the ancients as Buccina, and noted for a much prized variety of *Murex*. It is separated from the north-west coast of Sardinia by a channel two-thirds of a mile broad, and is three miles long by two-thirds of a mile broad. In 1836 King Carlo Alberto granted it to the Barteoloni family as a kingdom; but when King Paul I. died in 1882 a republic was proclaimed in accordance with his wishes. Women have a vote, and the president's term of office is six years. The population numbers 55.—*Mitt. der k.k. Geogr. Gesell. in Wien*, Bd. xxxviii. Nos. 9 and 10.

A commercial syndicate in London has raised funds to send out an expedition next autumn to **Victoria Land** for the purpose of whale and seal fishing. For a payment of £5000, Mr. Borchgrevink, with a party under his charge, will be taken out in the vessel equipped by the syndicate and landed at or near Cape Adare. The party will consist of about twelve persons. They will work towards the south Magnetic Pole, taking magnetic observations, will survey and sound fiords and bays, make collections, dredge, and record meteorological and pendulum observations. The party will leave England on September 1st.—*Geographical Journal*, February.

An interesting scientific expedition into **South Central Africa** has recently been organised by the Directory of the Field Columbian Museum of Chicago. The object of the expedition, which will be under the command of Professor Daniel G. Elliot (a Fellow of the Royal Society of Edinburgh), is to obtain for the Chicago Museum a collection of rare sub-tropical birds and mammals ; and the territory specially selected for exploration is Mashonaland and the country lying to the north and west. Professor Elliot hopes to start from Fort Salisbury, which will be his headquarters, in the beginning of June, and he expects that he will be occupied until December in obtaining his collection.

Another noteworthy undertaking on the part of a Chicago scientist is that of Professor Frederick Starr, of the University of Chicago, who is at present in **Mexico** engaged in investigating a dwarf race inhabiting the mountains in the neighbourhood of Lake Chapala. Professor Starr, in addition to his other anthropological studies, has paid special attention to the distribution and characteristics of dwarf races, and consequently the conclusions which he may arrive at with regard to this particular race must necessarily be of much importance.

Mr. R. C. Mossman has handed to us a copy of the first number of a *Bulletin*, which will contain a periodical report on **the Medico-Climatology of Edinburgh**. This number treats of the five weeks from December 28th to February 1st. The temperature was high and the sunshine above the average, while the rainfall was less than the usual amount for the period ; westerly winds prevailed. The death-rate was below the normal, being no less than 6·4 per thousand below the average, and the lowest for the period during the last eighteen years. The study of records of disease in conjunction with meteorological observations is of recent origin, and some time must elapse before any practical results can be obtained. A commencement has, at any rate, been made as regards Edinburgh.

The **Summer School of Art and Science** will be held, as usual, in the month of August. The course will be divided into two parts, the one lasting from August 3rd to the 15th, and the other from the 17th to the 29th, for either of which tickets may be procured, or for the whole session. The subjects are the same as in former years, and many former lecturers will take part in the meetings of this year. Among the new names we find Mr. G. F. Scott Elliot in the section of Geography and Geology, Professor Wenley in that of Philosophy and Social Science, and Professor Rein under Education. M. Elisée Reclus will again be present. A new and important advance is the formation of a permanent School of Social Science, with a Geographical Museum, Studio, etc.

A detailed prospectus, lately issued, may be obtained from the secretary, Mr. T. R. Marr, University Hall.

As announced in the last volume of the *Magazine*, a project is on foot in Belgium for an **Antarctic Expedition**. The promoter is M. Ad. de Gerlache of the

Belgian Navy, who has studied the question for several years, and last year made a voyage to the Arctic Ocean in a whaler. He intends to set out at the end of next summer in a steamship of 400 tons, and will pass two seasons in the Southern Ocean. In the first season he will steer for the lands situated to the south of Cape Horn, and in the second will make for Victoria Land, principally to ascertain more exactly the position of the southern Magnetic Pole. In the interval a cruise will be made in the Pacific.

The expedition will include a body of scientists, each a specialist in his branch of science, and will be provided with a complete set of instruments. It is proposed to make observations on the meteorology, magnetism, oceanography, fauna and flora of the regions visited.

The expedition is under the auspices of the Royal Belgian Society of Geography, which is now inviting subscriptions to the fund. £10,000 is the sum required.

Mr. C. G. Cash writes regarding the statement frequently made that Blaeu's Atlas of 1654 consisted of eleven volumes. He has himself the copy presented by Blaeu to Robert Gordon of Straloch, as appears from inscriptions in the work. It is a complete atlas in that it contains maps of all parts of the world, and is divided into five volumes, of which the first was published in 1644, the three following in 1645, and the fifth (Scotland and Ireland) in 1654. The original title of the fifth volume runs thus : Joannis Blaeu Theatrum Orbis Terrarum, sive Atlas Novus, Pars Quinta. But over this has been pasted the title "Scotia quæ est Europæ Liber xii.," to which again has been added in MS. "et Tom. Vtus Atlantis."

The maps relating to Scotland are forty-nine in number on forty-seven plates. On the inside of the cover still remains a fragment of Robert Gordon's book-plate. The *Privileges*, granted by "Ferdinandus Tertius," "Oliver P." and "De Staten Generael der Vereenighde Nederlanden," are all dated 1654. Besides maps by Timothy Pont and Robert Gordon there is a map of Fife by James Gordon of Rothiemay, son of the latter, and probably those of West Fife and East Fife are also by him.

There is an atlas of Blaeu in the British Museum of seven volumes bearing dates from 1649 to 1654. One volume, however, is merely an enlarged edition of the third ; the volume for Scotland is identical with Mr. Cash's. Mr. Coote, of the Map Department, informs us that P. A. Tiele in his *Nederlandsche Bibliographie van Land- und Volkenkunde*, Amsterdam, 1884, mentions an edition of 1640 in three vols. This seems to have been the first of John Blaeu's atlas, and to have been succeeded by that of which Mr. Cash possesses a copy, while the atlas in the British Museum is an emended issue of the latter ; the volume for Scotland does not appear to have been issued at all until 1654.

NEW BOOKS.

The Discovery of Australia: A Critical, Documentary, and Historic Investigation concerning the Priority of Discovery in Australasia by Europeans before the Arrival of Lieutenant James Cook in the "Endeavour," in the Year 1770. With Illustrations, Maps, Diagrams, References, Geographical Index, and Index to Names. By GEORGE COLLINGRIDGE, of the Royal Geographical Society of Australasia, etc. Sydney : Hayes Brothers, 1895. 4to. Pp. 376. *Price* 25s.

The History of Geographical Discovery and the science of Comparative Cartography have received a valuable contribution in Mr. Collingridge's *Discovery of*

Australia. With the exception of R. H. Major's *Early Voyages to Terra Australis,* it may be said to be the only critical and systematic investigation into the earliest discoveries of the great southern continent, yet that is by no means its special recommendation ; it is a thorough and scholarly piece of research, which may rank with the best work of the kind that has been attempted.

The whole question of early Australian maritime discovery seems to be so thoroughly enveloped in mystery that any attempt to fathom it must be set about with the greatest care as well as perfect impartiality of judgment. In his preface, Mr. Collingridge states that, so far as he is aware, he has read every book and examined every map of real importance on the subject. The present volume is to a great extent a digest of this material succinctly stated in chronological order. This is supplemented with copious extracts from original works, together with numerous facsimile reproductions of original maps and drawings, and presents to the student a full body of documentary evidence of the greatest interest. The sifting and weighing of the work of these early cosmographers is a task of no little difficulty, and any one familiar with such research must know how much collateral study is involved ; but Mr. Collingridge appears to have taken all due precautions to ensure accuracy, so that we may be fairly justified in accepting his conclusions with some confidence.

At the outset Mr. Collingridge is so careful that not even the earliest and most mythical views of Australia shall pass unnoticed that he takes us back to the first dawn of geographical knowledge, when it would appear that in remote Chaldean times, many centuries before the Christian era, Australia, and also America, were known to exist. But our geographical knowledge of later times may be said to begin with the Ptolemaic period, which, although one of retrogression, indicates what was known of the world before the dark ages of mediæval times obscured everything. In Ptolemy we find to the south of the Indian Ocean a land marked Terra Incognita, stretching from Zanzibar as far as Sumatra, which might be taken to represent Australia. This is about the extent of knowledge for centuries later until the great period of general renascence, which brought about a revival in geography as in other studies, and conjecture gave way to truth as navigators penetrated to the furthermost regions of the earth and rediscovered the world.

In some of the very ungeographical "wheel maps" of the Arabs about the first period of the Middle Ages, there is indication of a fourth continent corresponding to Australia. Their authority is, however, of little value, and we may start with the famous map of the world by Fra Mauro (1459), for which we are indebted to Prince Henry the Navigator. This map may be taken as representing the geographical knowledge of these times, and on it we find most of the East Indian islands but no trace of Australia, although at the place where it might be expected the perplexed cartographer puts a significant note to the effect that "in this oriental sea there exist many large and well-known islands, which he has not set down because he has no room!" According to Mr. Collingridge, the very earliest indication of Australia on any map is what appears to be a delineation of the western coasts of the continent on a chart drawn in 1480 by Bartholomew, the younger brother of Christopher Columbus, who was said to be one of the most efficient cartographers of that time. The map was prepared in London for Henry vi., and although the original is lost, a copy of it made in 1489 still exists in the British Museum. On Martin Behaim's globe of 1492 Mr. Collingridge also identifies the Australasian regions.

The first explicit claim to the discovery of Terra Australis has been found on a recently-discovered wooden globe, now in the Paris National Library. On it we find a southern continent distinctly outlined, and with the inscription, "Terra

Australis, recenter inventa anno 1499, sed nondum plene cognita." The last three words are amusing evidence of the imaginative powers of these early cartographers. No doubt, if space had permitted, we should have had a few lions and unicorns at the South Pole ! Without being able to question the genuineness of this globe, the date 1499 appears to us most improbable, for it appears to be based on the discoveries of Magellan (1521), and shows Tierra del Fuego as part of this southern continent. About this time we have the Hunt-Lenox globe (1506), Ruysch's Mappamundi (1508), Schoner's globe, and other maps of the same character, all more or less abounding with information regarding the Australian regions. Reviewing this early cartographical evidence as a whole, it seems to us that there is a tendency to over-rate its value. The maps were largely compiled from travellers' tales, and for the existence of the islands mapped all over the southern ocean we have to trust to the same authorities who are responsible for the mermaids which appear disporting themselves in the surrounding seas. It must also be borne in mind that from the earliest times there existed the general impression that a great continent lay somewhere in the southern seas, and accordingly discoveries of the northern coasts of islands, such as Tierra del Fuego, Java, or Timor, were supposed by the imaginative cartographer to be definite points of the coastline of this vast Terra Incognita.

With the advance of the sixteenth century we come to the stage of authenticated discovery, beginning with Magellan's circumnavigation of the world (1520-22), which for the first time revealed the vast extent of the Pacific Ocean. Sebastian del Cano, returning with Magellan's ship *Victoria*, touched at Timor, and proved the existence of an open sea to the south of Java. But rival claims and rival interests at this period caused many important discoveries to be kept secret. It seems probable that the Portuguese about this time discovered the north-western coasts of Australia, and that the first mapping of their discoveries appears in the Dauphin Chart of 1530. On this map the country is named Jave la Grande, and shown as immediately south of Java, only separated by a narrow channel, and without indication of the open sea discovered by Magellan. Other maps of the same school repeat this information, until, in 1542, Jean Roze prepared a map showing an extension of the coastline drawn in detail as far as 60° S. lat. But, notwithstanding this advance, we find that Gerhard Mercator in 1569, and Ortelius in 1570, appear to disregard all previous maps of these regions and fall back upon vague conjecture.

We now come to the famous Elizabethan period, when England and Holland began to contend with the declining power of Spain and Portugal for the right to trade with distant countries. With Drake begins the rise of the naval power of England. In 1580, on his voyage round the world, he sailed through the islands to the north-west of Australia, and eight years later was followed by Cavendish. The Dutch next appear on the scene, but as they seem to have surreptitiously possessed themselves of Spanish and Portuguese charts and documents, it is difficult to know to what extent they were the original discoverers. The name Terra del Zur on many Dutch maps would apparently indicate that it had been previously discovered, and so named on the Spanish and Portuguese charts, and that the Dutch rediscovered it by use of these charts. In a book by Cornelius Wytfliet, published at Louvain in 1597, Terra Australis is conjectured to be one-fifth part of the whole world, and is shown as separated from New Guinea by a narrow strait, and including all the South Polar regions.

In 1606, Spain is again represented by the discoveries of Torres, and in the same year a Dutch vessel, the *Duyfhen*, is reported to have sailed into the Gulf of Carpentaria. In 1616 Captain Dirck Hartog discovered the island bearing his name on the coast of West Australia, but as none of these discoveries is shown on a Dutch globe by Abraham Goos, published at Amsterdam in 1621, corroboration is wanting.

In 1622 the coast at Cape Leeuwin was discovered, or re-discovered, by the Dutch and named after the ship *Leeuwin*, and in the following year Arnhem Land in the north was discovered and named after the ship *Arnhem*.

A close rivalry now set in between the Dutch and English with regard to the newly developed East Indian trade. In 1624 we find Sir William Courteen, an enterprising English merchant, who was anxious to extend his trade to Terra Australis, petitioning King Charles the First for the privilege of establishing colonies there, but without success. In 1627 Pieter Nuyt's Land on the south coast was discovered, and named by a Dutch expedition. The next expedition is a most important one. It was sent out by Anthonie van Diemen, Governor of the Dutch East Indies, under command of Tasman in 1642. Sailing across the Indian Ocean from Mauritius, Tasman discovered what was supposed to be the southern portion of the Great South Land, and named it Van Diemen's Land ; he then discovered New Zealand, and, after visiting several islands in the Pacific, returned to Java without having touched the mainland of Australia. In 1644 Tasman made a second voyage in which he sailed into the Gulf of Carpentaria, named after Carpentier, a former Governor of the East Indies, but, failing to find Torres Strait, steered along the northern coasts of Australia and returned to Java.

About this time the power of Holland, then at its climax, began to decline and give way to England, and with it waned Dutch interest in Australian discovery. A map of the East Indies by Pieter Goos, about 1660, shows the northern, southern, and western coasts of Australia as "Hollandia Nova" at the close of the Dutch period.

We now come to the dawn of the English period, heralded by the arrival, in 1688, of M. Dampier, probably the first Englishman to set eyes on Australia, and whose name is given to the Dampier Archipelago, the first land discovered by him. In 1699 the Earl of Oxford was instrumental in again sending out Dampier, who in this voyage explored the greater part of the west coast, but his account of the country as "the barrenest spot on the globe" was not calculated to stimulate further enterprise. After this, nearly a century elapses before we reach the period of the development of English science and commerce, and find Captain Cook at the island continent on his course of circumnavigation of the globe. In discovering, in 1770, the eastern coast of Australia from Gipp's Land in Victoria to Cape York, he solved the problem of Terra Australis, and revealed a land rich in vegetation and minerals, and in every way desirable for colonial settlement. Thus was Australia opened up to British enterprise, and we may fairly say, that although it may have been known from remotest antiquity, it is to Captain Cook that the modern world is indebted for its re-discovery.

L'Archipel de la Nouvelle-Calédonie. Par Augustin Bernard, Docteur-ès-Lettres, Chargé de Cours à l'Ecole Supérieure d'Alger. Paris : Hachette et Cie., 1895. Pp. xxiv + 459.

It is not often that we have the pleasure of perusing a geographical work so thorough and so comprehensive as the volume before us. Indeed, to any one still disinclined to admit the claims of geography to an independent scientific status, M. Bernard's work should go far in refutation. It is geographical in the widest sense, the deduction drawn from any one branch of science being always carefully checked by comparison with others—for instance, conclusions derived from historical, biological, or economic phenomena are viewed, not by themselves, but in relation to the phenomena of meteorology, of climate, and, above all, of geology. M. Bernard, as we have said, is comprehensive. He also begins at the beginning : " What is an island ? " We are familiar with the categories of Mr. Wallace, which our author however criticises as too absolute in definition, and not therefore always fitting the actual facts of nature ; and he prefers as more philosophical and accurate the definitions of Baron von Richthofen.

The occurrence in New Caledonia of a long series of ancient sedimentary rocks down to the Cretaceous at once places this great island in the category of " Continental," as distinguished from islands of purely volcanic or coralline origin. Its flora and fauna, however, though much richer than those of these last two classes, are both very incomplete and much specialised. Why is this? The cause lies in long isolation. It thus becomes important to ascertain, first, with what other lands was there a former connection ? and secondly, the approximate date of the separation. The inquiry obviously takes the student far afield, a connection with Australia being shown to have terminated in Cretaceous times, while a connection with New Zealand is thought to have lasted a little longer, though not beyond the Cretaceous age. The flora is sharply divided ; there is an endemic flora, all ligneous, but this is gradually succumbing to age, and to the invasion of another flora, consisting mainly of herbaceous plants of wide distribution. Of the older, not only do some orders consist of one species only, but, so far as is known, only three or four specimens have been found.

To the structure of the great barrier reef which runs parallel to the shores of the island, bordering as it were the wide submarine plateau on which the island seems to rest, and extending beyond it on the north, the author gives full attention, perhaps the more because it is well known and often quoted as an example of the famous Darwinian generalisation. M. Bernard, while not denying that the region may be one of subsidence, adduces various causes, some as theory, others founded on observation, sufficient to account for the actual condition of things without the need of calling in subsidence at all. The whole of his reasoning on these great questions of physical geography, with the accumulated information on which it is founded, is both valuable and interesting; his work, besides being a monograph on New Caledonia, is really a treatise dealing with the chief physical peculiarities of all that part of the Pacific. The geology of the island is dealt with very fully, with special notice of the great and characteristic serpentine masses which chiefly yield the nickel and other minerals which constitute its most valuable economical resources. From climatic and meteorological data in relation to soil and elevation and exposure, he pronounces the island thoroughly healthy and colonisable in the fullest sense. Its climate is essentially a south temperate, with some tropical characteristics, and this is borne out by the varied vegetation, and in a less degree by the diseases of the country. M. Bernard speaks out freely as to what he considers the shortcomings of the colonial authorities. Coal might have been worked in a much more intelligent fashion, not merely for its direct value, but as a means of developing other industries. Nor are the forests developed as they might be, while wood and coal are imported from Australia, New Zealand, and the United States. There is no proper survey map of the surface of the country, which is especially necessary for intending agriculturists, owing to the broken and varied character of the surface. Various economic experiments have been carried out, but unintelligently, and usually in blind imitation of Australia, where the conditions are quite different. In 1885 there existed only 85 kilomètres of *routes carrossables*, which took thirty-two years to make.

In putting the Melanesian and Polynesian (the dark and fair races of the Pacific) on an intellectual level, M. Bernard runs contrary to most authorities. Speaking of their traditions, he says, "Quant à vouloir suivre, à l'aide de légendes, . . . ces migrations océaniennes, c'est peine perdue." But it is exactly in this matter of historical tradition that the Polynesians stand at an altogether higher level than the other race. The two races are doubtless akin. Profound linguistic affinities confirm this view; but when M. Bernard attributes the higher position always accorded to the Polynesians and their popularity with the early explorers to the easy virtue of their women compared with the stricter Melanesian practice, he perhaps hardly intends us to take him quite seriously.

We might demur to his statement that cannibalism in Fiji, which, as a matter of fact, was quite unbridled, depended simply on religious feeling; we might also question the perfect impartiality of his statement of the French claims to the New Hebrides. We question, also, whether it is only in Fiji, in the Western Pacific, besides New Caledonia, that an elaborate system of irrigation is found; but these are only small errors—if they are errors—in a work of so much merit as this. We should add that the work concludes with an elaborate description of the convict system in force in New Caledonia.

The Empire of the Ptolemies. By J. P. MAHAFFY, Fellow of Trinity College, Dublin, etc. etc. London : Macmillan and Co., 1895. Pp. xxv + 533.

Professor Mahaffy has made a special study of the Hellenistic world from the days of Alexander the Great till the time when the kingdoms of his successors were all finally absorbed into the Roman Empire. By his productions relating to that period he has taken high rank as a scholarly historian, and the work we have now to notice will tend both to confirm and exalt his reputation as such. It might be supposed beforehand that the subject he has here taken in hand is one in which the details, or, at least, all the more important, are known with unusual accuracy and fulness. The Ptolemies lived in a most enlightened age; they were themselves the most munificent patrons of learning in all antiquity, and attracted to their famous museum and library the best intellects of their time from all parts of the Grecian world. It might thence be inferred that ample materials must exist from which, without much need for searching inquiry or careful balancing of conflicting evidence, a full and faithful history of the Ptolemaic period might be constructed. But this is by no means the case. In spite of the importance and splendour of the dynasty, no connected account of it, if such ever existed, survived even to the days of Pausanias. A vast body of isolated facts is indeed at command, but to set these in order has been found an arduous task, requiring great patience and careful discrimination. The following is the Professor's description of the sources of his information : " We know the Ptolemies," he says, " through pompous hiero- glyphics, which were not intended to instruct us, through panegyrics, which were perhaps intended to mislead us, through episodes in the universal histories of Poly- bius and Diodorus. Recently we have added to these literary authorities a good many stray inscriptions, and a mass of papyrus fragments, which give us multitu- dinous isolated facts, seldom of public interest, but no connected history." As an example of the oblivion from which literature has failed to rescue even the greatest of all the military achievements of the Ptolemies, we may refer to the Asiatic expedition, by which the third Ptolemy, surnamed Euergetes, signalised the beginning of his reign. In this expedition, which was undertaken to avenge the murder of his sister, the young queen of Syria, he carried his victorious arms from the Nile to the Oxus, reducing, as he advanced, Syria, Cilicia, Pamphylia, Ionia, Hellespont, Mesopotamia, Babylonia, Susiana, Persis, Media, and, finally,

Bactriana. Yet, amazing as is the magnitude of this conquest, the only record that has preserved its memory is the copy of an inscription found on a tablet of basanite stone at Adule, in Abyssinia, in the sixth century of our era.[1] The copy was made by Cosmas Indicopleustes (then a travelling merchant), who afterwards inserted it in his *Christian Topography*, a work which was printed for the first time in 1706 from a Florentine ms.

Professor Mahaffy has made the most of the imperfect and perplexing materials with which he has had to deal, and out of them has framed a very lively and agreeable narrative. In his preface he explains that he has called his work the *Empire* of the Ptolemies, in order to emphasise the fact that they were not mere kings of Egypt, but sovereigns of a composite empire. Polybius has given a list of their dependencies, which shows that, when the fifth Ptolemy (Epiphanes) ascended the throne in 205 B.C., they comprised Cœle-Syria, Cyprus, most splendid cities, strongholds, and harbours all along the sea-coast from Pamphylia to the Hellespont, and the district round Lysimachia, Aenus also and Maroneia, and other cities still more distant. The possession of Palestine was a frequent subject of contention between Syria and Egypt; but after the defeat of Antiochus the Great in the battle of Raphia, 217 B.C., it fell to Egypt, which held it in undisturbed possession for a century afterwards.

The Ptolemies ruled Egypt from the death of Alexander in 323 B.C. to that of the celebrated Cleopatra in 30 B.C. During this period of well-nigh three centuries, the sceptre was swayed by thirteen sovereigns, of whom the first three—whose united reigns extended to nearly a century—were strong, sagacious, and altogether brilliant rulers. They showed great outward respect to the national religion, and by so doing conciliated the goodwill both of the priesthood and of the native population, by whom the old Persian rule had been abhorred because of its having pursued in this respect an opposite policy. The fourth king proved a degenerate scion of a noble stock. He is reprobated in history for neglecting business, and indulging to excess in the grossest and most debasing pleasures. Our author, however, judging from inscriptions, is inclined to think that he may not have been quite so black as Polybius and Plutarch have painted him. It was in his reign that the first of the great native rebellions, induced by the grinding oppression connected with the collection of exorbitant taxes, broke out. It raged for several years—a veritable civil war attended with the usual horrors.

Under the seventh Ptolemy, surnamed Philometor, the history of the dynasty declined in dignity and increased in complication, of which the cause is thus stated: "Hitherto, though it was usual to associate the queen or prince-royal in the government, there is no doubt about the reigning king. From henceforth we have almost constantly rival brothers asserting themselves in turn, queen mothers controlling their king sons, intestine feuds and bloodshed in the royal house, till the stormy end of the dynasty with the daring Cleopatra vi." Of the succeeding sovereigns, Ptolemy ix., surnamed Physcon on account of his obesity, was by far the most capable ruler, though, if history is to be believed, he was at the same time a monster of cruelty and vice. Be that as it may, he extended Egypt far south of the limits which his predecessors had accepted, and at his death, 117 B.C., after a long reign, he left his kingdom in a settled and safe condition, and free from Roman interference. The history of Ptolemaic Egypt might close with his death. Nothing of public interest followed till we come to the eventful reign of the sixth

---

[1] Our author inadvertently says the fifth century. He errs also in stating that the inscription was carved on a throne. The throne inscription was, indeed, that of an Axumite king of the third century after Christ. See p. 143.

and last Cleopatra. The closing scene in the great Ptolemaic drama shows us the chamber in the monument, where, arrayed in all her royal robes, and with a diadem encircling her brow, her queenly form lay stone-dead upon a bed of gold. Professor Mahaffy's sketch of her career is very effective, and forms the most interesting section of his important and valuable work.

Durch Afrika von Ost nach West. Von GRAF VON GÖTZEN. Two maps, some hundred Illustrations. Berlin : Dietrich Reimer, 1895. Pp. 416.

Although this is rather an unwieldy book, yet the brilliance and size of the illustrations more than compensate for the drawback. The author, accompanied by Drs. v. Prittwitz, Gaffron and Kersting, crossed Africa from east to west during the years 1893-94, and a most successful journey they had.

Geography, botany, geology, natural history and climatology, as well as medicine, are all enriched by the observations made by the expedition, all the members of which are to be congratulated upon the work they have done.

The chief discovery of the expedition was the active volcano, Kirunga, and the Kivu lake. The author's description of the ascent of Kirunga is exceedingly interesting. The party had the greatest difficulty in cutting their way through the dense forest which clothed the lower slopes of the mountain. After three days' hard work, they arrived at the summit at an altitude of 3470 mètres. The crater itself was estimated at from 200 to 300 mètres in depth. It looked like an oval amphitheatre, whose greatest diameter was about 2000 and smallest 1500 mètres. The floor of this amphitheatre appeared to be perfectly smooth, and in it were two sharply cut openings, from which steam was emitted, accompanied by loud rumblings from the interior of the mountain. The author is inclined to think that the activity of the volcano is passing away.

Lake Kivu is 1485 mètres above sea level, and it may be 80 or 100 or more kilomètres long, but the expedition was unable to explore its southern borders. The author believes that the Rusizi river is the only outlet of the lake, which, if his view is correct, would lead to Tanganyika. If this is the case, and if we take the altitude of Tanganyika as 810 mètres above sea-level, and the length of the river (90 kilomètres), we should get a fall of 1 in 133 ; and yet explorers of the Rusizi at its entrance into Tanganyika describe it as possessing a very sluggish current, a curious problem which has yet to be solved.

The book is well written and well printed ; the illustrations and the maps leave nothing to be desired.

The Key of the Pacific—the Nicaragua Canal. By ARCHIBALD ROSS COLQUHOUN. Westminster : Archibald Constable and Co., 1895. Pp. xvii + 443. *Price* 21s. *net.*

"The Key of the Pacific" is the Nicaragua Canal. In the splendid volume which bears this title, Mr. Archibald Colquhoun deals with the subject in all its bearings. First in importance is, of course, the "Engineering Problem." This the author discusses in detail with the help of diagrams and maps. He approves of Mr. Menocal's scheme, which, in spite of grave difficulties, he considers as quite feasible and presenting great advantages over the Panama route. Having made his observations on the spot, and being a practical engineer with much experience in tropical countries, Mr. Colquhoun speaks as an authority. It will cost more than the estimates show (according to Mr. Colquhoun about thirty millions in mere expenditure on the work), but it will have a greater traffic than is usually admitted. The time allowed for completion, six years, he considers sufficient. A feature telling strongly in favour of the future value of the canal is the country through which it passes. Nicaragua, with a fertile soil, great resources both vegetable and mineral, internal water communication, and a generally healthy climate, only awaits the immigrant and the capitalist to become one of the most prosperous States in America. This is the impression the reader receives from the author's interesting sketch of Nicaragua, which comprises chapters on the Social and Political Life, the Physical Features, and the Resources of the Country. Of course no private company or corporation could complete the canal, and it is certain that the project must be under the auspices of a strong Government, and without doubt this must be the United States ; but in the interests of the world the canal must be neutralised, and Mr. Colquhoun thinks that the United States and England could bring about the neutralisation. The benefit of the canal to European nations will certainly be great, as it will materially shorten the route to the Pacific littoral of the two Americas and to the South Sea Islands, but it will benefit America infinitely more, bringing Japan, Northern China, Australasia and part of Malaysia nearer to the Atlantic cities of the U.S. than they are now to England. Thus it will give an immense impulse to American trade and industries, especially cotton and iron, and " it will," our author concludes, "taken in connection with the vast changes occurring in the Far East, bring about the most serious rivalry to the commercial supremacy of Great Britain which she has yet had to encounter."

Algerian Memories: A Bicycle Tour over the Atlas to the Sahara. By FANNY BULLOCK WORKMAN and WILLIAM HUNTER WORKMAN. London : T. Fisher Unwin, N.D. Pp. xiv + 216.

This is an interesting book for cyclists who are in quest of fresh ground. Algeria, thanks to the French occupation, is a country where, with comparative safety, with wonderfully good accommodation, and with excellent roads, the traveller may see Oriental life to much advantage, still almost undiluted in some localities, enjoy the finest mountain scenery, and even have a short run into the desert. Most of the country visited by our cyclists can now be reached by railway ; but as they kept to the roads, sometimes across country, and took things leisurely, they saw much that a railway traveller must miss. Their route was from Oran eastward to the ruins of Carthage, with excursions to several outlying points in the south, their goal in that direction being Biskra, on the edge of the Sahara. The book is written in a terse, businesslike style, and some of the scenes of North African life are vividly depicted. " Tlemcen " is an excellent etching.

The most interesting, because the freshest, part of the book is the description of the Kabyles of the Djurjura, a picturesque mountain range to the east of Algiers. Their villages are perched on the summits of the *crêtes* or spurs which strike out irregularly from the main chain. They have ever been a brave, independent, and exclusive people, and can boast of never having been conquered by Roman, Moor, or Turk. They finally succumbed, however, in 1857, to the French under Mac-Mahon, who clinched his victory by building a very strong fort, Fort National, in the midst of their mountain fastnesses.

The best time for a cycling tour is from the middle of March to the end of May, as earlier the weather and roads are uncertain, and later the heat is too great for comfort. The writers complain of the want of road maps ; a simple sketch-map of their own route would have been a convenience to the reader.

NEW MAPS.

WORLD.

WORLD, Traveller's Route Chart of the ——, on Mercator's Projection. By J. G.
Bartholomew, F.R.G.S.   *Price* 1s.; *cloth,* 1s. 6d.

The Geographical Institute, Edinburgh.

The principal steamer routes are marked on this map, with distances. There
are also numerous insets, one of them showing routes in the Mediterranean, and
another routes from the British Isles.

ASIA.

INDIA. Indian Atlas. Whole sheets, 47, 65, 112, 121. Quarter sheets, 2 SE.,
6 NE., 15 NW. and SW., 28 SW., 31 NE., NW. and SW., 32 NE., 34 SW.,
42 NE. and SE., 49 NE. and NW., 60 SE., 67 SW., 70 SE. and SW.,
71 NW., 72 SE., 87 NW. and NE., 90 NE., 91 NW., SE., and SW.,124 SE.,
125 NE., 129 NE. and SE., 131 NW.

India, showing provincial and district boundaries (skeleton). Scale 1 in.=64 m.
India, showing railways. Scale 1 in.=80 m.
Telegraph map of India. Scale 1 in.=32 m.
India, engraved. Scale 1 in.=256 m.
Bombay Presidency. Scale 1 in.=32 m.
Berar. Scale 1 in.=24 m.
Upper Burma. Scale 1 in.=16 m.
    Do. (skeleton).   Do.
Central Provinces. Scale 1 in.=32 m.
The Nizam's Dominion. Scale 1 in.=16 m.
North-West Provinces and Oudh. Scale 1 in.=32 m.
Punjab and Kashmir. Scale 1 in.=16 m.
Skeleton Map of the Punjab and Surrounding Countries. Scale 1 in.=32 m.

PALESTINE, Raised Map of ——. By George Armstrong. *Photo,* 1s. *Collotype,*
2s. 3d. and 3s. 3d.   *The Palestine Exploration Fund, London.*

The Raised Map of Palestine on the scale of $\frac{3}{8}$ in. to one mile is represented in
the maps referred to. They give a general notion of the configuration of the
country.

AFRICA.

TRANSVAAL. Kaart van de Zuid Afrikaansche Republiek. Schaal 1 : 1,500,000.

J. Smulders & Co., den Haag.

A handy general map of the country. On a map published in 1896, however, Swaziland and Amatongaland should not be coloured as though still unannexed,
nor should the Bechuanaland railway stop short at Vryburg.

ATLAS

ZEICHENATLAS zum Gebrauch im Geographischen Unterricht. Herausgegeben von
E. Debes.   *H. Wagner and E. Debes, Leipzig.*

An earlier edition of this useful atlas was noticed at length in vol. ii. p. 318. The
maps are compiled with the assistance of Dr. A. Kirchhoff and Dr. R. Lehmann,
and are arranged for the middle and lower classes in schools. For the latter only
the general outlines of elevations and one or two chief rivers are given, while for
the more advanced pupils the details are more numerous. Each number has a
corresponding atlas containing only networks of latitude and longitude on which
the pupil has to draw the maps.

THE SCOTTISH

GEOGRAPHICAL

MAGAZINE.

THE INDUSTRIAL AND COMMERCIAL DEVELOPMENT OF JAPAN.

By JAMES TROUP, H.B.M. Consul at Yokohama.

(Read at Aberdeen, in February, 1896.)

THE Empire of Japan, as it exists at the present date, extends from the extremity of Formosa, in the south, to the farthest of the Kurile Islands, in the north—or, say, from about 22° to 51° north latitude; and from about 119° to 156½° east longitude. Consisting of a chain of islands fringing nearly the half, and that the temperate half, of Eastern Asia, its length may be taken at over 2300 geographical miles. Its greatest breadth, taking this from Tsushima on the west to the Bonin Islands on the east, extends to 800 miles; but the greatest breadth of the mainland of Japan is only about 220 miles. It is not my intention here to make more than a passing reference to the latest addition to the Japanese Empire—the islands of Formosa and the Pescadores. The remarks which follow will be limited in their application generally to the islands of the Japan group proper.

Like those of every other nation, the earliest sources of the civilisation of the Japanese are shrouded in obscurity. Ethnologists, while differing to some extent as to details, are agreed that the main stock of the race is Mongolian. It may therefore be assumed that this people, before the dawn of history, crossed over from the continent of Asia to the Japan Islands. Other races, however, were already in occupation of those islands. The principal of those were the Ainu, now reduced to a few thousands inhabiting the island of Yezo, but at one time to be found all over the Japan Islands, and opposing the advance of the Japanese.

From what I shall now proceed to state, the probability would seem to be that the ancestors of the Japanese, before they crossed from the continent, had made a commencement, at all events, at settled life ; had begun to lay aside the nomad, assuming that they had been nomads, for the agriculturist.

The earliest Japanese records which we have of their own history are contained in two compilations ; the first termed the " Kojiki," or " Record of Ancient Matters," which has been translated into English, and its contents analysed by Mr. Chamberlain ; the other the " Nihongi," or " Chronicles of Japan," now in course of being made similarly accessible to readers of English by Mr. Aston. The former dates from 712 A.D., the latter from 720 A.D. By the aid of the researches of Mr. Aston and of Mr. Parker in the Korean and Chinese records, and from songs, ancient rituals, and other sources, we are able to corroborate, correct, or supplement these accounts.

Although the Japanese records were not compiled until the above comparatively late times, they contain traditions of a much earlier date ; and it is of the materials thus supplied that we may make use, to arrive at some notion of what the early Japanese possessed.

From Mr. (now Sir Ernest) Satow's summary, in his account of Shinto ritual, it would appear that the better description of dwelling of the early Japanese, the " palace" of the noble, was constructed of a framework of wood tied together by cords of climbing plants. A raised bench, for resting and sleeping, ran round the side of the hut, the remainder being mud floor. Rush mats were used to sit or lie upon, as well as skins of animals. The roof was thatched, with a hole in it for the smoke. The doors were sometimes fastened with hooks. To this it may be added that the general structure of the Japanese hut probably finds something like its counterpart in the rude temporary shanty which we now sometimes see in the country, where to a framework of wooden posts is added some straw matting hung round to resist the weather.

Keeping now, in the main, to Chamberlain's analysis of the Records, we find that rice was the only cereal the possession of which went back with the Japanese to time immemorial. To their rice, which, according to the Chinese account, they ate with their fingers—although the Japanese account mentions the use of chopsticks—they added, for food, various roots, seaweed, fish, and the flesh of animals which they had shot or snared ; for Buddhism had not yet reached them, to prohibit, or at least to limit, their consumption of animal food. Their saké, or rice-beer, dates back perhaps to the same early period ; it is mentioned in the records, though as a rarity ; and their domestic utensils consisted of cooking-pots and cups and dishes of earthenware and leaves. The horse they possessed, and the barndoor fowl ; and they fished with the cormorant. There is repeated mention in these records, connected with the most ancient portion of the traditions, of spears, swords and knives, of iron, bows and feathered arrows, hooks for fishing, which were barbed with bone or iron ; traps and gins, for human as well as for the animal tribes ; the pestle and mortar, fire-drill, wedge, sickle, and shuttle. But no mention is made of the saw or the axe, or of any more advanced tool.

Navigation was in its most elementary stage. Rowing and punting are mentioned, and for these purposes water-proof baskets were used, with oars. Sailing, indeed, was little practised by the Japanese even long after the end of the ninth century.

Their clothing, which was made of hemp and paper-mulberry bark, and which they dyed with madder and other plants, consisted of both upper garments and skirts, trousers, and girdles. Skins were also used for clothing. They had veils, hats, and combs; and, unlike the later Japanese, wore bracelets, necklaces, and head-ornaments of stone. Their cloth was woven, but not sewed, for they had no needles. They had the straw rain-coat and broad-brimmed hat, which we see to this day. With the tendrils of creeping plants the warrior bound his sword around his waist. Among plants, besides rice, they had the bamboo, ginger, lilies, the vegetable-wax tree, the wild vine, the cherry, pine, and other trees. Iron was their only metal. They carved wood, or knotted cords, for records.

Such is a dim picture of the early economic state of the Japanese nation; and only a short time before the records which contain this picture were reduced to writing, a new and most important influence—one which, amongst other things, supplied the Japanese with the means of making such records—came and spread itself amongst that people. The indebtedness of the Japanese to China for so many of the arts and implements of civilisation is a feature of their history which on every hand forces itself upon our attention. Samuel Johnson, in his diary of his journey to the Hebrides, says that he was told at Aberdeen that the people there learned from Cromwell's soldiers to make shoes and to plant kail. "How they lived," says he, "without kail, it is not easy to guess. . . . When they had not kail, they probably had nothing." However this may be as regards our forebears, it is certainly not wide of the mark to say that, when the Japanese had none of the arts and none of the productions which they derived from Chinese sources, they had few or none of those things by the possession of which we have been accustomed later on to characterise them. Korea, however, was the stepping-stone by which those things came, and, in some cases, more than the stepping-stone.

Tradition, as well as the Chinese and Korean records, indicates intercourse between the continent of Eastern Asia and Japan in the early centuries of the Christian era. In the later period covered by the Japanese annals which I have just drawn upon, we hear of the existence of dogs and cattle in the country. Some Whittington must also have introduced the cat. The Chinese records of the third century mention the existence in Japan of towns and cities of from one thousand to thirty, fifty, and even seventy, thousand houses. The inhabitants travelled a good deal by boat, and laid in provision of grain, and they also held markets for the interchange of commodities; but the quantity of rice produced was insufficient for the inhabitants, the roads were mere tracks, and, judging from the time taken in travelling by water, the boats of the country must have been still of a primitive character.

Amongst the civilised arts which were first introduced from the continent was that of writing. The Chinese characters were not entirely unknown in Japan, probably, before the middle of the fourth century.

If we are, as seems probable, to adopt the chronology suggested by
Aston, the arrival of the first Korean teacher of the art was in the year
405 ; and the first notice which we have of the use of writing by the
Japanese relates to the early part of the same century.   But there is no
reliable evidence to show that the Japanese studied with any profit till
the time of the teaching of Buddhism in the sixth century.

Buddhism had been first received in China, from India, as early as
the year 65 A.D.   Korea received it, from China, in 372 A.D., but it was
not until 552 A.D. that it was first received from Korea into Japan.   It
may be interesting to note that St. Columba was removing from the
north of Ireland to Iona about a decade later.   From then onwards the
study of the Buddhist Scriptures began to be prosecuted in Japan, and,
with it, the art of writing, and, in time, the study of Chinese literature
generally.   Block-printing was known in Japan as early as the eighth
century, but was for long used only for the multiplication of Buddhist
books.   Movable types were known in China and in Korea before
the invention of printing in Europe, and copper types were founded
in Korea as early as 1403 ; but printing by means of such types does
not appear to have been practised in Japan until towards the end of the
sixteenth century, when the art was introduced from Korea.   To-day
there is in Tokio a large typefounding establishment ; and, not to speak
of books and the numerous newspapers published in the Japanese language,
numbers of books are printed in the country in European languages.   But
this is anticipating.

It would be impossible to attempt, within the limits of this address,
to trace with anything like completeness the progress of the influence of
this Chinese-derived civilisation in Japan.   Mulberry-trees, along with
the silkworm, would appear to have been introduced in the fifth century ;
and, without reference to precise order of dates, we may note also the
art of sewing, carpenters' tools, the abacus and musical instruments ; chess,
the game of " go," and dice ; the use of other metals besides iron ; wheat
and barley, different varieties of pulse ; the plough, water-wheels, the
balance ; the tea plant ; the manufacture of paper, porcelain, lacquer ware,
fans, umbrellas ; improved architecture;—all these, not to speak of medicine
and the study of astronomy, appear to have come from the continent.
The style of dress assigned to different ranks of society was settled in the
seventh century.   Cotton is said to have been introduced by a Hindu—
not from China—about 800 A.D., and to have been cultivated until the
twelfth century ; but, in the civil commotions, its cultivation was
neglected, and finally abandoned, until its reintroduction later, in the time
of the Portuguese intercourse.   The Chinese sugar-cane reached Japan
through the Lu-chu Islands, from which it was introduced into Satsuma.

The actual interchange of commodities between the continent of Asia
and Japan at this period, and for some time, could not have been otherwise than insignificant in amount.   What intercourse there was of this
nature with China was with the district of Foochow and Ningpo, and not
by way of Liaotung, although embassies passed by the latter route, and
through Korea, to and from the Chinese capital.   In the end of the
ninth and beginning of the tenth centuries, Chinese pirates are said to

have ravaged the coasts of Kiushiu, and the island of Iki. At a later period, the Chinese carried on considerable commerce with Japan, trading to ports in Kiushiu and on the Inland Sea, to Sakai, near Osaka, and to ports on the east coast as far as Nambu.

In the meantime the territory in the northern and eastern part of the Main Island had been reclaimed from barbarism, and more than one attempted inroad by Tartars from the continent—the last being that when the Mongol fleet was sent by Kublai Khan in the end of the thirteenth century—had been repulsed. The cultivation of rice in the early ages implies some irrigation and the construction of watercourses; but such works, with the construction of reservoirs, sluices, and the introduction of the irrigating water-wheel, had immensely improved and extended. Some development in the building of ships must also have taken place. In the third and fourth centuries we hear of periodical raids being made by the Japanese on the coasts of Korea. The emperor Ojin, subsequently deified as the god of war, is said to have built five hundred vessels near what is now the port of Hiogo. Later on, we learn that one of the sides in the sea-fight at Dannoüra, which took place in 1185, brought seven hundred vessels to the contest; and, still later, in the time of the civil wars, Japanese pirates ravaged the Chinese coast at Ningpo. Yet, a century later, the Japanese warships were much inferior to those sent against them by Kublai Khan.

Japanese material development is undoubtedly much indebted to the Buddhist priesthood. Those ecclesiastics did not confine themselves to the mere propagation of their faith; they built bridges, made roads, and cut through mountain passes, planted fruit-trees, constructed reservoirs and ponds, aqueducts and canals. They did carving, modelled and made castings in bronze, and used the brush both in caligraphy and painting. To the development of bronze-casting, the images of Daibutsu at Nara and Kamakura bear witness, and the bells at Kioto and Miidera.

During the period at which we have now arrived, much progress has to be noted in the arts of making porcelain, pottery, and lacquer ware, in engraving and inlaid metal-work. Defensive armour, which at first was probably of padded cloth, came, in the eighth and ninth centuries, to be made of leather and iron. Iron ores were worked from the tenth century, and in the twelfth we hear of the mines of Oshiu supplying iron for sword-making. In the sixteenth century defensive armour had come to be made of metal plates. The quarrying of granite is attested by the stone which we see in the construction of the castle of Osaka, and otherwise.

The most ancient specimens of architecture existing in Japan are the buildings of the monastery of Horiuji, completed in 607 A.D.; and those of the Hongwanji at Kioto, which date from 1591, are also of great interest. These, and such works as the building of the castle and city of Yedo by Iyeyasu, also in the end of the sixteenth century, imply a great development in handicrafts.

Buddhism had, as already indicated, restricted very largely the use of animal food, but in other respects the food of the people had, during those centuries, improved with the variety of food products. Among Japanese condiments we may well, for example, note soy, the basis of many of our own sauces.

Notwithstanding the civil wars of the fourteenth to the sixteenth centuries, during which much destruction was caused to towns and villages, monasteries and cultivated fields, the general mode of life had then become settled, and in the higher ranks was not without elegance. In those and in the centuries preceding, literature flourished, and the drama, painting, music, and the art of gardening.

But the advent of another wave of civilisation from another quarter was at hand. Marco Polo on his return to Europe, after his residence at the court of Kublai Khan in the thirteenth century, had made known the existence of a country in the Eastern seas called Zipangu. It was not, however, till the year 1542 that the first European who set his foot in Japan, Mendez Pinto, arrived there. This led to extensive commercial intercourse on the part of the Portuguese and Spaniards with Japan, but still more to the propagation of Christianity there. The Portuguese and Spaniards were, at the end of the century, followed by the Dutch and English. In the year 1600 the first Dutch ship arrived in Japan. Her two pilots were Englishmen, and one of them Will Adams. It is of the very greatest interest to read, in the accounts given of Adams, and in his account of himself—for we are fortunate enough to possess several letters written by him—how he obtained the favour of Iyeyasu and of his son, the Shogun of the day; how he was the means of negotiating for commercial privileges, both for the Dutch and English, obtained a grant of land, on which he lived, at the village of Hemi, close by Yokosuka (near Yokohama), what is now the principal naval port of Japan, assisted the Japanese to construct improved ships, and to navigate them; and, finally, how, notwithstanding his efforts to obtain permission from the Shogun to leave Japan and return to England, he died there. His tomb may be seen to-day on an eminence overlooking Yokosuka, the entrance to the Bay of Tokio, and the adjacent shores.

The Dutch and English factories were established on the island of Hirado; the former, under Jacob Spex, was opened about 1610; and the latter, under Captain Richard Cocks, who was left in charge there on behalf of the East India Company by his colleague, Captain John Saris, a year or two later. At this period Dutch and English ships had the privilege of visiting any port in Japan.

Through the fourteenth, fifteenth, and sixteenth centuries, especially the last, further developments had taken place in the art of shipbuilding in Japan; and we have details of what a Japanese warship had come to be in the sixteenth century, in the time of Hideyoshi. About that period the largest description of war vessel had 100 oars, with a complement of about 205 men all told, including 100 rowers and 100 fighting men, of whom 26 were armed with muskets, the rest being armed with bows and arrows, lances and spears. Such a vessel carried two guns, and had a capacity of somewhat under 190 tons. The merchant junk of that time varied from about 10 tons to as much as 400 tons burden, or thereabout. All were clumsy and ill adapted for sailing. Will Adams says of a junk which he saw at Shimonoseki in 1613, and which he describes as being of

800 or 1000 "tunnes" of burden, and "sheathed all with yron"—a transport, apparently—that "she was built in a very homely fashion, much like that which describeth Noah's Arke unto us."

Adams died in 1620; but before his death he had contributed to the improvement of shipbuilding in Japan, by providing a style of vessel better suited for navigation to distant parts.   In one of his letters he says he built two ships for the "Emperour," in one of which he went a voyage himself; and it is certain that at that period other ships were built on the European model, and employed by the Japanese in expeditions to the coast of China.

According to one authority, the Japanese had intercourse with the Indo-Chinese Peninsula as early as 1434; it is certain, however, that in the sixteenth, and the early part of the seventeenth centuries, Japanese traders had settled all over the Far East.   They traded to Cambodia, Siam, Formosa, the Philippine Islands, to the Eastern Archipelago and the Moluccas; and they even visited Mexico.   Japanese colonists settled in Siam and at Hué; and there, and in Tonkin, their descendants are said still to be found.   Japanese soldiers were employed by the King of Siam, and even by the Dutch and English, in their Eastern expeditions, and Japanese sailors came to England on board European ships.   The first mission to Spain and the Pope was sent out in 1582, the last in 1613, returning in 1620.

Among the imports brought into Japan during this period by Japanese merchants, are enumerated—silk and silk stuffs, damask, cotton stuffs, gums, spices, drugs, dyes, woods, varnish, wax, honey, sugar, sharks' skins, horns, skins, leather, gold, tin, lead, ivory, coral, camphor, birds and feathers; and among the exports—bronze and iron coin, porcelain, hempen and cotton cloth, sulphur, screens, fans, umbrellas, mats, urns, and chafing-pans.

Such was the state of intercourse of the Japanese with foreign nations, and of foreign nations with Japan, when, suddenly, it came by a great check.   It is beyond the purpose of this address to trace the spread of Christianity in Japan in those days, and its suppression in less than a century from the first advent of the Portuguese.   Suffice it to say that, in 1636, an edict was issued by the Shogun Iyemitsu, ordering the destruction of all ships built on the European model, the restriction of junks to a size rendering them unfit for voyages to distant countries, prohibiting Japanese from leaving their native land, and interdicting to them the study of any Western language.   Before the end of 1639, all Europeans, except the Dutch, were expelled from the country.   Japan's material acquisitions from the Portuguese and Spaniards were probably limited to the possession of tobacco and cotton, the samisen, a species of guitar, the art of making gunpowder and firearms, and a species of sponge-cake, known to this day as "kasutera" (= castilla).   Common potatoes were introduced by the Dutch.

As far as English trade was concerned, notwithstanding that Saris had obtained from Iyeyasu a commercial charter and treaty, and that he and Adams were strongly in favour of the East India Company carrying on a trade with Japan,—what between the difficulties imposed by the

Japanese Government, and the jealousy of the Dutch—the settlement at Hirado was abandoned. A subsequent unsuccessful attempt was made, in 1673, on the visit to Nagasaki of the ship *Return*, to re-establish English trade with Japan.

From the period at which we have now arrived, until past the middle of the present century, the intercourse between Japan and Western countries was restricted to that through the Dutch at Nagasaki, to which place their factory was removed in 1641. Among the articles which were imported by the Dutch, besides those imported by Japanese merchants as above mentioned, were steel ingots, woollen goods, holland and diaper, tortoiseshell, saltpetre, files, needles, glass and glass ware, quicksilver ; and, among the articles exported, in addition to those in the previous list, were copper, camphor, lacquer ware, bronze and other metal ware, silk goods, fancy work in horn, oil-paper, tinsel-paper, gold thread, rice, saké, soy, pickled fruits, tobacco, tea, gold and silver. The export of the precious metals was, however, prohibited in 1715, and other restrictions were then, or at other dates, placed on commerce. The annual value of the Dutch trade, each way, at different periods of their intercourse, varied from about £500,000 to £700,000 sterling.

Kæmpfer estimates the annual value of the Portuguese import trade, in the last years of their intercourse, when it was in its decline, at from £400,000 to nearly £800,000 sterling. In return for their imports, which were similar to those of the Dutch, they took away mainly gold and silver.

Chinese trade had been restricted to Nagasaki at the same time as Dutch. It was still, however, permitted ; and junks continued to arrive from Batavia, Siam, Cambodia, Canton, Formosa, and other places. As many as two hundred Chinese junks *per annum* came over, according to Kæmpfer, in the years 1683-84, having some ten thousand persons on board. Indeed, the Japanese, becoming apprehensive of the influx of Chinese into the country, made a regulation limiting their trade to a total of some £200,000 sterling per annum, the number of junks not to exceed seventy. But probably the value of the trade exceeded the nominal limit assigned to it.

Jealous guard against intercourse with all Westerns except the Dutch was thus kept for over two centuries. Nevertheless, Europeans were occasionally to be found in Japanese waters. In the eighteenth and nine-teenth centuries, surveys of the neighbouring coasts were made by navi-gators like La Pérouse, Broughton, and Beechey. In 1808, H.M.S. *Phaeton* paid an unexpected visit to Nagasaki, and even took soundings in the harbour, to the great trepidation of the officials, the chief of whom subsequently committed suicide for his remissness in not having succeeded in keeping the ship away. Unsuccessful attempts to establish intercourse were made by the Russians early in the century. These were followed by several Russian descents on the Japanese settlements in Saghalien and the Kuriles. No more curious account of the manner in which foreigners were treated by the Japanese in those days can be read than that of Captain Golownin, of the Russian navy, who visited Yezo in 1811, in the ship *Diana*, and was for some time detained a prisoner. An important relaxation was, however, granted in the eighteenth century in the matter of the study of the Dutch language by Japanese. Those who obtained this liberty were medical men; but through their medium some knowledge of what was transpiring in the outer world reached the secluded country.

The commencement of trade, under the conditions at present existing, between the West and Japan, dates, as you are aware, from 1859. With the history of the opening of the Treaty Ports I shall not trouble you; but it will be interesting, as illustrating the main subject of this paper, to give some indication of what have been the commercial and industrial results, or some of the results, of this opening of Japan to the commerce of the world.

The statistics of the foreign trade of the country during the earlier years after the opening of the ports are neither so full nor so exact as those of more recent date. I take advantage of the labours of Mr. Gubbins, of H.M. Legation in Japan, in applying certain corrections to those statistics in order to arrive at as exact a comparison as is, perhaps, attainable of the trade in different years, since the year 1873. The results of his calculations are given in an interesting article, published in the *Chamber of Commerce Journal* for October, 1894, from which I select the following:—

	Imports.	Exports.	Total Trade.
In 1873	£5,623,641	£4,325,895	£9,949,536
„ 1883	5,589,617	6,546,662	12,136,279
„ 1888	10,943,812	10,067,006	21,010,818

and, from the published report of H.M. Legation in Japan on the trade of 1894, I take the following:—

	Imports.	Exports.	Total Trade.
In 1893	£11,652,165	£11,397,720	£23,049,885
„ 1894	12,681,222	11,801,342	24,482,564

The above shows the increase as stated in gold values; but, stated in silver values, the currency of the country, the increase is still more striking:—

	Imports.	Exports.	Total Trade.
In 1873	$26,859,181	$20,660,991	$47,520,172
„ 1894	117,481,955	113,246,086	230,728,041

In other words, the foreign trade of Japan, stated in silver values, was almost five times as great in 1894 as it was in 1873; stated in gold values, it was about two and a half times as great. The figures lately received of the trade of 1895 are, for the value of imports, $129,260,839; and of exports, $136,112,165, or a total of $265,373,004. These figures are still subject to revision, but may be taken as a sufficient indication that there was no falling off in the trade last year.

I observe that both Mr. Hayashi, Japanese Consul-General in London, and Mr. Longford, the compiler of the Trade Summary for 1894, of H.M. Legation in Japan, have been led to make an interesting comparison between the foreign trade of Japan and that of China. Their results show that, although China proper has a population of, say, from eight to ten times that of Japan (which is now over forty-one millions), the foreign trade through the open ports of the former did not, in 1894, amount to twice that of Japan.

To enumerate the articles of this trade with Japan would be to enumerate nearly every article of civilised commerce; but the chief of them may be summarised as follows:—Imports : raw cotton and other fibres ; textile fabrics, including cottons, woollens, and linen ; cotton-yarn ; machinery and instruments, including cotton-spinning machinery, loco-motives, clocks and watches ; war material, provisions, drugs and chemi-cals, dyes and paints, books and stationery, rice and beans, hides and leather ; metals, including iron, steel, lead, tin, and zinc ; kerosine, and sugar.   And exports : raw silk, tea, camphor, rice, vegetable-wax ; marine produce ; mining products, including coal, copper, antimony, sulphur, manganese ; fabrics of silk and of cotton ; cotton-yarn ; wares of metal, wood, ivory, lacquer, paper, and other materials ; matches, matting, umbrellas in European style ; porcelain and earthenware ; provisions and drugs.

It is of interest to us to know that, of the total foreign trade of Japan, forty per cent. is with the British dominions, and of the imports which Japan takes, more than fifty per cent. come from the same.   Com-paring her again with China, we find that Japan is nearly as good a customer of the United Kingdom as is the larger Empire, the respective values of the imports taken from us by the two countries being in the proportion of forty-two to forty-five.

As to the shipping which carries this foreign trade of Japan, in 1894, there entered at Japanese ports, from foreign countries, a total of 2517 vessels of an aggregate of 2,689,781 tons.   Of this carrying trade, British vessels had a very fair share, the number of entries under the British flag being 974 vessels, of an aggregate of 1,614,112 tons, or, sixty per cent. of the total tonnage entered.

Since the opening of the ports, the Japanese mercantile marine, as well as the imperial navy, has undergone great developments.   At an early date, vessels on the European model began to be constructed, as well as purchased from abroad ; and, in order to assist this movement, the Government, in 1887, prohibited the building of junks of over ninety tons' burden, or thereby.   According to figures recently published, Japan now owns 517 merchant steam-vessels of an aggregate of 321,522 tons, of which 102 vessels, of 187,988 tons, are ocean-going steamers.   As many as 87 vessels, of an aggregate of 132,963 tons, are stated to have been added to the mercantile marine between the 30th June 1894 and the 30th June 1895, of which 46 vessels, of an aggregate of 77,515 tons, were purchased from abroad.   Practically the whole of the larger class of Japanese steamers were taken up by the Government to serve as transports during the war ; and their places in the coasting and other trade had thus to be temporarily supplied by European-owned vessels under charter.   By far the greater portion of vessels sold to the Japanese flag, as well as of those chartered, were British vessels. Japanese companies have established steamship lines from Japanese ports to Shanghai, to Korean ports and Vladivostock, and to Bombay; and vessels trade to other ports in the East. The number of Japanese sailing vessels of European form of construction was given, at 31st December 1893, as 749, of 44,967 tons. Large aspirations and projects are entertained by the Japanese for the extension of shipping lines to other ports of the world, and for the fostering of the ship-building industry. At the naval ports and at Nagasaki, fine graving docks exist, with all appliances for the building and repair of vessels; and further accommodation of this nature is in prospect at Kuré, at Yokohama, and at other ports.

Inland transport has been greatly improved by the development of the railway system. There are now considerably over 2000 miles of railway open to traffic, and extensions are being proceeded with under a system sanctioned a year or two ago by the Government. This will increase the mileage to some 3000. The telegraph and postal services extend throughout the country. In 1894 there were 3718 post offices and 716 telegraph offices open, including in each case 590 mixed post and telegraph offices.

An extensive system of banking, under regulations enacted by the Government, has, in recent years, been developed throughout the country. At the end of 1893, one hundred and thirty-five banks were thus in operation, having, besides their head offices, one hundred and sixty-six branch establishments. The Specie Bank has agencies abroad as well as in Japan. The total capital invested in those banks was then $62,916,100; reserve funds $24,275,268; sums deposited during the year $1,210,723,647; sums remaining on deposit at the end of the year $69,843,919; advances made during the year $545,624,365; and sums remaining on loan at the end of the year $127,653,852. The Bank of Japan had then redeemable notes in circulation to the value of $148,663,128; and, including this, all the banks together a circulation of $171,307,174.

The industrial development of Japan has, in the meantime, been proceeding apace. With the opening of the ports an immense stimulus was given to her production of silk. Improved filatures were introduced on the European model; and the silk export is now worth to the country nearly four and a half millions sterling per annum. At 31st December, 1892, the extent of land under mulberry plantations, including an estimate of that along the borders of fields and streams, amounted to 629,510 acres; and the production of raw silk of all sorts, but not including waste, is given by the statistics for the same year, the latest which I have at hand, as equal to 13,500,000 lbs. avoirdupois. About one-half of this product is consumed in the country.

Other produce, directly or indirectly, received a similar stimulus. The extent of land under tea plantations, in the year just named, amounted to 155,938 acres; and the production of tea for the same, to 64,766,000 lbs. There were, in 1893, 6,785,224 acres of land under rice, being an increase as compared with the previous years. The yield in that year was 184,510,328 bushels, giving slightly over 27 bushels per acre, or somewhat less than the average. Under barley, wheat, and "naked barley" (*Hordeum vulgare nudum*, incorrectly termed "rye" in statistics), there were 4,272,825 acres, being also an increase on that in previous years; and the combined yield of these crops amounted to 82,503,851 bushels, or an average of 19·3 bushels per acre, a fairly good yield as compared with that in other years. Under pulse, millet, maize, buckwheat, rape, potatoes, cotton, hemp, tobacco, indigo, and sugar-cane, together, there were 3,897,742 acres of land.

The production of coal for the year 1892-93 was about 3,163,000 tons. The principal coalfields of Japan are those in the north of Kiushiu, and in the island of Yezo; the ports of shipment for the former being Nagasaki, Kuchinotsu, and Moji, the last-named on the south side of the Straits of Shimonoseki; and for the latter, Otaru on the north-west, and Mororan on the south-east coast of Yezo. Japan coal is now very generally employed for steaming purposes all over the Far East. The existence of petroleum in Japan, particularly in the provinces of Echigo, Shinano, and Totomi, has been known of for many years; and, within the last twenty-five, various attempts have been made to develop the working of it. These have met with but indifferent success; but the industry is now receiving fresh attention, especially in Echigo. The latest figures of the total production in Japan which I have at hand are those for 1892. The production of refined petroleum had reached 1986 tons in that year, as against 1924 tons in 1891, and an average of 1169 tons *per annum* in the four previous years. The product is used in the districts where it is produced, and is also shipped for distribution to Osaka, to Yezo, and elsewhere in Japan. The production is but a small part of the oil required in the country, which is generally supplied with oil from the United States and Batoum. It seems still a question whether Japan will ever be able to supply her own wants in this respect; but her production of oil is on the increase in quantity, and the product improving in quality. The production of copper for the same year amounted to nearly 20,500 tons. The principal copper-mines are those of Ashiwo, near Nikko; of Akita, on the north-west coast; and of Iyo, in Shikoku. The antimony mines are also situated in the province of Iyo; and the production must have increased considerably within the last two or three years, seeing that the export in 1894 amounted to 1598 tons, of a value of £26,750. Antimony crystals of great beauty are extracted from those mines. The production of manganese, in the north of Japan, is also increasing. Sulphur is collected on various volcanoes, from Satsuma in the south to Yezo in the north. The present production of iron in Japan is probably only about 20,000 tons per annum. The principal seat of iron-mining is at Kamaïshi, to the north of Sendai. The pig-iron produced there is good for castings, but is too hard for purposes where machining is necessary, unless when it is mixed freely with soft imported iron. Most of the other iron at present made in Japan is obtained from iron sand. The charcoal iron made in Yamaguchi, Choshiu, is of excellent quality, but dear. A project is now on foot to develop iron-mining in the Northern Provinces, and for the establishment of steel works; but it seems as yet doubtful whether suitable and sufficient ores are to be found in the country for any considerable production. The production of gold in Japan is not large; for the year 1892-93 it is given at 21,540 oz. Of silver, the production in the same year was 1,916,549 oz. The principal gold-mines are in the island of Sado; silver, in Akita, Sado, Hida, Ikuno in Tajima, and elsewhere.

State forests cover 17,833,597 acres, the greater portion of which are situated in the island of Yezo. Among the trees growing in that part of Japan may be enumerated three varieties of elm, three of oak, two of birch, three of maple, five of pine and fir, two of poplar, willows, ash, beech, chestnut, larch, yew, juniper, mountain-ash, magnolias, and others. The acreage given above does not include woodland privately owned, which would probably go far towards doubling the above figures for forest land in all Japan. On the other hand, the state forests are mostly wild forest land, while woodland privately owned may generally be taken as cultivated forest. The woods ordinarily used for building are those known as " sugi," " matsu," and " momi,"—cryptomeria, pine, and fir ; the more durable but more expensive " keyaki," and such pine-woods as the "hinoki," are used for finer work, such as temple construction and decorative purposes ; while the former, with cherry-wood, is now used in furniture-making. Camphor-wood is used in the making of cabinets and nicknacks, as well as for the production of camphor; the "kiri" (Paulownia) for clogs ; the many varieties of bamboo for innumerable purposes, such as making fencing, scaffolding, measures, rods, water-pipes, flowervases, chairs, picture-frames, screens, fans, and ornaments. The elms and *Cladrastris amurensis* (Jap. *enjiu*), of Yezo, are used there in the building of railway carriages.

The value of the marine products of Japan, including dried fish, salt-fish, and seaweed, but excluding fresh fish, is given at about eleven million *yen*, or say, £1,200,000 *per annum*; and the value of fishing-boats, apparel, and nets at about twenty-five million *yen*, or say, £2,700,000. The industry affords employment to over two and a half millions of the population. About one-half of the above prepared marine products is exported to China.

But notable developments in other directions have taken place in the country since its opening to foreign trade. With the importation of Western fabrics and intercourse with Western people, changes have come about in the style and material of the clothing of important sections of the population. Not only are the people generally better clad than they were, say, thirty years ago, but their material condition is improved in other ways ; certain classes have changed their mode of dress. The uniforms of the army and navy and of civil functionaries have been remodelled in the European fashion, with a corresponding change of material. A similar change has taken place in the dress of a large proportion of private persons. While the ordinary population continue to clothe themselves mostly in cottons, in the old style of dress, a large part of the material for which is supplied from abroad, many have, in addition, taken to the use of blankets for wraps, and other woollen material.

The use of goods supplied from the West has prompted the desire of the Japanese to make such things themselves. Japan is now, more than ever, becoming a manufacturing country, and that not merely in the species of articles made by her in former times, such, for example, as silk

goods, which have of late shown an expansion. Considerable attention
has been drawn to the fact that she is now becoming, and is likely to
become still more, a competitor in supplying the East with those manufactures which have hitherto been supplied mainly from Europe and by
ourselves. We have seen, in the foregoing review, that she has profited
greatly by her intercourse with other nations in time past. If there is
one characteristic of the Japanese people which strikes one more than
another, it is, perhaps, their possession of the power of adopting and
assimilating foreign arts and industries, and frequently of improving
upon them. It was so in their intercourse with China and Korea in
former times,—witness their art industries in porcelain, in lacquer work,
in metal work; it was beginning to be so in their intercourse with
Europeans in the sixteenth and the early part of the seventeenth centuries.
This characteristic cannot better be studied, perhaps, at the present day,
than in one of the exhibitions of their own industries which the Japanese
are so fond of holding. There is hardly a manufacture of European make
but what finds a counterpart there. From manufactures in cotton and
wool, such as yarns, shirtings, blankets, and army cloth, carpets and rugs
of various fibres, canvas and cordage,—to matches, saddlery, and other
leather work, beer, glassware, patent safes, umbrellas, brushes, boots and
shoes, hats and caps, cotton and silken socks, knitted woollen work, soap,
chemicals, patent medicines, mathematical and surgical instruments,
photographic apparatus, spectacles, clocks and watches, even pianos and
violins—all are there represented. Not merely do the Japanese manufacture, to a large extent, arms for themselves, but arms have been made
in Japan for a British colony. The competition need not be limited
to supplying the East; Japanese porcelain and earthenware, wall-papers,
carpets and rugs are now exported to this country at prices which our
makers find embarrassing. It is true that some of the things above
mentioned are not quite up to the mark in quality; and unless this is
attained and maintained, permanent success in any line of competition is
not to be looked for. The national ambition of the Japanese to manufacture their own requirements has, in this respect, not always resulted in
success. Still, the effort is being made with more and more success.
There is an aptness in the people for industrial pursuits; and were the
wages of the Japanese mill operative, who is paid in free-coined silver,
to remain at a figure such as they were at a year ago—equivalent to,
say, from threepence to sevenpence a day—a strong additional ground
would be thus afforded for his success as against his European rival.
Combination between Japanese manufacturers and shipping companies
doing their carrying trade is also likely to second the efforts of the former
to extend their operations.

This competition has, as yet, been chiefly seen in cotton-spinning.
There are over 600,000 spindles at work in Japan, and their number is
likely soon to be materially increased, and that, partly, in the manufacture
of higher classes of yarns than those hitherto made. Competition in
other branches of manufacture has, however, as we see, commenced, or is
certain to come—in cotton textiles, some of which already, along with
cotton yarns, form an article of export to Shanghai, in blankets

and other woollens, in thread-making, in sugar-refining, and other things.

But there are certain indications that the conditions of this competition with the West are not likely to be so favourable to Japan as might have been anticipated a year ago. Money has become more plentiful, and food and wages are rising in consequence. There was also a certain dearth of labour during the war; and this appears to be becoming accentuated by emigration to Formosa. Should, again, there be any considerable rise in the gold value of silver, the discrepancy between the rates of wages of the operative in the East and in the West would be less striking.

And there is another side to the picture as to the future of our intercourse with Japan. She probably will, in time, largely supply herself with the textile fabrics which she at present receives from us; but, to do so, she requires from us the machinery to do it with. No doubt the Japanese are capable of making machinery too; they are already making much of their electric plant and other machinery. A considerable period must elapse, however, before they are likely to be able to construct that required for their cotton-spinning and weaving purposes; and, for the making of the machinery, iron and steel are wanted. As already noticed, the question of Japan being able to supply her wants in this respect from her own resources is, to say the least, problematical. It is hardly to Japan that we have to look for our most formidable competitors in such matters as these.

Then, again, with the growth of industries the needs of the nation will grow, and the means of satisfying them. The general wealth and capacity for trade will increase. Our trade, as trade always must, will have to modify itself as demands alter and opportunities change. It may be little consolation to the British manufacturer to be told that Japan imports raw cotton from British India, and wool from Australia, and is likely to do so more largely in the future. It is well, however, to face these facts, and to be prepared to adapt ourselves to changed conditions as they come. It is hardly to be supposed that the enterprise which has built up the fabric of British trade which we now see is not capable of adapting itself to new circumstances. While this enterprise exists, and the integrity which has been characteristic of our merchants—while, in a word, we are true to ourselves—we may hope to see new and further developments of intercourse with the interesting and enterprising country which has here claimed our attention. What, more than Japanese competition in the East, we have to dread, is misunderstandings between employers and employed at home, whether resulting from want of sympathy on the one side, or ignorance and mere self-will on the other.

A new epoch—I think we can hardly term it less—is in a few years, under the new treaties recently concluded, about to commence in the intercourse of Japan with foreign nations. In the past we have seen how the Japanese have profited by intercourse with Europeans,—by more than intercourse, by friendly co-operation. Much, no doubt, has been done by Japanese for themselves, and more will yet be done; but much also of what they have achieved is due to concert with Western people. Under the new conditions of intercourse is it not to be hoped that friendly co-operation will still further, and in other ways than at present, contribute to the development of the country and the promotion of her best interests ?

VENEZUELA.

Short Sketch of its History, Geography, and Industries.

On his third voyage Columbus sighted the continent of South America in July 1598, and it was part of the territory subsequently called Venezuela that he visited.   Having named the island of Trinidad, and taken possession of it for the Crown of Spain, he sailed through the Serpent's Mouth, the channel between Trinidad and the delta of the Orinoco into the Gulf of Paria, which he left by the Dragon's Mouth, the other channel between Trinidad and the mainland.   Whether Columbus was himself aware that he had discovered a new continent has been often disputed.   He appears to have named the land round the mouths of the Orinoco " Ysla Sancta," and the coast of the Gulf of Paria " Ysla de Gracia "; but in his own despatch he calls the latter " Tierra," and being struck with the large quantity of fresh water poured into the Gulf of Paria by the mouths of the Orinoco, speculates on the existence somewhere to the south of the Equator of the Earthly Paradise, which he supposes to be raised high above the globe like the thin end of a pear, and from which large quantities of fresh water might flow down and create the swift currents he noticed in these seas.   At any rate the river must, he affirms, proceed from a vast extent of land to the south hitherto unknown.   The natives also told him that the peninsula of Paria was part of the *tierra firme*.   But it is not clear that he considered it a new continent, and not, rather, part of Asia.

In the following year, 1599, Alonzo de Ojeda, Juan de la Cosa, and Amerigo Vespucci followed the coast westwards to the Goajira peninsula, and, penetrating into the Gulf of Macaraibo, found on its shore an Indian village built on piles, which they named Little Venice—Venezuela—and this name was afterwards applied to the whole of the adjoining country.   The exploration of the interior was commenced from Coro, which in the year 1527 was handed over to the Welser of Augsburg by the Emperor Charles the Fifth as guarantee for a loan.   The object of these merchants being to collect as much gold as possible, they sent their agents in all directions, and one of the governors of the town of Coro, Ambrosius Alfinger, reached Pamplona in the present Colombia.   Gradually also the more fertile eastern shores of the country were occupied, and the town of Caracas, called at first Santiago de Leon de Caracas, was founded in 1567 as a basis of operations against the tribe of the Caracas, and this town became the seat of government in 1578, Coro having been taken in 1558 from the Welser, who had abused their power.

The first European to penetrate into the Cordilleras was a certain Francisco Martin, a soldier of Alfinger, in 1531.   The mountains began, however, to be permanently occupied by the Spaniards in 1550, and within the next ten years they founded the towns of Trujillo, Merida, and San Cristobal. By the end of the century the conquerors had extended their authority eastwards over the whole of the country north of the Orinoco.

But, before this was accomplished, the river had been navigated by adventurers in quest of the fabulous El Dorado, which was the goal of so many unfortunate expeditions from Coro, New Granada, and Peru. In 1595 Sir Walter Ralegh sailed from England and burned the town of S. Joseph in Trinidad to revenge the murder by the Spaniards of some English sailors, taking prisoner the governor, Antonio de Berreo. Now this Berreo, maintaining that he had a claim on the kingdom of El Dorado, as heir of his uncle Jimenez de Quesada, had started from Tunja to make good his claim, and descended by the Meta into the Orinoco, and, following its course to the sea, had sailed to Trinidad, which he held to be part of his kingdom. Thence he sent Domingo de lbargoien y Vera to Spain, with objects in gold obtained from the Orinoco, to secure for him the possession of the country. Vera returning with five vessels entered the Orinoco, and solemnly took possession of Guiana and El Dorado in the name of Berreo. From Berreo Ralegh heard the fabulous accounts of the city of Manoa, and saw a copy of the narrative of Juan Martinez, who was supposed to have visited the city and given it the name of El Dorado. Dazzled by these descriptions, Ralegh entered the Orinoco, and ascended it as far as the mouth of the Caroni, and on his return to England handed to Lord Charles Howard and Sir Robert Cecil a full account of his discoveries and of the information he had collected from other travellers. In this narrative he speaks of a map which he was drawing. This map was long supposed to have been lost, until in 1878 J. G. Kohl found a manuscript map in the British Museum which was, doubtless, Ralegh's map, or an accurate copy of the original. It was reproduced by Herr L. Friederichsen in the *Festchrift der Hamburgischen Amerika-Feier*, 1892, and is very curious as showing the geography of the northern part of South America according to the ideas of the time. Between the Orinoco and the Amazons is seen an immense lake, with the city of Manoa at its western extremity.

Leaving these early explorations, which may be studied by English readers in such books as Hakluyt's *Voyages*, Payne's *History of the New World called America*, or in *The West Indies and the Spanish Main*, by James Rodway, the recently issued volume of the "Story of the Nations" series, we must pass on to the later history of the colony. In the seventeenth century Venezuela seems to have made steady progress; agriculture and cattle-grazing were rapidly extended, and the future prosperity of the country seemed to be secured. But in the following century its development, as well as that of the other Spanish colonies, received a serious check by the short-sighted policy of the home government. The colonies were compelled to trade with Spain alone, Seville, and subsequently Cadiz, enjoying the monopoly, and were not even allowed, except in a few exceptional cases, to despatch ships of their own. Consequently commerce declined, prices rose enormously, and the smuggling trade with the Dutch and English flourished. Another grievance was that the most remunerative posts in the administration were given only to born Spaniards. This raised enmity between the Creoles and the Spanish officials, who, on their part, treated the Creoles with gross injustice, and, aided by the ecclesiastics, oppressed them in every way. By the middle of the century the exasperation caused by these iniquitous proceedings had reached such a pitch that the country was ripe for revolution, and in 1749 Juan Francisco de Leon headed a rising in Caracas, the immediate cause of discontent being the proceedings of the Guipuzcoan trading company, which in 1728 had acquired the tobacco monopoly, and had gradually obtained control of the whole trade of the country. At first the insurgents were successful, and, a new governor having been sent from San Domingo, the affairs of Venezuela were for a time conducted satisfactorily, but in 1751 the Brigadier Ricardos supported the Guipuzcoan company and defeated Leon, who was shipped to Spain and died on the voyage.

The selfish policy of the Spanish Government with regard to its colonies was maintained until 1778, when the monopoly of Cadiz was abolished, and the trade thrown open to all Spaniards. But it was too late ; insurrections broke out repeatedly, stimulated by the ideas of freedom spread abroad on the outbreak of the revolution in France. At first, under Sebastian Miranda, and afterwards under Bolivar, the Venezuelans rose in revolt in the year 1797, and carried on the war with varying success until independence was secured by the final battle of Carabobo in 1821. In 1819 Venezuela was joined with Colombia and Ecuador, and Bolivar was elected President of the new republic, to which office he was re-elected in 1821. Then his popularity began to wane, for he was obliged to resort to strong measures to repress the ambitious men who thought to take advantage of the weakness of the new State. He was accused of monarchical aims, and was compelled to resign in 1829. The next year he died, and the Republic was divided into the existing States of Venezuela, New Granada (Colombia), and Ecuador.

From 1831 to 1847 Venezuela enjoyed a period of rest, but in the latter year new disturbances arose, and the country was involved almost continuously in civil strife until General Antonio Guzman Blanco restored order in 1870. To him are due great improvements in agriculture and commerce, the construction of roads and railways, and the establishment of schools and colleges. Under his influence peace was preserved until 1892, when Dr. R. A. Palacio refused to resign the office of President at the end of the term of two years, in accordance with the laws of the constitution, while a party in the Congress refused to prolong the term. General Crespo led the opposition, and brought the war to an end in seven months. For about a year he governed the State with the title of " Chief of the Executive," and in February 1894 was elected President for a term of four years.

Among the most noted travellers in Venezuela in later times Alex. v. Humboldt is the first that must be mentioned. It was he who first reported a connection between the Orinoco and the Amazons through the Casiquare. In 1886 and 1887 M. Chaffanjon made a journey to the sources of the river, and other travellers, English, French, and German, have described various parts of the country. But perhaps the most complete geographical description has been given by Dr. W. Sievers in a book entitled *Venezuela*, in *Die Cordillere von Merida* in Dr. Penck's *Geogr. Abhandlungen*, published in 1888, and in his *Zweite Reise in Venezuela in den Jahren* 1892-93, forming the twelfth volume of the *Mitt. der Geogr. Gesell. in Hamburg*, which has been recently issued.

Venezuela lies between 1° and 12° 26′ N. lat. and 60° and 73½° W. long., and contains an area of somewhere about 500,000 sq. miles, inhabited by only about two million persons. Its coastline, from the Columbian frontier to that of British Guiana, has an extension of 2000 miles. In the American continent there are three forms of surface configuration—lofty mountains, the Andes or Cordilleras on the west, on the east mountains of archæan rocks, chiefly granite, gneiss and slates, covered with sandstones of later dates, and extensive plains, called Llanos or Pampas. Venezuela, and Venezuela alone, contains large tracts of all three forms. The Llanos separate the Cordilleras from the older mountains, and extend along the Orinoco. To the south of the river the country is occupied by the older elevations of Guiana : to the north of the Llanos it is a lofty mountainous land. In the northern Venezuelan half of Guiana the comparatively low elevations take the form of series of small and large ranges of hills, without continuity or regularity of arrangement, among which numerous large streams wind along, falling in cataracts and rapids. On the underlying granite sometimes lie blocks and bosses of white and red sandstone falling steeply on all sides to the lower flanks of the hills. The most noted of these summits is Roraima on the British frontier. Northwards these heights diminish in elevation, becoming small hills on the southern bank of the Orinoco, which descends from the mountains by numerous fine cataracts, and, after receiving the large tributaries Ynirida, Guaviare, and Meta, bends at the confluence of the Apure towards the east, following, for the greater part of its course, the skirts of the Guiana highlands and entering the Atlantic by a broad delta. Few travellers, except Humboldt and Chaffanjon, have visited this southern part of Venezuela, and some of its rivers are very imperfectly known. The latter traced it upwards to where it was no more than a small torrent descending from the Sierra Parima, which has an elevation varying from 4000 to 4500 feet. During the dry season the water is low, and the bed of the stream is studded with banks of sand which render navigation difficult, if not impossible, while in the rainy season it fills its bed, 900 to 2700 yards broad, and inundates the surrounding country, in some parts to a distance of ten miles from its banks. In the dry season the current is sluggish at Bolivar, the only large town on the river, standing only 20 to 25 feet above the level of the Atlantic, which is 300 miles distant, whereas the large volumes of water carried down in the wet season lend extraordinary violence to the stream. The high tides ascend the river as far as to the island of Tortola, about 190 miles from the mouth. Since 1891 the Orinoco forms the boundary of Colombia and Venezuela along the part of its course between the tributaries Guaviare and Meta.

The elevated zone lying between the Llanos and the Caribbean Sea is the most populated part of the country, contains almost all the large towns, and the intelligence and industrial energy of the country. The Andes, entering Venezuela on the west, extend as far as Barquisimeto, where a deep hollow, through which flow the rivers Cojedes and Yaracui, separates them from the system denoted by Dr. Sievers under the name of the Caribbean mountains. The two systems differ both in outward configuration and in structure. The Caribbean mountains consist of two more or less parallel chains skirting the coast on the one hand and the Llanos on the other, and linked together here and there by transverse ridges. In the valleys thus enclosed formerly stood lakes, but these are nearly all dried up, and the only lake of any importance in the country is the Tacarigua or Lake of Valencia. This arrangement can be traced as far east as Trinidad, and is also perceptible in the island itself. The ranges are also broken in the transverse direction at the Bay of Barcelona and at the Gulf of Paria. The islands of Tortuga, Margarita, and perhaps Tobago, probably once constituted a northern chain of the same system, and the remains of still another line of upheaval may possibly be seen in the outer series of islands, Curaçao, Orchilla, Los Hermanos, etc.

The Andes in the west, from the mouth of the Yaracui to the frontier of the Republic, are known as the Cordillera or Sierra de Merida. Running from south-west to north-east, they rise above the snow-line, attaining, between the Llanos and the Gulf of Maracaibo, an altitude of 15,000 feet. They are characterised by groups of ridges radiating frequently from a definite point and afterwards converging again. Two main lines of elevation start from Pamplona in Colombia, of which the one runs up the peninsula of Goajira, now surrendered to Colombia, while the other is the Cordillera of Merida. The latter again breaks up into a series of ranges at Macuchies, and its north-eastern extension divides again at the Paramos of Jabon and Rosas, sending out the lower hills that occupy the districts of Coro and Barquisimeto. West of these hills lies the Gulf of Venezuela, with its entrance into the Gulf of Maracaibo blocked by a very inconvenient bar. Swamps and rank vegetation render the low-lying land on its shores almost uninhabitable.

A noteworthy point is the almost total absence of volcanic rocks in Venezuela. Actual volcanoes do not exist in the Cordillera of Merida, in the Caribbean mountains, or in Guiana. Comparatively recently, however, indications of volcanic activity having been at work in past ages, possibly as late as the Cretaceous period, have been noticed. Thus, in Guiana gold occurs in association with diabase and diorite, and south of the Caribbean mountains, near San Juan de los Morros, similar eruptive rocks have been discovered; the latter is, however, the only place north of the Orinoco where they are known to occur. The core of the mountains is formed of archæan rocks—gneiss, granite, mica-schists, chlorite-, talc-, and hornblende-schists. Above these are found in all parts of the mountain country of Venezuela deep deposits of white, red, and yellow sandstones, greyish-blue to dark bituminous limestones and marl, and clay which may be assigned to the chalk series. The Tertiary series is hard to determine, no distinctive fossils having been discovered. It probably forms the Cerro de Oro, a mountain 4300 feet high, south of San Cristobal, and skirts the flanks of the Cordillera. It is of rare occurrence in the Caribbean mountains, and whether it is present in Guiana is very doubtful. The Llanos were formed by Quaternary deposits laid down probably on Tertiary formations by a sea that then washed the foot of the mountains. The fossil remains of huge animals are found in this diluvial formation—the megatherium, glyptodon, toxodon, and varieties of the horse.

In the Tertiary period the sea covered the area of the present Llanos, and rose along the skirts of the Cordillera perhaps to a height of 700 to 1000 feet above the present sea-level. But the slope of the Llanos is not uniform in degree or in direction. It does not follow the direction of the Apure and Orinoco towards the Atlantic, but the lowest part of these plains is found along the line of the rivers Cojedes and Portuguesa, nearly along the boundary of the archæan rocks of the east and the sedimentary strata of the west, though the latter cross this line, extending to the Rio Camoruco. East of the Cojedes-Portuguesa line the Llanos stretch eastwards and north-eastwards, and reach the sea through the gap of the Caribbean mountains at Barcelona. Where the Orinoco makes its bend to the east, mesas—flat-topped elevations about 1000 feet in height—begin to appear and run north-north-eastwards, forming the watershed between the Orinoco and the coast of Barcelona and of the province of Cumana. Between Maturin and Soledad, at any rate, their composition is quite uniform. In the north they are of fine-grained reddish-brown sandstone, coloured and knit together by swamp iron ore, and in the south they contain small or large grains of quartz. The great abundance of iron points either to the formation of swamp ironstone in the stagnant waters of the plains, or to the denudation of the mountains of Guiana which are rich in iron. In the latter case it may be that the old formations of these mountains form the basis of the Llanos of Maturin, as they certainly do in the neighbourhood of Soledad. The western Llanos are well watered, and in the rainy season are flooded over large areas, but between the Rio Chive and Soledad lies a sandy desert, where the rainfall is small and the water is sucked up by the sand. According to Humboldt this desert reaches to the neighbourhood of Pao, and the *Apuntos Estadisticos* of Barcelona mention its occurrence in that district. Opposite Soledad huge coal-black rocks rise in the bed of the river, dividing it into two arms and forming the narrows, or Angostura, of the Orinoco; and along the course of the river, as far as the mouth of the Caroni, the dark rocks of the Guiana system appear on the north, as well as the south, side, rising in roundish, lumpy hills, mostly of granite, whilst the more level country is formed of gneiss and crystalline rocks. It is evident, then, that here the river cuts its way through the Guiana system, which is continued northwards underneath the Llanos.

The Caroni, 770 yards broad at its mouth opposite the long low island Fajardo, is noted for the beautiful falls, eight or nine in number, situated about two and a half hours' journey up stream.

Few meteorological records have been kept in Venezuela, and therefore the climate can be described only in general outlines. As in other countries of South and Central America, three zones are recognised—the *Tierra caliente*, extending from sea-level up a height of about 1800 feet; the *Tierra templada*, reaching up to about 7200 feet; and above this the *Tierra fria*. These zones have mean annual temperatures of 86° to 77° F., 77° to 60° F., and below 60° F. respectively. On the north coast the heat is excessive, owing to the trade-wind which blows across the hot Caribbean Sea. Maracaibo is said to be the hottest place on the coast, though La Guayra, situated on an exposed mountain flank and unsheltered by forest, is known as El Infierno de Venezuela; the mean yearly temperature is 85° F., and the minimum over 77°. But what makes the heat more felt here is the small difference between day and night temperatures, amounting even in the cooler months to only 5° or 6°. Caracas, lying at an elevation of 3000 feet above sea-level, and therefore in the *Tierra templada*, enjoys a cooler and more agreeable climate. Its annual mean temperature is 71° F. In the hottest months the thermometer ranges from 68° to 82°, and in the coolest from 71° to 52°.

In the Cordillera the changes of temperature are naturally very great, as the traveller may in a few hours ascend from the sheltered valleys to the bleak, wind-swept *paramos*, the high plains above the forest limit. It may be assumed that in general the temperature in the Cordillera diminishes at the rate of 1° F. for every 365 feet. But local conditions vary the changes, which are greater in the midst of the mountains than on their outer slopes. In the lower parts, though the heat is very great, it is not so extreme as in many other tropical countries, and is not so high on the northern slope as on the southern, in the Llanos, and in Central Venezuela, for on the north humidity moderates the temperature and the luxuriant vegetation wards off the burning rays of the sun. The highest readings of the thermometer have been observed in the Llanos: in Acarigua, south of the Portuguesa range, Dr. Sievers observed a temperature of 125½° in the sun and 89½° in the shade, in October 1885, at 2 P.M., while Sachs at Calabozo found the mean of the readings at the same hour during February 1877 to be 96°. As a rule the bare and dry tracts are warmer than the woodlands.

The rainfall is given by the Graf zu Erbach in *Wandertage eines Deutschen Touristen im Strom- und Küstengebiet des Orinoko*, after Villavicencio, as 70 in. at Maracaibo; 68½ in. at the lake of Valencia; 65 in. on the Caribbean coast; 63 in. on the Gulf of Paria; about 60 in. along the Orinoco; and 31½ in. at Caracas. In the Tachira valley and in Western Merida there is a lesser rainy season in spring, and a greater rainy season beginning at the end of July, while a short dry period intervenes, called the "little summer of St. John," when the sun is near the Tropic of Cancer, but in the highlands farther east, lying also in a more northern latitude, the short dry period disappears and the rainy seasons merge into one.

On the whole, the climate is fairly healthy, considering that the country lies near the Equator. Yellow fever frequently rages on the coast, especially where mangroves grow, and in the Llanos and the great forests of the lowlands, where decaying vegetable matter sends forth malarious exhalations. Occasionally, also, it visits the towns of the *Tierra templada*, but the higher mountains are quite free from it, and possess an extremely healthy climate. Here Europeans could live for a long time without any detriment to their health.

Closely connected with the climate is the distribution of vegetation. The coco-palms and cocoa are confined to the *Tierra caliente*; the banana, yucca, and sugarcane extend to the upper limit of the *Tierra templada*; and the *Tierra fria* is the land of wheat, oats, potatoes, beans, and peas. Cactuses, as well as mimosas, agaves, euphorbias, and thorny plants of all kinds occupy all the lower parts of Venezuela that are not well watered. The palm region extends upwards to about 3300 feet, and is succeeded by the fern forest, extending upwards to 6000 feet, and above this, up to 7800 or 8200 feet, is the cinchona forest, named from the frequency of different varieties of this tree. The upper limit of this forest varies, but may be placed in general at an altitude of 10,000 feet. Still higher, bushes and herbs are met with, and then come immense pastures where cattle are grazed, and then the region of the Frailejon (various kinds of *Espeletia* of the order of the *Compositæ*), extending up to the snow-line. Naturally these zones are not present in all parts of the Republic; on the Caribbean mountains the upper ones are absent owing to the lower elevation, while in the Cordillera the palm region does not exist, as the valleys have an altitude of 2600 to 3200 feet. But not only in a vertical direction does the vegetation change. It is also affected by the physical features of the country. Tropical forests and forests of hardier trees are found in the higher mountains, grass land predominates in the Llanos, and in the higher parts of the Cordillera meadow-lands are of most importance. And these surface features are by no means clearly marked out from one another; small savannahs skirt the streams in the Caribbean mountains, and unfruitful tracts occupy no small areas both in Central Venezuela and the Cordillera. The western Llanos also are better wooded than the eastern, where trees grow only in clumps, and agaves and cactus make their appearance in the drier parts.

A new division of the territory of the Republic was made in 1882, and there are now eight States, seven Territories, two Colonies, and the Federal District of Caracas. The States are Miranda (Guzman Blanco) Carabobo and Lara in the centre, Los Andes and Falcon in the west, Bermudez in the east, Zamora in the Llanos, and Bolivar in Guiana. The territories lie on the Orinoco and its tributaries and delta, and on the western boundary towards Colombia, while one of them includes the islands in the Caribbean Sea. The colonies are called Bolivar and Guzman Blanco; the former is near Guatire, not far from Caracas, and the other seventy-five miles south of Caracas in the Serrania del Interior, at an elevation of 5900 feet. These colonies are especially intended for immigrants.

The population of Venezuela at the census of 1891 was 2,323,527, and 2,075,245 in 1881. The increase is due to the excess of births over deaths, as emigration and immigration nearly balance one another. From the number of inhabitants must now be deducted the population of the peninsula of Guajira, some 60,000, and some 20,000 in Alto Orinoco and Amazonas, half of these territories and Goajira having been ceded to Colombia. For the same reason the area, 593,943 square miles, must be reduced by nearly 100,000 square miles. During the decade there was a considerable addition to the population of the Federal District, while in the States of Miranda, Carabobo, and Bolivar the numbers decreased. About half the population is concentrated in the States of Carabobo, Miranda, Bermudez, and the Federal District—that is, those east of the Yaracui, the combined area of which is only some 69,000 square miles, or about two-fifteenths of the total area of the Republic. The Llanos, the low lands, the sterile Coro, and the highlands of Merida are the most thinly populated, except the delta and the territories south of the Orinoco, which, occupying about half the area of the Republic, have less than 100,000 inhabitants.

The Indians form the groundwork of the population, and to these are added Whites and Negroes. These elements are mixed in endless varieties of proportion. The pure white race is in a small minority, forming only about one per cent. of the population, and pure Negroes have become rare. Of Indians few of pure breed are found in the central States, but there is still a large number in the wilds of the Orinoco and of Guiana, in the Cordillera and the eastern States. The half-castes, constituting the mass of the civilised inhabitants of Venezuela, consist of *mestizos*, half White and half Indian, mulattoes, and *zambos*, half Negro and half Indian. The last are rare, for the Indian has a greater repugnance to the Negro than the white man. The different elements have undergone further crossings, producing all shades of colour, though brown predominates owing to the larger proportion of Indian blood. The shade depends to a great extent on the situation, being darker in the low hot tracts, where the Negro finds a congenial climate, and lighter on the heights of the Cordillera. Here Indian blood prevails, and the inhabitants preserve the reserve and quiet proud bearing of their Indian ancestors. Little is known of the original inhabitants of Venezuela. They were, it would seem, at the time of the Spanish conquest, in a poor and primitive state. There are nowhere remains of large buildings, towns, or even of burial-grounds. The Indians were probably divided into a number of small tribes, without any close connection, or consciousness of a common origin. In the east and in the present central States lived Indians of Caribbean race, probably successors of the Arowaks, who are supposed to have previously lived on the coast of the Caribbean Sea. Something may, perhaps, be learned from the native place-names. At any rate, it is noteworthy that in the Cordillera the names are similar to those in Upper Colombia, most of them accented on the last syllable, while in the east and centre they are generally accented on the penultimate or antepenultimate. Again, the syllable *gua*, so common all over the country from Barquisimeto and the eastern shore of the Gulf of Maracaibo, does not occur in the Cordillera or on the west of the Gulf. At the present day, pure-blooded Indians live in small villages and towns in the Cordillera, and in the Llanos also there are villages with almost pure Indian inhabitants. This element tends more and more to prevail over the others, and on it depends the future progress of the country, for the Indian, however stupid, lazy, and indifferent he may be, has qualities that the Negro does not possess, while the few Whites that yet remain will soon disappear among the more numerous native tribes.

Agriculture is pursued in the valleys of the Cordillera and the Caribbean mountains, chiefly in the zone between 1000 and 6000 feet, for here grow the chief products of the country—coffee, sugarcane, maize, bananas, and cocoa. Some products rise or decline according to the state of the market, such as coffee and cocoa. Others have almost disappeared; indigo is now hardly cultivated at all. Cotton, at the time of the conquest the most important crop, has lately been planted again in some districts. The coco-palm is comparatively rare, but it grows on the coast in sufficient numbers to make the export of *copra* profitable. But this source of profit, as well as the export of bananas, is neglected, and rice is grown in such small quantities that ninety-nine per cent. of the large quantity consumed is imported from the United States and Europe. Maize, the only cereal the Indians possessed before the arrival of the Spaniards, is grown extensively, and in the *Tierra caliente* yields four crops a year; and yet up to the beginning of the eighties it was imported from the United States. The vine is not cultivated, though there are suitable sites in the mountains, and wild grapes are sometimes seen in the Cordillera. The cultivation of wheat has made the greatest progress, replacing maize in the higher regions. Its lower limit is in general the upper limit of the *Tierra caliente*, though it is grown even in the upper tracts of the hot region, where it comes to perfection in three months. Oats are used for fodder for horses and mules in the higher lands; in the low country, maize, sugarcane, and hay are used for the same purpose. The potato, with pease, maize, porridge, and wheaten bread, forms the staple food of the people in the upper parts of the Cordillera. It is grown in large quantities, and there are several kinds. Nor is it confined to high latitudes, but is found as low as the lake of Valencia, 1600 feet above sea-level. The yucca, celery, bananas, and yams share its importance as articles of food, and all kinds of vegetables are abundant. Tobacco can grow both in the *Tierra templada* and the *Tierra caliente*, but thrives better in the latter, where the soil is richer and moister. At one time the cultivation of tobacco was in a very flourishing condition in the Llanos of Zamora, but was thoroughly destroyed by the revolutions. Moreover, the people do not know how to make good cigars, and hence on the coast Havana cigars are smoked.

As already stated, the chief vegetable products of the country are the sugarcane, maize, bananas, coffee, and cocoa. There are plantations of sugarcane in all the valleys up to an altitude of 6500 feet. It thrives best where there is plenty of moisture in the air, water, and warmth, and therefore generally on the banks of streams in the warmer valleys. In the eastern part of Venezuela there are large factories for the extraction of the juice and the manufacture of rum, which is the usual drink of the

people, and the cause of the drunkenness so common in the small towns.
A brown, unrefined sugar is also manufactured, and is consumed in immense quantities.   A large number of people live almost exclusively on
it, adding at most a little manioc, banana, or cheese.   Maize and bananas
are important articles of home consumption.   The former grows up to
8000 feet, and the latter, in sheltered places, as high as 6500.   Cassava
bread, made from the yucca, is also a favourite food among the
Venezuelans.

The two most important export products are cocoa and coffee.   The
former grows up to a height of 1600 feet, the border of the *Tierra templada*.
It demands a small amount of labour, though it requires to be kept free
from insects, particularly ants, and regular watering, and the fruit is not
ready till seven or eight years after the plants are put in the ground, but
the profits generally compensate for the long waiting.   Of still greater
importance in Venezuela is coffee, the cultivation of which is the chief
industry of the country, for with the price of coffee rise and fall those of
all other articles of consumption, as well as the rate of wages, rent, etc.
Venezuelan coffee is not of such good quality as that of Java or Ceylon,
while the cocoa is the best in the world, but still there are very fine
kinds.   The tree grows as high as 6000 feet; it yields a fair crop in the
third year, and a plantation sometimes lasts for fifty years.   Coffee is
largely drunk in Venezuela, whereas, singularly enough, cocoa is very
little used.

Up to 1886, when the export of gold commenced, copper was the
only mineral exported in any quantity.   The best-known mines are those
of Aroa, between Tucacas and Barquisimeto.   They yielded largely for a
considerable time, but of late have been less productive.   Iron appears to
be abundant in the Cordillera near Trujillo, but is not worked.   The coal
of Barcelona gives great promise, as it lies so near the coast.   Probably
it is of the Cretaceous or Tertiary period, as the Carboniferous formation
does not occur in Venezuela.

But of course the great mineral wealth of Venezuela lies in the gold
fields of the Yuruari, the possession of which is disputed with Great Britain.
Gold was first discovered in this country by a Brazilian in the year 1842,
but no great progress was made until the discovery of the gold fields of
Caratal, by Friedrich Sommer, in 1856.   Soon a number of mines were
started, of which El Callao was the richest, yielding most of the gold
which came from Venezuelan Guiana.   In the last few years the output
has fallen off, and in 1893 El Callao was considered to be worked out.
There is, however, plenty of gold, no doubt, in the country, and improved
means of transport, and better methods of extracting the ore would probably render it profitable to work mines that have now been abandoned.

The trade of Venezuela has made great progress, though latterly it has
suffered severely from the effects of the revolution.   The chief exports are
coffee, cocoa, gold, hides, and copper, and the imports rice and manufactured articles, for the manufacturing industry of the country is very insignificant.   The chief ports are La Guayra, which receives a fourth of
the imports, Puerto Cabello, Maracaibo, and Ciudad Bolivar.   General
Guzman Blanco did much to improve the means of communication.   Up

to 1883 there was only one narrow-gauge line in the country, which ran from the mines of Aroa to the harbour of Tucacas. In that year a railway, constructed by English engineers, was opened, connecting La Guayra with the capital, Caracas. Though only twenty-four miles long, it is a great engineering feat, for it is carried over the coast range, attaining a height of 3000 feet. A line a little over four miles long connects La Guayra with the watering-place Macuto, and another thirty-four miles long runs between Puerto Cabello and Valencia. The latter is also in communication with Caracas by means of a railway 100 miles in length. Other lines bring the total mileage opened up to 186, while others are under construction or projected, and roads have been made and telegraph lines laid down over considerable distances.

The country has, therefore, made a good start on the road to prosperity, and, if it can avoid revolutions, bids fair, with its fruitful soil and varieties of climate adapted for various crops, to become one of the most flourishing of South American States.

THE SHOTTS OF ALGERIA AND TUNIS.[1]

THE question of flooding the Sahara to the south of Algeria and Tunis has long been discussed, especially by M. P. Vuillot, in a work entitled *L'Exploration du Sahara*; but the plan has had to be abandoned, and the reason is explained in a recent work of the same author.

The series of shotts in the northern Sahara may be divided into three chief groups, according to M. Vuillot. In the west lies Shott Melrir with its various branches and divisions, as Shotts Meruan, Raduan, Bejelud, El Hajel, Bu Shekua, Mujat Tajer, Hajele, Tuijin, Sellem, etc. Next comes Shott El Rharsa or Gharsa; and, thirdly, we have Shott El Jerid, with its eastern expansion called El Fejej, separated from the Gulf of Gabes only by a sand-covered stratum of rock at most 10 miles broad.

Between Shott El Jerid and Shott El Rharsa runs a chain of low hills starting off from Jebel bu Hellal in the south-west, while between Rharsa and Melrir there is a row of sand-dunes intermingled with low, swampy patches.

Several explorers have visited this country. In 1845 Violet d'Aoust found by barometrical observations that Shott Melrir was very little above sea-level, if, indeed, it were not below it. At once it was suggested that this depression was the site of the Lake Triton of Herodotus, which Jason was supposed to have navigated.

M. Vuillot refers briefly to the old geographical authors who have given information on this question—Scylax, Pomponius Mela, Ptolemy, Pliny, and Edrisi who lived in the twelfth century. Several traditions also are cited which tend to show that Shott Jerid was once filled by the sea, and

---

[1] From a paper by Paul Staudinger in the *Geographische Zeitschrift*, Jahrg. i. Heft. 12.

that Nefta on its western shore, where the remains of an ancient galley have been found, was once a harbour.

Ptolemy, however, does not mention a connection between the Triton lake and the sea, but speaks of a river flowing into it and of another lake, lying at some distance in the interior and called the Libyan or Turtle lake ; and other ancient writers say much the same. In modern times M. Tissot has written a learned treatise on the Triton lake. Another Frenchman also, named Pomel, combated the theory that the sea had once extended into the interior, relying more on actual facts than on tradition. The sea, he says, has never in the present geological period penetrated into the interior. A belt of shallow salt-pans, with a lower level than that of the sea, on the shores of which shells of extinct or partly extinct organisms have been found, has given rise to the mistake ; but all the molluscs which once inhabited these shells belonged to the fresh-water fauna. Duboscq and the Italian expedition Antinori-Fuchs came to the same conclusion.

Lastly, in 1884, Rouire, in a report to the Academy of Sciences, denied the connection of the shotts with the Triton lake on topo-graphical grounds. He believes that the Triton river of the ancients was the watercourse which is named on our maps Wed Bagla and, below Kairwan, Wed Zerud. The lake he holds to be one of the three to the north-west of Susa which the river passes through, the Bagla or the Kelbia or the *sebka* of El Mengel. (Lately he has fixed on El Mengel, having found an artificial connection with the sea. See vol. xi. p. 35.) This opinion he has defended by arguments drawn from Herodotus, Scylax, Pliny, and Ptolemy, etc. According to the last author the Triton river rises in the mountain Οὐσάλετον ; and the western branch of the Wed Bagla, called Wed Marguellil, has its source behind the Usalet mountain of to-day. From this statement, supported by other arguments, Dr. Rouire concludes that the Bagla lake was the Libyan lake of Ptolemy, and the Kelbia and Pallas lakes and the *sebka* of Mengel the Triton lake.

Yet another explorer appeared on the scene. M. Roudaire, charged by the Ministry of War with the triangulation of the meridian of Biskra, made a journey in the shott region, and for the first time established the fact that the bed of the Shott Melrir lies below the level of the sea. The result led him to the idea that it would be easy to connect the Algerian and Tunisian shotts by a canal with the Gulf of Gabes, and thus form an inland lake. Connecting the isolation of the Triton lake and the withdrawal of the sea from the Tunisian shotts with the increas-ing drought of the country and its present sterility, compared to its luxuriant vegetation in Roman times, he cherished the hope that the formation of a lake would restore its ancient fruitfulness. He thought also of the great advantages such a lake would present in giving easier access into the African continent and improved conditions for defence. His proposal was examined by the Ministry of Public Instruction, and in 1874-5 the Ministry of War gave him a new commission. The levelling operations he then executed on the Tunisian frontier convinced him of the necessity for further investigation ; and in 1877, in conse-quence of the report of a commission of inquiry appointed by the

Academy of Sciences, Roudaire was sent for the third time to the Tunisian shotts.

Numerous soundings and levels were taken, and it turned out that the bed of the shotts slopes down from east to west, and not towards the sea, and that the Shott El Jerid, which must necessarily lie in the lowest depression, if a connection with the Mediterranean be practicable, is considerably above sea-level. These facts, of course, entirely upset Roudaire's plans. Instead of simply a cutting through the ridge of Gabes, a canal in the bed of the Shott Jerid to conduct the water to the isthmus of Kriz was found to be necessary. This also must be cut through to admit the water into the Shott El Gharsa, and again another canal must be constructed through the Asluj ridge to fill Shott Melrir. Other technical and local difficulties cropped up; for instance, it was feared that the excessive evaporation of the sea-water would have an unfavourable effect on the healthiness of the country. Lastly, the cost was estimated by Roudaire at three millions sterling, a sum which would construct many a mile of the Sahara railway.

M. Vuillot's starting-point was Biskra, an oasis with 150,000 palm-trees. As far as Tozer the journey is more agreeable than along the latter half of the way to Shott Gharsa, as numerous towns and villages are passed through. Many persons imagine the Sahara to be a vast desolate sea of sand, through which the traveller wades with great difficulty. In reality, however, these sandy tracts are of small extent compared to the area of the Sahara, while hill and valley, stony plateaus, dry river-beds, and districts where a certain amount of vegetation is produced by the scanty rains, give some variety to the scenery. The oases are more numerous than is usually supposed, and many are still unknown to Europeans. M. Vuillot's marches were in general only about 12 or 13 miles a day, but during the first half of his journey he frequently came across villages, towns, and groups of habitations, even in that distance. The oasis of El Hamma reveals the wonderful fruitfulness of the Belad el Jerid, or Land of Dates; but even this is surpassed by the sea of verdure which encircles Tozer. Not date-palms alone, but also other fruit-trees and kitchen gardens are seen in an oasis; barley and wheat are often cultivated, and sheep are almost always to be found, so that a large oasis can not only support its own population, but can also export food. The following figures indicate the prosperity of Tozer.

Its area, including the sites of the villages, is 4416 acres, but the oasis alone covers only 2343 acres. The population numbers 10,000 souls. The official return for the purposes of taxation gives the number of palm-trees as 218,000. But this estimate is said to be too small, as also that of the produce, which is actually 21,600 tons of dates. Woollen fabrics are woven in the houses, and good *burnous* and *haiks* are made; the salt-pans are also worked.

The foundation of the town dates back to the dawn of history. The sound of the name suggests Tes-Hor (town of the sun), or Tuser (The Strong), the name of an Egyptian queen, daughter of Menephta, and points to an Egyptian origin. It appears as Τίσουρος in the table of

Ptolemy, and as Thuruso in that of Peutinger, and several Arabian savants visited the town. In 1307 the sheikh El Tijani writes of the numerous ancient monuments in the Jerid. Remains of extensive ruins, probably of an old temple, subsequently converted into a Christian basilica, are still to be seen, but most of the buildings have entirely disappeared. Shaw visited the place in 1730, and also found ruins. Since then several Europeans have been there, chiefly Frenchmen. In 1882 the town was occupied by French troops; in 1884 the administration of the Jerid was organised, and in 1887 Tozer was chosen as the seat of government. The Jerid no longer forms part of the province of Constantine, but belongs to South Tunisia.

Tozer was the most interesting point of Vuillot's route. The only other place worthy of mention is Nefta, with its many mosques, graves of saints, and fanaticism. Many points were found to lie above sea-level in contradiction to former surveys.

PROCEEDINGS OF THE ROYAL SCOTTISH GEOGRAPHICAL SOCIETY.

LECTURES DELIVERED IN MARCH.

Sir David Chalmers delivered in Glasgow, on 2nd March, the address on "British Guiana," which he had previously given in Edinburgh, Dundee, and Aberdeen. Sir Renny Watson occupied the Chair.

On March 5th, Mr. G. Seymour Fort read before the Society in Edinburgh a paper entitled, "South Africa of To-day, in relation to its Industrial Resources." The Chair was taken by Colonel Cadell, V.C. Mr. Fort read his paper the following evening at a meeting of the Aberdeen Branch, presided over by Councillor Fleming; also at Dundee on the 9th, when the Rev. Dr. Colin Campbell was in the Chair; and at Glasgow on the 10th, when Mr. W. F. G. Anderson presided.

On the 16th March, Mr. R. M. Routledge, Stipendiary Justice of Trinidad, delivered a lecture to the Aberdeen Branch on the Falkland Islands. Mr. John Clarke, M.A., presided.

On the 19th March, Mr. Arthur Montefiore, Hon. Sec. to the Jackson-Harmsworth Polar Expedition, addressed the Society in Edinburgh on the progress and aims of the Expedition. The Chair was taken by Dr. John Murray.

LECTURE IN APRIL.

Captain F. R. Maunsell, R.A., will address the Society in Edinburgh on Turkish Armenia, on April 9th, at 4.30 P.M.

Arrangements for other lectures are in progress.

CORRESPONDING MEMBERS.

The following gentlemen have been elected Corresponding Members of the Society for the current Session :—

Professor J. J. Stevenson,	*New York.*	Professor Dr. J. J. Egli, . *Zurich.*
„ Israel C. Russell,	*Michigan.*	„ Paul Chaix, . *Geneva.*
„ Dr. H. Wagner,	*Göttingen.*	„ Penck, . . *Vienna.*
„ H. Mohn,	*Christiania.*	„ Luigi Bodio, . *Rome.*

Mons. V. Dingelstedt,	*Geneva.*	Captain A. C. Yate,	*India.*
Dr. Johnston-Lavis,	*Monaco.*	Mons. de Margerie,	*Paris.*
Captain G. V. Tarnovski,	*Transcaspia.*	Colonel H. L. Wells,	*Persia.*
Mons. N. de Seidlitz,	*Tiflis.*	Professor Corstorphine,	*South Africa.*
Professor Carlos de Mello,	*Lisbon.*	Dr. Macdonald,	*East Africa.*
Clement L. Wragge, Esq.,	*Brisbane.*	Rev. W. C. Willoughby,	*Bechuanaland.*
Henry Gannett, Esq.,	*Washington.*	Dr. Hans Steffen,	*Chili.*
Dr. Robert Bell,	*Ottawa.*	Professor Pittier,	*Costa Rica.*
Michael G. Mulhall, Esq.,	*Buenos Ayres.*	Alex. Begg, Esq.,	*British Columbia.*
Baron F. von Mueller, K.C.M.G.,		Don M. Ballivian,	*Bolivia.*
	Melbourne.	Mark W. Harrington, Esq.,	
J. P. Thomson, Esq.,	*Queensland.*		*Washington State.*

GEOGRAPHICAL NOTES.

By The Acting Editor.

EUROPE.

Spitzbergen and French Explorers of the Seventeenth Century.—A recent number of the *Bulletin de Géographie Historique et Descriptive* contains a study of an interesting map in the possession of Mr. Cash, F.R.S.G.S. The map, measuring 35 by 24 inches, is drawn on fine white parchment in bright colours, and shows the north-east Atlantic, including Scotland, Norway, Jan Mayen Land, and Spitzbergen. It is remarkable as suggesting a French claim to the exploration or possession of the last two, for Jan Mayen Land is called "Y. de Richelieu," and Spitzbergen "France Artique" (*sic*); and the latter is decorated with a French coat-of-arms of the time of Louis XIII., and bears a· nomenclature pointing to the presence of French explorers in the north.

In the periodical referred to, Dr. E. T. Hamy, to whom the map was submitted through the kind intermediation of M. Ch. Maunoir, secretary of the Paris Geographical Society, follows two lines of inquiry, both of which lead to the same conclusion. First, he traces the history of exploration and whale-fishing at Spitzbergen, showing that, for some years after the discovery of the archipelago by Barentz in 1596, the English endeavoured to maintain by force a monopoly of the fishing, but were gradually constrained to admit the fishermen of other nations, especially Dutch and French. After the founding of the "Royale et Générale Compagnie du Commerce pour les voyages de long cours ès Indes occidentales, la pesche du corail en Barbarie et celle des baleines" in 1621, and while Richelieu was governor of Havre, an attempt was made to establish regular traffic between France and Northern Europe, and in connection with this enterprise a Norman gentleman of good family, Nicolas Toustain du Castillon, a pupil of the famous cosmographer Plancius, tried to make the North-East Passage. Of this attempt only too little record remains; but Mr. Cash's map is just such as would be drawn to illustrate and commemorate it, and in all probability such was its origin.

Dr. Hamy's second line of inquiry is directed to a comparison of the successive mappings of Spitzbergen. He shows from internal evidence that Mr. Cash's map is the work of a pilot who had actually visited the west and north-west of Spitzbergen, and that it is subsequent to, because more advanced than, the map of Jean Guérard published in 1628. But it is also earlier than 1631, because it shows no trace of the additional information given by Thomas Eyre, and published by Pellham in that year.

"La carte de M. Cash a donc été peinte vers 1629 ou 1630, ce qui correspond parfaitement à la date de l'expedition de Toustain du Castillon. J'ai déja dit que son style la rapproche des œuvres de la cartographie dieppoise, dont elle constituerait ainsi une des pages les plus modernes, ... un des monuments les plus intéressants de l'histoire des tentatives des marins de notre nation sous ces hautes latitudes."

The Population of France in 1894.—The report lately issued by the Office du Travail gives the following figures for the year 1894 :—Marriages, 286,662 ; divorces, 6419 ; births, 855,338 ; deaths, 815,620. Compared with the figures for 1893, only the divorces show an increase, namely, of 235. The diminution in the number of births was 2·2 per cent., and of deaths 6 per cent. In spite of the decrease in the number of births, they exceeded the deaths, though these were also fewer, by 39,768, as against an excess of 7146 in 1893. In the latter year the births exceeded the deaths for the first time. It appears that the improvement in 1894 was due to a smaller number of deaths, the births being 19,284 fewer than in 1893. The diminution, amounting to 1 birth for every 2000 of the population, is fairly equally distributed, seeing that it occurs in 70 departments, though the western departments—Brittany, Normandy, Maine, Anjou, and Poitou—show the largest proportion. There has been an increase in the south in the departments of Drôme, Vaucluse and Bouches du Rhône in the east ; and Ariège, Haute Garonne, Tarn et Garonne, Lot et Garonne, Gironde, and Charente Inférieure in the west. The average number of births has been 22·5 per 1000, varying from 14·5 in Gers to 32·6 in Finistère.

The general mortality in 1894 was 21 per 1000, Allier having only 14·8 per 1000 ; while the maximum, 26·9, occurred in the Bouches du Rhône.—*Revue Scientifique*, No. 6.

ASIA.

The Rainfall of the Malay Archipelago.—In the *American Meteor. Jour.* for February Dr. Voeikof discusses the tables for the year 1894 lately published in Batavia. These, he remarks, tend to discredit the long-prevailing belief that near the Equator the rains are always heavy, and of nearly daily occurrence. In the Malay Archipelago both some of the rainiest and some of the driest stations lie within 1½° N. and 1° S. of the Equator. The rainiest stations near the Equator are on the west coast of Sumatra and in the neighbouring mountains. The more level Eastern Sumatra and Western Borneo have less rain and also less marked contrasts. On the north-east of the peninsula of Celebes the rains are comparatively light, and there is a well-marked dry season. In general the rains are lighter in the east— *i.e.* in Eastern Java and on Bali, Sumbava, and Timor—than in Western Java and Sumatra. In the same direction the dry season becomes longer and more sharply marked.

The interior of Western Java, which in the number of rainy days and the equable distribution of rain does not yield to Western Sumatra, presents a striking contrast to the western coast and the south-western peninsula of Celebes. The rains are heavy but very differently distributed ; during the north-west monsoon they are excessive, especially in December and January, and the dry season is clearly marked. Such a large rainfall in a single month is not experienced even in Western Java and Western Sumatra, and is only approached in the Moluccas, where, however, the rainy season is in June and July, owing to the south-east monsoon impinging on the mountains of Ceram.

Out of the 137 stations from which records extending over no fewer than thirteen years have been received, only ten have a fall of over 24 inches in a single month. In Java the increase of rainfall towards mountains is clearly illustrated between Batavia and Buitenzorg, and on two other lines in the middle part of Java; and the action of forests in bringing down the rain on the lee-side is also perceptible at stations beyond the influence of mountains.

The greatest falls in one day are far from reaching those of India, or even of New South Wales and Queensland. Out of the 137 stations a fall of 12 inches or more in one day was recorded only at six in the years 1880 to 1887, and 1889 to 1893. The percentage of days with heavy rainfall is far from being everywhere in accordance with the yearly fall. Celebes has by far the greatest percentage, though its yearly fall is but slightly in excess of the average for the Archipelago; while at Preanger, where the rainfall is much above the average, the maximum in a single day during a period of twelve years did not exceed 8 inches.

Professor Voeikof gives a number of tables illustrating the points he comments on, and mentions that the rainfall observations of the East Indian Archipelago (with those of France) were the first published *in extenso*—a good example that has been followed by few countries.

AFRICA.

The Triangulation of Africa.—General E. T. Chapman, whose letter on the mapping of the continent was noticed in vol. xi., p. 367, has kindly forwarded an extract from a report on his proposals drawn up by the Surveyor-General of the Cape.

Cape Colony, the Surveyor-General maintains, has done its share, having carried out two large geodetic surveys since 1840, and recently a still more costly and expensive one in conjunction with Natal, from which, as well as from other sources, a map on the scale 1:800,000 has been compiled and will shortly be published. The maps of the Orange Free State and of the South African Republic are based on triangulations made from time to time for purposes of property and topographical surveys, and, though now disconnected, will prove useful when the general triangulation has been extended over these countries. In British Bechuanaland surveying operations were commenced in 1886 by Lieutenant Laffan, R.E., who measured a base line in the valley of the Hart River. The surveys that were subsequently executed furnished the means of constructing a map which needs only to be connected with the main triangulation of Cape Colony.

To complete the work already done towards mapping South Africa, the Surveyor-General considers three triangulations immediately necessary, viz. a chain of triangles across the Orange Free State, joining the Natal to the Kimberley series; a chain about 70 miles long to connect the eastern part of the Bechuanaland survey to the Kimberley chain; and one of 150 miles connecting the western part of the same survey with Sir Thomas Maclear's triangulation south of the Orange River.

Further extension of geodetic operations northwards must depend on the co-operation of the Governments concerned. Three chains of triangles should be pushed northwards,—one from the north end of Sir T. Maclear's triangulation through the German Protectorate to its northern boundary; a second from the surveys of Mr. Melville in Bechuanaland to the Victoria Falls on the Zambesi; and a third from the north end of the Natal chain through the South African Republic and Rhodesia to the Zambesi. These chains should be connected by a transverse chain approximately on the parallel of 18° S. lat. These proposals are illustrated by a map.

General Chapman suggests that an international committee should be formed, and that other officials engaged in survey work in South Africa should send in reports to the committee. An annual report might be published, and the work done by each country be plotted on a skeleton map.

The Future of Eritrea.—In a lecture delivered at the second Italian Geographical Congress in September last, Baron L. Franchetti dealt with the development of the Italian colony on the Red Sea. On Massaua and its neighbourhood he had not much praise to bestow. An expanse of bare land, gleaming white under a burning sun, studded with clumps of mimosa, which raise their contorted branches scantily clothed with leaves some four to six feet above the soil, does not give a favourable impression to the newcomer, and he is glad when the sixty miles of navigation along the so-called channel of Massaua are over. At first sight, the town with a row of neat-looking houses along the shore, some of them almost elegant, has an appearance of prosperity ; but a glance at the quiet port, where the glaring sunlight is broken only by a few Arab dhows no bigger than a fishing-boat, the guard-ship, and, perhaps, a steamboat, makes the stranger wonder how the inhabitants gain a living. Besides the native population, the soldiers and officials, there are about 1000 Europeans, of every rank and description, whose aim in life is to obtain directly or indirectly some share in the funds of the State. The commerce is small compared to the number of traders, and is chiefly in the hands of the natives and a colony of *banyans*, who spend little and transmit their profits to India. It is hoped that in time trade will be opened with Abyssinia and the Sudan ; but it would be imprudent to depend on it for the prosperity of Massaua, as the only means of transport must, for a long time to come, be baggage animals.

The neighbourhood is still more discouraging. A few houses of well-to-do natives are surrounded by small gardens, watered by the wells which the owners have had means to pay for ; and the brackish water, lying a few yards below the surface in the plains of M'Kullu and Arkiko, has enabled the inhabitants to lay out a few gardens, the produce of which yields a profit at the neighbouring market of Massaua ; but cultivation on a large scale has proved a failure, owing to the uncertainty of the winter rains in the region between the Red Sea and the plateau.

On the plateau, however, the conditions are totally different. Elevated about 6500 feet on the average above the level of the sea, it enjoys a healthy and mild climate. The thermometer rises above 92° F. on very few days in the year, and then for only a few hours, and never, even at 6500 feet, sinks to freezing-point. The greater part of these lands, gently undulating or hilly, is cultivable, and abundance of water is found at a depth of fifteen to twenty-five feet below the surface. On the productiveness of the soil Baron Franchetti spoke from the experience of four years. Cereal and leguminous plants yield crops equal to those raised on fairly good land in Italy ; the vine and the olive thrive splendidly, and a variety of wild olive, *Olea chrysophylla*, is one of the characteristic plants of the plateau and its slopes. Other fruits give a fairly good crop. The small native cattle, having strength proportionate to their size, must be yoked to the plough in greater numbers, but as their price is low the cost of traction is less than in Italy. On the whole, the climate, humidity, and soil on the plateau permit agriculture to be carried on profitably. The production of crops of warm climates is attended with greater difficulty. Besides tobacco, which being grown in Italy ought to thrive in the colony, Baron Franchetti experimented with cotton. At Asmara, 7800 feet above sea-level, the climate was too cold, and Cheren was considered unsuitable, because the fevers prevailing there at the time of harvest would be an obstacle to colonisation. In 1892 two experimental farms were started at Gura and Godofelassi,

6500 feet above sea-level. Tobacco gave here excellent results as far as the luxuriance of the plants was concerned; as to quality, no decision can be given owing to the want of men able to modify the cultivation and the preparation of the leaf in accordance with the local conditions. Egyptian and American cotton seeds produced small plants, but heavily laden with fruit. Coffee is cultivated profitably on the opposite coast of the Red Sea, up to a height of 6500 feet, and would no doubt thrive on the Ethiopian plateau. Experiments with this plant are being continued.

At the end of 1893 ten peasant families emigrated from Italy to Eritrea, and established themselves in the neighbourhood of the native village of Godofelassi, a few hundred yards from the Italian fort of Adi Ugri, in a fertile undulating country 6500 feet above sea-level, forming the district of the Sarae. Under a contract with the Government each family received fifty acres of land, on which it is bound to reside for five years and cultivate it. At the expiration of this period the land is to become the freehold property of the family. They received free passages and the means of putting their land under cultivation, as well as of support for the first year. This advance, in addition to interest at the rate of three per cent. *per annum,* is to be repaid in instalments of one half of the crop, beginning with the second harvest. The expense of transporting a family of seven to ten persons, and establishing them on the land, amounts to £160 on the average. Between November 1893 and July 1894 each of the families had cleared fifteen to twenty acres of ground and planted it, and in August the crops promised well, and in spite of exceptionally late mists yielded enough produce to support part of the family at least for the next twelve months. On this system Baron Franchetti believes that the extensive lands of the plateau, unoccupied by natives, may be successfully colonised, and that the peasants may make a comfortable living. Large undertakings, in which the capitalist employs labour, have not been successful as a rule. In some cases, plantations on free grants of land have been sublet to natives, who cultivate them with their own implements, oxen, and even seeds. Thus the concessionaire levies a tax on the native, which, if paid at all, should be paid to the Government.

As to the hot region, this is of course not suited to European colonisation, and here agricultural establishments under European supervision are the only means of utilising the land. Of the lowlands between the plateau and the sea not much can be expected, except through native cultivation and grazing during the winter rainy season. Another low tract lies to the north-west. The territory of Cassala is extremely fertile, but the disturbed state of the country will for the present prevent the influx of capital. There is also fertile land of great extent in the Baraka valley, where the rainfall, though not as regular as on the plateau, is generally sufficient. In all this region, however, fevers rage after the rains, just when the crops ripen and have to be harvested. But there is cause to hope that agricultural undertakings, maintained by sufficient capital and under competent management, may produce large returns, especially in tobacco and cotton.

Meteorology in the Congo State.—The *Mouvement Géographique* of February 2nd contains a note on observations made by Père de Hert at Kimuenza during the summer of October 1894 to May 1895. Kimuenza is a vast plateau situated in about 15° 22′ 30″ E. long. and 4° 29′ S. lat., some dozen miles south-south-east of Leopoldville. The altitude of the station, as determined by readings of a barometer and a boiling-point thermometer, a pillar erected by Captain Delporte in Leopoldville, where he had determined the altitude to be 340 mètres (about 1115½ feet), being taken as the point of reference, is 1568 feet nearly.

The maximum atmospheric pressure was recorded on May 26th and 27th, namely, 30·28 in. The minimum occurred on November 24th, 29·93 in. The mean of all the readings was 30·11 in. The mean monthly maximum, 30·15 in., occurred in May ; after that the barometer rises and remains high during the whole of the cool season. The monthly minimum was recorded in March—30·07 in. The greatest fall was 0·24 in., between the evening of November 10th and eight o'clock on the following morning. The barometer is independent of the state of the sky ; no predictions of weather can be drawn from its movement. Storms burst when the pressure is high as well as when it is low. But P. de Hert noticed several times that when the pressure was high in the morning, there was a storm at noon, and occasionally the barometer sank before rain.

On March 2nd the thermometer in the shade rose to its maximum, 94·8° F. ; the absolute minimum was 62·2° in the night between the 7th and 8th of January, and the range was therefore 32·6°. The mean for the season was 77·4°. The warmest month was March, with a mean of 78·1°, and the coolest November, with a mean of 76·5°. This small difference of 1·6° shows how very uniform the temperature is. On sixty days the thermometer rose to 86° or more, and on nineteen days only it sank below 68°. The most sudden fall of temperature on the same day was 19·6°, on March 16th. The highest temperature in the sun was 130·1°, on February 27th, and the minimum 70·2°, on November 2nd ; hence the range was 59·9°. The great falls of temperature occur on rainy days, or at any rate on the days following rain.

AMERICA.

The New British Route to the Pacific.—In February, Colonel J. Harris delivered a lecture at the Imperial Institute (*Journal*, vol. ii., No. 15), on the proposed route through Hudson's Bay, briefly noticed on p. 157. The capital of Manitoba, Winnipeg, now containing 30,000 inhabitants, will be connected with the new route by a railway, now under construction, to Sea Falls, and thence by a line to Churchill Harbour, and will thus be brought two days' journey nearer England. But the country that will derive most benefit from the new scheme are the vast North-West Territories, extending from about 25 miles west of Lake Superior to the Rocky Mountains, a distance of 1350 miles, and from the 49th parallel northwards to the watershed of the Saskatchewan River, an average breadth of 350 miles. Within these limits lies an area of 462,500 square miles, two-thirds of which have been proved to be capable of producing the finest wheat, while the remainder is admirably adapted for grazing and dairy-farming. In 1891 there was an area of 1,300,000 acres in Manitoba and the Territories which produced 25,000,000 bushels of wheat and 30,000,000 of barley and oats, while the cattle were estimated at about 600,000 head. At present the cost of transporting wheat from the centre of this region to Montreal, Boston, or New York is 35 cents per bushel, while the charge to Fort Churchill would not exceed 15 cents, and the saving of the carriage of live stock would be £3 per head.

To the north-west, again, of the Territories is a vast region embracing the Athabasca and Great Mackenzie basins, in which, as stated before a Select Committee of the Senate of the Dominion in 1887, there are 650,000 square miles suited for potato-growing, 407,000 square miles adapted for barley, and 316,000 for wheat; and at the head of the Peace, Liard, and Peel rivers there are 150,000 to 200,000 square miles which may be considered auriferous. There is also strong evidence that the Athabasca and Mackenzie valleys contain the most extensive petroleum field in America.

Hudson's Bay and Strait are remarkably free from rocks and shoals, and the depth of water is very uniform, averaging about 70 fathoms. Probably, as in the Gulf of St. Lawrence, the navigation will be confined to the summer, during about 4½ months. It does not appear that either the Bay or Strait is ever completely frozen over, but the difficulty is, as in the St. Lawrence, to enter harbours. If the produce of the North-West Territories can find an outlet through the Bay, the distance saved on land will be 1291 miles as compared with Montreal, and 1700 as compared with New York, and this without any increase in the length of the sea voyage.

The railway from Fort Churchill will be divided into three sections. The first to Sea Falls, 350 miles long, will be constructed without difficulty, as proved by surveys already made. It will place Fort Churchill in direct communication with the navigation of Lake Winnipeg, the Red River, and the Saskatchewan, which form highways for steamers over an aggregate distance of 1800 miles. The second section, to Prince Albert, will be 300 miles long, and will run through some of the finest wheat land in the North-West. At Prince Albert it will form a connection with the Regina branch of the Canadian Pacific. The third section will run for about 350 miles viâ Battleford to Calgary, through fine grazing country, and there join the Canadian Pacific. The saving of distance from ocean to ocean will be 1328 miles as compared with the route through Montreal.

Cuba.—Except in the barren islands of the northern archipelago there is, no island off the coast of America as large as Cuba. Its area is 43,316 square miles, or 45,880, if the smaller islands be included. It is, therefore, considerably larger than many European countries, such as Portugal, Bulgaria, or Greece, and occupies about half the total area of the West Indian islands. While from Cape San Antonio in the west to Cape Maisi in the east, it stretches across eleven degrees of longitude, extending in length to about 750 miles, its breadth is comparatively small, the maximum being only 100 miles. The farthest point from the coast, to the north-east of Bayamo, is not more than 37 miles distant, and the proportion of the average breadth to the length is 1 : 12. Cuba is in many respects the most similar island to Java, but even that is more compact. Consequently the interior is easily accessible to the conqueror or the agriculturist, and hence the speedy extermination of the native inhabitants, which seems to have been completed in a few decades. Where there are roads the island can be crossed in a day or two on horseback, while the train takes only as many hours to perform the journey, and the productions of the interior can be brought to the ports with little labour and expense. In the direction of the longer axis of the island the cultivation exhibits marked differences and contrasts. Pinar del Rio and Havana, at the western extremity, are the principal tobacco-growing provinces, the latter being also the chief commercial district ; then come Matanzas and Santa Clara, where sugar is mostly grown ; and lastly, Santiago de Cuba is the province of mountain and forest, where the cultivation of sugar, coffee, tobacco, fruit, the felling of timber and mining are carried on side by side, and where the population, interests, and political opinions are as varied as the industries.

The coast is indented to a remarkable degree, and scarcely any island can compare with Cuba for the abundance of excellent harbours. The bays of Havana, Mariel, Bahia Honda, Cienfuegos, Santiago, Guantanamo, Nipe and Naranjo can shelter ships of the deepest draught, while vessels of smaller tonnage can find access into a large number of smaller inlets. With a headwind it is indeed somewhat difficult for sailing vessels to enter or leave these harbours, for while they are often capacious and branch into fiord-like arms, their openings are narrow—at

Havana, for instance, 370 yards broad. This difficulty is especially felt on the north-east coast, which is exposed to the full force of the trade-wind.

A belt of coral reefs and islands encircles Cuba on all sides. Where deep water reaches close up to the coast, as in the south-east and north-east, and in the gulf of Cienfuegos, the belt is narrow, but elsewhere, notably on the north coast on both sides of Cienfuegos and north-east of Cape San Antonio, it covers a considerable area, sometimes 25 to 30 miles broad, and near Pinos as much as 60 to 70, constituting a great danger to navigation. The small islands or cays are begirt with breakers, and between them the currents and tides form a very complicated system, while the deep channel winds in and out among them without any regular course, and landmarks are completely wanting. It is, therefore, not to be wondered at that many vessels are wrecked or stranded on this coast, and that only the most experienced pilots can find their way through the labyrinth of islets and reefs. The most accessible stretch of coast is that which lies between Matanzas and Bahia Honda, and hence Havana has come to be the seat of the Government and the military centre of the island. The bays of the south-east coast are, indeed, easily accessible from the sea, but they have not such ready communication with the interior.

The mainland of Cuba rises on all sides from the sea as a cliff of coralline limestone 15 to 30 feet high, and then slopes somewhat steeply towards the interior frequently in the form of terraces, especially in the east and west, but on the whole attains no great elevation. High mountains—the Sierra Maestra and Sierra Cobre, culminating in the Pico Tarquino (8400 feet), the Gran Piedra (5210 feet), the Ojo del Toro (5190 feet), and the Lomo del Gato (3248), are found only along the south-east coast between Cape Cruz and Cape Maisi, bordering on the Bartlett deep, which here sinks to a depth of 3280 fathoms. Another line of elevation, or rather the steep margin of a plateau, the Sierra de los Organos, which attains a height of 2600 feet in the Pan de Guajabon, skirts the north-west coast from the Guadiana bay to the neighbourhood of Havana. Elsewhere the surface of the island consists of plains of greater or less extent traversed by low sharp ridges, or broken ranges and groups of hills. These elevations are most important in the centre of the island, where the Loma de Banao near Espiritu Santo rises to a height of more than 5000 feet (5510 ?), and gradually sink towards the west, corresponding to the narrowing of the land; indeed, every dip in the vertical elevation corresponds to a contraction in the breadth of the island. The principal depression is that between Jucaro and Moron, and here too is the greatest contraction, separating the island into two halves of equal area. The dip may be regarded as a part of the axis of the Yucatan deep, and it should be noticed that immediately to the north-east lies the narrowest part of the Old Bahama Channel, so that here meet two important geomorphic axes.

The predominating rock in Cuba is limestone—in the low coastal zone coralline limestone such as is now being formed round the coast, but in the higher and more inland parts Tertiary marine limestone, the coralline origin of which is doubtful. The soil—the famous red earth of Cuba—has been formed by the disintegration of this rock. At some points, namely, near the watershed and in the bases of many of the limestone hills, eruptive rocks occur—diorite, porphyry and granite—and together with azoic slates and gneiss compose the skeleton of the island, on which lie the Tertiary and more recent limestones and Mesozoic clays and marl.

Owing to the vertical and horizontal structure of the island, as already described, the rivers have no great length or drainage area. In the interior they flow gently between low banks, flooded in the rainy season, and descend to the coast through deep cañons with fine waterfalls and rapids. The largest, the Cauto, has a length of 200 miles and a drainage area of some 3900 square miles. Unlike the others, it carries down large quantities of detritus, and enters the sea through low lands mostly formed by its deposits ; it is also navigable for a distance of 75 miles.

Cuba is much better adapted for agriculture than Jamaica or Haiti, where mountains and unproductive lands are far more extensive. The tracts of red calcareous soil, often changed by an admixture of humus to a brown or black earth, are exceedingly fruitful ; but beside them are flats and slopes bare or covered with rubbish, which, owing to their permeability, are condemned to constant drought. Cuba presents an alternation of impenetrable woodland and sun-burnt heaths, savannahs and almost desert tracts, and the vegetation is distributed in oases. Where slopes are cultivated, the tropical rains, by washing the soil, tend to exhaust it.

Though Cuba lies close to the Tropic of Cancer, and even in its eastern part extends only three degrees towards the Equator, it has quite a tropical climate. Besides the Atlantic with its northern anticyclonic centre, two factors determine the meteorology of the island—the warm Central American sea and the broad plains of the continent ; the configuration of the island makes itself prominently felt only in the lofty south-eastern mountains, strengthening the effect of the above influences. Like the other West Indian islands, Cuba lies in the domain of the north-east trade-wind, which, however, between May and October, is often interrupted by calms and feeble changeable winds, particularly from the southern quadrant. In summer the trade-wind brings abundance of moisture from the Atlantic, which is in part condensed and falls on the island, but the southerly winds from the Caribbean Sea are those which really make summer the rainy season. The rain falls in deluges, and in Havana, which is by no means the most remarkable place in this respect, 8 to 12 per cent. of the rainfall of the year often falls in a single day, and as much as 33 per cent. in one month. In winter the trade is a dry wind, and in a still greater degree the northerly winds blowing from the plateaus of the continent, which cause sudden changes of temperature, so that in the interior of the island the thermometer often falls at night below freezing-point, ice is formed, and the country covered with hoar-frost. Snow never falls, as these winds are very dry.

The thermometer in Havana seldom rises during July and August as high as 97° or 99° F., and seldom sinks in January below 50°. The yearly mean is 77·4° (0·7 higher than in Calcutta), the mean for July is 82·4°, and for January, 72°. The annual rainfall amounts to 47 inches. Santiago de Cuba must be much warmer, especially in winter, and also receive a larger quantity of rain ; but exact data are wanting. For a tropical country the climate of Cuba is favourable to Europeans, owing to the frequency of sharp breezes from the north-west, and as a rule yellow fever does not carry off so many victims as diseases of the respiratory organs.

The population in 1895 may be taken in round figures at 1,700,000, or about 37 to the square mile. This population is distributed according to the cultivation. The province of Havana, with about 450,000 inhabitants, is by far the most densely populated, and the western department (the provinces of Pinar del Rio, Havana, Matanzas, and Santa Clara) contains four times as many people as the larger eastern department. The percentage of Negroes to the total population, which in 1841 was nearly 60 per cent., is now only 34·2 per cent. The economic condition of the country is now in a very depressed state. As regards the sugar industry, Cuba is the chief producing country in the world, but the quantity, as elsewhere, has decreased owing to excessive competition, especially with beet-sugar. The exportation of tobacco, the second chief product of Cuba, also shows a tendency to decline. Added to this, the pressure of taxation helps to check the progress of

the country. Scarcely ten per cent. of the land is at present brought under cultivation, and much might be done by irrigation, the construction of dams to afford
protection from floods, and of railways and roads, to increase the production.—
From an article by Dr. Emil Deckert in the *Geogr. Zeitschrift*, Nos. 2 and 3, 1896.

**The Temperature and Density of the Waters of the Gulf Stream and of the Gulf
of Mexico.**—In the years 1874 to 1882 numerous observations were taken by Commanders Sigsbee and Bartlett in the steamer *Blake*, but they have not till recently
been subjected to any thorough examination. The results will shortly be published
in the *Annual Report* of the U.S. Coast and Geodetic Survey. A preliminary
sketch is given by Herr A. Lindenkohl in *Petermann's Mitt.*, Bd. 42, No. 2.

The waters of landlocked seas are much more affected than the open ocean by a
variety of agencies—evaporation, precipitation, and the influx of water from river
mouths and currents. As regards evaporation, there are a number of data from
points on the northern coast of the Gulf and from the observations of Captain
Shufeldt on the Isthmus of Panama ; but these cannot be applied directly, as the
evaporation of salt water is less than that of fresh, and no doubt the evaporation at
a distance from the coast, where the winds are stronger and have free passage, will
be greater. Herr Lindenkohl has assumed that the annual evaporation amounts to
60 inches on an average, that evaporation is promoted by the prevailing north-east
trade wind, and that it is most effective along the middle of the Gulf from east
to west, corresponding approximately to the position of the mean barometric
maximum.

The precipitation may be estimated with greater probability of accuracy. There
are two zones of excessive rainfall separated by an almost rainless tract. On the
north-east coast, near New Orleans and Pensacola, the annual rainfall amounts to
59 inches more than in any other part of the United States, except the north-west
Pacific coast. The south-western part of the Gulf lies below the Tropic, and is
characterised by seasons of heavy continuous rain. At Vera Cruz the yearly mean
is 179½ inches. Between these zones lie the sandhills of Northern Mexico and
Southern Texas in the west, and the sandy wastes of Northern Yucatan in the east.
By the construction of curves of equal fall, Herr Lindenkohl has obtained 33 inches
as the average yearly rainfall over the Gulf. Thereby the volume of the water
would be increased by ·84 of a cub. mile daily, and the rivers which flow into the
Gulf add a daily volume of about ·685 of a cub. mile, of which over 70 per cent.
comes from the Mississippi. Against the loss of 1·54 cub. miles by evaporation
there is, then, a gain of 1·525 cub. miles. Though these figures may not be very
exact, they are near enough to the truth to show that the balance is very nearly
maintained.

The currents which affect the Gulf are the current through the Yucatan Channel,
flowing inwards, and the Gulf Stream, issuing by the Florida Channel. Of these
the former is the strongest current found in these seas during the 20 years' cruise of
the *Blake*. According to Lieutenant Pilsbury's measurements, made in 1887, it
occupies nearly the whole of the strait, developing its greatest velocity of 2½ to 5
knots an hour at its western edge—that is, on the shortest course to the Gulf.
Down to 130 fathoms the velocity diminishes rapidly with the depth, and it may
be assumed that at a depth of 200 fathoms there is no movement in the water.
After leaving the Yucatan Channel the stream expands, losing a great part of its
velocity. An approximate calculation of the volume of water which passes through
the Yucatan Channel in 24 hours gives the enormous quantity of 650 cub. miles,
which would be enough to raise the whole surface of the Gulf nearly six feet. The
outflow at the Florida Channel amounts to 431 cub. miles in the 24 hours, and the

difference of these volumes leads one to believe that there must be an undercurrent running back into the Caribbean Sea through the Yucatan Channel. The heaping up of the water by the wind in the north-western part of this sea is, no doubt, one cause of the current, but the circumstances seem to show that, even in the absence of wind, there would be a surface-current from the Caribbean Sea into the Gulf, and one below in the opposite direction, just as between the Mediterranean and Atlantic.

The fresh water which finds its way into the south-western part of the Gulf, whether as rain or the discharge of rivers, remains on the surface, but quickly absorbs salt from the strata beneath. Consequently, a thin layer of comparatively high temperature and salinity is found on the surface, while the water below is remarkably cold. From about 86° F. at the surface, the temperature falls to 44° at a depth of 250 fathoms. The fresh water which enters the Gulf on the north likewise floats on the surface, in virtue of its small specific gravity and in spite of its low temperature. It is only by slow degrees that it absorbs heat and salt from the lower strata. Strange to say, the Mississippi water does not flow direct to the Florida Channel, but turns westwards, making for the middle of the western part of the Gulf. At first Herr Lindenkohl was inclined to attribute this movement to the rotation of the earth and the prevailing winds, but finally came to the conclusion that the water sought the lowest part of the Gulf, the level of the eastern side being raised by the Yucatan current. Notwithstanding the action of this stream, the water of the northern and western parts of the Gulf makes for the Florida Channel, the Yucatan stream dipping here below the surface.

Forcing its way against the north-east or south-east trades, the water flowing towards the Florida Channel is much exposed to evaporation, by which its salinity, and consequently its specific gravity, is raised, so that it sinks, carrying with it a larger amount of heat than could penetrate to the depths by radiation or conduction. Hence, north of the western extremity of Cuba, a temperature of 59° F. is found at a depth of 250 fathoms, compared with 44° in the western part of the Gulf, and 46° in the Caribbean Sea. In accordance with the law that, where decided differences of temperature occur in the lower strata of parts of the sea closely connected with one another, undercurrents tend to restore equilibrium, warm and salt undercurrents flow westwards and into the Caribbean Sea. But it must not be assumed that hereby the Gulf suffers a permanent loss of salt ; the salt thus abstracted finds its way gradually (with the temperature) to higher strata, until at length, having reached the surface, it is carried back into the Gulf.

In the winters of 1889-90 and 1890-91 the steamer *Blake*, under Lieutenant Vreeland, was sent out to investigate the connection between the Yucatan current and the Gulf Stream. The latter first appears as a feeble current of 1½ miles an hour at the western entrance of the Florida Channel, with an axis in the direction of the Gulf, but occasionally displaced somewhat towards the Yucatan Channel. The facts of the case seem to be that when the Yucatan current develops great activity, the water heaped up in the north-western part of the Caribbean Sea overflows into the Gulf, part of it making its way direct to the Florida Channel— though no greater quantity, probably, than is found at the same distance in other parts of the Gulf. But the Yucatan current is very variable, and the Gulf Stream owes its origin in greater measure to the efforts of the Gulf to relieve itself of excessive pressure, by which even the Yucatan current is sometimes checked at its weakest points. The view, therefore, that the Gulf Stream makes the circuit of the Gulf is not correct, nor that it flows direct from Yucatan to Florida. Nor is it possessed at its commencement in the Florida Channel of such a store of heat as would suffice, according to the views hitherto prevalent, to moderate the climate of Western Europe. In fact it is no warmer than the North Atlantic at the same latitude, and moreover, lying on the surface, is liable to lose its heat by diffusion. It has also to contend against the cold and fresher water flowing from the Florida Bank and forcing its way to the centre of the Channel, and though, on arriving at Cape Florida it has gained the victory, forcing back the colder water to the coral reefs, yet it is so weakened by loss of heat and salt that it would be unable to maintain its existence were it not reinforced by warm salt water flowing through the Santarem Channel, and by still greater additions driven by the north-east trade wind to the eastern entrance of the Florida Channel.

The origin of high temperature and salinity that characterise the Gulf Stream in the Atlantic Ocean must be sought elsewhere than in the Gulf of Mexico. Evaporation, promoted by the winds, takes place in a high degree over the whole surface of the Atlantic between Bermuda and the Southern States of the Union, and the sinking water communicates to the lower strata a higher temperature and salinity than are found in any other ocean. This water, having a temperature of 60° to 62° at a depth of 250 fathoms, comes in contact, at a distance of 40 geographical miles from the coast, with the cold waters of the shallows, which, at the same depth, have a temperature of only 45°. An exchange of temperature and salinity takes place here and gives rise to what is called the Gulf Stream; for, though the movement of the water is originally in a vertical direction, it is probable that near the surface it passes into a horizontal movement, just as in the case of atmospheric circulation; and further, it must be assumed that the accumulation of heat and salt on the surface has an effect on the level.

GENERAL.

Wind and Ocean Currents.—Herr E. Witte remarks in the *Annalen der Hydrographie*, Nov. 1895, that, while no doubt is entertained as to the connection between wind and currents, there is great divergence of opinion as to the degree of influence exerted by winds on the waters of the oceans. While some hold that wind is of secondary importance in the production of currents, others consider the trade-winds the chief, if not the only, motive force to be taken into consideration. Equally divergent are the views regarding the effect of the rotation of the earth on the water.

The heating of the water can only act by causing the water to expand within the Tropics and thus rise above its normal level, so that the superficial layer flows off on both sides towards the Poles, thereby lowering the pressure in the lower strata, and causing a return undercurrent towards the Tropics. The wind acts by sweeping the surface water before it, and in the open sea this movement is propagated to greater and greater depths. But where the current impinges on the land, it heaps up the water against the shore, and the increased pressure forces the lower strata away from the land, or, under favourable conditions, there may be a flow of the surface water to the sides. Should, on the other hand, the wind blow off the land, it drives the surface water from the shore, and the underlying water wells up. Lastly, the rotation of the earth will, in the northern hemisphere, have the same effect as a wind driving the water against the shore, when the land is on the right of the current; and of a land wind, when the land is on the left. In the southern hemisphere the action will be reversed. Should the current-producing forces act in the same direction, the result will, of course, be intensified, and enfeebled when they are opposed.

Now, on the north and north-west coast of Norway, where the rotation of the earth tends to press the water against the coast, the winds which tend to drive the water from the coast are in winter by far the most prevalent as far south as Bergen, as appears in a table extracted from Dr. Supan's *Statistik der unteren*

Luftströmungen. And yet along the whole coast the cold ground water nowhere wells up, but the warm stream hugs the shore. Hence it would seem that, here at any rate, the rotation of the earth is a more powerful factor than the wind in producing movement in the water.

A second example is taken by Herr Witte from the southernmost part of the Peruvian current in about lat. 40° to 25° S. Here westerly winds are far more prevalent than easterly, and yet the warmer surface water is driven away from the coast, as though by the action of the earth's rotation, and the cold water from below wells up in its place.

These examples, which might be added to, show that wind is not the only, and perhaps not even the most important, factor in current-production. And the same conclusion may be arrived at in another way. The air currents are a consequence of the unequal heating of the air by the sun. Now, in a cloudless atmosphere, such as prevails in the region of the trades, the greater part of the solar rays pass through to the land and sea, while only a smaller portion warms the air directly. Indirectly the air is warmed from below, and this rise of temperature produces air currents which in turn cause movements in the surface water. And only a part of the energy of the air currents sets the water in motion by friction, while the rest is probably dispersed among the air currents flowing rapidly in different directions. Thus, under all circumstances, it is the sun's heat that causes ocean currents, either directly or indirectly through the air, and it is, at least, improbable that the first, direct, influence should be less effective than the other, or disappear entirely.

Other facts seem to point to the same conclusion. The trades have a tendency to heap up the warm surface water about the Equator, especially in the western parts of the oceans. It might therefore be expected that near the Equator there would be a zone of warm water of exceptional depth, whereas it is well known that the opposite is the case, and that here the temperature falls most rapidly with the depth. And, secondly, an undercurrent ought to flow from the Equator towards the Poles, whereas the movement of the lower layer is undoubtedly in the opposite direction. Certainly the surface currents on the west coasts of the continents in the region of the trades indicate that the wind is the prevailing force in these parts. And this exercises its full influence as long as the water can move freely, but when it is brought up by the east coast of a continent the heating effect of the sun asserts its influence, driving the upper layers towards the Poles. As it flows off, this water draws the air with it, so that in the western parts of the ocean a trade-wind cannot come into existence, but over a space of 20° of latitude the air currents follow those of the water. According to this view, it is the warm currents on the east coasts of continents that determine the winds, and the low atmospheric pressure is a consequence of the highly heated water.

The editors of the *Annalen* wish it to be understood that they do not, by publishing this suggestive article, signify their agreement with the author on all points.

Deep Soundings in the Pacific.—In vol. xi., p. 650, reference was made to a sounding of 4900 fathoms made by H.M.S. *Penguin* in the Pacific. The *Geographical Journal* for March states that Captain Balfour, having put a new wire on the machine, sounded again and found the bottom at 5022 fathoms in lat. 23° 39′ S. and long. 175° 04′ W. He then steered 50 miles eastwards, and sounded 3100 fathoms. Sailing south parallel to the line formed by the Tonga and Kermadec Islands, and finding depths between 2200 and 4400 fathoms, he struck a deep hollow 100 miles east-north-east of Sunday Island, and recovered a specimen of the bottom from a depth of 5147 fathoms; and again, 100 miles east of Macarthy Island, a sounding of 5155 fathoms was obtained.

German Balloon Ascents.—Dr. Assmann, whose experiments were alluded to in vol. xi., p. 32, has given a short account of the principal results obtained in the *Meteorologische Zeitschrift*, of which the *Quarterly Journ. of the R. Meteor. Soc.* contains a summary. The temperature above the height of 12,000 feet has been found to be decidedly lower than inferred from previous ascents ; this is due to improvements in the instruments. The rate of decrease of temperature is uniform, but possibly increases with the altitude, probably owing to an increase of temperature in the zone of cloud formation between 6000 and 12,000 feet. Above the level of 15,000 feet a nearly constant temperature prevails throughout the year. The inversion of temperature in winter and at night is apparently regular up to 3000 feet, and is usually connected with differences in the direction of drift of the different strata. The formation of cumulus in the neighbourhood of a depression extends up to an unexpected height. The upper surface of a cloud area showed the same thermal and electrical relations to the air stratum above as are noticed at the earth's surface—a confirmation of a prediction by Professor von Bezold. The exchange of air between anticyclones and cyclones has been thoroughly explained, and is found to obey a simple law. It is nearly certain that the electric potential does not increase with the height, and that it tends to approach a constant value at considerable altitudes, whence it would seem that the earth is the source of all the electricity of the atmosphere. Lastly, the humidity is extremely low at moderate heights, but great variations occur between different cloud strata.

MISCELLANEOUS.

The erection of a meteorological station at Ushuaia, the seat of government in **Tierra del Fuego**, has been decided upon by the Argentine Government.—*Deutsche Rundschau*, Jahrg. xviii. No. 6.

Between Angoulême and Limoges immense caves have been discovered. They contain long passages and extensive halls with stalactites and stalagmites of all kinds. In one of the rooms have been found a skull, potsherds and stones which appear to form a musical octave.—*Deutsche Rundschau*, Jahrg. xviii. No. 6.

The largest continuous forest occurs in **Siberia**, extending for a distance of 3000 miles, with a breadth of 1000 to 1700, from the Obi valley eastwards to the Indigirka, across the valleys of the Yenesei, Olenek, Lena, and Yana. The trees are chiefly conifers, sometimes attaining a height of 150 feet. Even the most experienced trappers do not venture into these wilds without marking the trees.—*Deutsche Rundschau*, Jahrg. xviii. No. 6.

Major F. M. Von Donat has published in English a lecture he delivered to the Berlin Geogr. Soc. in 1892 on the drainage of the **Pontine Marshes**. In the time of the Volscians, 500 years B.C., this tract was a well-drained and flourishing country, with, it is said, twenty-six towns. After the conquest by the Romans, the drainage system was allowed to perish, and the many schemes that have been since proposed either were not carried out or failed in their object. The extensive works constructed 120 years ago by Pius VI. were based on a bad plan, but would be of great service if incorporated in a new system. Major von Donat estimates that his scheme would entail an expense of £40,000—in itself rather a large sum, but easily recouped if the result were a success, and the greater part of the 120 square miles reclaimed for agriculture. Now unworkable, and spreading malaria over the surrounding country, this tract would, if relieved of its excess of water, be more productive than any other land in Europe. Professional engineers only can judge of the practicability of Major von Donat's scheme ; but what was done 2400 years ago can surely be done now, and one would expect that the Italians would readily provide the money to restore this and other districts to their healthy and fertile condition in ancient times.

NEW BOOKS.

A Naturalist in Mid-Africa. Being an account of a Journey to the Mountains of the Moon and Tanganyika. By G. F. SCOTT ELLIOT, M.A., F.L.S., F.R.G.S. Pp. xvi + 413, 49 figs., frontispiece, and 4 maps. 8vo. London : A. D. Innes and Co., 1896. Pp. xiii + 413. *Price* 16s.

Mr. Scott Elliot was led by "a most inconvenient love of botany " to undertake explorations in the Victoria region, Ruwenzori, and the central watershed of Ankole, Karagwe, and East Tanganyika. One of the results of his hazardous journeyings is the book before us—a record full of interesting facts and suggestions, both scientific and practical, and spiced with adventures and mishaps which some-times came perilously near putting in a full stop. In this last connection, as else-where, the author's dry humour is exquisite.

There is much in the book that we cannot possibly deal with, for if there be such a person as " a mere naturalist," Mr. Scott Elliot certainly is not he. Indeed, if we except missionaries, for most of whom his admiration is too great to allow him to do much more than thank them, there seems scarce any relevant subject which this naturalist in Mid-Africa does not deal with.

He discusses the possibilities and impossibilities of development in British Central Africa, the relative merits of proposed railways and routes of water-transport, slavery and the Arab question, the proper method of dealing with the natives, the most practical outfit, besides geology, meteorology, natural history, and much more. No one can accuse him of being a stylist ; he does not charm the senses like "the naturalist of La Plata," but if he is sometimes brusque, one feels that his grip is firm. He may leave one doubting, but never as to his meaning ; witness the charming footnote on Weismannism, p. 209. The last word suggests the occasionally obvious lack of accuracy in proof-correcting, for Weismann does not spell his name with two *ss*, and the famous author of *Pflanzenleben* is Kerner, not Körner. But who expects a travelling naturalist to care for spelling ? In the same cavilling spirit we may notice that few of the figures could be called works of art, but it is probably better to have conscientiously reproduced photographs than pretty pictures conjured up by an artist who never saw the realities. We venture to cite a few samples which may convey some impression of the interest and manner of the book :—

Thorny plants are protected by their thorns from antelopes and giraffes ; but the reason of the thorns is to be found in the action of the intense heat, "which, by transpiration, makes the walls of the cells very thick and hard, and thereby produces a cure for the evil which it itself brings about."

The main advantage of long necks—to Clarke's gazelle, ostrich, and giraffe, etc., which are found where the bush is rather thick and short—is the " power of obser-vation, which renders them extremely difficult to approach, and not the ability to crop trees."

"It gives one a curious impression to march day after day over lovely grass plains covered with zebra, hartebeest, and other antelopes, past beautiful lakes where geese, ducks, and other water-fowl almost cover the water ; then, perhaps, through a dense virgin forest with magnificent timber, and all the time to see no human beings whatever. Yet the country is healthy and in every way suited for Europeans, while we have hundreds of people in England who do not know where to turn for employment."

"Then at last I struggled up to the forest, and managed to get to about 7000 feet, where I found myself in a new world. There was the common English sanicle, a beautiful meadow rue, a cerastium, etc. . . . The difficulty is to understand *how they got there.* I have never seen birds either swallowing or carrying about on their feet the seeds of the three plants named (I could also mention the forget-me-not, the willow herb, the St. John's wort, etc.), so that one must hold it as proved that there was once a continuous European climate from the original birthplace of most of our genera (the Garden of Eden—vegetable department) to Ruwenzori, Kenia, and Kilimandjaro on the one side, and Ireland on the other." Chapter xii. is devoted to a serious consideration of the problem.

."On the rotting sticks and leaves, in this rainy and hot climate, fungi were curiously abundant. I have gathered perhaps seventy, all different, in as many minutes. One form has the most curious distribution I have ever heard of. It is only known from three places—Texas, Japan, and Ruwenzori !"

" The conditions in these Ruwenzori valleys are very much those of the Carboniferous age. They may be roughly described as a 'dim, hot steaminess.' Although it is apparently fanciful, it is none the less literally true, that there has been a continuity of climate and physical conditions from that ancient date to the present, though the climate and conditions may have migrated over a whole continent in the interval."

" When I asked Makwenda how they punished a man who had killed somebody, he replied, 'They never did anything bad now they knew Europeans'—a delicious piece of humbug."

"It will scarcely be credited that these huge beasts (rhinoceros), after running wildly to and fro for a minute or two, fled precipitately, with little Bobby barking courageously after them for a quarter of a mile."

" My men were attacked in only three places during the whole of my journey of some 2500 miles. The first was in Ankole, just at the spot where Captain Langheld killed thirty-five people in his fight with Antari. The second was at Tenge-tenge's, almost on the site of Emin and Stuhlmann's camp. .The third was in Urundi, just where I entered the particular district crossed by Dr. Baumann. The conclusion is obvious." Impartial as the reader may be, it will, we think, be difficult for him to avoid concluding from Mr. Scott Elliot's narrative that the Germans are *etwas zu übermuthig.*

But it is dangerous to begin quoting from a book which abounds, as this does, in interesting facts and acute suggestions. We are not competent to discuss the author's corrections of Mr. Stanley's geography, nor his conclusions on the distribution of plants, nor his practical advice as to the development of the country, but we would very cordially recommend the work as a solid, unpadded record of travel, rich both in dramatic and scientific interest. Nor can we leave it without expressing our admiration of Mr. Scott Elliot's "pluck" and self-controlled respect for life.

A Breath from the Veldt. By JOHN GUILLE MILLAIS, F.Z.S. London : Henry
	Sotheran and Co., 1895. Pp. 236. *Price £3, 3s. net.*

This magnificent volume is of exceptional interest as a work of Art, as a contribution to Natural History, and as descriptive of the country and character of the Boers, so much in evidence at present. The author's father, the eminent President of the Royal Academy, contributes a frontispiece, but the remaining numerous and beautiful sketches are by Mr. Millais himself. Nowhere have we seen more lifelike or interesting drawings. Such a picture as that of the "Locust Storm" enables one thoroughly to realise this dire visitation. Again, Mr. Millais' pictures of the Lesser Green cormorants fishing and rhinoceros birds feeding on a Koodoo cow, are among the most curious we have ever seen. What particularly impresses any one examining the pictures in this remarkable volume is Mr. Millais' truthful, although necessarily unconventional, drawings of wild animals; his minute observation of their habits and appearance; and his graphic representation of certain of their attitudes not generally known, or at least not previously depicted. The illustrative part of the volume is thus thoroughly original, and naturalists owe a debt of gratitude to Mr. Millais for leaving the beaten path of conventional art, and for giving to their science the benefit of his exact representations of the wild animals of the Veldt. To sportsmen, also, the volume must be of surpassing interest, as it casts so much new light on the habits of big game.

As for the literary part of the volume, it is as unconventional and charming as the sketches. At the same time, Mr. Millais writes with great precision on zoological subjects, as such a description as the following shows:—" I should like to add a few notes from personal observation of the rhinoceros bird (or 'rhinaster vogel,' as it is called by the Dutch) which, to my mind, is the most interesting bird I have ever seen. In size, it is about equal to our common starling, which it also resembles to some extent in habit, being fond of resting on the backs of animals and relieving them of the tics that infest most creatures, domestic or wild, in the bush country. To aid it in these forays after insects, nature has endowed the rhinoceros bird with a tail of horny feathers for support, and claws of such extraordinary strength and sharpness that they can cling on securely while performing feats of acrobatism most amusing to witness. The prehensile power of the claws is, as I found by experience, so great that when a dead bird which had grown stiff was thrown on to the back or sides of an ox, so that the feet touched the animal's hide, the claws held fast at once and could not be easily withdrawn. It is most interesting to notice the way in which a party of these birds will move about on the body of a horse or ox, searching every part of his skin as they run or hop over it in the most lively fashion."

With regard to the Boers, the following paragraphs are worth quoting:—" I like the Boer, and admire his free, independent spirit; and though amongst a certain class of Englishmen and Afrikanders in South Africa it is almost looked upon as being unpatriotic to vent such opinions, I can certainly say that the Boer is, and has been, much wronged. There are thousands of Boers at the present day who are wishing to extend the hand of friendship and comradeship to the English and South Africans, but their fears of being swindled and imposed upon hold them back, their experiences in the past having lain only amongst Englishmen of an inferior type. All this is of course only the result of ignorance. Some Boer farmer has been badly bitten by a travelling loafer or rascal calling himself an Englishman—for till quite recently men of this stamp swarmed in South Africa. What is the result? The Dutchman naturally includes the whole nationality of that rascal in a sweeping condemnation, and hates them accordingly . . . I do not wish to make out the Boer as a sort of hero by any means. Most of them are very ordinary creatures—stupid, conceited, and lumpish to the last degree; but at the same time, I must say it is enough to make any man, of whatever nation he may be, wild to hear of a race of fine, hardy pioneers and farmers, as they are, abused by a lot of third-rate potmen, bar-loafers, and counter-jumpers, who know as little of the real Boer as they do of the solar system, and judge of them only by the class of anti-English Boers which they themselves have done so much to create."

Peru.   Von E. W. MIDDENDORF.   III. Band : *Das Hochland von Peru.*   Berlin :
   Robert Oppenheim (Gustav Schimdt), 1895.

This, the third and last volume of Dr. Middendorf's great work on Peru, lacks
none of the merits by which its predecessors were characterised. Having in the
previous volume described the Peruvian Lowlands extending from the Pacific to
the foot of the Cordilleras, which run parallel to the coast-line, he now conducts us
to these mountains themselves, and makes us familiar with their scenery, their
passes, valleys, plateaus, lakes, and rivers, as well as with their towns and villages,
and their inhabitants. He has taken what we conceive to be the most attractive
and effective method of setting forth the facts with which he deals. Led by his
love of mountaineering and his zeal for antiquarian exploration, he visited, in a
series of journeys beset with many hardships, difficulties, and dangers, all parts of
the Peruvian Highlands from Lake Titicaca in the south to Chachapoyas, the
capital of the department of Amazonas, in the north. Each of these journeys
forms the groundwork of a narrative in which the author relates with great liveli-
ness and picturesque effect all the incidents he met with on his way. The work,
although it contains some 600 pages, and these of more than ordinary length, never
tires the reader, so interesting and varied is the information it conveys.

Dr. Middendorf's antiquarian researches were partly of a philological nature,
and in his preface he presents us with a summary view of his conclusions regarding
the indigenous languages and their relation to each other. In the coast region, as
one might expect, these languages are dying out, or are rather extinct, except in a
single locality in the north—in the town, namely, of Eten. In the hill-country again,
they still survive, and, though much corrupted and mixed with Spanish words, form
to this day the general medium of intercourse. The Quechua speech which was introduced by the Incas, is that which is still commonly used in most of the provinces
which pertained to their old kingdom, while the still older Aimará is now only
heard in the country to the south of Lake Titicaca. The Aimará race, however,
must have at one time extended far in the opposite direction, for in the north of
Peru there are mountains, rivers, and places with names which have no meaning in
the Quechua tongue, but which are explicable in the Aimará. The condition of the
Indian races as depicted by our author is deplorable. On his railway journey from
La Paz to Cuzco he saw, near one of the railway stations, a great crowd of Indians,
mostly women, wearing the round Quechua hat, and among them not a few Aimarás,
all of them pinched with hunger, sad and sorrowful, ragged and dirty ; and of
those whom he saw in their villages during his travels, he has no better account to
give. The Aimarás, however, notwithstanding their present degraded and pitiful
condition, had of old attained a degree of civilisation which in some respects surpassed that of the Incas by whom they were conquered. This is shown by the
superior skill and taste displayed in their architecture, of which remains are still to
be found, as well as by their dexterity in some of the arts which they still practise,
one of which is weaving.

Dr. Middendorf introduces us to his subject by presenting a general view of the
mountain system of South America from the Straits of Magellan up to the Isthmus
of Panama, dwelling more particularly upon the features which characterise it in
Peru. When seen from the banks of Lake Titicaca, the snowy peaks of the
Cordillera do not appear so high as they are in reality, being only 12,000 feet above
the lake level. Their stupendous magnitude can only be realised by one who views
them from the eastern valleys. The summits of the coast-chain are so remote that

they cannot be seen from the lake, but northward from it they gradually approach the eastern chain until their bases touch each other behind the small place Santa Rosa, and form a deep valley, which after a long ascent culminates in the famous pass called El Paso de la Raya, where the Huillcanota has its source. The heights on the left side of this river belong entirely to the coast-range, but are not its main summits, whereas the mountains to its right consist of an unbroken chain of loftier altitude. This forms the continuation of the Bolivian *Cordillera real*, the proper Cordillera of the Andes. Our author here remarks that the name of Andes which is now applied to the whole of the South American high mountain-chain, is in Peru limited to this section of it which has from of old been so designated. The coast-range is called simply La Cordillera, and is distinguished into the eastern and western Cordillera. The people living to the north-east of the capital, Cuzco, are called *Antis*, their territory *Anti-suyu*, whence this name was applied also to the lofty mountains which separate this territory from the river Huillcanota.

In the regions between the Pass of the Raya on the one hand, and Ayacucho and Huancavelica on the other, the mountain system of Peru attains its greatest breadth. Here the summits present themselves in such a confused mass that it becomes no longer possible to trace the direction of the chains, and the traveller on his way from Cuzco to Ayacucho often reaches elevated points from which he sees himself surrounded on every side with snow-clad peaks. Four rivers have their sources in this knot of mountains, and these when united form the Ucayali, which, after running parallel with the more western stream of the Marañon, joins it to form one of the mightiest rivers in the world—the Amazon.

It is well known how rich the Peruvian mountains are in minerals—especially silver and gold. Our traveller visited several of the mines. One of these was situated in the range of the Cordillera Negra in Middle Peru, at an elevation of some 13,000 or 14,000 feet. The proprietor, who gave Dr. Middendorf a friendly welcome, belonged to our ubiquitous race ; he was *ein Schotte Namens Bryson*. A picture is given of his hospitable but solitary place of abode.

One of the most interesting visits is that which our author made to the island of Titicaca, in the lake of the same name. It is famous in Peruvian history ; for here, according to the native traditions, the first Inca, Manco Capac, and his wife Mama Ocllo, made their appearance from behind a rock, and announced that he and his wife were children of the Sun, and were sent by that glorious presence to instruct the simple tribes. How the Incas came to subjugate these tribes, who were as far advanced as themselves in culture, is attributed in this work to the advantage they derived from their political unity, to their well-ordered government, and good system of laws, to their success in promoting unity of speech, and facility of transit throughout their dominions, and to their suppression of the worship of gods in favour of the Sun-culture. It would appear from the fact that they introduced their own religion and brought several races under subjection, that the Incas have not always displayed the mildness and forbearance for which their eulogist, Garcilaso de la Vega, so often extols them.

The three volumes are all well printed on good paper, and are also copiously and beautifully illustrated. An English translation of the work is much to be desired.

Tafilet. The Narrative of a Journey of Exploration in the Atlas Mountains and the Oases of the North-West Sahara. By WALTER B. HARRIS, F.R.G.S., author of "A Journey through the Yemen," "The Land of an African Sultan,"

"Travels in Morocco," etc. Illustrated by MAURICE ROMBERG from Sketches and Photographs by the Author. Edinburgh and London : William Blackwood and Sons, 1895. Pages xii + 386. *Price* 12*s.*

Mr. Harris has given a most interesting narrative of exploration in the Trans-Atlas region of Morocco, which has very rarely been visited by Europeans, and of which very little was known. The late Mr. J. Thomson in a lecture to the Scottish Geographical Society (reproduced in this *Magazine* for 1889, pages 169-180) told how, crossing the Atlas range, he reached the imposing castle of the Kaid of Glawi or Glawa on the southern slope, but was then compelled to turn back. Mr. Harris not only crossed the Atlas, but, passing Glawi, travelled for a considerable distance along the valley at the foot of the Anti-Atlas, and rounding the range penetrated as far as the district of Tafilet, from which come the well-known dates. He travelled disguised as a Mussulman, running the risk of certain death if his true character were discovered, and disregarding his sufferings from want of food, intense cold at night, and the exhausting toil of long marches, often on bare foot, through deserts. The privations were so great as to produce a serious illness, which would have ended in death, had he not on his arrival in Tafilet, where the Sultan was encamped with his army, been so fortunate as to obtain the assistance he needed. Altogether his journey is one of the most notable pieces of adventurous exploration that have been accomplished by any of our countrymen for many a year, and although Mr. Harris lacks the literary faculty, and his sentences are not infrequently bald and awkward in construction, wanting in ease and grace, and showing at times an imperfect apprehension of the meaning of words, yet in spite of this the narrative is one of engrossing interest. He never wastes time or space on unimportant or irrelevant matters, but has an eye for and notes what is really worth telling, and makes the reader feel at home in the country traversed, among the wild Berbers, whose character and customs are described, and above all with the companions who shared his trials, and are praised in the most generous spirit. To any one who enjoys reading the perilous adventures of an enthusiastic explorer the book can be most strongly recommended. Mr. Harris generally does not fail somewhere in his book to explain the vernacular terms he is compelled to use, but it would have been convenient if they had all been grouped together in a separate glossary. The book is furnished with two maps, the larger of which is a detailed map of the route between Marakesh (Morocco city) and Tafilet, taken from the map published by the Royal Geographical Society. There is, it must be noted, a remarkable discrepancy between this map and Mr. Harris's own description of the country traversed in regard to the position of the portion of the Trans-Atlas range which is known by the name of Jebel-el-Kebir. According to the map, Jebel Saghru (Anti-Atlas) and Jebel-el-Kebir are one continuous range, and Mr. Harris passed along the northern slope of the latter, and rounded it before he turned south to Tafilet. But, according to the text, Jebel Saghru and Jebel-el-Kebir are separated by a gap, and Mr. Harris passed along the valley between them, and rounding only Jebel Saghru turned south to Tafilet, Jebel-el-Kebir lying farther to the east. Mr. Harris, telling what he actually saw and heard, is probably right, but, if so, the map needs correction.

From Far Formosa : The Island, its People, and Missions. By G. L. MACKAY, D.D. Edited by the Rev. J. A. Macdonald. Edinburgh : Oliphant, Anderson, and Ferrier, 1896. 346 pp. *Price* 7*s.* 6*d.*

"Mackay of Formosa" is a name familiar to all who have any interest in the Far East. More than that of any other single man his name is associated with this fruitful island, which has recently passed under the rule of the Japanese Emperor. A quarter of a century spent amid its relaxing airs and trying climate is an experience which few could stand, and none could possibly desire. It is, therefore, · matter for congratulation that his friends have prevailed upon the eminent Canadian missionary to give to the world a permanent record of much that he knows of a little-known land. The early chapters are biographical, and tell how the young enthusiast, after pursuing theological studies in Toronto, Edinburgh, and Princeton, finally found his way to China. There then follow concise and instructive descriptions of the island of Formosa, its geography, history, geology, flora, fauna, and people. The Chinese immigrants and their descendants are gradually pushing the aboriginal tribes, for whom they have a great contempt, inwards towards the mountains. The mode of government described by Dr. Mackay, happily superseded now by the more wholesome Japanese rule, is the usual Chinese system of legalised robbery, bribery, and injustice. Into the mountain fastnesses, however, the unscrupulous pig-tail has not penetrated; and here "savage life can be seen in all its lights and shades." These Malay hillmen live, as their ancestors have lived for unknown centuries, on forest game, big and small. What little farming is done, is done chiefly by the women. Two particularly interesting chapters give a glimpse of this savage life, as witnessed by Dr. Mackay and Captain Bax of H.M.S. *Dwarf*. They narrowly escaped death because, in their innocence, they began to sketch some of the huts. The great ambition of these mountaineers is to bring in one Chinese head ; and a man is (theoretically) not allowed to marry till he has proved his merit as a "head-hunter." But Chinese heads are not so easily secured, and a chief may grant special dispensation to one who has made a reputation in hunting the deer and wild boar. On the return of a successful head-hunting party from their murderous raid, the whole village holds a wild bacchanalian *fête* for days. The skull is finally hung as a trophy on the wall of the brave's hut who did the deed. A considerable part of the book is naturally taken up with an account of Dr. Mackay's missionary labours ; but there is hardly a page which does not touch upon some interesting custom or superstition, more or less peculiar to the inhabitants of Formosa. The editor, to whom was intrusted the assortment of the material, has done his work well, having evidently succeeded in his aim of preserving not only the substance, but the crisp, vigorous style of the author. Several maps, and nearly twenty illustrations, reproduced from photographs, enrich an eminently readable and instructive book.

The Geology and Scenery of Sutherland. By H. M. CADELL, B.Sc., etc. Second edition. Edinburgh : Douglas, 1896. Pp. 104. *Price 4s.*

In this revised and enlarged edition, Mr. Cadell has brought his interesting little work up to date. He has contrived to pack into small compass a very considerable amount of geological information, which he sets forth in so clear and entertaining a manner that no one can fail to follow his interpretation of the stony record of Sutherland. The book is illustrated with a number of excellent sketches by Mr. G. Straton Ferrier, Mr. B. N. Peach, and the author himself, and with orographical and geological maps of Sutherland and Caithness, and accompanying sections. No visitor to Sutherland who wishes to have an intelligent appreciation of that wild region, should be without this excellent guide to its geological structure. Mr. Cadell shows, as only a geologist can, how intimately dependent on that structure is the [physiographical aspect of the country, and how much [the scenery gains for us in interest when we can read in the fabric of mountain and glen the strange story of their origin.

The Place-Names of Fife and Kinross.   By W. J. N. LIDDALL, M.A., Advocate. Edinburgh : W. Green and Sons, 1896. 8vo., pp. xiv. + 58.

The author's introduction is most interesting as indicative of the information to be found in place-names, and deserves attention.   The most surprising result as to the Fife place-names is that, with the exception of Kilwinning, all those beginning with *Kil* are derived from *coille,* a wood, rather than from *cella,* a church.   They number altogether over twenty, and of most of them there can be little doubt about the origin ; respecting Kilconquhar, Kilduncan, Kilmany, Kilnynane, and Kilrenny, some may perhaps differ.   The nomenclature is purely Goidelic in origin and free from Brythonic admixture, or nearly so.

This small volume is a scholarly and important contribution to the subject of place-names and their etymology, and will throw important light on names quite outside the Fife area ; to natives of the "Kingdom" it will be specially interesting.

Die Liparischen Inseln. Fünftes Heft—Filicuri. Prag : Heinr. Mercy, 1895.

The small island described in this number is less than four square miles in area.   The coast presents many picturesque views, such as the great blocks of the Sciara, which are reproduced here.   The island consists of a volcanic cone, some 25,000 feet high, surrounded by several lesser elevations.   It is almost treeless.

Australien und Ozeanien.   Eine allgemeine Landeskunde von Professor Dr. WILHELM SIEVERS.   Mit 137 Abbildungen im Text, 12 Kartenbeilagen, und 20 Tafeln in Holzschnitt und Farbendruck.   Leipzig und Wien. Biblio-graphisches Institut, 1895. Pp. 521. *Price* 16 M.

This is the fifth and concluding volume of Dr. Sievers's valuable and most attractive series, which now forms a complete compendium of *Erdkunde* treating the various geographical aspects of the continents in a systematic and scientific manner.   The series was originated and planned by Dr. Hans Meyer, the well-known publisher, and explorer of Kilimandjaro, and it may safely be said that, but for the exceptional resources of the "Bibliographisches Institut," such an elaborate and richly illustrated work could never have been produced at such a cheap price.   Australia and Oceania in the present volume are treated in a similar style to the other continents already dealt with.   The author begins by sketching the history of the discovery of Australasia and the islands of the Pacific, leading on to an account of the inner exploration of the Australian continent.   Then, after a general preliminary survey of the whole region, he takes up each of the special aspects of geographical study.   Starting with an investigation of the structural geology and surface relief, he reviews separately the phenomena of climate, the varied animal and plant life, the different races who people these scattered islands, their political and social problems, and finally sums up the present development in trade and industry, even to the gold mines of Coolgardie.   A supplementary chapter summarises all that we know of the great south Polar land of Antarctica, and the fact that it includes the results of Borchgrevink's discoveries shows that it is up to date.   Well selected illustrations and maps, representative of the most charac-teristic and special features of interest are liberally supplied throughout the work. They are beautifully executed, although some of the chromolithograph plates would be more true to nature if less highly coloured.

In such a compendium a feature of great value would have been numerous references to original authorities and their works, but that is altogether wanting ; without it these volumes are of much less value to the student who would pursue his research further.

Ritters Geographisch-Statistisches Lexikon. Achte Auflage: Unter der Redaktion von JOHS. PENZLER. Zweiter Band, L-Z. Leipzig, 1895. 1202 pp.

The first volume of this new and enlarged edition of Ritter's *Lexikon* has been reviewed in a previous number of our Magazine. Of this, the second and concluding volume, it will therefore be sufficient to note some of the inaccuracies which it contains, confining ourselves, as seems appropriate, to localities in Scotland. It is a little strange that a gazetteer so comprehensive, and so full of minute detail, should so often lapse into error. It is true that the errors are for the most part of a trifling nature; but, for that very reason, they might have been avoided altogether with a little more care in revision. Examples of these are:—" Feebles " (*s.v. Selkirk*) for " Peebles "; " Men Lomond " (*s.v. Stirling*) for " Ben Lomond "; "Locharron " (*s.v. Lochalsh, Lochcarron, Scourie*) for " Lochcarron "; " Hangshaw," " Kilvin," and " Cartwater " (*s.v. Lanark*) for " Hagshaw," " Kelvin," and " White Cart Water "; and Stornaway (*s.v. Lewis*). And although Sutherland is correctly spelt in its own place, it appears as "Southerland " in other places, as, for example, under *Lairg, Lochinver,* and *Scourie.* Nairn is carefully mis-pronounced " Naïrn "; Loch Sunart masquerades as "Laffuart" (*s.v. Shiel, Loch*); while the spellings "Musselburg," "Edinburg," and "Ferthgolf," all occur within the few lines descriptive of Musselburgh. It is somewhat disconcerting to find the Firth of Forth figuring under so many disguises. At one time it is *Forthbai,* at another *Forthgolf,* and again it is " die Mündung des Forth "—probably the best German equivalent of them all. Under "Tranent" it appears simply as the "Firth of Forth "; and this, of course, being the local name, is absolutely correct. It cannot be expected, however, that a gazetteer written in German should employ anything but German names for physical features. In English gazetteers, the words used to denote the mountains, lakes, and rivers of foreign countries are almost invariably the recognised English words, and not their foreign equivalents.

Other errors are—" Orchill " (*s.v. Strathmore*) for " Ochil "; " Pyrawall " for " Pierowall " (*s.v. Westray*); " Wighton " (*s.v. Whithorn*) for " Wigtown " or " Wigton "; " Rhoe " and " Rhove " (*s.v. Rhoe* and *Shetland*) for the " Meikle Roe " in Shetland; and " Nyan " for " Ryan " in the loch of that [name beside the Rhinns of Galloway, under which last title the spelling is " Rhyns " in the *Lexikon.* Tobermory appears as " Tobermorry," Gareloch as " Loch Gare " (*s.v. Row*), and the Moray Firth is always " Murraygolf." Of course, " Murray " is quite a correct spelling, though out of fashion. It may be noted, however, that there is no mention of the *county* of Moray or Murray. Indeed, there are several important omissions. The districts of Menteith and Stormont are not mentioned, although Morvern and Mearns both occur. The united parish of Whitekirk and Tyninghame, in East Lothian, does not appear under either of its names, neither is Yester parish, in the same county, mentioned. But, for no special reason, the neighbouring parish of Whittinghame is duly entered—although with the error that it is situated in the county of "Huntingdon." There is no mention of the parish of Nenthorn in Berwickshire; the island of Pabay does not occur, in spite of the fact that many other of the Scottish isles, of no greater magnitude, are carefully entered; and although mention is made of a " Ruthven " in the United States, there is no word of the several places of that name in Scotland. The parish of Walston in Lanarkshire is also ignored. One curious inaccuracy is that Marykirk, correctly placed in Kincardineshire, is at the same time stated to lie to the south-west of *Berwick*—a mistake evidently due to a confusion between Marykirk and Ladykirk. Lochmaree is said to be in the *county* of Dingwall.

From a consideration of these various inaccuracies, it will be seen that, after making all allowance for printers' blunders which it is impossible to avoid altogether, the greater number are due to a want of careful revision. Their occurrence is all the more to be regretted as the *Lexikon* is, on the whole, very reliable, and includes much information of the most minute description.

The Natural History of Plants: their Forms, Growth, Reproduction and Distribution. From the German of ANTON KERNER VON MARILAUN, Professor of Botany in the University of Vienna. Translated and edited by F. W. OLIVER, M.A., D.Sc., Quain Professor of Botany in University College, London, with the assistance of MARIAN BUSK, B.Sc., and MARY F. EWART, B.Sc. With about 2000 original woodcut illustrations and sixteen plates in colours. Vol. i., Biology and Configuration of Plants, pp. xiv. + 777 ; vol. ii., The History of Plants, pp. xiv. + 983. London and Glasgow : Blackie and Son, Limited, 1894 and 1895.

The completed translation of Kerner's *Pflanzenleben*, which would have been reviewed sooner had it been less interesting, forms an impressive couple of volumes, well deserving to share the popularity of the original.

Professor Oliver has re-written much of the systematic part of the book, and has used editorial discretion both in adding and in omitting, while always conserving the intention of the author. That such a book should be in every detail up to date would mean that botany was a finished science, and we shall pay the work no such foolish compliment. Nor shall we go to the opposite extreme of hunting for passages where, to our particular knowledge, the text seems to lag behind discovery. The important fact is that the book is *facile princeps* of its kind.

The qualities which give it this high place may perhaps be summed up in the word *vitality*, by which we mean that the plant is never a dead dried "specimen." in this book, but is always a living creature—feeding and feeling, growing and reproducing, struggling, varying, evolving. It is true that many a botanical naturalist has seen more or less of this vision ; it has been Kerner's credit to reproduce it so vividly that his readers must see it with him. He has been able, moreover, to base his natural history of plants on a broad foundation of observation and experiment, to sweep in the results of analysis so as to build up a living whole, and to illumine this with the light of evolution.

We confess to a very strong admiration for this book, particularly for those parts of it which deal with the relations of plants to one another, to animals, and to their surroundings. Where the author deals with the internal physiology of the plant he is not always so effective—largely, of course, because vegetable physiology is a bit backward.' Moreover, the systematic part—to have been quite in tone with the rest—should have been more of a real history of the pedigree of plants, for which, however, the botanist's data and insight are still insufficient.

The translation is admirable, and the book is beautiful. The illustrations are probably the best that have ever been published in a work on general botany. In short, we think that author, editor, translators, and publishers have much reason to be proud, and the public to be grateful, on account of *The Natural History of Plants*.

Handbook for Travellers in Algeria and Tunis. By Sir R. LAMBERT PLAYFAIR, K.C.M.G. Fifth edition, thoroughly revised. London : John Murray, 1895. Pp. 363. *Price* 10s. 6d.

This new edition brings a most excellent handbook up to date. It has been thoroughly revised, and to a large extent re-written, several new maps and plans having also been introduced. The result is most satisfactory. A capital account is given of the new winter-resort, the oasis of Biskra ; and the author has noted the latest archæological discoveries at Timegad, Tebessa, and Tipasa. He makes a few guarded remarks upon the creation of the new French naval base at Bizerta, and gives a short description of its surroundings ; those who wish fuller information can find it in the *Annales de Géographie* (July 1895).

Cassell's New Geographical Readers. Fifth Book—Europe. Sixth Book—The British Colonies and Dependencies. Seventh Book—The United States. London, etc. : Cassell and Co., 1895. *Prices* 1s. 6d., 1s. 6d., 1s. 9d., *respectively.*

The earlier books of this set have already been noticed from time to time. The information is very full—perhaps almost too detailed for *Readers*—and the illustrations are exceedingly helpful. Besides the descriptions of countries, there are brief explanations of the seasons, latitude and longitude, currents, etc.

The Statesman's Year-Book. Statistical and Historical Annual of the States of the World for the Year 1896. Edited by J. SCOTT KELTIE, Assistant Secretary of the Royal Geographical Society, with the assistance of I. P. A. RENWICK, M.A., LL.B. London : Macmillan and Co., 1896. Pp. xxxii + 1164.

The present issue is as excellently edited as its predecessors. The latest political and other changes are, as far as we have at present been able to test the book, been recorded. Perhaps the division of the North-West territories of Canada into provinces should have been mentioned, though at present of little significance.

A noteworthy innovation is the illustration of recent territorial changes by means of maps. Four clear and well-executed maps are given showing the boundaries on the Pamirs, the Siamese boundaries, the British and Venezuelan claims in South America, and the recent changes in Bechuanaland, respectively. Most readers will look for the note on the disputed territory in South America under British Guiana, not under Venezuela where it is placed.

The Australian Handbook, Shippers' and Importers' Directory and Business Guide for 1896. London, Melbourne, etc. : Gordon and Gotch. Pp. 632.

The *Handbook* has been repeatedly noticed in these pages, and therefore there is nothing much to be said of the new issue except that, as usual, it seems to have been brought well up to date in all its divisions. It is a most useful book of reference, not only for the business man, but also for the geographer, in all that relates to Australia and the neighbouring colonies.

Beobachtungen der Russischen Polarstation an der Lenamündung. I. Theil. *Astronomische und Magnetische Beobachtungen.* 1882-4. Bearbeitet von V. FUSS, F. MÜLLER, and N. JÜRGENS. St. Petersburg : 1895.

This is the record of important scientific work done, at the mouth of the Lena, by an expedition organised by the Imperial Russian Geographical Society. The letterpress is in both Russian and German. The volume consists largely of tables of numbers. Of general interest, however, is the account of the experiences of the expedition, with the difficulties and hardships which are inseparable from a winter life in Arctic regions. The great value of the observations taken lies in their continuity, which reflects great credit on the scientific members of the staff.

NEW MAPS.

EUROPE.

EUROPA, Schulwandkarte von ——, nach dem Entwurfe und unter der Leitung des Vincenz v. Haardt ausgeführt. Massstab 1 : 4,000,000.
Verlag v. Ed. Hölzel in Wien, 1896.

A good wall-map, showing the mountains and rivers clearly, and with the principal towns in large type. A very useful map for schools.

PELOPONNES, Vegetationskarte des ——, von Dr. A. Philippson. Massstab 1 : 625,000. *Petermanns Mitteilungen, Tafel 18*, 1895.

ISLAND, Geologische Karte des südöstlichen ——, von Th. Thoroddsen. Massstab 1 : 1,000,000. *Petermanns Mitteilungen, Tafel 19*, 1895.

AMERICA.

PERU, Mapa del ——. Por A. Raimondi. Escala de 1 : 500,000. Fojas 1-20, 22 y 24. *Presented by the Geographical Society of Lima.*

Good maps of South America States are rare, and therefore this large map compiled by the well-known Peruvian geographer, Sr. A. Raimondi, is most welcome. It gives a large variety of information, forests, marshes, railways, bridges, ruins, passes, and many other details, having their distinctive signs. Much of the country is yet unexplored, and here the map does good service in showing the situation of the unknown tracts. There still remain twelve sheets to be issued, some of them of very interesting parts of the country, as Lake Titicaca, for instance.

The execution, for which MM. Erhard F^res, of Paris, are responsible, is good, and the colours agreeable and not too dark. The mountains are a little stiff, perhaps, and railways and roads would have been more conspicuous with rather a darker shade.

AUSTRALIA.

QUEENSLAND, Geological Sketch Map of Part of the Eastern Margin of the Artesian Water District of ——. By R. L. Jack and A. Gibb Maitland. *Annual Progress Report of the Geological Survey for the Year* 1894.

ATLAS.

HANDY ROYAL ATLAS OF MODERN GEOGRAPHY. By the late Alexander Keith Johnston, LL.D., F.R.S.E., etc. With additions and corrections to the present date by T. B. Johnston, Geographer to the Queen, F.R.S.E., etc. Price £2, 12s. 6d.
W. and A. K. Johnston, London and Edinburgh, 1896.

This is a selection of maps from the Royal Atlas, some on the original scale, others reduced. Since the larger atlas was produced several small changes have been made in the maps to bring them up to date. The *Atlas* is of a convenient size and well bound. We take this opportunity of suggesting that a good map of Siam and Indo-China should be added to the Royal Atlas when a new edition is issued. There is none larger than that in the general map of Asia.

THE SCOTTISH
GEOGRAPHICAL
MAGAZINE.

EASTERN TURKEY IN ASIA AND ARMENIA.

By Captain F. R. MAUNSELL, R.A.

(Read before the Society in Edinburgh on April 9th.)

EASTERN Turkey in Asia has been the scene of some of the most stirring events in the world's history. The tide of conquest has ebbed and flowed across it many times, the last wave being that of the Osmanli Turk, and since his advent, now some centuries back, it has lain fallow, and its geography and natural features become quite forgotten by the Western world. Recent events have revived the interest in the country, and a description of its very varied topography, and of the races that form its inhabitants, will be useful at the present time.

Under the stress of continual wars, the political boundaries have been subject to frequent variations; the Osmanli are now the governing race, but the remaining sections of the population are composed of numerous discordant elements—rival races with rival interests and religions. Of the Christian races, the Armenians are much scattered through the northern portion of the country, principally in the large towns and villages on the plains; but in no district, except, perhaps, in the mountains of Zeitun, is the population exclusively Armenian or Christian, but there is always an admixture of more warlike Moslem races. Thus, there is no definite district that can be called Armenia because its inhabitants are solely, or even principally, Armenians.

I propose in this paper first to deal with the high plateau enclosed in the quadrilateral, Erzerum—Mush—Bashkala, and Bayazid, to which the name of Turkish Armenia is often applied, and which once formed part of Greater Armenia, when Armenian kings ruled the land; and then to direct your attention to the south-east—to the great mountain districts of Kurdistan, which follow the frontier of Persia, and overlook the wide plains watered by the middle Tigris and Euphrates. The whole of North-east Turkey in Asia is an elevated plateau, diversified by many lofty ranges, which in the north abut steeply on the Black Sea, and in the south on the Tigris valley and Mesopotamian plain. On the west it merges into the somewhat lower plateau of Anatolia; on the north-east it continues across the Russian frontier until it overlooks the valleys of the Kur and Rion; while on the east, it connects with the plateau of Central Persia, slightly lower in elevation, but with many similar charac-teristics. In the quadrilateral portion of the plateau I have just mentioned, three of the great rivers of the country have their origin; the Murad Su, one arm of the Euphrates, rises near Diadin, and flows out past Mush; the Frat Su, the other arm of the same river, rises just north of Erzerum. The Aras, the ancient Araxes, rising in the Bingeul, flows just north of the Russian frontier range, and so to the Caspian.

The general level of the plateau is 5500 feet above sea-level, the town of Erzerum being 6300 feet, the lake of Van 5360 feet, and the plains of Alashgird and Bayazid 5500 feet above the sea.

The formation is probably due to some great volcanic upheaval, as most of the mountains bear traces of former volcanic action, while severe earthquake shocks are not uncommon. On the north shore of Lake Van is the Subhan or Sipan Dagh, a splendid conical peak with a crater at the summit, which rises nearly 7000 feet above the lake, or 12,000 above sea-level. It is streaked with snow throughout the summer, and forms a striking landmark, especially when seen from the north. The name of Subhan is short for Subhanullah, "Praise be to God," which the pious Moslem is supposed to ejaculate on seeing so noble a mountain.

Nimrud Dagh, on the west shore of the lake, has a very fine crater some five miles in diameter, which contains numerous hot springs and small lakes. The Bingeul Dagh, or "Mountain of the Thousand Lakes," has a crater, the outer rim of which from a distance looks like the walls of some enormous fort commanding the long gentle slopes of the mountain. As the name implies, there are countless streams coming from springs near the summit, which cut their way in deep basaltic ravines to form tributaries of the Euphrates, or the head-waters of the Araxes. In the Aladagh range is the Tandurek or Oven mountain, having a large crater said to shows signs of activity at intervals, and on whose northern slope are extensive deposits of sulphur.

Hot fountains or sulphur baths are not at all an uncommon feature in the country, the largest being Ilija and Hassan Kale, near Erzerum and near Diadin, while there is a fine mineral spring just south of Bitlis.

With the exception of the isolated mountain masses just mentioned, and a few other minor ranges, the country is mainly open, rolling downs or steppes, not difficult to traverse in any direction. As the traveller reaches the summit of a pass the view is a dreary one, and usually the same, the distance being filled in by wave upon wave of bare, brown hills, a confused mass like the waves of some vast sea suddenly petrified, absolutely treeless, and altogether most desolate-looking; in the far distance, towering above the rest, but with every feature sharply defined in the clear summer air, is the isolated cone of Sipan, the long grassy slope of Bingeul, with its Kar Kala or Snow Castle at the summit, or the historic peak of Ararat itself.

Probably not a sign of human habitation will be visible, for the villages are collections of flat-roofed dwellings of mud and stone, some on the plain, which are scarcely to be distinguished from the surrounding country, others hidden in the ravines to escape the rigour of the winter blizzards.

The country, although its general aspect is so forbidding, is by no means a desert. It is full of perennial streams and springs, a most important feature in the East, while there are extensive plains, such as Alashgird and Passin in the north, and Bulanuk Khanus and Mush in the south, of rich black soil and very productive. Along the Persian frontier are the plains of Tusi, Abagha, and Serai at an elevation of about 6500 feet, the population of which are mainly Kurds who never cultivate the soil to any great extent, but use the plains for grazing their flocks in summer.

Around Lake Van the climate becomes more propitious, and the Van gardens, with those of Aganz and Akhlat, and other points on the shore, produce grapes, apricots, pears, peaches, melons, and cucumbers in the summer months. The lake water is undrinkable; carbonate of potash is extracted from it in drying pans near Van, and is used to manufacture soap. On the southern shore the scenery is strikingly grand, as there the mountains dip sheer into the water at several points, the whole being more like a sea-coast than that of an inland lake.

Abundant fresh water from streams and springs is obtainable all round the shore, and the rich gardens of Van, Aganz, and Artemid derive their fruitfulness from the many canals which carry the water in every direction.

Of lines of communication across the plateau, the principal is the great trade route from North-west Persia, which comes from Tabriz and Khoi, enters Turkey near Bayazid, traverses the Alashgird plain, which it leaves by the Delibaba defile at the west end, passes along the Passin plain, and over the Deveboyun pass to Erzerum, and then reaches the sea-coast at Trebizond.

The great road from Russia, traversed by the invading armies in the last two campaigns, lies through Kars, over the lofty rounded ridges of the Soghanli Dagh, into the Passin plain, and so to Erzerum.

From the direction of Erivan is another line of invasion, which has been utilised to reach Bayazid, the best routes crossing the frontier range between Ararat and the Baluklu Lake, and leading south towards Lake Van. The hills along the Persian frontier are of rounded outline, with several valleys leading to the lower plateau round Lake Urmia; communication is by no means difficult between Van, Urmia, and Khoi.

North of Erzerum, towards the Black Sea, the plateau changes into a strikingly indented and contorted country, intersected with deep, chasm-like valleys, the principal being the impracticable gorges of the Chorokh. The coast ranges are extremely rugged, and the northern slopes and spurs are densely wooded.

As one ascends to the plateau by the Trebizond-Erzerum road the change of scenery is very striking. At first, the road lies along a narrow valley, with well-wooded spurs on either hand, where the rhododendron and azalea may be found in bloom, and where collections of scattered wood chalets, with red-tiled roofs, of the Greek settlers along the coast, are prominent, until, near Baiburt, the dreary view over the plateau supervenes, and mud hovels take the places of bright cottages.

Erzerum first comes into view some fifteen miles off, as the traveller ascends the broad valley of the Euphrates, appearing as a blackish patch on a dark grey mountainside. Its houses of black basalt rising above the fortifications, and narrow streets with an open drain down the centre, do not improve on closer acquaintance.

Notwithstanding this, its position as a great fortress has always been deemed of the highest importance. Away to the north is the rugged, difficult country about the Chorokh valley; to the south are lofty ranges commencing close to the town, and connecting up with Bingeul and other ranges in that direction. Consequently, we find that all routes from east to west converge towards this important gateway, and must pass over the Deveboyun or Camel's Neck ridge just east of the town. Thus the great roads from Kars, Van, and Mush, as well as the great Persian trade route from Bayazid, come in on the east, and the Trebizond cart road and roads to Erzingan and Sivas branch out to the westward.

The problem of the formation of Van lake, eighty miles long and forty broad, elevated 5360 feet above the sea, is a very interesting one, and although ancient writers aver that it has an underground outlet it is difficult now to discover any traces of one. The most acceptable theory as regards its formation is that at one time the Mush plain extended eastward to Van, but that an overflow of lava from the Nimrud Dagh blocked the outlet in that direction. But also on the north shore the Murad valley at Melasgird is 250 feet below the lake level, and here again we can only suppose that a lava overflow from Sipan is answerable for this dividing ridge. That the ridge between Aganz and Patnotz is volcanic is obvious from the numerous pieces of obsidian strewn about on it. Mr. Ainsworth has suggested the great spring of Norshen, which forms the source of the Kara Su, as the true underground outlet, and this is very probable; but there are in the neighbourhood a number of very large springs which have not their origin in the lake, such as Erishat and Meshingird, and a very large one in the Guzel Dere at Sakh, issuing from the base of a cliff, and forming at once a stream fifty yards wide and eighteen inches deep, which flows south towards the Tigris, so that Norshen may be but one of many others. The watershed between the lake and the Tigris is a very slight one on the south shore, and at Ziwa Bay a single ridge some 600 feet high rises from the water's edge, and all streams to the south of this find their way through an enormous rift in the mountains, called the Kesen Dere, into the Bohtan, a tributary of the Tigris.

The lake level has been known to fluctuate, and the village of Arjish, near Aganz, was some years ago overwhelmed, and its ruins still remain half under water. A still more curious feature is that in the centre of the lake a large spring of fresh water exists, which in calm weather forces its way to the surface. A more careful survey of the lake with soundings will have to be made before the theory of its formation can be satisfactorily settled.

Of the population of the plateau no reliable statistics are available; nothing is very definite in the East, and least of all statistics, which in these regions vary according to the religion or politics of the person questioned; about one-half Armenians, and the remainder Kurds and Osmanlis is perhaps correct. The Armenians are principally in the towns, especially in Van, which has a population of 14,000 Armenians and 11,000 Moslems, and in large compact villages on the plains.

They are industrious and good cultivators of the soil—far better than the Kurds, whose instincts are mainly pastoral. The Armenian towns-people are remarkable for their aptitude in business as bankers, merchants, and tradesmen of all kinds, and they have most of the trade of the interior in their hands. The effect of the higher education of the Armenian by the American missionaries is often not fully realised, though it has been in progress since 1819; colleges and schools have been established in most of the large towns with the most excellent results, raising the standard of education among the Armenians much higher than that of the Moslem races with which they come in contact.

The habitations of both Kurds and Armenians are half underground, with flat, earthen roofs; and cattle as well as human beings find shelter inside during the long winter months. The usual living-room has a small pit used as an oven in the centre of the earthen floor. Over this is a hole about two feet square to let out the smoke and let in as much light as it can. There may be a few very small windows covered with oiled paper. The stable has a raised dais at one end for persons to sleep on, so that the general temperature may be kept above freezing-point by the addition of animal warmth.

At harvest time there is a busy scene round the villages; the corn is brought home and stacked, a threshing-floor is made in a suitable level place by forming a circular floor of clay, watered and flattened with a roller, and then left to harden in the sun. Sheaves are placed round the circle and oxen harnessed to a spiked roller are driven round and round until the corn is threshed out. Winnowing is simply done by pouring the grain from a sieve held shoulder high and allowing the wind to do the rest. This is the primitive process in use from the earliest times and still universal throughout Turkey and Persia and the East generally. The barley and chopped straw are stored in deep underground pits lined with clay and closed by a large stone. The wheat is stored in large earthenware jars ranged round the walls of the principal room of the house. Hay is cut in the autumn in the valleys and piled on the roof or stacked close by.

Want of fuel is severely felt in the long winter on the plateau, and although coal is known to exist just east of Van, it is not worked. Wood fuel for Van has now to be brought from the far shore of the lake, and oak shrubs, roots and all, are dug up and sold. *Tesek* or cow dung cakes are the usual fuel, but any small shrubs, and even large thistles, are gathered also.

A few poplars and willows are usually found near the villages, and willows and pines in secluded valleys in the mountains, which proves that it is possible for trees to survive the winter, and that the general treeless-ness may be due to the great demand for winter fuel, and not entirely to the climate.   The winter climate of the plateau is of extraordinary severity, and the temperature falls sometimes to 15° below zero.

The blizzards or *tipi* are very dangerous to the traveller, or to cara-vans if overtaken without means of shelter on one of the numerous passes. Small sledges are then used, and even in summer, on Bulanuk plain, I found the Kurds using sledges to bring home their harvest, although the ground was quite practicable for wheels.   Also at several places on the steep mountain slopes south of the lake the corn is drawn home on sledges.   The summer weather is clear and dry, and excellent for travel-ling.   The great elevation tempers the sun heat, and the various autumn fruits are very acceptable to the thirsty traveller.

The Kurds always retain their original organisation, which is tribal and pastoral, some tribes being at constant feud with each other.   In summer it is customary to abandon their villages altogether and live in black goat-hair tents pitched either close by, or on some neighbouring pasture belonging to the tribes.   Grass is to be found all through the summer on the *yailas* or grazing grounds on the Bingeul Dagh and the rounded summits and elevated valleys of some of the other ranges, and thither the flocks are taken for grazing.

The Kurds on this plateau were drawn from the mountains farther south in the fifteenth century, and settled here, the country being found wasted and almost depopulated by long wars, and thus this country has scarcely the same right to be known as Kurdistan as the great mountains farther south, which have been the home of the Kurds from earliest history.   Look-ing now to the southern edge of the plateau, we find below Lake Van a far more mountainous and intricate region with a complete change of scenery.   Along the south shore is a long rocky wall, the summits of which are bare sheets and pinnacles of rock, but whose lower slopes enclose beautiful wooded valleys watered by the many mountain torrents.

The districts of Karkar and Karchikan, containing many Armenians, are among the most fertile portions of the Van province, and form a fringe between the main range and the lake.   To reach Bitlis from the lake, I followed one of these valleys known as the Guzel Dere, or Beautiful Valley, and it certainly justified its name, especially to a traveller who had just accomplished the dreary treeless journey from Erzerum.   Up the slopes, densely wooded with low oaks, were many small Kurd and Armenian villages, while high overhead were the bare rock summits, the valley itself having a carpet of brilliant green meadow grass through which flowed a fine mountain stream, while walnut, elm, ash, and poplar grew by the banks.   Three stone bridges were passed, almost in ruins, but they testified to a former more prosperous era, the Kurdish bridge of the present day being but a sorry construction of wickerwork. Elsewhere, also, there are evidences that this fine mountain region of the Bohtan supported a much larger population, and it constituted formerly the rich province of Zabdicene.

The wall of mountains continues beyond Mush, on the south bank of the Murad to near Palu, where it gradually dies away in the Kharput plain. The principal break in this long chain is the narrow valley of the Bitlis river, on which is situated the Kurdish town of Bitlis. The watershed between this stream and Lake Van is but a slight ridge, while from the foot of the pass roads branch out west towards Diarbekr and the Mediterranean port of Alexandretta, and south towards the Tigris valley and Mosul. Consequently this pass forms the great highway between the Armenian plateau and the Tigris basin.

East and south-east are the various ranges of Central Kurdistan, which enclose the rugged difficult valleys of the Bohtan and Great Zab, the summits of which rise to 12,000 feet, culminating in the Jelu Dagh, south of Bashkala, at a height of 17,000 feet. These mountains sink towards the Tigris in a gradual series of zones or levels, which can be fairly well defined.

The plain country between the Kurdish hills and the left bank of the Tigris is traversed by many tributaries of the main stream, which supply water in abundance—the life-giving element in these regions, if they are to be productive and fruitful.

The plains east of Mosul, those of Zakho, Shemamlik, Erbil and Kifri, produce quantities of corn, maize, and millet, with cotton and rice also.

The alluvial portion between the great rivers Euphrates and Tigris commences not far north of Baghdad, and extends to the Persian Gulf.

From the earliest history this region of Assyria has been one of the most fruitful in the world ; now, however, the extensive irrigation canals which were the source of its greatness are only traceable by their ruins, and both Tigris and Euphrates at flood-time overflow their banks uselessly and convert the country into a marsh. It only needs to restore an efficient control over the water, and a proper canal system, for the country to regain its former prosperity and population, and outrival Egypt in the abundance of its produce. The hill zone, of from 2000 to 4500 feet altitude, is by far the most beautiful to the eye of the traveller. In it are the outliers of the main ranges, comprising several parallel ridges, usually of limestone, the slopes well wooded with oaks, giving rise to many springs and streams, and enclosing long narrow valleys and plains extremely well watered and of great fertility.

The upper zone comprises the mountain summits, in many instances great rolling downs, at a level of from 9000 to 10,000 feet, from which the snow never entirely disappears, and which constitute the summer grazing ground or *zozans* of the nomad Kurds. These mountains are the true home of the Kurds, the Kardu of the Assyrian inscriptions, the Carduchi of Xenophon, who occupied this region in far ages back.

In Central and Southern Kurdistan the tribes have easy access to the plains on either bank of the middle Tigris, and a large proportion are true nomads, living in tents all the year round. These nomads are often looked on as a set of idle vagabonds who should, at all costs, be induced to settle, if the country is to be kept at peace ; but this is, I think, a wrong view of the case. They do a large trade in wool and goat's hair; the summer heat on the plain drives them to seek the grass on the mountains, which springs up as soon as the snow melts; the intense winter cold and snow drives them down again to find abundant grass on the so-called desert east of Mosul and Jezirc, which appears with the autumn rains and continues until early spring; so that if they are to retain the pastoral instincts they have had for countless generations, they must, perforce, be nomad. Even the Kurds settled in villages among the mountains usually forsake them in the summer heat for tents or booths of oak branches on some grazing-ground on a neighbouring mountain.

It is most interesting to watch the scene at the Jezire boat-bridge in May and June, when innumerable flocks of sheep and goats, with their Kurdish owners, are to be seen filing across to reach the mountain pastures. They have to pay the sheep-tax before they can cross, and this is almost the only point in their migrations where they come in touch with Turkish officials. I spent a fortnight among the Hartoshi tribe on the high watershed range between the Bohtan and Great Zab rivers, soon after they had arrived there for the summer. The temperature in the Tigris valley at Jezire a week previous had been 110° in my tent, while now, at their first mountain camp at Maidan Jashush, at an elevation of 8200 feet, there was hoar frost outside in the morning and a noon temperature of 63°. This was in the middle of June, and several considerable drifts of snow still lay about. The soil was of marvellous richness; a thick carpet of grass appeared directly the snow melted, in which were studded numbers of flowers of great variety and brilliance. Among others were a blue Alpine gentian, hyacinths, the Star of Bethlehem, and various kinds of iris and polyanthus. Many small streams ran from the melting snow or from springs; the black Kurdish tents were scattered along the banks of the main stream, with the horses and sheep feeding not far off. When on the move it took a very short time to strike camp and advance in a long straggling column, the sheep grazing as they went; the men rode along fully armed, either with sword, dagger and long lance, or more usually with a Martini, Peabody, or some home-made rifle. The meanest shepherd always had a gun of some sort with him. The women had to do most of the work, and had to trudge along with, perhaps, the huge iron family cooking-pot strapped on their backs, or gain a seat among the tents and furniture on a pack bullock or donkey. They are never veiled, and are accorded more freedom than their Osmanli neighbours. The Kurd is very fond of bright, showy dress and accoutrements, and longs for the possession of the best rifle or horse he can get hold of.

The distinctive dress of the Hartoshi Kurds, and of those who spend the summer in the same region, is a short pelisse of goat's hair, braided with gold lace and ornamented with green tassels and open in front, displaying a red waistcoat, wide trousers, often bright red or a curious patchwork pattern much prized, and boots dyed some showy colour. A white conical cap is worn with a turban of black and red cotton wound round it. Most of the cloth is made from goat's hair. It is very strong and durable. The shirt-sleeves are wide open and hang down to the ground when walking.

On the summer pastures a low prickly shrub called *gavvan* is used for firewood, stalks of thistles which grow to a height of five feet, and stalks of several kinds of umbelliferous plants like fennel or wild carrot. The leaves of the latter are sometimes stored for fodder, and are called *karnoch*. These plants are not found below about 5500 feet. The organisation of the Kurds, both nomad and sedentary, is still tribal, although the semi-independent *begs* who used to rule in Amadia, Rowanduz, and Sulaimania have now disappeared, and the present *aghas* are much restricted in their power. There is a certain number of *shaikhs* or holy men, who possess a considerable influence over their ignorant following.

In the religion of the Kurds, especially of the nomads, there exist, under the outward forms of Mohammedanism, traces of Pagan doctrines, which probably descended to them from very early ages. Mosques are never seen in the villages. Much reverence is paid to the tombs or *ziarats* of their local saints. Sometimes inside the low wall surrounding the tomb are crowded together various propitiatory offerings which the nomads have placed there as they passed by. I have noticed entire tents, tent-matting and cloths, brightly painted cradles, and various other articles, all of them of considerable value among these simple people. On the branches of the trees which overhung the shrine were numerous rags and scraps of clothing which passers-by had tied there, imagining that by so doing they were leaving their ailments behind them. Around the shrines is often a fine grove of oaks which are never lopped for firewood. These isolated groves are often met throughout Kurdistan, and are always treated with veneration and the wood is left untouched.

Usually there is no control of tree-cutting, and much of the country which would otherwise be well wooded is almost denuded by reckless cutting and burning.

A heap of stones to mark the spot where a murder was committed is also a roadside incident in some places, and each passer-by is supposed to add an additional stone to the pile.

The Kurds of the Armenian plateau are far more fanatical towards Christians than those of the mountains. While passing up the valley of the Murad north of Melasgird, it was impossible to halt in any of the villages of the Sipkani, and a twelve and a quarter hours' ride with tired baggage-animals had to be undertaken to get through to a more hospitable village. It was always far different in the south, and there the rugged outline of the country assists the people to maintain a state of semi-independence, and the Kurds are a frank and hospitable race of mountaineers with many good qualities, not fanatical or cruel by nature, although addicted to inter-tribal raids and not averse to robbery on occasion.

The borderland between Turkey and Persia, with its ill-defined frontier, is always the scene of many raids. The Shekak tribe will raid over into Persia from Albak, the Zeza into Ushnu plain, the Bilbass into Central Kurdistan, the redoubtable Jaf and Hamawand of Sulaimania keep that border in a constant ferment, and the Lur chief, Hassan Ghuli

Khan, sometimes essays to defy the Turks or harass the Arabs of the Bani Lam.

When crossing the border from Bashkala to Urmia, the village I had halted in was alarmed about midnight by some raiders of the Shekak. I had pitched my tent in a sort of village green in the centre of the houses. In a moment the whole was a scene of the wildest confusion; outlying cattle were driven into the shelter of the houses. Every villager grasped his gun and rushed forth to discharge it into the darkness, to the common danger, to show thereby that the village was ready to meet all comers. Horsemen soon began to arrive from Bawis, a little Kurdish village close by which had not been so well on the alert, and we heard that 500 sheep had been lifted, and were now well on their way over the Persian border.

Bashkala, the "Castle of the Head" as it is called, from its old castle standing sentry at the head of the valley of the Great Zab, forms, with Julamerk, the two great centres of Turkish authority in the Kurdish mountains. It is 7900 feet above sea-level, and looks out eastward over the extensive upland basin of Albak, through which the Zab flows in a wide, grassy valley bordered by low cliffs. A few miles below this the Zab enters the mountains, which it traverses in a series of wild, impracticable gorges, until it emerges again near Amadia. Down stream it has still to traverse several ranges of the lower outlying hills, and does not reach the plain until a short way west of Rowanduz, so that for quite three-quarters of its course it is a mountain torrent in a valley affording the grandest scenery which the traveller could desire. South of Bashkala comes the Jelu Dagh, the highest mountain in Kurdistan, with two great peaks called the Tura Daouil and Tura Bermarghi, rising to 17,280 and 16,300 feet respectively. The view from Diza, itself 6000 feet above the sea, is most striking; in the foreground is the level plain of Gavvar, obviously the site of a former lake, still marshy in the centre, with black, loamy soil, having Kurd and Christian villages dotted about it. Beyond is the splendid mass of Jelu, rising 11,000 feet above the plain, and in July, for quite a third of its height, covered with perpetual snow, the summit a vast line of cliffs and pinnacles with glistening masses of snow in their clefts. The summit of this mountain still awaits the explorer.

Were the plain of Gavvar still a lake, the view across it of these two snow-clad peaks, towering high above the rest of the Kurdish mountains, would be one of the grandest in the world.

The watershed range between the Bohtan and Great Zab is deserted by the Kurds on the first snow falling in November, and remains so until next spring. The Bohtan valley may be taken as the southern limit of Armenian villages, the Great Zab valley being occupied by Nestorian Christians.

There are, nevertheless, a few villages on the summit of the range, for instance, Merwanen, an Armenian village, at 8095 feet above sea-level, which is occupied throughout the winter. It has the usual flat-roofed houses, but here the street, so to speak, is covered over as well, and the whole village has one common roof. The interior is like a rabbit warren, and it is difficult for the uninitiated to grope about and escape being knocked down in the semi-darkness by cattle being driven in and out. On one side of the house-passage a door opens into a cow-house, on the other into a living-room, which, besides the inevitable oven sunk in the centre of the floor, contains a great vat for storing water in winter to prevent its being frozen.

The Assyrian or Nestorian Christians are an important race in this region, who number about 200,000, and inhabit the basin of the Great Zab, part of the Mosul plain, and many villages near Urmia in Persia. The Great Zab valley is a very difficult, mountainous region, comprising a succession of deep, narrow valleys radiating from the watershed, each of which contains villages of Nestorians and Kurds, each Nestorian tribe taking its name from the valley in which it resides.

The original Nestorians of the Mosul plain and of Diarbekr seceded in the seventeenth century from their own Church and joined that of Rome under the title of Chaldæans. A few years ago part of these reverted and formed a new sect, electing their own patriarch and calling themselves New Chaldæans.

The mountain Nestorians are a manly, warlike race, difficult to distinguish from the Kurds of the same region; they are savage and untutored to a great extent, but an English mission, with headquarters in Urmia, is at work among them, and among those on the Urmia plain. The Chaldæans are crushed by centuries of oppression, but are most industrious agriculturists, and their villages on the Mosul plain are amongst the most thriving there.

South-east of the Great Zab, towards Rowanduz and Sulaimania, are found no Christians, and the population consists exclusively of Kurds, who extend into Persia, as far as Sihna and Kermanshah, and south until they meet with Luristan or the country of the Lurs.

This frontier district is perhaps the most beautiful part of Kurdistan, and is abundantly watered by the Lesser Zab and its tributaries. High up the slopes of the hills, often in the most inaccessible places, are perched the little Kurdish villages, their flat roofs overgrown with grass and scarcely distinguishable against the hill-side. Round each of them, probably irrigated by some spring, are vineyards and gardens, laid out in terraces, growing melons, cucumbers, apricots, pomegranates, and other fruits, and often a tall group of walnut- and mulberry-trees by the stream bank half hides the houses from view. Low oaks, usually of the gall-bearing kind, are the commonest trees, with poplar, ash, and sycamore by the streams.

Some of the tribes here are entirely nomad, the principal being the Jaf, who move south of Sulaimania and cross to the mountains over the Persian frontier for the summer, returning in winter to the plain between Kifri and the Diala, and leaving a train of international difficulties behind them.

Rowanduz was once a great Kurdish stronghold, and occupies a remarkable position of great natural strength. The Rowanduz river, a mountain torrent from the Persia frontier on its way to join the Great Zab, passes through a ridge, which here bars the valley by a narrow rocky gorge 500 feet deep at one point, and narrows to about ten yards wide opposite the town, where it is crossed by a rickety bridge of wooden beams and fascines.

A tributary stream from the south-east passes the ridge in another similar but smaller gorge, and on the hill-slope between the two great rifts is built Rowanduz. A line of round towers in addition made the place secure in former times. Space is limited, and the houses are piled one on the other in a curious haphazard fashion, and to find a way down the narrow stair-like streets one has to pass perhaps over the flat roof of a building, where care must be taken not to enter by the chimney instead of the door, or else through a dark tunnelled archway on a level with the basement. In summer, every one sleeps on the roof in these villages, a sort of screen of wattle or bushes being erected round the edge of the roof. High ranges, wooded almost to their summits, border the Rowanduz valley, up which leads the principal trade route from Mosul into West Persia.

Sulaimania is another typical Kurdish town, and may be called the capital of Southern Kurdistan. It contains 2500 houses, nearly all low huts, with few buildings of any size. The narrow winding streets of the bazaar with the stalls on either hand, with their eager salesmen, are shaded from the sun by an arrangement of oak boughs stretched across. The articles in the stalls reflect the tastes of these warlike Kurdish mountaineers, always fond of something bright and showy, and of service-able weapons.

The principal manufacture of Sulaimania is saddle and horse furniture, shoes, and leather work generally, the leather being cleverly dyed in brilliant colours by natural dyes made from oak galls.

A curious assortment of flintlocks and various guns, swords, knives, and daggers of all shapes, round shields of bullock hide, belts with a row of pouches for powder and bullets, and felt saddlecloths, embroidered with coloured worsteds, made a very interesting display. A fair quantity of Manchester piece-goods, cotton kerchiefs, etc., could be seen; but to a Kurd a Martini rifle or a good horse, looted from some Arab on the plain, is of more value than very many yards of cloth. The universal Kurdish drink is a kind of curdled milk known as *yaurt*; cheese flavoured with a kind of wild garlic, raisins, dried plums and apricots, piles of thin, flat, circular cakes of wheat and rye bread, are the eatables. There are numbers of vineyards on the hill-slopes close by, and a good red wine is manufactured; every village has its fruit orchard, which produces fruits in great profusion, such as pears, apples, peaches, apricots, plums, figs, and pomegranates. Numbers of mulberry-trees are also grown, and a quantity of coarse silk is produced, which is used by the Kurds in the manufacture of clothing, and which if under proper control would probably yield good results. Walnut-trees are very plentiful in the frontier ranges; wheat, rice, and tobacco are cultivated in little terraced gardens. Valonia oaks, which produce galls for dyeing, are common on the hill-sides, as are also pistachio-trees, from the berries of which a kind of soap is extracted.

Sulaimania stands near the north-west end of the Shehrizur plain, watered by the Khaujiru stream, a tributary of the Diala, and is some thirty miles long and about eight broad. It is well watered and has an extremely fertile soil, and is dotted at various points with tumuli, the sites of villages in ancient times; but now the population is very small in comparison to the productiveness of the soil, and only a few Kurdish villages can be seen down its length. As a general rule the inhabitants do not care to waste their labour in producing more than suffices for their bare maintenance, when any surplus may be taken from them by the tax collectors, who collect in kind tithes of all the produce.

The Arab race supersedes the Kurd after the last outliers of the mountains are left behind and the great Mesopotamian plain is entered. The population of Mosul, Baghdad, and Busra is principally Arab, while along the lower Tigris and Euphrates are many Arab villages, but a larger proportion of Arabs are nomad Bedawin, who are practically independent of the Turkish Government. To turn now to the communications or main trade routes through this part of the country, we find the two great centres are Baghdad and Mosul. Baghdad, which contains 100,000 inhabitants, is most favourably situated for commerce. The principal route from Teheran and Central Persia to the head of the Gulf reaches Baghdad through Hamadan, Kermanshah, and the Tak-i-Girra pass; this avoids the several lines of rugged hills which cut off Central Persia from the sea at Bushire; the Tigris is navigated by two British steamers and a Turkish Company from Baghdad to Busra, to which point on the river ocean steamers can ascend.

Near Baghdad are the great shrines of Kerbela, Samarra, and Nejef, which attract a great number of pilgrims from Persia. To the north the main Mosul road passes through Kifri, Kirkuk, and Erbil, and routes follow the Euphrates to Deir, and cross the desert to various points in Syria.

Mosul, another great trade centre, is a town of 40,000 on the right bank of the Tigris, opposite the ruins of Nineveh. Roads branch from here through Rowanduz into West Persia, through Jezire and Mardin to Diarbekr, and from there either to the Mediterranean at Alexandretta, or through Kharput and Malatia to Sivas and Samsun.

To the north a track runs through Jezire and Sairt to Bitlis, and thus connects with the Armenian plateau.

The connection with North-west Persia by Rowanduz is an important one, as round Lake Urmia are some of the richest districts of Persia. The streams from the Turkish frontier hills are utilised to irrigate numberless fruit gardens, orchards, and vineyards before they finally lose themselves in the great salt lake of Urmia, and the plains of Salmas, Urmia, Baranduz, and Sulduz on the west shore are thickly populated and of marvellous fertility.

As regards roads in the country, some few in the Sivas and Erzerum provinces have been well laid out and completed, but in South-east Turkey in Asia there is no effort at improving or making roads, which take the easiest natural line from one point to another, and in time, by the constant passage to and fro of caravans, of pack animals, or strings of ox-carts, are rendered passable.

With the exception of the ox-carts of the Sivas and Erzerum

provinces, caravans of pack mules, horses, or donkeys are used for commerce. Camels are used in the Tigris valley and on the great road through Mosul to Aleppo and Alexandretta, and also on the Persian trade route through Erzerum to Trebizond.

The Tigris is much utilised by large rafts from Diarbekr to Baghdad. These consist of about 150 inflated sheepskins tied underneath a light wooden framework, and can carry cargo or travellers, who fit up tents on them. This use of skin rafts is of great antiquity, as sculptures at Nineveh show that similar vessels were employed when that city was at its prime. They cannot ascend the river, but are broken up on arrival at Baghdad and the skins taken back by caravan. The concession for two British steamers to navigate the Tigris permits their use only from Busra to Baghdad. Although a Turkish company was talked of in 1892, which was to employ steamers to navigate the river from Baghdad to Mosul, nothing has been done as yet. There are the remains of an ancient stone dam across the river some twenty miles below Mosul, made for irrigation when Senacherib ruled in the adjacent city of Nimrud, which would require clearing away.

The Euphrates in not navigated at all owing to numerous rocks, rapids, and remains of ancient dams, but can be ascended in the flood season, April to August, to Meskine. The higher mountains, particularly those of the Bohtan and in the basin of the Great Zab, are never entirely denuded of snow in summer; the first melting causes the spring floods, and its gradual disappearance later on keeps the water in the Tigris and its tributaries high until August. The low season then commences and continues until the November rains.

The winter climate of the plain country is not at all severe, although heavy rains occur, especially in January and February, and occasional frosts at night, while in spring the Mesopotamian plain is a carpet of green grass, and water is to be found in all the valleys. The summer heat is undoubtedly scorching on these treeless plains, but is quickly mitigated on the ascent of the outlying hills.

Turning now to the communications of the Armenian plateau with the Mediterranean, we find from the Gulf of Alexandretta routes inland directed on Kharput and Diarbekr, and the foot of the Bitlis defile.

The Syrian ports are all small and exposed to the west, and Alexandretta, which is sheltered in the gulf of the same name, is the great outlet for the trade of the Aleppo, Diarbekr, and Kharput provinces. The Aleppo carriage road from here crosses the coast hills by an easy pass at Bailan, and caravan routes lead through Killis to the Euphrates ferry at Birijik, and from there over open country by Saverek to Diarbekr. Another way leads from Adana, which is connected by a short railway with the roadstead of Mersina, through Marash, Malatia, and Kharput to Diarbekr. This keeps to a wide depression in the hills after leaving Marash, and crosses low outliers from the Taurus ranges, well wooded, with numerous streams. From Diarbekr there are easy routes to the foot of the Bitlis pass, to Jezire and to Mosul. Diarbekr, a town of 30,000 inhabitants, still surrounded by the old walls of black basalt, which held out against

Tamerlane, occupies therefore an important position as a road junction on these routes. From Kharput up the Euphrates valley to Erzingan are only rough paths across the Dersim mountains, and north-east towards Erzerum there is much difficult mountain country to traverse.

The mineral capabilities of the country have never been properly exploited; but the belt of hills between the Bohtan and the Great Zab is commonly reputed to contain minerals.

For my own part, without leaving the track to search, I have passed outcropping seams of coal near Shernakh, north of Zakho, probably part of other seams that have been slightly worked near the latter place; also several seams in the Great Zab valley near Zibar.

Magnetic iron ore was picked up near the foot of the Bitlis pass, and nodules of hematite iron ore at the gorge of the Great Zab at Berdin, where it finally enters the plain, and also on the Persian frontier east of Mendali.

In the Tigris valley are many springs of petroleum and bitumen, which if properly exploited would probably prove of immense value. The principal springs are at Al Hadhr, Hammam Ali, Kirkuk Erbil, and Mendali, which cover a wide field. Erbil is the ancient Arbela, which used these very springs to illuminate the town in honour of Alexander's victory over Darius. · At El Fatha, some thirty miles below the junction of the Lesser Zab with the Tigris, the river for about five miles is covered with long threads of bitumen which ooze from a low rocky ridge, which follows the left bank for that distance.

Wheat, barley, maize, millet, rye, and rice are the principal crops. Wool is collected principally at Jezire and Mosul from the flocks of the nomads; the extensive oak forests produce galls for dyeing. A species of wild silk called *gez* is found in the hills near Jezire, and the mulberry-tree is common throughout Kurdistan.

The whole of these regions suffers greatly from want of communication with the outer world and profitable markets; the population is at present insufficient, and much of the country lies idle and almost depopulated; but the soil, if need be, is capable of a greatly increased production. Railways are undoubtedly the best means of developing the country, and would at the same time bring Western civilisation more in contact with the various interesting races of the region. Projects have been put forward at various times for a trunk line through the country to connect up with Indian lines and give through communication with the Mediterranean.

The Euphrates valley line was the first of these, which was to start from Alexandretta, pass through Aleppo down the Euphrates valley to Baghdad, and then descend the Tigris basin towards the Gulf, turning along the coast to Bushire. By keeping to the coast difficulties would be encountered in the hills and in the barren tracts along the coast, while the traffic of the interior of Persia would be quite untouched by such a line, as between the Gulf coast and the central plateau there is a continuous belt of mountain country difficult even for mule caravans to traverse. Again, the section of the line in the Euphrates valley between Aleppo and Hit would be through entirely unproductive country. At no time in history have there been any large towns on this portion of the river, as it flows in a valley some 200 feet below the general level of the desert, which is crossed by numerous ravines ; and irrigation canals, the great source of wealth in the alluvial plains farther down, would here be impossible.

A more favourable line would avoid the desert country altogether by going north of it. The line would start from Alexandretta, pass by Aleppo, and cross the Euphrates at Birijik, following the same general line as the caravan route at present. From Birijik the line would skirt the north edge of the desert through Urfa Veranshehr and Mardin to Jezire, where the Tigris could be crossed. The left bank of the river could be followed through Mosul, Erbil Kirkuk to Kifri. The ascent to the Persian plateau could be begun near the Tak-i-Girra, or Gates of Zagros, which is the route followed by the present Baghdad-Kermanshah road, the great outlet of the trade of West Persia. The central plateau being reached, the line would pass eastward through Kermanshah and Ispahan, thus traversing some of the most fertile districts of Persia. At no portion of its length does this line pass through unproductive country. Branch lines could be made to Diarbekr and into the provinces of Kharput and Van, and would connect up with the line to Constantinople when it is prolonged to the Gulf of Alexandretta. Such a railway would not have to depend entirely for its success on the through traffic with India, as the Diarbekr, Mosul, and Baghdad provinces, with those of Central Persia, would benefit to an immense extent by such communication with Western markets, and trade would be developed in these regions as well.

Such in outline are the geographical features and characteristics of the various races of this remarkable country, and there are a few points I should like to recapitulate before I conclude. The country has from the earliest ages occupied a prominent position in the world, and was the meeting-ground of East and West at the time of the Persian and Roman empires. The fertile basin of the two great rivers, with their vast system of canals, maintained the kingdoms of Assyria and Chaldæa, the ruins of whose capital cities, Babylon and Nineveh, are still among the wonders of the country, and the rich valleys among the Kurdish hills supported populous provinces.

The situation of Eastern Turkey in Asia is a most favourable one as a half-way house between Europe and Asia, as on the south-east it opens on the Persian Gulf, and through there on the Indian Ocean ; on the west it has access to the Mediterranean ; on the north to the Black Sea ; while on the north-east it can command the Caspian and routes through North Persia from Central Asia, while the lofty Caucasus range shuts it off from Europe and makes it Asiatic.

Particularly important within this area is the Armenian plateau or quadrilateral which projects south like a strong bastion overlooking the Tigris basin and Central Persia.

Through Erzerum pass all routes going west to Anatolia ; through the short pass of Bitlis is another open door which would lead to the Mediterranean on the one hand, and give access to the open basin of the

Tigris and the Persian Gulf on the other. From the east of the plateau the way is easy into Azerbaijan, one of the richest provinces of Persia, and from here another way opens into the Tigris valley by Rowanduz and Keni Sanjak. These considerations, I think you will admit, show the great geographical importance of the Armenian plateau apart from the sorrows of its much troubled Christian inhabitants.

THE FALKLAND ISLANDS.

By R. M. ROUTLEDGE, Stipendiary Justice of Trinidad.

(*Read before the Society at Aberdeen, March* 1896.)

WHEN I had the honour of being asked by the Committee of this Society to give an address on some part of those foreign lands where my lot has been cast during the past few years, I did not hesitate long in making up my mind as to the part of the world I would ask my audience to accompany me to this evening. Two reasons prompted me to choose the Falkland Islands: the first, that I considered the West Indian Islands, where I am now employed, really so well known, that I dreaded lest I should be unable to give my hearers much information beyond what is already known to many, either by hearsay from relatives, or from the many books which have been written from time to time regarding those colonies; and the second reason why I selected the Falklands was that the two lecturers who have just preceded me have been treating you to a description of two countries around which, within the last few months, a large amount of interest has been centred, diverting our thoughts for the time being from other colonies. For these two reasons, then, I considered I should best assist the Society to-night in its object, if I asked' you to allow me to be your guide and instructor in a part of our British possessions which, I think I may venture to say, without any injury to the feelings of the great majority of my audience, is almost unknown, and therefore I have undertaken to tell you what I know of the tight little group of 200 islands or so, known as the Falklands, and lying near to the South Pole at a distance of nearly 8000 miles from this country, 400 miles east of Patagonia, 500 miles north-east of Cape Horn, and 250 miles from the mainland of South America, in S. latitude 51°-53°, longitude W. 58°-61°.

The first question, then, which I found myself daily called upon to answer between the time I had orders to take up my residence in the Falkland Islands and sailing was, "And where are the Falkland Islands?" The reply I have already given you. That question being answered, was usually followed by another. "And how do you get there?" To which the answer is that, in order to reach the Falklands, there are two courses open, the easiest being to take one of the German steamers belonging to the Kosmos Line, calling formerly at Dartmouth but now at London, which touch first at St. Vincent in the Cape

Verde Islands, and then at Monte Video, and from the latter port sail to the Falkland Islands *en route* for Central America by the Straits of Magellan. Another route is taken by the Pacific Line, which perhaps, on the whole, is preferable to a Briton, seeing that these boats, calling as they do at so many ports on the French, Spanish, and South American coasts, have more passengers on board, and one does not find himself, as it was my misfortune to do when making for the Falklands, the sole occupant of the first-class saloon. The hindrance to passengers travelling by this latter route is the transhipping, either at Monte Video or at Punta Arenas, otherwise known as Sandy Point, in the Straits, into the Kosmos boat, which alone makes for the Falklands, as that Company holds the mail contract. The voyage occupies either way from five to six weeks. Mails are now forwarded to the islands by a three-weekly service, which is a decided improvement on the system in practice in my time. The islands are entirely dependent on the arrival of the mails for information as to the ongoings in the outer world, not being connected by cable. Cablegrams are used, however, being sent on to Monte Video by boat, where they are cabled by responsible parties to their destinations, and are received in the same way by the residents in the islands.

From the Cape Verde Islands to beyond Monte Video, the temperature is everything that can be desired, but soon after quitting the harbour of Monte Video one realises that there are other parts of the world colder even than the north-east of Scotland, though not so far removed from the Equator ; and during five days' sail from Monte Video to the islands was the only time when the sea showed us any sign of those mountainous waves which cause ships to roll " gunwale under" and lessen the comfort at any rate of the landsman accustomed to *terra firma*.

At 8 A.M. on the morning of Saturday, June 27th, 1891, we sighted Cape Pembroke lighthouse at the entrance to Port William, the outer harbour of the Falkland Islands. This harbour is from seven to eight miles long, and at the head of it on its southern side lie what are known as " The Narrows," through which one sails into the inner harbour known as Port Stanley. This harbour is landlocked, and also extends a distance of seven to eight miles from end to end, and is very wide and of great depth in some parts. When you hear that vessels of the size of H.M.S. *Sirius*, and H.M.S. *Royal Arthur*, of 3600 and 7700 tons respectively, have no difficulty in anchoring and in putting about in it, you can have some idea of its size. Were it not for the vast quantity of kelp which abounds in it, it would be a perfect harbour.

The object which first catches the eye of the stranger is not one of a very cheery kind, for it is none else than the cemetery, which lies at the east end of the town or settlement.

Sailing up the harbour to the usual anchorage ground of the mail steamer, one is struck with the numerous hulks lying in mid-channel, some of them in the harbour, with their masts and rigging all struck. They are used by the merchants as store-houses. In mid-channel is to be seen one which by its size naturally causes the stranger to inquire what it is. This is none other than the *Great Britain,* which nearly half a century ago was launched under royal patronage at Birkenhead, and was then considered one of the wonders of the shipping world. After many changes, this noble vessel passed into the hands of the company known as the Falkland Islands Company, and is at the present time used as a coal and wool store.

One disembarks from the steamer and is carried to the shore by a steam launch or boat which the depth of water permits to lie alongside one of the many jetties running out into the harbour.

The whole settlement of Port Stanley, as the principal town is named, or to give it really the name which to you will best describe it, the "village," lies to the south side of the harbour. The main road, and the only road in the islands, with the exception of a few small side roads through the village, skirts the south side of the harbour for about a mile and a half or so, coming to a terminus at the west end at a house known as Sullivan House, and at the east end at the hotel, very fitly named the "First and Last." All along this road are situated the principal dwelling-houses and places of business, such as the stores, carpenter and engineering works, the gaol and printing office, which are all in one, the gaoler discharging the double duty of gaoler and Government printer.

At the extreme west almost of the main road lies Government House, situated in a paddock or compound of several acres in extent and surrounded by gardens.

Let the eye travel as far as it may, yet you cannot exclaim, "I see a tree"; none grows in the Falkland Islands. The cultivation of trees has been attempted times without number, but whenever they reached to any height above the shelter provided for them when planted, they refused to grow any more. This sad want is said to be due to the extremely cutting winds which prevail almost daily in the islands, and when you look at the chart and recollect what I told you of their position, knowing as we do that wind is ever blowing somewhere, you cannot wonder at their refusing to face the blasts.

At first sight this absence of trees gives a sad and somewhat melancholy feeling, but ere long one begins to look with pleasure upon the grand peaks of the many mountain ranges rising one above the other as far as the eye can carry, and especially when illumined with the most glorious sunsets it has ever been my privilege to behold; under these circumstances one, I say, begins to look upon such scenery as possessing charms which make up in some measure for the want of the verdure of a tree-clad country.

There are several notable ranges of mountains in the island. In the East Island is a range known as the Wickham Heights, whose elevation is from 800 to 2000 feet above the sea. They cross from east to west, and are composed mostly of quartz rock. The highest mountain is Mount Adam in the West Island. It is 2315 feet above sea-level.

Numerous valleys there are, and several lakes. Very few rivers run through these islands; the San Carlos in the East Island is the largest known, but there are what are known as "Streams of Stone," consisting of myriads of cubical fragments of quartz rock of varying size, some large enough to afford shelter to a man.   It is said water can be heard here, flowing many feet below ground.

Near to Government House stood the old Government Offices, now removed to a more convenient situation in the town, and adjoining these are the Court-House, the Treasury, the Secretariat, and Post Office.

The Falkland Islands is a Crown colony, having as its official head a Governor, who, along with the Colonial Secretary (who is also Colonial Treasurer) and the Colonial Surgeon, forms the Executive Council. Besides this Council, there is the Legislative Council, composed of the above members of the Executive, with the addition of other official members nominated by the Governor and appointed under Her Majesty's warrant.   To this last body, as to the House of Commons, attaches the duty of passing Ordinances, as 'the laws are called, voting supplies, and such like.

In Stanley, as through the whole of the islands, the houses are, for the most part, built of wood ; stone, though to be found in abundance, being used only for repairing roads.

In the settlement are churches of different denominations.   There is Christ Church, otherwise known as the Cathedral, the Bishop of the English Church in South America taking his title from the Falkland Islands.   The Rector of Christ Church is Dean of the Diocese, as well as Colonial Chaplain.   Christ Church is built of material almost entirely obtained from England, and has been raised at a cost of about £8000 or more.   The rebuilding of this church was necessitated owing to the destruction of the former building through what is known as a peat-slip, by which a large part of the town was also carried away.   The other churches in Port Stanley are the Roman Catholic and the Baptist, and attached to the first-named is a neat little Presbytery, famed for its simplicity externally and for the kind-hearted geniality of the priests who from time to time have inhabited it.   Formerly there was a Presby-terian Church in Darwin, a settlement on the western side of the East Island, but this has been closed and is now used as a school.   Through-out the entire islands, then, with the exception of the town of Stanley, the inhabitants are dependent for religious services on the periodical visits which the clergy of the different churches pay to the camp, as the country is called.   The farmers gladly welcome the clergy, and lend them every aid, permitting services to be held in their private dwellings. No clergyman, doctor, or in fact any one, who is a bad traveller on horse-back or by sea need think of venturing to these islands, for there is no other mode of conveyance by land than horseback, and none by sea but small schooners, not exceeding 100 tons as a rule.

As for the education of the young, it must be confessed that outside Stanley and Darwin it is a thing to be dreamt of merely, for so large are the sheep-runs throughout the islands, and so scattered from each other are the dwellings of the inhabitants, that it would be impossible, even were the sheep farmers and other owners of the lands willing to engage teachers for the young of their individual stations.   I say it would be impossible for the children to travel through the camps, especially in winter, so great are the distances, and so impassable are the camps during that season. The education of the young in Stanley and its neighbourhood is well provided for by two Government schools, a Roman Catholic school, and another attached to the Baptist Church. There is also a private school, attended for the most part by infants.

Close to the Government school lies the Government dockyard, in which are the Government stores, over which the chief-constable has charge as storekeeper. In the dockyard there is also a battery of four guns, one of which is discharged on a fixed day, at noon, to signal the correct time throughout the settlement. This time is taken from the chronometer kept in Government House, which is checked from time to time when any ships of war visit the island.

There used to be three hotels in Port Stanley, but only two remained at the time of my departure, owing to the total destruction of one by fire in the end of 1893. Public-houses abound, and from these a good revenue is derived, for beer is largely consumed, and not much blame attaches to the people, for the water is for the most part collected from off the houses after rain.

The engineering works and carpenter shops employ many hands, and the wages are high, especially when vessels are in the harbour for repair— work paid for at a special rate of wages. This is not surprising, for, with the exception of beef and mutton, no article of food is produced on the islands, but everything has to be imported, either from America or Europe, and is, in consequence, very dear when retailed to the consumer. A large number of the people in the settlement are employed at these trades, for it is seldom that there are not two or three, and sometimes more, "lame ducks" (as the vessels are termed which make for Stanley for repairs) lying in the harbour, and on which the sound of the hammer is to be heard from day to day. I have myself seen the same vessel put back into Stanley for repair two or three times ere it could make the passage round the Horn, each time coming back in a worse condition than before, and, to add to the misery of the captain, the men would begin to take a dislike to the craft, and in many cases desert.

In these workshops machinery of the most advanced types is managed by workmen of much skill and ability, and you can learn from them much about the home counties of Scotland, for most of the principal hands, as I soon perceived, talked in an accent familiar to me.

I trust I have succeeded in some measure in picturing to you what may be termed the capital of the Falklands, or the seat of the Government; and now, before taking you into the camp life and giving you an account of the industry of the islands, from which arises their financial success, I had better say something about the inhabitants.

It will, I have no doubt, be information to many to learn that throughout these islands there exists not a human being possessing any distinctive mark of nationality, such as would draw attention to him were he walking down the principal street of this town, with the slight exception which will be referred to immediately.

I have often been amused by the question, "And what are your natives like?" and I have felt I was causing much disappointment when

I was unable to picture them as resembling the Patagonian, as we read of him or see him in pictures (for there alone do these gigantic figures exist). Well, all I have to say of the islanders is, they are just the same as you or myself, not more nor less European-looking.

In a small island called Keppel, on the north coast of the West Falklands, is to be found a colony of natives from Tierra del Fuego, which lies to the extreme south of Patagonia. They have been brought to this island by the energy of the Bishop and his clergy, in the hope of civilising them and educating them. There are, or were at the time of my visit to the Mission, about thirty to forty boys, and everything was being done to inculcate into them the habits and customs of civilised States, for good schools with competent teachers and superintendents were provided.

With this exception, the people of the Falkland Islands are for the most part of British origin, with a sprinkling of Germans, Norwegians, and Spaniards, and number, according to the latest return, 1902. (There are ordinances providing for the naturalisation of foreigners.) When you hear this, you must be naturally led to think that large tracts are still lying waste, but that is not so. Every acre is taken up, either on lease or as the freehold of the owner. From this source arises the principal part of the revenue of the settlement, which can boast of having had for many years the annual balance of its accounts on the right side. The other sources of revenue arise from the duties imposed on spirits and tobacco, and certain minor taxes.

The inhabitants are for the most part law-abiding. A few police there are, who also discharge the duties of officers of Customs. Judges and magistrates have an easy time of it, as far as the inhabitants go. Where their labours come in is when vessels have to put into Stanley with mutinous crews, or when the crews of vessels run riot when allowed to come ashore from the vessels lying at anchor.

There is a very salutary law in these islands, which I do not hesitate to say accounts in some measure for the happy absence of crime or criminals—a law which, alas, I fear, can only be workable in very small communities. It is a law protecting people in their own interest against over-indulgence in intoxicating liquors. When any one is brought before the magistrate on the charge of drunkenness, the magistrate has the power to order the supply of intoxicating drink to such person to be stopped. This prohibition is published in the *Weekly Government Gazette,* and at the same time all the licensed houses receive notice under the magistrate's hand that such order has been made. Any one supplying any person thus prohibited during the time for which the order runs renders himself liable to a heavy penalty. I have known people, before I had made up my mind to carry out the law thus far, request me of their own free will to place them under prohibition, and thus secure them as far as possible against the temptation.

Perhaps if there is some ardent supporter of the temperance movement in my audience to-night, he may be able to devise some such scheme of prevention as may prove workable in large communities.

The social life throughout the islands though simple is hearty, and it is considered a great slight if a traveller rides through a camp without inquiring after the well-being of the householders, nor does any one avoid doing so, if only to get a rest, for it is no easy task to ride over the rough camp, through bogs into which one's horse often sinks up to the girths. The houses, on account of the large size of the holdings or runs, are of course very far apart.

In Stanley there is a Social Club, which meets during the winter on Saturday evenings, to hear and discuss papers. All reference to local politics is forbidden; and to this is due, no doubt, the harmony which prevails. Government officials generally contribute a paper or two during the session, on subjects best known to them. To this club is attached a reading-room, in which the members find the leading papers and magazines of the day, and can also engage in the healthful games usual in such meeting-places.

The inhabitants of Stanley are given also, as those of other parts of the world, to more frivolous amusements. There is an excellent assembly room, in which as a rule a couple of balls are held every month during the winter, and an occasional performance of theatricals, for a very complete stage, with all its accessories, occupies one end of the room. Concerts are very frequent, especially during visits from ships of war. But if there is one thing more than another in which the Falklander excels, it is in piling on the "bull's eye" at the rifle-range. To my knowledge, only once has the Stanley rifle team been beaten on its own range. It is customary for this local club to challenge each ship of war in turn, and, in 1893 I think it was, the Stanley lads received their first defeat from the men of H.M.S. *Sirius*. Of course, as is usual, some reason had to be assigned for it, and the reason the landsmen gave was, "That for once in the Falklands the wind was at rest"; and this latter remark leads me to say a word or two on the climate of these islands.

Taking the latitude and longitude, which I have already given, and looking for the same latitude and longitude to the north of the Equator, one would be inclined to say, "What a nice climate you must have, for you are in about the same latitude south as Yorkshire is north." That is so, but I cannot say that the climate exhibits much kinship to that of its *vis-à-vis*. In the Falkland Islands it seldom snows to any great extent. I spent three winters there, and I never saw snow lying as thick as the sole of my boot, and I never saw any of the ponds covered with more than a thin coating of ice. The thermometer is hardly known to have sunk below 29°. But, notwithstanding, the cold is sometimes intense, especially when the wind blows from the SE., and it was no uncommon thing for the news to run through the settlement, "There is an iceberg to be seen floating past, if you go on the look-out." This outlook is on the top of the hill, just behind Stanley, from which approaching ships are signalled.

Beautiful as the harbours are around, as also the many creeks, and suitable as they appear to be for sea-bathing, yet none ventures to try the luxury. It is a strange fact, and still unaccounted for, that the cold of the water is so intense in these harbours, that the best of swimmers have been known to get in but never come again to the surface alive. The temperature of the water is nearly at freezing point. I cannot help thinking that the great quantity of kelp in the harbour, combined with the salt of the water, must in some measure account for it.

Not a day passes, summer or winter, without high winds, either in the morning or evening, and owing to these the inhabitants have acquired a somewhat peculiar gait by beating against them, and accordingly have gained the name of "kelpers," which is sometimes used as synonymous with "natives." I have myself been told, when walking with friends and quite innocent of having assumed this gait, " Routledge, you are getting into the kelpers' walk."

Summer weather in the Falkland Islands, on the other hand, may be compared with the weather usually experienced in this country in May, the temperature seldom being above 76°. I have seen the temperature at 90° on Boxing Day, but that was only once, in 1891.

With a climate such as I have described, one would be apt to say, "This is no place for invalids." I fear I am not sufficiently up in medical science to be able to tell you how it is that, with all these apparent disadvantages of climate, the Falkland Islands are a good place for people of weakly constitution. There is an absence of all pulmonary diseases. Of course, the usual infantile complaints prevail, and so does that complaint common to every part of the world, I believe—rheumatism. You will best understand the healthfulness of the place, when I tell you that the latest statistics show the death-rate to be less than 6 per thousand. The climate is undoubtedly stimulating and encourages people to work, and to this is due, in a great measure, the happy result that there is no poverty in the Falklands ; at the date of my departure there was not a single person receiving aid from the Poor Fund, which has been kindly formed and is administered by one of the oldest lady residents.

The Government Savings Bank, opened in 1888, has proved the thrifty habits of the people, the shepherds and labouring classes being its customers. The latest balance-sheet shows an average of £131 odd standing to the credit of each depositor.

I have informed you already that skilled labour is highly paid in the town. Other labour is also paid on a liberal scale. The wages of labourers, shepherds, and servants in the camps is very liberal. Of course there is a very limited demand for domestic servants, who therefore do not venture to emigrate in search of situations. Almost every servant, male or female, who is not a native of the islands, has at one time or another found his or her way thither under a special contract of service.

One very fruitful source of income to the labouring class is the cutting and preparing of peat for fuel, which is abundant and is universally used, except in the kitchens of the more wealthy, and in the workshops I have before referred to. If I am not mistaken, peat used to be looked upon in this country as a cheaper article of fuel than coal, but it is not so in the Falklands, notwithstanding that every ton of coal has to be imported, no seams having, as yet at least, been found in the islands. The daily wages paid to the cutter and stacker, together with the heavy charges for cartage, make it a dear commodity, but the quantity of coal imported annually is not nearly commensurable with the consumption of fuel required to counteract the cold of the climate.

You must surely wonder how it is that these islands flourish, as I have endeavoured to show you they do. The answer is simple. *On sheep*, or, to be more correct, *off sheep*. I have up to the present directed your attention more or less to the settlement or town of Stanley, where the larger number of the inhabitants live. But I will now take you into the camps, and tell you about the wool and other products, which, as they have to be carried into Stanley for transhipment to Europe, give much employment to many of the inhabitants of the town. The whole area, then, outside Stanley, consisting as it does of 6500 square miles, or 4,160,000 acres, is devoted to sheep farming and is leased off to farmers, many of them gentlemen of private means, who, or their forefathers, have from time to time taken up lands throughout the islands. Others are men who by their industry and carefulness have saved sufficient to enable them to secure lands for themselves.

The islands are marked out in what are known as sections. Each section contains 6000 acres. A station may contain any number of sections. Most of the lands are held on lease, but several of the holders have from time to time purchased their entire holdings. It is one condition in the leases that the holders must purchase the freehold of a certain number of acres in each section. The length of lease is usually twenty years. It is from the rents paid by the tenants for their holdings that, as I have before remarked, the greater part of the revenue of this colony is derived.

Some of the many outlying islands, such as Lively, Speedwell, Weddell, and many others all round the group, are held by one owner, and are fortunate in having the sea as a boundary fence against all the others, for this secures them against the risk of infection from "scab," that most destructive of pests, which is the dread of all sheep farmers throughout the world. Steps have had to be taken to stamp out this horrible scourge, and I have just learned, since my return from Trinidad, that a practical inspector has been got from one of the large colonies on the other side, who is likely to successfully drive it out. The introduction of sheep into the Falkland Islands is comparatively of recent date. At the time of our occupation, there was abundance of wild cattle inhabiting them. The landholders having acquired the lands, soon also acquired the right to remove these animals, and with the exception of a few which still exist, and may be said to be kept more for the pleasure of sport than for any other purpose, sheep have taken their place. It is still considered, however, a treat to those who are fond of the exciting sport to have a day with the gun and lasso and bolas amongst these wild cattle.

Sheep have now, as I have remarked, taken their place, and the success which has attended the industry of sheep farming on these barren islands, where nothing grows but a scrubby grass, almost warrants one in saying that nature intended them for no other purpose than sheep grazing.

The number of sheep on the islands, according to the latest statistics

I have been able to procure, viz., at the end of 1894, was over three-quarters of a million. The quantity of wool exported was 3,808,475 lbs., not quite so much as in the previous year, when the quantity nearly touched 4,000,000 lbs. A large number of carcases are now annually shipped home to the London market; and I can assure you that if the Bill becomes law, which I see it is proposed to introduce into Parliament, insisting on the labelling meat with the name of the country from which it comes, no one need hesitate in accepting Falkland Islands mutton. It is excellent, and contrary to one's usual experience of cheap goods, namely, that they are nasty, is cheap but delicious. Mutton is only 2½d. per lb., and beef 3d. per lb.

The exportation of the mutton from the islands is carried out as follows :

A large vessel, containing refrigerating chambers of enormous size, goes out to a station called San Carlos on the north-west of the East Island. To this port all the farmers purposing to ship home their mutton drive their sheep. Here they are slaughtered and prepared by an experienced butcher in charge, sent out by the firm at home, for transference to England. After loading up at the East Island the vessel proceeds to a port on the West, and the same process of consigning goes on.

Since many farmers on the large farms have now taken up land in Patagonia, another resource has been opened to the sheep farmer in the supplying cargoes of live stock with which the farms in that enormous continent are being stocked. These are shipped in schooners from the Falklands to the nearest port on the Patagonian coast. In addition there are small exports of sheepskins, tallow, and a very trifling quantity of sealskins. Seals are not found in any great numbers around these islands.

In 1892 an expedition of several vessels visited this southern region in the hope of securing whales. The expedition paid a visit to the Falklands, both going and coming from the whaling ground, and during their absence we were hopeful that their success would be so satisfactory as to induce the promoters of the scheme to adopt the Falklands as a station for storing and equipping. But on their homeward visit they gave but a doleful account of their doings.

Cattle are still, of course, to be found on the farms in the Falkland Islands. These have to be kept in order to meet the requirements of the market, and according to the last return the number was 8192.

Horses, many of them bred on the islands, are of first-rate quality, and can be had at a very reasonable price. £20 is considered a high figure for a horse. Many of them are imported from the continent. Every shepherd has his horses, two or more, and almost every other inhabitant also. Horseback is the only mode of conveyance through the islands. No carriage is to be seen even in Stanley. In Stanley there are a few carts, but throughout the camps they are rare. All the carrying is done by the horses.

The farmers themselves own large numbers of horses, some as many as a hundred; and a sight worth seeing is the gathering every morning into the *corral*, as the enclosure is called, of the troop of horses, so that any one going on a journey can select his animal or animals for the day.

Travelling through the islands is, as I have remarked, a difficult matter. No one unused to the camp thinks of venturing any distance from Stanley unaccompanied by a guide or *vaquero*. One can but keep an eye on the peaks of the mountains, which are capital guide-posts for the stranger. What are known as tracks appear to be creatures of the imagination; and in order to make for, as one must do, certain gates in the fencing surrounding the huge farms or stations, it is necessary to be under the guidance of an experienced hand.

The only other mode of getting about the islands is by schooner, of which many regularly ply from Stanley to the different ports, and it is by such means that those living on the West Island reach Stanley, or send their wool there for shipment home. If a schooner arrives in Stanley when there happens to be no steamer in the harbour ready to receive the wool, it moors alongside the *Great Britain*, of which I have spoken. The wool is stored there, and then shipped home by the Kosmos boat on its first visit.

I have shown you that to one species of animal is due much of the success of the inhabitants of these islands. Of other animals I have little to say, for they are few indeed. There are a few birds of different kinds to be found, such as the snipe, the teal, and a few of the more common kinds. The robin-redbreast is a very large specimen with brilliant breast. There is an abundance of geese, the upland geese being specially good. The penguin still swarms over parts of these islands in millions; and at one time a good revenue was derived from them, but for some time back there has been no demand for their oil, and consequently the licences for killing them have not been much in demand for many years.

In some parts rabbits are very numerous, and afford good sport to the men of the ships of war on their annual visits. Hares have been introduced, and with care are likely to do well. As regards insects and reptiles, I am sure some of my hearers will envy the Falklander, and exclaim, "What a Paradise!" when I say that none exists, and it is only of recent years that a bluebottle fly has been seen in the islands, and the ships calling for the frozen mutton have been blamed for its introduction. Fish are plentiful, the kinds most relished being the mullet and smelt.

Such is a sketch of the life and habits on those little islands, which, according to history, John Davis discovered in 1592. The name of Falkland was given to the group by an English navigator of the name of Strong in 1690, when he sailed through and anchored in the sound dividing the two islands.

Between 1706 and 1714 the islands received also the name of Les Malouines, and yet they are sometimes called the Malvinas, into which the Spaniards corrupted the name.

In 1764 Captain John Byron, who had been sent by George III. on a mission of survey to the islands, hoisted the Union Jack, fired a royal salute, and took possession of them in the name of his sovereign.

In the same year Mons. de Bougainville attempted to found a settlement under the authority of the French Government at his own expense. Spain then laid claim to the islands on the ground that they formed a part of her American possessions, and she induced France to abandon the further prosecution of the scheme. This was done, and on 1st April 1767 the standard of Spain was hoisted. Some of the settlers remained, others returned to Europe.

At the same time the English Government were establishing a colony at Port Egmont, and for three or four years an English and Spanish settlement, each ignorant of the presence of the other on the same archipelago, were engaged in prosecuting similar objects.

In the year 1770 Spanish men-of-war, with 1400 troops on board and artillery, reached the harbour where the British flag floated. The English for a time withdrew, but it was not for long, for in 1771 the Spanish Government was compelled to make satisfaction for the violence, and agreed to restore the islands. In 1774 the settlement was abandoned, and for forty-six years nothing much is known of the Falklands. At times they were visited by whalers, who found them inhabited by wild cattle and horses left by the Spaniards. In 1829 the representative of the Sovereign of England at Buenos Ayres protested against the unwarrantable occupation of the Falklands by the Buenos Ayres Government, and in 1831 the commander-in-chief on the South American station received orders to remove any force belonging to the Argentine Government. In 1833, on the 2nd January, the British re-occupied the islands in earnest and hoisted their standard, whereupon the Argentine garrison quietly withdrew. In 1834 a lieutenant in the royal navy, with a cutter and crew from one of the men-of-war of the South American squadron, was stationed at Port Louis; and in 1842 a lieutenant-governor and a small establishment was appointed by Her Majesty's Government to form the colony which I have endeavoured to describe to you as now existing. Though far removed from the mother country, the breasts of all the inhabitants, be they British-born or naturalised subjects, harbour an ever keen and, I am sure, lasting love and regard for the home country, whose flag, the Union Jack, they are proud to float, as occasion may arise, from the staff in the Government dockyard.

GEOGRAPHY IN THE SCHOOLS.

THERE is a great want in this country of works treating of Geography as a subject of study in schools and the method of teaching it. The Germans have devoted much more attention to the question, and among the publications which have recently appeared is one by Dr. S. Günther and Dr. A. Kirchhoff,[1] in which the latter treats of General Geography.

---

[1] *Didaktik und Methodik des Geographie-Unterrichts (Mathematische und Allgemeine Geographie)* von Dr. Siegmund Günther und Dr. Alfred Kirchhoff. C. H. Beck'sche Verlagsbuchhandlung, München, 1895.

The following is the substance of his introductory remarks regarding the sphere, division, etc., of Geography, and its usefulness in education.

Dr. Kirchhoff defines Geography, perhaps not very exactly, as the knowledge of the Earth and of the mutual relations of the Earth and Man. As a natural body cannot be explained scientifically, that is, as the result of causation, unless the history of its development be traced out, Geology is necessarily inseparable from Geography. Botany, Zoology, and History are also handmaids to Geography, but Geology is far more indispensable, since, without its assistance, the formation of the Earth's surface can only be treated descriptively. Specially important to the pupil is a certain amount of geological knowledge in the study of his native country, that he may understand the structure of the surface by referring to its origin. Similarly, without a knowledge of history, the territorial divisions of a country may certainly be learned as an exercise of memory, but their true significance will be lost sight of.

Ethnology stands in a peculiar relation to Geography; it is not, like Geology, an integral part of the science. As only the mutual relations between the Earth and its inhabitants—plants, animals, and man—should be the subject of geographical study, Ethnology, as well as Industrial Economy and Statistics, have in their full extent as little connection with Geography as Botany or Zoology. Since, however, scientific Geography borrows only from the latter the necessary conceptions and facts, the Geography of the schools must itself provide what it wants, for there is in schools an independent system of instruction regarding flora and fauna, but no separate scheme of ethnology, industrial economy and statistics. The teacher, then, must exercise his judgment in transgressing the bounds of geography proper, in so far as circumstances may demand. He will instruct his pupils in the elements of the classification of races, and introduce, as occasion may occur, facts of economic importance—free ports, the difference between registered and measured tonnage, etc. On the other hand, he may cast aside without hesitation much of the material of geographical instruction that is to be found in the old system—everything, for instance, relating to the work of man that has no immediate connection with the Earth. Historical details should not find a place in geographical teaching, except such as are needed to explain facts connected with the world—the deeds of a Columbus, a da Gama, a Magellan, or a Cook. So, too, the inheritance of the county of Katzen-ellenbogen by the Hessian landgraves must be mentioned to explain the almost unique example of a State extending across the boundary between North and South Germany. On the other hand, the birthplaces of Schiller and Goethe, or the field where the battle of Marathon was fought, are matters for the history class. Heraldry, and the names of reigning princes, or the minutiæ of the constitution and administration, are not admissible. Some few points may be touched on in treating of our own country, and certain subjects—universities and faculties, torpedoes, brass, damask, etc.—though not strictly within the domain of Geography, are hardly to be avoided in the description of towns.

Classification is of more importance in Geography than in any other science. During the present century the time-honoured division into

Mathematical, Physical, and Political Geography has done much mischief. It was considered as correct and natural as the saying that man consists of a body and soul, and, as almost all men deduce from the latter that the soul may be separated from the body, so it was held that Physical Geography, like Mathematical, might be differentiated from Political, although the latter was understood to mean the geography of countries, which could not be thought of apart from the natural configuration of the land. Indeed, the view became familiar that Political Geography comprised the historical side of the science of the Earth, which was accordingly divided into Physical and Historical. That this mistake carried confusion even into the highest scientific circles, was shown, in Vienna, for instance, by the establishment of separate Chairs for Physical and Historical Geography, which caused, indeed, little inconvenience, as the lecturers did not confine themselves to their arbitrary limits. In Prussia, also, in the regulations of 1891 for geographical instruction, it was laid down that the political geography of Germany should be kept apart from the physical, and, most singularly, the former was to be taken in the Lower Third class and the latter in the Upper Third—the cart before the horse.

As early as the middle of the seventeenth century Bernhard Varen clearly recognised the difference between General Geography and the geography of countries (*Länderkunde*). These are the two divisions into which Geography naturally falls. The former treats of the Earth as a heavenly body, its atmospheric envelope, its seas, the general formation of its land masses, the characters of its rivers, glaciers, and lakes, the elements of the distribution of the fauna and flora, the general relations between the Earth and Man (Anthropogeography). The practicability of making a division into general Physical Geography and into Anthropogeography, containing essentially historical elements, is at once apparent, whereas the geography of countries, or Special Geography, as it may be called, cannot be separated into natural and historical subdivisions. It describes countries and depicts their characteristics on the basis of terrestrial laws learned from the study of general geography. Thus, countries, except the uninhabited Polar lands, must be regarded from two sides—that of their natural features, and that of the inhabitants and the effects of their activity on the lands they occupy. A country may be examined with regard to its natural conditions, as is done by the geologist, climatologist, hydrographer, etc.; or it can be investigated and described exclusively from a historical point of view, with regard to its settlement, the industry of its inhabitants, and the effect that the latter has exercised on the character of the people. But these two ways of considering a country are only fragmentary when isolated; a living whole can only be produced by their coalescence. An inhabited country, with its complication of surface divisions, coastlines and roads, canals, towns and villages, is a microcosm, the peculiar character of which depends on a host of natural and human conditions determining one another. No naturalist, or historian, or economist can thoroughly unfold its nature; that is the work of the specialist in the geography of countries. He that finds a pleasure in demonstrating that no geography is an independent science may amuse himself by analysing General Geography, and declare that its individual parts—Mathematical Geography, Climatology, Oceanography, etc.—are sciences in themselves, and that Anthropogeography properly belongs to the domain of the historian and economist. But let him not touch Special Geography. The complete knowledge of what the conception of a country involves—whether the country be as large as India or as small as Germany—can only be collected in its Special Geography; and this does not solely consist of physical and historical data placed side by side, but as a science it unfolds the causal connection of natural circumstances with those that are called into existence by human agency. Certainly the best results, even in a very small area, may be obtained by distributing the work among specialists in the sciences concerned; but it remains the task of the geographer to form the materials supplied by the geologist and botanist, the historian and linguist, into an organic whole. The case of General Geography is to a great extent similar, for though solid progress is obtained through the accumulation of material by the independent work of specialists, a controlling force is needed to prevent the knowledge acquired from falling asunder into an endless chaos of disconnected data.

Through its central position among all branches of objective science, Geography is of great importance as a subject of education. Herbart's dictum that Geography is an associating science is strikingly true. If it be erased from the time-tables of schools, the mathematical and natural science branches of instruction lose all connection with the linguistic and historical.

If the co-ordination of knowledge furnishes a good guarantee for its continuance, an indispensable condition for profitable comparison, though terrestrial phenomena may in the scholastic syllabus find their places under distinct sciences, Geography, the only study that forms a connecting link between almost all other school subjects, should not be treated as a Cinderella, and, above all, should not be banished from the upper classes of schools; for here, where the historical knowledge of the pupils on the one hand, and the scientific on the other, receive their finishing touches, Geography is best able to give the full benefit of its associating powers. Our schoolmen, in placing ancient or modern languages, mathematics, or natural science in the foreground in the arrangement of the time-table, and assigning a meagre share of the time to Geography, or eliminating it altogether, as though it were of no more importance than botany, or, at most, history, show a want of appreciation of the true value of the subject as a branch of education. The time allowed for it must, of course, be regulated according to the whole time at disposal and according to the importance of the principal subjects, but it should not be relegated to any spare time that may be left over, perhaps only an hour a week, regardless whether the prescribed course can be completed in this one hour. The utmost restriction, or even the total abolition, of geographical teaching is in these days too lightly excused by the remark that a due regard for the principal subjects will not permit of any other arrangement. It is forgotten that Geography is the one subject which is always important and indispensable, whereas the others are, with respect to it, really subsidiary. A time-table which in any class leaves hardly any room for Geography deserves censure under any circumstances, for it injures the whole by weakening the cohesion of the constituent parts.

The value of history should certainly not be despised. But at the present time most German teachers are inclined to over-estimate its value as compared with that of Geography. Besides exercising a centralising power in a greater degree, Geography possesses other advantages. To Immanuel Kant, the first German High School teacher of Geography, we owe the apothegm, "Nothing is better calculated to awaken the healthy human intelligence than Geography." And this holds good from the sixth to the first class, if the teacher sets about his work in the right way, if he teaches his pupils to think instead of cramming them with names and figures. Even in the lowest class it is possible, particularly in the knowledge of their own neighbourhood, which should naturally come first, to rivet the attention of pupils by the attractions of true science, which searches into the reason of things. Without too great a demand on their powers of comprehension, they may be taught out of doors to recognise the creative terrestrial agencies, and the dependence of man on his environment, and, again, the effect of man's industry and thought in improving the fields. What other subject learnt in the sixth class can so easily and forcibly lead from perception to thought in reference to the most familiar phenomena, which, however, point the way to the highest conceptions of Nature and Man? This is the grand characteristic of geographical instruction up to the highest classes, that it teaches the most simple and homely intellect to exercise itself in searching out the connection of cause and effect in matters which are of striking importance in the practical life of peoples and in the history of mankind.

Moreover, the imagination guides the intellect. The views and descriptions of distant lands stimulate the pupil to consider why everything in other countries has such a different appearance from what he sees at home. And again, in the geographical class the creative faculty is exercised alternately with the receptive, knowledge develops into power, as soon as the freehand drawing of maps receives its due attention.

And how thoroughly Geography combines material and intellectual interests! Through it alone the pupil learns to understand what the activity of head and hand accomplishes in the world's industry, to read the productive capacity of its inhabitants inscribed in gigantic letters on the face of a country. History shows in a few cases that there is an inherent justice in the destiny of nations, while Geography demonstrates this truth as applied to all mankind. It alone makes the whole world homelike, leads to an unprejudiced estimate of peoples in all parts of the world, and enables a man to understand his native country.

PROCEEDINGS OF THE ROYAL SCOTTISH GEOGRAPHICAL SOCIETY.

LECTURES DELIVERED IN APRIL.

On April 9th Captain F. R. Maunsell, R.A., delivered a lecture in Edinburgh on "Eastern Turkey in Asia and Armenia." Major-General Sir R. Murdoch Smith presided.

Another address was given in Edinburgh on April 30th by M. Lionel Dècle, the title being "Through Africa from the Cape to Uganda." Mr. William C. Smith took the Chair.

LECTURES IN MAY.

Major C. Barter will lecture in Edinburgh on Ashanti, in the Society's Hall, on May 7th, at 4.30 P.M.

ELECTION OF MEMBERS.

During the Session the following 174 ladies and gentlemen have been elected members of the Society :—

Adams, D. S.
Adams, J. F.
Aikman, John.
Ainslie, Miss J. K. M.
Alexander, William.
Anderson, Rev. Andrew.
Anderson, John.
Anderson, J. Mitchell.
Anderson, W. Simpson, M.A.
Andrews, Ernest.
Anstruther, Henry T., M.P.
Bain, William.
Balfour, Mrs. Alexander.
Barrie, Charles.
Bernard, Professor Augustin.
Black, S. D.
Boyne, James.
Briggs, General James.
Bruce, William P.
Campbell, Archibald.
Charteris, Rev. Prof. A. H., D.D.
Clarke, John.
Coats, P. Herbert.
Colville, John, M.P.
*Cook, Henry.
Croll, George.
Crombie, Theodore.
Cumming, David.
Cunningham, James.
Cunningham, Miss L. L.
Cunninghame, J. C.
Dalkeith, The Earl of, M.P.
Dalrymple, The Hon. Hew.
Dalziel, T. Kennedy, M.B., C.M.
Davidson, George.
Davidson, J. Mackenzie, M.B., C.M.
Davidson, John.
Davidson, R., C.S.I.
Denny, Colonel, J. M., M.P.

Dewar, Colonel, J. C.
Dick, Adam.
Dickson, Leonard, W., C.A.
Dickson, W. S., Advocate.
*Don, Robert B.
Downie, Alexander A. E.
Drummond, Sir James, Bart.
Dundas, Captain Robert.
Edmondston, Thomas.
Edwards, John.
*Elphinstone, The Right Hon. Lord.
Fergus, A. Freeland, M.D.
Ferguson, Fergus.
Ferguson, Mrs. M.
Ferguson, R. M., Solicitor.
Finlay, Sir Robert B., Q.C., M.P.
Fleming, John.
Fraser, William.
Galloway, M. P.
Geddes, George N.
Gibson, Campbell.
Gilmour, Captain R. Gordon.
Gray, Baxter.
Grierson, Prof. H. J. C.
Gunn, John, M.A., D.Sc.
Gunn, Neil Ballingall, F.I.A., F.F.A.
*Hamilton, His Grace the Duke of.
Hardie, Prof. W. R.
Harrison, G. C.
Hay, His Excellency Sir James S., K.C.M.G.
Hay, Councillor Robert.
Henderson, Miss E. F.
Hill, Major T. A.
Home, Miss G. Milne.
*Hope, Captain Thomas.
Horn, William.
Horne, James.
Hozier, James, M.P.

Jamieson, Claude Auldjo.
Jarvie, William.
Jenkins, Thomas.
Johnston, Brigade-Surgeon, Lt.-Col. Wm.
Jolly, J. Keith.
Kinnaird, The Right Hon. Lord.
Leadbetter, James G.
Lithgow, Wm. Todd.
Logan, Mrs. C. B.
Lyde, Lionel W., M.A.
Macdonald, A. Spence.
Macdougall, Colonel J. W.
*Macgillivray, Evan J.
*M'Iver, Charles.
Mackay, Miss.
Mackay, Prof. J. Yule, M.D.
Mackie, David, M.A., M.D.
M'Kie, Thomas, Advocate.
Mackinlay, P., M.A.
Mackintosh, Miss A. B.
MacLellan, John.
Macmillan, Archibald.
M'Neill, Duncan.
Macqueen, John Otto.
Marshall, John, LL.D.
*Maxwell, Sir J. Stirling, Bart., M.P.
Mearns, Daniel, Lord Provost of Aberdeen.
Mess, John, C.A.
Miller, J. W.
*Miller, Rev. Alexander, B.D.
Milligan, David M. M., Advocate.
Mitchell, Andrew, J.P.
Mollison, John Croll.
Monro, Captain David.
Muirhead, George, F.R.S.E.
Munro, Thomas R.
Murdoch, Robert.
Murison, William, M.A.
Murray, Patrick.
*Murray, Wm. Hugh, W.S.
Oliphant, E. H. Lawrence, M.D.
*Orr, Robert, of Kinnaird.
Orr-Ewing, C. L., M.P.
Pearson, The Right Hon. Sir Charles J., M.P.
Pearson, Dalziel, W.S.
Philip, J. Bentley, M.A.
Philip, Joseph.

*Pirie, Captain Duncan V.
Pirie, Mrs. M. C.
Prothero, Prof. G. W.
*Pullar, Frederick P.
Pulsford, G. Marsom.
Reid, James R.
*Reid, Prof. R. W., M.D.
Rettie, William, J.P.
Robinow, Miss E. F.
Rodger, Anderson.
Russell, David.
Russell, Thomas.
Rutherford, Alastair M.
Ryley, Rev. H. Buchanan.
Saintsbury, Professor.
Sanderson, A. N.
Service, George W.
Sharp, F. B.
Sharp, R. B.
Shearer, J. Elliot, F.S.A.Scot.
Sibbald, Rev. S. J. Ramsay, B.D.
*Simpson, J. Young, M.A.
Smith, Robert.
Smith, Thos. Henry, M.A.
Somerville, R. Anderson.
Spence, Miss.
Stevenson, Prof. J. J.
Stewart, Sir Mark J. M'T., Bart., M.P.
Stewart, James.
Stitt, Charles Alfred.
Sym, Colonel J. Munro, C.B.
Thomson, William, J.P.
Thorburn, Robert.
Tod, Mrs. C. C.
Tod, Frederick.
Trotter, Alexander E. C.
Troup, James.
Walker, James, of Dalry.
Wallace, Hugh R.
Wallace, James, M.A., Advocate.
Ward, Miss L. L., B.A.
Wauchope, Colonel A. G.
White, J. Martin, M.P.
Williams, James.
Willis, Mrs. N. H.
Wilson, Professor J. Dove.[1]
Wilson, Miss J. T.
Wilson, Major R.
Wilson, William.
Wylie, Alexander, M.P.

The asterisk denotes Life Members.

GEOGRAPHICAL NOTES.

By The Acting Editor.

EUROPE.

The German Census.—The results of the census of the German Empire taken in December 1895, as at present worked out, are given in *Petermanns Mitt.*, Bd. xlii. No. 3. The total population numbered 52,244,503, and the increase during the five years, 1890-95, was 5·7 per cent. Of this number 31,847,899 belonged to Prussia, where the increase was 6·3 per cent. Bavaria contained 5,797,414 inhabitants, the increase having been 3·6 per cent., and for Saxony the number was 3,783,014, with an increase of 8 per cent. The growth was greatest in Brandenburg and Westphalia, being 11 and 11·2 per cent. respectively, while in Hohenzollern alone was there a decrease, namely of 1·5 per cent. East Prussia and Alsace-Lorraine also show small percentages. The movement of the population is greatly affected by the attraction of large towns and mining and industrial districts, and the influence it has exerted will be apparent when fuller details are published.

The Mouth of the Volga.—Astrakhan, the centre of the shipping traffic on the lower Volga, has become, during the last few decades, cut off more and more from the Caspian Sea. In consequence of the silting up of the mouth of the river, the navigable channel is less than ten feet deep when the water is high, and is therefore closed to the Caspian fleet. Consequently the ships have to anchor in open roadsteads at a distance of twenty to thirty miles from the coast, exposed to the wind and waves, and there the goods have to be transferred to vessels capable of navigating the Volga. This work keeps 10,000 men employed, and entails an expense of £300,000 to £400,000 yearly. To remedy this inconvenience, the Russian Government have, since 1891, been investigating the practicability of regulating the mouth of the Volga, and restoring a navigable channel for vessels up to Astrakhan. It has been decided that the work can be accomplished, and at the end of last year the preparatory works were commenced. Astrakhan will also, before long, be connected by a railway with Tsaritsyn, and thus with the Russian network.—*Globus*, Bd. lxix. No. 7.

ASIA.

Railways in India.—The following abstract has been kindly forwarded by Professor P. Chaix :—

I borrow from the *Times* (January 31st, 1896), the following quotation from the Marquis of Lansdowne's speech, delivered at a meeting at the South-West London Polytechnic Institute :—" The success with which we have in India accomplished the great task of infusing new life and vigour into an Eastern society, with a civilisation quite different from our own, is such as every Englishman may well be proud of."

The more deeply we study the progress of the English rule in India the more do we agree with the opinion of the late Viceroy, and the more are we impressed with the high price at which such a result has been obtained. Few people on the continent are aware of the number of those devoted servants of their country—*Rulers of India*, as they are styled—who have lost life or health in the pursuit of that noble cause, the sad case of the Earl of Mayo being the most striking example.

Lieutenant-Colonel Gracey's Report on the Railways of India affords us the means of seeing at work one of the most powerful engines of English civilisation in the East. We are told of distances exactly measured on the banks of the Nile by the pace of the *Bematists* in the service of the Ptolemies, of runners or state-messengers instituted by Darius in his vast monarchy, of richly laden caravans from Mazaca to Gangra, Tavium Sardis, and Ephesus ; but in none of those cases are real roads mentioned. The pompous court of the Persian kings had to fight or to negotiate its way across mountains and through mountaineers on its yearly migration from Babylon to Ecbatana ; and Professor Ramsay, in his learned study of Asia Minor in early times, shows that the caravans found their way somewhat at random from one large city to another across the steppes of Phrygia and Lycaonia. In road building we see one of the most imperishable features of Roman greatness, and, to the shame of so many centuries which followed it, roads did not survive the disturbances of the Middle Ages.

The few faithful companions who brought the corpse of Otto the Third back from Italy carried it on their shoulders, and fought their way over the Brenner Pass. Across the St. Gothard the difficulties were such as to require the help of Rudolf, Count of Habsburg, to carry the Archbishop of Mayence to Italy, which earned for him the vote of that powerful churchman when canvassing for the imperial crown. To Shah Abbas the Great is due the credit of having built across the Caspian range roads which, in their present dilapidated state, still bear his name. But, notwithstanding the pomp and vast treasures of the Mogul Emperors, there were no roads in their States except the broad avenue between Agra and Delhi ; and the regular migrations of their retinue to the valley of Cashmir were sometimes attended with fatal accidents in the mountains. To the British rule was reserved the merit of initiating, in 1849-50, a new and more beneficial era for Hindustan. In 1853 twenty miles of railroads had been built, and the yearly progress went increasing by rapid strides till 1884, when 1177 miles were opened in a single year. The yearly progress has since slackened down to 451 in 1893, and 360 in 1894-5.

The first lines were of course opened in the most populous provinces of Bengal, Behar, the Doab and Carnatic. Sections, with double tracks, are to be found from Calcutta to Benares through Patna, and in the neighbourhood of Madras and Bombay, 1150 miles in all. The most important lines are (1) from Madras to Calicut, with a branch to Bangalore ; (2) from Madras to Bombay through Poona ; (3) the Great Indian Peninsular from Bombay to Allahabad and Benares ; (4) from Bombay to Baroda and Ahmedabad ; (5) from Bengal to Nagpur ; (6) the East Indian from Calcutta to Patna, Benares, and Allahabad ; (7) from Allahabad to Cawnpore, Agra, Delhi, Lahore, etc. ; (8) the North-Western, from Kurrachee to Mooltan and Lahore, with branches from Mooltan and Bukkur on the Indus to the Bolan Pass and Quetta, and from Attock to Peshawur. The line of the Bolan Pass, reaching Quetta at a distance of 536 miles from the port of Kurrachee, is a purely military road, and very expensive ; Pishin can be reached by a loop from Quetta and a side line. The gradients average 1 in 33, with a maximum of 1 in 25, and a minimum of 1 in 55. The dryness of the climate has necessitated the building of a tank capable of holding 120,000 gallons of water supplied from springs and conducted through pipes of wrought iron.

The aggregate length of the seventy-three railways open on March 31st, 1895, was 18,855 miles, of which 10,895 were built on the standard gauge of 5 feet 6 inches, and 7,712 on the metre gauge. The balance is made up of lines with special gauges. A short branch of ten miles was completed, in 1894, from Bowringpet station to the Mysore goldfields of Kolar. In countries where the population is not very thick, with indifferent prospects of trade, the metre gauge has been deemed sufficient. A few lines of that size will impart life to the venerable states of Oodeypore, Chitor, Jayselmer, Jodhpur, Bikaner in Rajpútana, of chivalrous fame. The absence of running water on the line Rajpútana-Malwa has led to the building of two tanks containing 30,000 and 40,000 gallons of water respectively, while the abundance of rains and frequency of floods in Burma has made it advisable to raise two feet above the level of the plains the lines from Toungoo to Mandalay and farther northwards, also built on the metre gauge.

The East Coast railway is actively progressing, and has been opened as far as to Khurda Road, while on the north of the gulf a line is under construction from Chittagong to Makum. A connection will be made with the Burmese system, and with this object surveys have been made east and south of Chittagong, but it is now believed that the Pakhoi range will offer the best route. Among the most remarkable works executed on that extensive mileage may be noted a bridge at Shadara on the Ravi, and large bridges of ten spans of 100 feet each on the line Bareilly-Rampur, sometimes threatened by heavy floods. A bridge of boats has been constructed across the Mahanaddy at Sambalpúr in place of the ferry of former times, as well as bridges on the Jhelam in the Punjab, and on the Gogra. A few branches are being built to keep the valley of the Ganges in touch with the foothills of Rohilkund and Oudh. The isle of Ceylon has but one single line, from Colombo to Kandy and the foot of Pedro Talla Galla. A large extent of country between Poona, Hyderabad, Nagpúr, and Orissa has not yet been tapped by railway lines.

The heavy rains of India are a fruitful source of damage, and every year a list is given in the report of bridges carried away, embankments breached, and tanks burst by floods and torrents.

The progress of railway development has made it necessary to search for fuel; wood is used on a limited scale, and coal is found in many places. The best collieries worked are in Sylhet, Eastern Bengal, in Tirhoot, at Kurhurbaree and Serampore, at Mokum and the Sado valley in Assam, at Warora, at Singareni in the territories of the Nizam, at Umaria and Mohpáni along the Nerbudda. The coal is in general an excellent fuel for railways, and is much prized for use in foundries. The deepest pit at Kurhurbaree is 650 feet deep. The Warora colliery is worked by the use of compressed air. The north-western region on the river Indus suffers from less favourable conditions in the quality and the abundance of the output; mines are opened at Sharigh, Khost, Dandot, and Killa-Ali-khan. The Sor range, fifteen miles from Quetta, yields coal of poor quality, which is made into briquettes with the produce of the neighbouring petroleum borings at Khattan in Baluchistan. Petroleum has also been found at Makum in Assam, and at Digboi. Borings have been carried to a depth of 474 feet, striking oil at 158 feet at Makum, and to 1120 feet at Digboi. The yearly output of the coal mines in India rose, during the fourteen years, 1880-1895, from 1,019,793 to 2,529,853 tons. The quantity of English coal used on the railways was 28.62 per cent. less in 1894 than in the preceding year.

The total capital expenditure on Indian railways, up to December 31st, 1894, amounted to Rs. 25525,30,392, whether on state railways, railways of native states, subsidised or guaranteed lines, etc. The total length of railways worked during 1894 was 18,694 miles as compared with 18,207 worked during the previous year. The gross earnings showed an increase of 5.9 per cent., the passenger earnings being 3.85 per cent. greater, and the goods earnings 6.78 per cent. The passenger unit-mileage rose 5.15 per cent., and the ton-mileage of goods 9.87 per cent. The number of passengers carried per mile was 315,085 in 1894 as compared with 307,661 in 1893, of which the lower classes contributed 97.43 per cent.

The Temperature of Siberia.—The following figures are extracted from the Siberian Calendar for 1896, published in Tomsk. They refer to the agricultural districts. The mean annual temperature in 31.6° F. in East Siberia, 32.6° in West Siberia, and 32.5 in European Russia. The summer temperatures are 61.7°, 63.5°, and 62.6° respectively. The mean winter temperatures are −0.4°, 1.4°, and 11.3°. Lastly, the mean temperatures during the time the corn is growing, from May 1st to September 15th, are 57.2°, 59°, and 59° respectively. Thus, though the annual temperature in Western Siberia is considerably lower than in European Russia under the same latitudes, the summer in the agricultural districts of Western Siberia is as warm as in Central Russia.

The following table shows the temperature of the most important places on the east coast of Siberia in order from south to north:—

	Annual Mean.	Summer Mean.	Winter Mean.	Mean of the Corn Season.
Vladivostok, . .	40·1	64·4	10·4	60·8
Diu (Saghalin), .	32·9	61·7	5·0	53·6
Nikolaievsk, . .	27·3	59·0	− 7·6	52·9
Petropavlovsk, .	35·6	55·4	17·6	51·1
Aian, . . .	24·8	51·8	− 4·0	46·4
Okhotsk, . .	23·0	51·8	− 3·1	46·4

The difference between the temperatures at Nikolaievsk and Petropavlovsk, nearly on the same parallel, is very striking. The milder climate of the latter is due to the warm current flowing up from the south. In the forest zone of the Government of Yakutsk the mean temperatures are 17·6° for the year, 59° in summer, − 27·2° in winter, and 51·8° during the corn season.—*Globus*, Bd. lxix. No. 70.

The Area of the Dutch East Indies.—Hitherto there has been a great want of reliable data regarding the area of the individual islands of the East Indian Archipelago. Now, the Government of Insulinde has caused to be executed a planimetric measurement. These measurements, which have been executed with considerable care on the best maps available, four persons being engaged in the work, are as follows :—

	Sq. Miles.
Borneo, ⎫	213,253·4
Sumatra, ⎪ With the smaller islands belonging to them,	175,913·2
Celebes, ⎬	73,317·1
Java, ⎭	50,774·1
The Moluccas, ...	36,896·7
Dutch New Guinea, ...	149,117·2
West Flores and Sumbava, with adjacent islands, . .	7,628·8
Residency of Timor and its Dependencies, . . .	17,781·7
Residency of Riouw and its Dependencies (except Indragiris),	3,157·0
Residency of Bali and Lombok, ...	4,062·8
Residency of Bangka and its Dependencies, . .	4,489·7
Sub-Residency of Billiton, ...	1,869·5
	———
	738,261·2

Globus, Bd. lxix. No. 15.

The Toba Lake.—Dr. J. F. Hoekstra gives, in the *Tijdschrift van het K. Nederl. Aardrijkskundig Genoots.*, Deel xiii. No. 1, the latest information about the Toba Lake and the peninsula Samosir. The areas of the lake and peninsula differ considerably, as given in Droeza's *Kaart der Battaklanden en van het eiland Nyas*, and by the mining engineer, N. Wing Easton, who visited the country three years ago in search of bismuth. Probably the latter's estimates are more reliable, though perhaps only approximate. It may, then, be said that the area of the lake itself is about 530 square miles, and that, if the peninsula be added, the total area becomes about 790 square miles. It is therefore about two and a half times the size of the Lake of Constance, and far larger than the other lakes of Sumatra, Singkarak, the next in size, having an area of only 43 square miles.

The peninsula Samosir has as a rule a narrow coast strip, which gradually passes into hilly and mountainous country, while the interior consists of a high tableland which only here and there—at the north, east, and south-east—reaches the lake in the form of a high steep shore. All the streams are small, and even the largest, on the south coast, cannot be navigated by small boats. The surface is treeless, in some places cultivated, in others covered with grass or bare. The steep banks of the Toba Lake consist to a large extent of quartz trachyte, an eruptive stone of very recent origin. The eruptions were accompanied by the deposition of large quantities of tuff, which now form the tablelands on Samosir and at the southern extremity of the lake and elsewhere. The trachyte forms a huge wall along the east coast of Samosir and on the bank of the lake just opposite, whence Verbeek and Easton independently have inferred that the peninsula was once connected with the eastern shore of the lake. Trachyte also occurs in many other places in the neighbourhood, and some of the products of the Pusuk Bukit, the volcano which, according to Verbeek, connects Samosir with the western shore of the lake, are trachytic. Unfortunately Easton was unable to examine the mountain thoroughly and ascertain whether there is a crater on its summit. The plateau in its rear was visited during one of the military expeditions, and it was said that a crater form could not be distinguished, but it is very unlikely that the plateau commands the whole of the summit of this mountain, 6580 feet high. At any rate, warm springs and sulphurous exhalations show signs of volcanic action.

On Samosir also solfataras occur. Recent volcanoes, built up of loose andesite so common in other parts of Sumatra, are absent. The older rocks are mostly covered with the volcanic deposits. On the south shore of the lake clay schists crop out, and perhaps on the eastern shore. On the south breccia also occurs, rising in huge walls, sometimes to a height of 650 feet, and in some parts of Samosir it is largely developed, rising to a height of 1000 feet above the surface of the lake.

Easton did not visit the north, west, and east shores of the lake, but he believes that the lake basin is not a collapsed crater, but an enclosed arm of the sea, referring to the regularity of the tuff beds which must have been deposited in water, and to other facts. At present, however, the question must be considered unsolved.

The Kangean Archipelago.—This is a group of about thirty islands lying to the east of Madura and north of Bali, in about 7° N. lat. The chief islands are Kangean, Saobi, Paliat, Sapeken, Saseël, and Sepanjang; the other and smaller islands are uninhabited. Kangean is hilly, its highest summit, Gunong Batupotih, rising in the centre of the island to nine hundred or a thousand feet above sea-level. The hills are formed of limestone, and contain caves, in some of which edible bird's nests are found. Sepanjang, next in size, is covered entirely with forest. The inhabitants are true bushmen, and live by collecting bast and hunting. Sapeken, though small, has a large population, about 6000, and a large trade and fishery. Yet it is nothing but an unproductive sandheap resting immediately on coral. The population lives on the shore, while the interior is a wilderness, in the midst of which are one or two plantations of palms.

The islands are surrounded with reefs which are very dangerous to vessels, especially as their exact positions are unknown. There are no rivers or lakes; fresh water is obtained from wells. The climate of the main island is good, but cases of *beri-beri* are not uncommon. Fevers and cholera are more prevalent on the smaller islands. The chief products of Kangean are coco-nuts, paddy, and other field crops, which after a good harvest are exported in considerable quantities. The natives (about 15,000 in number) have little aptitude for trade, so that the export is chiefly in the hands of the people of Madura, Bali, and Sapeken. Trade is carried on by barter—linen, dried fish, earthenware, petroleum, etc., being exchanged for rice, paddy, or coco-nuts—for coin is scarce. Fowls, jungle-fowl, and *bakisars*, a cross between the two, are also exported, but cattle and buffaloes, though plentiful and cheap, to a very limited extent, owing to the cost of transport. From the smaller islands are exported tortoiseshell, fish, sea-cucumbers, fibres, and dried venison.

It is singular that wild boars and squirrels are not to be found on the Kangean Archipelago. The woods contain several kinds of valuable timber and bamboos. The inhabitants of the island are principally Madurese, with a considerable admixture of Makassaren, Buginese, and, probably, Balinese.—*Bijd. tot de Taal-, Land-, en Volkenkunde van Nederl. Indië,* Deel xlvi. afl. 1.

AFRICA.

Ivory and Elephants in Africa.—Geheimrat K. Mobius gave the following statistics at a meeting of the *Gesells. Naturforschender Freunde* in Berlin on February 18th:—11,650 tons of ivory were offered for sale in 1895 in London, Antwerp, and Liverpool, besides 1570 tons of old stock. From the Sudan came 1140 tons, much of which had probably been collected by Emin Pasha. German East Africa and Mosambique now yield less than formerly—only 1830 tons in 1895—and Cape Colony hardly any. 6680 tons came from the Congo, chiefly of inferior quality, 668 from the Niger and Benue, and 727 from Gabun and Cameroons. As an African elephant yields, on an average, about 30·8 lbs. of ivory, the 11,650 tons must have been obtained by the slaughter of some 42,360 elephants. At the present day, the regions of Africa inhabited by elephants stretches from the Tropic of Capricorn northwards to 13° W. lat.—*Globus,* Bd. lxix. No. 75.

Phosphates in Algeria.—The first information of the existence of these deposits was given by Ph. Thomas, who, in 1888, was entrusted with the geological survey of Tunisia. The richest beds occur in strata of the Eocene period, and particularly in an old beach towards the edge of the Sahara. Besides extensive accumulations of oyster shells, there are found remains of fishes and reptiles in extraordinary quantities, among them huge rays and crocodiles. Of these remains, consisting of bones and coprolite, the layer of phosphates, 25 to 30 feet thick, is in great part composed. As the distance from the old shore increases, the layer becomes harder and contains fewer fossils. It is now ascertained that these phosphates extend for a distance of 430 miles from the south of Tunis, through Constantine, to the centre of the Department of Algiers. Attempts to exploit these deposits failed, until some Englishmen succeeded in starting extensive works in the neighbourhood of Tebessa. Tebessa lies close to the ruins of the ancient Thevesta, on a tableland 2913 feet above sea-level and nineteen miles long. The southern edge of this plateau is bordered by a steep range of hills, on the slopes of which are the phosphate works, 300 to 2000 feet above the plain. The most productive and the oldest, though opened only two years ago, is Crookston on Jebel Dyr, about nine miles from Tebessa. The land conceded by the Government has an area of 6900 acres, and is estimated, by the mining engineer of Bona, to contain one to two hundred million tons of valuable material. And it surpasses all other beds yet known, not only in extent, but in its richness in phosphates, of which it contains 60 to 68 per cent. on an average, with a maximum in some places of 83 per cent. About 300 Italian workmen are employed in the extraction, and over 550 tons of phosphate are forwarded daily to Bona for shipment.—*Revue Scientifique, Jan.*

Timbuktu.—A few notes are given in the *Tour du Monde* for April on the climate, commerce, etc., of Timbuktu and its neighbourhood. The temperature is one of the highest in the world, though not quite attaining to that of the Central Sahara. During the greater part of the year the thermometer ranges between 86° and 104° F. in the shade, and in the sun marks about 18° more. In December and January the nights are very cold, the thermometer falling to 43° or 44°, though it stands as high as 86° about noon. Between June and October storms are numerous, and the country near the river is deluged with rain ; during the rest of the year the climate resembles that of Algeria, but is somewhat warmer. The winds have a great influence on the climate ; the simoom, which blows from the east for about twenty days in every hundred, is very scorching, and raises up clouds of sand, so that it is impossible for people to leave their houses or tents.

Beyond the limits of the inundations of the river begins the desert and the "Azauad," the dreary Saharan forest, which extends some seventy miles north-wards to Arauan. From the north of Lake Fanguibine it stretches, it is said, to Lake Tsad. Colonel Joffre, however, as reported in the *Bull. du Comité de l'Afrique Française*, limits it to Taudenni on the north, Mabruk on the east, and El Akela on the west. This forest is composed exclusively of mimosas, jujube trees, and other thorny plants. Though terrible even to cross, the forest is the home of Tuaregs and Moors. The latter are divided into several tribes, one of which, the Berabiches, occupies the neighbourhood of Timbuktu as far as Arauan, and carries on a large caravan trade, exchanging the salt of Taudenni at Timbuktu for the millet of the Sudan, and dealing also in manufactured goods from Morocco and Cape Juby. Besides camels to carry their merchandise, the Berabiches own flocks of sheep and herds of goats.

The Tuaregs are divided into three classes—Tuaregs of pure race, forming the warrior nobility, vassals, and slaves. Each of the noble families has a number of vassals, who occupy themselves with cattle-grazing and dealing, and in time of war are obliged to follow their masters. The slaves are for the most part born among the Tuareg, and are most devoted to their owners. The wealth of the Tuaregs consists of cattle and sheep, and they are obliged to move about in search of fodder. When in want of provisions they extort blackmail from their neighbours or plunder them ; sometimes they rob one another.

The town of Timbuktu contains 7000 to 8000 inhabitants, composed of very different elements. The bulk of the population consists of the Ruma, descendants of the Moorish conquerors, and Haratin, descendants of the subjugated Songhay negroes. Its importance as a commercial centre has much diminished since the Tuaregs became masters of this region nearly a century ago. Placed at the northern limit of the Niger, and accessible by the large native canoes in time of flood, and close to the countries of the Sahara, it has naturally become the-*entrepôt* for traffic between these and the Sudan. The inhabitants produce scarcely any-thing, being occupied almost exclusively with the transfer of merchandise between the north and the south. European textiles, chiefly English, are brought from Morocco, as well as sugar, spices, glass ornaments, and weapons and other goods in copper and iron. Tobacco and dates from Tuat and Tafilet, and tanned hides and leather from Tenduf, are also seen at Timbuktu. Besides the Berabiches, who deal chiefly in salt from Taudenni, the nomads of Tajakant, Tuat, and Tafilet are engaged in this trade. Timbuktu is in communication with Mossi, Kong, and the whole country to the east as far as Gogo, importing from these districts millet and rice, vegetable butter, cotton, raw hides, ground-nuts, etc. At present it is impossible to estimate the value of the trade owing to the disturbance caused by the occupation by the French.

AMERICA.

The Currents of the Gulf of St. Lawrence.—Mr. W. Bell Dawson, C.E., in charge of the Tidal Survey, has sent us his report on the observations made last year. The investigations of the previous season, to ascertain whether a current existed across the Gulf to the Strait of Belle Isle, pointed rather to a current from the Gaspé coast to the Cabot Strait (vol. xi. p. 369), and accordingly it was decided to examine the current in the Gaspé region, and to endeavour to trace it across the Gulf to Cape North.

The entrance to the St. Lawrence river between the Gaspé coast and Anticosti lies on the line of a deep channel connecting it with the ocean. Between the Grand Banks and the Banquereau and Misaine banks on the west it is 40 miles broad, with a depth of 250 fathoms, and retains these dimensions across the Gulf, and even at the mouth of the Saguenay, 420 miles from Cabot Strait, it has shallowed only to 150 fathoms. The temperature in summer ranges at the surface from 50° to 65°, falls to 30° to 34° at a depth of 40 to 50 fathoms, and then rises again. In the Gaspé region, also, the coldest water forms a layer between the depths of 30 and 40 fathoms, as well as at the Strait of Belle Isle, and as the same condition has been found at the few points within this triangle where observations have been taken, it seems probable that this cold layer extends over all the Gulf area. The deeper water of the channel referred to above has a temperature at 100 to 200 fathoms of 38° to 41° in the Cabot Strait, and the same temperatures were observed last season between the Gaspé coast and Anticosti. This deep water seems to be entirely quiescent, and therefore to have little connection with the currents in so far as they affect navigation.

The surface water in the eastern part of the Cabot Strait, along the west coast of Newfoundland, and in the Strait of Belle Isle, has a density ranging from 1·0233 to 1·0245 (reduced to 60° F.), or much the same as in the open Atlantic, while water from depths of 100 to 150 fathoms, both near Gaspé and in the Cabot Strait, has a density of 1·0254 to 1·0261. It is owing to this higher density that the colder water at 50 fathoms is able to float upon the warmer water. In order to trace the current across the Gulf, the density of the water was taken along a series of sections at various depths down to 50 fathoms, the results of which had not been worked up when the report was written.

Mr. Bell Dawson also gives some notes on the variations, velocity, etc., of the Gaspé current, which will be useful in the navigation of the channel.

Seriland.—A party sent out by the Bureau of American Ethnology visited in November last the Sierra Seri and its neighbourhood, the home of the fierce Seri Indians. The country lies on the Gulf of California, opposite the island of Tiburon. The district including Seriland slopes down like a roof from the Sierra Madre to the Gulf, but the slope is broken by outlying ranges and buttes, and in Seriland these ranges are of exceptional magnitude, and extend beyond the coast-line, forming the largest island in the Gulf. The general outline of the coast would not be much altered, but only moved a little inland or seawards, if the plain of Sonora were to sink or rise a few hundred feet. A rise of 100 feet would connect Tiburon with the mainland, while a fall of 200 feet would convert the whole province into two great islands; and even if Sonora were to sink 3000 feet or more, Seriland would still rise above the water.

Most of the vapour from the Pacific passes over the country till it comes in contact with the Sierra Madre, and therefore the region is extremely arid. Between the Colorado and the Yaqui, which rises in the Sierra Madre, no stream reaches the sea. The precipitation is greater on the outlying ranges than over the plains, but it is so slight that the country is a semi-desert, and little moisture in any form reaches the Desierto Encinas, the natural boundary of Seriland.

The topography of Seriland is marked by the great ruggedness of the ranges, which rise from great foot-slopes or plains. The abrupt change from jagged cliffs to smooth slopes conveys the impression that the foot of the mountains is buried in vast torrential deposits, but the geologist finds that the expanses between the mountains are simply planed rock strata, with a thin veneer of alluvium. Hardly less striking are the strong local features of the topography. Tiburon is but thirty miles long, yet it contains several ranges, and the Sierra Kunkaak is remarkably rugged, almost to its summit, 4000 feet above the sea. Sierra Seri is an imposing assemblage of peaks, precipices, and gorges, rising to fully 5000 feet, though its breadth from the desert to the Infiernillo strait is but ten miles. Very curious, too, is the Desierto Encinas; freshets seem to flow from all directions into its broad basin, and yet it is never filled, and rarely even wetted. It is partly shut off from the Gulf, and lined in its lower levels with recent marine shells, which show that it must comparatively recently have been an arm of the sea. The Infiernillo strait, also, is evidently an old valley that has not long been invaded by the sea.

The flora of the country is meagre. Cacti are the most conspicuous forms. The trees are little more than shrubs. *Acacia greggii, Parkinsonia microphylla,* and the creosote bush (*Larrea tridentata*), are the most common. The fauna includes the bighorn and the bura (a large, sluggish deer) in the mountains, the antelope, peccary, and black-tailed deer on the plains, and the jackrabbit and coyote everywhere. Ground-squirrels and kangaroo-rats are common. Turkeys are said to live in the woods, and the Californian quail is constantly seen. Rattle-snakes, scorpions, tarantulas are to be found, and multitudes of ants, but mosquitoes, gnats, and other small pests are unknown. The green turtle forms the chief fare of the Indians, and fish and crustaceans swarm in the Gulf, the rocks and islets of which are frequented by wild-fowl. The fierce Seri Indians have always preserved their independence. Intermarriage with other peoples is a capital crime among them, and to slay aliens is considered meritorious. They have an excellent physique, and great endurance.—Article by W. J. M'Gee and W. D. Johnson in the *National Geographic Magazine,* April 1896.

OCEANS.

Specific Gravities and Oceanic Circulation.—In the *Challenger* reports, *Summary of Results,* 2nd Part, a chart of the surface densities was published, the densities being reduced to a uniform temperature of 60° F., and the standard density being that of distilled water at 39·2° F. (4°C.). Thus the salinity of the water was approximately shown, which is one of the most important factors in the physics of the ocean. But the movement of the waters depends on the specific gravity at the temperature observed, and this Dr. Buchan has dealt with in the *Transactions of the Royal Society of Edinburgh,* vol. xxxviii. part 2.

A table was drawn up showing the mean annual specific gravity at the mean annual temperature at the latitude of observation. It was then found that the mean of the specific gravities, at the 640 points where observations had been recorded, was 1·0252, and lines were drawn on a chart showing where the specific gravities differed from this average by one, two, or more *promilles* in excess or defect. As regards the specific gravity at various depths below the surface, the observations are not sufficiently numerous for the compilation of a shaded chart, as in the case of surface temperatures. For depths of 100, 200, 300, 400, 800, and 1500 fathoms, therefore, the differences from the mean at the depth are marked on Dr. Buchan's charts in red or blue, according as they are above or below the mean, at each point of observation. For other depths than those mentioned there are not sufficient data to indicate the specific gravity of the oceans, even over limited areas. There are still enormous blanks in the charts of specific gravity at depths of 100 fathoms and over.

A table of the mean specific gravity of the ocean at the depths mentioned and reduced to 60° F. shows that the specific gravity (and the approximate salinity) is greatest on the surface and steadily diminishes down to 800 fathoms at least, while at 1500, 2000, and at greater depths it increases again. On the other hand, the actual specific gravities at the observed temperatures, which alone determine movement, steadily increase from 1·0252 at the surface to 1·0280 at 2000 fathoms and below. This increase is caused, as far down as 800 fathoms at least, by decrease of temperature, as shown by the maps in the *Challenger* reports. But at 1500 fathoms and at lower depths there is also an actual increase of salinity, which is an important factor in the distribution of low temperatures over the bed of the ocean from the Polar and adjacent regions.

The distribution of salinity over the surface of the ocean is determined by the prevailing winds and their relative dryness, the upwelling from lower depths where the prevailing winds blow off the land, and the amount of rainfall. The prominent features of the circulation of the atmosphere may be understood from an examination of the prevailing winds of January and July. In the former month the circulation over the Atlantic north of lat. 35° is determined by the low pressure near Iceland and the high pressure areas over Eurasia and North America. Hence the prevailing winds at this season are north-westerly over the eastern parts of America and south-westerly over Western Europe, and drive the surface water from the American coast and the West Indies, across the Atlantic, round the north of Norway, and along the coast of Siberia. Hence a strong current brings warm water to the western shores of Europe and drains it off from the shores of America, causing water of lower temperature to well up from greater depths. In the North Pacific the wind system is controlled by the low pressure in the north of that ocean, causing north-westerly winds in the east of Asia, and south-westerly and westerly winds in the west of North America. The great dryness and low temperature of the north-westerly winds have a great effect in lowering the temperature.

In the southern hemisphere, south of lat. 35°, a broad ring of low pressure about 30 degrees of latitude in width encircles the globe, in which strong westerly or north-westerly winds blow nearly all the year round. The surface currents produced by these winds bring enormous volumes of warm and heavy water from lower latitudes, which, on reaching those parts of the southern ocean where heavy rainfall, icebergs, and melted snow lower the temperature and salinity of the surface, sink below the lighter waters of the Antarctic. These currents of warm water mitigate the cold of the southern Polar regions and confine the ice-clad area to its present limits. At this time of year restricted areas of low pressure occur over the land of the Southern Hemisphere, and hence the winds and surface currents are directed towards the land.

In July the distribution of pressure in Australia is exactly the reverse, the highest pressure occurring in the basin of the Murray, and the lowest near the north coast, and this diminution of pressure is continued northwards to the low pressure system of Asia. The high pressure is continued westwards in the same latitudes through the Indian Ocean, and southerly winds sweep northwards into

Asia, precipitating large quantities of rain on the coasts of India and Burma, which retard the downward circulation of the ocean in those parts. On the other hand, the rainfall in the north and west of the Arabian sea is very small, and consequently the salinity is increased and the vertical circulation greatly accelerated. Similarly, in the Pacific, the prevailing summer winds pour a heavy rainfall on the coasts of Asia and the outlying islands. The summer winds of Europe are determined by the high pressure of the Atlantic in relation to the low pressure areas of Asia and Africa. In Spain and north-west Africa they are northerly, in France and the British Isles south-westerly, and westerly and north-westerly on the coast of Norway. In North America the centre of lowest pressure over the States is about Utah, from which pressure rises all round, but chiefly to the south-east and west. Consequently the winds blow inland from the Gulf of Mexico, and on the Pacific coast are north and north-west as far as Vancouver.

The general result, as regards salinity, is that the salinity is high in tropical and sub-tropical regions, where temperature and evaporation are high, and rainfall small. In the anticyclonic areas it is large, because the descending currents of dry air promote evaporation. Thus the salinity is high on the east coast of South America and on the west coast of Australia, whither the winds blow directly from anticyclonic areas during a large part of the year. The most remarkable region of low salinity is that which extends from India eastwards to long. 143°, and stretches across the Equator to lat. 9° S. This brackish water is driven northwards and then eastwards across the North Pacific by the prevailing winds, and accounts for the low salinity of that ocean.

One of the most important factors of oceanic circulation is the position of the line of lowest mean atmospheric pressure, for towards it flow the prevailing winds and the currents produced by them. In the Atlantic this critical line lies at all seasons north of the Equator, and consequently the surface currents generated by the south-east trades render the temperature and salinity in the North Atlantic higher than in any other ocean. A great contrast is exhibited by the western part of the Pacific, where the line of lowest pressure lies for eight months in the year to the south of the Equator, and northerly winds and currents advance southwards to 15° S. lat.

But when specific gravities at the observed temperatures are dealt with, it is found that the conditions are almost reversed. The lowest specific gravity is intertropical and the highest extra-tropical. The absolutely highest, 1·0277, occurs in the North Atlantic and is found along a line running from the Färö islands north-eastwards and then eastwards midway between Norway and Spitzbergen. It is not due to abnormally low temperature, for the temperature here is higher than in other seas at the same latitude, but is mainly the result of high salinity caused by the longest current in the world, starting from the northwest coast of Africa, and passing through the West Indian islands and then from Florida to Spitzbergen. This affords the strongest proof that the great surface currents of the oceans are produced by the prevailing winds.

The absolutely lowest specific gravity, 1·0222, is found in the Bay of Bengal and the western part of the Chinese sea. This is the most extensive region of unusually heavy rainfall ; indeed, at all seasons the rainfall is heavy. The abnormally low specific gravity is, therefore, occasioned by the low salinity of the water as well as by the high temperature. Further, as the prevailing winds blow for a large part of the year from the northern part of this region over the North Pacific, the low salinity of this ocean at all depths is probably to be traced to this rainy area. The Arabian sea presents a strong contrast, being one of the driest regions of the world, and, as the winds blow off-shore, causing an upwelling of

cooler water from below, the surface temperature is low. The Red Sea, in particular, presents some interesting features, which Dr. Buchan discusses at length. We have not space to give an outline of this part of his paper, nor to deal fully with specific gravity and salinity at various depths. As at the surface, so at all depths down to the bottom, the North Atlantic, between lat. 20° and 40°, is characterised by a salinity higher than that of any other ocean, and it seems to be a rule that the high or low salinity of any area is continuous through all depths, indicating that a vital connection is maintained between the surface and bottom, for otherwise through the diffusion of salt the salinity of the lower strata would be the same everywhere.

Dr. Buchan concludes that the prevailing winds are the most powerful agents in oceanic circulation, driving the water before them, so that on the eastern shores of continents are accumulated large masses of water of high temperature and, owing to evaporation, of high salinity, causing colder water to well up on the western coasts of the continents, where they start on their course. The ice-cold water occupying the bottom of the ocean in all latitudes comes from the Polar regions, and this slow bottom current is caused by the reduction of the surface water in inter-tropical regions through evaporation, by the prevailing winds in extra-tropical regions blowing Polewards, and by the greater specific gravity of the water in high latitudes. To explain fully the vertical movements of oceanic waters, additional observations on a large scale are necessary.

MISCELLANEOUS.

Captain Lugard, accompanied by his brother, Lieutenant Lugard, is about to lead an exploring expedition to **Lake Ngami.**

Père C. de Deken, who accompanied Prince Henri of Orleans across Tibet, died at Boma, on the Congo, on March 3rd. He was only forty-four years of age.

Professor W. H. Dall, who has visited Alaska to examine the coal deposits, has brought back fragments of the flesh of a mammoth. This is the first time that remains of this animal, other than bones, tusks, and teeth, have been found in America.

Signor G. Boggiani has published in the *Memorie* of the Italian Geographical Society an account (published also in pamphlet form) of the **Caduvei** of Matto Grosso and Paraguay. It contains the chief ethnological details of the book reviewed in vol. xi. p. 205.

In January last Dr. Hans Steffen started on another expedition to the cordilleras of **Chile.** He intended to ascend the Rio Manso, a tributary of the Rio Puelo, and if possible to reach the Tronador.—*Verh. der Gesell. für Erdkunde zu Berlin*, Bd. xxiii. No. 2.

Sir W. Martin Conway intends to visit **Spitzbergen** this summer. He will be accompanied by Mr. Trevor-Battye, and will try to cross overland from the Icefiord to the Foreland Sound. Should there be time to spare, it will be devoted to an exploration of the southern part of West Spitzbergen.

NEW BOOKS.

The Mameluke or Slave Dynasty of Egypt. By Sir WILLIAM MUIR, K.C.S.I., LL.D., D.C.L., Ph.D. (Bologna). London : Smith, Elder, and Co., 1896.

This work is a very welcome addition to the existing stock of English historical literature. The period of Egyptian history which it traverses, though marked by very striking and quite unique characteristics, has not hitherto received from our countrymen the separate treatment to which its importance well entitles it. Germany, as usual, has been in the field before us, and has thus pioneered the way along which the learned Principal of our University here conducts us. He states in his preface that for his material he has been mainly indebted to the last two volumes of the great work, *Geschichte der Chalifen,* written by the late Dr. Weil, whose authorities were mostly accessible only in rare Arabic MSS., for which he had to ransack several of the great European libraries. · As the Principal is himself a distinguished Arabic scholar, and has written several works of standard authority on the history and tenets of Islam, we can have no misgivings as to the general accuracy of the history of the period he now calls us to survey, especially when we learn that this history "is supported by a singular concurrence of contemporary writers." These authorities are chiefly Abulfeda, Noweiry, Ibn Batuta, Macrizy, Abul Mahâsin, and Ibn Ayâs. The contents of the MSS. examined by Dr. Weil have brought to light events and narratives which were for the most part previously unknown, and they thus supplement, and occasionally correct, the information derived from the published sources.

The historian of the Decline and Fall, with a few strokes of his graphic pen, outlines the general character of the Mameluke sway to which Egypt was for several centuries subjected : "A more unjust and absurd constitution," he says, " cannot be devised than that which condemns the natives of a country to perpetual servitude under the arbitrary dominion of strangers and slaves. Yet such has been the state of Egypt above five hundred years. The most illustrious sultans of the Baharite and Borgite dynasties were themselves promoted from the Tartar and Circassian bands. . . . With some breathing intervals of peace and order, the two dynasties are marked as a period of rapine and bloodshed ; but their throne, however shaken, reposed on the two pillars of discipline and valour : their sway extended over Egypt, Nubia, Arabia, and Syria ; their Mamelukes were multiplied from 800 to 25,000 horse ; and their numbers were increased by a provincial militia of 107,000 foot, and the occasional aid of 66,000 Arabs." As the Mameluke rule was established about the period when Egypt was assailed for the last time by the arms of the Crusaders, our author introduces his subject by giving a succinct review of the leading incidents of the various crusading expeditions. Then, proceeding to his proper subject, he offers some remarks descriptive of the strange Mameluke race of whom he has to write. Their name, he says, signifies a slave, and comes from the same root as *Mâlik,* an owner or king. They were first employed by the caliphs of Baghdad in the declining days of that caliphate, both as body-guards, and also as a force for countervailing the overweening insolence of their Arab soldiery. These slaves, becoming conscious of their power, fomented riots and rebellion, and having finally made themselves masters of the Court, assumed, like the Roman Pretorians, the right of deciding the succession to the throne. The example of the Baghdad caliphs was followed in Egypt, both by the Fatimide caliphs, who became its masters in 970, and fixed their seat at Cairo, and also by the Eyyubite dynasty, founded by the famous Saladin in 1170.

The Mamelukes were not only prisoners of war, but were in many cases children purchased by Syrian slave-dealers from Turcoman, Mongol, and other eastern hordes, and sold by these, sometimes at enormous prices, to sultans and emirs. They occupied in Cairo a position elsewhere unparalleled in the annals of slavery. They lorded it with a high hand over the weak and subservient native population. The great mass of them pursued, no doubt, a low life; but the favourites of the emirs, and especially of the sultans, were well educated and trained to arms, and as pages gradually rose to the position of their masters. It, indeed, not seldom happened that he who was to-day a slave found himself on the morrow the commander of an army, or even the occupant of the throne itself. One of the many striking characteristics of the Mamelukes was that they did not intermarry with the races, whether Coptic or Semitic, in the midst of which they lived, nor did they form with them any ties, either domestic or social. Another was, that the slave succeeded his slave-master, and became the protector, or perhaps the murderer, of the lawful family of his master. Their principal occupation was fighting—either with themselves or with the people of the country to hold them down, or with foreign enemies. They were the supreme rulers of Egypt for the space of 267 years—from the year 1250, in which they slew Turan, the last of the Eyyubite sultans, till the year 1517, when Selim I., the Sultan of Turkey, who had defeated Toumanbeg in the battle of the Pyramids and entered Cairo without opposition, incorporated Egypt in the Ottoman Empire. Sir William Muir, however, curtails by ten years the duration of the Mameluke supremacy, dating its commencement from the year 1260, when Beibars, the commander of the Egyptian army, was unanimously elected sultan. His dynasty, called the Bahrite,[1] ruled Egypt till 1382, when it was replaced by the Burjite dynasty, the members of which belonged chiefly to the Circassian race.

Some of the Mameluke sovereigns were men of distinguished ability, shining in the arts alike of war and of government. They long strenuously maintained their authority over Syria and their subject states in Asia Minor. Towards the close of the fourteenth century Barkûk, who belonged to the Circassian dynasty, had the audacity to slay a messenger sent to him by Tamerlane, and he even scornfully defied that terrible conqueror as an "angel of the Evil One destined for hell-fire." In general, however, the Mameluke sultans were men of profligate life, perfidious, and remorselessly cruel. Their history, as presented in the volume before us, is a dreary and melancholy record of wars, assassinations, usurpations, rivalries, and brutalities. The author would have enlivened his narrative had he told us more of the life, manners, and condition of the people, and of their progress in art and literature, which was considerable during the Mameluke period. The book is illustrated with a dozen beautiful pictures, chiefly of remarkable buildings, and with a map which we have not found quite satisfactory.

Hausaland; or, Fifteen Hundred Miles through the Central Soudan. By CHAS. HENRY ROBINSON, M.A. London: Sampson Low and Co., 1896. Pp. 304.

Hausaland is that great African territory extending from lat. 8° N. to 14° N., and from long. 4° E. to 11° E. It is approached either by sailing 300 miles up the Niger, and then proceeding overland, or by crossing the Great Sahara from the Mediterranean coast. The latter route is, however, at present closed to Europeans, owing to troubles in the vicinity of Lake Tchad, caused by the French occupation of Timbuctoo. In Mr. Robinson's opinion, "the Hausas are superior, both intel-

---

[1] The Eyyubite Sultans settled their Mamelukes, who were chiefly Turks or Mongolians, on an island in the river Nile. *Bahr* is the word for *a river*, and hence their name. The Burjites lived in the *Burj*, *i.e.* the citadel or other fortified places.

lectually and physically, to all other natives of Equatorial Africa." Their language serves as a sort of *lingua franca* over practically all Africa north of the Equator and west of the valley of the Nile. Within Hausaland itself it is spoken by fifteen million people. Owing to the importance of the language and people of Hausaland, "the Hausa Association" was established in 1891 to provide "for a thoroughly scientific study of the Hausa language, with a view of promoting the higher interests of that people, and of translating the Scriptures and other appropriate literature into their tongue." An earnest student of that language, Mr. John A. Robinson, died on the river Niger, and a few months later this Association was founded in his memory. The author took up the work of his deceased brother, under the auspices of the new Association, and in this interesting work details his travels through Hausaland and the Central Soudan in 1894-95.

"Slave-raiding and the traffic in slaves," says Mr. Robinson, "is the great overshadowing evil of the Central Soudan." In Hausaland alone five million persons, or one-third of the population, are in a state of slavery. Yet Hausaland, where slave-raiding and trading are flourishing, is "within the British sphere of influence, having been definitely recognised as such by the treaty of Berlin." The practical result of the slave-trade here is that "the country is subject to nearly all the horrors of perpetual civil war. There is no real security for life or property anywhere. . . . Any who attempt to resist are massacred on the spot." Yet all this is done within the British "sphere of influence"!

We commend this book to all interested in the Central Soudan, and particularly to those who can help to remove the terrible scourge of slavery under which the active and intelligent population of Hausaland has so long groaned.

Terre d'Espagne. Par René Bazin. Paris : Calmann Lévy, 1896. Pp. 336. *Price 3 fr. 50 c.*

Nobody could write better than a Frenchman such a light and readable book of traveller's notes as this. A German would have encumbered it with science and an Englishman with statistics. M. Bazin describes Spanish scenery, leading sights, and men and manners in 1894, and generally follows the well-known route through Spain. As a specimen of his style, take the following account of Gibraltar :—

"A truly English welcome (rain). But really, am I not in a port of the great island ? The first man I see is a policeman, phlegmatic and polite. The first roof is one of corrugated iron made in Sheffield. I enter the town—after duly written authority—and I meet the same soldiers as one sees in Malta, Jersey, London, or India, in red coats, small caps, cane in hand, ruddy and well-fed. The windows are *à guillotine.* The pictures in the corridors represent steeplechases or foxhunting. Little bottles of sauce stand on the sideboard of the dining-room. Ladies are chatting in the ' Ladies' Room.' Midshipmen are reading the *Times*, or quaffing port-wine, in a room reserved for gentlemen. Outside—for the rain has ceased—and the streets, rocks, and whole island smoke like a corner of Florida at sunset— English soldiers and sailors walk stiff and grave, and seem as numerous as the civil population, which is varied and mixed, half Spanish, half Jewish. There is not a street without its barracks, its artillery magazine, and its guard-house and sentinels at their posts. Where are the tennis courts ? There are few in the town, but I discover one. Where is the English parson ? Ah, there he comes, on horseback, from his parochial duties. The rosy English children are no doubt at home, but their mothers and sisters are beginning to walk towards the Alameda to enjoy the cool of the evening. They have the same round figures, the same short dresses, the same energetic and athletic appearance which they have in all climates. England is there in all her entirety, with her habits, her fashions, her masterful air, her constant activity. Degrees of latitude may change, but she changeth not. The very sun fails to injure the charming complexions of these young Misses who watch the crowd of passers-by from the windows of their cottages surrounded by climbing jessamine.

"This corner of Spain resembles Spain so little, it has been so powerfully influenced by its masters, that the first sentiment we feel is one of veritable admiration for that power which possesses such creative force. Memories and regrets may intrude themselves ; we may hope, knowing what this mutilation costs the Spaniards, that Gibraltar may one day become Spanish again ; but the impression received from the first is that it is pleasant to live on English soil."

M. Bazin has none of the petty jealousy of Britain entertained by Parisian *boulevardiers.* He warmly praises the British officers for their politeness in showing him over Gibraltar ; and he records his sense of the extremely obliging behaviour (not experienced by him for the first time) of English gentlemen towards a stranger who had asked their good offices without any formal introduction.

Naturalists will be interested to know, on his authority, that, after being decimated by smallpox, the monkeys on the rock numbered fifty in October 1894.

The Gold Diggings of Cape Horn: A Study of Life in Tierra del Fuego and Patagonia. By JOHN R. SPEARS. New York and London : G. P. Putnam's Sons, 1895. Pp. x + 319.

If the reader expects to hear from Mr. Spears good news about rich and unexplored gold deposits at Cape Horn, he will be sorely disappointed. The working of what gold there is is evidently beset with so many difficulties that the game seems hardly worth the candle. We are not surprised, therefore, that after the first chapter, the subject which furnishes the principal title of the book gives place to the Study of Life in Tierra del Fuego and Patagonia, the literary outcome of a two months' cruise along those inhospitable shores. The reader whose acquaintance with those parts is derived from accounts of English or European travellers will find much that sounds strange to him in these sketches, written by an American newspaper correspondent. The Yaghans of Tierra del Fuego, *e.g.*, whom we are in the habit of regarding as specimens of the lowest type of humanity, are asserted by Mr. Spears to have been skilled in various arts, and are credited with the possession of a vocabulary of forty thousand words ! Their present deterioration and almost complete extinction he ascribes to the civilising efforts of English missionaries. So also the Patagonian Tehuelches seem to have been very unfairly dealt with by former observers. Welshmen may justly feel proud when reading of the Welsh settlement of Chubut, which, after many years of hard struggle, is now in a very prosperous condition. Lovers of animal nature and sportsmen will find much to interest them in the accounts of the guanaco, the puma, called out there "the friend of man," and other "beasts odd and wild." The sheep-farmer's life appears to have a singular attraction for many even highly educated people, and nobody could deny that the cow-boy, the gaucho, becomes a most interesting individual under Mr. Spears' treatment. One would hardly expect to come across tramps in a country like Patagonia, and yet we are told they are plentiful even there. We gladly indorse the author's advice to the Yankee : "if he can afford to go away and see some other part of the world, let him travel out of the way to Patagonia and Punta Arenas instead of Paris " ; but we fear not many will follow it if, like him, they have to make the voyage on board an Argentine naval transport. Readers need have no fear of finding Mr. Spears' book dull.

Parts of the Pacific. By A PERIPATETIC PARSON. London : Swan Sonnenschein and Co., 1896. Pp. 388. *Price* 10s. 6d.

The writer of this amusing yet instructive book wandered through the less frequented regions of the Pacific—North Queensland, New Caledonia, Fiji, New Zealand, Hawaii, and the coast of North America—and gives in its pages a colloquial description of strange lands, touching upon such matters as Emigration, Missions, the Kanaka Labour Question, and the General Treatment of Natives. The book abounds in curious and interesting passages, such as the following, upon the " making of young men " among the aborigines of Australia :—

" When a good number of youths are ready for initiation, a large gathering is held, at which the festivities and solemnities last for a week or so. The youths are instructed in the existence of the Father and in the code of morality—the poor women being always excluded from any theological instruction, for there are no New Women among the Blacks. Amongst other things, the youths are placed close to a hot fire to test their endurance—sometimes, indeed, hung by the heels over it—and one of the front teeth is next knocked out with a stick and mallet, amid much excited dancing, after which they are introduced to a mud figure o Daramulun, the Great Spirit—the Master, the Father. The figure is afterwards destroyed.

" The youths are then turned out for several days to forage for themselves, being forbidden certain foods. Upon their return to the camp they are considered full-blown men, and are at liberty to take wives unto themselves."

Folk og Natur i Finmarken. Af HANS REUSCH. Kristiania : i Kommission hos T. O. Brögger, 1895. Pp. 176. *Price* 3 Kr.

The northern part of the Scandinavian peninsula is especially interesting as inhabited by people still in an almost barbarous state, though living so close to civilised Europe. Dr. Reusch describes the country between the Alten and Varanger fiords, the greater part of his remarks referring to the people. The Laps are a feeble race, both in their muscular and nervous systems. Their senses are not less acute than those of other men, but they lack energy, courage, and endurance. This is perhaps both the cause and the result of the semi-starvation they suffer during a great part of the year. They cling to their old customs, and refuse to adopt improvements in agriculture, fish-curing, etc., which would improve their condition. The schoolmaster has a hard task to teach children speaking three different languages, and his exertions have not as a whole been successful. The Kvæns of Vadsö send their children willingly to school, and are anxious that they should learn Norwegian, but in most places the priest, the schoolmaster, and th constable are looked upon as useless parasites.

Dr. Reusch describes the huts and occupations of the people, the trade with the Russians, and also gives some notes on the geology and natural features. A fact not generally known, is that the North Cape is not the most northern point of Europe, but Knivskjærodden, another promontory of Magerö, low and inconspicuous.

National Geographic Monographs, prepared under the auspices of the National Geographic Society. New York, Cincinnati, Chicago : American Book Company, 1895. Vol. i., large 8vo, pp. 336. *Price* $1·50.

This volume consists of a series of ten monographs by various experts, who are well-known expositors of physiographic science. Major Powell leads off with three monographs, entitled respectively, *Physiographic Processes, Physiographic Features,*

and *Physiographic Regions of the United States.* Iu the first of these the action of
the geological agents of change is clearly and succinctly set forth ; the second gives
an admirable *résumé* of the leading characters of plains, plateaus, mountains, valleys,
hills, cliffs, and other physiographic features ; while in the third we are presented
with an instructive outline sketch of the mountains, plateaus, and plains of the
United States.   Prof. I. C. Russell discourses on the *Present and Extinct Lakes of
Nevada,* of which he has a tale to tell which will appeal to geographers and
geologists alike.   Professor N. S. Shaler describes the *Beaches and Tidal Marshes
of the Atlantic Coast* in that clear and graphic manner which marks all that comes
from his active pen.   *The Northern Appalachians* are treated of by Mr. Bailey
Willis, who first delineates the topographic features of that mountain-tract, and
thereafter shows how the region has acquired its configuration under the influence
of denudation.   In subsequent sections he considers the genesis of the Appalachian
type of mountains, and describes the influence of the Appalachians on settlement.
The Southern Appalachians are dealt with in another monograph by Mr. C. W.
Hayes, who treats his subject in much the same way.   Mr. G. K. Gilbert discusses
that perennial question—*Niagara Falls and their History.*   His monograph is not
only well written, but gives the reader a much clearer notion of the whole subject
than he will obtain from any other source.   As regards the age of the Niagara
gorge, there have been, as every one knows, many estimates.   Dr. Pohlman's
estimate was 3500 years, while Dr. Spencer puts it at 32,000 years.   Mr. W.
Upham, with the same data for computation, thinks 7000 years a more reasonable
estimate.   Mr. F. B. Taylor, on the other hand, while considering the data altogether insufficient for the solution of the problem, is of opinion that Mr. Upham's
estimate should be multiplied by a number consisting of tens rather than units.
One is not surprised that so cautious an observer and clear thinker as Mr. Gilbert
should come to the conclusion that no estimate yet made has great value, and that
the best result obtainable may perhaps be only a rough approximation.   Not the
least important monograph in this volume is one on *Mount Shasta, a typical
Volcano,* by Mr. J. S. Diller.   Like all the other articles, it comes from one who is
thoroughly familiar with the subject he discusses.   Professor W. M. Davis describes
the *Physical Geography of Southern New England,* and shows how a clear perception of surface-features can only be gained by tracing out their development.
This, indeed, is the lesson taught by all the monographs of this instructive volume.
Professor Davis insists that only by a knowledge of the evolution of land-forms can
we obtain that appreciation of the facts which are the foundation of all thorough
geographical study.   No mere study of the facts themselves without regard to
their meaning or development will suffice to place them clearly enough before the
mind.   That these monographs will give an impetus in America to the methods of
geographical study advocated by their authors, can hardly be doubted.   They can
be commended also to the attention of students and teachers of geography in this
country.   The examples given, it is true, are all American, but they are so well
related and admirably illustrated that the lessons they convey can hardly fail to be
appreciated.   It need only be added that the work is beautifully printed, and that
each monograph can be had separately for a few pence.

Elementary Physical Geography. By RALPH S. TARR, B.Sc., F.G.S.A.   New York :
    Macmillan and Co., 1895.   Pp. xxxii + 488 ; illus. 267 + plates 29.   *Price
    7s. 6d. net.*

This work is apparently intended as a text-book for secondary schools and
colleges, and confines itself strictly to Physical Geography, and does not present us
with an *olla podrida* of the sciences such as we find in so many books purporting

to be elementary text-books of physical geography. The matter is well arranged, the text concise and easily understood, and errors seem to be few. As deep-sea deposits, for instance, we find Globigerina Ooze and Red Clay mentioned, but neither Blue Mud nor Diatom Ooze (p. 164). Neither does the moon rotate on its axis in 29½ days, the period of its revolution round the earth (p. 14).

One very prominent feature of the book is its illustrations, consisting, besides maps, chiefly of reproductions of photographs. Many of these photographs, which are chiefly of examples from the American Continent, have been rendered almost quite useless in reproduction. Some indeed are simply unintelligible even with the aid of their legend, e.g. figs. 50, 89, 177. Others are quite spoiled by excessive reduction, some being no larger than an ordinary visiting card and giving blurred representations of complicated natural phenomena, as of a tornado (fig. 52), cloud forms (fig. 57) ; also figs. 121, 202, 209, 214, 231. As to the maps, they are in general mere sketches, the only lines of reference being the equator, and the tropical and polar circles. In plate 10 there are no lines of reference whatever, which, in this case, seems to be due to sheer carelessness, for on the opposite page we have the same map for a different season of the year *with* a few lines of reference. Other maps on a very small scale seem to have been photographically reproduced from larger ones, as figs. 151, 185, and are of little use ; and indeed there is hardly a map which does not suffer from some infirmity or another. With these exceptions the book seems well suited to its purpose, and a greater measure of care in the preparation of the maps and illustrations will greatly enhance the value of any future edition.

-work, Present and Past. By T. G. Bonney, D.Sc., LL.D., F.R.S., etc., Professor of Geology, University College, London. London : Kegan Paul, Trench, Trübner and Co., Limited, 1896. 8vo, pp. 295. *Price 5s.*

This is a disappointing book. The writer tells his readers that he first saw a glacier in 1856, and that his earlier geological papers dealt mainly with ice and its work. For nearly twenty years, he says, he has written mostly on petrological subjects, but has never obeyed the well-known dictum and wholly cast off the old love. Nevertheless his pages yield abundant evidence to show that he is but inadequately acquainted with the work which has been done in the realm of glacial geology within the last twenty years. In dealing with the interpretations of glacial phenomena proposed by others, he points out where these appear to him to be strong and where weak. He remarks that he has endeavoured to follow the example of a judge rather than an advocate : that is, to sum up the evidence on each side of a case, and leave the verdict to the jury. " Like any such official," he continues, " I have my own view as to what that verdict should be, and this will doubtless be disclosed to those who can read between the lines." If by " the jury " the author means the average reader, we think that the latter would have preferred a frank statement of the " judge's " own views ; for how can one who has had no training in geology estimate the weight of the evidence, and why should he be expected to grope between the lines in search of the author's opinions ? So far will such readers refrain from thanking Professor Bonney for his reticence, that they will be more likely to conclude that he either does not know his own mind, or lacks the courage of his convictions. If, on the other hand, " the jury " be one of experts, they will feel that " the judge " has presented a feeble and inadequate outline of the evidence—that he often misunderstands the interpretations which he attempts to criticise, and not infrequently advances objections apparently in ignorance of the fact that they have been long ago met and answered.

Ethnology.   In Two Parts.   *I.—Fundamental Ethnical Problems; II.—The*
    Primary Ethnical Groups.   By A. H. KEANE, F.R.G.S.   Cambridge : At the
    University Press, 1896.   Pp. xxx + 442.   *Price* 10*s.* 6*d.*

The geographer who admits that the study of the human inhabitants of the
earth falls within the range of his science, must feel the necessity of making himself acquainted with what can be ascertained as to the origin and evolution of the
present types of mankind—the special domain of the science of Ethnology.   But
hitherto a comprehensive English work on the subject, reflecting the present state
of the science, has been wanting.   The editor of the *Cambridge Geographical Series,*
therefore, deserves our thanks for issuing as the first publication a treatise on
the subject.   For writing such a book, no man is better qualified than Mr.
Keane.   While making use of nearly everything that has been written on the
subject up to the present day, he has not produced a mere compilation, but an entirely original work.   We have no doubt that many of his views will provoke
violent dissent ; yet every fair critic must acknowledge that Mr. Keane does not
indulge in wild speculation, but that he always endeavours to support his reasoning
by the established facts of natural science.   Unfortunately the arrangement of the
matter is not such as to make the book an easy introduction to the science for those
who have no previous knowledge of the subject ; and, well illustrated as it is, a few
more diagrams would greatly help the tyro, especially in the chapter on the Physical
Criteria of Race.   Mr. Keane is a monogenist and a thoroughgoing evolutionist.   All
the present varieties of man he traces back to one pliocene precursor who originated
somewhere on the now submerged Indo-African and Austral continents.   From
thence man migrated by overland routes to every part of the eastern hemisphere,
and thus arose in the new centres of evolution the several pleistocene groups,
whence are derived the present primary divisions of the human family.   Having
discussed the origin of these primary divisions in the first part of his book, he
devotes the second to a more detailed study of each.   He illustrates the relationship of the many minor divisions by means of family trees, and perhaps the reader
will be surprised to discover Polynesians, Teutons, and Semites on the same
Caucasian tree.   With regard to Homo Caucasicus, we expect many ethnologists will
be rather startled on learning that he was evolved not in Asia, nor in Europe, but in
North Africa.   There is no room for an Aryan race in Mr. Keane's scheme : Aryan
is not a racial, but only a linguistic, term, and the Aryans were but a community of
Caucasic type and of primitive Aryan speech, evolved in some Eurasian region.

On the whole, Mr. Keane seems to have made sure of his facts.   But there is
one statement in the book which we cannot allow to pass unchallenged.   It professes
to record a most remarkable instance of the influence of environment.   Some
Suabian Germans, who " had originally fair or red hair, light or blue eyes, and broad,
coarse features," settled near Tiflis in 1816, and within the course of two generations
became in appearance like the dark native Georgians, "and these changes were due
entirely to the surroundings, no instance of crossing with Georgian natives being
on record."   Mr. Keane even notes the successive changes : " In the first generation
brown hair and black eyes began to appear, in the second black eyes and hair
became the rule, while the face acquired a noble, oval form."   Mr. Keane's only
authority appears to be Reclus, whose words he quotes in the footnote (p. 203).
But Reclus does not say that the Suabian immigrants were all fair-haired and
blue-eyed ; in fact (vol. iii. p. 629 of his *Géogr. Univ.*), he expressly declares,
when speaking of the inhabitants of Suabia, that the majority have black hair.
We may, therefore, reasonably suppose that among those Suabian immigrants
there were a good many black-haired ones, and we should not be surprised to find

that now the majority of them are dark. But this might result from marriage between fair and dark Suabians in any environment.

Bibliographie du Congo, 1880-1895. *Catalogue méthodique de* 3,800 *Ouvrages, Brochures, Notices, et Cartes.* Par A. J. WAUTERS. Bruxelles: Administration du *Mouvement Géographique*, 1895. Pp. 356.

This volume is practically a *résumé* of the history of the Congo State from 1800 to the present day, with the outstanding events which have happened shortly noted. A list of books from then until now is also given, and the author has done good service to geography and the allied sciences thereby.

The bibliography is not quite complete, but in such an undertaking it is most difficult to get together all the works and papers published on the subject, the author, however, has not, so far as we are able to judge—and we have looked the work carefully through—missed very much which should have been incorporated in its pages.

Postal Directory of the Madras Circle, showing the Name of each Village in the Madras Presidency and of the Post Town through which it is served. First edition. 3 vols. folio, pp. 1506. Madras, 1893 [not for sale].

Apart from its official use, this work is of geographical value, for it is the first complete list of all the towns and villages in the Madras Presidency, Mysore, and Coorg, numbering close upon 93,000. Each village is located by Táluk and District, and coupled with the post town through which it is served, while post towns themselves are printed in capitals. It is thus an index or gazetteer on an exhaustive scale. The spelling—the most perplexing feature in all Indian place-lists—is generally most satisfactory: occasionally the *oo* takes the place of *u*; and sometimes a "district" or "post town" is wrongly given in the 3rd and 4th columns, as "Pudmanabhapuram" and "Martandam" on p. 965, the latter not to be found in the list and the former elsewhere spelled Padma-; Kanigiri (p. 533) is put in Vellore district instead of in Nellur, etc. These may be corrected in a second edition.

Pacific Line Guide to South America, containing information for Travellers and Shippers to Ports on the East and West Coasts of South America. London: Simpkin, Marshall, and Co.; Liverpool: Rockliff Brothers, 1895. Pp. 157. Price 2s. 6d. net.

Of course this book contains full information regarding the Pacific Line steamers, but it also gives very useful, though short, notes on the ports and the excursions that can be made from them. The list of books recommended is rather deficient: De Bourgade de la Dardye's *Paraguay*, and Mulhall's *Handbook of the River Plate*, should have been included. The work contains maps and illustrations, and rather too many advertisements.

Cocoa: All about it. By HISTORICUS. London: Sampson Low, Marston, and Co., 1896. Pp. 99.

In this small book are given extracts from old writers, illustrating the history and use of cocoa from the time of the discovery of America, and its varieties and cultivation are described, as well as the modern processes by which it is prepared for use. A chapter on vanilla concludes the work. The volume contains a number of illustrations, some of them taken from old books, connected directly or indirectly with the history of the plant.

NEW MAPS.

AFRICA.

AFRIQUE, Carte d' ——. Échelle 1 : 10,000,000. 1895.
Société de Geographie de Paris.

This map, we are informed, is intended to show the political divisions of Africa according to French ideas. With current politics we, as a geographical society, have nothing to do, but we may point out that one or two accomplished facts have been passed over. The boundary agreed to between Great Britain and Germany in the Lake Tsad and Benue country is not marked, and Swaziland is not included in the Transvaal. Nor is the southern boundary of the Somaliland Protectorate shown, as defined by the Anglo-Italian Agreement of 1894.

NUBIA AND ABYSSINIA, Map of ——, to illustrate Dongola Expedition, 1896. Scale 1 : 2,854,868 = 45·057 miles to an inch.
W. and A. K. Johnston, Edinburgh.

AMERICA.

BRITISH COLUMBIA, Map of the Province of ——, 1895. Scale 20 miles to an inch. 2 sheets.

KOOTENAY DISTRICT, Map of the West Division of the ——, and a portion of Lillooet, Yale, and East Kootenay, 1896. Scale 8 miles to an inch. Compiled by direction of the Hon. G. B. Martin, Chief Commissioner of Lands and Works.   *Presented by the Government of British Columbia.*

These maps contain a large amount of detail not hitherto published in general maps. The map of British Columbia, compiled by Gotfred Jörgensen, C.E., shows the mountains, roads, railways, mining settlements, etc. The map of the Kootenay District embraces the country from the Kootenay river on the east to the Shushwap lake on the west, and from the U.S. boundary northwards to the boundary of the Province. Both are clearly executed by the Sabiston Litho. and Pub. Co. of Montreal.

ATLAS.

L'ANNÉE CARTOGRAPHIQUE. Supplément Annuel à toutes les Publications de Géographie et de Cartographie, dressé et rédigé sous la direction de F. Schrader. Cinquième Supplément.   *Hachette et Cie, Paris,* 1895.

The latest number of this very useful publication contains three sheets of maps relating to Asia, Africa, and America. Besides new boundaries, the routes of Dutreuil de Rhins in Central Asia, and of Dr. Donaldson Smith and other travellers in Somaliland are shown. The American sheet represents the results of the explorations of F. J. Vergara y Velazco in Colombia, and Drs. Steffen and Stange in the cordilleras of Chile. These sheets are of the highest importance to geographers.

NEW PHOTOGRAPHS.

Richmond River, N.S. Wales.
Clearing of " Brush " Timber, Experimental Farm, Richmond River.
Superintendent's Residence, Experimental Farm.
    Taken by WALTER S. CAMPBELL, Esq.
Presented by Baron Ferd. von Mueller.

THE SCOTTISH
GEOGRAPHICAL
MAGAZINE.

BRITISH SOUTH AFRICA.

By G. SEYMOUR FORT.

(*With a Map.*)

(*Read before the Society in March* 1896.)

WHEN you did me the honour of asking me to address you, I was at first, I must confess, very reluctant to do so—mainly because, although I had lived several years in South Africa, I had never devoted myself to geographical research, nor been to places not generally visited by travellers. However, your secretary very kindly suggested that I should frame my address on more general lines, giving some account of the present state and prospects of the country, and on these conditions I gladly accepted your invitation. I was the more pleased to do so because I deem it a great privilege, as an Englishman, to address a Scotch audience on matters connected with the expansion of our Empire.

I have spent many years of my life in Australasia as well as in South Africa, and have also visited New Zealand, Tasmania, and New Guinea ; and I have always been much impressed by the success of Scotch people as colonists, and by the fact that so many of our prominent men of mark in these new countries are Scotchmen. I have always maintained before English audiences that Scotchmen are the backbone of our Colonial Empire, and I feel it a great privilege to do so face to face with a Scotch audience.

I have divided my address into three heads. In the first instance, I will give some account of Rhodesia, that vast tract of country, some 2000 miles in length and 1000 in width, which lies to the north and north-west of the Transvaal, and stretches north of the Zambesi nearly to 8° S. longitude, and which is now partly under the Chartered Company, and partly under Imperial administration. I will then give a sketch of the native problem in South Africa, more especially in reference to the question of native labour supply, and, finally, I propose to say a few words about the Boers and the present crisis in the Transvaal.

At the outset, however, I would draw your attention to the political geographical conditions of South Africa. Within this area, as you will observe, there are no natural boundaries separating one set of homogeneous people from the other. The boundaries of the Cape Colony, Orange Free State, and Transvaal are merely political and artificial. In all these three States Dutch and British live side by side, and there is no natural or geographical reason why the whole of South Africa, from Cape Town in the south to the Zambesi river on the north, should not be confederated under a political and industrial system.

I have chosen to begin with Rhodesia, not merely because I have lived there for some years past, but mainly because it is, with the exception of Natal, the most British possession in South Africa. Cape Colony is self-governing under our Colonial Office, but with a Dutch population almost equal in numbers to the English, and with a solid Dutch vote exercising a most important influence in its Legislative Assembly. In the Orange Free State you have a pure Dutch republic, liberal and elastic in its constitution, but legislating entirely in the interests of its Afrikander burghers. In the Transvaal, some twenty-four Dutch farmers, elected by about 16,000 Dutch burghers, legislate for, and control the destiny of, the 80,000 British and foreign unenfranchised subjects living in this territory. My object in drawing your attention to these facts is to indicate how closely interconnected are the industrial and political factors, and I shall hope to show you later on in what way the Dutch element, both at the Cape and in the Republic, is opposed to industrial progress and the development of the agricultural and mineral resources. In Rhodesia, however, although there are a few Dutch scattered amongst the settlers, they form no bar to progress. Every effort is being made to test the natural wealth of the country, and every advantage is given to British trade and interests. This is a most important point to note, for, as I shall hope to point out later, the Boer Government directly hamper British, and favour in every way possible foreign, and especially German, trade.

In area, Rhodesia is seven and a half times as large as Great Britain, and comprises the district of Manica in the south-east—bordering on Portuguese territory—Mashonaland in the centre, Matabeleland, and an area across the Zambesi of many thousand square miles. As you are aware, eight years ago in the whole of this vast district there was hardly a single white resident—the land was practically in the possession of the barbarian and wild game. To the north and north-west the military despotism of the Matabele under Lobengula prevailed, and the whole region of the south was subject to the powerful chief Gungunhana. In 1893 the power of the Matebele was practically broken, and Gungunbana's raids have ceased for years past. The natives throughout Rhodesia[1] are peace-

---

[1] Since this lecture was delivered the Matabele outbreak has occurred. Throughout the rest of Rhodesia the natives, who number considerably over 1,000,000, have remained peaceful and loyal.

ful and settled, and towns have sprung up—in Manica the town of Umtali with some 600 white inhabitants; in Mashonaland, Salisbury with 1500; in Matabeleland, Buluwayo with 2000. Roads, along which coaches frequently ply, connect these towns, and in the surrounding sub-districts are numerous small townships in constant communication with these centres. Some 1500 miles of telegraph line runs from Port Beira on the east through Umtali to Salisbury, from thence to Buluwayo and onwards through Tuli down to Cape Town; moreover, Mr. Rhodes's trans-continental telegraph is completed across the Zambesi and to beyond Blantyre. There are twenty-two telegraph offices in the country. As an instance of the usefulness of this system of telegraphic communica-tion, I may mention that in August last a desperate convict escaped from the gaol in Salisbury, and after hair-breadth escapes succeeded in forcing his way across the Zambesi to Blantyre, a distance of 500 or 600 miles. News of his escape had, however, been telegraphed there, and he was immediately arrested.

On the east, a railway runs from Fontesvilla on the Pungwe inland for nearly 120 miles. This line will shortly be completed to the coast, and afterwards continued into the interior. On the west, railway extension is steadily progressing through Bechuanaland towards Buluwayo, so that before very long there will be direct railway communication between this town and the Cape. Thus, to a very large extent, the initial difficulty of communication—a difficulty which is always felt in new countries—is being mastered.

People living in old civilised countries can hardly conceive what an immense amount of energy and labour is spent in these new countries simply in travelling from one place to another—in fighting with space. The distance between Buluwayo and Umtali is about the same as from London to Edinburgh—400 miles—but this journey would take at least twenty days to accomplish in a bullock-wagon, in a light cart with mules twelve days, and in a mail-coach travelling night and day about six.

Let us now turn to the country itself, and first to its agricultural resources. In many places, especially on the high plateaus north of Umtali and in Matabeleland, the conditions for successful farming are assured. The climate is healthy, and, although the sun is powerful, there is always a fresh breeze; the soil is fertile, vegetables of all sorts, and fruits, including the ordinary English fruits, growing luxuriantly. I have seen splendid potatoes and crops of wheat and oats; cattle thrive well and are always fat. Wherever the farms are near a market, large profits are made: butter fetches 6s. a pound, milk 6d. a bottle, potatoes 4d. a pound, eggs 6s. to 10s. a dozen, and forage for horses 1s. 6d. a bundle.

In Gazaland, a healthy district south of Umtali, some three to four hundred families, principally Afrikander, have settled and devoted their attention to farming. Hitherto there has been no road into the district, and this community has been almost entirely self-sufficing. They have grown their own wheat and tobacco, tanned their own leather, used honey for sugar, and the juice of the indiarubber trees for candles. As soon as the road connecting Umtali with Salisbury is completed, their farms will be very valuable.

Briefly, the conditions for success in agriculture are all present. In a few years Rhodesia should be able to grow all its own produce, and breed its own horses and cattle. The one thing necessary to make agriculture profitable is a market, and that depends upon population; and whether the country becomes comparatively densely populated or not depends upon the extent and richness of the gold mines there.

It is a question often asked, Is Rhodesia going to be a second Rand? but in order to answer this, it is necessary to compare briefly the conditions under which gold is found in the two places. On the Witwatersrand goldfields you have an almost unique geological formation—the "banket" formation—and the payable ore is practically concentrated within some twenty miles of Johannesburg. It is because the payable gold area is so concentrated that the gold industry has given such rapid results, and that as an industry its life and value are so accurately known. In Rhodesia, however, this element of concentration is absent, and the development of its gold industry must necessarily be prolonged. Throughout Rhodesia the gold belts or mineral areas are scattered at considerable distances apart. Within these areas lie the quartz veins which contain the gold, and the work of the past five years has been to examine and sink down in these various quartz veins, in order to ascertain which of them contains the greater quantity of gold, and is likely to prove a payable and permanent mine. This state of experimental work is, however, now over. A large number of very rich reefs have been discovered, and their richness proved to a very considerable depth and length. There are thousands of tons of rich stone already exposed and only waiting to be crushed, and machinery is already on its way for that purpose. In the meantime, the short stretch of railway between Port Beira and Manicaland is rapidly being completed. At present machinery and goods have to be trans-shipped at Beira on to a lighter, then proceed up the river Pungwe, and be again transferred to the train about fifty miles up the river. All this handling is necessarily expensive and risky, and directors are wisely waiting until their batteries can be taken direct from Beira inland. Personally, I am convinced that as soon as the batteries can be got to work in the country, the richness and permanency of the gold industry will be proved, and that in a few years the output of gold from Rhodesia will equal, if not exceed, that from the Rand. In the meantime, as communication improves, the country becomes more and more fit for European settlement, and the demand for skilled labour of all sorts increases. Blacksmiths, carpenters, masons, can all earn very high wages. Farming for the present is better undertaken by men born in South Africa; but any young man with a small capital at his back and of average intelligence will find many opportunities of making money. I knew personally a young fellow who arrived with £300. He at once got a small appointment, and shortly afterwards invested £200 in the purchase of a wagon and oxen, and when I saw him he was getting £2 *per diem* for the hire of them. It can safely be stated that there is hardly any place in the world where the possibilities of making money quickly are so great as in Rhodesia.

In connection with this sketch of Rhodesia, it is impossible to omit a brief reference to the two men who have been mainly responsible for adding this territory to our Empire—namely, Mr. Rhodes and Dr. Jameson. It is only when we remember how indifferent to Imperial interests in South Africa have been our English statesmen in the past, and how eager Germany has been to take advantage of this indifferentism and to unite her western and eastern possessions, that we can realise the enormous debt we owe to Mr. Rhodes for securing this vast territory for British colonisation and trade. For years past the genius‘ and energy of this great man have been devoted to the maintenance and extension of British interests in South Africa, and history will not fail to recognise in him a great patriot and a great statesman. But Mr. Rhodes would never have been able to secure and develop the resources of Rhodesia if it had not been for the loyal and phenomenally able assistance of Dr. Jameson. It is difficult to describe in a sentence the ability and many-sidedness of this remarkable man, and if I once began to tell you stories about him I fear I should never stop.

We must now pass on to a consideration—necessarily hurried—of the native question.

It is not generally recognised that the black and yellow population in South Africa outnumber the Whites by six to one. There are roughly half a million Whites in the midst of three million natives. These comprise every grade of native from the wizened, ape-like Bushman to the educated Transkei native, who reads a newspaper edited in his own language. There are in Cape Colony, Bushmen, Hottentots, Red Kaffirs, Galœkas, Tembus, Fingoes, and farther north Basutos, Swazis, Zulus, and a large Kaffir population in the Transvaal. On the east coast is a dense native population generally called Mhimbawes, and in Rhodesia alone it is estimated that there are some two million natives—Shangaans, Mashonas, and Matebele. Like the white population, these are all very much mixed—in fact, I venture to think that no house in South Africa contains persons all of one nationality. Let us take a farmhouse in the Cape Colony as an illustration of this : the father is perhaps English, the mother Afrikander, with either Dutch or Huguenot blood in her veins, the children are a blend, the cook is a Zulu, the stable-boy a Basuto, the herd-boy comes from the Zambesi, the boy who drives the wagon is a half-caste.

This is more or less a concrete instance of the intermingling of white and black which prevails. But the most important point to note is, that despite the excess of black population, and despite the fact that this population is rapidly increasing, there is throughout South Africa a scarcity of native labour. From Cape Town to Zambesi one hears the same cry—the difficulty of obtaining reliable native labour : the whole success of the mining industry depends upon the reliability of this labour. In Matabeleland some 10,000 natives are employed in the mines, in Johannesburg about 60,000, and large numbers in the Kimberley diamond mine. These natives are recruited from all parts of South Africa, and I believe that recently natives have even been brought from the Congo Free State. As one drives into Johannesburg it is a curious sight to see the two streams of natives either entering or returning from the city. The former, lean and gaunt, with scarcely a rag to cover their nakedness, some of them having travelled hundreds of miles; the latter, fat and beaming, staggering under the purchases they have made, which include every possible article and utensil under the sun. It, however, too frequently happens that the natives seeking work are not sufficient, and as one goes through the mines one sees the stamps hanging up idle for lack of quartz, or white men sitting idly by their huts waiting for the necessary hands to enable them to get on with their work. It is an admitted fact that the native is essentially lazy. He will only grow sufficient food for his immediate wants, and his principal induce- ment to work at all in the mines and elsewhere is to obtain sufficient money to purchase a wife, and then remain a domestic loafer for the rest of his days. He is perfectly content to stay at home, to drink beer and talk scandal, and occasionally make excursions to other villages. If daughters are born to him, so much the better; they will always be a source of cattle and income.

The great object of British policy in Cape Colony and Rhodesia has been to raise the wants of the native—to impose a hut tax, and to try every method which will compel him to become useful as an industrial factor. Some of the measures adopted may seem harsh from a purely English point of view, but the natives throughout South Africa generally recognise the sympathy that lies behind English legislation, and look to the Queen as their Sovereign protectress. But this is emphatically not their feeling towards the Dutch, and especially the Boers in the Trans- vaal. The Boer is, himself, disinclined to labour and content with a low standard of comfort; his first care is to intimidate the native and afterwards to treat him as a mere machine. In many essential respects the native suffers great injustice at the hands of the Boer Government.

Marriage rights are not recognised. A native Christian woman properly married in the Cape Colony loses all conjugal rights, even the position of wife, as soon as she goes to reside in the Transvaal. The hut tax is ruthlessly enforced, even though through drought or other reasons its payment means starvation. If more than five families settle on any piece of ground, they are broken up practically and handed over as slaves to the surrounding farmers. The ordinary rights of justice are denied to natives. A few months ago some sixty of them were thrust into prison; their trial, however, was so long delayed, and the food given them so insufficient, that thirty of them died. Again, a Dutch farmer was fined by the courts for grossly ill-treating a native employé, and was condemned to pay a fine; but President Krüger practically overruled the judgment of the court and ordered the fine to be paid out of the public funds. Finally, the natives, especially those working in the mines in Johannesburg, are able to buy drink. All these factors, and especially the sale of drink, tend to increase the difficulties of obtaining native labour, and thus seriously hamper the progress of the mining industry. But this brings me to my third heading, namely, some account of the present position of the Dutch in South Africa.

I cannot attempt, at this stage, to enter into any historical details, but will confine myself mainly to the influence of the Dutch element upon the industrial progress of the country. In entering upon this question, I wish at once to state that I have no wish to stir up race feeling or frame any indictment against them as a nation. On the contrary, I recognise that the Boer has many fine qualities. He is intensely patriotic, and, in many respects, is not unlike the English landed proprietor of two or three hundred years ago. But the radical difference between Dutch and British is, that the former are entirely anti-industrial and anti-progressive, and would like, if possible, to lock up the resources of the soil in the interest entirely of their own pastoral instincts and habits. When we compare the exports of Australia and Cape Colony, we are struck with the fact that, although the former is nearly twice as distant from Britain, and suffers from a less equable rainfall, yet from Australia we get our wool, our frozen meat, our butter, and our fruit, whereas Cape Colony has to import its wheat, its butter, and its meat; and quite recently protective duties were placed on Australian mutton and flour. Undoubtedly, one of the main causes of this difference is the solid anti-progressive vote exercised by the Dutch farmers in Cape Colony. They are opposed to immigration, to irrigation—in fact, to every factor that makes for progress. The Scab Act is an emphatic instance of this. The scab is an infectious disease which affects sheep. For years past, the Governments in Australia have kept it down by compulsory precautionary methods. Last year, with great difficulty, a modified form of this Act was introduced into the Cape Colony; but the agitation against it by the Dutch farmers was so keen, that it was never put in force. In other words, these farmers would rather see their flocks diminish and become riddled with disease, than allow any outside interference in their own interest. While this is the attitude towards agricultural progress of the comparatively progressive Dutchman in Cape Colony, the Boer in the Transvaal has not been behindhand not only in hampering to the utmost the progress of the mining industry in Johannesburg, but also in placing British trade and imports under every possible disadvantage. As you are aware, when we gave the Boers the Transvaal in 1881, the terms of the convention implied that all British subjects should have, after a reasonable term of residence, political rights. Since that date, however, President Krüger has steadily legislated against these rights being obtainable by any but Dutchmen, until at the present time no stranger can acquire a vote until he has resided ten years in the country. And there are other restrictions which make it practically impossible for a British or foreign resident to obtain the franchise. The consequence is, that you have a population of some 15,000 or 16,000 Boers legislating for a population of some 100,000 British and foreign residents. But this is not the worst feature in the matter, for these Dutchmen, who are quite incapable of carrying on the departmental work of government, have called in to their aid, for this purpose, a large number of Hollanders, who form the official class, and carry on the entire work of the Government.

Of the many evils of the Hollander-Boer administration, I will mention those more directly affecting the gold industry; and it must be remembered that it is the working of the gold by the stranger population which has raised the Transvaal from an almost bankrupt condition into the most wealthy and powerful centre in South Africa. In 1887 gold was first washed in Johannesburg, and the output last year amounted to nearly £8,000,000. Of this, the Government secured nearly one-half by taxation. And this brings me to the first evil—taxation. In the first place, a far greater revenue is raised than is required, and an enormous secret service fund is maintained. It has been stated that every fifth man in Johannesburg is a Government spy; and certainly the Uitlander population, who have no representation, are taxed to supply their enemies with all the sinews of war—even to voting £20,000 to give the Dutch farmers a trip to Delagoa Bay. Again, taxation is distinctly of a class nature; one instance is the Stamp Receipt Law. Only the commercial man—*i.e.* the Uitlander—is compelled to stamp a receipt; the Boer escapes. Thirdly, the necessaries of life are severely taxed by means of import duties. Flour pays seven shillings and sixpence per 100 lbs.; mealies five shillings per bag; bacon, ham, and butter one shilling per lb. These taxes, and the absence of State-aided education and the English language, press heavily upon the working and professional classes, and prevent them from bringing their families to the Transvaal.

The next evil is the concessions. These are monopolies granted to individuals for the manufacture or sale of certain goods. A concession was granted to a German to manufacture dynamite; he, however, was proved to be actually importing dynamite manufactured elsewhere, and the Government took over the concession. The effect of this is, that dynamite, which is very extensively used, costs the consumer ninety shillings a case instead of thirty shillings, which would be the cost if there were an open market. It is estimated that this concession alone costs some £600,000 per annum, and that, if it were abolished, the profits on the working of the mines would be increased nearly five per cent. Besides this dynamite concession there are many others, but this more especially affects the profitable expenditure of capital on the mines. In fact, Krüger has himself stated that there is too much gold in Johannesburg. These are some instances of the manner in which President Krüger—and he more than his Executive is responsible—has pandered to the anti-industrial instincts of his race, and has used the wealth which British industry has placed in his hands to strengthen his despotic position. There is also another side to his policy, namely, the advantages he has given to German trade and influence. In other words, he has used the independence granted him by Great Britain to deny her subjects political rights and undermine her position, even to the extent of using British money for the purpose of giving trade advantages to Germany. Into the details of this anti-British policy it is impossible for me at this moment to enter. But I have said enough to show you that although we are the paramount power in South Africa, yet the agricultural and mineral resources of that country are at present mainly in the hands of the Dutch, whose policy it is to lock up this vast wealth, as far as possible, in the interest of their own anti-progressive and pastoral habits.   The attempt on the part of the Johannesburg people themselves to compel the Boer Government to grant them some measure of justice and fair play resulted, as you all know, in a failure.   But, at all events, British subjects in Johannesburg did obtain from her Majesty's Government an understanding that their grievances would be redressed; and they are now anxiously awaiting the fulfilment of that promise.   In the meantime, however, President Krüger, despite his protestations, is rapidly increasing the strength of his military position, by building forts and arming his people, by importing German officers and men, and by holding out promises of further trade concessions to alien powers.   It is impossible to foresee the political outcome of the situation; but this at all events it is possible to say, that it would be a grave national disaster if British subjects in the Transvaal should ever be shaken in their belief in, and loyalty to, the Empire, and should, under pressure of circumstances, be induced to give their support to a confederation in South Africa not based on Imperial lines, nor pledged to advance in every way British trade and interests.   This has been for years past the main plank of Mr. Rhodes's scheme and outlook for South Africa.   On these lines he has added Rhodesia to our Empire, which, as I have already explained to you, is at present our most Imperial possession in South Africa.   It is in my opinion most important that the eyes of people in Great Britain should be opened to the silent industrial and political conflict that is now taking place in South Africa.   On the one hand, President Krüger, who is a great statesman, is endeavouring to mould the lines of a future confederation, in which the dominating influence shall be Dutch and German, and to which our disheartened colonists will be compelled, more of necessity than of choice, to give an unwilling adherence.   South Africa will cease to be an integral part of the Empire, and become an independent Dominion, with possible hostile tariffs and railway rates.   On the other hand, Mr. Rhodes and his followers wish to bring about a union of the Dutch and British within a confederation which shall make South Africa an integral part of the Empire, with no bar to British progress or to British trade.

I fear that in thus advocating the policy of progressive Imperialism I have ventured dangerously near the deep waters of politics; but, as I have already pointed out, the industrial and political affairs of South Africa are inseparably connected.   So long as our Dutch fellow-colonists endeavour to place obstacles in the way of British energy and enterprise, no real industrial progress can take place.   Living, as we do, side by side with them, with no geographical barriers to keep up differences of race and interests, it should be the aim of Dutchman, Afrikander, and British Colonist alike, to unite in developing to the utmost the agricultural and mineral resources of this most valuable portion of the Empire.

AËRIAL NAVIGATION.

By A. SILVA WHITE, Hon. F.R.S.G.S.

To navigate the atmosphere as the mariner navigates his ship has been the dream of countless enthusiasts. Records reach us from the Middle Ages, and even earlier times, of flying men and flying-machines. But, from the classic fables of Dædalus and Icarus and "the dove of Archytas" down to the adventurous attempt of the Italian alchemist who, in the early part of the sixteenth century, endeavoured to fly from the walls of Stirling Castle, it is difficult to discover where fact ends and fiction begins. The problem of aërial navigation has, however, at last passed into the hands of competent investigators, who are confident that its solution will be successfully accomplished before the close of the current century.

The art of aërostation has a curious history. The year 1783 is the *annus mirabilis* in the science of aëronautics : it records the construction at Annonay, near Lyons, of the first balloon. The brothers Montgolfier, sons of a famous papermaker at Annonay, were the discoverers. It occurred to them that, if they could obtain any vapour lighter than air and enclose it in a large bag, the same natural law which sustained the clouds in the heavens would raise their simple contrivance to similar heights. So, taking a leaf from the open book of nature, they constructed a paper balloon with their skilled hands. In order to reproduce "clouds," they lighted chopped straw and caught the smoke. And the balloon rose into the air !

This successful experiment created the greatest enthusiasm throughout Europe. It was repeated on the 19th September 1783, by the Montgolfiers, before a brilliant assemblage of the French Court at Versailles. On this important occasion the balloon rose 1500 feet, and descended after being in the air for eight minutes. Attached to the balloon was a cage containing the first aërial travellers—a sheep, a cock, and a duck.

The brothers Montgolfier believed it was the smoke that raised their paper balloon; and it was only subsequently recognised that the ascensional power was due merely to the lightness of heated air as compared with an equal volume at a lower temperature. The difficulty of generating heated air was overcome by fixing a brazier containing ignited matter under the neck of the balloon, which was thereby sustained as long as the fire lasted, and was enabled, besides, to carry human freight.

In spite of the Montgolfiers' discovery of the fire-balloon, perhaps greater credit was due to Professor Charles, of Paris, who, shortly after the experiments at Annonay, used hydrogen gas (then known as "inflammable air") as the ascensional power. Professor Charles constructed a balloon which differed very little from those in use at the present day : it was made of silk, varnished over with an elastic gum, and had not only the netting and hoop as now used, but also the escape-valve at the top.

Both in this gas-balloon of Professor Charles and in the fire-balloon of the Montgolfiers several personal ascents were made; and they were imitated all over Europe. The first personal ascent in Britain was made from Comely Gardens, Edinburgh, on the 27th August 1784, when Mr. Tytler rose in a fire-balloon of his own construction; but, as the balloon was not provided with fuel, it descended half-a-mile from the place of starting. This achievement Mr. Tytler modestly called "a leap into the air"; and he very courteously surrendered the credit of "the first personal ascent in Britain" to Lunardi, whose feats a few days later created the utmost enthusiasm in London.

To Mr. Charles Green was due the substitution of coal gas for hydrogen gas. Since his first ascent on 19th July 1821, this gas has almost invariably been employed for balloons.

All kinds of balloons have since been constructed; but, saving the improvements of a few minor details in their gear, no real progress has been made. We have, it is true, a Balloon Society, at whose meetings every subject under the sun—except balloons—is discussed; and every year a large number of patents for flying-machines are registered: but, until quite recently, practically no progress has been made in solving the problem of aërial navigation. The balloon itself has been used chiefly as a toy to fascinate the public mind. Sometimes it has been put to practical use, as at the siege of Paris, when balloons were the only means by which communication with the outside world could be maintained, and by which some of the besieged, notably Gambetta, effected their escape. Moreover, in nearly all the more recent wars, a balloon corps has rendered important services by spying out the movements of the enemy: consequently, nearly every European army, and especially the French, maintains a carefully-trained corps equipped for this service. For scientific purposes, too, the balloon offers incalculable advantages; but, though many useful contributions have been made to meteorology and the physical sciences by competent observers and aëronauts, the cost of conducting a series of systematic synchronous observations is too great to popularise the employment of this agency.

For the practical purposes of aërial navigation the balloon has been proved, over and over again, to be quite inadequate. So-called "navigable balloons" have frequently been constructed, more or less on the same lines as *La France*, the most successful of all the inventions in that direction. The shape of *La France* resembled that of a fish or a cigar, and it was propelled by a screw, the driving power of which was derived from an electric motor. The advance in our knowledge of electricity has, in fact, opened out new prospects to inventors of flying-machines, for it has long been recognised by engineers and scientific experts that the solution of the problem of aviation, or flying in the air, rested on the discovery of a motor which should have "sufficient energy in proportion to its weight." "This motor," Mr. Hiram Maxim now asserts, "has been found, its power has been tested, and its weight is known." But it has been proved to demonstration that the ascensional power must come from the machine itself, and not depend on the lifting capacity of the balloon in any shape or form: since it is impossible to drive an inflated envelope against the impact of the wind. This principle may be made clearer to the reader if

I describe some of my own humble experiments in the utilisation of air-currents in ballooning.

In the present state of aëronautics it would be of some value to know how far it is possible for a balloon to utilise air-currents in order to reach a definite goal. Clumsy as is the balloon for the purpose of aërial navigation, I know that much may be accomplished by a capable steersman. It is useless to attempt to guide, by any contrivance whatever, the balloon as it is constructed at the present day. Immersed totally in the elastic fluid of the atmosphere, it is not possible to overcome the lateral resistance of the wind. Vertical gravity is overcome by an imprisoned gas, lighter than the surrounding fluid, which ascends until it establishes an equilibrium—that is to say, until the weight of the floating body is equal to the weight of the fluid atmosphere it displaces. The inherent force or contrivance for resisting the impact of the wind has, therefore, still to be perfected, as no doubt it will be, judging from the tentative experiments with Mr. Maxim's aëroplane. Hence, flying-machines and not balloons will be the future navigators of the aërial ocean.

Anybody who has witnessed a balloon filling during the prevalence of a high wind must have felt convinced of the impossibility of any conceivable agency being capable of driving such a light body against it; and even if this were in a measure possible, the light material of which the balloon is necessarily constructed would not stand the friction. Those cases in which a certain measure of success has been reported, of "navigable balloons" going within a few points of the wind, are more likely attributable to the influence of the cross-currents that everywhere are met with in the higher, and often in the lower, strata of the atmosphere. The surface wind prevails sometimes no higher than 500 feet; but it has also been found constant at as great an altitude as four miles. An intelligent use of the air-currents, therefore, would enable the aëronaut, within certain restrictions and under favourable conditions, to guide his balloon.

Formerly, in the early stages of scientific inquiry, it was thought that the atmosphere surrounding our globe was of no great depth, and that it might be possible to raise a body to its uppermost limits—as a cork floats to the surface of a basin of water—where, having two mediums instead of one, the atmosphere might be navigated like the ocean. But in the light of established science, the analogy between the navigation of the ocean and that of the air—an inelastic and an elastic fluid—has been shown to be false. A balloon floating in the air is at the sport of the wind, and the only way of navigating it is to fall in with favourable air-currents. It is true that our knowledge of air-currents is most imperfect; but meteorology is a young science, and aëronautics is still in its infancy. Only by constant and comprehensive experiments, therefore, can we hope to discover reliable "sailing directions" for the balloonists of the future.

As far as I am aware, little or no attention has been given to this aspect of ballooning. This is the more surprising, because one would have thought that a single ascent would be sufficient to impress on the mind of the aëronaut the great part played by air-currents in the guidance of his balloon. At least, this was my experience; and I immediately set myself the task—though in a very limited way—of ascertaining how far air-currents might be relied on.

Taking advantage, ten years ago, of the visit to Edinburgh of Mr. Dale— who, it will be remembered, was killed in a balloon accident at the Crystal Palace—I conducted a series of experiments, some of which yielded decisive results. Mr. Dale's balloon, the *Edina*, was a new one, though with a capacity of only 28,000 feet of gas. It had a diameter of thirty-eight feet, and measured with the car sixty feet in height. Thus, our scope for experimental ballooning was extremely limited, owing to the small size of the balloon and its consequent want of power. The *Edina* was able to carry only two persons and a very small supply of ballast.

My first ascent with Mr. Dale was on September 26th, 1886. The wind was blowing from the south-west, veering occasionally to the south ; but the surface current, as shown by the ultimate course taken by the balloon, and ascertained beforehand by a small "pilot-balloon," was due south. The *Edina*, on being released, rose perpendicularly to 500 feet, where she obtained an equilibrium, and was driven by an under-current in a north-easterly direction. By easing her of the weight of some ballast, the balloon rose to 1500 feet, where she met the true and constant surface wind from the south. We therefore commenced the passage of the Firth of Forth in a northerly direction. About half-way across, the draught drawing down the Firth slightly altered the balloon's course, setting her about a mile to leeward, but the danger of being blown seawards was easily obviated by raising the balloon to 2000 feet, and ultimately to 2500 feet, above the influence exercised by the physical formation of the land. The passage of the Firth was accomplished under a heavy shower of rain in fifteen minutes, at a speed of twenty miles an hour. Skirting the Fife coast, a landing was effected two miles north of Kirkcaldy.

I have given some particulars of this ascent in order that the reader may note two things—the action of air-currents in determining the course of the balloon, and the ease with which, under favourable circumstances, these may be utilised by an intelligent aëronaut. Of course it may happen that the prevailing wind may be constant for a considerable altitude ; but, still, it is rare that a divergent or even contrary current may not be found by a balloon of large capacity. I remember reading of a French aëronaut who, ascending from Calais during a high wind, was blown straight out to sea, but returned to his starting-point by a contrary wind, which, happily, he found by mere accident at a great altitude.

By a careful study of the weather, and the means which the aëronaut has of ascertaining the direction of the wind both before and during an ascent, it is quite possible to place a fair amount of reliance on the direction which the balloon may ultimately take. In several of the ascents made by Mr. Dale at Edinburgh, I was able to ascertain beforehand the approximate course of his balloon, and also the places at which a landing might be effected. The most successful of these experiments was on the occasion when he reached Kinross, twenty-one miles distant from Edinburgh, in three-quarters of an hour. This spot I had fixed on several hours before the ascent was made : and it was gained only by seeking out the most favourable air-currents, for the prevailing wind would have taken the balloon considerably to the east. A steady breeze from the south-east, with an area of low-pressure skirting our north-western shores, causing the upper air-currents to veer between south-east and south-south-west, offered to the aëronaut the choice of a number of places in Fifeshire located within a vast triangle (the apex of which was the starting-point) at which a landing might have been effected. Owing to the short distance which a balloon of such limited capacity as the *Edina* could safely accomplish, the easiest spot was chosen for a goal : for to manœuvre the balloon requires a large expenditure of its very life-blood, gas and ballast, and only very large balloons have much of either to spare.

In the other ascents I made with Mr. Dale (partly with the purpose of making physical observations), untoward circumstances prevented any adequate tests being applied, but their partial failure was not without instructive results.

On one occasion we arranged a race between two balloons. Mr. Dale and I were in one—the *Sunbeam*—and Captain Taylor was alone in the *Edina*. The race caused a considerable amount of public interest, because, I believe, it was the first time on record of an attempt being made by two balloons to race towards a *fixed goal*. It failed because the pressure of gas at the meters where the balloons were filled was insufficient to inflate both of them at the same time, the wind being high and a good deal of gas escaping in consequence. The result was that the balloons were released with only sufficient gas and ballast to keep them above the house-tops ; and, as the wind was blowing directly out to sea, it would have been impossible for the balloons, thus crippled, to have accomplished the passage of the Firth. A landing was therefore made in the town. A very rough landing it was too ! The *Edina* was "rushed down" perpendicularly, in order to "fetch" the only available anchorage, on a small plot of ground between some houses. Unfortunately, we "carried away" the telegraph wires and wrecked the balloon, which was subsequently trampled to shreds by the mob.

Perhaps my most interesting ascent was on the 19th October 1886. Our intention had been to make for Carstairs, a town in Lanarkshire, about thirty miles distant ; but, on account of circumstances I shall refer to, we were able only to pass within three miles of this spot, landing at Carluke. The weather was grey and stormy when the *Edina* was released, with Mr. Dale and myself in the car. Before the balloon had passed well over the town, we had already entered the clouds, at 2300 feet ; and 300 feet higher we emerged into brilliant sunlight. The thermometer, which before starting had registered 60° Fahr., then stood at 43°, and the temperature continued to decrease until, at an altitude of one mile, it was at the freezing-point, 32°. Here there was not much wind ; and we had leisure to witness the shifting cloud-scene over which we hung suspended. We even ventured on a cigarette, at the risk of an explosion. The heavy cumulous masses, which entirely shut out the earth from our view, except here and there when they parted for a moment immediately under the balloon, formed towards the east a sea-scape : great crested waves breaking (as they appeared) against gaunt and jagged rocks, and rolling on to the distant horizon. The upper layer of these clouds was, except for these wave-forms, on a sharply-defined plane, and no vapour apparently intervened between them and the high cirrus clouds above.[1] To the west, on the other hand, a very different cloud-scape was observed. There the scene was purely Arctic in character—an illusion that was heightened by the perfect stillness, and made very real by the relatively low temperature—snow-clad mountains of every shape and form casting shadows from the sun, and a rocky ice-bound coast, with the familiar ice-foot and "leads" of blue running water. The colouring of this Arctic scene, instead of resembling the low tones in the High North, partook of the splendour of an ideal landscape : pearly greys, azure or deep indigo blues, and warm-coloured fleecy snowdrifts enhanced the beauty of a scene which, in the Arctic regions, except under a refracted light, typifies the stern reality of death. At other points of the compass many of the most characteristic land-sculptures were represented in cloud-forms; and the scenes were constantly shifting, though imperceptibly to the eye. For a few moments the balloon cast the curious double shadow—that of a balloon inverted on another balloon—so often witnessed in aërial navigation. The *Edina* descended of her own accord into a nimbus, or rain-cloud, to avoid passing through which some ballast was thrown out; but another swift downward movement shortly afterwards took place, until, at 3800 feet, the clouds parted and the earth again became visible. We were at that moment over Cobbinshaw reservoir; and at the same mean elevation we passed at a great speed over the marshy, inhospitable moorlands to the west of the Pentland Hills. A landing was effected with much difficulty. The strong wind prevalent at the time carried the balloon, after her first contact with the earth, rapidly over two fields; but at the second hedge the car, instead of leaping over or through it, as in the first instance, caught a stout paling which held it captive until, with the assistance of several men, the balloon was safely secured. Another leap would have taken us into the midst of several blazing furnaces; and it was this consideration which induced me to hang on to that providential paling "like grim death," in spite of Mr. Dale shouting to me, "Let go, or you will break your arm!"

On this trip we laboured under the disadvantage of having the valve-rope (which leads to the top of the balloon to release the gas) twisted into a knot in the very centre of the inflated envelope, so that it could be reached only by means of an umbrella attached to a stick. We had, in fact, been "sent up" in a hurry, the very car having been slung awry. Consequently, we were at the sport of the wind and were most careful

---

[1] This "sea of clouds" resembled in every particular that which I viewed for a week on end whilst camping out in the old crater of Tenerife. In the North-East Trades, the clouds round the Peak form every day with extreme regularity : one can almost tell the time by them.

not to jeopardise our frail hold on the valve-rope by using it, except on an emergency; and so we permitted ourselves to ascend or descend as the varying temperatures affected the gas in the balloon : in the end, however, the balloon had to be ripped open in order to avert a catastrophe.

In attaining high altitudes, the fact of passing through several hundred feet of rain-cloud into the warm sunlight above, causes the gas to expand and to escape through the open neck of the balloon, thus lessening its carrying-power. The result is that, in again passing through the clouds towards the earth, condensation of the gas increases the velocity of the falling balloon to such an extent that only a liberal disposal of ballast can check the rapid descent.

Although with only these partial successes to look back upon, I am convinced that much might be done to navigate the balloon, within a few points of the prevailing wind, by the judicious use of air-currents. At the same time it must have been evident to the reader who has troubled to peruse my descriptions so far, that under no circumstances can the balloon be relied upon as offering the best mechanism for navigating the air. My own limited experiences show that, in order to accomplish this end, a flying-machine, heavier than the air, and capable of withstanding the impact of the wind, is the only promising air-ship of the future.

Inventors and discoverers, who are unable to circumvent a natural law, endeavour to utilise it. I have dwelt on the obstacle to aërial navigation presented by the impact of the wind. And it is precisely this apparent obstacle which Mr. Maxim has turned to account in the construction of his aëroplane, which, more than any other known mechanism or principle, appears to me to offer the best prospect of solving the problem of aërial navigation. This is, in the words of Professor Langley, "a mechanism designed to secure artificial flight by taking advantage of the inertia and elasticity of the air."

The problem of aviation, as the French call it, is now attracting considerable attention; and many of our leading inventors—especially in America and France—are turning their minds to its solution. Mr. Maxim's experiments have, in the first instance, been directed towards the discovery of general principles, " to ascertain the amount of energy required for flying, and also to ascertain what influence, if any, the factor of size has upon flight." The leading principle of the aëroplane has been lucidly described by Professor Langley, who says: " If, in such aërial motion, there be given a plane of fixed size and weight, inclined at such angles and moved forward at such speeds that it shall always be just sustained in horizontal flight, then the more the speed is increased the less will be the power required to support and advance it, so that there will be an increasing economy of power with each higher speed, up to some remote limit not yet attained in experiment." This principle, though directly contrary to that which obtains in all transport by land and water, is partly illustrated by the action of a skater on thin ice, which is too weak to bear his weight except he move over the surface at a great speed.

As soon, therefore, as the means of directing this ingenious mechanism (the aëroplane or aërodrome) has been found, it may be quite possible to "navigate the air." The question of air-currents, to which I have purposely alluded at some length, does not, so far as I have ascertained, appear to have entered into the calculations of Mr. Maxim and his coadjutors: consequently I am curious to know how it is to be met. For, should the flying-machine soar up into a contrary or even diverse air-current, it is evident that the very principle of its flight will be the cause of its instant destruction, unless it be provided with some automatic mechanism which will enable it, *without altering its direction*, to adjust itself to the changed conditions.

We have heard a great deal within the last few months of Herr Andrée's bold project of reaching the North Pole by balloon. In the discussion which followed the reading of his paper at the late International Geographical Congress he received a certain amount of popular support, because such an attempt as he proposes to make naturally appealed to the sympathies of an English audience. For my own part, I did not deny that Herr Andrée might, under the most favourable circumstances, reach the Geographical Pole. Given a strong southerly wind at starting he might attain his goal from Spitzbergen in a few hours. But how would he return? Our knowledge of the meteorology of the North Polar regions is so slight that all we can say is, that southerly (SW.-SE.) winds may possibly prevail during the month of July, when the attempt, this year, is to be made. But if this hypothesis is favourable to his reaching the Pole, it necessarily militates against his leaving it. His proposed use of a number of guide-ropes (to retard the progress of the balloon, in order to utilise a sail) might, if the wind be light, enable him to deflect his course by a few points; but in a strong wind this contrivance would not act. It would, moreover, be an added danger. Herr Andrée cannot expect to find an open sea at the Pole. The Palæocrystic ice, which he is likely to encounter, will afford the worst possible anchorage, and present many obstacles to the free use of his guide-ropes. In a strong wind, too—in spite of the probable uniformity of temperature, comparatively speaking, and the absence of night—he will find great difficulty, though he preserve his guide-ropes, in maintaining the low mean altitude necessary for their use. Long balloon voyages are difficult enough in Temperate regions, over well-known countries, but a voyage towards the North Pole is a flight into the unknown. The longest voyage ever made in a balloon (Paris to Sweden in 1871) lasted fifteen days; but Herr Andrée has to provide for one that may last, at least, thirty days—though the distance to be covered, 2000 to 2500 miles, can, of course, be traversed in less time, given ideal circumstances which are not likely to occur. Aëronauts have been totally lost in attempting the passage of the English Channel; but Herr Andrée must penetrate an uninhabited region, inimical to human life, with no prospect of returning to civilisation, should an accident occur to his balloon or should he run short of gas and ballast, except he and his companions manage to make their way back over the ice to an Eskimo or Danish settlement. Altogether, considering the unknown factors against which he will have to

contend, and the totally inadequate capacity of the balloon for long
aërial voyages, Herr Andrée's project, much as we may applaud it, must
be characterised by experts as foolhardy in the extreme, even suicidal.
Nobody can deny that there is an element of chance in his favour; but
this is so slight that the adventure does not, in my opinion, warrant
the support of scientists.

THE METEOROLOGY OF CENTRAL ASIA.

THE meteorological observations made by N. M. Przhevalski during his
four journeys in Central Asia have been worked up by Professor A. J.
Voiekof, and published in a large volume by the Imperial Russian Geographical Society.[1]   To these are added observations of M. M. V.
Pievtsof, and a review of the climate of the region by the Professor himself.   We can only touch on some general questions.

The first of these is the extension of the summer rains of the Asiatic
monsoon towards the north, and the limits of its principal regions—the
Indian and the Chinese, or East Asiatic.   In his *Climate of the Monsoon
Region of Eastern Asia,* Professor Voiekof, on the ground of M. Przhevalski's observations during his first journey, included the Eastern Nan-
shan in the domain of the rainy Chinese monsoon, and stated that
Northern (more correctly North-Eastern Tibet) had the climate of the
monsoon, that is, received rain chiefly in summer.   On his third journey
Przhevalski formed the opinion that the summer moisture at the sources
of the Whang-ho, at Kuku-nor, and in North-Eastern Tibet, was brought by
west or west-south-westerly winds, and that in all probability the monsoon
of Western India carried this moisture from beyond the Himalayas, while
in the Eastern Nan-shan the summer rain came from China.   Yet, on the
other hand, he remarked that neither the Chinese nor the Indian monsoon
reaches the Western Nan-shan, the Altyn-tag of Lob-nor, and the country
south as far as the Kuen Lun, for there, notwithstanding the great altitude, the air is remarkably dry throughout the year, rendering the country
a desert.   When Przhevalski spent the summer of 1884 in North-Eastern
Tibet, rain and snow were frequent and abundant, and he was confirmed
in his opinion that the precipitation was carried thither by the south-
west monsoon of the Indian Ocean.

With this opinion Professor Voiekof is unable to agree, for the winds
in the great plain of the Ganges, in the Eastern Punjab, and on the slopes
of the Himalayas, blow from the east-south-east, proceeding, therefore,
from the Bay of Bengal and not from the Arabian Sea.   Hence the west-
south-westerly winds of Tibet cannot be a direct extension of the Indian
monsoon passing over the Western Himalayas.   Moreover, if such were
the source of the precipitation of North-Eastern Tibet, it would be diffi-

---

[1] *Scientific Results of the Journeys of N. M. Przhevalski in Central Asia.   Meteorological
Section.*   Edited by A. J. Voiekof.   St. Petersburg, 1895.   (In Russian.   Remarks accompanying the meteorological tables in French also.)

cult to account for the fact that in North-Western Tibet the air in summer is dry, and the snow that falls is quickly evaporated, as appears from the reports of Pievtsof. Here the climate resembles that of the lower slopes of Central Asia, as described by Przhevalski—great dryness of the air, strong winds, storms not infrequent in the daytime, frequent but not abundant snowfalls, swept by the wind and soon evaporated or melted by the sun. This is very different from the weather of North-Eastern Tibet, where snow and rain fall in abundance and the air and soil are moist.

Professor Voiekof then sets forth his own theory. Towards the east the elevation of the Himalayas becomes considerably less, the summits in Bhotan being only 17,000 to 18,000 feet, and the ridge under 15,000, while a little farther east the Brahmaputra and two of its chief tributaries cut their way through the range. Now in the northern part of the Bay of Bengal and the Eastern Himalayas is the rainiest region in the world, and three-fifths of the precipitation falls in the three summer months (the excessive rain of the Khasia mountains is well known), whereas the rainfall of the North-Western Himalayas is only a fraction of the amount observed here. It is also an important fact that the winds in the Eastern Himalayas blow from the south-south-west, or even from the south-west—being turned aside by the mountains. Evidently, then, a current may pass northwards, not only through the ravines but over the crest of the range, which is not much higher—in some places even lower—than the highlands of North-Eastern Tibet. In summer, too, the clouds are usually not lower than 3300 feet above the surface of the ground, and consequently from the warm, damp countries to the south of the Himalayas a large quantity of vapour may be carried in summer to the north. This vapour causes a large rainfall not only in North-Eastern Tibet, but also in South-Eastern, as shown by the observations of the French missionaries. It seems certain, then, that the rains from the Bay of Bengal extend to North-Eastern Tibet, which has the characteristic monsoon climate—a damp, cloudy, and rainy summer, and a dry winter.

The year after he ascertained that the summer in North-Eastern Tibet was very rainy, Przhevalski spent twenty-five days in the mountains of Keria, when rain fell almost continuously, with calms or feeble northerly winds. These mountains adjoin on the north the uplands of Eastern Turkestan, of no great elevation, which are one of the driest parts of the world, and they also constitute the northern border of the Kuen Lun. Przhevalski accordingly added this country to the domain of the Indian south-west monsoon, and M. Venukof expressed his opinion that this monsoon held sway over a broad zone of Central Asia, stretching from the sources of the Amu Daria and Tarim to the neighbourhood of Lanchau on the Whang-ho, and attaining the 36th parallel between Keria and Khotan, and the 40th at its eastern limit. But Przhevalski expressly says that the rain came with feeble northerly winds or calms. Moreover, why in North-Western Tibet, which is higher and nearer to India, is the summer dry? The Western Nan-shan and the Altyn-tagh are also dry. In the former, Przhevalski remarks that in summer the air was filled with dust borne by north-west winds from the adjacent deserts. Very different weather prevails on the eastern slopes. It should also be noticed that in the south of Eastern Turkestan, east of the Cherchen river, not one important stream issues from the mountains, and there is not a single oasis. Professor Voiekof believes that the rains noticed in this country proceed from the vapour that rises from the irrigated fields and the woods of poplars on the uplands of Eastern Turkestan, which, carried by the northerly winds mentioned above, is condensed on the mountain slopes. Very probably the summer in which Przhevalski visited the mountains of Keria was exceptionally rainy there.

It appears from Przhevalski's journal of his four journeys, as well as from his meteorological records, that in all the parts of the highlands of Central Asia he visited westerly winds prevailed, except in Eastern Mongolia, with Ala-shan and Ordos and the Eastern Nan-shan, where in summer light south-east winds were predominant, and in Eastern Turkestan, near Lob-nor, where strong north-east winds blew in winter and early spring. In the greater part of Turkestan westerly winds prevail, as noticed by Pievtsof. And the greater the elevation, the more general does this direction of the air-currents become. Professor Voiekof regards these currents as part of the thermal circulation of the atmosphere, set in motion by the heat of the Tropics and diverging towards the east under the influence of the Earth's rotation, as they pass towards the Poles.

Central Asia is a country of vast extent and of very varying altitude; the climate therefore may be expected to exhibit great variety. It is in a high degree continental, with a very large annual range of temperature, both absolute, and as compared with that of other climates in the same latitudes. The reason of this is that Central Asia is not only remote from the ocean, but separated from it by lofty mountains. Very probably the winter is cold in Central Asia for the latitude, and the summer warm, so that there are two factors present to cause a large annual range of temperature. Continental climates of high latitudes are usually marked by cold winters, and those of low latitudes by hot summers. Consequently, the mean annual temperature, reduced to sea-level, is usually in high latitudes lower than that of the oceans and higher in low latitudes. Siberia and the northern part of North America are examples of the former case, the Sahara and Northern India of the latter. Central Asia is situated in the middle latitudes, and there is every reason to believe that on the whole the winter there is colder and the summer warmer than in coast lands in the same latitudes and at the same elevation.

That the daily range is large in Central Asia, and greater than on plains and in low valleys in the same latitudes is known from the observations of Przhevalski during his first journey, and from those of Severtsof on the Pamir. The summer months, however, and often also the end of spring and the beginning of autumn, present exceptions to the rule in those parts where at this season there is much cloudiness and abundant rain. This is especially the case on the boundary of China and Tibet and the neighbouring mountains—the Eastern Nan-shan and others. In winter and late autumn, however, the weather is clear and the winds feeble, and then the daily range is large, notwithstanding the low meridian altitude of the sun.   The extensive highlands of North-Eastern Tibet have elsewhere also a larger daily range in winter than in summer.

A considerable part of Central Asia is so high that the temperatures reduced to sea-level differ largely from the actual.   It is impossible to calculate the elevation by reference to the temperature on the coast, where the climatic conditions are so different, and therefore the discovery of the low basin of Lukchun, in the very heart of the continent, is of the utmost importance, and the observations recorded there under the auspices of the Russian Geographical Society are very valuable, not only to the science of meteorology, but also as furnishing a basis for barometrical observations of altitude.

The difference of elevation is so great, even outside the mountain ranges, that it frequently has a more marked effect on the temperature than the latitude.   There is no doubt that, as regards the yearly mean and that of the seven warmer months from April to October, the high plains of Tibet are the coldest part of Central Asia.   In the middle of summer this country has a lower temperature than Siberia on the Arctic Circle, and only the coastal region within seventy to a hundred and thirty miles from the sea, which is exposed to the influence of the ice, is colder.

As regards the winter temperature the question is more intricate.   If the most northern part of Mongolia, around the sources of the Yenesei and its affluents and the affluents of the Selenga, be excluded from Central Asia, as being Siberian in its natural characteristics, there are three districts where, according to the observations recorded, the winters are very severe.   These are the uplands of Tibet, the uplands among the ramifications of the Kuen Lun system, south of the Altyn-tagh (the valley of Zaisan-Saitu, for instance), and the centre of the northern part of Mongolia, a little south of Urga.

The warmest districts of Central Asia are undoubtedly in the less elevated parts of Eastern Turkestan, but their exact position is still unknown.   If the whole year and the warmer months be taken, the choice rests between the more southern, but somewhat higher, desert of Takla-Makan and the hollow of Lukchun that descends to below sea-level.   In winter, and even in all the five months from November to March, Lukchun is certainly the colder.

We turn next to atmospheric pressure.   This is (reduced to sea-level) high in winter and comparatively low in summer—in other words, exhibits a continental or thermal character.   It might be expected that, in accordance with what is known of the phenomenon in other parts of Asia, the highest winter pressure would be found in the eastern parts of Central Asia, but that the gradients from east to west would be very small—in other words, that south-east or east winds would prevail in Eastern Turkestan ; but such is not the case.   It would also seem that the pressure in the north would be greater than in the south, but in reality the gradients are so slight that north-easterly winds do not prevail anywhere in the middle of winter.   In general these winds occur only in the central and eastern uplands of Eastern Turkestan at the end of winter and in spring.

The winter north-west and west winds of China fully confirm the opinion that at this season the pressure is higher in Central Asia than on the shores of the Pacific, while the summer south-west and south winds of China and Eastern Mongolia show that then the distribution of pressure has been reversed. Where precisely the centre of low pressure is to be found is not known. Professor Voiekof at one time supposed that it was in the neighbourhood of Lob-nor, but certain data furnished by Pievtsof seem to indicate that it may be in the desert of Takla-Makan. A transference of air between the inner part of Central Asia and the coast of the Pacific Ocean possibly takes place, and the prevailing winds serve as a guide in the questions concerning the atmospheric pressure.

As to India, it is separated from Central Asia by such lofty mountains that there can be no interchange of air in the lower strata, but the winds blowing over the crests and passes of the Himalayas are determined by the distribution of pressure at greater heights where it may, or rather must, be different from below. The winds which constantly blow from the south through the passes clearly show that at these heights the pressure is greater over India than over Central Asia, and this circumstance is a consequence of the greater rarity of the air over India, in consequence of the high temperature, and the abundance of vapour. In winter the former factor predominates, the difference of temperature between the two countries being strikingly great, while in summer the air of India is laden with a larger amount of moisture. Hence the prevalence of westerly winds in Tibet both in summer and winter.

In Central Asia there is in general a marked freshening of the wind in the middle of the day, with calms or very gentle breezes at night and in the early morning; in the afternoon there is frequently a strong wind, or even a gale, a characteristic phenomenon of continental climates in the middle latitudes. These winds are most piercing in spring, the season *par excellence* of strong winds and storms. Then are most commonly observed the dust-storms, which, it seems, are nowhere so frequent and violent as in Central Asia; they also prevail in Northern China, Northern India, the steppes of the Aral and Caspian Seas, and, crossing over into Europe, invade South Russia. Indeed, stormy springs, with sudden changes of temperature and dust-storms, and comparatively mild, clear autumns, are characteristic not of Central Asia only, but are found far to the west of it.

It is singular that in the western parts of Central Asia, outside the domain of the Asiatic monsoons, a rapid increase of cloudiness takes place from the middle of winter to the middle or end of spring. And this is the more remarkable because at the same time the humidity of the air and soil diminishes and dust-storms prevail. In Northern and Central Europe, in higher latitudes in general near the sea, and in countries having a climate intermediate to the continental and oceanic, the cloudy season is very different, the cloudiness being greatest in late autumn or in winter and rapidly diminishing in the spring. The minimum cloudiness occurs in Central Asia in November and December, and extends beyond its limits almost over the whole monsoon region of Southern and Eastern Asia. In these parts also the cloudiness begins to increase in

January, but not equally, for in Northern India it attains a second maximum in February, and falls again to April.   But, as in Central Asia, the cloudiness is greater in April than in November and December, though in these months the humidity of the air is comparatively great and very small in April, and, while the spring is the season of winds and dust-storms, November and December are the calmest months in the year.

Over all the vast expanse of Central Asia few rainfall records have been kept, and almost all that is known of the precipitation of the region is from the observations of travellers taken on the way and their remarks about the character of the vegetation, absence or presence of lakes and rivers, etc.   From these it may be gathered (1) that the greater part of Central Asia receives little rain, and that consequently the vegetation is poor and dust rises into the air ; (2) that some districts of no great extent receive abundant rains in summer and even in the six months from May to October, but that the winter there is dry ; and (3) that, except in these well-watered districts, more rain falls in summer than in late autumn and winter.   The last circumstance sharply distinguishes Central Asia, even its western parts beyond the range of the monsoons, from Turkestan adjoining it on the west, where, on the contrary, the summer is the dry season and rain falls in the cold part of the year.   And this is the case in the western part of Asia in the Mediterranean basin, where the rainless summer is more and more prolonged towards the south and the more rain falls in the winter months.   Summer rains are, indeed, characteristic of a continental climate, and occur in Asia, Africa, North and South America and Australia, while dry summers and rain in the colder part of the year are the rule in oceanic climates of the lower middle latitudes, but only in the Old World are they found far inland, and even here they do not extend to the very heart of the continent.

PATAGONIA AND CHILE[1]: THEIR OROGRAPHY AND GEOLOGY CONTRASTED.

THE unequal division of the southern extremity of the South American continent by the Andes is at once apparent on the map, and it is natural to look for some explanation of this fact.   Between Santiago and Buenos Ayres the Argentine Republic is about 740 miles broad, whereas Chile in its broadest part, between the parallels of 37° and 39° S. lat., measures only 130 miles across.

As regards its orographical configuration, Chile may be divided into four parts.   The most northern, extending down to about Copiapó, is a fairly even plain, falling steeply to the sea and rising gradually to the Bolivian plateau, 12,000 to 14,000 feet above sea-level, over which the traveller may wander for days almost at the same level before reaching

---

[1] From an article by Dr. R. A. Philippi, Director of the National Museum in Santiago, in the *Zeitschrift der Gesell. für Erdkunde zu Berlin*, Bd. xxxi. No. 1.

the descent to the La Plata basin.   Here and there is a terrace escarpment, and in some places mountains of considerable size rise above
the plateau, but there is no continuous range, and the numerous
volcanoes, one of which, the Llullailaco, surpasses Chimborazo in height
by some 100 feet, stand completely isolated from one another.   No
eastern and western cordilleras exist here.   Of this the railway is a
proof, which runs from Antofagasta to Oruro in Bolivia, for in the whole
of its great length there is not a single tunnel, not a single large cutting,
no great embankments, and no zigzags.

The second division of Chile, from Copiapó to Santiago, is marked
by a number of transverse spurs running from the cordilleras to the
ocean, and separating the river valleys from one another.   In travelling
from north to south, these spurs must be crossed by passes which are
often very steep.

In the third section, the mountainous coast-lands are separated from
the cordilleras by a longitudinal valley extending uninterruptedly from
the transverse ridge of Chacabuco, which separates the valley of the
Aconcagua from the basin of the Maipu, to Puerto Montt (officially
styled Melipulli) on the bay of Reloncavi, and, continued under the sea,
cuts off the island of Chiloe from the mainland.   It extends from lat.
33° S. to 41° 30', and gradually sinks from an elevation of about 2300
feet near Chacabuco to the sea-level at the bay of Reloncavi.   It is this
part of Chile that Dr. Philippi especially deals with in his comparison
of Chile and Patagonia.

South of Puerto Montt the cordilleras advance close to the sea, and
the country consists of nothing but the slopes of the mountains and a
strip of country lying to the east between the foot of the cordilleras and
the watershed between the two oceans, which often runs at a considerable distance from the main chain.   It is this strip which has been the
subject of such prolonged dispute between Chile and the Argentine
Republic.   The old boundary treaty defined the boundary as the watershed formed by the high cordilleras.   Consequently the Argentines,
when it was discovered that the watershed did not coincide with the
line of greatest elevation, claimed the cordilleras as their frontier, while
the Chilians claimed all the country west of the watershed.   This
question is now likely to be settled amicably.

The long coast of Chile rises nearly everywhere steeply from the sea,
and the land extending at the back inland to the longitudinal valley
already mentioned is called the Coast Cordillera.   It is not, however, a
range, but rather a much diversified tableland which at some points
rises to 3300 feet, but is on the average much lower.   It consists of
archæan rocks—granite and mica-schists.   In parts it is skirted by
Tertiary deposits, which at several points extend far inland.   Older
sedimentary rocks do not occur in the coast-region, except a narrow strip
of chalk, often only a few steps in width, skirting the shore.   Chiloe
and the smaller islands to the south are a continuation of the Coast
Cordillera, for they have the same orographical and geognostic characteristics.

The longitudinal valley was originally a huge cleft which was gradually filled up by detritus washed down from the Andes and the Coast Cordillera. This layer is fully 330 feet deep; no well has yet been sunk through it. Between the rivers Itata and Renaico, that is, between 36° 50′ and 37° 50′ S. lat., the valley is covered with a deep layer of sand. A microscopic examination by Dr. Pöhlmann has proved that it is of volcanic origin; perhaps it is derived from the volcano Antuco.

The chief rivers of Chile all run from the Andes straight to the sea through openings in the Coast Cordillera. It is, however, very singular that their principal tributaries all flow from south to north, in spite of the general fall southwards of the great longitudinal valley. This curious fact was first noticed by Dr. Peter Möller.

The chain of the Andes is composed not only of volcanic products, but also of upheaved strata of the older Cretaceous and Jurassic formations. Older rocks also crop out here and there as, for instance, in the Cordillera of Talca, where syenite was observed by Pissis. Ammonites and other fossils belonging to the Jurassic formation are found at a height of 13,800 feet, where the road from Santiago to Mendoza crosses the crest of the range, and the valley of the Tinguiririca river is full of Cretaceous fossils.

The Andes continue straight to Cape Horn, forming a labyrinth of fiords, almost all of which start from glaciers, islands, and peninsulas often joined to the mainland by very narrow necks. This configuration resembles that of the Norwegian coast, and the coast of North America north of lat. 50°, and the resemblance, Dr. Philippi believes, cannot be the result of chance, but must have some connection with the rainfall.

A striking characteristic in the third section of the Chilian territory is the existence of large lakes at the western foot of the Andes. There are the lakes of Antuco (Laguna de la Laja), Villarica, Riñihue and Lacar, the Ranco lake, and the Puyehue, Llauquihue and Llanquihue, to which succeed on the eastern side of the Andes the Todos los Santos and Nahuelhuapi lakes. The Ranco and the Llanquihue lakes are much larger than the Lake of Constance.

The points to which Dr. Philippi directs attention are, (1) that nowhere in the Andes of the second and third sections of Chile have any traces been found of the older Secondary system, or even of Palæozoic rocks, the Carboniferous series, for instance; (2) nor has there been found anywhere in the Andes upheavals of Tertiary formation; (3) the volcanoes—at any rate from Talca southwards—do not lie on the crest of the range, but on its western side—sometimes they are isolated, rising from a low plain, as the Antuco, Villarica, Osorno, and others; (4) the watershed between the two oceans in the southern part of Chile lies to the east of the Cordillera and is very low, being in the neighbourhood of the lake of Villarica, for instance, only 1600 feet above sea-level. Many rivers rise to the east of the Cordillera, and for a time run north or south until they find an opening in the range through which they can reach the ocean. They almost always form a number of rapids, so that it is very difficult to reach their upper course, even by land, from the Chilian side. From the Patagonian side it is a very different kind of journey to this country. The traveller gradually ascends with his ox-wagon, and often crosses the watershed without being aware of it.

The orographical configuration of Patagonia is very different. The whole wide country is a plain, gently descending in terraces to the coast, and intersected by narrow river valleys running parallel to one another and perpendicular to the coast. The geognostic formation is equally uniform; it is everywhere an Eocene clay. Florentino Ameghino divides this formation into two; the lower, laid down in the sea and characterised by the presence of *Ostrea Bourgeoisii*, he calls the *Formation patagonienne classique*. In other places, however, he speaks of *Ostrea Patagonica* as occurring in it. The upper strata he calls the *Formation Santa Cruzienne*; it is due to the action of fresh water and wind (*Löss*), and contains enormous quantities of bones of extinct mammals never found elsewhere—a quite peculiar local fauna. It may, perhaps, be inferred that Patagonia was in the Tertiary period completely cut off from the land to the north. The one deposit followed immediately after the other, and therefore in the lower strata of the Santa Cruzian formation both *Ostrea Bourgeoisii* and the remains of mammals are found. This may be explained by the assumption that at such places lay a sea-shore to which the remains of the land fauna were swept down by rivers. Only at one spot, on Mt. Observation north of the Rio Coyle, could Carlo Ameghino observe the Santa Cruzian formation undoubtedly overlying the Patagonian. The latter may be seen near the coast of San Julian lying immediately on reddish sand belonging to the Cretaceous system and containing numerous remains of Deinosaurus and fossilised wood. According to Darwin, the Tertiary foundation rests on plutonic and metamorphic rocks.[1]

Darwin says in his *Geological Observations* that the plains or large terraces have been built up by the denudation of the old Patagonian Tertiary strata and the deposition of a quantity of well-rounded gravel on their surface, which at the coast varies in thickness from ten to thirty-five feet and grows deeper inland. This detritus is often covered with a thin layer of sandy soil. The *débris* must have been spread over the whole surface by streams descending from the Andes. Unfortunately, nothing has been said of the petrographical structure of the drift, whence some information might be obtained of the nature of the rocks on the eastern flanks of the old Andes, whether they were purely plutonic or formed of upheaved strata of the Cretaceous system. The sudden appearance of these water-borne pebbles needs explanation, as well as their absence during the long period when the Patagonian and Santa Cruzian formations were being laid down.

Darwin states further that the plains rise gently, scarcely perceptibly to the eye, from one talus to the foot of the next. Between Santa Cruz and Puerto Deseado, a distance of 150 miles, there are at least seven such steps, one above another, and on the three lower, with

---

[1] Dr. Philippi asserts that Darwin mentions no definite spot where the Tertiary formation rests on porphyry. He seems not to have noticed the allusion to claystone porphyry near Puerto Deseado.—Ed.

elevations of 100, 250 and 300 feet, shells of existing species are extremely abundant, either scattered about the surface of the ground or imbedded in the thin upper layer of sandy earth. He adds that undoubtedly the upper terraces up to a height of 950 feet near San Julian and 1200, probably, along St. George's Bay, were formed by the same agency, whatever that may have been, and that the raised deposits of shells seem to point to the sea as the agent which moulded the surface during the stages of upheaval.

It is, therefore, evident that at the beginning of the Eocene age the whole of Patagonia was a sea bottom, which was afterwards upheaved and constituted the *Formation patagonienne* of F. Ameghino. After it became dry land, birds and five hundred species of various mammals, monkeys among them, lived upon it in considerable numbers. How this singular fauna, confined within a small area, was developed all at once in the Cretaceous period is hard to say. There the animals certainly were, and in course of hundreds, or rather thousands, of years their bones were covered by the silt of rivers and wind-borne dust, and a huge layer was formed above them—the Santa Cruzian. During this period there must have been a more luxuriant vegetation to feed the herbivorous fauna than now, but it must have been a steppe vegetation, like the present, without large trees; or coal seams, however thin, and leaf impressions, etc., would certainly occur. On the other hand, the presence of monkeys may indicate the existence of woods locally, though, of course, there are monkeys which do not live in forests. It is also hard to explain the absence of land and fresh-water snails in the immense pampas formation.

Later occurred a fearful catastrophe. The whole of Patagonia sank below the waves and all the land animals were drowned. Not a single one of the five hundred species of mammals, not a single bird had time to escape to the lofty mountains in the west, and where they had swarmed now lived molluscs of the present day. But their time was short, or a layer of shell conglomerate would have been formed. However that may be, it seems certain that in a time geologically very recent, when these molluscs already existed, the Patagonia of the present day rose out of the sea. In the Eocene period, then, Patagonia was raised twice above the surface of the sea and sank twice beneath the waters.

Of the thickness of the Tertiary strata of the Patagonian pampas, Dr. Philippi has no definite information,[1] but it must be very considerable. It is undoubtedly the product of the weathering of older rocks which can have been situated only in the west, that is, on the Andes.[2] The breadth of Patagonia diminishes from north to south, and therefore the ancient Andes must have supplied more material

---

[1] Dr. Bodenbender observed a depth of 160 to 200 feet in a well near Cordoba, where the bottom was not reached. This is probably a maximum, as it is in a deep depression.—*Petermanns Mitt.*, 1893, p. 233.

[2] At Nuevo Golfo, Puerto Deseado, and Santa Cruz, Darwin found the strata containing *Ostrea Patagonica* associated with a pumiceous mudstone and overlaid in the valley of the Santa Cruz by a basaltic lava of great thickness. These volcanic products must have come, he believes, from the now dormant southern craters of the cordilleras.—Ed.

in the north, whether because the weathering of the strata was less
rapid with the decrease of temperature, or that the exposed surface
was greater in the north, just as now the height of the mountains
lessens towards the south.

On the western side of the cordilleras there is nothing similar to
the formation of the Patagonian pampas. The surface of the old Andes
must certainly have shed off denudation products during the thousands
of years of the Eocene period, when the pampas of Patagonia were
formed. The detritus was certainly washed down, but it had to fill
up the deep-sea channel between the Andes and the Coast Cordillera before it could contribute to the building up of dry land. And
owing to the steep declivity the fragments set free by superficial
fissuring would roll down as boulders to the sea, losing only their
corners and edges. But still it is singular that no Tertiary fossils are
found at the foot of the Andes, for the old sea-shore must have had
sheltered recesses where organic life might have been developed. The
detritus from the granitic islands has furnished most of the material
for the Tertiary deposits of the Chilian coast, which apparently correspond
in age to the older submarine and the later fresh-water deposits of
Patagonia as well as to the sub-aërial formation. They contain, in many
places, coal seams more or less thick, some of which are worked and are
of great value to the country, for the coal is good, though inferior to
that of the Carboniferous series. Near Lota and Coronel the seams
extend under the sea, and headings are driven below the water. Pieces
of fossilised wood are frequently met with, and near Coronel has been
found a clay-slate with well-preserved leaf impressions. Evidently a
rich vegetation and dense forest grew on the shore of this sea in which
the Tertiary formation of Chile was laid down. Accordingly, on the
Chilian side also upheavals and subsidences have taken place. Near
Leon there is a blue clay, rich in sea shells, interposed between seams of
coal. After the lower seam was formed it must have sunk, and in the
mud the molluscs must have lived and thrived.

No bones of land animals have been found in the Tertiary strata of
Chile, though some of them must have been formed on land. Probably
this is because the western side of the Andes was covered with dense
forest. It was in the steppe regions of South Africa, not in the thick
forests, that the numberless herds of antelopes, giraffes, and ostriches
found food. There was then in Chile no animal life to compare with
the plentiful fauna which lived contemporaneously in Patagonia and has
left the innumerable bones of the Santa Cruzian deposits. At a later
period, when the estuary of the La Plata river was formed, and when,
perhaps, Patagonia was first connected with Brazil, long after the
Tertiary fauna of Patagonia had passed away, an exceedingly rich fauna
sprang up on the pampas—mastodons, megatheriums, scelidotheriums,
mylodons, glyptodons, the singular *Macrauchenia*, with nostrils in the
middle of the head and very small nasal bones, toxodons, etc., which
spread far to the north and south. In Chile, on the other hand, Dr.
Philippi has found bones of only two old-world forms—teeth of the
Equus curvidens and bones of the *Mastodon Chilensis*, Ph., distinguished

from all other varieties by its chin. The latter must have been common in Chile, though a complete skeleton has not yet been obtained.

The Chilian volcanoes existed already in the Tertiary period. This is proved by two specimens of shale conglomerate from the neighbourhood of Navidad, near the mouth of the Rapel river, where Tertiary fossils are abundant, and by a specimen of sandstone from the Hacienda La Cueva, about twenty-three miles south-east of the river's mouth. Sections of the two former, when examined by Dr. Pöhlmann with a microscope, showed grains of quartz, mica, plagioclase, orthoclase, hornblende; secondly, fragments of augite-andesite, pumice, obsidian, and ashes; and thirdly, a cement of carbonate of lime with crystals of aragonite. No section could be cut from the sandstone, but many of the grains proved to be andesite.

A petrographic examination of the drift stones which covered the surface of the pampas above the Santa Cruzian formation, and a microscopic examination of the clay and harder rocks, would show whether the mountains which then bounded Patagonia on the west differed from the Andes of the present day, and explain how it happened that denudation yielded such different products on the eastern and western sides of the continent.

Dr. Carl Ochsenius, through whom Dr. Philippi's article has been published, states that a sea could not have existed in the great longitudinal valley in Tertiary times, for near Coronel andesite pebbles, as large as the fist, are contained in the lowest coal-bearing strata. These rounded pebbles could only have reached the sea-shore from the flanks of the Andes along torrent beds.

THE HISTORY OF GLOBES.

A SHORT account of the history and progress of globe-making, translated from the Italian of Matteo Fiorini by Dr. Günther, is noticed by Dr. Alois Bludau, in the *Geographische Zeitschrift* for April.

The earliest steps in the art are, as with many other things, involved in darkness. Naturally, no earth globes could have been constructed until the globular form of the earth was recognised. The Pythagorean school were the first to maintain that the earth was spherical, and Aristotle gave strict proofs of the correctness of this view. In the second century B.C., Krates of Mallos, a contemporary of Hipparchos, made the first terrestial globe, and set it up in Pergamon. But it was not a globe in the modern sense; probably it only showed the distribution of land and water, according to the knowledge of the time in a conventional manner. Both Strabo and Ptolemy discuss the construction of globes, the latter implying in his concluding words that he possessed one. He recommends the insertion of meridians and parallels, and would extend the usefulness of globes by the addition of a meridian circle.

There is more information about celestial globes in classic times. Apparently they were known as far back as three centuries B.C. A statue

of Atlas, preserved in Naples, bears on its shoulders a globe six and a
half feet in circumference. From the positions of the constellations
relative to the points of intersection of the Equator and Ecliptic, Heis has
referred its construction to about the year 300 B.C. Hipparchos is said to
have made a celestial globe, and Ptolemy gives copious directions for the
construction and the drawing of the constellations.

In the Middle Ages, the Arabs were the great students of geography
and astronomy. Of terrestrial globes, however, they make no mention ;
to the preparation of celestial globes they devoted more attention. Eight
of Arab manufacture are known to exist, the oldest of which was made
about 1080 in Valencia. Others are to be found in Velletri, London,
Paris, St. Petersburg, and Dresden. The one at Dresden dates from the
year 1289, is made of bronze, is 5·7 inches in diameter, and is provided
with horizontal and meridian circles.

In Christendom, the Middle Ages were a period when much even of
the knowledge of geography gained by the Ancients was lost. It is not
surprising, therefore, that no terrestrial globes of these centuries have
been preserved, and that there is no record of their construction.
Celestial globes were known, and the Venerable Bede, Notker Labeo,
and Pope Sylvester II., were well acquainted with these instruments.
The Emperor Friedrich II. and King Alfonso X. of Castile also showed
great interest in them.

The age of the great discoveries rendered necessary a more exact
knowledge of the form and extension of the earth's surface, and aroused
a desire to be able to study the surface on a solid representation. Moreover, through the invention of printing, together with wood and copper
engraving, drawings could be multiplied mechanically with accuracy,
which gave great assistance in the construction of globes. Until then, a
globe had been a work of art, which could only be reproduced by
repeating the same processes throughout. Still, at first, hand-drawn
globes were more common. Of these, the most famous is that of Martin
Behaim of Nuremberg. He was in Portuguese service, but spent much
of his time in his native town during the last ten years of the fifteenth
century. At the desire of his fellow-countrymen, Gabriel Nützel, Paul
Volckamer, and Nikolaus Groland, he constructed, in 1491-92, the
celebrated globe which is still in the possession of his descendants, and
was examined by Professor Hermann Wagner in 1893. Many copies
exist, which are, however, inaccurate. The sphere has a diameter of a
little more than twenty-one inches, and is made of pulp covered with
plaster; on this is laid the parchment inscribed with the delineation of
the earth's surface. There is no network of meridians and parallels, but
only the Polar circles and Tropics, the Equator, divided into 360 degrees,
and the Ecliptic. On the South Polar area, of which nothing was then
known, are some armorial bearings. The globe turns on an iron axis, and
bears an iron meridian and a brazen horizontal circle, added eighteen
years later. The reception this globe met with was such that the Town
Council invited Behaim to give instruction in the making of globes, and
Nuremberg soon acquired a reputation for their manufacture.

The globe of Laon, a metal sphere six and a half inches in diameter, dates from about the same time. It seems to have formed part of an orrery, and has a network only on the Northern Hemisphere. The initial meridian is drawn through Madeira, which indicates a Portuguese origin. It was found at Laon, and is now in Paris. A celestial globe now in the Lyceum at Constance was made by Johannes Stöffler, Professor of Mathematics in Tübingen.

The discoveries of Vasco da Gama, Columbus, Amerigo Vespucci, the Cabots, Magellan, and others, gave increasing practical importance to globes, and the fairly numerous globes of the early part of the sixteenth century furnish, as well as the maps, an interesting insight into the notions then entertained of the extension of the New World, and its relation to the eastern side of Asia.

Among the hand-made globes of this period may be mentioned the Lenox globe in New York, engraved on copper. "America" first appears on a globe in 1513, as the name of the newly discovered continent. In the twenties and thirties, Johannes Schöner made several globes, working first at Bamberg, and afterwards in Nuremberg. Some of these were covered with hand drawings, others with printed slips. Four are known, made in 1515, 1520, 1523, and 1533 respectively; of the third and fourth, the printed slips only have been preserved. Other globes of this period are the Nancy globe (about 1530), over six inches in diameter, with the land surface gilt, and the water in blue enamel; the copper globe of De Bure, and the Eruy globe; an Italian globe, now in Paris, of about 1535, made of wood, with a plaster covering, and eight inches in diameter; and the globe of the Minorite friar, Franciscus of Mechlin, dating from the same period.

With the introduction of the arts by which they were enabled to multiply copies of their works, the globe-makers found themselves confronted by a problem, the difficulties of which were only fully understood by slow degrees, and which is still discussed in the present day. As cartographers have had to contend with the difficulty of representing a curved surface on flat paper, so globe-makers have had to seek the best means of adjusting their drawings to the convex surface of their globes, for the strips of paper when applied to the globe must crease or split. To avoid this distortion entirely is, of course, impossible, and all that can be done is to reduce the imperfection as far as possible. The method of dividing the paper into rings bounded by parallels of latitude was not found to answer, and has been little used. Gores extending from Pole to Pole were used from the first, and since then the aim of globe-makers has been to find the most suitable size and form for these gores, so that they may meet closely when laid on the globe and not overlap. As the paper cannot adapt itself perfectly to the convex surface, it must evidently have a greater area. The stretching of the paper during the printing and by the moisture of the paste, and its subsequent contraction on drying, have to be guarded against. Originally the gores were bounded, when laid out flat, by circular arcs, but it was gradually discovered that curves of sines were better forms. Peculiar difficulties attend the proper adjustment of the gores at the Poles, and very soon the device was adopted of covering the space round the Poles with circular discs, instead of carrying the gores quite up to the axis. With all the resources of mechanical improvements, aided by experience and theory, the globes of the present day are by no means perfect.

Of globes thus produced there are no specimens extant from the earliest period. Some are known only from detached remarks in geographical authors, while of others the segments, or some of them, have been preserved. The well-known cosmographer, Martin Waldseemüller, seems to have possessed such a globe, and one constructed by Gemma Frisius (1508-15) is known only from an illustration. Nordenskiöld discovered at Rome, in a copy of the edition of Ptolemy of 1525, the twelve segments printed from wood for a globe of four inches diameter. About the same time Kaspar Vopell of Cologne (1511-61) and Johann Honter, a Transylvanian Saxon, actively pursued the art. Peter Apian appears to have done nothing in this direction, but his son Philipp in 1575-76 made a celestial globe and a terrestrial globe on which the demarcation line of Pope Alexander v. is shown as the line of no magnetic deviation.

Gerhard Mercator was not only the first cartographer of the latter half of the sixteenth century, but was also among the first globe-makers. From the gores still remaining, he appears to have made eight globes. Finished copies are preserved in Vienna and Weimar, in Italy, the Tyrol, and Styria. The most noted is one completed in 1541 and dedicated to the Cardinal Granvella.

In the seventeenth century the great endeavour was to make globes of large dimensions. They were often provided with rotating mechanism. To these belonged, though dating from the preceding century, the globe of the astronomer Tycho Brahe, six feet in diameter, which was burned in the palace of Copenhagen in 1738. W. J. Blaeu and the Italian Marco Vincenzo Coronelli were among the noted globe-makers of this century. The latter made for Louis XIV. in 1683 two huge globes, each fifteen feet in diameter, and rotated by machinery. Another large globe (the Gottorp) was made by Andreas Busch between 1656 and 1674 for the Duke of Holstein. It was eleven feet in diameter, and was a terrestrial globe on the outside and a celestial globe inside.

Most of the ordinary globes of the sixteenth and seventeenth centuries were constructed according to the precepts of Glareanus (1488-1551). Then Antonio Floriani of Udine introduced new methods in the preparation of the gores, in which he was followed by Varenius. While Glareanus had given his gores a circular edge, curves of sines now came into use. The theoretical study of these questions was continued in the eighteenth century, particularly by Kästner and Tobias Mayer, while Gerhard, Valck, Moll, and Senex, and lastly the French cartographer, Delisle, were distinguished in practical construction. No important improvements as regards the construction and the form of the gores have been made in the present century. Möllinger and, above all, Fiorini have studied these questions more thoroughly than any one else, and the results obtained by the latter will soon find a practical application. The technical processes have been considerably improved, and embossed and other globes unknown in preceding centuries are manufactured.

PROCEEDINGS OF THE ROYAL SCOTTISH GEOGRAPHICAL SOCIETY.

LECTURES DELIVERED IN MAY.

On May 7th Major C. Barter delivered an address in Edinburgh entitled "Notes on Ashanti." Colonel Cadell, V.C., presided.

The Rev. W. Campbell, of Tai-nan-fu, lectured in Edinburgh on Formosa on May 30th. The Chair was taken by Dr. George Smith, C.I.E.

ELECTION OF MEMBERS.

At a meeting of Council held on May 28, the following gentlemen were elected members of the Society :—Robert Allan ; A. Cameron Corbet, M.P. ; J. A. T. Ross Cormack ; B. Bentham Dickinson, M.A. ; David Reid.

GEOGRAPHICAL NOTES.

By THE ACTING EDITOR.

EUROPE.

Some Place-Names between Loch Lomond and Loch Long.—At a recent meeting of the Helensburgh Naturalist and Antiquarian Society, Mr. Donald Maclean, postmaster, read a paper on the probable derivation of certain names in this district. The etymology of those he dealt with, except in the case of *kil*, representing the Latin *cella*, was entirely Gaelic, but many had been so corrupted that the original meaning was hard to find. After referring to a few cases of mistaken interpretation, Mr. Maclean began his explanations with the name Camis-Eskan, the first part of which means bay, as in Camstradden, "bay" whence roads "branch off" (*Rathadean*=branch), and Camsail, "beautiful bay." Eskan is probably from *Iasgairean*, "fishermen." From Benbuie *dh* has been dropped, *buidh* meaning "yellow." Bun-a-chara, as the name ought to be spelt, where the river almost doubles back, means the "foot of the turn," and the old fort on the top of the adjoining height may be the "Dun-fine" or tribal stronghold, either of the Colquhouns or their predecessors. Sith'mor, a contraction of "Sithean-mor," is a large fairy mound, and the low hill here may be the Cnoc Elachan which appears in the arms of the Colquhouns, *elachan*, or more correctly *dealaichan*, meaning a partition, and referring possibly to the small island in the Fruin below Daligan, whence this place also derives its name. Auchentulloch-na-moan is pure Gaelic, and spelt almost correctly ; it signifies an upland field of moss or peat-bogs. The "natre" of the other Auchentulloch has suffered much corruption ; it may be derived from *na threib*="under tillage." Blairnyle appears to be *Blar na feill*, a market stance, and Duchlege, a common name in the district, is from *achlais,* a hollow or armpit, referring to the angle made by the two roads that meet here. Luss may be from *lois*, "groin," or *leas,* "thigh," terms not inapplicable to the outline of the shore and the hill contours, and Edentaggart—*Aodann t-sagart*—means "priest's field," and preserves the memory of the holy men who officiated at the chapel of St. Michael, which formerly stood a short distance to the east of Glenmolachan farmhouse.

Several names near Tarbet begin with *stuc,* a precipice or cliff, such as

Stuckendepert, where sacrifices were made (*iobairt*), and Stukgown (*gobhain*, a smith), though in the latter and other cases, *stuck* may be akin to the German *Stück*. Tyvechtan is the watchman's house (*Tighvechtan*) near Stronafyne (*Sron a finneadh*), where the clan usually mustered when the fiery cross was sent round. Portincaple near Finnart, and Portachaple, opposite Inversnaid, both mean a landing-place, horses having been ferried across here. Arrochar, Mr. Maclean interpreted as *Dara car*, "second turn," referring to the windings of the road.

Turning to Glen Fruin, Mr. Maclean remarked that the principal point of interest was the derivation of the name of the glen itself. Dr. Murray, in his *Old Cardross*, mentioned that about the middle of the thirteenth century the third Earl of Lennox granted land in "Gleanfreone" to Donald Macynel, which disposed of the derivation of the name from the encounter of the clans Gregor and Colquhoun in 1603. *Fraon*, "shelter in a hill"; *freumhean*, "roots"; and *fodh sron*, "below the ridge," or nose, are possible sources, the last satisfying all the topographical features. Dairland (Durland) seems to be *Dara linne*, the second pool, the first being below Chapel schoolhouse. In Inverlauren the latter part of the name is a corruption of *loorean*, the cloven-hoof-shaped bends formed by the river at this point.

On the Gareloch we have Balernock (Ballernick) probably Baile-air-cnoc, "a dwelling-place on a knoll," while *Sithean-dun* "fortified fairy mound," has been converted into Shandon. Rahane, on the other side the loch, is composed of *rath*, a circular entrenchment, and *hane*, which stands for *teine*, "fire," and indicates that a beacon was lighted on the rath. Analogous are Ardnahane, "beacon height," and Ardentinny, on the Cowal shore.

Clynder is probably *Cille an tir*, the church on the shore, or the church of the district (or from *an tir chli*, the left shore), and Kilcreggan, the church beside the rocks, while Rosneath is *Ros 'n fhe*, the point of the marsh, which described it well in former times. For Peaton, formerly written Peitoun, Mr. Bain of Cove suggests a derivation from *bidean*, a point or pinnacle. Blarnachtra (Blarnachra), he says, is a cultivated plain or dale ; Craigrownie, a modern compound of "rowan," Cursnoch (*Coire's cnoc*) corrie on the brow or side of a hill, while Letter is "half," being one of the two farms on the Peaton estate. Mr. Turner of Clynder interprets Stroul to mean water trickling over a rock, and Barremman may be from either *Barr meadhon*, middle point, or *Barr na moine*, the edge of the moss, or a point where peats were shipped.—*Helensburgh and Gareloch Times*, May 6th and 13th.

Sunshine and Wind in Edinburgh.—In the *Journal of the Scottish Meteorological Society*, Nos. xi. and xii., is a short paper by Mr. R. C. Mossman on the duration of sunshine in relation to winds from various quarters. A table is drawn up from observations made in Edinburgh during five years from 1890. The sunshine in the four seasons is recorded, in percentages of the possible, under the eight principal points of the compass, and the annual means are obtained by dividing the gross totals by the number of observations. In the annual means the north-west wind appears as the most sunny, after which follow the west and north, while the most sunless days are those when the wind blows from an easterly quarter, north-east to south. During spring and summer anticyclonic systems to the north predominate, and the sky with the easterly winds is then much clearer than in autumn and winter. Westerly winds do not exhibit nearly so large a seasonal drop.

Another table shows the percentage of the days when each wind blew on which no sunshine was recorded. Not a single calm day during the five summers was sunless, while in winter half the days without wind were also without sunshine. During winter nearly all the sunshine is registered with westerly winds and a low barometer, and if pressure be high, and there be a calm or easterly wind, cloudy weather predominates. On the yearly average one day out of ten on which the north-west wind blows is sunless, while the frequency of cloudy days with south-east winds is four times as great.

Mr. Mossman has contributed to the same journal papers on the Diurnal Range of Temperature Variability at the Summit and Base of Ben Nevis, Lady Franklin Bay, and Hongkong, on the Number of Auroras observed over the north-east of Scotland from 1773 to 1894, and on the Frost of 1895 in Scotland.

Lake Onega.—During the past twenty years a survey of this lake has been executed by the Russian Government, of which an account is given by M. F. K. Drizhenko in the *Izvestiya* of the Russian Geographical Society. The lake lies in the Government of Olonets, which contains as many as 2000 separate lakes, covering, together with the swamps and rivers, nineteen per cent. of the surface. The longest axis of the lake, from the Black Sands on the south to the river Kumsa in the north, measures 140 miles, and its greatest breadth, from Logmozero to the parish of Pudosh, fifty-six miles ; its height above sea-level is about 128 feet. Its southern half has the form of a broad open expanse of water, while the northern portion branches off into a number of long narrow bays running north-west, or more exactly north, 32°-35° west, which are crowded with islands of all sizes, for the most part, lofty, stony, and covered with woods, firs predominating. The shores of the lake are mostly high on the north and west, are covered with dense forest, and are formed of ranges running parallel to the direction of the bays. The south shore of the lake is low and swampy, especially near the river Vytegra. The eastern shore in its southern and middle parts is sandy, with promontories of reddish granite, while farther north it is high and stony.

Round the lake, and on the rivers of its basin, are situated nearly all the centres of population of the Olonets Government. In the absence of railways and satisfactory highroads, the lake has always afforded a convenient and cheap means of communication. Its importance, strategically as well as commercially, was long ago recognised. In 1702 Peter the Great caused two frigates built in Archangel to be brought from the Gulf of Onega, partly along the rivers and the lakes Onega and Ladoga, and partly by haulage, to St. Petersburg, where they were added to the fleet of galleys which, appearing in the rear of the Swedes, decided the issue of the war. From this time the question of an uninterrupted waterway between St. Petersburg and Archangel was often discussed, and even now has not been forgotten, though nothing definite has been the result, and, in spite of the import-ance of Lake Onega, with its area of 4200 square miles, for navigation, its hydrography was hardly investigated at all till a quarter of a century ago.

In 1870 Adjutant-General Posiet directed attention to the want of a chart of the lake, and the first rough survey was made in that year under his directions. The inhabitants of the neighbourhood being thus awakened to the advantages to be derived from a thorough survey, a memorial was forwarded to the Ministry of Marine, with the result that in 1874 a systematic survey was commenced, which was completed in the autumn of 1894. Before this, what was known of the hydrographic features of the lake was largely conjectural. In a map accompanying Lieutenant Drizhenko's paper are shown the outlines of the lake as previously drawn and as ascertained by the survey, and the discrepancies are seen to be in some parts very considerable. The lake was also said by the fishermen to be very deep, sinking to 1400 feet, whereas the bathymetrical chart shows that in the middle of the sea is a hollow running parallel to the bays, which gradually sinks northwards to a maximum depth of 476 feet. Another hollow lies at the end of the most northern bay, with a maximum depth of 315 feet.

Lieutenant Drizhenko describes the methods employed in the survey of the lake, which, during the twenty-one years it was continued, was directed by several different officers, and lastly by himself. Among the points he notices is the local attraction, which was found to cause an error of 2″, or 196 feet, in the distance of points fixed astronomically. Levelling operations were also executed, both to connect the lake with the White and Baltic seas, and to ascertain the mean level of the surface, and it was found that the fluctuation of level deduced from the means for the six summers these operations were carried on was two feet, while the extreme range was six feet. Observations of the magnetic elements were also made, but it does not appear that any investigations were instituted into the temperature of the water. The final result of the work will be an atlas of thirty-four sheets, some of which are already finished, and a sailing directory.

In conclusion, Lieutenant Drizhenko refers to the economical products of the Onega basin. Its mineral deposits are very rich ; its iron was exploited in the time of the Great Novgorod, and Peter the Great employed the private foundries to cast cannon and other war material. At the present time the Russian Admiralty has four foundries in the Olonets mining district. Copper also is found in several places. Marble of many varieties is worked, and has been· used in the building of churches and palaces, and the tomb of Napoleon in the Invalides at Paris is of porphyry from the quarry of Shoksha. Then, again, the forests contain abundance of timber suitable for building ; the waters swarm with excellent fish, and wild fowl and fur-bearing animals supply articles for export. Bread, groceries, and piece goods are the imports, while berries, mushrooms, and dairy products are consumed locally.

The Molluscs of the Mediterranean.—The great uniformity of the molluscan fauna in the deep waters of the Mediterranean, from about 220 fathoms downwards, announced by Fischer, and proved by the results of the *Travailleur* expedition, has been confirmed by the dredgings of the *Pola* in 1894. Professor F. Brauer stated further, at a meeting of the Akademie der Wissenschaften in Vienna, that the specimens obtained supported Dr. N. Marenzeller's view that a distinct deep-sea fauna does not exist in the Mediterranean. Many of the deep-sea molluscs of this basin are identical with those of the Atlantic and North Atlantic species, and are also found in the Tertiary formations of Sicily and Italy. Their migration into the Mediterranean must have taken place when there was freer communication between this sea and the Atlantic. One hundred and twenty species of gastropods, scaphopods, and lamellibranchs were dredged up by Dr. R. Sturany. More were found in shallow waters than in deep, for, the fauna being so uniform, there is a decrease in the number of species towards the bottom. The deepest spot where mollusc shells were found was 1320 fathoms, north of Alexandria. Here nine varieties were dredged up, of which one had only been known before from North Australia and South Africa.—*Globus*, No. 18.

ASIA.

The Ob and the Yenesei.—The leader of the expedition for the exploration of the coasts of the Arctic Ocean, M. A. J. Vilkitski, made a report in February last to the Russ. Geogr. Soc. on the work accomplished in 1895. There are two passages out of the Yenesei into the ocean, on either side of the Siberiakof island. In 1894 the eastern passage had been examined, when it was found that there was a good navigable channel with a depth of fifty-six to seventy feet and no bar. In

1895 the expedition left Yeneseisk on June 15th, to explore the western passage. The first attempt to sail out to the sea failed because of the blocks of ice which closed the passage, and it was not till August 3rd that the expedition was able to leave Golchikha and sail past Siberiakof island to Cape Mate Sale. Here a large sandy island, which is to be called Vilkitski, keeps the ice off the coast, so that there is a convenient passage between the island and the mainland. The boats continued their way along the east shore of the Gulf of Ob, this being less known than the western, and on August 26th entered the Khamanchau Ob over a bar covered with ten feet of water. On September 30th the expedition entered Tobolsk, having accomplished in the reverse direction the voyage made by Lieut. Ovtsyn in 1737. The Gulf of Ob is considerably narrower than has hitherto been supposed, and does not run in a straight course from north to south. In 72° N. lat., the eastern shore shoots out suddenly westwards, and is forty-five miles from its position on maps; at 71° it turns back eastwards and then runs south, thirty to thirty-five away from the position indicated on maps. The first forest on the Ob river is met with three degrees farther south than on the Yenesei. Along the whole distance of six hundred miles to the mouth of the Ob proper, there is no human habitation. The first inhabited place, Obdorsk, has eight hundred inhabitants, and the first town, Beresof, five hundred. The chief occupations are fishing and hunting, the former being more productive than in the Yenesei.—*Verh. der Gesell. für Erdkunde zu Berlin*, Bd. xxiii. No. 3.

The Transcaspian Railway.—The prolongation of this line has been alluded to in previous volumes. One branch is to run from Samarkand to Tashkent, and subsequently to be connected through Vernoie with the Trans-Siberian line. Another will open up Ferghana, passing through Khojent and Marghelan, and terminating at Andijan on a tributary of the Syr Daria. The line from Samarkand to Tashkent, 205 miles, is already marked out, and the first section, from Samarkand to Jizak (sixty-two miles) has been under construction since the middle of 1895. The time allowed for its completion is two years, but it is possible that it may be finished in the present year. Three routes have been marked out between Jizak and Tashkent, of which one crossing the Syr Daria at Chinaz will probably be adopted. The Ferghana line will branch off at Jizak and run to Khojent through Ura-tiube. A branch from the Transcaspian line from Dushak to Serakhs is also contemplated.

At some future time a line is to be constructed from Tashkent to Orenburg, 1370 miles long, following the valley of the Syr Daria. A project for one from Uralsk to Bokhara has not much chance of success, the country on the route being a desert.

The port of Krasnovodsk, the future terminus of the line on the Caspian sea, is approaching completion, but the railway has to be prolonged to Krasnovodsk from Uzun-ada. An earthquake on June 27th, 1895, caused subsidences in the latter port, in consequence of which some steamers are now able to enter the harbour which could not do so before; but another earthquake would be necessary to render the harbour accessible to all vessels.—*Revue Française*, May.

Mr. Littledale in Tibet.—In February last Mr. Littledale gave an account to the Royal Geogr. Soc. of the journey he accomplished last year, accompanied by Mrs. Littledale and his nephew, Mr. W. A. L. Fletcher, across Tibet from north to south (see *Geogr. Journal* for May). To gain the borders of Tibet he took the route by the Transcaspian railway and through Kashgar and Khotan. On March 19th he reached Cherchen and marched up the Cherchen valley to the Tokus

Dawan. On April 28th a pass was crossed, and the travellers thought they were then to the south of the Akka-tagh and on the Tibetan plateau. But on climbing a ridge to the south, they saw in front of them seven extinct volcanoes, with summits covered with snow and ice, and found that the Akka-tagh lay farther to the south with a very rough country between. A passage was therefore sought in another direction, and on May 15th the Akka-tagh was crossed by a long high pass where several donkeys and horses were lost. The party was at last on the Tibetan plateau, with lakes and low mountains in front, and the Akka-tagh with fine glaciers and snow-fields behind. By angles taken from different points, Mr. Littledale measured the height of two peaks close together, which towered above the rest, and found it to be 25,340 feet. The plateau is about 2000 feet higher than the Pamirs, and has, in its northern part especially, a very small rainfall. The drainage finds its way into the lakes, which, having no outlet, are salt. Nearly the whole of this district lies above the timber-line, and it is probably possible to travel from the Pamirs nearly to the border of China proper without seeing a tree or shrub higher than an umbrella.

The party then travelled through a volcanic country where the herbage was, at that time of year, poor, and water scarce. Antelope were fairly numerous and very tame. Short ridges and peaks traversed the country from east to west, but no continuous mountain range was seen between the Akka-tagh and the Ninchen-tangla, south of Tengri-nor. Between 36° 50' and 33° 50' N. lat. numerous un-mistakable volcanoes were visible, three of them being considerably over 20,000 feet high. South of the latter limit no volcano was noticed until the Tongo was reached three months later. On June 26th the first Tibetans were seen, and then the expedition took every precaution to avoid detection, often travelling by night. They constantly crossed from one lake basin to another, the gradients becoming less and less steep. The Zilling-tso (Captain Bower's Garing Cho) was passed on the eastern side. At last the party was discovered by the Tibetans ; they were friendly, and it was evident that the common people bear no ill-will to strangers, but that all the trouble springs from Lhasa. After this Mr. Littledale pushed on towards Lhasa as fast as he could, continually watched by the Tibetans. Crossing a low pass, he came in sight of the Tengri-nor, known locally as the Nam-tso (Great Sky Lake). On the south it is fringed by the Ninchen-tangla, a succession of snow-clad peaks and glaciers, partly hidden in clouds and vapour, while high above all rises the great precipitous peak of Charemaru, 24,153 feet high, which farther on the way was seen to be a sharp ridge with a peak at either end.

In spite of the opposition of the Tibetans, Mr. Littledale and his party still marched on, passed the Ninchen-tangla by the Goring-la (19,587 feet), and arrived within forty-three miles of Lhasa. Here negotiations with the Tibetan authorities were prolonged for many days, Mr. Littledale insisting that his party should be allowed to spend two days in Lhasa, and then travel to India by the Jelap La and Sikkim, while the Deva Jung insisted that they should return by the way they came. A Tibetan trader informed Mr. Littledale that the Dalai Lama was then twenty years old, and was to come of age in November. The last two Lamas however, had died between the ages of eighteen and twenty, and so the Rajah of Lhasa, who acts as regent, had held the office for forty years.

Mr. Littledale would probably have succeeded in forcing his way to the capital, had not Mrs. Littledale become seriously ill in the encampment in the damp Goring Tangu valley, 16,600 feet high. It then became necessary to move in one direction or another, and the travellers had reluctantly to consent to march west-wards. On August, the long march of 1200 miles to Kashmir was commenced, the route chosen running to the south of Nain Sing's and Captain Bower's, and passing Rudok. Ladak was entered on October 27th, at the village of Shushal, and here Mr. Littledale brought his traverse of 1700 miles from Cherchen to a conclusion. From April 26th to October 16th the route was never lower than 15,000 feet, and for four weeks the party camped above 17,000 feet. Almost all the lakes of the country between the Tengri-nor and Ladak, Mr. Littledale states, have greatly decreased in size and are still falling. Lines of gravel show where the surface formally stood, and marks high up on the rocks, as much as 200 feet above the present water-level, were occasionally noticed. In most places there was capital grazing and the country was easy to traverse, being a great contrast to the region south of Cherchen.

AFRICA.

The Source of the Niger.—The Anglo-French Commission for the delimitation of the frontier of Sierra Leone left Freetown in December last, and reached Tembi-Kunda on January 13th. This place was found to lie much farther west than had been supposed, and its latitude proved to be 9° 5′ 20″ N. From here the Liberian frontier runs eastwards and the Anglo-French frontier northwards to the 10th parallel on the Scarcies. In Tembi-Kunda, on French territory, in a deep, wooded ravine, rises the Tembiko, the head stream of the Niger. The ravine is regarded with superstitious awe by the natives, and they believe that any one who looks into it will die within a year. North-west of Tembi-Kunda lies a range of mountains difficult of access. As the frontier is here to follow the watershed, it must take a very zigzag course ; Boria or Bogoria falls within the French territory, and Nerekore, Kulakoya, Samaindu, Yalaukolia, Yerdia, and Mussadugu belong to the British.

The highlands of the *Hinterland* traversed by the Commission are composed of granite, with summits rising 2000 feet above the plain. The climate is much healthier than that of the coast. The thermometer frequently marked only 57° F. in the early morning, and the rainfall was less than on the coast. The soil is very fertile, and offers good prospects for cultivation when communication with the coast is improved.

Lake Rikwa.—The *British Central Africa Gazette* contains an account of a journey made by Mr. Nutt from Fwambo, a station of the London Missionary Society, south-east of Tanganika, to the above lake. He skirted the eastern side of the Fipa plateau, and came to the lake not far from its southern extremity. At first he traversed a forest, interrupted here and there by the rocky beds of rivulets, and then entered a more mountainous district intersected by streams flowing to join the Saisi, the chief affluent of Lake Rikwa. The upper part of the Fipa plateau is bounded by a slope of 1300 feet. A chain of hills intervenes between this slope and the lower terrace of the plateau, and the plain commences 2900 feet below the upper surface of the plateau. Crossing the Saisi, Mr. Nutt entered the country of the Uwanda. He found abundance of game on the south of the river, while there was none on the north. Near the banks of the lake the ground is flat and regular in outline ; there are no bays, sandbanks, or rocks. The water is shallow near the banks, and apparently very few streams enter the lake. Mr. Nutt speaks of abundance of hippopotami, crocodiles, and rhinoceros, whereas Dr. Kerr Cross did not notice any (vol. vi. p. 292). The mountains approach closer to the shore on the east than on the west. On this side they end abruptly to the south in the mountains Kiumba and Ibonga, and on the north in the Mbutwe.

West Madagascar.—In the *Verh. der Gesells. für Erdkunde zu Berlin*, Bd. xxiii. No. 3, Dr. A. Voeltzkow publishes a few notes on the west of Madagascar and its inhabitants, the Sakalava. These, as stated by Dr. Hamy (see p. 42), are of African origin with an admixture of Hova, that is Malay, blood. Dr. Voeltzkow says that the name, according to the Sakalava of Menabe, considered to be the original home of the people in Madagascar, is derived from *misaka*, meaning "to cross over," and *lava*, "far," and he believes that the advents of both the Hova and Sakalava took place at about the same time, the former arriving, possibly, a little earlier. The Sakalava spread over the west coast and were powerful enough not only to subjugate the aboriginal tribes, but even to make the Hova tributary. It was only in the beginning of the present century that the latter, under Radama I., threw off the yoke of the Sakalava, took Mojanga (Majunga), and forced the Sakalava to recognise their supremacy. Another element of the population are the Makua from Mozambique, who are held in bondage. Those in Hova territory received their freedom by the treaty with Great Britain in 1877, but elsewhere slavery is still a recognised institution.

A small steamer calls at the ports along the coast—Moronzanga, Majunga, Mainterano, Morondava, and Tullear—every four weeks, conveying mails from and to Nossi Bé in connection with the vessels of the Messageries Maritimes which run between Marseilles and Mauritius. Majunga, the largest town on the coast, has a population of about 10,000, composed of Hova, Sakalava, Makua, Hindus, and Talantra, the last being the offspring of Arabs and native women. The importance of the town lies not only in the possession of the only large harbour on the coast, but also in its position at the mouth of the Betsiboka, the largest river in Madagascar, and navigable by small boats for a distance of ninety miles. Boats of ten to one hundred tons are employed in the coasting trade, and for personal conveyance and fishing *Laka fiara*, boats twenty to twenty-six feet long by two or three broad. They are made of a tree trunk, the sides being raised by planks, which are fixed on with pegs of hard wood. They are provided with outriggers and carry a sail.

The trade is in the hands of Hindus who act as agents for Europeans, and also import goods themselves. Printed calico, blue and white cottons, beads, rings, gun-flints, etc., are exchanged for the products of the country—ebony, hides, caoutchouc, tortoiseshell, wax, etc.

The west coast is almost entirely in the hands of the Sakalava, who are divided into numerous tribes. Those of the north, particularly in Nossi Bé, are Christians; in Majunga and the other coast towns they are Mohammedans, and the rest are pagans. The northern districts are partly under Hova rule, but south of Majunga begins the independent territory, though here also are a few Hova posts. South-western Madagascar has never been entered by Hova.

The Sakalava are exceedingly superstitious and are fetich-worshippers. They also use ordeals to test the guilt or innocence of a man charged with crime. Their mode of disposing of the dead is somewhat peculiar. The corpse is buried in the forest until the flesh has disappeared from the bones, and then the skeleton is deposited with others of the same family in a large boat sunk in a lake. Another custom worth noticing is that the mother is named after the child, and in some countries the father also—just as occasionally among the Arabs—Abu Bekr, for instance. The Sakalava feed on rice, maize, pease and beans, bananas, sweet potatoes, etc., to which are sometimes added beef, fish, or other meat. They do not know how to make butter or cheese, but they extract the juice of the sugarcane and distil rum.

It is well known that the fauna of the island is peculiar, many of the forms common on the mainland being absent. The Aye-aye (*Cheiromys Madagascariensis*) and the extinct *Æpyornis maximus* are noted for their singularities. The only dangerous animal is the crocodile (*Crocodilus Madagascariensis*) which kills many persons every year. The trees of the island are also well known, especially the Raphia palm and the Traveller's tree (*Ravenola*).

The climate of the west coast is comparatively healthy. The temperature is never excessively hot ; in summer the thermometer at Majunga stands at 86° to 88° F. at noon, and falls four or five degrees lower at night.

One of the chief causes of the value of Madagascar is its wealth in cattle. These are of the humped breed so common in East Africa. There are no sheep, but goats are seen in all the native settlements. 'Rice is grown in large quantities and is exported to Mauritius, the Comoros, and Zanzibar. Raphia fibre, timber, ebony, and caoutchouc are the chief exports. The mountains are rich in minerals. There is coal in the northern districts, and gold, it seems, occurs in several places.

The Congo Railway.—The engineer Goffin, writing from Matadi on February 26th, stated that during the nine months from the beginning of June 1895 more than sixty kilomètres (about twenty-seven miles), from kil. 102 to kil. 163, had been constructed, and he expected that by this June kil. 193 would be reached, and that fully ninety kil., or fifty-six miles, would have been completed within the year. The country presents less formidable obstacles than along the first part of the route, and accordingly the progress has been much more rapid than in the earlier years of the construction. The railway was commenced in 1890, and in the past twelve months nearly as great a length has been constructed as in the previous five years. A trial trip was made from Kimpesse to Matadi, a distance of ninety-eight miles, and the run was accomplished in eleven hours, from 7.15 A.M. to 6.15 P.M. The central station, which was to have been erected at Kimpesse, will be constructed at Tumba, 116 miles from Matadi, so that the journey may be made in one day, and Stanley Pool be reached in a second day. During the eight months from June 1st to January 31st fully 7000 passengers were carried as far as Tufu (kil. 80, or about fifty miles), and 18,000 tons of goods, materials of construction, tools, and provisions.—*Le Mouvement Géographique*, April 5th.

AMERICA.

The Public Lands of the United States.—*The Public Lands and their Water Supply* is the title of an essay by Mr. F. H. Newell, being an extract from the sixteenth *Annual Report of the Geological Survey*, 1894-95. There are still open to settlement nearly 630 million acres, or one-third of the total area of the United States. This land is partly utilised, a large proportion of it feeding herds of cattle and flocks of sheep. No rent is paid by the owners of the cattle and sheep, and the land is open to all. Consequently grazing has been developed to such an extent that many ranges are overstocked, and in many parts valuable grasses are threatened with extinction. The system also tends to retard settlement and the building of houses. Within the vast extent of public lands are also some scores of millions of acres of valuable forests and woodlands, which are unprotected and subject to depredation on every side and, still worse, to the ravages of fire kindled by careless or malicious persons.

The first step necessary towards the classification of these lands is the drawing of a good map. To compile it from the thousands of ledgers wherein the records of all the various parcels of land are contained would entail an enormous amount of labour. Accordingly, at the suggestion of Major Powell, the more important tracts have been marked off on the maps of each State, and the result is a sufficiently accurate general map, a copy of which accompanies Mr. Newell's essay. The table below shows the areas of the various classes of land in the territory of the United States (exclusive of Alaska).

	Sq. miles.	Acres.	Per cent.
Vacant Public Lands, . .	980,337	627,415,680	33·01
Indian Reservations, . .	131,689	84,280,960	4·43
Forest Reservations, . .	30,445	19,484,800	1·03
Lands disposed of, . .	1,107,044	708,508,160	37·27
Area of older (19) States, .	458,195	293,244,800	15·43
Area of Texas, . . .	262,290	167,865,600	8·83
Total area of United States,	2,970,000	1,900,800,000	100·00

Of the 627,415,680 acres of vacant lands, 26,822 square miles, or 17,166,080 acres, are in the eastern part of the States, and the remainder is in the West, mainly within the arid regions. The 131,689 square miles under the head of Indian Reservations includes Indian Territory and the military reservations of the West. Portions of these are every year allotted in severalty or thrown open to settlement. The Forest Reservations are a new institution, dating from 1891 ; the area given includes the Yellowstone National Park. Of the 1,107,044 square miles of lands disposed of, 671,728, or 60·68 per cent., are in the eastern part of the country, and 39·32 per cent. in the West.

The State which contains the largest extent of vacant public lands is Montana, where there are 114,000 square miles still open, while the largest proportion, 95 per cent., is found in Nevada. California comes next with 90,000 square miles, or little more than half the area of the State, and Wyoming and New Mexico follow. In Kansas less than two per cent. of the land is still in the hands of the Government, but this is largely due to waves of popular excitement, and, the settlers having found it impossible to make a living, a great part of the country is still vacant.

Another map shows the forests, woodlands—by which term is denoted tracts where the wood is small or scattered—and the irrigated areas, of the western States. It appears that of the 990 millions of acres in these States, there are in round numbers 147 millions of forest, 167 of woodlands, and 605 millions treeless. The greater part of this last vast area is covered with herbage which affords excellent grazing, though the water-supply is too small to admit of cultivation, except in so far as plants adapted to arid conditions may be introduced. The extent of country absolutely desert and worthless has been found in the progress of settlement to be relatively small. It is impossible to discriminate between lands which for a few weeks or months in the year produce a scanty growth sufficient for grazing purposes and those which are always barren. It has been estimated, however, that there are about 70,000 acres of such desert lands. Irrigation has not to any considerable extent diminished this area, and probably not more than one per cent. of the arid lands is now irrigated and tilled. Some parts, even within the arid region, are, however, cultivated each year by what is known as " dry farming." By taking advantage of local conditions of rainfall, thorough cultivation, and the selection of crops which will stand considerable drought, or mature before the heat becomes extreme, it has been found possible to raise profitable crops on such lands. Nevertheless, the farmers are not as a rule successful, the profits of a favourable year being swallowed up by continued losses at other times.

Mr. Newell also discusses the sources of water-supply and the local conditions of the lands in each of the western States.

The Olympic Country.—This peninsula is situated in the north-west of the State of Washington, west of Puget Sound, and extends for ninety miles along the south side of the Straits of Juan de Fuca, and for one hundred miles along the Pacific coast. Along its southern border there is deep water for thirty miles in Gray's harbour and along the Chehalis river, a neck only twenty-five miles across connecting the peninsula with the mainland. The interior and western parts, previously little known, have been explored by Mr. C. S. Gilman, C.E., recently deceased, and a description from his pen is published in the *National Geographic Magazine* for April.

The whole peninsula has an area of 5700 square miles, of which probably 3000 are occupied by the Olympic mountains. The main watershed of this group begins at Cape Flattery, and runs parallel with the Fuca Strait at a distance of about twelve miles as far as the meridian of Port Angeles, where it turns abruptly to the south, passing the east end of Mt. Olympus, and, after six miles, deviates to the south-east as far as Pyramid Peak, twenty miles distant ; thence it runs south-west, gradually turning to the west, for thirty miles, when it reaches Mt. Frances at the head of Quinault lake ; finally, running in a south-westerly direction for about eighteen miles, and considerably increasing in altitude, it reaches its termination. From the main line many spurs and ranges run out, some of them of greater height, and, branching off again in their turn, almost completely cover the country.

Mount Olympus, the culminating point, 8150 feet high, lies about twenty miles south of Freshwater bay, and to the south of the main crest, with which it is connected by a short sharp ridge. It is a cluster of sharp jagged peaks projecting upward through an ice-cap four miles long by two broad. At the close of summer the ice seems to be in some places as much as 500 feet thick, and the winter storms add many feet to the depth. The view from one of the many accessible peaks is remarkably fine, extending over Hood's Canal and Puget Sound and to the wooded slopes and snow-clad summits of the Cascade range beyond.

Lakes Cushman, Crescent, and Quinault are of considerable extent and depth. The western shore from Gray's harbour, twenty-five miles northward, a stretch of broad, smooth, and hard sand beach, extends to Point Grenville, beyond which to Cape Flattery bluffs 100 to 250 feet high skirt the ocean. On the mountains, above 4000 feet, the timber is very scrubby and rare, owing probably to the barrenness of the soil and the great snowfall as much as to the elevation. Below this limit there are fine woods of Alaska cedar, hemlock, spruce, fir, balsam, alder, cottonwood, and other trees. It is a great country for moss, and contains many small and delicate ferns. There are also thousands of acres of fine grazing lands, well-watered and fanned by the breezes of the ocean, and free from mosquitoes and other troublesome insects. Of course these would be habitable only from June to December. Among the rocks at the upper limit of these grass lands grow berries in profusion, which attract large numbers of black bears.

Between the mountains and the coast are about 1300 square miles of comparatively level valley lands, of which about 225,000 acres are rich bottom lands along the streams. The timber is generally heavy and not easy to clear, but it will be valuable in a few years, and a help instead of a hindrance in establishing a home. Coal is found in some places, several varieties of iron ore are scattered over the peninsula in small quantities, and traces of gold have been seen. In the valley of the Solduck is a group of hot springs of undetermined medical value, and springs heavily charged with iron and sulphur are very numerous. Alum occurs in springs on the coast, just south of the Queets river.

The climate of the western slope of the peninsula is somewhat different from that of other parts of western Washington. Owing to its proximity to the ocean, there is more wind and much less foggy weather. The rainfall is in excess of the average for the Sound country, but it comes in short, sharp showers, and leaves a greater proportion of fine weather. In summer the nights are cool, but not too cold for tomatoes and corn, which ripen perfectly, as they do not elsewhere west of the Cascades. Except on the mountains, ice and snow are seldom seen and never remain long.

Dr. Nordenskiöld in Tierra del Fuego.—In the beginning of December Dr. Nordenskiöld arrived at Paramo on the bay of San Sebastian, whence he followed the coast to the Popper river. He then made his way far into the interior by paths never before trodden by a European, and discovered a tributary of the river which he named the Candelario. This country is designated on many maps "impenetrable forest," but in reality it consists of broad valleys where travelling is easy, and woods that may be traversed even on horseback. From a hill rising 1600 feet above the surrounding country the explorer obtained an extensive view on all sides, embracing about a fourth of the Argentine territory of Tierra del Fuego. To the north and west lay forests and ranges of hills intersected by fruitful valleys. The mountain lake Solier lay to the south-east, while to the south the panorama was closed in by mountain ranges, which at one place permitted the traveller to catch a glimpse of the lake Tagnano. The whole country is, in Dr. Nordenskiöld's opinion, adapted for grazing and cultivation, and at the mouth of the Popper river, or Rio Grande, is a good harbour. Crossing over the ranges skirting the Beagle Channel, the party found about forty persons washing gold at San Sebastian and obtaining about thirty-five ounces a week. The natives living to the south of the Rio Grande probably number 2000 to 2500. The botanist of the expedition is well pleased with his collections, but the zoologist, Dr. Ohlen, has not obtained such good results.—*Deutsche Rundschau*, Jahrg. xvii. No. 8.

MISCELLANEOUS.

The French occupied Timbo, the Capital of **Futa Jallon**, on March 18th, and a Resident will control the government.—*Revue Française*, May.

In order to preserve the **kangaroo**, threatened with extinction by sportsmen, the Government of South Australia has appointed a close time from November 1st to April 30th.

The Senate at Washington has passed a bill for the laying of a telegraphic cable between the **United States and Japan** past Hawaii. The American Post Office will pay an annual subsidy not exceeding £34,000. The cost of the undertaking is estimated at £726,000.—*Le Mouvement Géographique*, May 10th.

Not long ago a volcanic eruption, which lasted two days, took place in the island **Stenosa** (or Denusa), one of the Naxos group in the Grecian Archipelago. On the north side of the island a crater was formed which threw out stones and lava. Forty persons lost their lives and a large number of cattle were destroyed.— *Deutsche Rundschau*, Jahrg. xviii. No. 8.

A **high-level observatory** has been established on the Bielašnica, one of the chief peaks on the range between Sarajevo and Mostar. This, the first high-level station in the Balkan peninsula, is situated on a "true peak," and the observations are likely to be of extreme interest. The observatory is connected by telephone with Sarajevo, 5350 feet below.—*Geographical Journal*, May.

By the treaty of Simónoseki between the Chinese and Japanese the following **Chinese ports** have been opened to foreign trade :—Su-chau-fu in Kiang-su, Hang-chau-fu, and Shao-hsing-fu in Che-kiang, in the centre of the silk-producing district ; Sha-shi (Hu-pei), commanding the routes through the two Hu and the road to Sze-chuen; Heng-chau (Hu-nan), situated on the trade route between Canton and the middle course of the Yang-tse, in the centre of the cotton district and near to coal mines; and Chung-king-fu in Sze-chuen. The Si-kiang, or river of Canton, has also been opened at the demand of Great Britain.—*Revue Française*, April.

The island of **Tristan da Cunha** was visited last year by the Governor of St. Helena. This volcanic island, discovered in 1506, has an area of only forty-four square miles. The inhabitants, numbering sixty-one, of whom eighteen are men, twenty-three women, and twenty children, were in excellent health; they asked for a schoolmaster but not for a doctor. They cultivate the soil and occasionally make an excursion to Inaccessible Island to the south-west in order to hunt seals, the fat of which supplies oil for their lamps. The Governor found on the island six hundred head of cattle, some donkeys and pigs, a few hens and numerous geese. At one time the albatross lived in great numbers on the island, but it has been exterminated by the wild cats, which have multiplied exceedingly.—*Globus*, No. 18.

The subject of **Acclimatisation** is of increasing interest, and Professor W. Z. Ripley has written a very compact, and at the same time valuable, article on the subject in Appleton's *Popular Science Monthly* (March, April 1896). For those working at the question the appended bibliography will be of use. After studying the subject from all points of view, Professor Ripley concludes that immediate colonisation of the Tropics is impossible, and "Let it be understood that a colonial policy in the Tropics means a permanent servile native population, which is manifestly inconsistent with political independence, or with any approach to republican institutions."

There are two radically different policies which may be adopted in attempted colonisation: the first, and the only one Britain can adopt, is by sending a large body of men—the more the better—to a new habitat at once, from which, by elimination, a few fortunate variations may survive. "Thus after a long time, and enormous sacrifice of life, a new type, immune to some degree, becomes established. All that the State need do, therefore, is to keep up the supply of immigrants long enough, and leave the climate to do the rest."

The other policy is to imitate the operations of natural ethnic migrations; to utilise the natural aptitudes of various nationalities—until finally a great drifting movement toward the Equator takes place. "In other words, the peoples of the Mediterranean basin, learning of their aptitude for a southward migration, would perhaps move to Algeria, displacing the people of the Soudan and the Semitic stocks toward the Equator. To fill the place thus left vacant, the people of northern France slowly drift to the Rhone Valley and Provence for a generation or two, and their place is taken by Germans and Belgians."

But in any case "great problems for science remain to be solved before the statesman can safely proceed to people those tropical regions of the earth so lately apportioned among European States." All interested should certainly read the paper.

The Geographical Association.—We are glad to report that this Association is prospering. It has now sixty-three members, among whom are thirty-four representatives of secondary schools, several of the first public schools in England, the Edinburgh Academy and Fettes College being of the number.

It has often been pointed out in these pages that Geography cannot receive proper attention in schools until the study receives encouragement from the boards of public examiners, and on p. 103 it was announced that the Association proposed to send a memorial to the various examining bodies requesting that the papers set in examinations should be such as to promote a more scientific system of geographical instruction, and that the study of the subject should be encouraged by a larger proportion of marks. This memorial has now been drawn up, and, having received the approval of the Geographical Societies (including the Royal Scottish Geographical Society), the Education Committee of the Teacher's Guild of Great Britain and Ireland, and other bodies, has been sent to the Delegates for Local Examinations of the Universities, to the London and Victoria Universities, the Scottish Education Department (leaving certificate) and Scottish Universities Preliminary Examination Joint Board, etc.

The following is the text of the memorandum accompanying it :—

1. That the main Principles of Physical Geography should form the basis of Geographical teaching at all stages, and should be fully recognised in all Examinations in Geography.

2. That a general knowledge of Geography, based on Physical Principles, should be required, together with a special study of some selected region, *e.g.*, India, a group of British Colonies, South America, Central Europe.

3. That it is desirable that all Public Examining Bodies, such as the Civil Service Commissioners, the Universities (in their Local and Certificate Examinations, and London Matriculation) and the College of Preceptors, should recommend a course of instruction in accordance with the ideas suggested above. This would stimulate Geographical teaching in schools, ensure that Geography should be systematically taught throughout the school, and do away with the need for separate classes to prepare candidates specially for the various Public Examinations in Geography.

4. That in the Examinations above referred to Geography and History should be dealt with in separate papers, and that the maximum of marks should be approximately the same for each.

B. BENTHAM DICKINSON (Rugby),
Hon. Sec.
J. ROBINSON (Dulwich).

NEW BOOKS.

Problems of the Far East. By the Right Hon. G. N. CURZON, M.P. London : Archibald Constable and Co. Pp. xvi + 444. (New and Revised Edition.) *Price 7s. 6d.*

This is the second edition of a work which, in its first form, appeared on the eve of the great war between China and Japan. Partly for this reason, but chiefly because of the political and geographical reputation of the author, the book attracted much attention. The views put forward by Mr. Curzon were not such as generally commend themselves to men who, partly from ignorance, partly from early bias, thought of China as practically invulnerable, and of Japan as an insignificant race of artists. Events have proved that Mr. Curzon was right and his critics wrong. With remarkable political insight he realised, what certain diplomatists and military men did not, that China was rotten to the core, and her people lacking in all true patriotism, but that Japan, in her confidence and energy, had all the makings of a Great Power. The new edition now before us contains a good deal of additional matter, bearing chiefly on the results of the recent war. It is satisfactory to find that in their general character Mr. Curzon's opinions have needed no revision, but simply amplification. The last two chapters, on "The Destinies of the Far East" and "Great Britain in the Far East," are particularly interesting, and worthy of the pen of the author who dedicates his book "to those who believe that the British Empire is, under Providence, the greatest instrument for good that the world has seen, and who hold, with the author, that its work in the Far East is not yet accomplished." All who wish to get a broad, sane view of the problems of the distant Orient will find in Mr. Curzon a safe and sure guide. Nearly fifty illustrations, chiefly of Korean scenes, add to the interest of the book, which is also provided with two maps.

Rambles and Studies in Bosnia-Herzegovina and Dalmatia. By ROBERT MUNRO, M.A., M.D., F.R.S.E. Edinburgh and London : Blackwood and Sons, 1895. Pp. xx + 395. With numerous Illustrations. *Price 12s. 6d. net.*

"The *raison d'être* of the book," explains Dr. Munro in his preface, "is to give an abbreviated account of the attractions—scenic, social, and scientific—of a portion of the Balkan peninsula, which, till lately, was almost inaccessible and unknown to the people of Western Europe. On the success or failure of my efforts to combine the popular and scientific elements, probably, hangs its fate. From the current opinion that archæology is a dry, uninteresting study, and incompatible with the gay and pleasurable side of life, I dissent *in toto*."

There will always be some people who resent the idea of making science "popular," but that is not the attitude of the present reviewer, in whose opinion such a book as this of Dr. Munro's ought to be heartily welcomed by every lover of archæology. For surely the writer who arouses an interest in any scientific subject in the minds of those who have previously held aloof from such studies is doing a great and lasting service to science in general, and more definitely to his own particular branch of it? And there is every reason to believe that this will be one of the results of Dr. Munro's recent experiment of combining the genial, gossipy narrative of the modern traveller with the authoritative statements of the archæologist.

The Congress of Anthropologists and Archæologists held at Sarajevo in August 1894, in which Dr. Munro took a leading part, appears to have been the bait which drew him to the Balkan peninsula at that date. But he also gave himself time to visit many places of interest in Bosnia-Herzegovina and Dalmatia. For the details of his various experiences one can only refer to the book itself, the text of which is admirably supplemented by its many illustrations.

Of much interest and value are Dr. Munro's observations on "Interlaced Ornamentations" (pp. 280-295). The interlaced style of ornamentation used to be regarded as specially and exclusively Celtic, but "there is now a consensus of opinion that it originated in the Byzantine school, which arose upon the downfall of the Roman empire, and thence spread throughout Europe, acquiring greater intricacy and significance as it reached the outward limits of Christendom." This doctrine is very strongly enforced by the designs, from various parts of Europe, which are introduced at this portion of the book.

By an unfortunate accident, the headlines of Chapter x. announce that that chapter relates to "Bosnia-Herzegovina in *P*rehistoric Times," instead of "in Historic Times." This mistake has been rectified, as far as possible, by an erratum slip, inserted at the beginning of the chapter before binding the book. But it may be asked whether the word "prehistoric," so much in use at the present day, might not also have been dispensed with in some of the preceding chapters? The fact is that that word is very commonly used by modern writers to denote races or objects of whose place in the world's history they know nothing positive, but who, for aught these writers can say to the contrary, may have been quite "historic." Thus, in Dr. Munro's book, the cemetery at Glasinac is styled "prehistoric"; but the reasons for giving it this title are not very apparent. Even distinguished archæologists like Dr. Munro are still much hampered by the misleading term "age," which (itself signifying *a period of time*) has been and continues to be used to denote *a condition of culture*. Thus, we have "the Bronze Age," "the Stone Age," and so on. People of "the Stone Age," for example, by which is meant people employing weapons and implements of stone, are assumed to have been "prehistoric." But the unprejudiced reader opens the latest volume of "the Saga Library," and therein he reads (pp. 40 and 494-5) that the Northmen employed flint and other stone weapons in the time of Edward the Confessor. Now, the average antiquary who picks up a flint weapon in Scandinavia would dub it and its user "prehistoric," although it now appears that it may have been used, and fashioned, by an absolutely historical man of the eleventh century. Who can say that such "prehistoric objects" as the perforated stone hammer figured (No. 122) in Dr. Munro's book were not the property of those eleventh-century Norsemen, or perhaps of the equally historic Huns of the fifth century?

These, however, are questions of detail. The book as a whole will undoubtedly stimulate the apathetic into an interest in the past and the present of the Balkan countries. The author himself is so genuinely interested in his subject, be it modern country life or archæological discussion, that he carries his reader with him. And this is the secret of all successful books; in which category we hope and believe this work of Dr. Munro's will be ranked.

The Finding of Wineland the Good: The History of the Icelandic Discovery of America. Edited and translated from the Earliest Records by ARTHUR MIDDLETON REEVES. To which is added *Biography and Correspondence of the Author*, by W. D. FOULKE. With Phototype Plates of the Vellum MSS. of the Sagas. London : Henry Frowde, Oxford University Press, 1895.

This handsome volume contains the best and most trustworthy account yet given to the world of the Icelandic discovery of North America by Leif Ericsson in 1000 A.D. It is trustworthy, because it consists mainly, that is to say fundamentally, of transcriptions of the Sagas by an accomplished Icelandic scholar, accompanied with prolegomena and copious notes. Unfortunately, it is a post-humous work. The author, after completing his editorial labours in Berlin in 1890, returned to America, and was killed in a railway accident at Hagerstown, near Richmond, on February 25, 1891, at the early age of thirty-four. The prefixed memoir by Mr. Foulke, which contains copious extracts from the author's corre-spondence, is extremely interesting. It gives the impression of an amiable, clever, sensitive, and high-toned character—of one whose premature death deprived the world of letters of an industrious and intelligent craftsman. He was born in 1856 at Cincinnati, where his father was a prosperous merchant. Education and natural tendency made him an accomplished linguist, so that he mastered German, Swedish, Danish, Icelandic, French, Italian, and Spanish. In the Norse languages he was an enthusiast, and in Icelandic, in particular, he was an acknowledged philological authority. As was natural in the circumstances, he travelled widely, his range including Iceland and the chief countries of Europe, especially England, Scotland, and Germany. His letters are vivacious and interesting, and are distinctly stamped with the character of the man.

The preliminary chapters are extremely valuable. They deal with the early fragmentary references to Wineland, with the Saga of Eric the Red, with the Wineland History of the Flatey Book, with the Wineland of the Icelandic Annals, with the fictions that have gathered around the historic narrative, and with the circumstances in which the discovery of the story of the Sagas was published. Regarding the supposed site of Wineland, Mr. Reeves does not commit himself to a definite opinion. He refers to the theories of Rafn and Gustav Strom on the subject, and he seems to accept the view of the latter, which places the northern limit of the Icelandic settlement approximately at 49° N. lat.; and he thinks that it could not have been much farther south than that limit. Professor Horsford's speculations (see vol. x. pp. 101-103 of this Magazine) were directed to proving that the site of the settlement was on the river Charles in Massachusetts, which is between 42° and 43° N. lat. Horsford's work was not published till after Reeves' death, so that he knew nothing of his theory, or of his supposed discoveries. But Reeves says distinctly that "there is no suggestion in Icelandic writings of a permanent occupation of the country"; and as Horsford's theory, based on the alleged discovery of solid stone foundations, necessarily implies permanent occupation, it must be assumed that it would have been rejected by Reeves as fictitious.

Undoubtedly, the most valuable feature of the present volume is the reproduction in phototype of the vellum manuscripts of the Sagas, the pages being interleaved with transcriptions in modern typography. This must have been a work of great labour, implying exceptional knowledge and skill; and it has been carried out with complete success. The only thing the unlearned may be excused for desiderating is an interlinear translation of the texts. With that the volume would have been complete.

The West Indies and the Spanish Main. By JAMES RODWAY. ("Story of the Nations" Series.) London: T. Fisher Unwin, 1896. Pp. 371. *Price 5s.*

In this interesting volume the author (who is well known as the historian of British Guiana) gives the story of the acquisition of the West Indies and adjoining South American coasts by Europeans. The Spaniards were the first to arrive, and their conquest was marked by shocking cruelties perpetrated upon the unhappy Indians. Very different was the treatment of the latter by the English under Sir Walter Raleigh, "the father of English colonisation," who sailed up the Orinoco in 1595, having received Letters Patent from Queen Elizabeth. He informed the Indians that his Queen was an enemy to the Spaniards, had freed other nations from their oppression, and had now sent to rescue them. To confirm his statement, he gave each cacique (or chief) a coin, so that they could possess the Queen's likeness, and these were treasured and even worshipped for a century afterwards." The fabled "El Dorado" was believed to exist in Guiana, but it has been left to the present century to discover it, for the well-known Caratal gold-diggings occupy its site. Raleigh's beloved Guiana became the parent of the West Indies, for, after various attempts to colonise it, the British abandoned Guiana for the West Indies, leaving the Dutch and French in possession. The country lying between the Essequibo and Orinoco was considered in 1613 to belong to the Dutch; and as in 1814 the British succeeded to them, they legally acquired that territory, as the recently published Blue Book shows.

Whilst British Guiana (both in letterpress and illustrations) bulks largely in this volume, no reference is made to its most celebrated explorer, Sir Robert H. Schomburgk. Yet not merely does Mr. Rodway adopt, in his map, the "original Schomburgk line" as the boundary between British Guiana and Venezuela, but also he gives, as one of his illustrations, a representation of the "Victoria Regia" lily, which Schomburgk was the first to discover whilst exploring British Guiana.

Excellent pictures (from photographs) of Negroes and British Guiana groups are interspersed, but some more ideal illustrations taken from old books, such as "Suicides," "Voyage of the Sable Venus," and "Europe supported by Africa and America," are rather curious than instructive.

Die Insel Tenerife: Wanderungen im canarischen Hoch- und Tiefland. Von Dr. HANS MEYER. Leipzig : Verlag von S. Hirzel, 1896. Pp. viii+328. Mit 4 Originalkarten und 33 Textbildern.

Although this excellent work appeals primarily and for the most part to the scientific traveller, it nevertheless contains many details as to mode of travel and choice of residence, which will be appreciated by every intending visitor to Tenerife. Indeed, Dr. Meyer, having realised the lack of a complete and minute guide for any one desiring to make himself thoroughly acquainted with the island, has endeavoured to furnish such information as will supply this want. Accordingly, he has given a careful account of the various means of access, the kind of clothing one ought to wear, the scientific instruments which the more serious traveller should take with him, and many other references equally useful and interesting.

But, of course, Dr. Meyer did not himself visit Tenerife in the character of the ordinary traveller, and to him the real attraction of the place was undoubtedly its famous " Peak "—" a king among mountains, like its distant East African brother-volcano, Kilimanjaro." And one reads with a peculiar interest the words of the conqueror of Kilimanjaro when at length he had attained the summit of the Peak of Tenerife. " The bird's-eye view which we had was of a vastness so immense that never from any of our own Alpine summits, never even from the top of Kilimanjaro itself, have I seen anything to equal it. . . . On no other mountain have I had so strong a feeling of being lifted up above the underlying landscape as on the Peak."

The scheme of the book is as follows. After the preliminary instructions already referred to, Dr. Meyer gives an interesting description of Tenerife, past and present. Then follows his itinerary :—From Santa Cruz to Taganana, by Laguna and over the Anaga ; from Laguna to Orotava and Icod ; from Icod to Santiago over the Teno ; from Orotava, by the Pedro Gil, to Guimar and Santa Cruz ; from Orotava across the Southern Cañadas to Vilaflor ; from Vilaflor, by Arona, to Adeje and Guia ; thence to the Pico Viejo and Icod ; and finally, his ascent from Orotava over the Montaña Blanca to the summit of the Peak. It will thus be seen that he made a very thorough exploration of the island, all the while placing on record the various scientific facts ascertained by him as he went along.

The four coloured maps are finely executed. All represent the island on the same scale. The first is an ordinary *carte du pays*, while the three others show respectively the geological formation, the natural vegetation, and the various crops cultivated on the island. The numerous illustrations in the text are also of interest, many of them—probably most—being from photographs taken by Dr. Meyer.

Of the Guanches and two other coeval or earlier races a good deal is said ; and Dr. Meyer brought home with him forty-five skulls ascribed to one or other of those races. These have received a careful description in an Appendix by the eminent Berlin scientist, Dr. F. von Luschan, who, it is interesting to note, points out that the practice of trepanning followed by those early races is another link connecting them with the Berber Kabyles.

One statement of Dr. Meyer's is perhaps open to question, viz. that (p. 56) Santa Cruz is the largest town in the whole archipelago. A few years before his visit in 1894 Las Palmas was credited with a larger population than Santa Cruz.

Herbsttage in Andalusien. Von GEORG WEGENER. 2te Auflage. Berlin : Allgemeiner Verein für Deutsche Litteratur, 1895. Pp. 322.

Students of comparative literature will be interested in comparing this book with its French counterpart, *Terre d'Espagne*, by Réné Bazin (*rev.* p. 273). As, in reviewing the latter, we gave a Frenchman's account of Gibraltar to-day, we cannot do better than translate a portion of Herr Wegener's German description of the same scene by way of comparison :—" Our carriage rattled, through strongly fortified gates, over a rough causeway to the Royal Hotel, which is considered the best in the town. The hotel was not on a large scale, although its bill was. But the latter fact was still unknown, so we cheerily walked through the streets, darkened by the dusk, and were glad to stretch our legs after our long voyage. From inside, Gibraltar has anything but a foreign appearance, the English having given it quite the look of a little northern town. The narrow, badly-lighted streets with poor houses, the grocery stores, the drapery establishments with shockingly dressed windows, the little confectioners' shops, etc.—all this was on as homely a scale as in one of our small towns. Did we not see the red-uniformed soldiers at every step, and were not the gigantic rock visible at every street corner, towering to the sky in the grey moonlight, we could perfectly well believe that we had been suddenly transported to some village in Pomerania or Silesia. A military band played in an open place, and gay, idle groups lounged through the streets. At half-past nine thundered a gun. That was the signal that the gates were shut. Whoever arrives too late must remain outside. In the morning a similar signal announces the opening. The hour of these signals changes somewhat with the season of the year. Thus, Gibraltar is, in a way, always on a war footing as regards Spain, and the whole life of the town is subject to strict military rules. Strangers require formal permission for an extended stay. Spaniards arriving to purchase provisions must leave the town that very night. On the other hand, the latter do their best to outwit the odious part-owners of their peninsula. It is not merely because of the stain on their honour, nor because their sea-power is so diminished by the loss of Gibraltar, but also because the latter furnishes such opportunity for smuggling into Spain. Gibraltar draws her supply of victuals, not from Spain, but Tangiers, and broad, flat vessels may daily be seen carrying cattle over the straits to Gibraltar, so that the Englishman may not want the beefsteak he cannot do without."

The Riviera, Ancient and Modern. By CHARLES LENTHÉRIC. Translated by CHARLES WEST, M.D. London : T. Fisher Unwin, 1895. Pp. 464.

Not as a guide-book, but "as a companion to the intelligent traveller" (who, by the way, usually carries a guide-book), Dr. West says he presents this celebrated work of M. Lenthéric to English readers. It tells them of Grecian and Roman colonies placed along the Riviera ; it points out to them the trophies, monuments, and works of public utility erected by these colonists; it recounts the arrival of the Saracens, and the final triumph of the Cross. But, more than this, it pictures the glorious scenery of maritime Provence with a master's hand.

Examining one of the admirably clear maps given by Dr. West, we observe that the "Aurelian way" which led from Italy into France passed along the Corniche route *via* Mentone, Nice, and Cannes to Fréjus, proceeding thence in a westerly direction *via* Vidauban and Tourves to Aix, where a branch road led to the north, and another southwards to Marseilles. A Roman "secondary way" led from Vidauban to Toulon, but no road led along the coast from Fréjus *via* Toulon to Marseilles.

Monténégro, Bosnie, Herzégovine. Par H. AVELOT et J. DE LA NÉZIÈRE. Paris : Henri Laurens, 1896. Pp. 248.

Thanks to the Austrian police and Viennese tourist clubs, the southern territories of the Austrian Empire, which now include Bosnia and Herzegovina, form safe and attractive fields for the tourist who desires to get off the beaten path of travel. Such a work as this under review ought to give an impetus to new tourist excursions, for it is written with that delightful mixture of erudition, artistic taste, and gaiety in which educated Frenchmen excel, and it is illustrated with four coloured plates and two hundred admirable original designs by the authors. Our travellers started from Trieste, and first investigated Istria. They next "did" the Quarnero, after which they traversed Dalmatia. From the Bocche di Cattaro they made their way into Montenegro, winding up with a visit to Herzegovina, Bosnia, and Croatia. Wherever they went they filled their sketch-books, and the charming results given in this volume make it one of the most artistic books of travel ever published. Artistic in everything, the authors impart to their descriptions that local "bit of colour" which is so sadly wanting in the ordinary "guide-book," which describes every country in exactly the same dry tone. Take the following on Bosnian gypsy musicians : "In a corner the gypsy musicians preserve admirable stolidity, although blows are raining around them. Among them are young girls, very pretty little savages, with dull hair and hard eyes ; under an embroidered chemise one sees their copper-coloured necks, and their wrists are encircled with most tasteful bracelets ; to-day lovely, to-morrow they will be repulsive ; for among these creatures of the south a few years suffice for the transition from youth to decrepitude." A map showing our friends' exact route would have been useful, but might not have been considered artistic.

Through the Buffer State : A Record of recent Travels through Borneo, Siam and Cambodia. By Surgeon-Major JOHN MACGREGOR, M.D. London : F. V. White and Co., 1896. Pp. 290.

An officer of the Indian Medical Service, the author wrote what he calls "this random narrative" after an interesting tour, during which he visited Bangkok, Mount Phrabat, Pechim, Wattana, Siam-Reap, Lake Tele-Sap, Penhom-Penh, etc. As a lively, if somewhat rambling, account of a journey undertaken in a region attracting much present attention, the book is welcome, although a more careful revision would have been desirable. The author has, however, found time to revise in a footnote his assertion "that the highest mountain in the world, Mount Hercules, has quite recently been discovered in New Guinea, and is 32,000 feet high," an assertion which Sir Wm. MacGregor, Governor of British Guinea, afterwards informed him " sprang from a romance written some years ago by some Captain Lawson."

Marcel Paisant—*Madagascar.* 51 gravures et cartes. Paris : Larousse, N.D. Pp. 142.

This work commences with a few pages of geographical description of the long double range of mountains with the great upland valley of the Maningory ; the East coast with its more abundant rainfall and luxuriant flora ; the absence of harbours—excepting always Diego Suarez at the extreme north—compensated by the long chain of lagoons within the reef, which might without much difficulty be rendered continuous ; the West coast much more *accidenté*, with numerous rivers, but comparatively desert and unhealthy. There are then a few notes on the ethnology and customs, but two-thirds of the volume are occupied by the history. That of the earlier period is perhaps the more interesting, and also impartial. The writer attributes, we believe rightly, the failure of the early French settlements to the hostility and jealousy of their own countrymen, the authorities of the Isle of France—a spirit not unknown in recent French colonial history. He refers to the story of the adventurer Benyowski, and attributes his destruction to this cause rather than to any fault of his own.

In the more recent history the main source of confusion is *la perfide Albion* and her missionaries, one of whom, *le révérend* Shaw, was accused of an attempt to poison a company of French soldiers. The truth of the accusation, repeated by the writer, may be gauged (if this is necessary for English readers) by the fact that Mr. Shaw received £1000 as *solatium* from the French Government.

Wolfe. By A. G. BRADLEY. ("English Men of Action.") London : Macmillan and Co., 1895. Pp. viii+214. *Price 2s. 6d.*

Wolfe will occur to most simply as the striking hero of a single notable achievement, and after reading this sketch of his life one is led to think that the biographer has felt a difficulty on that very account. He has had to cover a certain space, and yet he has had to keep a certain proportion—to maintain some regard for perspective. This he has done only with an effort : thus, he inserts a chapter on our American Colonies as they were at the time of Wolfe's arrival before Quebec ; and, even then, of the two hundred pages, considerably over half has had to be given to Quebec itself, its preliminaries and consequents.

It is a very readable volume and one that deserves to be read. Many, too, formerly uninformed, will find in the doings of the young Wolfe as here presented the earnest of the final triumph, and they will not readily find a chronicle more lucid and absorbing. As to the matter of the book, we have only one fault to urge : extracts might very well have been given from the despatches of both Wolfe and Montcalm, and some material apparently forced into the work might well have been sacrificed for the purpose. In style some unexpected inaccuracies appear, and one or two details in questions of fact are also loosely represented. " He opened the ball " (p. 34) is a really suggestive mode of introducing a fight, but is of doubtful taste, and unworthy of repetition. There is a grammatical mistake on p. 50, l. 6 ; on p. 90 there are two (l. 1 and l. 3). The peace of Aix-la-Chapelle (p. 51) did *not* "restore everything to its original position," in respect of two of the leading combatants, Austria and Prussia. "Townshead" (p. 79) is a printer's slip for "Townshend." The list is not complete ; but the blemishes are petty, and, in spite of the lack of due proportion, Mr. Bradley has carried his verdict on Wolfe : "The promise with him had already been fulfilled, for in every branch of a soldier's duty, in peace and in war, he had shown the highest capacity."

Modern Civilisation in some of its Economic Aspects. By W. CUNNINGHAM, D.D. ("Social Questions of To-day.") London: Methuen and Co., 1896. Pp. xvi+227. *Price 2s. 6d.*

This work, drawn from history and from theory, has many of the characteristics of the author's previously published writings : smooth narrative, regard for historic fact, recognition of possible economic fiction, and others ; even (by way of particulars) the familiar hit at the rigid Free Trader and at the champion of a golden age for the fifteenth-century labourer. Though written with hardly sufficient force, it is an interesting book—a favourable instance of applied economics. It deals with the fundamentals of political economy, and applies concurrently actual social life and ethical and other maxims, often with consummate judgment. Thus, we are told (p. 205), "Romance, art, religion, and everything that appeals to emotion" have their "economic use" ; and, yet (p. 211) it is "not altogether easy to see how any minister of education is to evolve public spirit, even if he inspires the songs of the people with the help of the music halls" : political sermons, State warnings, in little.

Grundzüge der Physischen Erdkunde. By Prof. A. SUPAN. Second Edition. Leipzig : 1896. Pp. x+706 ; illus. 203+maps 20.

This edition is considerably enlarged from the first edition of 1884, containing over 200 more pages and over 60 more illustrations. A conspicuous improvement from an English reader's point of view is the substitution of Roman for German characters. The text has been completely revised and the matter re-arranged, being now divided into five sections, viz.: The Atmosphere, The Sea, The Dynamics of the Land, The Morphology of the Land, The Geographical Distribution of Plants and Animals. The maps have undergone considerable alteration, embodying the results of recent research ; the maps 6 and 7, for example, of the first edition, showing the isabnormals for January and July, have been incorporated into one, and a new map added of Temperature Zones. A very valuable addition are the copious references to the literature of the subject at the end of each section.

Gazetteer of the Rawalpindi District. Revised Edition, 1893-94. Compiled and published by authority of the Punjab Government. Lahore, 1895. Pp. x+268+lviii.

The "Settlement" of an Indian district, a comprehensive operation which is undertaken at long intervals for the purpose of ascertaining the remunerative value of village lands so as to secure the settling of an equitable taxation between the cultivating occupiers and the Government as landlord, affords opportunity to the "Settlement" officer of prosecuting other statistical and historical inquiries, and renders the time of a Settlement most favourable for the compilation of a Gazetteer.

Mr. F. A. Robertson, whose "Settlement" operations were completed in 1887, submitted thereon an independent report, leaving the first edition of the Gazetteer untouched. Now, he has extended the Gazetteer to a second edition by embodying in it the report of his Settlement, also bringing it otherwise up to date of publication (July 1895), and rendering it replete with information concerning the history and present circumstances of a district which presents many features of exceptional interest. Historically, it is shown that the names of Alexander the Great, Mahmud of Ghazni, Baber, and Tamerlane are closely connected with the district, and the remains of Taxila, a large and wealthy city at the time of Alexander's invasion, have been identified. There are also many remains of Buddhist *stupas* and monasteries.

The present rights of landed proprietors and the occupancy rights of tenants, the determination and due recording of which form a very important part of a Settlement, are necessarily fully stated and explained. Many particulars are also given concerning the occupations of the people and their industries; concerning their social life and superstitions; concerning prices and wages; concerning the various crops and the system of husbandry. A reference to these few matters is a sufficient indication of the comprehensive character of this Gazetteer.

Insect Life: a Short Account of the Classification and Habits of Insects. By FRED. V. THEOBALD, M.A., F.E.S. ("University Extension Series.") London: Methuen and Co., 1896. 8vo. Pp. xi + 235. 53 Figs.

This is a condensed account of those insects which are of economic importance. The subject is fascinating, but the author has failed to do it justice. His work is marred by an irritating looseness of grammar and by inaccuracies which might readily have been avoided. What does the author mean by saying, " At the present day we are taught not to consider the useful insects, but those that do us injury"? Has it really come to this? We do not understand the sub-title of the book, for the author is certainly not strong either on classification or habits. It must be said, however, that Mr. Theobald has packed together much useful information, and has avoided a too free use of technical terms. The book is also cheap, but a glance at the illustrations, which are mostly bad, shows that this is not an unalloyed advantage.

The Riviera; or, the Coast from Marseilles to Leghorn: Ninth Edition. *Florence and Environs*: Second Edition. By C. B. BLACK. Adam and Charles Black, 1896. In 1 vol. *Price 3s. 6d.*

Looking to the elaborate contents, the numerous excellent plans and maps, and the low price of this admirable guide-book, the tourist would be hypercritical indeed who did not value it. The reviewer's function, besides unstinted praise, is limited to very small points indeed. Why does Mr. Black call the town known in English as Mentone by its French name "Menton"? and why does he not call Marseilles "Marseille," and Lyons "Lyon," if he follows French rules? Again, in describing the island of St. Honorat, one of the Lérins near Cannes, he misses a fact of interest, viz., that in its ancient monastery St. Patrick was a monk before becoming the famous missionary to Ireland. For the benefit of gamblers, Mr. Black furnishes, with a diagram, ample details regarding the roulette table at Monte Carlo.

Un Mois en Phrygie. Par Prof. H. OUVRÉ, Bordeaux. Paris: Plon, 1896. Pp. 269.

The French have always had an affection for "l'Orient," and M. Ouvré assures us that Phrygia is as full of picturesqueness and interest as the Far West of America. He visited Turkey in Asia during July and August 1893, and graphically describes his tour in this little book, which is illustrated with fourteen photogravures. His closing words are practical, for he bids his French compatriots show less delicacy in exploiting the Orient. "The English and Germans have no delicacy. They know how to speak and even to scream, if necessary. They know that Eastern countries resemble the bazaar at Stamboul, where every one goes armed with a stick. Less refined and more clever than the French, they don't play at who loses wins, and don't believe that they will gain wholesale what they lose piecemeal."

NEW MAPS.

EUROPE.

RUSSIE D'EUROPE, Carte Hypsométrique de la Partie Occidentale de la —— et des Régions Limitrophes de l'Allemagne, de l'Autriche-Hongrie et de la Roumanie, par le Lieutenant-Général Alexis de Tillo. Échelle de 1 : 1,680,000.
Annales de Géographie, April 15th.

AFRICA.

IL GIUBA ed i suoi Affluenti, esplorati dalla Spedizione del Cap. Vittorio Bottego inviata dalla Società Geografica Italiana, Settembre 1892—Settembre 1893. Scala di 1 : 4,000,000.
Memorie della Società Geografica Italiana, vol. v., parte 2.

BENGUELLA E MOSSAMEDES, Carta Dos Districtos de ——, 1985. Escala 1 : 1,000,000. Four Sheets. *Commissão de Cartographia, Lisbon.*

A considerable amount of new material is incorporated in this map, and the latest information has evidently been collected. Centres of population are marked, military posts and mission stations, as well as the routes of travellers.

AMERICA.

ALTO PARANA, Plano de la Parte Media del ——, desde el Yabebiry hasta el Ibitorocay, con datos ethnográficos de J. B. Ambrosetti. Escala 1 : 150,000.
Boletin del Instituto Geográfico Argentino, Tomo xvi., Cuadernos, 9-12.

OCEANIA.

NEW ZEALAND. Geological Sketch Map of the S.W. part of Nelson and Northern Portion of Westland. By Alexander Mackay, F.G.S., Mining Geologist.

This map of the north-western part of Middle Island is compiled to accompany Mr. Mackay's report on the Geology of the District published in *Papers and Reports relating to Minerals and Mining*, 1895.

ARCTIC REGIONS.

ARCTIC REGIONS, The ——, with the Tracks of Search Parties and the Progress of Discovery. Compiled from the latest information, 1896. *Price* $1.00.
The U.S. Hydrographic Office, Washington.

The map issued by the U.S. Hydrographic Department is an exceedingly useful one. The routes of a great number of navigators are given, and the stretches of coast either visited or observed by the various expeditions. The only doubt is whether too much has not been attempted, the colours and shades being so numerous that they are often hard to distinguish. For some reason exceedingly difficult to divine, Franklin's explorations along the north coast of the American mainland are not shown, but his discoveries are assigned to later travellers. This is rather a serious error, but in other respects the map is highly to be commended.

CHARTS.

INDIA. Chart (on Mercator's Projection) of the Tidal and Levelling Operations of the Survey of India Department, 1858-1895, showing approximate Cotidal Lines. *Price* 1 *Rupee*. *Survey of India Offices, Calcutta.*

MOZAMBIQUE. Reconhecimento Hydrographico da Barra do Rio Licungo (M'Gondo). Escala 1 : 10,000. 1895. *Commissão de Cartographia, Lisbon.*

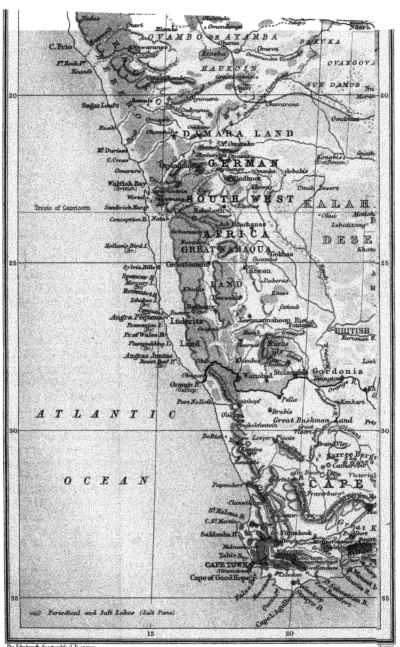

MAP OF SOUTH AFRICA SHOWING LAND SURFACE FEATURES

Scale 1:12,000,000 or 189.3 Miles to an inch

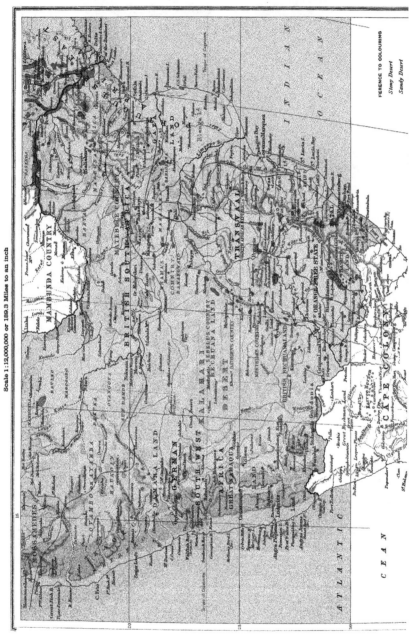

FERENCE TO COLOURING

Stony Desert

Sandy Desert

THE SCOTTISH
GEOGRAPHICAL
MAGAZINE.

RECENT EXPLORATIONS IN CELEBES.

THE interior of this singularly shaped island, though the deeply indented outline of the coast seems to render it easily accessible, has been hitherto little explored. Messrs. P. and F. Sarasin,[1] who have been travelling about the island during the last three or four years, have therefore been able to add considerably to the knowledge of its geography. The travellers crossed the island in several places, and found a remarkable number of lakes.

Their first journey was in the northern peninsula, to the west of Minahassa. Leaving Amurang, on the northern coast, the travellers ascended the Rano-i-apo river, and reached Karoa, a tobacco plantation on the frontier between Minahassa and the independent principality of Bolang-Mongondo, on the 25th. It lies at an altitude of 870 feet. The path then led through thick forest, which covers the mountains forming the boundary between Minahassa and the adjoining territory on the west, following the river, and crossing a number of tributary streams.

Underwood was scarce, owing probably to the intense shade thrown on the ground by the spreading crowns of the trees. Animal life, also, was remarkably rare, even monkeys being absent. Rhinoceros birds were heard, and beautiful butterflies hovered round the brooks. At a height of 3120 feet the forest suddenly came to an end, and the travellers found themselves on the banks of a small lake, called by the guides Mokobang. It consists of two basins united by a narrow passage, the larger being some 200 yards in diameter. The guides stated that it was drained by the river Rano-i-apo, but more probably its waters find an outlet into the Poigar, which was struck soon after. Crossing this, the

---

[1] *Zeitschrift der Gesell. für Erdkunde zu Berlin*, 1894, 1895, 1896. *Verhandlungen*, Nos. 4 and 5, 1896.

travellers marched over a rich loamy soil, decked with small begonias here and there and bird's-nest ferns, until at length they stood at the edge of the plateau, and looked down on the cultivated lands of Mongondo.   This plateau, drained by the Poigar river, has a mean elevation of 3300 feet.   The river is said to rise in a large lake called Dano, lying probably to the south-east of the Mokobang.   On the south side lies a group of volcanoes, one of them apparently still showing slight signs of activity, which have probably built up the plateau with their ejectamenta. Rich soil, abundant irrigation, and a splendid temperate climate, distinguish this tract, which is, nevertheless, totally uncultivated and covered with very dense forest.

Mongondo is a tableland, or broad hollow, surrounded by forest-clad mountains.   The villages, some of which are enclosed in a bamboo fence, consist of houses elevated on low piles, as in Minahassa; in front is a verandah, which gives entrance to the living-room, while a small apartment at the back is used as a kitchen.   The inhabitants are well dressed, some of them almost in European style.   In some villages they are of pure Malay race, while in others they exhibit an admixture of Chinese or Japanese blood.   Many of them are neither Mohammedans nor Christians, but devil-worshippers.   The gardens are carefully fenced in, and the crops seem to thrive well, especially the large fields of maize.   Coffee is also grown, and the arenga palm for its sago.   Pigs are the chief domestic animals; the young ones are fed on ripe papaya fruits, which the people despise as food.

Having visited Kottabangon, the chief place in the district, the Messrs. Sarasin turned their steps towards the northern coast, being unable to continue their journey westwards to Gorontalo, as they wished, owing to the difficulty in procuring rice.   The path, though it is the main thoroughfare between Mongondo and Bolang, is beset with obstacles —fallen trunks, swamps, etc.   Where it began to descend to the valley of the Ongkag, two solfataras were passed, and immediately afterwards the forest was entered.   The river is a rushing stream, with numerous windings.   On its banks the travellers came across sedimentary rock for the first time; it was a grey clay-slate, with a strike approximately north-west.

Parts of the journey described above had already been traversed by Schwarz, Wilken, De Lange, De Clercq, and Riedel, but now the travellers started from Bolang into a part of the interior never before visited by a European.   They followed the coast at first, passing the estuary of the Lombagin, which is formed by the union of the Ongkag and Dumoga.   It swarms with crocodiles, and the natives living near have to construct a bamboo fence around their huts, to keep these reptiles out when they leave the stream at night.   The interior was gained by the valley of the Lolak, the party passing thence into the plain drained by the Dumoga, where to their surprise they found themselves almost at sea-level again.   On reaching the river itself they found it a full rushing stream.   It flows in a deep bed, and is sometimes confined in a ravine among basaltic rocks which occasionally exhibit a regular columnar formation.   For some distance the ground was not more than 200 feet above sea-level. Where the river is joined by a fairly large tributary, the Mau, it becomes navigable with prauws and bamboo rafts, and, after passing through a narrow cañon, it flows quietly through level country, in which is situated Dumoga-besar, a large well-kept village. By difficult paths through the forest, frequently swampy, the travellers reached Dumoga-Ketyil, and crossing the river Dumoga, here only knee-deep, came to the village of Duluduo, in the kingdom of Bintauna. The elevation of this place is not more than 560 feet above sea-level, so that the Dumoga flows to the Celebes Sea across a very gently-sloping plain.

Here again it was impossible to procure sufficient rice or sago for a journey overland to Gorontalo, direct or through Bintauna on the northern slope. Many days were spent in palavers and inquiries among the natives as to the road to Bintauna and the distance, but provisions in sufficient quantities being unattainable, the Messrs. Sarasin had after all to make for Malibagu, on the southern coast. The road ascended rapidly, soon reaching the watershed at an elevation of 1150 feet. Here the travellers perceived that they had been crossing the peninsula along a line of low elevation, while lofty mountains rose to the east, north, and west. The eastern mass is probably volcanic, and the origin of the basalt observed along the bed of the Dumoga. These Mongondo mountains, as the travellers named them, mark the south-western limit of the volcanic region occupying the north-eastern extremity of the peninsula. To the north rose a massive ridge with bold contours, which the natives called Huntuk-Buludawa; it constitutes the north-easterly extremity of the region of archæan rocks which lies to the west of the Dumoga. Lastly, to the west lies a group to which the name Bone mountains were given, and which was visited soon afterwards.

As to the Dumoga river, it appears to be formed to the north-west of Duluduo by several streamlets, of which some descend from the Buludawa and others from the Bone mountains. At a very recent geological period its valley, according to all appearances, must have been covered by the sea.

The low ridge connecting the Bone and the Mongondo mountains and forming the watershed falls steeply to the ravine of the Malibugu river. Lower down the valley widens out into a plain, on which is situated the village of Malibugu, and thence to the sea the path traverses a large dirty swamp, where nipa palms grow. The coast westwards is fringed with mangrove swamp and partly with raised coral reefs. Occasionally a rocky spur descends to the sea, an offshoot of the Bone mountains. While on the northern coast signs of an advance of the coastline in recent geological times were often observed, such a displacement could not be verified on the southern coast. On the contrary, the waves have eroded the rocks, forming caves and overhanging vaults, which indicate that either the sea-level is stationary, or that the sea has recently gained on the land. Monkeys, rhinoceros birds, and pigeons abound in the woods, and a babarussa was killed.

Having been unable to traverse the country from Duluduo to Gorontalo, the Messrs. Sarasin endeavoured to make the journey in the opposite direction, from Gorontalo up the Bone river. The valley for a short distance rises gently, and then suddenly contracts, and is covered with dense forest. Here and there it narrows to a dark ravine, and then the path has to leave the river and pass over the wooded hills. In three days the party reached Pinogo, 790 feet high, the chief village of an extensive plain covered with cultivated fields, which is named Bawangio. The mouths of the Monoti and the Bulawa were passed, above which the Bone can be forded without much trouble, and at length the bed of the river, though encumbered with smooth boulders, was the only available path. Even this failed at last, the stream losing itself in a narrow cañon with perpendicular walls, whence was heard the rushing of a waterfall. There was nothing for it but to cut a way through the wood, clambering over the rocky ridge through which the cañon had been eroded. Wild boars, snakes, and the interesting dwarf wild ox of Celebes (*Anoa depressicornis*) were seen on the journey. The Bone grew smaller and smaller, and presently divided into two brooks, where the travellers began to ascend the ridge of the mountains, and then crossing summits and ravines, ascended the Gunung Bulawa, 4760 feet high, which is probably one of the principal peaks of the Bone mountains. These, rising in several dome-shaped summits, attain a height of 4600 to 4900 feet, or perhaps more. To the north-east is seen the sharp outline of the Huntuk-Buludawa ridge, separated by a saddle from the Bone group. The Sinandaka, Panega, and other rocky spurs rising immediately from the coast are probably direct continuations of the latter. The unexplored mountain tract to the south-west may also merge in the Bone mountains, as well as the range skirting the Bone river on the north. These, then, are the central elevations of the broad section of the peninsula between Gorontalo and Duluduo and the gathering ground of the waters of the most important rivers, feeding the Bone on the one side and some of the tributaries of the Dumoga on the other. A whitish-grey granite forms the core of the mountains, but here as well as at the foot of the Sinandaka were observed remains of a gneiss casing that originally covered the granite. In the upper regions the boulders and the trees were covered with moss, and ferns were common. As in all the forests of Celebes, the large number of succulent plants was very striking, especially among the epiphytes. Leeches were very troublesome, and therefore the travellers were curious to know how the wild oxen managed to exist; it was found, however, that the hide of one that was killed showed no signs of the attacks of these pests, being perhaps tough enough to withstand them. The temperature in the mountains was at night as low as 59° F., and the cold was felt severely.

On the descent a south-easterly direction was taken, down the valley of a stream reaching the sea at Negeri-lama. The country was of the same character. Nibong palms, pandanus, wild betel, and an abundance of tree-ferns were the trees which covered the ground, and here and there stood clumps of casuarina. At 2000 feet the first sago palm appeared, and at 1600 the first bamboo thicket. A warm spring with water at 112° F. was passed, and here, as at a spring on the way up from Gorontalo, eggs of the maleo (*Megacephalon maleo*, Fem.) were found laid in the sand to be hatched by the heat of the spring. On arriving at the coast the Messrs. Sarasin proceeded by sea to Kama.

The next journey was across the peninsula west of Gorontalo, from Buol to the Gulf of Tomini. Strange to say, the fauna and flora of this part of the island exhibited striking differences from those of Minahassa. For instance, the cockatoos, so plentiful in Buol, and the land crab (*Testudo forsteni*) are unknown in Minahassa, and similar remarkable instances might be mentioned among the molluscs, centipedes, land crustaceans, etc. As the climate is the same, the cause of these differences must be explained by the geological history of the island.

The Buol district has been for centuries noted for its wealth in gold. Mines have been opened at Palele, to the east of Buol, which seem likely to yield good returns. The gold occurs in veins intersecting the eruptive rocks of which the mountains are composed.

The journey inland was commenced at Matinang, immediately south of which rises a lofty range culminating in the imposing Gunung Matinang and Gunong Timbulon. To the west it seems to have no continuation, while to the east rises the Palele mountain, which, however, is probably a separate elevation. The mean height of the range is about 6500 feet. Two hours' march from the coast, forest took the place of the cultivated fields, and soon the river was heard rushing through a narrow glen. At the height of 650 feet numerous ferns quite new to the travellers were met with. The forest was very poor in animal life, small squirrels and a few birds appearing only occasionally. One of the summits of the Gunung Matinang was ascended; it proved to be 6730 feet high, and one or two of the others may be three or four hundred feet higher. At the camp, about three hundred feet below the summit, the thermometer fell during the night to $54\frac{1}{2}°$ F.

From the Matinang range, which forms the watershed between the Sea of Celebes and the Gulf of Tomini, the travellers descended by the valley of the Panu, perhaps one of the headwaters of the Molango. On this side the forest was marked by the presence of damar trees, from which the natives collect resin. They are straight, majestic trees, with quite round trunks; one of them measured fully twenty-four feet in circumference.

South of the Matinang range lies another, the Oleïdu, much lower but still of no small height. Beyond this again lies the Oleïdu-kiki (Little Oleïdu), a summit about 3850 feet high. It is probably part of the Oleïdu range, but that the travellers could not determine beyond doubt as the wood never allowed them to obtain a distant view. The Oleïdu chain is composed of red clay-slates, whereas the Matinang consists of eruptive rocks of ancient origin; they have probably burst through the clay slates and crumpled them up to form the Oleïdu and the chains farther to the south. No recent volcanic rocks occur in this part of Celebes.

The path remained at nearly the same height for some distance, passing over another summit, the Gunung Bontula, and then descended to the valley of a large stream, the Mangkahulu, only 590 feet above sea-level. For three days' journey the bed of the stream was the only

path.   Narrow ravines alternated with broader stretches, where was
seen the spoor of wild oxen, deer, and wild boars.   After the Buhu
joined the main stream the valley became wider, and the village of
Randangan was reached, the first met with after ten days' march.
From this point the journey was continued by water.   After leaving
the plain of Randangan the river, now called the Butaio-daa, wound
between forest-clad hills, forming small rapids where fallen trees and
drift-wood dammed up the water.   Here and there villages appeared
surrounded by fields.   Gradually the country became more level, but
was still hilly, and the river flowed quietly in endless windings, being
fifty-five yards broad, and in places as much as ninety.   A herd of black
baboons showed themselves on the bank, and large crocodiles basked on
the sand.   Near the coast the river divides into two arms and enters the
sea through a belt of mangroves.   From Marisa the Messrs. Sarasin
returned in prauws to Gorontalo where they took ship to Minahassa.

The explorations of the travellers had hitherto been in the great
northern peninsula ; they now turned their steps to Central Celebes, and
early in 1895 landed at Palapo in the Gulf of Boni, a dirty little place
situated on the banks of an insignificant stream and its branches, with a
very unhealthy climate.   Behind rise lofty mountains with lower hills in
front, from which Palapo is separated by a plain an hour's march in breadth
and extending southwards and westwards for several hours' journey.   Along
the little river Toka and farther up in the mountains gold is found in
considerable quantities.   The march inland was commenced at Borau to
the north-east of Palapo.   To the north of the village stands the Tampoke, its culminating point rising in a dome to a height of, perhaps,
4900 feet.   The route followed at first the Borau river, passing over
level swampy ground covered with grass, huge ferns, and low brushwood,
and then traversed a forest marking the boundary between the coastal
zone inhabited by Buginese and the territory of the Toraja tribes.
Forest alternated with fields and bamboo thickets, and the travellers
crossed several small streams flowing into the Kalaëna, which enters the
sea at Wotu on the east.   From the bank of this river long high ranges,
without any prominent peaks, were seen to the north and north-west,
bounding the Kalaëna plain, while to the west steep rocky mountains
formed a prolongation of the Tampoke.   From the Kalaëna valley the
party crossed a ridge 2000 feet high into that of a tributary of the river,
and, crossing to the other slope, came to the village of Lembongpangi,
situated in a picturesque hollow 1600 feet above sea-level.   The village
was quite empty, as the people had gone to their plantations in the
mountains.   The guides here warned the travellers that the range they
were now approaching was held sacred by the Torajas, and that they
must not collect plants or animals on it or break off specimens of stone
—at least they must not be seen doing it.   Nor must the men shoot or
sing, lest they should disturb the spirits.

Having crossed some outlying spurs, the party began to ascend the
main range named Takalekajo.   The ascent was very laborious, the path
winding between rugged blocks of stone, and sometimes so narrow that
the porters could scarcely force their way through with their loads.

The rocks were bluish-black, hard, and crystalline; at one place mica-schist was seen *in situ.* The height of the crest of the ridge was found to be 5480 feet; the culminating summits rise a little higher. In the night the thermometer at the camping-place, about 300 feet down the northern slope, fell to $55\frac{1}{2}°$ F.

After a toilsome descent the party came to a precipice affording a fine view of the Poso lake. The deep blue sheet of water lay a day's march distant in a north-westerly direction. About the middle a pro-montory shot out far into the lake from its western shore. Nearer, at the south end of the lake, a large plain with woods and fields showed that the lake had once extended farther in this direction. The Takal-ekajo range, clothed with unbroken forest and preserving the same height throughout, skirted its western shore, falling steeply to the water's edge, while the east bank sloped up to low hills with cultivated fields. A river, the Kordina, drains the country to the south of the lake, and on one of its tributary streams is situated the little village of Tamakolowe, at a height of 1770 feet above sea-level. The night spent here was remark-ably cold, the thermometer falling to $54\frac{1}{2}°$ F.

The Poso lake lies at an altitude of 1640 feet. Its longer axis, lying about south-south-east and north-north-west, measures some twenty-five miles, and its maximum breadth is a little over nine. Its principal feeders are the Kordina, already mentioned, and the Kaiya on its western side. Its waters abound with fish and crayfish, and the quantity of molluscs is extraordinary. The eastern shore is fairly well peopled by natives living in scattered houses. From its surface the Takalekajo range could be seen stretching south-eastwards in a chain marked by many sharp peaks, which probably runs down into the south-eastern peninsula of Celebes. To the north-west it skirted the lake, giving occasionally glimpses of parallel ranges in the background, and continued to the north of the lake in high mountainous country. No traces of cultiva-tion could be seen on its slopes, and the western shore appeared to be occupied only by a few fishing settlements. Soundings were taken at the northern end by means of a rope of lianas weighted by a stone, and depths down to 754 feet were measured; in the middle of the lake, in the direction of the promontory on the western shore, the rope, 1023 feet long, failed to reach the bottom. The bed of the lake is not a crater-basin, but a tectonic cleft of great depth; indeed, volcanoes were never met with on this journey.

The Torajas, who inhabit this country, clothe themselves principally with bark cloth, in the preparation of which they are great adepts. The finer cloths are as thin as paper, and are frequently dyed red and adorned with grotesque figures. Sometimes the Toraja wears only a loin-cloth. To this may be added a *sarong* of bark thrown over the shoulders; jackets of the same material are also seen now and then. When travel-ling across the country, the Toraja wears a long apron of skin hanging down behind so that he can sit comfortably on the damp ground. A certain fancifulness is exhibited in the head-covering. Sometimes it consists of a simple piece of bark cloth, occasionally dyed red, or more seldom of many colours, and arranged so as to form two horns. The half of a gourd-shell and caps of rattan are also used, and these, again, are covered with skins and ornamented with civet cat's tails, plumes of heron feathers, tines of deer, etc. Bangles of shell, horn, iron, and bronze, necklaces of glass beads, or threads adorned with the bill of the Celebes cuckoo (*Phoenicophaës calorhynchus*), the cuirass of the rhinoceros beetle, and other things, are worn as ornaments. The Toraja never goes unarmed; he always carries a spear with a well-wrought iron point, and is seldom without a shield of rattan coloured in patterns, or of wood ornamented with inlaid bone or shell and red and white goat's hair. His principal weapon, however, is his *klewang*, often an heirloom; its handle and sheath are carved, and the latter is further adorned with hair and feathers, or the head of a rhinoceros bird.

The country between the Poso lake and the Gulf of Tomini contains numerous villages, usually perched on the summit of a hill. From the top of a ridge about 1900 feet high the travellers could survey the country for some distance. To the north the gulf was visible through gaps in the hills; in the direction of the lake the country was full of rounded hills; to the west and north a multitude of ridges, with some very imposing peaks, stretched from north to south, forming a prolongation of the Takalekajo range; while to the east a long level ridge stretched away in the distance. The route of the travellers lay sometimes along the Poso river, which drains the waters of the lake into the Gulf of Tomini, but more frequently over hills 1200 to 1400 feet high; one of them was cleared for a new plantation, and proved to be composed of grey clay capped by coral. On the north shore of the Poso lake coral was observed at a height of 1600 feet, and nearer the coast it was again met with. The grey clay was seen also on the south side of the watershed, and probably it was formed at a period when large fresh-water lakes were numerous in the island. Afterwards the island sank, and coral reefs were built up on it.

From Mapane the explorers sailed to Gorontalo, skirting at first the Tojo coast eastwards, and passing the Tanjong Api, or Fire Cape, a small truncated cone connected with the mainland only by an isthmus of low land. Its form is that of a volcano, though the mountains which rise behind it in a succession of ranges must be of a different character. It owes its name, according to the missionary A. C. Kruijt, to the gases which rise out of the ground along the coast and burst into flame.

On March 10th the Baliohuto chain north-west of Gorontalo presented a very fine picture. It is probably the highest elevation in the northern arm of Celebes. Its many-pointed outline does not indicate a volcanic origin, which has often been assigned to it. Both east and west of Gorontalo, however, masses of breccia, which are more than probably of volcanic origin, form the shore hills, and perhaps these coast volcanoes may lie on a prolongation of the line of volcanic energy running from Cape Api to Togean and Una-Una, and be connected on the other side by the warm springs of the Bone river with the recent volcanoes of Bolang-Mongondo, and finally with those of Minahassa.

In the southern peninsula of Celebes, inland from the point where the western coast begins to trend westwards, lies, according to the unanimous statements of the natives, a lake enclosed by mountains and of considerable area and depth. This lake has already appeared on some maps of the island under the name of Kariangung. The Messrs. Sarasin were desirous to ascertain the truth of these statements, and also, if possible, to continue their journey across the peninsula to Palapo. Several Government officials had traversed parts of the route, and they prophesied a successful result of the journey, though they warned the travellers against the wild marauding Torajas of the interior.

At the end of July the two travellers took passage on a steamer to Pare Pare, an unhealthy place in spite of its cleanliness, where the thermometer marked $90\frac{1}{2}°$ inside the house. Thence they travelled in boats up the coast to Maroneng, at the estuary of the Bungi river, so called from a settlement on the banks. The shallow lake basins of Tempe and Sidenreng, near Pare Pare, are situated in a hollow but little above sea-level, to the south of which the mountains begin to rise, forming still farther south several chains and mountains of varying geological formation, and attaining their greatest height in the peak of Bonthain. Northwards there are also mountain ranges, and from Maroneng a small independent mass is seen between the Sadang river and the Bungi, of which the highest point, Mount Tirasa, does not exceed 1600 feet. North of the Bungi a high massive ridge stretches into the Letta country. Between these lay the route into the kingdom of Enrekang in the interior.

The rocks observed on the route were grey clay and sandstone, while the Loko rock at the south side of the pass was volcanic. It is not, however, a production of recent eruption, but, like all the other mountains in the neighbourhood, has been upheaved by folding. The volcanic rock probably underlies the clay and sandstone, which were originally covered with recent limestone now continued seawards by the living reefs. The height of the pass is 1360 feet, and the Loko rock may be 300 feet higher or thereabouts. At the watershed the expedition entered the territory of Enrekang and descended into a rich cultivated plain watered by the Sadang. The village Enrekang lies at the confluence of the Kalupini with the Sadang. The bed of the stream is 218 yards, only half of which was covered with water, as it was then the dry season. The height above sea-level is only about 160 feet.

The expedition was led past Enrekang, hidden among fruit-trees, which it was not allowed to enter, and through the rice and maize fields that surrounded it, reaching at length a longitudinal valley running northwards between two parallel chains. Here also cultivated fields, carefully fenced in with stone walls, were frequently passed. The rice was nearly ripe, and, to keep off the swarms of finches, clappers were placed in different parts of the field, which could be worked by strings running to a hut in the middle, or set in motion by the wind. Presently coralline limestone was observed, and for some days it was always present. The landscape had a desolate appearance ; innumerable blocks of limestone lay scattered over the grassy hills. To the left was seen a fantastic pile of rocks, the summit of the mountain Bambapuang, fully 3300 feet high, and a landmark conspicuous for a long distance. At the foot of the mountain a small tarn, the Lura lake, lies on the summit of a pass. It is shallow and covered with reeds, and measures about 330 yards in diameter. Flocks of ducks (*Anas arcuata* Horsf.) swarm among the reeds. The altitude was 2100 feet.

From the height of the pass the travellers looked down into two valleys, that of the Walida river on the left, and that of the Kalupini on the right. They descended into the latter and came to the village of Sosso, near which they encamped at a height of 1940 feet. A grey tone prevails in the landscape; a greyish-green hard grass, growing in tufts as high as a man, covers the valleys, hills, and mountain flanks. A tree with dark-green crown or a bush is seldom seen. The ground is grey clay or sand, with light-grey limestone rocks here and there, and the higher ranges are composed of the same limestone. Low green bushes with shining leaves are sparsely scattered over the rocks. This dreary country, a grassy desert, is a striking contrast to the damp, unbroken forests of the north of the island. Certainly this was not the original condition of the country. Here also unbroken forest covered the surface, but in course of time it was cleared by the axe and fire, and after the ground had been cultivated for a short time the grass took possession, to the exclusion of all other vegetation, and changed the character of the landscape. Only orchids, some of them with stalks exactly like the grass, relieve the monotony.

No satisfactory information could be obtained in Sosso regarding the position of the lake, and therefore the expedition moved on to Kalosi, in spite of the opposition of the King of Sosso, who had received orders from the King of Enrekang to prevent their further progress. For the second time on the journey slaves were seen fastened together with chains and rings round the neck. The trade still continues in spite of the efforts of the Dutch Government to repress it, and Torajas of the mountains are led down to Enrekang and Sidenreng by Buginese and Arabs, and are even exported to Borneo. A rajah, head of a small village near Kalosi, informed the travellers that a good riding-path led to the lake, which could be reached in two days, and also that the route to Palapo presented no difficulties. However, all further exploration was stopped one morning by the appearance of 500 armed men, mounted and on foot, and the expedition was compelled to return by the way it came.

A letter from the Messrs. Sarasin, dated March 30th of this year, when they were about to return to Europe, contains a few particulars about their last journey, across the south-eastern peninsula of Celebes from the Gulf of Boni to the Gulf of Tomaiki. The chief result of the journey was the discovery of two large and beautiful lakes, the Matanna and Towuti. The former is twelve nautical miles long by three broad, while the latter exceeds the Poso lake in area. The Matanna lake is very deep; in the middle the bottom was not reached at a depth of 1575 feet. A village on piles, of true prehistoric style, stands in the water at its southern shore. Earthenware and bronze utensils, similar to those found in European lake-dwellings, are manufactured by the inhabitants. The Towuti lake extends north and south over half a degree of latitude,

with a great breadth, especially in its northern part. In the middle of
this part an island, named Loëha, rises like a mountain from the water.
On the way to the island a depth of 499 feet was sounded, but probably
the water is deeper towards the south. The lake of Matanna stretches
east and west, and pours its waters into the Towuti, which in its turn is
drained into the bay of Ussu. The surface of the Matanna lies about
1300 feet above sea-level, and that of the Towuti at 1150. The natives
of the interior are Torajas, in about the same state of civilisation as the
Dyaks of Borneo. They are pure Malays.

In these excursions the Messrs. Sarasin have obtained a large amount
of fresh information, not only as regards the physical features of the
island and its geology, but also relating to its fauna and flora and the
habits and customs of the natives. Their narrative shows how difficult
it is to explore Celebes, though at first sight its comparatively narrow
peninsulas seem to offer great facilities. Sometimes the character of the
country, covered with dense forest where no villages are found over long
distances, necessitated the carrying of large quantities of provisions,
and these, again, were not always obtainable. At other times the
natives opposed the advance of the expedition, or pretended that the
route was impassable, or the guides professed ignorance of the country.

THE ELEVENTH SWISS GEOGRAPHICAL CONGRESS.

By V. DINGELSTEDT, *Corr. Member.*

THE Swiss Geographical Societies are in the habit of holding a Congress
every three or four years, for the purpose of exchanging ideas and for
social intercourse. The National Exhibition being held this year in
Geneva, and attracting a great number of visitors, not only Swiss, but
also foreigners interested in industrial, artistic, and scientific progress,
the eleventh congress has been organised in this town, under the
auspices of the municipal authorities and the direction of the Geneva
Geographical Society, the oldest of the Swiss geographical societies.

The Congress was opened on May 25th, and was closed by a reception given by the cantonal and municipal authorities on May 27th.
Some hundreds of visitors, including several ladies and delegates of
foreign societies, attended the Congress, which was fruitful, not only in
speeches and amusements, but also in geographical papers and communications of a solid and interesting nature. During the three days five
general meetings, each lasting not less than three hours, were held for
the reading of papers and for oral communications, as well as three
social gatherings of still longer duration, at which there was an almost
uninterrupted succession of speeches.

On Monday morning the Congress was opened in the University
Hall by the Geneva State Councillor, M. Dunant, with an appropriate
speech of welcome and a discourse on the importance of geographical
studies. The subsequent meetings were presided over by M. Arthur de

Claparéde, the President of the Geneva Geographical Society. In all there were twenty-one communications, dealing with all branches of geography—physical, zoological, geological, climatological, cartographical, geodetical, etc.—and some interesting accounts were read of travels and explorations in unknown or little-known countries.

The most noteworthy communications were by Prince Roland Bonaparte on the periodical variations of glaciers in France; by Dr. F. Forel, the author of the remarkable monograph on the Lake of Geneva in two large volumes, who treated of the same phenomena as connected with Swiss glaciers; by Colonel Lochmann and Dr. J. Graf on ancient and modern map-drawing; by M. Henri Moser on the present condition of Bosnia and Herzegovina; by Mr. Ed. Neville, the famous Egyptologist, on the excavations conducted by him with such excellent results at Deir el Bahari; by an African traveller, M. Bertrand, on a recent journey in the Barotse country (Central Africa); by Herr Bieler and Dr. Keller on the geographical distribution of the more common domesticated animals in Africa and Switzerland; by Professor Pittard on the distribution of animals in general; and, last not least, one by the eminent physicist, Raoul Pictet, on the solar heat in Egypt, sand-spouts, and irrigation by solar action.

We cannot enter into a detailed account of all these communications, but we will give a *résumé* of M. Moser's paper on the Balkan provinces under Austrian rule, and a short account of the new and bold ideas set forth by M. Raoul Pictet in his lecture on irrigation by solar energy.

The progress of Bosnia and Herzegovina, which were occupied by Austria-Hungary, with the consent of Europe, in 1878, seems to be truly remarkable, and is the more interesting and instructive because the Austro-Hungarian Government has had to deal with a population containing a large proportion of Musulmans notorious for their fanaticism and turbulence, and considered, not without some reason, to be incapable of improvement and civilisation of any kind. Governor-General Kallay, however, to whom the difficult task was entrusted, has, nevertheless, succeeded, in eighteen years, in establishing public security, and reconciling the Christian and Musulman populations, who formerly lived in constant feud. He has constructed roads and railways, and now there are five hundred miles of railways, and roads of about ten times this length, through country which previously had no such means of communication. The commercial traffic has been doubled, and the number of travellers, owing to the erection of several hotels and the safety of the road, have increased in a still greater proportion. In addition to these improvements, public education has been cared for, all creeds found in the country—Catholic, Musulman, and Orthodox Greek—have been placed on an equal footing, local usages and customs are respected, justice is meted out impartially to all, and the people are instructed in agriculture, cattle-breeding, and other industries. The number of primary schools has risen from five to three hundred. High schools, commercial and technical schools, have been opened in all the principal towns, and agricultural schools, with model farms attached, have been established for the purpose of illustrating the best systems of cultivation. Nor has art been neglected by Governor Kallay, who has energetically promoted schools of industry and art, in which every effort is made to develop the æsthetic taste and natural aptitudes of students, Moslem art being taken as the basis.

The progress already made under the supervision of the Austro-Hungarian authorities by the two Balkan provinces, and their future advance, are of considerable importance to the whole of Europe, not merely on account of the peace and security established in this hitherto disturbed country, but also because the commerce seems capable of considerable extension, and rich cultivable soil is here awaiting colonists, and travellers can find healthy resorts and beautiful scenery. In order to attract foreign artisans and agriculturists, the Government offers gratuitously ground, motive power, and fuel for the establishment of new industries, and also exemption from taxes for several years. In consequence of these measures, the population has increased by 32 per cent. in seventeen years, and now numbers 1,565,357. In recent years, 533 families, mostly German, have come to settle in the country as agriculturists.

The interest of M. Raoul Pictet's paper is of another order. Having demonstrated that the sun is the main source of all terrestrial energy, the lecturer declared that this is especially the case in Egypt, where there are more sunny days than in most other places. The persistence of high temperature and the constancy of weather conditions allow observations of great precision and value to be made in this country. Observing the sand-spouts (*trombes de sable*) the learned professor conceived the possibility of utilising in a not remote future the enormous quantity of solar heat poured down over Egypt for the irrigation of the country, and even of the whole Sahara. Near Cairo, as in other places in Egypt, dark columns of sand may be observed to rise early in the morning, especially above small elevations of the ground. These columns assume a conical form, about thirty feet in diameter, where most contracted, and sometimes rise to the enormous height of 10,000 or 13,000 feet. The professor, with several companions, made his way in the early morning to some spot where the phenomenon was likely to occur, and scattered around various light objects, such as coloured feathers, small scraps of paper, etc. He also placed maximum and minimum thermometers at intervals on a circle of about 1600 feet radius, which indicated a progressive elevation of temperature. It was soon ascertained that the temperature rose far more rapidly on the east side of the hillock of sand forming the base of the spout than on the opposite side. The difference between the temperatures of the sand and the air was very marked, the temperature of the air being only 72° F., while that of the sand was 82° to 86°. The temperature rises very rapidly, for by four o'clock or half-past in the morning, the time at which sand-spouts usually form, the thermometer on the eastern side marked 113°, or even 122°. Sand-spouts were observed at this time of day in Mexico by M. Henri de Saussure.

All light objects lying on the ground within a radius of 1000 or 1300 feet begin to move round in a circle, gradually drawing nearer together until they are collected at the centre of the disturbance. Then the gyratory movement increases in rapidity, and suddenly all the objects are raised from the ground amidst clouds of sand, attaining to greater and greater heights. At this stage the observation of the phenomenon becomes very difficult owing to the sand, dust, and gravel; yet the observer succeeded in ascertaining that the temperature of the sand column is from 100° to 122°, that the sand is raised in the space of an hour to a height of 13,000 feet, and that the funnel-shaped top of the spout has a diameter of 1300 to 2000 feet. Small objects raised up with the sand become so small that they are scarcely visible even through the strongest telescopes, and, when the column is dispersed, they are found scattered about to a distance of fifteen to eighteen miles. The column expands into a funnel-shape only at the height of five hundred feet. Sometimes the phenomenon continues through the entire day. All sand-spouts have the same appearance, and pass through the same phases of development. The movement is always from below, and the maximum temperature is reached between twelve and three o'clock, when, in the months of May and June, the sand not infrequently is heated up to 167° F. The column rises to its greatest height about two o'clock. Sometimes eight, ten, or even twelve of these spouts are visible at once.

The lecturer sought to determine, by means of an ingenious apparatus, the amount of heat received by the sand in a definite interval of time, and found that it is $5\frac{1}{2}$ units of heat in a minute—that is, the heat absorbed in that time would raise the temperature of a litre of water $5\frac{1}{2}°$ C. In the winter the quantity of heat is sometimes as much as $6\frac{1}{2}$ units.

M. Pictet looks forward to the time when this enormous quantity of heat will not be wasted as at present, but will be employed in raising the water of the Nile for the irrigation of Egypt. This, he pointed out, might be effected by covering a large area of ground with blackened plates of sheet iron, under which water would be conducted. Thus a huge boiler would be constructed capable of heating the water by the solar rays alone to a temperature of 150° F., the water of the Nile being at 68° to 77°. Such a boiler, with an area of 1 hectare ($2\frac{1}{2}$ acres), would develop two thousand horse power, by which the water of the Nile might be raised to the necessary height and poured in abundance over the desert lands. The properties of the Nile water on the one hand, and of the sand on the other, are such that these two agents need only to be brought together to produce luxuriant crops.

Evidently this bold scheme can only be realised by a combination of great enterprise and enormous capital.

THE TERRITORIAL GROWTH OF STATES.[1]

IN political geography not only must the territory over which a nation holds complete sway be considered, but also the extension of its rule over the adjacent seas, and those various rights which encroach on the domain of one state to the advantage of another. Such is the acknowledgment of the Caspian Sea as a Russian lake, contained in the Russo-Persian treaties of 1813 and 1828. The inclusion of Luxemburg in the German Customs union, and the control of marine and sanitary regulations by Austria-Hungary along the coast of Montenegro, are also important in political geography. For modifications such as these usually occur at the borders of the domains of a state, and are the signs of an advance for which they prepare the way, or the remnant of a former extension. The territory of a state is no definite area fixed for all time—for a state is a living organism, and therefore cannot be contained within rigid limits—being dependent for its form and greatness on its inhabitants, in whose movements, outwardly exhibited especially in territorial growth or contraction, it participates. Political geography regards each people as a living body extending over a portion of the Earth's surface, and separated from other similar bodies by imaginary boundaries or unoccupied tracts. The peoples are constantly agitated by internal movements, which are transformed into external movement whenever a portion of the Earth's surface is occupied afresh, or one formerly occupied is surrendered. Thus, in a metaphor employed by more than one writer, a people resembles a fluid mass slowly ebbing and flowing. It has seldom occurred in history that such movements have taken place over unoccupied areas; as a rule they take the form of encroachment and usurpation, or small territories, with their inhabitants, are annexed to larger ones. Similarly these larger states fall to pieces, and this union and disintegration, expansion and contraction, constitute a great part of those historical movements which geographically are represented by a division of the surface into greater or smaller portions.

The extension of the geographical horizon, a consequence of the corporeal and mental efforts of numberless generations, has continuallv provided new domains for the territorial growth of nations. To obtain political control over these, to fuse them and keep them together, has demanded ever fresh forces, which could only be developed with the slow growth of civilisation. Civilisation constantly supplies new foundations and bonds by which the sections of a people may be built up into a connected whole, and increases the number of those who are knit together by a consciousness of their interdependence. Ideas and material possessions spread out from small centres, and gradually extend their domains. We see the close connection between religious and political

---

[1] Abstract of *Die Gesetze des räumlichen Wachstums der Staaten*, by Prof. Dr. F. Ratzel, in *Petermanns Mitt.*, Bd. xlii. No. 5.

expansion, but this is far surpassed by the immense influence of commerce, which now gives a mighty impulse to efforts of expansion. And
all these motive forces derive new energy from the increased population,
which must find room for itself, and leads to expansion after it has
helped on the march of civilisation by its concentration.

Though the most civilised peoples have not always been the greatest
founders of states—for state organisation is an application of the forces
of civilisation under particular conditions—still, all the great states of
the past and present have been those of civilised nations. This is
evident at the present day, for the great states are situated in Europe
and the European colonial territories. China is the only state of vast
dimensions that belongs to a non-European civilisation, and of non-
European civilisations the East Asiatic is the most highly developed.
At the commencement of our own civilisation we find the largest states
in its cradle around the Mediterranean Sea, where the lands, however,
owing to their form and situation in a steppe zone, could not give rise
to states of continental dimensions. Only when several of them were
merged into the Persian Empire, did a state spring up, the area of which,
some 2,000,000 square miles, could compare with European Russia.
Egypt, with its desert lands, contains not more than 150,000 square
miles, and the inhabited tracts of Assyria and Babylon not more than
50,000. During the short period of its greatest extension, Assyria ruled
over a country only about three times as large as Germany. Of all the
earlier world-empires only Persia deserved this title. The kingdom of
Alexander the Great (1,700,000 square miles) and the Roman Empire
(1,300,000 square miles at the death of Augustus) did not attain to its
truly Asiatic dimensions. The empires of the Middle Ages were merely
fragments of the Roman. The feudal system favoured the formation of
small states, for the land was divided and subdivided into private estates,
and hence the general dissolution of states in which the last remnant of
the old Roman occupation passed away, after its other creations, science
and commerce, had already perished. On the ruins new organisations
arose, and in other continents, first in America and Asia, dominions
sprang up in conjunction with the commerce, faith, and civilisation of
Europe, which occupied twice and three times the area of the largest states
hitherto founded. The rapid advance of geographical discovery allowed
these new states to spread in three hundred years over America, Northern
and Southern Asia, and Australia, and the practically continuous growth
of population in Europe during the past two hundred years, together with
the invention of new means of communication, furnished ever new material
for, and inducements to, further expansion. The British Empire, the
Russian Empire in Europe and Asia, the United States of America,
China, and Brazil, are of a size never attained in the past.

Since the areas of states grow with their civilisation, people in a low
state of civilisation are naturally collected in very small political organisations, and the lower their condition the smaller are the states. Before
the Egyptian occupation Schweinfurth counted thirty-five—and probably
there were more—in the Sandeh country, within an area estimated at
about 53,000 square miles. In Junker's time a large state possessed a

territory scarcely as large as a third of Baden (Ndoruma's covered some 1900 square miles), while most of them measured one to five square miles, being indeed merely villages with their lands. Hardly less was the subdivision in Roman times among the Rhætians, Illyrians, Galls, and Germans.

Peoples also of more powerful organisation, who threatened with their locust-like swarms the infant colonies in North America and South Africa, founded only small states. They laid waste wide areas, but could not retain possession of them and weld them together. At the annexation Basutoland contained 12,000, Zululand 8,500 square miles, and, but for the interference of the Whites, these territories would have been still further disintegrated. The confederation of the Six Nations in the Alleghany country, which for a century was the most dangerous enemy of the young Atlantic colonies, held sway over, perhaps, 20,000 square miles, only partially inhabited, and in 1712 could send 2150 warriors into the field. The empires of Montezuma and of the Incas were not great states in respect of the size of their territories, nor were they firmly consolidated. The Incas at the arrival of Pizarro had not extended their conquests over as large an area as that of the Roman Empire in the time of Augustus, and their dominion was but a conglomeration of conquered states, barely a generation old, and beginning to fall to pieces even without the help of the Spaniards. Before Europeans and Arabs planted great states in America, Australia, Northern Asia, and Africa, these vast areas lay politically fallow. Politics, like agriculture, learned by degrees the resources hidden in the soil, and the story of each country is that of the progressive development of its geographical conditions. The attainment of political power by the union of small territories was introduced as a new idea into the divided lands of primitive peoples, and the struggle that naturally arose between the policies and needs of large and small state organisations is one of the chief causes of the deterioration of these peoples since they have come in contact with Europeans.

Accordingly the size of states diminishes with their age. Of the present great empires only China can be called old, and it has acquired the greater half of its territory within the last hundred years (Mongolia and Manchuria, Tibet, Yunnan, Western Sze-chuen and Formosa). All the others—the Russian Empire, Brazil, the United States, British North America, and Australia—have sprung up during the last three-quarters of a century on the territories of small native states. On the other hand, Andorra is more than ten centuries old, and Liechtenstein and some other small principalities of Germany are among the oldest in their part of the world; compared to them Prussia and Italy are in their first youth.

Certain developments have been alluded to which progress more rapidly than the state and prepare the way for it. Ranke lays stress on community of life among men. This lies in the ideas and wares which are passed on from one people to another. It is seldom that a state has been able to set limits to their wanderings; as a rule, indeed, they have drawn the states after them. Animated by the same impulse and following the same path, are frequently found together ideas and wares, missionaries and merchants, drawing peoples closer and assimilating them, and thus preparing the ground for political approach and union.

All old states and those in a low state of culture are theocracies. The spiritual world here not only controls the life of the individual, but also determines that of states. Every chief has priestly functions, every tribe its holy place, every dynasty prides itself on a divine origin. Over the universal political decay of Europe hovered the Church, preparing the way for new and greater states, while Islam undertook the same task in Western Asia and North Africa, and in Africa of to-day the powers of Islam and Christianity preside over political separation, while between lies paganism, with its smaller divisions.

Primitive states are national in a very restricted sense ; their development is brought about by the removal of the barriers, and then they become again national in a more extended sense. The states of primitive peoples are family states, but their earliest growth is often due to the advent of strangers. Thus people of the same origin may indeed be brought together, as far as the territory of the race extends, but the union is not national, though the community of language and customs produced by intercourse independent of politics may facilitate political connection. In times of greater intellectual development these common possessions produce a feeling of nationality, and exercise an attractive and cohesive force. This feeling of nationality, however, does not spread with the rapidity of religion and commerce, and, therefore, comes sooner into conflict with the territorial extension of the state, which has always gained the final victory ever since the Roman Empire first strove after a cosmopolitan character. But the state recognises the value of national feeling, and seeks to convert it into political patriotism by helping on the fusion of peoples, and to use it for its own ends—Panslavism, for instance. The modern state, embracing a large area of territory, but truly national, is the peculiar product of this process. Between this and the restricted state of early times lie the numerous states of the past and present, which lack a civilisation sufficiently powerful to weld together the heterogeneous elements of the ethnographical foundation.

Commerce and traffic hurry on far in advance of politics, which follow in their footsteps and are never sharply separated from them. Peaceful intercourse is the condition of the growth of a state. Its primitive network of routes must be laid down beforehand. Knowledge of a neighbouring territory must precede its political annexation. When the state has commenced to expand, it shares with traffic in its interest for means of communication, and may even take the lead ; the made roads of the Iranian and of the old American states owed their origin to political aims rather than commercial. Every route prepares the way for political influence, every waterway is a natural agent for state development, every confederation entrusts its traffic arrangements to the central power, every negro chief is the first, and if possible the only, trader in his land. The advance of the customs boundary precedes that of the political boundary ; the customs union proclaimed the coming of the German Empire.

The connection of the enlargement of the geographical horizon with political expansion is too evident to need much discussion. Even in the present time the greatest results have been obtained through geographical exploration, of which the Russians in Central Asia have afforded the most brilliant examples.

The growth of states proceeds through the annexation of small territories to amalgamation, while at the same time the attachment of the people to the soil becomes ever closer. Out of the mechanical union of territories of varying size, population, and stages of culture, an organic growth is started by the approachment, mutual intercourse, and intermingling of the inhabitants. Growth which never goes beyond mere annexation creates only loose, easily dismembered conglomerations, which are only temporarily held together. The Roman Empire was constantly threatened with disintegration till the first century B.C., when the military organisation necessary to hold it together was established, and the commercial supremacy was won for Italy, which made the fortunate peninsula in the middle of the Mediterranean the centre of a region traversed by excellent trade-routes.

This process of fusion of various tracts also indicates a closer connection of the people with the soil. The expansion of the state over the Earth's surface may also be accompanied by a downward growth which helps to fix it to the soil. It is not a mere metaphor to speak of a people taking root. A people is an organic body which in the course of its history is fixed more and more firmly in the soil on which it lives. As the individual contends with the virgin soil until he has converted it into cultivated land, so a people struggles with its territory and makes it ever more and more its own by shedding its sweat and blood on its behalf, until the two cannot even be thought of apart. We cannot think of the French apart from France, or of the Germans apart from Germany. But this union was not always so close, and there are states which even now are not so intimately associated with the land. There is a historical succession in the relation of states to their territory, as there is in their dimensions. We never meet with that complete severance from the soil which, according to the theories advanced by many speculators, is the mark of a primitive state of existence. But the further we go back, the looser becomes the connection. Men are more scattered, and their cultivation is more superficial and is lightly transferred from one field to another, and their social relations bind them so firmly together that their relation to the soil is comparatively weaker. And as the small states of this stage of culture are isolated from one another by uninhabited marches, etc., not only does a large tract, often more than the half of an extensive territory, lie politically fallow, but the emulation is wanting to develop what is politically of value in the country. Thus the large rivers are not used by Indians and Negroes as boundaries or waterways, while they immediately acquire inestimable value as soon as they are reached by Europeans.

Hence on the whole there is a decrease in the appreciation of the political value of the ground as we go back from the newer to the older states. It stands in close relationship to the decrease of political areas. The earlier observers of African life have remarked that the constant petty wars lead to no extension of territory, but are merely a means of obtaining slaves. This fact is of the greatest import in the history of the Africa of the Negroes; the slave raids decimate the population, and at the same time hinder the development of states. The state never rests, and the continual excursions across the frontiers makes it a centre for expeditions of conquest surrounded by a zone of depopulated and desolated country. The frontiers are not clearly defined, and evidently depend on the energy with which excursions are made. As soon as this energy declines the territory shrinks. There is no time for the people to take root on a particular area. Hence the usually short duration of these powers, of which examples may be found in south-east Africa, from the Zulus to the Wahehe. In the more advanced states of the Sudan this region of conquest, or rather of raiding expeditions, occupies only a part of the state; the position and extent of the Fulah states, in Bornu, Baghirmi, Wadaï, Darfur, remain for a long time unaltered, but vary continually where they come in contact with the subject heathen lands, that is, chiefly on the southern side. Nachtigal in the north, Crampel and Dybowski in the south, have shown how indefinite the position and extent of Wadaï is in those parts. Still more indefinite is the political value of the soil. The land-hunger of the conquering states of antiquity, especially of the Romans, so often mentioned, does not stand out clearly by any means. The acquisition of territory is only a secondary feature in the great political upheavals of antiquity. Power, slaves, treasure, are the prize, especially in the wars of the Asiatics; hence the ephemeral character of their growth. In Rome, after the wars with Pyrrhus, a struggle with the necessity of acquiring new territory may be observed, in which, as empire was desired, the system of alliances and of keeping one power in check by others, had to give way. Cæsar's greatness lies in his having first given the state, together with extension of territory, a defined and secure boundary.

Territorial growth is effected on the periphery of the state by the displacement of the frontier. A state which aims at the possession of certain tracts sends out spurs which are filled with more vigorous life than the rest of the periphery. The outposts of Peshawur and Western Tibet, of Merv and Kokan, show even to those who do not know their history, that in these directions British India and Russia are pushing on with peculiar energy, endeavouring to possess themselves of all the advantages of the intervening lands, just as Rome, by the conquest of Gall, made haste to oppose the advance of the Germani. On its German and Italian frontiers, for centuries the scene of especially vigorous growth, France concentrates her forces with the object of regaining lost ground. The marches of Germany, as it expanded eastwards, were fortified and colonised bit by bit as they were conquered, and the same process was repeated in the United States of America and in Argentina, where in a few years large towns sprang up on the site of the primitive block-houses of the Indian frontiers. In the crowded state of Europe, such parts of the periphery are the most exposed to danger, and the most strongly fortified; the wounds they may receive are the most to be feared.

In the most primitive states the frontiers are so indistinct as to vanish altogether. The attempt to apply our notion of a frontier, as a well-defined line, to states which occupy no sharply delimited tract, has led to the worst misunderstandings, both in the Indian policy of the American states and in Africa. Lichtenstein spoke of the futile attempts to fix a definite boundary for the Kaffir territories, which neither party should cross without the special permission of the chiefs. Not lines, but positions, are the essential features in these cases. Contact is avoided, and the state draws itself together, surrounding itself with a politically empty zone. But if its people thrust themselves forward beyond these limits, they insinuate themselves in among their neighbours rather than displace them. The proprietary rights of the chiefs among primitive peoples are inextricably mixed up. Though the confusion has caused much perplexity to colonial offices, it has afforded conquering and colonising powers, bringing with them different notions of what a frontier is, excellent opportunities for interference and aggression. Combined with the different appreciation of the value of land politically, it has wonderfully accelerated the dispossession of the natives. Their policy was like their trade, for they recklessly gave away their most valuable possession, because they had no idea of its value. Long before, the isolation of one small state from another had made its disastrous consequences felt in causing stagnation, which on the arrival of Europeans turned into decay. At a higher stage—in the Sudan and Further India—the frontier is defined in many parts, mountains and watersheds being made use of, though the system of unoccupied marches is still retained. China had separated herself, till a few years ago, from Korea by such a frontier zone, which was, however, in contrast to the African and Further Indian, clearly defined.

The state in its growth selects the geographically advantageous positions, occupying the good lands before the bad, and, if its growth is accompanied by encroachment on the territory of another state, it takes possession of the important points, and then advances towards the less valuable parts. In new lands (colonies), the history of which is fully known, the new political organisations spread up from the sea, beside the streams and lakes, on the fruitful plains, while the older political organisations are thrust back into those parts of the interior which are difficult of access, and offer little temptation to the aggressor, into steppes and deserts, mountains and swamps. This is what has happened in North America, Siberia, Australasia, and South Africa. Pioneers of the same civilisation have on the whole the same notions with regard to the value of the land, and hence the harmony in the expansion of all the European colonies of the last century. But in other times different notions prevailed. The old Peruvians did not descend to the Amazons, but spread themselves over a small strip of plateau 2500 miles in length. The old Greeks did not seek for large fertile inland tracts, but, like the Phœnicians, islands and peninsulas, while the Turks occupied the elevated steppes of Asia Minor. Beside the stage of civilisation the force of habit plays an important *rôle*, and that is why political expansion proceeds as far as possible over regions where the conditions of life and work are the same. The Phœnicians settled on the coasts, the Dutch on islands, the

Russians on river-banks. How much the landlocked character of the Mediterranean Sea contributed to the expansion of the Roman Empire was well known to the ancients. To both Greeks and Romans these lands were a most happily situated colonial sphere, where they could everywhere feel themselves at home, just as the Europeans of Central Europe do in America between 45° and 35° N. lat. The enclosure of politically advantageous points often determines the configuration of states. The extension of Germany along the North and Baltic Seas, the incorporation of the Meuse north of Sedan in France, and England's possession of the Channel Islands, are cases in point. Chile's northern frontier across the apparently worthless desert of Atacama, in lat. 24°, was pushed forward to the 23rd parallel as soon as the guano deposits in the bay of Mejillones were discovered, and the advance of Great Britain across the Orange river followed the discovery of diamonds on the Vaal river in 1867. At a lower stage states show a preference for trade-routes and the neighbourhood, as may be remarked in the Sudan and in Central Africa.

In many cases the growth of states preserves for long periods the same direction, with the object of obtaining political benefits, for there is an advantage in following the line of country most favourable to the movements, or rather succession of movements, by which this growth is effected. So people make for the coasts, pass along the rivers, and spread over plains. Others push their way up to the limits of the country accessible to man, the incentive being the advantage of filling up a territory with natural boundaries. Rome spread out in North Africa and West Asia along the desert. 222 B.C. it had reached the southern foot of the Alps, but it was not till two centuries later that it progressed further, when it had already spread eastwards and westwards far beyond the Alps. Bohemia filled its hollow before any neighbouring state had acquired fixed boundaries, and when it expanded further, it was towards Moravia, in the direction of the opening in its hollow. To the same group of cases belongs the growth in the direction of least political resistance. The spread of the states of Central Europe in the east, beginning with the first division of Poland, was a reflux of the political energy long in vain directed westwards.

A simple political body, if left to itself, renews and multiplies this body continually, but never creates another. The family is renewed in its offspring and creates new families. From the family tribe or the race another family tribe branches off, and so on. All these corporations become states through connection with the soil. As they increase, no larger state arises out of the small one, but a number of states of the same size. In order that the customary limit of growth may not be transgressed, the number of the population is kept down by every possible means, some being of the most ghastly description. Thus the state may be kept within manageable limits and be held in the grasp of one hand. As far as we are acquainted with the states of primitive peoples, their growth has always been brought about by external influence. Men from countries where larger ideas of space prevail, carry the conception of larger states into those where the ideas of space are more restricted. The stranger is superior to the native, who is only acquainted with one state, by the fact that he knows at least two. If we think of Africa before the foundation of European colonies, we find the large states all along the line where Negroes are in contact with Semitic and Hamitic peoples, and scarcely any where Negro states lie side by side or border on the sea. Where, moreover, Negro states exist in the interior, their foundation is always traditionally associated with a stranger. All states in Africa are due to conquest and colonisation. History exhibits again and again the quiet immigration and spread of a people which, at first only tolerated, has suddenly come to the front as rulers. Such has been the course of European colonisation in almost every case, thus the Chinese established their kingdoms in Borneo, and at the dawn of the Roman Empire we find, veiled indeed in mythical obscurity, the strangers whose advent gave to Rome, advantageously situated for trade and navigation, a preponderance over the other towns of Latium. In the whole of Melanesia at the advent of Europeans there was only one state formation, through immigrant Malays on the north-west coast of New Guinea, and similar cases may be found in America.

Where Europeans have not brought in the conception of an extensive dominion into the region of small states, its introducers have been sea, desert, and steppe peoples—Hamitic and Semitic, Mongols and Turks. If we further inquire whither the search after its origin leads us, we arrive at the shore of the eastern Mediterranean, where fruitful lands are situated amid extensive steppes. Egypt and Mesopotamia, Syria and Persia, are large oases favourable for the concentration of the people on a narrow area, and surrounded by regions which tempt the inhabitants to expand. As Lower Egypt has extended over Upper Egypt, China from its löss districts in all directions, all these regions have supplied men for armed incursions and slow conquest by colonisation. The political organisation of these masses, and the wide supremacy which knit together the individual lands, were the creations of the steppes, whence sprang the founders of large states in Egypt and Mesopotamia, Persia, India and China, and the African Sudan. The absence of pastoral peoples who once ruled the greater part of the Old World, deprived pre-European America of an ever-active political ferment, and hence is partly due the weakness of its state organisation.

The influence of nomad pastoral tribes on agricultural and industrial peoples is only one side of a deeper contrast. This is the fundamental fact which underlies the state foundation of the seafaring nations—the Phœnicians, Normans, and Malays—and of the most recent European colonies. We find it also in the universal tendency of settled peoples, and especially agriculturists, to withdraw or cut themselves off from political affairs. All purely agricultural colonisation, of the Achæans in Grecia Magna, the Germans in Transylvania, and the Boers in South Africa, tends to stagnation, and the success of Rome lay in stimulating a sturdy peasantry by more mobile and experienced elements.

There is a difference in the historical progress of mankind. Some remain stationary, while others push on, and both conditions are favoured by the nature of the country, wherefore from seas and steppes, the regions of movement, state formation advances into the forest and arable lands, the regions of fixed settlement. In a stationary condition weakness and decay creep in, whereas aggression demands organisation, which among the Tartar hordes, as in the ships of the Vikings and Malays, accomplished great results with small forces. The most extreme cases exist in Africa, where a martially organised people, like the Zulus, is found side by side with a people that for generations has been falling to pieces, like the Mashona. They form part of a single whole, for the one people exists at the expense of the other.

As the appreciation of the political value of land becomes greater, territory becomes to a greater degree the measure of political strength and the prize towards which the efforts of a state are directed. That the relative areas of Austria-Hungary, Germany, France, and Spain may be represented by the figures 100, 86, 84, and 80, those of the Netherlands and Belgium by 100 and 90, of the United States and British North America by 100 and 96, is the result of slow development and a balance struck after many struggles. From the smallest beginnings of growth up to the mighty states of the present age we see the same struggle of small states to raise themselves to an equality with the larger, and of the larger to push themselves to the front. The impulse has been as strong in the village states of Sandehland as in the huge states which occupy half a continent. We see it now at work in continental Europe, which it has awakened to the necessity of combination, in commercial affairs at least, against the giants, Russia, North America, and the British Empire. And the same law has prevailed no less in the most recent colonial enterprises ; in Africa there has been an eager competition among the powers for the acquisition of territory, and the east of New Guinea has been divided between Great Britain and Germany in the proportion of 125 : 100.

The end has been attained in very different ways. A small state has taken enough land from its neighbours to make it equal, or nearly so, to the largest of them. Or states develop side by side or in succession, the later comer appropriating as large an extent of land as the first has occupied—the United States and British North America, for instance. Again a state may split into two of nearly equal areas, as the Netherlands and Belgium. A state that has been reduced in area may annex in another direction enough land to make good the loss, as when Austria acquired 19,730 square miles in the Balkan peninsula to compensate for the loss of 17,110 square miles in the Apennine peninsula. Russia and China, finding themselves in a huge territory where it was impossible to remain stationary, have become the masters of Central Asia, the former having similar problems to solve with regard to the Turkish peoples, as the latter in dealing with the Mongols.

Naturally the efforts of nations are not confined to extension of territory. Even large states come into close contact at length. Neighbouring states share in the advantages of position and natural resources, and hence arises a uniformity of interests and modes of activity. Beside the lines of communication between the Atlantic and Pacific in the United States, Canada has constructed her Canadian Pacific railway, and the navigation of the great lakes is made available by canals on both sides. Throughout America the constitution and political life of the United States serve as a model, just as in the Sudan one form is observable through all the Mohammedan states, whether their founders were Fulahs or Arabicised Nubians.

In peaceful competition, as in armed contests, the rule is that the aggressor must advance on to the ground occupied by his opponent, and in overcoming him he assimilates himself to him. States bordering on steppes in their struggle with the inhabitants of the steppes must acquire so much of the same character as will enable them to avail themselves of the advantages the steppe affords. Russia and Central Asia and France in Algeria illustrate this principle.

PROCEEDINGS OF THE ROYAL SCOTTISH GEOGRAPHICAL SOCIETY.

CLOSING OF THE SOCIETY'S ROOMS FOR CLEANING.

THE Council has approved of the Society's Rooms being closed for cleaning from Monday, 31st August, to Saturday, 19th September, both days inclusive.

GEOGRAPHICAL NOTES.

By THE ACTING EDITOR.

EUROPE.

The Climate of Italy.—The *Annuario Statistico Italiano* for 1895 contains a mass of valuable material collected by various departments of the Government. The first part is devoted to the climatology of Italy, the tables referring to the year 1893. As regards climate, Italy may be divided into four regions—the northern, eastern, western, and southern. The first, coinciding with the valley of the Po, has a short, warm summer, and large daily ranges of temperature. Places situated on the Alpine slope have a mild winter relatively to their elevation and to those lying along the axis of the valley. Towards the Adriatic both the annual mean and the winter mean increase. Very low temperatures have been recorded at Padua, Milan, Turin, and Alessandria, and the annual range, from the greatest heat at Milan (99.5° F.) to the greatest cold at Alessandria (0.1°), marks the climate as distinctly continental.

Central Italy is divided by the Apennines into two great parts, the eastern and western. The extremes of temperature diminish towards the south, and the temperature rises on both slopes towards the sea. On the eastern slope, the mean annual temperature does not exceed 57°, and the absolute range is only 93°. In the western zone the annual mean is 57.7°, and the range 91°.

Calabria, the Basilicata, Apulia, Sicily, and Sardinia form the last division, in which the annual mean varies from 61° to 64°, and the difference between summer and winter is only 25°. The higher mean in this zone is due to the greater warmth in January, February, March, November, and December, and therefore it does not appear that the summer heat increases in Italy towards the south. August is warmer than July in this southern region.

A comparison of recent records with those of Schouw (*Tableau du Climat d'Italie*, 1839), referring to the early years of the century, shows that the climate has undergone no marked change.

In the Alpine region, particularly in the valley of the Po and the larger valleys of the Apennines in Central Italy, there are two maximums and two minimums of nebulosity. The maximums occur in spring and autumn, that of autumn being the highest. The lowest minimum occurs in summer. The average number of clear days is above 150 in Venice, Pavia, and Ancona, and less than 120 in Turin, Moncalieri, Alessandria, Urbino, Leghorn, Siena, and Perugia.

In distribution of precipitation Upper Italy differs from Central, and still more from Southern, Italy. In the south there are two seasons, a wet and a dry, whereas North Italy has two maximums and two minimums, the rain falling abundantly and being accompanied by storms, especially in May and June. The provinces of Udine and Belluno are especially rainy, the fall amounting to about sixty inches in the year. Genoa and the province of Vicenza follow, while the Capitanata, the Salentine peninsula, and Sardinia are the driest parts of the kingdom. For the rest, the distribution of rain is so irregular that the amount received by adjacent places may differ considerably.

In summer Italy is not affected, directly or indirectly, by the cyclones of the Atlantic, and the numerous storms are due to local winds. In spring and autumn also, though the great cyclones pass from west to east at lower latitudes, they seldom impinge directly on Italy, but the danger from secondary disturbances is greater when anticyclones bar the passage of a cyclone to the east, or when great barometrical depressions lie over the land or seas of Italy. The storms of winter sometimes, and these the most violent, pass southward from the British Isles, impinging on the Alps between Mont Blanc and the Maloja pass; others descend from the Baltic; others, again, reach the western coast from the Gulf of Gascony and Spain, while storms from the Straits of Gibraltar visit the southern parts of Italy.

The Plitvitz Lakes.—Dr. A. Gavazzi has examined seven of these lakes, lying to the east of Kl. Kapella, and obtained a large amount of geological information and data of the temperature, colour, and transparency of the water, the constitution of the bottom, etc. The following are the altitudes, areas, and depths of the Lakes:—

	Altitude in feet.	Area in acres.	Maximum Depth. in feet.
Prošće, . . .	2110	179	92
Eiginovac, . . .	2093	19	36
Okrugliak, . . .	2073	14	26
Batinovac, . . . ca.	2034	2.5	13
Galovac, . . .	1909	31·6	367
Gradinsko, . . .	1824	16	20
Koziak, . . .	1758	185	134

—*Geographische Zeitschrift*, Jahrg. xi. Heft 4.

New Explorations in the Karst.—The demand for water in Trieste, not only for drinking, but also for industrial purposes, now that the town is becoming a seat of industry, is growing more and more urgent. It has been proposed to conduct water from the Feistritz springs, thirteen miles off, which issue from limestone strata and are of excellent quality. After ten years, however, the scheme is no nearer fulfilment, and therefore it has been taken up by private speculation. In connection with the undertaking, the subterranean hydrography of the neighbourhood has been investigated, and a number of so-called blow-holes have been discovered, none of which is accessible. Independently of this undertaking, the cave explorers have been busily engaged. Herr Masinitsch has obtained entrance into the chasm of the Kačnajama, 700 feet deep, by fixing ladders in it, and has found vaults of great beauty, adorned with elegant stalactites, which descend to a depth, as far as at present investigated, of 1000 feet. The rise and fall of the flood water in October corresponded so exactly with that of the Reka that there are now hopes of penetrating into the hitherto inaccessible part of the subterranean channel of that river. The Reka is supposed to be the main stream into which the others pour their water underground. It falls into the Reka cavern twenty-five miles from the Timavo mouth, which is held to be its outlet. The Karst hydrography is not yet fully known, and the Societá di Scienze Naturali, under the direction of Dr. Marchisetti and Dr. Valle of the Museo Civico, are turning their attention to its investigation.—*Globus*, Bd. lxix. No. 20.

The Rainfall in the Caucasus.—Lying between the Black and Caspian Seas, and traversed by a continuous lofty chain running from east to west, the Caucasian territory presents singular and interesting meteorological conditions. The means of precipitation during the course of the year attain their maximum of 200 mm. (7·87 inches) to the north and south of Sukhum-Kale, near Socha and Batum. From the former point they diminish towards the north-east, and from the latter towards the east, not forming separate regions but one area of decreasing fall towards the Caspian Sea, so that the line of 3·94 inches (100 millimètres), beginning somewhat south of Novorossiisk runs parallel to the crest of the range as far as a point a little east of the road from Vladikavkaz to Tiflis, where, turning sharply to the west, it follows, north of Tiflis, the southern slope of the mountains and south of Kutaïs bends suddenly southwards ; then passing north-westwards past Ardagan it reaches the Black Sea again close to Batum.

The curve of 1·97 inches (50 mm.) begins in the north at Anapa, extends eastwards as far as 42° E. long., passes southward to the medicinal springs of Piatigorsk, and thence following the direction of the mountains, and farther on running parallel to the Caspian Sea at no great distance from the shore, reaches Shemakha. Here it makes a sharp angle to the north-west, and runs in this direction to Tiflis, and here turning still more sharply to the south-east, strikes the Caspian Sea on the right bank of the Kura at about 39° N. lat. To the south of Lenkoran, the zone of 100 mm. again makes its appearance on the Caspian shore. The steppe skirting the Caspian Sea north of the mouth of the Kura receives less than 25 mm. (·98 inches), and also a narrow zone stretching Elizabetpol south-eastwards to the Caspian.

The winter rainfall (December to February) is very similar to, but heavier than, the annual. In isolated patches near Batum and Socha it amounts to 600 mm. (23·62 in.), diminishing rapidly towards the north-east, so that the curves lie close together. The curve of 200 mm. embraces an unbroken region, beginning at Anapa, and, running across the Vladikavkaz and Tiflis road in the mountains, follows thence a south-westerly direction to the Chorokh river south of Batum in 41° N. lat. The curve of 100 mm. (3·94 in.) runs near and parallel to the former, but forms a small island near Stavropol, and encloses a small strip of the Caspian coast from Kizliar to about 41° N. lat. The middle of Daghestan around Khunsakh, and the lower course of the Kura are enclosed in curves of 50 mm. (1·97 in.). The 25 mm. line first appears west of the mouth of the Volga.

The curves of the spring rainfall (March to May) are very similar to those of the year. The extremes are in the same position, with 400 mm. at Socha and 300 at Batum. But two ovals occur on the road from Vladikavkaz to Tiflis of 400 and 300 mm. respectively. The lower course of the Kura and the coast of the Caspian are very dry, and west of the Volga mouth is the driest district of all. Summer and autumn also exhibit the same general characteristics. In summer (June to August) the extremes are 500 mm. (19·69 in.) north of Batum and 400 mm. (15·75 in.) near Socha, and the curve of 300 mm. starts from the Black Sea at 44° N. lat. and ends to the south of Batum, making a wide sweep into Daghestan. The driest zone lies near Baku and is very small.

The autumn rains are much heavier, being 800 mm. (31·5 in.) over a district round Batum and 500 mm. round Socha. The curve of 300 encloses a zone along the Black Sea, beginning to the south of Novorossiisk and passing by Kutaïs to Artvin. To the north-east the country is dry, especially south-west of the Volga mouth. On the Caspian Sea lies a small strip with only 100 mm., and the country round Erivan receives only 50 to 100 mm.

The results show that the rainfall centre lies in the heart of the mountains round Kazbek, beginning south of Vladikavkaz and ending half way on the road to Tiflis. The next zone surrounds the former and covers a broad tract extending to the Black Sea and from Kuban to the upper course of the Araxes. Isolated strips occur by the Gokcha lake, round Stavropol, and east of Piatigorsk. The third zone embraces the rest of the territory exclusive of the country north of the Kuma, round Elizabetpol, and on the Lower Araxes. The driest summers occur in the steppe north of the Kuma, on the Caspian Sea between Derbent and Baku, on the Middle Kura from Elizabetpol upwards, and on the Middle Araxes. Dry winters are the rule north of the Lower Kuban, in a narrow strip stretching eastwards to the Kuma past Piatigorsk, and Vladikavkaz to Baku.

On the whole the central parts of the mountains and their southern slope are the wettest parts of the Caucasian province, and the larger, north-eastern district, with the slope to the Caspian Sea, the dry part. This is in harmony with the lower limit of snow on the south-west slopes of the mountains.—*Deutsche Rund-schau*, Jahrg. xviii. Heft 9.

ASIA.

Meteorology in Mysore.—Observations were started in the Mysore province three or four years ago, at the request of the Government of India. Bangalore, Mysore, Hassan, and Chitaldroog were chosen as suitable stations, and were supplied with sets of instruments by the Meteorological Department of India. The Director of Meteorology in Mysore, Mr. John Cook, has sent us his *Meteorological Results for the Years* 1893 *and* 1894.

As regards atmospheric pressure, the similarity between the records at the four stations is very marked, and even the variations at Madras, 190 miles to the east, are singularly close. In the period of the north-east monsoon the mean daily atmospheric pressure is above the yearly mean, while, from April to September, the period of the south-west monsoon, it keeps below the yearly mean. The monthly mean temperature in the shade falls below the yearly mean from October to January, which may be said to constitute the cold season. The coldest month in Mysore, and Madras also, is December, but January is very slightly warmer. There is a great contrast between the daily range in Mysore and at Madras. On the Mysore plateau the range reached nearly to 34° at Hassan, in January and February 1894, and 26° at Chitaldroog, while in Madras it never rose as high as

23° in 1893, averaged 25° for six days in January 1894, but rarely touched 20° during the rest of the year. The mean annual rainfall for the four Mysore stations was 31·68 inches on 110 days in 1893, and 30·59 inches on 104 days in 1894. October was the rainiest month, with an average of 8·1 inches. The greatest falls in one day were 3·26 inches at Chitaldroog, and 3·23 at Bangalore.

AFRICA.

The Geographical Relations of St. Helena.—This isolated volcanic block rising from the depths of the South Atlantic, 1000 miles from Africa, and 1800 from South America, has lately acquired increased importance for zoologists and geologists, because Ihering regards it as the remains of a connecting link, the Helenis, between Africa and America, which existed in Mesozoic times. The island possesses no indigenous mammals, reptiles, or freshwater fishes, while the only sea-bird peculiar to it, the *Ægialites Sanctæ Helenæ*, is closely related to an African variety, and may at some time or other have found its way thence to the island.

In order, then, to form some conclusion about the former connections of the island, Dr. W. Kobelt has to fall back on the molluscs, plants, and insects. As regards the last, the great majority, according to Mr Buchanan White (*Proc. Zool. Soc.*, 1871), may be referred to the Palæarctic region. The land molluscs, Dr. Kobelt holds, may with a high degree of probability be considered part of the molluscan fauna of a submerged southern continent extending to Polynesia on the one hand, and to South America on the other. A connection with Africa could not have existed, and the connection between the continents inferred by Neumayer from the distribution of Jurassic ammonites, must, if it existed, have been situated farther northwards.

The flora of St. Helena, on the other hand, is totally different from that of the American continent. The indigenous fauna has, indeed, to a large extent disappeared. However, a migration from the supposed southern continent is not impossible, and the abundance of ferns points in this direction.

Snails and plants, therefore, lead to a conclusion diametrically opposed to that deduced by Mr. White from a study of the insects. But there is no proof that the migrations of animal forms and plants all took place in the same period. Plants and snails may have been introduced in the Mesozoic period, while the insects may date from a later age. Now, the currents round the island set in from the direction of the cape, but Dr. Croll remarked that in the Glacial period the prevailing winds and currents over a large part of the Atlantic Ocean probably flowed from the north, as a maximum atmospheric pressure must have been formed over the great polar land-ice just as now over Russia and Siberia. The position of St. Helena could form no hindrance to the acclimatisation of extra-tropical forms, for its climate is very moderate, the temperature seldom rising above 61°. As regards marine molluscs, Smith (*Proc. Zool. Soc.*, 1890) mentions forty kinds which are carried with the sea-horn from the Cape, but none of these can establish itself in the warmer waters. Of the living varieties half are West Indian, while nearly a third occur in the Mediterranean. Consequently, these also indicate a very different arrangement of currents from that now existing.—*Geogr. Zeitschrift*, Jahrg. xi. Heft 4.

AMERICA.

Lieut. Peary in Greenland.—The chief particulars of this explorer's work in 1894 and 1895 have already been reported. A few interesting notes may, however, be gathered from his paper in the *Bull. of the American Geogr. Soc.*, vol. xxviii.

No. 1. Lieut Peary remarks on the regularity of the winds of the inland ice. He sojourned on the ice for more than seven months, besides making visits to it of shorter or longer duration, and always found that, except during unusual atmospheric disturbances, the winds blew from the interior outwards, in a direction perpendicular to the nearest coast-line. These constant winds must, he believes, tend to prevent the ice-cap from increasing in height by continually transporting loose snow to the periphery.

The appearance of the northern land strip in the neighbourhood of Independence Bay is much less forbidding than that of the Whale Sound strip. The northern shores of Whale Sound are almost everywhere bold, and the plateau above the cliffs is almost completely covered by tongues of the inland ice, or detached ice-caps on the islands, whereas the Independence Bay land, though elevated, is rolling and bare of ice-cap, broad areas of the red and dark brown land being exposed to view. The geological features are the same at both places. The lateral moraine at Independence Bay contains the same rocks, and the level tops of the high mountains and ridges show the same hard, compacted surface. Dark granite and gneissose cliffs and trap dykes may be seen as about Whale Sound, and wave-marked slabs of red sandstone, similar to some found at Bowdoin Bay, were observed 3000 feet above the sea at Independence Bay.

Lieut. Peary was much struck by the greater abundance of flowing water, not only on the land, but along the edge of the ice-cap in this latitude. In 1895 the slopes of the inland ice for several miles from the moraine were free from snow, while at Bowdoin Bay deep snow partially covered even the moraine. On the land small lakes and pools are numerous, and rushing brooks are everywhere, and sharply marked tumuli and embankments of morainic material, miles in advance of the present edge of the ice, indicate more strongly than anything to be seen in the south that the ice is undoubtedly retreating from a considerable area. The smoothness of the bay ice, the absence of icebergs, and other circumstances render it probable that the ice is driven every year out of the bay by the wind. The mainland of Greenland ends between the 82nd and 83rd parallels, and beyond it lies an archipelago of unknown extent, stretching north and north-eastwards.

A map of Whale Sound, drawn from the surveys of the expedition, accompanies the paper.

The Mexican Census.—According to the census taken at the end of 1894, the total population of the country is 12,542,057, giving a density of 16·3 to the square mile. In 1890 the total was 11,490,830, and, therefore, there has been a considerable increase in the four years. The capital has 339,935 inhabitants, Puebla 69,676, Guadalajara 83,870, San Luis Potosi 69,976, Monterey 56,835, Merida 56,702, Pachuca 52,189, Durango 42,166, and Zacatecas 40,026. The increase in San Luis Potosi, Monterey, Pachuca, and Durango has been strikingly large, chiefly in consequence of the construction of new railways. It must, however, be taken into consideration that the older figures were merely the result of calculation, not of actual enumeration. Hence the decrease in the population of the capital, which was given in 1890 as 350,000, may be only apparent.—*Geogr. Zeitschrift*, Jahrg. xi. Heft 4.

The Future Capital of Brazil.—The project of establishing a new capital on the central plateau of Brazil, and the expedition of Dr. L. Cruls to examine the country, have already been briefly alluded to in these pages. The *Tour du Monde* of April contains fuller details on the country and the proceedings of the expedition.

The reasons that induce the Brazilians to remove the capital are that Rio de

Janeiro, though possessing an excellent harbour, is too exposed to the attacks of a hostile fleet, and that, placed at the extreme edge of the country, it tends to keep the commerce in the coastal region instead of attracting it into the interior. The climate also is warm and damp, and yellow fever rages at times with great virulence. In 1890, therefore, it was decided to mark off an area of about 5560 square miles on the central plateau for a new federal district, and in 1892 the commission, under Dr. Cruls, started for the interior, with orders to take accurate observations of positions, and thoroughly investigate the orography, hydrography, climatological and hygienic conditions, the nature of the soil, the quantity and quality of the water supply, timber, and other materials of construction, and any other matters bearing on the fitness of the district for the purpose intended.

The central plateau occupies a large part of the States of Rio de Janeiro and Minas Geraes, and a smaller proportion of the State of Goyaz, and sends off narrow belts into the State of Bahia, and to the west of the Rio São Francisco to where Goyaz adjoins the State of the Maranhão and Piauhy. It was to the central part of this plateau, in the neighbourhood of the Brazilian Pyrenees in the State of Goyaz, that the commission directed its steps. Leaving Uberaba on June 29th, the party marched towards Pyrenopolis, and were much surprised at the remarkable fall in the temperature. The inhabitants of the country asserted that there had never been such cold during the previous ten years. At Mariano dos Cazados, 1600 feet above sea-level, the thermometer marked 28·2° F. in the night between the 12th and 13th of July, and everything was covered with a coating of ice. This frost extended for about two hundred miles, from the Rio Paranahyba to the Pyrenees, and caused great damage to the plantations.

On August 7th a detachment left Pyrenopolis to ascend the Pyrenees, situated some dozen miles to the east-north-east of the town. Around the range the plateau attains an elevation of 3300 to 3900 feet, and hence the four culminating peaks appear to be of small elevation. On the highest summit the instruments indicated a height of 4577 feet, or much less than that usually assigned to this range. From the top Dr. Cruls could survey the country for a long distance, and noticed that the numerous sources of the Rio Corumba all lay to the north of the Sierra, while the river itself, passing round the chain, flowed southwards to join the Paranahyba. On the other hand, the Rio das Almas, which farther down becomes the Maranhão, and finally the Tocantins, rises a little to the east of this range, and follows its southern slope before turning to the north-west. Accordingly, the watershed between these streams describes a curve somewhat like an S.

The day following, Dr. Cruls and his party returned to Pyrenopolis, and, having made their preparations, left in two divisions towards the end of August, meeting again in Formosa on September 14th, with sufficient information for the delineation of the federal district. The climate appears to be temperate, the mean being estimated at 67°, while that of Rio de Janeiro is 74°. Severe frosts were experienced in July, with temperatures from freezing-point down to 26·5°. The winter, from April to September, is not damp—much drier than the summer, which is the rainy season. Drinking-water is abundant, and, though the forests are not extensive, there is fair quantity of large timber suitable for building purposes, and stone, lime, and brick-clay in abundance.

The commission being free to choose any form they pleased for the federal district, they decided on a quadrilateral, 160 kilomètres by 90 (about 100 miles by 56), bounded by the parallels of 15° 20′ and 16° 8′ S. lat., and the meridians of 3h. 9m. and 3h. 15m. east of Greenwich, and they separated into four parties to fix the angular points of this quadrilateral, two being exactly north and south of Formosa, and the others to the west of it. No particular spot within this area has been chosen as yet for the site of the capital. The topography of the greater part of the zone delineated, in which are plains intersected by hollows with gentle slopes, is admirably adapted for the site of a large city, both from an æsthetic point of view and as regards salubrity, water-supply, and facilities for drainage.

Lago Argentino.—On a journey to Santa Cruz at the end of 1894 it occurred to Señor Clementi Onelli to look for a waterway connecting the Andine lakes of the extreme south with the Pacific. He did not think that the enormous volume of water collected in these lakes, especially during the thawing of the snow, could be drained off by the river Santa Cruz alone. In the early days of March he entered the Lago Argentino and sailed across it westwards, arriving at length at the foot of the first wooded mountains, where the lake makes a great turn to the south between precipitous cliffs. To the south-west he entered a channel which he followed for two days, coming to the foot of a ravine down which a stream descended. Leaving the boat fast moored and, as he supposed, sheltered from every danger, he continued his journey by land, through wood and up a hill amid snow and wind. After passing the night in the open air, he descried to the north-west the deep waters of the Pacific, to which he marched. On the return journey Señor Onelli found that the stream which discharges its waters into the outlet of the lake is fed by brooks on the west side of the hill at a height of 1300 feet above the Pacific. Following the hollows of the mountain it turns to the south and then to the east into the channel proceeding from the lake. As its discovery is likely to complicate still further the question of the frontier between Chile and Argentina, Señor Onelli named the stream "Gordiano."

The boat, which had been left in a quiet bay, was found in a small blue lagoon fast frozen in among enormous blocks of ice, but a cartridge of dynamite, judiciously fired, served to free it, and the journey to Santa Cruz, with a west wind and a current of eight or nine miles an hour, was made in six days, whereas the outward journey had taken thirty-five days of hard labour. The position of the pass into Chile is 50° 34′ 7″ S. lat., and 73° 25′ 2″ W. long.—*Boll. del Instituto Geogr. Argentino*, Tom. xvi. Cuad. 9-12.

GENERAL.

The Polar Libration.—On this subject, which has been once or twice alluded to in the *Magazine*, Herr F. K. Ginzel, of the Royal Observatory in Berlin, contributes a long article in *Himmel und Erde*, vol. viii. p. 7. After referring to the Precession of the Equinoxes and Nutation, which can only change the direction of the axis in space, but cannot affect its position with regard to the figure of the Earth, he mentions Euler's essay, published in 1758, on the rotatory movements of a solid body in space, in which it is shown that when the two moments of inertia lying in the equatorial plane are unequal, the axis of rotation will not coincide with the principal axis of inertia, but will undergo deviations having a period of ten months. Euler's theory received little attention at the time, for among the earlier astronomers and mathematicians the identity of the axes of rotation and of figure was accepted as an axiom, and the diameter of the circle described by one pole round the other would be too small to be detected by the instruments then in use. The geographical positions determined in the fifteenth and sixteenth centuries by means of the quadrant, astrolabe, and Jacob's staff were liable to errors of several minutes of arc. Tycho Brahe succeeded in reducing the probable error to one minute, while the introduction of Hadley's sextant and the substitution of mural circles for the astrolabe, rendered still greater accuracy possible. Still, Bouguer in 1740 estimated the limit of error with the instruments used in his time at five seconds. Consequently it was then impossible to test Euler's theory by actual observation. But with the present transit and other instruments, the errors of which have been carefully examined and can be allowed for, even small displacements of the Pole can be detected, and the fact of their existence proved.

The first to notice these movements was Bessel, who in 1844 wrote to Humboldt that he had observed a variation of 0·3″ in the altitude of the Pole, which he was inclined to ascribe to a change in the Earth's mass ; but he does not appear to have followed up the investigation. At length, when a large number of observations of latitude had been accumulated in the observatories of Europe, doubt again arose of the constancy of the Earth's axis, and in 1883 Fergola proposed that the supposed movement should be investigated by the co-operation of several observatories. In 1888, the permanent Commission of International Geodesy determined to carry out this proposal. The method adopted was the Horrebow-Talcott, in which the altitude of the Pole is deduced from the difference of the zenithal distances of two stars, culminating, the one to the north, and the other to the south, and their declinations. Observations made in 1889 at the observatories of Berlin, Potsdam, Prague, and Strasbourg failed to detect any unmistakable variation of the altitude during the first six months ; in the third quarter a gradual augmentation commenced, followed by a diminution which continued till January 1890, then reaching the considerable amount of 0·5″ to 0·6″. When the same variations were reported from the Pulkova observatory, no doubt about the existence of changes in latitude could any longer be entertained.

The question to be solved next was whether these alterations in latitude were due to a displacement of the Earth's axis of rotation relatively to its mass. If this were so, the same changes in latitude must take place in the western hemisphere, but in the opposite direction. To ascertain this Dr. Marcuse was sent from Berlin, and Mr. Preston by the U.S. Coast and Geodetic Survey, to Honolulu, nearly 180° west of Berlin, and they made observations at Wakiki, near the town of Honolulu, from the end of May 1891 till May 1892. A programme was carefully drawn up and groups of stars so chosen that errors of position would not affect the results. The same stars were observed at the co-operating observatories in Europe, and a comparison of the results showed that the variations in the western hemisphere corresponded exactly to those in the eastern, being, however, in the contrary direction.

The next point of interest was the period of these variations. Euler's theory fixed it at ten months or, more exactly, 306 days, but the observations at Honolulu and the German observations seemed to indicate a much longer period, lying apparently between 380 and 400 days. Several different periods have been given by other observers, and Chandler has sought to prove, from 33,000 observations recorded during the years 1837-1891 at seventeen observatories, that the period is not constant ; between the years 1863-1885 it may have been 427 days ; while in Bradley's time it was perhaps a year. It is possible also that these periods may be included in a larger cycle of seven years. Chandler also attempted to express the variations in formulas, but the observations on which his theories are based are of differing degrees of reliability, and it is evidently too early to venture on definite hypotheses concerning the nature and variableness of the periods. It can only be said that the movement of the axis of rotation takes place from west to east, as demanded by the theory. It seems from the observations of 1891 and 1892 to take an elliptical spiral form, and the amplitude, as far as has been at present ascertained, varies between 0·10″ and 0·55″, corresponding to a displacement of the Pole of about 3¼ to 56 feet. A remarkable fact is that the plane of maximum displacement has turned through a considerable angle within a very short period.

In 1891-92 it lay 40° from the meridian of Pulkova, while by 1892-94 it had moved 90° farther west.

Herr Ginzel discusses lastly the possible cause of these polar displacements. Euler's period was based on the supposition that the earth was a rigid body with moments of inertia no longer liable to alteration, and the difference between his period and the longer one lately observed may be due to the fact that the earth is still to a certain extent elastic, containing down to a certain depth cavities and channels full of movable matter and fluid. Meteorological phenomena—winds, rain, the melting of the polar ice, etc.—concentration of volumes of water at certain points under the influence of atmospheric pressure, the transport of detritus by rivers, upheavals and subsidences of land surfaces, and other possible causes, have been discussed by Thomson, Schwahn, G. H. Darwin, Haughton, Gylden, and others, and an account of their hypotheses and calculations forms a large and interesting part of Herr Ginzel's paper. But all explanations of the phenomenon must at present be largely speculative, and its causes can only be ascertained by long-continued investigations in many branches of science. The astronomers will continue their labours in order to obtain fuller information on the amplitude and periodicity of the polar libration. For this purpose Professor Foerster, Director of the Berlin Observatory, proposed at the Berlin Geodetic Conference that observations should be taken at stations on the same latitude, as far as possible equidistant from one another, but the Conference contented itself with entrusting the arrangements to Professor Helmert.

The Aurora Borealis.—A new and ingenious theory of the causes of the Aurora has been propounded by M. Adam Paulsen of the Meteorological Institute at Copenhagen, and is explained by M. Durand-Gréville in the *Revue Scientifique*, May 2, 1896. A relation between these displays and the variations of atmospheric electricity were long sought for in vain. M. Andrée, member of the Swedish Mission, which sojourned at Cape Thordsen, Spitzbergen, in 1882-83, was the first to institute regular observations on this subject. He found that the potential sensibly diminished at the commencement of the displays, soon returning to the normal. On one occasion it became negative, and that with a clear sky. M. Paulsen has obtained more exact data than these isolated facts. He confined his attention to the form of aurora consisting of a single curtain, almost vertical, for, at Godthaab in Greenland (lat. 64°), the inclination of the needle approaches very closely to the vertical; this aurora consists of a sheet of rays parallel to the direction of the needle, so thin that, in passing over the zenith, it appears as a luminous streak. At M. Paulsen's request, M. Vedel observed the effect of the aurora on the magnetic needle at Denmark Island, on the east coast of Greenland, in the season 1891-92. When the curtain appeared to the south the needle deviated slightly towards the north-west; when it moved to the zenith, the needle oscillated about its original position, and slightly turned to the east as the aurora moved northwards. Consequently, the auroral draperies are the seat of electric currents directed from below upwards. In other words, whereas the positive potential of the air increases as a rule with the altitude, it follows the contrary law within the luminous sheet of the aurora. That this fact does not confirm the view that auroras are of the same nature as the illumination of the air in Geissler tubes by the electric current M. Paulsen has good reasons to produce.

As all important auroras are in the form of rays, they must be the result of an energy that is propagated by radiation, and the source of the energy must lie at one of the extremities of the radiating sheets. It cannot be below, or how can it shoot out instantaneously rays hundreds of miles in length upwards and not send any downwards ? The source, then, must be situated in the higher regions of the atmosphere, and yet within it, for the position of the sheets, always sensibly parallel to the inclination of the needle, shows that they share in the terrestrial movements of translation and rotation. But, while the source of energy is at a great altitude, the illumination often commences only in strata relatively low, for it is only when the electric current traverses the denser atmosphere that its energy is gradually converted into light until the whole is absorbed.

The next point is to inquire whether any phenomenon is observed in the laboratory analogous to this radiation. Such is the radiation from the negative pole in tubes filled with highly rarefied air. These rays possess the property of propagating themselves in straight lines without their course being influenced by the position of the positive pole ; of producing fluorescence by impact on the sides of the tube ; and of being transformed into light in passing through air at ordinary pressure. It is true that these " cathodic " rays exercise no influence on the magnetic needle, and this fact seems out of harmony with the perturbations observed by M. Vedel. But experience proves that the cathodic rays in passing through the air produce ozone, and M. Lenard has proved that the absorption of cathodic rays increases with the density of the air, and that air rendered fluorescent by their absorption is a very good conductor of electricity. The air, thus rendered a good conductor through strata several hundred miles in thickness, forms a communication between points of different potentials, and currents arise which are the result, and not the cause, of the auroral display. These currents soon establish equilibrium, and, therefore, if the aurora remains stationary, the needle quickly returns to its normal position, but if it changes its position new currents are produced. It is known that, other circumstances being the same, the magnetic perturbations are more intense during rapidly moving auroras. The great objection still remains that cathodic rays have never been obtained except by interrupted or alternating currents, and it is not easy to understand how rapid breaks like those contrived in laboratories can be produced in the upper regions of the air. A specialist in electricity suggested, however, to the author the analogy of a lightning flash in which the discharge is oscillatory. It may be, then, that the auroral rays are due, not to complete discharges, but to variations in the charges of the condensers.

The intimate relations which subsist between the periodicities of auroras and sun-spots leave no doubt as to the origin of the energy which produces the former phenomena. M. Paulsen remarks, moreover, that the intensity is greater during the earlier part of the night, and diminishes towards morning. The energy must, therefore, be stored up during the day, and not be due to a direct and immediate action of the sun similar to that of a magnet or electrified body, a view confirmed by the fact that the two annual maxima of frequency of auroras, corresponding to the equinoxes, occur some weeks after these epochs. M. Paulsen explains the absence of auroras in the equatorial regions by supposing that the insolated molecules mutually repel one another, and that they lose their energy only when they come in contact with non-insolated molecules near the Polar regions.

M. Paulsen has also accounted for the formation of vague and diffused auroras. When rays have a direction oblique to the lines of force, they will be diverted along a helix having the line of force as an axis, and the nearer the angle between the rays and the lines of force, the greater the diameter of the helix, and then the ray parts with all its energy in transit, and cannot descend to the lower regions of the atmosphere. In this case innumerable helices are superposed, and the result is a vague and diffuse aurora covering a considerable part of the sky. The almost vertical direction of the lines of force in the Polar regions facilitates the descent of the rays, but in the intertropical regions they cannot descend towards the earth, and the auroras become diffuse. They are also so feeble as to escape notice, but Mr. Wright has found that the greenish-yellow line of the spectrum, 557, characteristic of auroras, and formerly supposed to proceed in tropical regions from the zodiacal light, is produced in the atmosphere, and is due to diffuse auroras detected by the spectroscope in all parts of the sky, even at very low latitudes.

Lastly, M. Paulsen explains the cloud-like appearance of many auroras. These forms are, he says, clouds of frozen particles lighted up by the aurora, and this explains many of the circumstances connected with their appearance—their orientation, identical with that of the aurora, the rapid pulsations of the light which they emit, etc. The presence of water vapour, at altitudes of thirty to sixty miles, M. Paulsen accounts for by the total absence at that height of the dust necessary for precipitation, in consequence of which the air is supersaturated. M. Paulsen's theories, though needed to be supported in several points by confirmatory evidence, account so clearly and completely for all the phases of the phenomenon, that they merit the serious attention of all who are interested in the question.

MISCELLANEOUS.

Professor Sollas, accompanied by Professor Gardiner, is shortly to conduct an expedition to the South Seas for the purpose of examining the coral islands.

During the first two weeks of this year the heat in Australia was phenomenally great. The thermometer did not fall below 90° F. during the whole fortnight, and in some places rose to 122° in the shade.

The Essequibo in British Guiana is blocked by dangerous falls. To obviate the difficulty presented by them a railway is being constructed from Wismar on the Demerara river to a point above the falls. Nine miles of rails, or more than half, have been laid, and by the end of the year the whole line will be completed.—*South American Journal*, May 16th.

The Green Lake in Colorado is one of the highest in the world. Its surface lies at an elevation of 10,250 feet above sea-level, and its shores are perpetually covered with snow. The water is clear as crystal, and large trunks of petrified trees may be seen at a depth of 100 feet. Its maximum depth is 230 feet.—*Bull. de la Soc. R. Belge de Géographie*, No. 2.

The culminating point of the Riesengebirge has hitherto been held to be the Tafelfichte, the extremity of the Iserkamm in Austria. But the Austrian trigonometrical operations have shown that its height is only 3681 feet, and therefore it must give place to the Hinterberg, which according to the Prussian measurements of 1895, is 3696 feet high.—*Deutsche Rundschau*, Jahrg. xviii. Heft 9.

The population of Russia in 1892 numbered, according to the reports of the Medical Department, 119,288,804. Of those 100,251,510 belong to European Russia, 7,864,202 to the Caucasus, 4,856,902 to Siberia, and 6,316,190 to Russian Central Asia. The annual increase has been 540,261 in European Russia, 16,724 in Siberia, and 18,492 in Russian Central Asia, while the population of the Caucasus decreased by 2992 in the year 1892.—*Deutsche Rundschau*, Jahrg. xviii. Heft 9.

Not long ago we noticed M. Lancaster's report on rain in Belgium, the distribution of which is so remarkable. Since then we have received *Le Climat de la Belgique en* 1895 by the same author, containing tables of all the meteorological phenomena for the year. The temperature, which, after seven consecutive cold years (0·7° below the normal), had been rather warmer during 1893 and 1894, again fell, being 0·2° below the normal. The rainfall was about the average, but was very irregularly distributed. Storms were frequent, accompanied locally by deluges of rain. On June 10th a remarkable example occurred, which is described in an appendix to M. Lancaster's work. In some places it was accompanied by very heavy rains, while in others scarcely any fell. At Uccle the fall was 2·59 inches, the most remarkable during the past sixty years.

A new mammal lately obtained by Mr. Oldfield Thomas from Bogota, now in the British Museum, greatly strengthens the case for a former land connection between Australia and America. The existence of polyprotodont marsupials in Australia and South America is not of great importance, for fossil forms are found in the Northern Hemisphere, and as long as F. Ameghino's fossil Diprotodonts only were known, which he referred to the Eocene period, there was still no very strong argument in favour of the land connection. But the marsupial obtained by Mr. Thomas, *Cœmolestes obscurus*, renders such a connection much more probable. Mr. Thomas has shown that it belongs to the Epanorthidæ, one of Ameghino's fossil Diprotodont families. The Diprotodonts, therefore, both living and fossil, are confined to Australia and South America.—*Natural Science*, June.

The Rev. W. Campbell of Taiwanfu writes to draw attention to an error regarding the name of **Mount Morrison**, the highest point of Formosa. " Years ago the general impression was that the mountain was named after the first Protestant missionary to Formosa, and this is shown by allusions in several books, notably in *Our Mission in China in* 1866, by Donald Matheson, Esq. Mr. G. James Morrison, C.E., of Shanghai, seems to have been the first to assign a different origin to the name. He visited Formosa in the spring of 1877, and in the November Number of *The Geographical Magazine* he wrote :—' Mount Morrison is not named after the well-known missionary and Chinese scholar, but after the captain of one of the early vessels trading to Taiwanfu.'

"Mount Morrison was so named in 1844, when most, if not all, the vessels sailing to Formosa were engaged in the contraband opium traffic, and one wonders who this captain was, and what he had done to have his name handed down to posterity. The mistake would be scarcely worth noticing had it not been copied into the last edition of the *Encyclopædia Britannica* (vol. ix. p. 415).

"The true state of the case may be learned from the *Proceedings of the Royal Geographical Society*, vol. viii. p. 25, in which the late Admiral Collinson is reported to have said that he gave to the highest point of Formosa the name of Mount Morrison, 'a name which, he believed, all those who were acquainted with our original connection with the Chinese would acknowledge ought to be perpetuated throughout all ages.' Letters still extant show that Collinson, during the time he was engaged in his survey work, was still mourning the death of Robert Morrison as of a dear friend.

"It is, therefore, certain that the mountain was named, not after a merchant captain, but after the missionary who first translated the Scriptures into Chinese, who was asked to accept the Fellowship of the Royal Society, and whose pioneer *Chinese-English and English-Chinese Dictionary* was published by the East India Company at a cost of £15,000."

NEW BOOKS.

L'Homme devant les Alpes. Par CHARLES LENTHÉRIC. Paris: Plon, 1896. Pp. 480. *Price* 9 *francs.*

After writing well-known works on the Mediterranean coast and Rhone valley, the author—"Ingénieur en chef des ponts et chaussées"—was sent on a special mission to the Alps by the French "Ministre des travaux publics." The result is a popularly written scientific volume, based partly on the author's own observations, partly on the researches of others. From French and German authorities he obtains the theory of the leading mountain-chains of the globe being the result of three successive wrinkles in the earth's crust. The first wrinkle (*ride*) occurred in Silurian times and formed the "Caledonian" chain, represented to-day by the Scottish Grampians, the Scandinavian chain, and the Green mountains of North America. During a subsequent epoch occurred the second wrinkle, forming the "Hercynian" chain, represented to-day by Brittany, Cornwall, the French central plateau, the Ardennes, the Hartz and Bohemian mountains, the Ural mountains, and, in America, the Alleghanies. The Alps belong to a third wrinkle which formed the "Alpine" chain, represented to-day by the Alpine region of the Mediterranean and Central Europe, the Caucasus and Balkans, stretching in Asia to the Himalayas, and westwards by the Pyrenees to the Atlantic, and probably extending to the West Indies.

Theories of elevation and subsidence were enunciated by the Pythagorean School of Philosophy five centuries before Christ, and are repeated by Ovid in his *Metamorphoses.* Yet M. Lenthéric (perhaps to spare French feelings) does not remind his readers that Voltaire had the audacity to declare that the shells found at high levels on the Alps did not indicate the previous presence of the sea there, but might have been dropped by the innumerable crowd of pilgrims (wearing shells in their hats) on their way to Rome from every quarter of the Christian world! As our Scottish philosopher, Professor Playfair, retorted, "We can excuse in a poet and a wit that ignorance of the facts in mineralogy which concealed from him the extreme absurdity of his assertion."

A good map by H. Rollet, Paris, shows that old Roman roads crossed the Alps at the following passes:—

Alpis Summa (Mont Agel).
Mons Matrona (Mont Genèvre).
Mons Graius (Petit St. Bernard).
Summo Pennino (Grand St. Bernard).
Tarvesede (Splügen).
Vipiteno (Brenner).

Roman roads also crossed the Julier and Toblach passes, and the heights near Adelsberg.

M. Lenthéric derives the term "Alps" from the radical *alt*, *alb*, *alp*, signifying elevation and steepness, and consequently the coldness and dazzling white colour of snow. He points out that Snowdon in Wales, the White mountains in America, the Sierra Nevada in Spain, the ancient Haemus (the modern Balkans), the Imaus of Ptolemy in northern Scythia, and the Himalayas, all similarly derive their names from being snowy or brilliant mountains, whilst the monarch of the Alps is termed "Mont Blanc." M. Lenthéric does not however explain why the Alpine mountaineers themselves apply the term "alp," not to the mountains but to green pastures on the mountain-slopes, whither they drive in summer their cows and goats. Did the mountains acquire their title from these green alps or from the snow which covered some of their peaks ?

A chapter on Transalpine Railways closes this erudite volume, and the author declares that " after the triple experience of the Mont Cenis, Saint Gothard, and the Arlberg, all technical difficulties (for railways crossing the Alps elsewhere) have been removed."

Persian Life and Customs. By the Rev. S. G. WILSON, M.A., Fifteen Years a Missionary in Persia. Edinburgh and London : Oliphant, Anderson, and Ferrier, 1896. Pp. 333. *Price* 7s. 6d.

The general scope and nature of this book are thus indicated by the author in the Introductory chapter. "The preparation for writing this present volume was a residence of fourteen years in Tabriz, Persia, as a missionary under the Presbyterian Board of (American) Foreign Missions. The prosecution of mission work in its usual departments as preacher and evangelist, principal of the Memorial Training and Theological School, mission treasurer, superintendent of buildings and of legal affairs, has brought me into contact with Persian life and character among those of low and high degree alike. Itinerancies and mission business have led me to visit many localities, and given me wide opportunities to observe the manners and customs of the people. I have read nearly every work in English on Persia and Mohammedanism, yet have recorded, for the most part, only my own observations and experiences, and what I have heard from the people regarding themselves."

The work is divided into seventeen chapters, as follows :—Coasting the Black Sea : Georgia and Ararat : Tabriz : Maragha : Circuit of Lake Urumia : The Kurdish Raid : Mountain Armenians and Nomads : Teheran : Hamadan and Takht-i-Suleiman : The Conditions and Needs of Persia : The Sacred Year of the Shiahs : Religious Life and Morals : Among the Ali-Allahis : Social and Family Life : Village Life : Business Life : and Modern Missions in Persia. Of these the most interesting is the chapter on the Kurdish raid into Azerbaijan under Sheikh Obeidalleh in the autumn of 1880, of some of the episodes of which Mr. Wilson was himself an eye-witness. As he lived for fourteen years in Tabriz, the residence of the Vali Ahd, who has just succeeded to the throne on the death of his father by assassination, his summary of what is known of the character of the new Sovereign of Persia is of interest. "The Vali Ahd or Crown Prince of Persia is Mazaffir-ed-Din, the eldest son of the Shah by a royal mother. He has never been tried by the great responsibilities of government, so that his capacity is largely unknown. One marked characteristic of the Prince is his religious disposition. He is a man of faith, free from the rationalistic tendencies of some of the ruling class. He is devout in his fulfilment of religious rites, and a total abstainer from intoxicants. He has a humane and sympathetic nature. The mother of a condemned man can appeal to him with large assurance of mercy. He is a friend of education, taking a great interest in the Government school at Tabriz, having European tutors for the education of his sons, and encouraging his interpreters to translate standard works into Persian. He is much interested in telegraphy, photography, and like practical arts. He has a fondness for flowers and new and rare plants, and enjoys tent life and the pleasures of the chase, often camping with his retinue on the mountains."

The book, although containing little that is new, is pleasantly and modestly written, and is, on the whole, a valuable edition to the bibliography of modern Persia.

The Yellowstone National Park, Historical and Descriptive. By HIRAM MARTIN CHITTENDEN, Capt. U.S. Army. Cincinnati: The Robert Clarke Company, 1895. Pp. 397. *Price* $1·10 *net.*

The wonders of the Yellowstone are now tolerably well known in Europe, and have been already described in the pages of the *Scottish Geographical Magazine.* Since the terrible eruption of Mount Tarawera in 1886, when the renowned pink and white terraces of Rotomahana were wiped from the map of New Zealand, the Yellowstone area probably stands unrivalled among the regions of hydrothermal activity known to geographers.

In the volume before us the author has endeavoured "to collect the essential facts, historical and descriptive, relating to this region, and to place them in form for permanent preservation." The first part of the book deals with the history of the park, and here it may be interesting to note that "the park to-day stands on a firmer basis than ever before." The author is not here speaking from the geological standpoint, nor does he refer to geological processes when he tells us at page 19 that "the park was created, March 1, 1872." The purposes for which the park was "created" (by an Act of Congress) are :—

(1) The preservation of its natural curiosities, its forests and its game.
(2) The reservation of its territory from private occupancy, so that it may remain in unrestricted freedom "for the benefit and enjoyment of the people."
(3) The granting of such leases and other privileges as may be necessary for the comfort and convenience of visitors.

These excellent intentions have only lately been adequately given effect to, and until 1894 "wanton vandalism, destruction of game, or burning of forests, could be visited with no heavier punishment than ejection from the park and confiscation of 'outfit.'" For many years the park was left by Congress wholly without the means for its improvement or protection, and this vast domain, which now covers an area of about five thousand square miles, was administered by a series of governors who were generally either unable or unwilling to protect it against poachers and unscrupulous adventurers bent on exploiting it for their private advantage. Until the government of the park was put into military hands, and the corrupt civil administration was abolished, police regulations could not be enforced ; but it is pleasing to learn that during the last two years the superintendent has received much more extensive power, and that the condition of the reservation is now in an eminently satisfactory state.

The second part of the book describes the topography, geology, fauna, flora, and administration of the park, and concludes with detailed descriptions of a series of tours through it. In the third part the author does well to point to the future of this magnificent heritage of the American people. It is surprising to learn that the original patriotic intentions of Congress in setting apart this delightful region for the benefit of all lovers of nature should be constantly in danger of frustration by private schemers with greedy ends in view. For the last twenty years no session of Congress has been free from attempted legislation hostile to the best interests of the park. Railway promoters have been among the worst of these enemies, and it is fortunate that hitherto all attacks from such quarters have been repelled. The presence of a railway would not only destroy the amenity and "splendid isolation" the park enjoys, but would also have the effect of banishing the native fauna to which the park is intended to offer a secure sanctuary. In the Appendix is given a list of the mountains, geysers, streams, and other natural features with their altitudes and characteristics. The volume is enriched with abundant illustrations, a map, and a copious index. The remarkable colouring of the magnificent cañon of the Yellowstone river is perhaps the most striking feature in the scenery of the park, and had the author, instead of giving so many monochrome illustrations, supplied a coloured view of the cañon from Mr. Haynes' beautiful selection, the value of the book would have been considerably enhanced. The addition of a geological or orographical map would also have been an advantage, but to the general reader this work supplies in an attractive form a mass of valuable information about a region of the greatest geographical interest.

The Burman, His Life and Notions. By SHWAY YOE, Subject of the Great Queen. London : Macmillan and Co, 1896. Pp. xii + 603. *Price 12s. 6d.*

The author of this book has evidently an intimate acquaintance with Burmese life and manners. He describes very fully all the customs of the people connected with births, marriages, festivals, deaths, and other events, their religion and superstitions. This part of the work is, perhaps, rather heavy reading for all except ethnologists. The chapters on cultivation, manufactures, etc., will appeal more strongly to the general reader, and some, such as those in which King Theebaw is discussed, are not wanting in humour. A chapter on the character of the Burmese in general, and their probable future under British rule, would have been acceptable from an author who seems so well qualified to give an opinion on these questions.

The work was first published in 1882, and has not been materially altered. In spite of the verbal alterations mentioned in the preface, there are many passages that imply that King Theebaw still rules over Northern Burma. Some of the illustrations that adorn Signor Fea's work, noticed below, would have greatly assisted the text in making the reader acquainted with the architecture, manufactures, and dress of the people.

Quattro Anni fra i Birmani e le Tribu Limitrofe. VIAGGIO DI LEONARDO FEA. Milano : Ulrico Hoepli, 1896.

The object of Signor Fea's journeys in Burma was primarily zoological, and his perseverance and industry in the collection of specimens met with a large share of success, many rare, and some new, species being found. The traveller, however, did not fail to take note of other objects of interest, and describes the pagodas of Rangoon, Mandalay, and many other towns, the priests, the dress of the people, and their manufactures, especially musical instruments and lacquer-ware. The numerous illustrations represent a great variety of objects—pagodas, bells, native costumes, scenery, animals, plants, insects, fishes, etc.

The most interesting part of the book to geographers and ethnologists is, perhaps, that in which Signor Fea narrates his experiences among the Karens. His ascent of Mount Mulai and sojourn among the Karens of the Dona range was noticed at length in vol. iv. p. 486—for the publication of his book has been delayed. He also visited the independent Karens to the north of Tenasserim. These are in a rather more advanced stage of civilisation than the tribes of Tenasserim, but also less peaceful, and in many villages Signor Fea received no very friendly reception, as he was suspected of being an agent of the British Government, sent to spy out the country and report on its means of defence.

Geschichten und Lieder der Afrikaner. Ausgewählt und verdeutscht von A. SEIDEL. Berlin : Schall und Grund, 1896. Pp. 340.

Up to very few years ago the impression widely prevailed that intellectuality was an undoubtedly wanting factor in Africa. The investigations of the past twenty years show that such is not the case, and our author has culled from numerous publications a well-selected store of facts to show the thoughts, feelings, and intellectual endowments of the inhabitants of that continent.

Traditions, stories, poems, and fairy tales are given in this book, gathered from all parts, from modern Egyptian fables in the north to Hottentot stories in the south. Reading the translation of these—and the translation is kept as near the original as possible—the curious reader can form an opinion as to the comparative mental development of the various races and tribes, and can see how near they come to, or how far short they fall of, like literary productions of Europeans. Mr. Seidel has done his part very well, and given, we think, a very fair sample for judgment. Ethnological and Folklore students will be much indebted to the author for this most interesting and instructive volume.

Drei Jahre im Lande Heinrich Witboois. Schilderungen von Land und Leuten. Von F. J. von Bülow. Mit zahlreichen Abbildungen nach photographischen Aufnahmen und zwei Karten. Berlin : Ernst Siegfried Mittler und Sohn, Königliche Hofbuchhandlung, 1896. Pp. 365.

The author claims, and with justice, to give a readable and accurate account of the geographical, political, and economic position in the German Protectorate in South-west Africa. He writes of what he has lived through and seen himself, and trusts that his strong personal opinion may be credited to his wish for the well-being of the colony.

Lieutenant Von Bülow went to South-west Africa in 1881, and remained there for three years, returning in consequence of an accident whilst hunting, which rendered him blind. He made no very remarkable journey, but travelled all over the German colony, and describes both the journey and people with accuracy and care. He pays considerable attention to the economic and political questions of the colony, and believes that Cecil Rhodes' advice should be followed—to do what-ever will pay. He thinks that it is most important to send officials who will have a real interest in the country and not hamper its development by too much red tape, and advises emigration on a large scale. He saw something of the German war against Witbooi, and describes the fighting which took place during the first year. The illustrations are good, and the book can be recommended to those who have any interest in the region with which it deals. It is written in a lively strain and in a very interesting manner.

Aus dem Lande der Suaheli. Theil i. *Reisebriefe und Zuckeruntersuchungen am Pangani.* Von Gustav Meinecke. Vegetationsbilder von Dr. Otto Warburg. Mit 40 Illustrationen und einer Karte im Text. Berlin, SW. ; Deutscher Kolonialverlag (G. Meinecke), 1895. Ss. 194.

This brochure consists of a reprint of letters which the author wrote to the Berlin *Post*, and a description of the vegetation in German East Africa, written by Dr. Warburg. The illustrations give a very good idea of the scenery met with. The concluding chapter deals with the food supply obtainable in the coast towns. The author's description of the manufacture of curry is very amusing.

Although well written, there is nothing very new in the author's description of the method of reaching East Africa or of the various ports and settlements. It is pleasantly written, and shows careful observation. On the other hand, the chapter on "Zuckeruntersuchungen am Pangani" is new, and gives in some detail informa-tion as to the districts which are cultivated by sugar-planters, the possibility of extending the cultivation of the sugar-cane, etc., as well as a description of the sugar manufacture. It is unnecessary to enter into detail, but the author's con-clusions are that the sugar-cane is very good and cheap, that the plantations might be very much improved, as they are at present carried on in a very primitive fashion, and that under competent supervision it would be possible to compete with Indian sugar. Labour is cheap.

Dans la Haute-Gambie. Voyage d'Exploration scientifique par le Docteur André Rançon, Médecin de première classe des Colonies, Chevalier de la Légion d'Honneur, 1891-1892. Paris : Société d'Editions Scientifiques, 1894. Pp. 592. *Price 10 frs.*

The true gutta-percha of the East Indies is, as is well known, gradually disappearing, because the Malays cut down the trees instead of simply tapping them. For some purposes, such as telegraph cables, this material is indispensable, no kind of india-rubber having been found to answer the purpose of an insulator, and therefore the French Ministry of Commerce, Industries, and Colonies, at the suggestion of Professor Heckel, who has written an introduction to the above work, sent out two expeditions, one to Guiana, under M. Geoffroy, for the purpose of investigating the properties of the *Mimusops Balata*, and the other under Dr. Rançon to the French Sudan. M. Geoffroy succumbed to the effects of the hard-ships he had undergone before he had time to write his report, but his work gave a great impulse to the exploitation of the forests of French Guiana, and several companies have been formed to develop these resources of the colony. Dr. Rançon's journey was likewise attended with great success, and he has been able to give an account of the valuable observations he made on the flora of the countries he visited.

Ascending the Senegal to Kayes, Dr. Rançon travelled southwards to the Gambia and, passing through the numerous districts which lie on or near the banks of that river, finally returned to the Senegal at Boufoulabé. He treats of each district separately, describing its geographical features, geology, fauna, and flora, its inhabitants and commerce. Among the places he visited are some in British territory, such as the island MacCarthy and Yabuteguenda, and his remarks on them will be of special interest to British readers. The work is intended rather for the student than the general reader, and is a mine of information about the natives, their customs, cultivation, musical instruments, etc. Naturally the flora is treated of most fully, and we find descriptions interesting to botanists of a number of plants yielding food-stuffs, dyes or other products used in manufactures, poisons, or antidotes. Some of these are accompanied with illustrations.

Of the value of the country to France, Dr. Rançon's remarks do not give a very high estimate. Some tracts are fertile, and will, in time at any rate, give rise to an active trade, but of too many districts Dr. Rançon says that no commerce is to be expected with them. Nor does the French authority seem to be by any means firmly established. In many parts raiding still goes on unchecked, while of others we are told that they are too remote from the military posts to have close, if any, relations with the French Government.

The Province of South Australia. Written for the South Australian Government by JAMES DOMINICK WOODS, J.P. ; with a Sketch of the Northern Territory by H. D. WILSON. Adelaide : C. E. Bristow, Government Printer, 1894. Pp. 446.

Mr. Woods has compiled a handy general account of the colony, its history, national resources, political constitution, etc. He sketches the progress of discovery from the visit of Nuyts in 1627, and, particularly, details the discoveries of Flinders and Kent, Eyre and Sturt. The land laws, agriculture, and commerce are also dealt with, and a chapter is devoted to the aborigines. The book will be useful for reference to the principal facts of the past and present of the colony.

The chapter on the Northern Territory, of which comparatively little is known in this country, is a welcome addition to the book. This land, covering an area of 523,620 square miles, was granted to South Australia in 1864, John M'Douall Stuart, who crossed the continent in 1862, having brought back a favourable report of the country. "Visitors from Java, Singapore, Ceylon, and China . . . have spoken in the highest terms of the suitability of the soil and climate of the Northern Territory for the growth of sugar, coffee, tobacco, rice, indigo, and other tropical products." Gold is worked in several places, and silver mining has been carried on for some years, and copper and tin are also found. With these natural resources, it is surprising that the territory has not made more rapid progress. The chief obstacles seem to be the climate, which will not allow Europeans to work in the fields, and the difficulty of procuring cheap native labour to enable the planters to compete with other tropical countries. The development of the mines has been hindered by want of capital.

Die Erde. Eine allgemeine Erd- und Länderkunde. Von DR. FRANZ HEIDERICH. Mit 215 Illustrationen, 143 Textkärtchen und 6 Karten in Farbendruck. Wien, Pest, Leipzig : A. Hartleben's Verlag, 1896. Pp. xii+876. *Price* 30 M.

This is a compendium of geography in one volume, intended for the use of schools as well as for the general reader. We have no hesitation in saying that it is one of the best books of its kind we know. The author has succeeded in condensing into one by no means very formidable volume as much geographical knowledge as a well educated person need aspire to in these days. It is a popular book in the best sense of the word ; based only on the best authorities, thoroughly reliable in facts and figures, and very pleasant to read. It presents some features which are sure to commend it to the general reader. The author has wisely refrained from encumbering it with dry political and statistical details, and he has used the space thus rendered available to give more prominence to the description of the physical features of the earth and their evolution than is usual in books of the kind. Ethnology also has been treated very fully. We notice that the author sides with Max Müller rather than with the evolutionists in some fundamental ethnological questions. His remarks on commerce and industry are short and to the point ; perhaps a little more information regarding internal and international means of communication would be welcome. The mathematical part, with which the book opens, states in clear and simple language all the general reader requires to know. In the physical geography we are glad to note that the latest authorities especially Suess and Penck, have been consulted to excellent purpose.

The book is well printed, profusely and beautifully illustrated, and excellently bound. Altogether it is a production on which both author and publisher must be congratulated.

Austria, including Hungary, Transylvania, Dalmatia, and Bosnia. By KARL BAEDEKER. Leipsic, 1896. Pp. 468. *Price* 7 *Marks.*

This eighth edition, remodelled and augmented, with twenty-eight maps and twenty-five plans, will furnish travellers to the Austro-Hungarian empire with the most ample and reliable information at a very moderate expense. The lesser known territories of the empire are now being opened up to tourists by Austrian and other tourist agencies, and form delightful excursions.

Black's Guide to the Isle of Man. Edited by M. J. B. BADDELEY and E. D. JORDAN. London : Adam and Charles Black, 1896. Pp. 144. *Price* 1s.

An extremely cheap handbook to a very interesting island, once a fief of the King of Scotland, and whose lighthouses are still under the Commissioners of Northern Lights, Edinburgh. Excellent maps by Bartholomew elucidate the text. The chapter on the geology of the isle is not very satisfying, and the authors should refer to a standard paper on the subject by Mr. John Horne of the Geological Survey of Scotland (*Trans. Edinr. Geol. Soc.*, vol. ii. p. 323.)

Svenska Turistföreningens Årsskrift för År, 1896. Stockholm : Wahlstiöm und Widstrand. Pp. xxxiv+309.

One of the great attractions of this *Year-Book* is the large number of illustrations, some of them very clear. There are also many descriptions of various parts of Sweden which will be useful to intending visitors. One contributor enlarges on the advantages of a new route to Stockholm from Copenhagen, through Helsingborg and Jönköping. When the railway between Markaryd and Vernamo is completed, that is, probably, in 1898, the journey will be accomplished in forty-eight hours, and at a considerably smaller expense than by the Göta canal. Another article, *Esperanto som Turistspråk*, gives a few details concerning an international language, to extend which a society has been founded in Upsala. The advantages of such a language to tourists and other travellers are undoubted, and there is no difficulty in compiling a language easy to learn, as Esperanto certainly is ; the difficulty is to ensure its general adoption.

Views of Sweden. A River Voyage through Northern Sweden. Guides of the Swedish Tourists' Club, Nos. 12 and 13. Stockholm : Wahlström und Widstrand, 1896.

The Swedish Tourist Club publishes a number of small handbooks intended to make known to foreigners the attractions of the country. The latter number deals with the rivers Indals and Angerman. The scenery, though not grand, is beautiful, wood and water combining to produce charming landscapes. The views are taken all parts of the country, and represent both buildings and scenery.

Illustrated Europe, No. 167. *The Hungarian System of the Royal and Imperial Southern Railway Company.* By EDMUND STEINACKER. Nos. 168 and 169. *The Baths of St. Moritz.* By Pastor CAMILLE HOFFMANN. Zürich : Orell Füssli, N.D.

The former of these little books deals with a large extent of country, and therefore is very general in its descriptions, particular places being described in other numbers of the series. St. Moritz is well known to British tourists ; it is well described by Pastor Hoffmann, and the views are good.

Summer Tours from Leith to Norway, Denmark, and Germany.

Messrs. James Currie and Co. have again issued their useful little Guide, containing sketches of tours which may be easily made from Leith. Norway occupies a large part of the book, twelve routes being described. As before it is accompanied by a good map, and this year several plans of towns have been added.

Summer Tours in Scotland and England, 1896.   Caledonian and North-Western Railway.
Official Tourist Guide to Scotland.   North British Railway.   *Price* 6d.

On the approach of the tourist season the great railways have issued their guide-books as usual.   They contain notes of the places most worth visiting, and time-tables, etc., of the chief tourist routes, and are also well provided with maps, plans, and illustrations.

Summer Tours in Scotland : Glasgow to the Highlands.   Glasgow.   *Price* 1s.

The book published this year by Mr. David Macbrayne is a useful guide to the west coast of Scotland.   It is illustrated with some very fair views, and has four sectional maps.

A. Hartleben's Kleines Taschenbuch über alle Länder der Erde.   Nach den neuesten Angaben bearbeitet von Professor Dr. Friedrich Umlauft.   Pp. 98.   A. Hartleben's *Statistische Tabelle über alle Staaten der Erde.*   Wien, [etc. : A. Hartleben's Verlag, 1896.

The present issue of the *Taschenbuch* is the third, and the *Tabelle* is now issued for the fourth consecutive year.   We need not say much of these convenient publications as they have been noticed in former years.   Evidently the latest data have been collected ; for instance, the population of Germany according to the census of 1895.   The compiler in this case had to put up with the nearest approximation he could obtain at the time, which happens to be rather too small.

Bartholomew's Pocket Guide to Edinburgh and Neighbourhood.   Second Edition. The Edinburgh Geographical Institute, 1896.   Pp. 87.   *Price* 1s.

This is a new issue of the excellent little guide-book noticed in vol. viii. p. 395. Three new maps have been added, one of which is De Wit's *Plan of Edinburgh*, 1647.

Nell' Harrar.   Del Ing^{re} L. ROBECCHI BRICCHETTI.   Milano : Casa Editrice Galli, 1896.   Pp. viii+409.   *Price* L.7.50.

Signor Robecchi has made himself a name in Europe as an African traveller, particularly by his exploration of the *Hinterland* of the coast between Obbia and Berbera.   In the present volume he describes his experiences during an earlier journey which he undertook to Harrar in 1888.   This town, famous as a commercial centre, has been visited by several European travellers, and is the residence of agents of commercial firms, a Greek colony, etc.   It is, therefore, a place which has frequently been described with more or less detail.   Nevertheless, Signor Robecchi's book is welcome, for he spent several months in the town and in excursions to places in the neighbourhood, so that he is able to describe very fully the character, customs, and occupations of the many different races to be found here—Arabs, Abyssinians, Galla, and Somali.   He also made interesting geological, ethnological, and botanical collections.

Signor Robecchi met several men of note at Harrar—Count von Teleki and Captain von Höhnel, Dr. Nerazzini, and the engineer Ilg, who has acquired great influence in Abyssinia.   He has also some amusing incidents to relate, such as the building of an Abyssinian church, of which he was appointed architect by Ras Maconnen ; some of the circumstances attending it were rather ludicrous, and are humorously portrayed by the author.

Norway: its Fjords, Fjelds, and Fosses. By JOHN BRADSHAW. London : Digby, Long, and Co., N.D. Pp. 217. *Price* 3s. 6d.

The title of this book promises rather too much, for the author describes only the parts of Norway he visited in a couple of tours. He has put together his impressions of the scenery and towns he passed through in an easy readable style, but there is nothing in the book that is new or striking. To those who have not been in Norway it may give some notion of the kind of travelling, and the usual incidents of the road.

The Log-Book of a Midshipman. By Captain BASIL HALL, R.N. London : Blackie and Son, N.D. Pp. 224. *Price* 1s. 4d.

To the geographer the Log-Book is only interesting as being the work of Captain Basil Hall, whose narratives of voyages and discoveries are well known. This work contains discussions on a variety of subjects, and a great many amusing incidents, and, if not geographical, is very readable. It is well printed and neatly bound.

A Directory of the Chief Industries of India. With which is incorporated a List of Tea Companies and Gardens, Indigo and Coffee Estates, with their Factory Marks, and a Directory of Tea, Coffee, Cinchona, Cardamom Estates in Ceylon. Calcutta : Thacker, Spink, and Co.; London : W. Thacker and Co., 1896. Pp. 195.

The scope of this book is sufficiently described in the secondary title, for the volume is little more than a directory. A bibliography at the end is very useful to those who want to know something of the cultivation, etc., of Indian products.

Aegypten, 1894. *Staatsrechtische Verhältnisse, wirthschaftlicher Zustand, Verwaltung.* Von A. FRHR. VON FIRCKS. Zweiter Theil. Berlin : Dietrich Reimer, 1896. Pp. 279 and Bibliography.

A short time ago (see p. 108) we noticed the first part of this work in the *Magazine* with approval, and we are glad to see that the author has maintained the same high level of accurate observation in this, the second and concluding, part of his work. Although detail is needful, it is not overdone, and Herr von Fircks has aimed as far as possible at clearness and brevity. Internal government, laws, finance, taxation, religion, education, commerce, and communication are all dealt with in a masterly manner, and we obtain almost at a glance a comprehensive view of each subject. The tables are well chosen, and the bibliography is excellent, all necessary works being mentioned both concerning Ancient and Modern Egypt.

The progress which Egypt has made during the past few years is very remarkable, and is conclusively proved in these pages.

The enormous increase in the post and telegraph work shows the progress made by the people. The number of telegrams sent now is three times greater than in 1874. The number of telegraph offices in 1873 was 77 ; in 1894 it was 222. The value of money transmitted by post in 1878 was £E8,926,000 ; in 1894 £E14,200,000. The foreign post has increased from 672,000 letters and postcards in 1876 to 4,107,000 in 1894 ; and the local letters and postcards from 1,536,000 in 1876 to 10,060,000 in 1894. In every way prosperity seems to be on the increase. People travel, too, far more than they did. In 1877, 2,265,377 passengers were carried by rail, in 1894 9,827,813.

We can with every confidence recommend the study of this book.

NEW MAPS.

EUROPE.

ENGLISH LAKELANDS, The Waterways of ——.　Scale, 4 miles to 1 inch.
　　　　　　　　　　　　　The Geographical Journal, June 1896.

DEUTSCHLAND, Karte der winterlichen Sonnen-Aufgänge und -Untergänge in
——, für mitteleuropäische Zeit.　Entworfen von Heinrich Vogt.　Massstab
1 : 3,700,000.　　　　　　　　　　*Petermanns Mitteilungen*, Tafel 8.

ASIA.

SHAN STATES, Sketch Map of the ——.　Scale 1 : 2,000,000 (31·56 m. = 1 in).
　　　　　　　　　　　　　The Geographical Journal, June 1896.

AFRICA.

AFRICA, Political Map of ——.　By J. G. BARTHOLOMEW, F.R.G.S.　*Price, cloth 3s.*
　　　　　　　　　　　　　Geographical Institute, Edinburgh.

This map is after the style of that published in the *Magazine* in 1890.　The
results of the latest discoveries are shown, and the boundaries are drawn according
to the most recent treaties.　Several inset maps are added.

THE SUDAN, Bartholomew's Special Large Scale Map of ——, with General Map
of North-east Africa and Enlarged Plan of Khartum.　*Price 1s.*
　　　　　　　　　　　　　Geographical Institute, Edinburgh.

The map of the Sudan, on the scale of 1 : 2,000,000, or 31·5 miles to an inch,
embraces the country from below Wady Halfa to above Khartum and Abu Haras
on the Blue Nile, and extends eastwards to Massowah.　It therefore shows all
parts of the Sudan which have any connection with the military operations both
of the British and Italians.　The general map on the scale of 1 : 12,000,000 extends
from the Mediterranean to Lake Tanganika.

DER CONGO-STAAT nach seiner Districts-Eintheilung.　Nach den neuesten
Quellen.　Massstab 1 : 8,000,000.
　　　Deutsche Rundschau für Geographie und Statistik, Jahrg. xvii. Heft 9.

SÜDAFRIKA, Karte der politischen Eintheilung von ——.　Zugleich zur Übersicht
der Verkehrsverhältnisse, Goldfelder, &c., nach den neuesten Quellen entworfen
von Dr. B. HASSENSTEIN.　Massstab 1 : 10,000,000.
　　　　　　　　　　　　　Petermanns Mitteilungen, Tafel 7.

AMERICA.

UNIONS STAATEN, Die kulturgeographische Gruppierung der —— nach Dr. EMIL
DECKERT.
　　　Zeitschrift der Gesell. für Erdkunde zu Berlin, No. 2, 1896, Tafel 4 u. 5.

CHILOÉ, Derrotero de las dos espediciones del P. Fray Francisco Menendez á la
Cordillera al Este de ——, trazado segun Menendez, D. Francisco Vidal
Gormaz y otros autores por Francisco Fonk, 1895.
　　　Bol. del Instituto Geográfico Argentino, Tomo xvii. Cuadernos 1, 2, y 3.

THE SCOTTISH

GEOGRAPHICAL

MAGAZINE.

THE ISLAND OF FORMOSA: ITS PAST AND FUTURE.

By Rev. W. CAMPBELL, F.R.G.S., of Tainanfu.

(With a Map.)

ABOUT twelve months ago the island of Formosa formed the nineteenth, and only insular, province of China; but, soon after, it came under control of its own republican government, while to-day it ranks as an integral part of the Empire of Japan. Even from the geographical standpoint rapid changes of this kind carry with them a good deal of interest, but when one thinks of the effect throughout China and elsewhere of those further changes which the Japanese have already commenced in their newly-acquired territory, it will be felt that no apology is needed for submitting the following brief remarks on this important and very productive island.

Roughly speaking, it stands about a hundred miles out from the Chinese mainland, the south-eastern part of the channel being occupied by the small Pescadores group, which formerly made up a *Ting* or sub-prefecture of Formosa. A careful calculation from the latest Admiralty chart makes out the area of the island to be 14,982 miles, its length from north to south 245 miles, and its greatest width 80 miles, the whole extent being thus larger than Holland, and about one-half the size of Scotland.

It is now well known that the eastern side of Formosa is very mountainous, that its long bisecting middle range attains a height of nearly 13,000 feet, and that several spurs away to the north-east form the steepest sea-cliffs in the world; their lofty masses, rising some 7000 feet sheer from the water's edge, and wooded to the summit, present a sight which the beholder will probably never be able to forget.

The great drawback to Formosa is the want of good harbour accommodation, and the fact of its rivers and streams being navigable only to a very limited extent. With the exception of the small inlets at Saw-o and Black-rock Bay—which are suitable only for junks and incapable of extension—the entire line of the east coast is rock-bound, shelterless, and impracticable. Kelang harbour, to the north-east, no doubt possesses good depth of water and is open at all tides, but it, also, is too narrow, and too much exposed during the winter monsoons, to be a considerable centre of trade. A somewhat similar remark applies to the north-western port of Tamsui, as that is the mere estuary of an insignificant river, with a shifting troublesome bar which ocean-going steamers do not attempt to cross.

With all its disadvantages, the lagoon at Takow on the south-west coast is likely to become the headquarters of the import and export trade of Formosa. Its complete shelter, and good holding-ground in a part of the world where coral reefs abound, are important; while the very easy process of widening the entrance, and dredging out several sandbanks within, would secure an enormous extent of shipping accommodation. Consul Hurst's recently issued Report remarks on this subject as follows:—"A natural 'lagoon' exists at Takow, about seven miles long, by an average breadth of half a mile, and divided from the sea by a narrow 'spit' of coral throughout its whole length. The dredging of this lagoon and its conversion into a splendid harbour would be a simple and not very costly operation; but the Government have not yet been induced to take any steps in this direction, beyond ordering a preliminary survey some five years ago." It ought, however, to be added here, that any lack of harbour-room in Formosa is amply provided for at the Pescadores. Steamers reach the safe and spacious anchorages among those islands in about four hours from the port of Anpeng, and the recent victors knew well what they were about when insisting that the Pescadores also should be ceded to Japan.

One other remark under this head is, that the only fresh-water lake of any size in Formosa lies among the high mountain ranges some three days' journey south-east from the city of Chiang-hoa. The first European visit was paid to it twenty-three years ago by the writer of these notes, who named it Lake Candidius, in memory of that Dutch pastor who began Protestant missionary work in the island, about 1624. Regarding the river system, it should be remarked that, according to the geological formation of the island, most of the streams on its western side flow from the eastern mountain region across the plains, without much winding to the north or south, on their way to the sea. Of course the rainy season brings down great volumes of water, which oftentimes rush with so much force as to be quite impassable by boat or catamaran. That considerable river which debouches at the market-town of Tang-kang in the south has a direction more from north to south than any of the others, its general course being down through several valleys near Lau-long and La-ku-li, and farther on, till it flows out into the more level country of the Hong-soa county. As a matter of fact, large quantities of rice, sugar, camphor, rattan, charcoal and other products are brought to market over the streams and rivers of Formosa by means of very long bamboo rafts. These rafts draw only a few inches of water, and admit of great loads being placed upon them. It may be added here that the general lie of the land both to the north and south of the western plain region is such that much use may be made of the streams in Formosa for purposes of irrigation; and were more effort made for storing up water near the base of the mountains, and thence distributing it over the level fields of the west, the result could not fail to be a satisfactory one. Several such undertakings have already proved both easy of accomplishment and highly remunerative to the originators.

There is nothing very special to remark about the climate of Formosa. From October till March the weather is mild, often bracing, with occasional showers in the north, but very few throughout the southern regions. The highest temperature and the heaviest rainfall are to be met with between June and September, although even then the thermometer seldom registers more than ninety degrees in the shade. Severe storms sometimes occur during midsummer, but those terrible typhoons which start in the China sea and travel northward, usually slant off at South Cape to drive with full force across the low-lying Pescadores, or over the islands of Botel Tobago and Samasana to the east of Formosa. There are no active volcanoes in the island, and only slight earthquakes are of frequent occurrence, especially in the neighbourhood of Tamsui and Kelang.

It is safe to say that Formosa is an exceedingly rich island, because the alluvial plains of the west, far stretching and well watered, offer simply illimitable opportunities for raising sugar-cane, rice, sweet potatoes, ground-nuts, indigo, ginger, turmeric, and suchlike. Moreover, its present fruit production is sufficient to show what abundant increase would follow the introduction of any kind of general and systematic method of cultivation. Large beautiful pine-apples can be had almost for the lifting, and probably not a more palatable and wholesome fruit could be found anywhere than the Sai-le loose-skinned oranges, or those juicy and delicious little *pumelos* from Bantan. During 1894 twenty-one million pounds of tea, and over forty thousand hundredweights of camphor, were shipped from Formosa. It has been ascertained that rich coalfields exist, not only in the north—where they have long been worked in European style—but in the A-li-kang region, twelve days' journey farther south. The petroleum wells at Toa-kho-ham, and the sulphur deposits near Tamsui, could also be turned to great account; while it was owing to the recent war that negotiations were broken off between the Government of Formosa and the representatives of a Chinese syndicate for conceding exclusive right to work the gold mines in the island. In short, the wealth of this still undeveloped country may be seen on considering that, during 1893, trade to the value of four and a half million pounds sterling passed through the ten or twelve European houses doing business there.

Within the limits of this paper it is not possible to make anything like a detailed statement regarding the history of the island. Traders from the opposite mainland began to visit it about the middle of the fifteenth century. On nearing it, the sight presented was a wide level shore with lofty mountains rising range upon range into the interior, and this suggested to them the Chinese name which it still bears, that of Tai-wan or Terrace-beach. It was the same sight a hundred years later which led some Portuguese adventurers to shout out, Ilha Formosa! Beautiful Isle! another descriptive name which has now become current all over the world. At this early time, the island was found thickly peopled by an aboriginal race, or rather a collection of non-Mongolian tribes widely differing from each other in their appearance, language, and customs. Of course no strictly accurate account of those tribes can be looked for in the notices of this period, and one has surely some right to complain of the extent to which theorising is carried by writers like the late Professor Delacouperie, who, from such meagre data, say all sorts of things about the dwarfs, the black giants, and even the tailed men of Formosa.

Like a great many other good things, our earliest knowledge of the island comes to us from the Dutch. Wishing to share with the Spaniards and Portuguese in the lucrative trade of the Far East, their East India Company effected a settlement on the Pescadores in.1622, but the resident and provincial Chinese authorities strongly objected to this, and did not cease their opposition till the new-comers removed to the little-known, but much larger, island of Formosa. Dutch rule lasted there from 1624 till 1661, and during those thirty-seven years small military establishments were set up, and authority exercised from Long-kiau in the south up along the western sea-board, and on as far as what is now the north-east county of Gi-lan. Civil affairs were administered by a Dutch Governor with the members of his Council, who had all to report to colonial headquarters at Batavia; and one interesting feature of their work was, that not only were efforts made for the furtherance of trade, but also for bringing the natives of the island to the knowledge and obedience of the Christian faith. No fewer than thirty-seven ordained pastors came from the home land to engage in this latter service; who, besides attending to their more official duties, superintended the labours of the Dutch schoolmasters, and reduced at least three of the aboriginal dialects to a written form.

Indeed, the very success of the colony began to awaken the envy and covetousness of people living under less favoured conditions. China was then passing through that epoch-making crisis which resulted in the overthrow of the Ming, and the usurpation of the present Manchu-Tartar, dynasty; so that the unsettled times led many of the Fokien Chinese to cross the channel and try their fortune under the rule of those western barbarians, of whose influence and generosity they had been hearing so much. True, the Formosans were represented to be a warlike race, but it was believed that sharp-witted refugees like themselves would be sure to hold their own against people who were still looked upon as mere savages.

All this, however, was but the beginning of trouble to the thriving little colony at Taiwan, for that great Chinese patriot or pirate, the chieftain Koxinga, came himself to be so hardly pressed by the invading Manchus that he, too, began to think of Formosa as a place of rallying, if not of probable possession. His first move was to send over repeated messages from the Pescadores with the view of fixing a quarrel on the Dutch; but all pretence was set aside when he placed himself at the head of an immense flotilla of war-junks, both sea-worthy and well-provisioned, as well as manned by thousands of daring outlaws, who thought less of the fight than of the idleness and plenty they hoped to find in this newly-discovered retreat.

Meanwhile, Governor Coyett was shut up in anxious consultation with his councillors at Fort Zeelandia. Frequent were the warnings and earnest the appeals they had sent to Batavia; but other influences were at work among the higher officials there, so that reinforcements which ought to have come never reached Formosa, thus compelling the comparatively small garrison to find shelter within the castle walls, in which position both soldiers and officials were found when Koxinga appeared to demand their unconditional submission.

The sturdy Hollanders held out for nine long weary months, during which time they made several damaging attacks on the enemy; who, however, retaliated by perpetrating the most shocking cruelties on such Dutch people as were scattered throughout the island, their very clergymen being tortured to death, either by impalement or by crucifixion. Contemporary records unite in singling out the case of Rev. Antonius Hambroek, who was sent by Koxinga into the castle, under a flag of truce, to propose terms of surrender, and told to back these up with threats of most terrible vengeance. Mr. Hambroek was forced to leave his wife and two children (one of them described as a sweet and comely maiden of eighteen) in the invader's camp as pledges, which sufficiently proved that any failure of his undertaking would be a most ominous signal for those poor defenceless ones. And yet, this noble man was so far from persuading the garrison to surrender, that he encouraged them to continue the defence by hopes of relief, assuring them that Koxinga had lost some of his best ships and soldiers, and began to be weary of the siege. When his speech was ended, the Council left it to his own choice either to stay with them or return to the camp, where he could expect nothing but instant death. He had also two daughters within the Castle, who hung upon his neck, overwhelmed with grief and tears to see their father decided to go where he knew he must be sacrificed by the merciless foe.

But he reminded them that having left his wife and his other two children as hostages, death would be their certain fate if he returned not; so, unlocking himself from his daughters' arms, he exhorted them all to a resolute defence, and cheerily said as he left the castle gate, that God might yet make use of him in bringing deliverance to his poor fellow-prisoners.

Koxinga received his answer sternly, and without further delay issued an order for the massacre of all Dutch captives, and of every native who persisted in the profession of Christianity; Hambroek himself was put to death by decapitation, and the before-mentioned daughter compelled to become a member of his murderer's harem.

At length, worn out with disappointment, fatigue, and famine, the little garrison was compelled to surrender, all the public property falling into the hands of the enemy, and the brave but heavy-hearted defenders being allowed to depart in their only remaining ship.

Strong feeling was shown by the home authorities over the loss of so rich a colony; and, therefore, on arriving in Batavia poor Coyett was arrested, and a long trial afterwards ended in his being banished to the desolate island of Pulo Ay. A vindication ought yet to be made of the character of this noble but deeply-wronged man. As for Koxinga, he died a miserable death after having been king of Formosa for little more than twelve months.

Ching-keng-mai succeeded his father on the throne, and reigned for about twelve years, but was often in trouble through his trading ships being attacked by subjects of the now dominant Manchu ruler of China. It was in the hope of fortifying himself against this enemy that royal circular letters were sent out to European merchants frequenting those seas, in which tempting facilities were held out if they would only come and open warehouses in the neighbourhood of Taiwan.

Such an offer from Ching-keng-mai would be scarcely worth referring to here, were it not that the only response to it was made from a very unexpected quarter. It certainly is not generally known that during the latter half of the seventeenth century, the English East India Company had one of their factories on Formosa for a number of years; but there can be no doubt about the matter, because the old yellow documents which were courteously produced for my inspection at the India Office a month ago, abundantly prove it. One of these letters is dated 1670, and is addressed to the King of Tywan, its opening sentences running thus: —" Charles, by the grace of God, King of England, Scotland, France, and Ireland ; having most graciously licensed severall of his Merchants to trade into all the habitable partes of the World, amongst whom Sir Wm. Thompson, with some other Merchants, are, by the said most gracious King, authorized Governors of the Merchants to trade into these Eastern parts. Now for the directing and overseeing their Affaires at Bantam and partes adjacent, they have appointed mee (Henry Dacres), Agent. The said Henry Dacres, therefore, on behalf of the said Sir Wm. Thompson, Governor, sends greeteing unto your most Excellent Majesty ; and having seen your most gracious Letter directed to all Merchants in Generall, inviting them to trade into the partes under your Majestie's Jurisdictions, has, without delay, sent this small Ship and Sloope with Mr. Ellis Crispe, Capt., to acquaint us with the Merchandize desireable to bee Imported, and of Merchandize proper for us to Exporte, and when wee shall bee acquainted therewith by him, and have the permission of Friendship and Affection of your Majesty (which wee moste humbly desire) wee shall requeste the said Sir Wm. Thompson's leave to solicite your Majesty ; and because we would have your Majesty know that wee are Englishmen, and a distinct Nation from Hollanders (some people of which Nation about ten years since were driven out of your Land by his Majesty your Renowned Father), we have sent on this our Shipp Capt. Sooke, with eight other Chinamen, who have for long time traded and been acquainted with us and our Nation." There follows a long table of conditions for the settlement of the factory, while subsequent letters report concerning the reception of the supercargo, and of a very favourable commencement having been made. It would appear, however, that the king's desire to benefit himself out of this trade soon gave rise to a great amount of friction, and led ultimately to the following peremptory mandate from the Court of Directors to their representatives at Bantam, the date of it being 28th Feb. 1682:—"As to the Trade of Tywan, we hereby expressly require you, that if you have made no better earnings of it before this comes to your hands, you do order our Factors to desert the Place, and bring off what they can with them. To which purpose we have written a menacing Letter to the King, and probably may send a Ship to be with you in March or April next, to go down to Tywan to fetch off our Servants; and after that to use some forcible means for our satisfaction of the debt he owes us." The "satisfaction" was obtained, with interest.

At this stage it remains only to add that Koxinga's grandson was very young when he succeeded his father in the government of Formosa, and that his officers found it increasingly difficult to maintain their independence alongside the growing success of the great Manchu authority. Accordingly, the young Prince ended the trouble by tendering his submission in 1683, thus bringing Formosa under direct control of the Emperor at Peking.

And now begins a long period in the history of the island during which it emerges only now and then in such a way as to attract the notice of Western nations. On the cessation of its own monarchical government, it and the Pescadores were joined on as a prefecture to the opposite Province of Fokien, the insular officers—both civil and military—having all to report to their departmental superiors at Foochow. For administrative purposes, that portion of the western region occupied by the now numerous Chinese was made up of what might be called the Metropolitan County of Terrace-beach—where Tainanfu now stands—the County of Phœnix-Hill or Hong-soa immediately to the south of it, and Varigated-Net-Hill or Tsu-lo-san adjoining it on the north. It may be remarked in passing that this Tsu-lo-san is a mere Sinicised form of the old native Tilosen; and that, on the occasion of a civil outbreak there, when most of the people remained faithful to the Imperial cause, the Emperor graciously changed the name to the one of our present-day maps—that of Kagi or Established Righteousness.

Within the limit, then, of those three counties, the population was divisible into two great classes: (1.) The Chinese themselves, by far the bulk of them being immigrants or the descendants of immigrants from the Chin-chew and Chang-chew regions of the Fokien Province; (2.) Those agricultural aborigines, who rendered a general allegiance to the Chinese, conformed to many of their customs, and knew more or less of their language, but who lived somewhat independently in small townships or hamlets of their own.

There was very little intercourse between this population and the unsubdued tribes who inhabited the remoter parts of the island; certainly much less than existed between those tribes and the earlier new-comers from Holland. The Chinese did not dare to venture amongst them, because long years of oppression and trickery on their part had quite appropriated that western region where the native was wont to hunt and to fish, and where many of his little villages nestled in comfort and security; whereas, although sometimes acting towards them in a very high-handed way, the Dutch had come to adjust inter-tribal quarrels, to act fairly, and to prove an unspeakable blessing to the aborigines of Formosa.

The Chinese-Formosan annals of this time, and for long after, contain much that is very dry reading, being chiefly made up of vague topographical details, with an account of official appointments, clan fights, rebellions, and disasters; to which is added any number of wonderful stories about the inhabitants and productions of the island.

A valuable monograph has come down to us from one of the Jesuit Fathers, who spent some time in Formosa during the first half of last century. De Mailla writes in a sober and very interesting way about what he saw, and cordially testifies to the traces of Christianity which survived from the period of the Dutch occupation. On this subject he says:—"Before leaving Amoy, we had been informed that there were Christians in Formosa. Accordingly, we made inquiries, and certainly there are none among the Chinese; but there are traces as if Christianity had been known among the aboriginals from the time when the Dutch were in possession. We met several who were able to speak the Dutch language, who read Dutch books, and who, in writing, used Roman letters. We even found among them fragments of our five books (probably the Pentateuch) in Dutch. Those natives worship no idols as the Chinese do, and have a horror of anything approaching to such an act; and yet they perform no religious rites, nor recite any prayers. We spoke to several who acknowledged a God, Creator of heaven and earth—a God in three Persons, Father, Son, and Holy Ghost. They told us that the first man was called Adam, and the first woman Eve; that these, having disobeyed God, had drawn forth the divine anger upon themselves and all their posterity, and that it was necessary to have recourse to baptism to efface this stain; of which rite, too, the very formula is remembered to this day."

Coming down to more recent times, we find that certain events which took place in Formosa during 1842 attracted the notice of many Europeans at home and abroad. The first Anglo-Chinese war had just been closed by the Treaty of Nanking, when two British ships were driven, by stress of weather, on to the north-west coast of the island, the wretched persons who came ashore being brought to await instructions from the local Chinese officials. After consultation, those civil and military underlings concluded that the opportunity was a very favourable one for enriching themselves and obtaining substantial promotion in their respective services. They commenced proceedings with a wholesale plundering of the two ships, and by inflicting as heavy fines as possible on any private individuals who had been found looting on their own behalf. Those miserable sailors and passengers who escaped the dangers of the sea were then marched down the island to Tainanfu, were most scandalously treated during four months of imprisonment there, and were at last led out beyond the Great North Gate of the city, where they suffered decapitation to the number of one hundred and ninety-seven.

The report sent up to Peking quite ignored the cessation of the war with Britain, represented that two of the enemy's battleships had attempted to land troops on Formosa, but that during a severe engagement the Imperialists proved equal to the occasion, attacking and vanquishing the foreign barbarians with great loss of life. Trophies of war were also forwarded in the shape of articles which had been stolen from the two ships, while marks of the Imperial favour were besought for those who had conspicuously distinguished themselves in securing this most glorious victory.

Of course, there was a terrible outburst of feeling amongst Europeans in China when the real facts of the case came to light. Some would have at once proceeded to overthrow the dynasty, and few dissented from the proposal that the services of the still present squadron should be made use of; but wiser counsels prevailed in the end, for in the face of a profound expression of regret from Peking, the condign punishment of many Formosan officials, and the peace which had only recently been proclaimed, it was felt that this question was one we could not well take out from the region of diplomatic treatment.

An incident of much greater fatefulness to the island took place during the autumn of 1860. Once again there is war between Great Britain and China, and this time hostilities are terminated by the provisions of the Treaty of Tientsin. It is in accordance with Articles 8 and 11 of this Treaty that two ports in Formosa are declared open to foreign trade, and British subjects professing or teaching the Christian religion are free to go anywhere beyond, if supplied with passports counter-stamped by the local officials.

The result was that Consular offices were speedily opened at Takow and Tamsui, sub-offices being also provided at Tainanfu and Kelang. Before long, too, European warehouses and bungalows were established there, with accommodation for the European officers who were to administer the Customs Service. To meet the religious wants of the people, a Dominican Mission had already been at work for a few years, but in 1865 the English Presbyterian Mission broke ground at Tainanfu, to be followed seven years later by the only other Protestant Mission in the island, that at Tamsui from the Canadian Presbyterian Church.

These various concessions furnish conclusive evidence of the progress which had been made since the time when only a part of western Formosa was included in the Chinese Empire. The very fact that they were possible in a land then so much overrun by savages, shows that the Chinese population must have steadily increased, and is a testimony to their possession of at least some qualities which cannot be undervalued. No doubt instances did occur where the aborigines adapted themselves to the new order of things; but, generally speaking, how was it possible that such tribes could increase and thrive in daily touch with a shrewd, industrious, and plodding race like the Chinese?

What threatened to be a very serious interruption to the later prosperity of Formosa took place towards the close of 1874. About two years previous to that, a boat, from the Loochooan part of Japan, was wrecked on the east side of the island, and its crew murdered by the Bau-tan savages there. Soon after, the Japanese authorities presented a claim for compensation against the Government of China, but it was met in a very evasive way, and the interchange of several plainly-worded despatches was followed up with a curt intimation from China, that she refused to be held responsible for the action of savages inhabiting an extra-territorial region like the east of Formosa.

The response called forth was a sufficiently startling one; for the people of Japan itself were ignorant regarding the destination of a certain warlike expedition which left their country under sealed orders about this time. In a word, Japan accepted the ultimatum from Peking, caused the mysterious expedition to be landed on South Formosa, and very soon succeeded in chastising those whose cruel treatment of shipwrecked people had become a byword.

Hereupon, however, China entirely changed front, and made loud complaints of what was described as sending military forces into the territory of a friendly Power. All right, said the Japanese, the forces will be at once withdrawn when China acknowledges her responsibility, and meets the expenses of sending them to Formosa; which concessions were ultimately obtained only through mediation of Sir Thomas Wade, the British Minister at Peking. The incident must have cost China millions of money, for everywhere along the coast preparations were made as if war with Japan was all but certain.

One outcome of these proceedings was the desire shown by the Chinese Government to adopt measures for a more thorough development and defence of the island of Formosa than had ever been contemplated; and, accordingly, instructions were issued to Tang Tih-chiang, Governor of the Fokien Province, an officer who was known to be as just and capable as he was free from the superstition and hide-bound conservatism of his class.

No wonder, then, that under this administration, and especially during His Excellency's repeated visits to Formosa, a great amount of progress was made, and many more reforms decided upon. For example, roads were opened across the island, several of the aboriginal tribes subdued or pacified, coal-mining undertaken, telegraphs and railways introduced, and Formosa raised to the rank of a separate province of the Empire. Indeed, had his life been prolonged and Imperial help been forthcoming, there is some likelihood that this island might still have remained a rich Chinese possession, and a strong defence against any hostile foreign fleet.

The next occasion on which Formosa emerged from obscurity was during the Franco-Chinese war, twelve years ago. Many stirring events took place then, including the bombardment of Kelang and Tamsui, with that strict blockade which French men-of-war kept over the island for a period of nearly six months. A very outstanding figure all through the struggle was Liu Ming-Chuan, who directed the operations against the French, and who afterwards became the first Chinese Governor of Formosa. .

Whatever effect this war had elsewhere, it left China with one more opportunity for acting in a generous and enlightened way while legislating for Formosa.   And it cannot be denied that the Governor's forward policy did receive a certain amount of encouragement; but the man was evidently far in advance of those puissant old reactionaries who control everything at Peking.   They ought to have abundantly backed up their really capable representative, and who knows but Formosa might have been acknowledging his authority to-day?

It was under Liu Ming-Chuan in 1885 that that fresh and all-inclusive division of territory took place which still holds good.   According to this scheme, the whole of Formosa and the Pescadores are made up of four prefectures; these, again, being subdivided into eleven counties or districts and five sub-prefectures, two of the latter including all the eastern side of the island, and one of them the whole of the Pescadores group.   Their native names are as follows :—

1. The northern prefecture of TAI-PAK, made up of the three counties of Sin-tek, Tam-sui, and Gi-lan; with the sub-prefecture of *Ke-lang*.

2. The middle-western prefecture of TAI-WAN, made up of the four counties of Hun-lim, Tai-wan, Chiang--hoa, and Biau-lek; with the more eastern sub-prefecture of *Paw-li*.

3. The south-western prefecture of TAI-NAN, made up of the four counties of Heng-chun, Hong-soa, An-peng, and Ka-gi; with the sub-prefecture of *Phe-aw* (the Pescadores).

4. The eastern prefecture of TAI-TANG, made up of the two sub-prefectures of *Pi-lam* and *Hoe-leng-Kang*, with headquarters at the middle-eastern centre called Tsui-boe.

If it be objected that this distribution must have embraced a large extent of country occupied by independent non-Chinese tribes, the reply is that Liu Ming-Chuan did everything in his power to make it a reality and not a mere name ; for his efforts were unceasing to bring those head-hunting savages within the restraint and protection of the common law. Nor was he satisfied with simply issuing orders for the accomplishment of this ; for, on at least one occasion, the writer of these notes was an eye-witness of the Governor's self-denial and pluck in directing operations against savages whose ceaseless midnight attacks had depopulated one of the inland valleys.   At that time the Governor had for months been living this life of hardship, and next year the *Peking Gazette* was able to report that 478 villages, containing an aggregate population of 88,000 aborigines, had already given in their allegiance.   It should be added that by far the majority of these made voluntary submission, severe measures being resorted to when all other expedients had failed ; for the Governor was determined that, no matter what might be done or left undone during his term of office, the hurtful and most scandalous practice of head-hunting should be completely stamped out.

Another matter which had much attention given to it under this administration was that of providing Formosa with railways.   The proposal was to have one main line all down the west side of the island, adding on branches as they came to be called for, and there was no interruption till all the surveying part of the work had been finished.　Raising of the necessary capital caused some delay, but railways in Formosa are now an accomplished fact.

The general terminus of the two lines, which have been working during the past five or six years, is in the town of Twa-tiu-tia, a short distance up the Tamsui river.　One of the lines crosses the country for about twenty miles over to Kelang, while the other comes down the island for forty-four miles to the county town of Sin-tek.　Of course, it is beyond all question that the completion of this work would be a very great boon to every one.

And now, it is quite time to say a few words about still more recent and more sweeping changes than any that have been yet referred to. I happened myself to be travelling through Japan when that culminating point in the war was reached—the fall of Port Arthur.　The appearance of the lounging, well-fed Chinese prisoners who were then at Osaka was noticeable, and those regiments of tight little fellows who were in marching order for the field of action seemed fit for anything.　There could be no doubt as to what was coming, for every one believed that the proud, unwieldy, and traditional foe of the country would soon be suing for peace at any price.

At that time it was surely an insult to Japan, and the very height of folly, for the Chinese to take the initiative in this direction by sending over Mr. Detring, a foreigner in their employ, to try and arrange matters with the *Eh-law*, or dwarf slaves, as the Japanese are often called in China.　The officials at Hyogo wouldn't speak to him, but simply gave orders that he should be shadowed by policemen till he left the place. Nor was the next commission much more successful.　It was made up of several high-class mandarins—including the Governor of Formosa— but no proper credentials had been given them, and negotiations were not even entered upon.　The third attempt was made by the mighty Li Hung-Chang himself, whose full powers at once led to the Treaty of Shimonoseki, according to which peace was restored by China consenting to pay a war indemnity of thirty-five million pounds sterling, and ceding to Japan the southern half of the Liau-tung peninsula, with the island of Formosa and all its dependencies.

There is reason to think that the proposal to occupy Liau-tung was a piece of mere diplomacy on the part of the Japanese; because, so far as natural resources are concerned, the place is useless, while huge warlike establishments would have been necessary to retain it.　It was otherwise with Formosa, for that is a country rich in coal and agriculture, one which completes the line of islands reaching up so easily through Majicosima and the Loochoos to Japan itself; and one, especially, whose occupation had long been a cherished aim of the subjects of the Mikado. Wishing, therefore, to make sure of the island, an additional demand was made, about which discussion was certain to take place; and so Japan gave up the Liau-tung peninsula on the European Powers guaranteeing an additional seven and a half million pounds of indemnity, no objection at all being made to the cession of Formosa.　The whole transaction may become more intelligible to us on remembering that the Oriental mind is very fond of working in curves, its method of obtaining any desired object being to say a great deal about something else.

The Treaty of Shimonoseki was signed on 17th April of last year, but three weeks previous to that the Pescadores had already been bombarded and taken possession of, and it is at this point the Japanese were much blamed for their long delay in carrying out those clauses of it which refer to Formosa.   Chinese rule in the island quite ceased on its being formally handed over to the Japanese at the end of May, but it was months after that before anything was seen of the new authority throughout the region south of Chiang-hoa, and it is easy to understand how this position of things should have given rise to an amount of lawlessness which brought suffering and death to many a home.

No doubt local order of a kind was maintained through the establishment of a short-lived republic in the walled city of Tainanfu, the capital of the island; where, it may be added, the English Presbyterian Mission has its headquarters, the few missionaries being the only Europeans resident there.   This effort at self-government was chiefly due to thousands of soldiers belonging to the Black Flag division of the Chinese army, who had been sent over here during the earlier stages of the war.   They conferred with some of the leading citizens, and both agreed that the Brigadier-General Liu Yung-fu should be proclaimed president of a republic, and that all should unite in doing whatever they could to thwart and drive away the invaders of their land.

As might have been expected, however, the whole movement ignominiously collapsed when the Japanese army halted within a day's march of the city to prepare for the attack.   Many of the Black Flags ran off to the hill region, while Liu Yung-fu saved himself by escaping in the disguise of a woman carrying a baby.

After this, the four great gates were closed, and an ominous silence brooded over the city, people going about carefully as if walking on the thin crust of a volcano.   There seemed no way to avert the approaching doom, for the Japanese soldiers were irritated at having to fight every inch of their way over this newly ceded territory, and every one knew that terrible reprisals would be made in the place where the Black Flags had entrenched themselves.

It was on the Saturday afternoon of the President's flight that a deputation of leading inhabitants sought out the missionaries and pleaded with them, for God's sake, to go and bring the Japanese into the city in peace.   The undertaking was anything but free from risk, because it was difficult to ascertain how far this request indicated the general wish of the people, and native Christians in different parts of the island had been cruelly murdered on a charge of being in collusion with the Japanese. However, another and even more influential deputation came forward to say that they would put their request in writing, so that any one might see where the responsibility lay, and that this service was being rendered by the missionaries at the people's own urgent desire.

The sun was just setting when all the needful preparations were made, but not an hour was to be lost, and therefore, taking the stamped document with them, my two colleagues went out from the Great South Gate on their errand of mercy.   Nineteen unarmed Chinamen accompanied them, but they plodded along in silence.   The stars were shining brightly, and stillness reigned everywhere, till the party was suddenly stopped by the *ping* of a rifle, and the loud challenge of a Japanese sentry. Signals were made, but they were immediately surrounded and led to the presence of the General, who consulted with his officers, and afterwards told them of the acceptance of their invitation, and that the army would begin to move before daybreak, having Mr. Barclay with the nineteen Chinamen in front, and Mr. Ferguson with several officers proceeding somewhat in the rear. It was also plainly stated that, on the slightest show of treachery or resistance, the soldiers would open fire, and the whole city be burned to the ground.

The time occupied by that long march back again was, indeed, an anxious one; and as the missionaries drew near, and saw the city closed, their hearts sank within them lest some fatal interruption had taken place. That sound, too, seemed something more than the mere barking of dogs. Could it be possible that the roughs of the city had broken out at last, and were now engaged in their fiendish work? They looked behind, and saw only a wall of loaded rifles; in front, but there was no hopeful sign; and the strain was becoming almost insupportable, when the great gates were swung wide open. Hundreds of gentry were seen bowing themselves to the ground, and in a minute more the flag of the Rising Sun was waving over the city.

It would be out of place to say much at present about the future of Formosa under the altered condition of things, and only a few remarks are necessary on changes which have already taken place, and others which are almost sure to follow. Among the former may be noted—(1) that the Mandarinate has now left the island, bag and baggage. Now, it is no part of our duty to speak evil of dignities, or of anybody else, but twenty-five years' observation leads to the conclusion that there are tremendous difficulties in the way of regarding Chinese officialdom with anything like feelings of confidence and respect. No doubt some members of the class are capable (from the native point of view), unselfish, diligent, and really helpful to the people. Generally speaking, however, this countless host, from the viceroy down to the lowest *yamen*-runner, goes on the fundamentally pernicious principle that the country was made for the mandarins, and not mandarins for the country. (2) The influence of the so-called literary class is now gone for ever in Formosa. These are the gentry who swear by Confucius and all his opinions. They are held in high esteem, the οἱ πολλοί looking upon them as dungeons of learning, and as very fortunate in being able to make potsful of money at teaching and in every low kind of pettifogging. About seventy per cent. of their learning is a mere fraud, and consists in the power of memorising the classics and keeping close to the traditional comments which have been made upon them. Their anti-foreign tendencies are well known, and it would be difficult to find anywhere a prouder or more narrow-minded and impracticable body of men. (3) The Japanese authorities in Formosa issued a proclamation last February forbidding the importation of opium, except for medicinal purposes. This action has not attracted much attention, although it is a very significant one. The first Chinese anti-opium edict appeared in 1729, having also been directed against the use of the drug in Formosa, and ever since the island has had an evil reputation in this respect. The importation during 1893 (the last year unaffected by the war) was 5680 cwts., valued at £419,839. But everything is to be changed now, for the Japanese say that the whole thing must be stopped, and a clean sweep made of the opium. The proclamation is very suggestive reading after the voluminous report of our own late Royal Commission on this subject.

But it is unnecessary to make further enumeration of changes already accomplished in Formosa, and as for those which are still to come, one may forecast a little by considering, on the one hand, what Japan itself now is; and, on the other, the expressed determination of its rulers that Formosa, body, soul, and spirit, must be made a part of their empire. Connecting these two things, then, it goes for the saying that, before long, good roads will be all over the island, that the railway will be carried down from north to south, harbours opened, and a proper currency introduced, with parliamentary representation, upright officials, skilled native doctors, newspapers, and cessation of work every seventh day in all Government offices. Of course, too, there will be things to vex the soul of the European merchant and the ardent Christian missionary, but patience must be exercised, and great things still be expected from such a people as the Japanese have proved themselves to be. The movements of populations under the new order of things will be interesting, and be likely to appear in (a) the departure of many Chinese from Formosa: (b) steady increase in the number of Japanese immigrants; and (c) in the result of civilising influences brought to bear on all the non-Chinese-speaking tribes. These and other matters make up a problem of first-class importance, and one cannot but accompany Japan with gentle wishes through this critical, yet very hopeful, period of her history.

BRITISH UNITY.

By Arthur Silva White, Hon. F.R.S.G.S.

It is a commonplace to speak of the British Empire as being the greatest, the most powerful, and the most pacific that the world has ever seen. Ours is an eclectic epitome of all preceding empires, and therefore signalises a genuine advance in the development of national life. It owes its origin, its growth, its power, not to state-craft, nor to state-aid, and relatively little to aggressive wars. It is not of artificial expansion through the selfish nursing of monopolies; still less is it the sequel to an ambitious Imperial policy. It is the People's Empire—bone of their bone and flesh of their flesh—the creation and the heritage of a freedom-loving race. In a word, it is a political organism, fulfilling definite national functions.

Again, the British Empire makes for peace and not for war. Peace, it has been said, and said with truth, is the greatest of British interests: its preservation, therefore, is the first duty of our statesmen. Peace at any price we cannot afford to accept; but peace at a high price is merely a sum in arithmetic to a commercial State such as ours. By peace we gain time to develop along natural lines; and a large measure of time, judged by human standards, is required for so complex an organism as the British Empire to attain to full maturity as a responsible world-Power. To ensure peace we must, however, be prepared for war; and in these days of mammoth armaments this necessarily involves a great tax upon national industry, though, properly speaking, it amounts to merely an annual premium on war-policy.

The ocean unites the scattered members of an empire whose cohesion is dependent on sea-power. If, therefore, it can be shown that this cohesion is no artificial or imaginary bond, but one growing out of the natural co-ordination of its parts, and that the circulation of all vital interests is of necessity stimulated by nourishment from within and pressure of opposition from without, the organic unity of the Empire becomes a demonstrable fact. If sentiment were the only tie—as is too often asserted—between the Mother-country and her Colonies, the Empire would run grave risks of disruption. Sentiment, however, plays merely a subordinate part; and it is entirely overruled by self-interest, which forms the best possible basis for an imperishable union.

Let us consider for a moment how this great Empire of ours has grown up beyond the seas; what it really is, and what are its national functions. To establish my thesis, I must follow out two lines of argument—the historical and the geographical. Take first the historical, in which the late Sir John Seeley is an admirable guide.

When Elizabeth, the much-courted Queen of England, finally decided against the Habsburg system of royal marriages, which in her day were the object of the highest statesmanship and controlled the dynastic fortunes of States no less than their territorial aggrandisement, she was impelled by her love of peace—national rather than domestic. As she herself said, she was wedded to her people. This fortunate choice, isolating her to a large extent from the battle-fields of European rivalry, in spite of her championship of the Reformation, led, by a curious combination of circumstances, to the foundation of the British Empire. She gave us our insularity, not consciously, nor with any prophetic views of empire, but through the irony of negation. She was anxious to secure peace almost at any price : and the secret of her policy was " masterly inactivity." In the second half of her reign, however, she was forced into a war with the Spanish Monarchy—the united sea-Powers, who, by virtue of a Papal Bull, claimed the undiscovered Ocean-world as theirs by Divine Right; and the issues of that war revolutionised Europe. England rose from the sea as a commercial and maritime State, and entered upon a course of legitimised piracy in the New World. When the seventeenth century dawned, the two Protestant States, England and Holland, stood forth as the coming sea-Powers; and the two Catholic States, Spain and France, faced each other in fixed rivalry for military ascendency on the Continent. The Dutch Republic and England pursued a common policy, at the expense of the Spanish Monarchy; though the former was at first the more enterprising in her efforts to create emporia beyond the seas. England, on her part, was impelled by a blind instinct to seek the line of least resistance towards commercial and political independence. In fact, it was only under James, who united the English and Scottish crowns, that Great Britain embarked upon a conscious policy of colonisation, properly so called. Cromwell clearly foresaw the true destiny of England, and interpreted it by his Navigation Act, which aimed a vital blow at the Dutch monopoly over the carrying-trade, as well as by the capture of Jamaica; but under Cromwell we were an aggressive Power—the foremost maritime and military State in Europe. It required a foreign Prince, William of Orange, to unite the dynastic and commercial interests of England and Scotland on the one hand, and of England and the Netherlands on the other. The battle of the Boyne laid the foundations of the present European system; and the war of the Spanish Succession, carrying with it the question of monopoly in the New World, finally settled for us our character as a Commercial State.

Elizabeth, Cromwell, and William III. were, therefore, the three rulers who mainly contributed to the growth of British policy. The first withdrew us from European entanglements and alliances, and wedded us to the Ocean. The second adjusted, though in a very summary and inadequate fashion, the differences between the two kingdoms and Ireland, laid the basis of our commercial policy, and suggested the future greatness and scope of the Empire. The third gave us the rudiments of a durable Constitution, freed England and Europe from the French ascendency, and fixed the main lines of British policy as it exists at the present day. England sought the Ocean for a wider and a fuller life. She found this in America, and, as time went on, in other parts of the world. Sturdy emigrants, driven from their Mother-country by religious persecution; bold adventurers, to whom the illimitable horizon made an irresistible appeal; and genuine colonists who followed in their path: each and all bore with them their birthright of freedom, and created new homes in a New World.

The great English Revolution, whilst it laid the basis of our national policy, at the same time ushered in that series of wars with France, or chiefly with France, which came to an end only with the fall of Napoleon. This traditional rivalry imposed its law upon Europe during the eighteenth century. It was for us a period of incessant strife, during which the Empire grew—not by peaceful settlement, as for the most part it had done during the previous century, but by conquest. In the Napoleonic era the leading land-Power was pitted against the leading sea-Power: and in the end victory remained with the sea. Finally, the Victorian age has been characterised by unexampled Colonial expansion, chiefly through peaceful settlement, and by an unprecedented advance in national life and industry. Our inventors and scientists have revolutionised the conditions of international life and have been universal benefactors. The Ocean-world has been Anglicised.

We thus see, that in our search for a national policy we have oscillated between a pacific and an aggressive attitude : opposite poles, marking the extreme swing of the pendulum. And, if I may be permitted to interpret such conflicting evidence, historical facts appear to me to indicate, that the British Empire finds the widest scope for its dynamic development and expansion under conditions which provide for peace, unfettered commercial intercourse, and unchallenged naval supremacy. The latter, it is true, may be provocative of war; but it is an essential condition of our insularity, upon which the inviolability of our national rights depends, and, as such, counterbalances the immense standing armies of Continental States. The fundamental principles of British policy, may, consequently, be embodied in three watchwords— Conciliation, Amalgamation, Consolidation : conciliation towards the outside world, amalgamation of interests within the Empire, and the consolidation of its component parts on the basis of Free Trade, free institutions, and the freedom of the sea. The first may be secured by diplomacy, the second by statesmanship, and the third by naval supremacy permitting full play to our national genius.

Whilst the historical argument in favour of our withdrawal from Continental rivalries may, perhaps, fail to carry the conviction I have sought to establish, much greater confidence may be placed in the geographical argument of insularity and organic unity. That, of necessity as well as by deliberate choice, we do occupy a position of isolation (or, as we would say, of independence) in international politics, in consequence of our world-wide interests and the rivalry they excite, is a fact universally recognised ; and, in view of the comparatively recent revival of Colonial enterprise on the part of European States, it is evident that this unfortunate rivalry must increase as time goes on, thus necessitating on our side a corresponding activity in strengthening the bases of British Unity. Not so self-evident, however, is our hypothesis of the organic unity of the Empire.

A glance at the map is apt to convey a precisely contrary signification, if only to those who suffer from geographical astigmatism. Portions of the Empire are situated on every continent and in every sea : they girdle the earth and penetrate every zone of temperature. Representatives of every race and religion, and examples of almost every form of government, find a place in this political microcosm. Nevertheless they converge to the point of geographical unity. We are not, in fact, a simple marine organism, but a highly developed maritime State : not an undifferentiated mass, but a self-conscious and responsible unity with co-ordinated powers of adaptation.

Dominant characteristics may be classified and examined under two divisions : internal and external. With the first we associate all Colonial interests, under the second we group all Foreign relations affecting the Empire as a whole.

The British Empire is chiefly constructed out of islands and peninsulas. Although Canada, British South Africa, and India are connected with large continental masses, politically speaking they turn their backs upon the three continents to which they are attached, and face the ocean. Australia is an island-continent. New Zealand is antipodean only as regards her geographical position : in all else she closely approximates to the Mother-country. The minor British Possessions are all islands, portions of islands, or peninsular lands, with few exceptions, of which the most notable are the African Dependencies and two small American Colonies. The Empire may therefore claim to be, physically and politically, an ocean empire.

Other interesting homologies may be referred to. Great Britain lies off the continent of Europe and is flanked by Ireland. Australia is situated near the coast-line of Asia and has its satellite-island in Tasmania. New Zealand is composed of two islands, and its territorial waters are studded with islets. Canada, united physically with what were once British colonies, has its island-pendants in Vancouver and Newfoundland; India, the peninsula-continent, has Ceylon.

Since, therefore, we find some physical correspondence between the components of the Empire, we may be prepared to discover a sympathetic and complementary relationship in regard to their political institutions and national life. Canada, Australia, and New Zealand are self-governing colonies with a predominant white population—colonies of settlement, built up chiefly by immigrants from the British Isles, which have undergone a natural process of expansion: consequently, their political institutions are in close affinity with those of the Mother-country. The same, too, may be said of their national life, under the modified conditions of climate and circumstance, in conformity with the universal law of transformation. The most striking characteristics of our race—free institutions, great industrial activity, and individual commercial enterprise—which contribute to the promotion of British Unity, find unfettered scope in congenial climates and undeveloped lands.

It is quite otherwise when we penetrate the Tropics or enter sub-Tropical countries, where the prodigality of the soil offers an easy livelihood to the indigenous inhabitants. Northern Europeans cannot establish permanent homes in a climate so dissimilar to their own, and in a country where they form a mere fraction of the population. Their superior attainments and resources enable them, however, to act as rulers and taskmasters. Consequently, in all Tropical and sub-Tropical countries under the British flag we find colonies of exploitation and not of settlement—Dependencies, in fact. Of such are India, the African Colonies and Protectorates, and the West Indies. True, Europeans have established permanent homes in the West Indies; but their descendants have undergone inevitable deterioration by partial assimilation with local elements: moreover, insular climates are less deleterious to European constitutions than are continental climates under the Tropics. In India, owing to the advance of civilisation and the introduction of Western institutions and home-comforts, Europeans have been able to live for prolonged periods; but even India is a Dependency in the strictest sense, and can never be a colony of settlement. Again, Cape Colony and Natal, situated in the Temperate zone, afford a suitable base for European political expansion towards the Tropics, along a relatively healthy high-plateau: but, as yet, European settlers form a mere handful of the population. On the other hand, British South Africa is undoubtedly the most suitable area for experimental attempts to colonise

Tropical lands, and in that respect represents an exception very difficult to classify among our Colonies.

Apart from minor distinctions and the many qualifications evident in so complex a structure as the British Empire, we are justified in adopting two broad geographical classifications: Colonies of Settlement and Tropical Dependencies. Under the first we include Canada, Australia, New Zealand, and, to a modified extent (owing to the predominant native populations), Cape Colony and Natal, all of which lie in the Temperate zones, except the northern part of Australia. Under the latter we group India, all Crown Colonies, Protectorates, and the numerous emporia or commercial settlements, islands, naval and coaling stations lying within the Tropics. So that when we speak advisedly of British Colonies we mean only the self-governing Colonies, which, in truth, are really an extension of the Mother-country beyond the seas : the remainder (with the exception of isolated commercial settlements, forming a distinctive class of their own) are, properly speaking, appanages of the Empire. Other classifications, such as that of the Colonial Office, based upon degrees of self-government, for purposes of administration, are equally true ; but, from our present point of view, regarding the Empire as an organic unity, we must rigidly adhere to the biological relations between political aggregates.

Thus, though our Tropical Dependencies are less intimately associated than are the great self-governing Colonies with the internal development of the Empire, they play an important part in its organic functions. The economical relations between the Tropics and the Temperate zones—the one complementing the other as regards raw and manufactured products— is a natural bond of union. Vital circulation between them, or the interchange of commodities, is controlled by the universal law of Demand and Supply. Intercolonial, like international, commerce must flow along the lines of least resistance. From the Mother-country to the Colonies this vital circulation is to a certain extent impeded by tariff-barriers, which, as in foreign countries, are raised to protect local industries from being flooded by the enormous volume of British exports. The Colonies themselves are not sufficiently advanced to absorb this pulsating stream of British origin into their undeveloped systems. Consequently, the bulk of British exports (72%) goes to foreign countries, and chiefly to those from which we derive our raw materials and food-supplies—exports paying for imports—though India, within the Empire, is one of our best customers. On the other hand, from the Colonies to the Mother-country, as also from foreign countries to the United Kingdom, no fiscal barriers are placed in the way of imports, except a few trifling imposts, for revenue purposes only, because of our dependence on the world at large for the greater portion of our food-stuffs and for nearly the whole of our raw material.

The juxtaposition of coal and iron leading to the creation of active manufacturing centres, Great Britain, by her fortunate possession of these natural advantages, has become one vast workshop and warehouse from which other nations, as well as our own Colonies, have been compelled to purchase, in the absence of local competitive industries or cheaper markets of the same kind. This is one, if not the chief, reason why British Colonies take the bulk of their imports from the Mother-country,[1] though, partly for fiscal purposes and partly on account of a constitutional inconsistency to which I shall presently refer, they do not discriminate between her and foreign States. The most notable exception is Canada, whose commercial intercourse with the United States is naturally stimulated by geographical contiguity and by the complementary conditions of their respective resources. But, in these days of steam and electricity, mere mileage is not the chief controlling factor. For practical purposes, cheapness of production and cost of freight are, other things being equal, the true measure of contiguity between centres of demand and areas of supply.

These fundamental conditions of British trade are, however, undergoing a change which is not generally recognised. In some respects commercial intercourse between Great Britain and her Colonies is improving, if not in the same ratio as that between the Colonies and foreign countries. For instance, out of the twenty-eight million quarters of wheat annually consumed in this country, twenty millions are imported from abroad. Formerly this supply came from the Baltic, the Black Sea, and the United States; but now it comes in an ever-increasing degree from North-west Canada—the Indian quota being strictly limited. Canada, in fact, has the largest undeveloped area in the world for raising wheat, and cattle also, of which we take one-fourth of our imports. Tea, too, affords another example of Demand and Supply coinciding within the Empire: three-fourths of our imports, which fifteen years ago came from China, now come chiefly from British India and Ceylon. Again, of the five areas of production from which we obtain our woollen fibres—Australia, New Zealand, India, South Africa, and South America—four are under the British flag, whilst the fifth is bolstered up chiefly by British capital; and it costs little more to convey a bale of wool from these countries to London—owing to their areas of production being near the seaboard—than it does to subsequently carry it to our manufacturing centres in the North. Finally, cotton, which comes largely from the United States, is being produced in increasing quantities by India and Egypt, and might be grown profitably in other parts of the Empire, especially in Australia and Africa: in fact, the same may be said of most of our raw products and food-stuffs, which now we are obliged to import so extensively from foreign countries. Though artificial stimulus is inevitably followed by reaction, and bounties are an abomination, it is perfectly true that State aid—or, perhaps, State organisation only—is sometimes necessary for the stimulation of commercial intercourse along natural lines, the least pernicious form of which is a Government guarantee. Thus, there cannot be any doubt as to the important results to Intercolonial and British trade that must inevitably ensue from the proposed trans-Pacific cable, and a faster trans-Atlantic service of steamships to Canada, when these are finally established.

---

[1] Colonial imports from foreign countries amount, collectively, to about 30%. Colonial exports to foreign countries reach 40%.

On the Debit side, we have to note the competition that is growing up within the Empire. India, which exports more Tropical produce than any country in the world, has benefited in a greater relative degree than the United Kingdom from the creation of the Suez Canal, apart from the increased volume of commerce falling to the carrying-trade, of which we enjoy the world-monopoly.[1] Indian products, which formerly went round by the Cape, now find a new distributing-centre (or, rather, their old one) in the Mediterranean. Indian manufactures—*e.g.* coarse yarns, cotton-stuffs, and jute—enter into rivalry with those of the United Kingdom ; and, under the changed conditions of International commerce in the Far East, we may now expect to find India taking full advantage of her geographical position and great industrial development by opening up new markets for her produce. Australia, too, which thirteen years ago exported all her wool to London, for distribution, now possesses an important market in Germany ; and it is instructive to note, that this direct market was created by the institution of a steamship service between the two countries, thus leading to a steady increase of commercial traffic and greatly benefiting German exports.

These illustrations have been selected in order to illustrate the community of commercial interests existing within the Empire, and the danger to which this is exposed by the competition of foreign States. The ties of sentiment, of a common language and of methods of business, play an important part in international commerce ; but they are not in themselves sufficient to attach the Colonies to the Mother-country. Canada and Australia are naturally gravitating, not only towards each other, but towards every open market in which they can dispose of their products. It is true that the aggregate of our trade with the Colonies represents only a small proportion of our external commerce, and that hitherto it has varied within narrow limits. No less true is it, that the bulk of Colonial trade is with the Mother-country and the Empire, thus proving the connection to be more valuable, commercially speaking, to the Colonies than to ourselves. But our community of interests is so vital, so capable of expansion in directions that may eventually prove necessitous to the Mother-country, that we cannot afford to see the Colonies one by one achieving their commercial independence. Should they ever, indeed, attain to this, the political tie would be still further weakened. Consequently, no rational measures should be left untried to identify our respective interests, both as regards Intercolonial and International commerce : and it is satisfactory to note that Mr. Chamberlain, unlike most of his predecessors at the Colonial Office, is prepared to adopt business principles, which hitherto have been so sadly lacking, in the management of our Colonial relations. We are much in need of an English Colbert.

A striking instance of Intercolonial co-operation is now being decided,

---

[1] Three-fourths of the carrying-trade of the world falls to British shipping, which is paramount in Southern and Eastern Asia, and in South America. Seventy-five per cent. of the tonnage passing through the Suez Canal is also British. In other directions we are maintaining our lead, though with difficulty.

happily in a liberal and statesmanlike spirit, by the Home Government. Canada and Australia, our most advanced Colonies, are entering into reciprocal trade-relations, and have inaugurated a trans-Pacific steamship service. They now request the Home Government to aid them in laying a cable across the Pacific; and Canada urges a faster trans-Atlantic service, so as to give the Mother-country a new route and an independent line of communication, entirely through British territory, to the Far East. All the Colonies interested—in particular, Canada, the chief beneficiary—are prepared to contribute a fair proportion of the initial cost. On her part, the Mother-country will greatly benefit, strategically and commercially, by this new channel of intercourse. Two-thirds of the cables of the world have been manufactured and laid by Great Britain, and are now owned by British capitalists, who have over twenty-six millions thus invested. Telegraphs and cables having added a new nervous system to the world, national life within the Empire now beats with a single pulsation. But at the present time our telegraphic connection with the Far East—so vital to us—passes through foreign countries. To carry all land-lines through British territory, and to lay a cable along the bed of a deep and wide ocean—uniting the East with the West—would, therefore, give the central executive a strategic advantage that might prove to be of supreme importance in time of war, quite apart from the commercial facilities to be derived in periods of peace. Not only should we obtain comparatively secure communications with our naval bases in the Far East, but a considerable economy of effort would result to the fleet whose duty it is to protect our sea-borne commerce from the traditional *guerre de course*. Moreover, measures of self-protection and co-operation such as this will do more to promote British Unity than any cut-and-dry scheme of Imperial Federation.

Into the subject of Britannic Confederation, as I prefer to define it, we need not enter at any length. Enough has been said to convince all but hardened sceptics that the organic unity of the Empire is already an accomplished fact; and sufficient evidence has been brought forward to indicate the lines of its natural development. Peace and time, I repeat, are the essentials of our organic growth as an Empire. But since peace cannot be assured by a simple expression of the national will, and as our preparations for a great maritime struggle are deficient in some essential details, a feeling of general uneasiness exists in this country, that our Colonies and sea-borne commerce are singularly open to attack by a powerful and enterprising enemy. Thus, the subject of Britannic Confederation has given place, to some extent, to the more immediate need of Imperial Defence. It is argued that the Colonies and Dependencies, who in the aggregate contribute about 6d. in the £1 towards Imperial Defence, might reasonably be expected to bear a more adequate share of the cost to the Mother-country. But for reasons—chiefly strategic and administrative—into which it is unnecessary to enter, the balance of expert opinion is not in favour of direct Colonial contributions towards a war-fund. The Mother-country, who made a free gift of the fee-simple of her possessions beyond the sea to the few handfuls of Colonists capable of self-government, who asked in return no share or relief in the

burden of paying off a colossal National Debt, who set up and now
maintains at her own cost the world-embracing Diplomatic and Consular
Agencies by which the political and commercial affairs of the Empire are
controlled, does not propose to invite Colonial co-operation in the
relatively slight sacrifice involved by the annual outlay, or premium,
on war-policy. But as she has dealt generously and confidingly with
her offspring, she may reasonably expect reciprocal treatment. In granting
to all the rights, the privileges, and the protection of British citizenship,
she reserves to herself the Crown prerogative of making peace and war
—for obvious political and strategic purposes—the exclusive duty of
contracting commercial treaties, and a veto on Colonial legislation, which
is rarely exercised. All these conditions are essential to British Unity.
Where, however, the Colonies might reasonably be expected to bear a
share in that Imperial policy which is the guarantee of their existence
as semi-independent States, is in removing the restrictions that at present
interrupt the spread of British commerce and colonisation. Let them
lower their fiscal barriers, so soon as the obstructive treaties with Belgium
and the German *Zollverein* are denounced,[1] or otherwise enable them to
claim their natural rights; and let them open their doors wider to
emigrants from our shores, whom for selfish reasons they are anxious to
exclude, but who are surely entitled to share in the advantages over
which a generous Mother-country has delegated her control.

The fulfilment of these two conditions would, in my opinion, do
more to relieve British taxpayers and Home industries than any direct
contribution from the Colonies for purposes of national defence. It
would not be wise for Great Britain, whose trade with the self-governing
Colonies amounts to only twelve per cent. of the volume of her external
commerce, to initiate a preferential tariff in their favour. Such a policy
would be contrary to the principles of Free Trade (though those principles
need not necessarily be a national *fetich*), and would injuriously affect
vested interests in some of the most important British markets—notably
those of the United States and Germany. But the Colonies are now
sufficiently robust to suffer temporary loss of revenue, equivalent to a
contribution in kind, by lowering their protective tariffs against the
Mother-country. Moreover, once Australia is united as a Federal Com-
monwealth—a step that must necessarily precede that wider Britannic
Confederation which shall eventually include even British South Africa
—Free Trade within the Empire, growing out of Free Trade between the
individual units of the Empire, must become a vital condition of British
Unity—that is to say, in its ultimate expression.

---

[1] These treaties (dated 1862 and 1865 respectively) do not prevent differential treatment
by the United Kingdom in favour of British Colonies, nor do they prevent differential treatment by British Colonies in favour of each other, but they do prevent differential treatment
by British Colonies in favour of the United Kingdom—a logical inconsistency, if the Empire
possesses that organic unity which we claim for it, and one not to be upheld by international
law. Canada and Australia are no less a part of the Empire than is Algeria a department of
France. Our Empire is an Ocean Empire; and our lands beyond the sea are theoretically
contiguous, so long as we uphold our maritime supremacy. By permitting the Colonial
clause to remain in our commercial treaties, we practically deny the Unity of the Empire.

The other fundamental condition affecting British Unity is that concerning emigration and colonisation. The greatest of British products is strong men and virtuous women, of whom there is a drug in the Home market. Some of our Colonies refuse to receive our surplus population, for fear of increased competition and the expense of maintaining unsuccessful immigrants. Australia, perhaps, is the chief offender in this respect. She is not, of course, unreasonable in refusing to receive paupers and loafers; but under an adequate system of emigration, Her Majesty's Government might do much to remove this cause of complaint and to stimulate the outflow of population from congested districts at Home towards lands the original title to which, though alienated, should be the birthright of every British citizen. There is room to spare in all our Colonies for a much greater influx of population from the Mother-country. Canada, in particular, possesses vast undeveloped lands awaiting the advent of the settler to turn to profitable account; whilst Queensland and Western Australia stand much in need of colonists. That there is not a healthier stream of migration is partly the fault of the Home Government and of its traditional objection to State aid—an objection which, though right in theory, can be carried too far in practice. Instead of the small extra-mural institution at Westminster, whose resources and staff are quite inadequate, the State might properly undertake, with the co-operation of the High Commissioner and the Agents-General, to devise and carry out a comprehensive scheme of emigration, by which the Colonies no less than the Home-country would benefit in the long-run. An attitude of give-and-take would rapidly overcome the initial difficulties. In return, say, for the assistance demanded of Her Majesty's Government for the improvement of Intercolonial communications, some concessions might be made by the Colonies towards the promotion of a steady flow of immigration from our shores. We are active just now in acquiring new markets in Africa—markets that will take generations and an enormous outlay of capital to develop up to the point of profitable investment, if ever they reach that stage—whilst, at the same time, we are starving our most profitable markets in the Far East and are wholly neglecting the development of our Crown estates. Surely there is some inconsistency in this policy, or lack of policy? All the Colonies require cheap money and more colonists; and these they can obtain only from Home, under what amounts to an Imperial guarantee. From more intimate and responsible ties of relationship Great Britain and her Colonies would derive mutual benefits, by which the organic and political unity of the Empire would be immeasurably strengthened. But under the present one-sided arrangement the self-governing Colonies enjoy all the advantages and the protection of British citizenship, whilst bearing practically no share of its charges and responsibilities.

A glance at the map illustrates this apparently artificial equilibrium. What are the facts?

An Empire embracing the world in its every-day national functions, depending for its existence on the Command of the Sea and on the safety of its sea-borne commerce. An Empire which, it is estimated, has cost 5740 millions sterling, and is not yet paid for—the National Debt now

standing at £660,160,000.   An Empire whose capitalists have fertilised
every civilised country in the world: to whom the world owes 2000
millions sterling.   An Empire whose commercial fleets monopolise sea-
borne commerce.   An Empire which enjoys a comparative monopoly—
i.e. over fifty per cent. of the imports and exports combined—of the
external commerce of China, Japan, Persia, Siam, Morocco, Tripoli, Egypt,
the South African Republic, the Orange Free State, the United States of
America, Peru, and Chili; and which has a predominant share in the
external commerce of several other countries.   An Empire with official
representatives in every civilised land, and at every commercial centre—
a marvellous organisation and a costly one; with fleets of warships
dominating every strategic sea-area, naval bases, and a chain of coaling-
stations girdling the earth.   An Empire with an army the expense of
which is out of all proportion, judged by the Continental standard, to its
limited numbers: but an effective army, sufficient for a maritime State,
whose first line of defence (strategically speaking) is the sea-frontier of
the enemy.   Finally, a strong centralised Administration: an army of
Home officials, whose work is largely Imperial.   These are the chief
attributes of our Empire, for which the Mother-country pays the bill.
The Colonies (unlike India) contribute nothing, or next to nothing.   Yet,
though at any time they may invoke British protection, civil or military,
at any time they are at liberty to secede from their political allegiance;
and though our Diplomatic and Consular officers are their representatives
also, they place British imports on the same footing as foreign imports.

   Surely this is an anomalous condition of affairs calling for some
adjustment?   There is, of course, no talk of secession, any more than
there can be a question of forcible interference in the event of such
secession; nor are the Colonies sufficiently developed to embark upon a
policy of intercolonial Free Trade, though each geographical unit of
the Empire must necessarily make Free Trade the basis of its political
unity and economical intercourse.   But the fact remains; and it is a fact
not sufficiently recognised by our Colonial friends, who demand so much
of the Mother-country and are prepared to do so little in return.

   This unrecognised aspect is, however, only the common-sense or
business-like view of the relations between Great Britain and her Colonies.
It is scarcely necessary to add, that it is not the attitude of Her Majesty's
Government, who regard the Colonial connection otherwise than in a
spirit of commercial expediency.   Great Britain has need of her Colonies:
she has need of them chiefly on account of their strategic advantage to
her as a maritime Power.   As British lands, they are valuable bases of
supply in time of war: as neutral States their ports would be closed to
us, in so far as regards contraband of war.   So that there is a *per contra*
side, the value of which is not to be assessed by actuaries.   Moreover,
in spite of their indirect expense to Great Britain as a maritime Power,
all the self-governing Colonies pay their way, costing the Mother-country
nothing for local administration; and the same may be said of our richest
dependency, India, as well as of most of the Crown Colonies.   Even
Egypt pays handsomely for British tutelage and Protection.   Every other
European Power has, on the other hand, to contribute a large annual sum

of money towards the maintenance of their Colonial establishments. So that, in the sense that Colonies are established for the benefit of the Mother-country, the British Empire may be said to be the only prosperous Colonial Power of our day. Spain and Portugal, the earliest colonising States of modern times, still retain costly vestiges of national glory. Holland, who preceded us in the path of colonisation, continues to struggle with a fluctuating balance of profit and loss in the Dutch East Indies—perhaps on account of the poet's stricture:

> "In matters of commerce, the fault of the Dutch
> Is giving too little, and asking too much."

France, our traditional rival, has of recent years entered upon a course of ambitious colonial expansion, and has many possessions beyond the seas —playfully called *France d'outre-mer*—which are maintained only at a heavy annual deficit. Finally, Germany, now serving her apprenticeship as a Colonial Power, has had to pay heavily for her indentures.[1] Only Great Britain has learnt the secret of making Colonies pay for their up-keep and be prosperous at the same time.

We thus see that, as regards the internal economy of the British Empire, the balance of profits and loss, judged by a world-standard, is on the side of the Mother-country. At the same time it has become evident, in the course of our inquiry, that some adjustment of our Colonial relations and a certain amount of reorganisation in our administration are the necessary outcome of the organic growth of the Colonies from a position of apprenticeship to the dignity of co-partnership. To meet this requirement it has been suggested that, in return for the guarantees, present and prospective, of the predominant partner, the Colonies on their side should enter into more intimate, responsible, and co-operative relationship with the Mother-country, since, if they share profits, they should be prepared to face losses arising from the inauguration of an extension of business, so to speak. But since the Empire, as a going concern, is the only Colonial establishment that pays dividends to its shareholders, it is argued that all the partners will derive profits from its solvency.

We are now in a position to glance at the relations between the British Empire, as an organic unity, and its environment of friendly and hostile States. In periods of peace this relationship is controlled chiefly by economic laws, in time of war by the survival of the fittest.

The complementary functions between the Tropics and the Temperate Zones coinciding with the actual geographical distribution of our Dependencies and Colonies, it might at first sight appear possible for the Empire to gradually attain to the position of a self-contained State with regard to a dependable supply of food-stuffs, if not also of raw materials; but this ideal consummation, for reasons that are obvious, does not come

---

[1] "Greater Germany" costs the Mother-country eleven million marks annually. It is colonised by some 700 Germans, of whom 250 are officials. The "thousands and thousands" of whom the Emperor boasts are chiefly in the United States and British Colonies.

within the sphere of practical politics. Our Diplomatic relations with
foreign States are so intimately associated with our commercial connections, that the two cannot be separated. From their conjunction arises
that inevitable rivalry, friendly or otherwise, which is as characteristic of
national as it is of individual intercourse. There is little or no room for
the play of affinities and for community of sentiment in International
politics, because the family of nations is composed of units more diverse
in structure and temperament than is the family of individual units. For
the same reason, when communities spring from a common ancestry—as
in the case of Great Britain and her self-governing Colonies—greater
correspondence of structure and functions is a natural result. But even
consanguinity is no guarantee of peace. Family quarrels and family feuds
inevitably arise from incompatibility of temper and hasty actions. Hence
the secession of the United States; hence also the incompleteness of any
Britannic Confederation that does not include the greatest of English
Colonies.

Taking things as we find them, however, the British Empire is
singularly united, in spite of marked diversity among individual members
and of the vast size of the aggregates. Its mission—if it have a mission
—is to weld together the nations of the world by that community of
interests which arises spontaneously from free commercial intercourse :
consequently, its policy of Free Trade is an instrument well adapted to
achieve this end, theoretically speaking. Its *pax Britannica* is to the Old
World what the Monroe Doctrine is, or should be, to the New World.
Thus, Great Britain and the United States may be said to have developed
a common national policy, which finds its most eloquent expression in
the mutual desire for pacific arbitration on all points in dispute between
them, and its fittest instrument in the English language, the *lingua
franca* of the world and the mother-tongue of no less than 120 millions
of people.[1] It is therefore devoutly to be hoped that the United States
will develop sufficient strength to support its share in the burden of promoting universal peace.

Universal peace being, however, a synonym for the Millennium, its
consideration does not come within our purview. We are not all sure,
though we sincerely hope, that we are the fittest to survive : hence the
interaction of warring forces which are only held in check by that frail
leash known as the Balance of Power. The Balance of Power in
Europe is a calculable factor. It gives its law to the world ; and,
for the present at least, the European Hegemony is not seriously
threatened. We may therefore regard it as the controlling factor of
International relations. The old civilisations, which in Asia still enjoy
a large measure of temporal power, are without coherence ; whilst

---

[1] The Chinese language is spoken by an even greater number (360 millions) of people :
but it is strictly localised, and is broken up into numerous dialects, the sole point of union
being their common ideographs, or graphic characters. On the other hand, our own
language is familiar to many non-English people, and is essentially the medium of common
intercourse between the seamen of all nations. The populations of the British Empire and of
the United States amount in the aggregate to about 573 millions, or nearly one-third of the
total population of the world.

modern civilisation is crystalising through the affinity of world-intercourse and is developing greater and greater solidarity. Even an Asiatic people like the Japanese can now assimilate it. The Ocean is, in fact, the great amalgamator. All highly developed nations are pressing towards it: cutting canals from their chief inland emporia towards their nearest sea-board, piercing isthmuses that obstruct free maritime intercourse, building trans-continental railways and uniting oceans—thus opening up new routes, by rail and steamship, for International commerce. It is the age of maritime canals. In time, no doubt, Berlin and Paris will be seaports, as Manchester is now a seaport. And, above all, when the Nicaragua Canal is built, a new stage of International development will be reached: namely, that in which the Pacific shall vie with the Atlantic for supremacy in human interests, in the same way that the Atlantic competed with the Mediterranean after the discovery of America. The course of Empire moves ever westward, and with it the centre of gravity of world-power.

As the opening up of new channels of commerce gives rise to the in-evitable displacement of trade-centres, it follows that, by the inauguration of an inter-oceanic canal in Central America, a new nodal point will be created, by which International intercourse will be profoundly affected. The United States will then be forced to take up a definite and pro-gressive, as against a negative and retrogressive, attitude towards the unsolved problems of American politics. Commercially, she will benefit more than ourselves from this new ocean-highway; but, strategically, she will suffer seriously by admitting the navies of Europe into the very heart of her military system. Even if the canal escape neutralisation, which, in view of the Clayton-Bulwer Treaty, is unlikely, political domination over the Isthmus—then the true path for sea-power—cannot vitally affect the nation that holds the Command of the Sea. It is true that, though we hold Jamaica, dominating the Atlantic entrance to any trans-Isthmian canal, we have no naval base in the South-east Pacific; but this is a strategic defect that might possibly be remedied by the purchase of the Galápagos Islands from Ecuador. Though the Monroe Doctrine may operate against us in taking this very necessary measure of pre-caution, it will not in itself save Central and South America from the pacific invasion of Europe, any more than the Papal Bull of Alexander VI. prevented the spoliation of Spanish and Portugese colonies in the sixteenth century. Men are made that way: they listen best to the brutal Maxim Doctrine. There are Americans, of course, who regard the Monroe Doctrine as a kind of *Code Napoléon*; but the educated classes—and certainly those entitled to be regarded as statesmen— are lacking in this comfortable conviction, which, after all, is merely an expression of the national conscience, and finds no sanction in Inter-national Law. For, to follow that dogma to its logical conclusion, the United States must be prepared to guarantee the integrity of the corrupt Republics of Central and South America (with whom they have nothing in common, except in name), by creating adequate military and naval establishments, or to adopt the alternative of admitting some of them into the Union. Moreover, many of the South American Republics

repudiate the moral protection of the United States.   For instance, two
of the most advanced of these States—Chili and the Argentine Republic
—have, by a Protocol dated 17th April 1896, appealed to Queen
Victoria to act as arbitrator in the final settlement of their frontier
disputes; and Her Majesty has graciously consented to serve, if necessary,
in that capacity.   It is difficult, therefore, to escape the conviction that,
since the Nicaragua Canal must eventually be built, to meet the pressing
demands of International commerce, the best interests of the United
States would be guaranteed by falling into line with the Power that holds
the Command of the Sea and which already possesses equal, if not larger,
vested interests in the American Continent.

No further arguments are required to prove—at least to my mind
—the paramount necessity of promoting more vital and rational relations
between the Colonies and the Mother-country, if ever the Empire is
to achieve consolidation, nor to establish the obvious desirableness of
limiting our Foreign responsibilities to those which, as a member of the
European system and as an Oceanic Power, we are bound in honour
and by necessity to accept.

GEOGRAPHICAL EDUCATION.

By A. J. HERBERTSON.

I. NOTES ON SOME RECENT WORKS DEALING WITH HOME GEOGRAPHY.

IT is difficult to write on the subject of geographical education without
first of all discussing the scope and functions of geography.   There is no
doubt that ignorance or misapprehension of these explains much of
the opposition to the progress of geographical teaching in our schools and
colleges.   The views of many English authorities were expressed in the
discussion on Mr. Mackinder's stirring address to the Royal Geographical
Society on "The Scope and Methods of Geography"; [1] and Mr. Arthur
Montefiore has quoted these and others in a paper on "Geographical
Methods." [2]   In this *Magazine* there recently appeared a very full abstract
of the views of Professor Kirchhoff of Halle [3] on the meaning, divisions,
and educational significance of geography; and a summary of other
German definitions will be found in a paper by Professor Neumann of
Freiburg i. B. [4]

It is impossible to discuss in detail the subdivisions of geography in
the present article.   The antiquated mathematical, physical, and political
trio is certainly an unsatisfactory classification; that into physical
and historical geography even more so; the more recent threefold
division into physical, biological, and anthropological geography is more
rational than either.   Some authorities make a fourth subdivision of
mathematical geography, separate from physical geography and of equivalent rank; but the term is not satisfactory, and the division hardly
necessary although occasionally convenient.

---

[1] The numbers refer to papers in the list at the end of this article.

One of the most difficult questions for the teacher is how to begin geography—from the general or special point of view. Is he to teach general geography (*Erdkunde*) or the descriptive geography of different countries—for which the Germans have the convenient term *Länder-kunde*—beginning with his own? Most geographers say : Neither; he must begin with home geography; he must use the world visible to his pupils for their first lessons.

Here it is necessary to remark that home lore (*Heimatskunde*) is not the same as home geography, but includes it. There is a difference between home knowledge, "including all facts of the home, zoological, botanical, historical, etc.," and geographical home knowledge "forming a special preparation for future geographical work," as Matzat has pointed out.[5] It is important to notice this, as many opponents of geographical teaching are constantly complaining of the vagueness of the subject, and assert that many of its advocates make it an *omnium gatherum*. However convenient and advisable it may be to make no distinction between the two in practice in the school, it is well to do so here.

Home lore is inexhaustible and should be studied throughout school and college life as well as later. It is not merely a simple subject good for a beginner, but one complex enough to puzzle the most learned to interpret. The trained geographer can find plenty work to do in studying home geography. To do so at most stages of his progress is essential. Even the geographer who is a great traveller will find it invaluable. For the aim of geography is not only to gain, but also to interpret, and to apply, our knowledge of the earth's surface. We can do this most effectively in our own neighbourhood, of which we are forced to have some sort of interpretation in order to do anything in it.

1. In the School.

The geography of the school region should first be studied. This does not mean that a county geography should be given to the children from which to learn lessons. The children should not be troubled with such artificial restrictions as county boundaries. The geography of Lanarkshire is not the home geography of Glasgow, nor is that of Midlothian the home geography of Edinburgh. Home geography is limited by what a person can see for himself, and for schools it is the geography of the region which the children can see and more particularly visit in ordinary circumstances. For a town situated like Stirling or Edinburgh the area visible is very extensive and varied in character, and home geography can be studied ideally. But for a town in some inland valley, hemmed round by hills, it is more restricted, yet far more adequate than in the case of a city in a wide plain far from mountains or seas. Round the worst centre there is plenty material for studying home geography. The sun, moon, and stars rise and set, the clouds pass, the rain falls, and the seasons succeed each other, yielding impressive lessons.

How are we to begin the teaching of home geography? Assuming that the child already knows a fair amount of home lore, before having systematic lessons in home geography, the question is : How is he to be trained to look at his surroundings from the geographical point of view? The actual order of teaching will depend not only on the nature of the region but also on the pupil and on the teacher.

In what follows I shall be guided mainly by Professor Kirchhoff's paper, of which this article is in part a summary, with running comments and some additions.

One of the first requisites is to teach children to measure—and first of all to measure distances by paces. Afterwards heights should be measured, and later, when the child is far enough advanced in other ways, areas. At an early stage the notions of direction and of time should be taught.

These things are done in our schools, but not always thoroughly enough. The children are not sufficiently practised in estimating distances or heights, in finding out directions, or in judging time by sun or stars and not by a clock. Indeed, many of their elders are often puzzled by such things, and much amusement may be derived from the attempts of a company of ordinary intelligent people to estimate heights or slopes, and especially the angular distance of a star above the horizon. They were not taught such things at school. Let us hope their children will be.

Without this concrete idea of the value of our units applied to a small area, the figures used in geography, which should be given always in the form of round numbers, are only sounds, and convey no real impression to the student. This concrete idea, which is of immense practical importance, can only be gained by constant experience. To use Dr. Kirchhoff's illustration: How much would the shooting of our soldiers improve were they better trained to estimate distances at school?

As it is with distances and heights, so is it with other geographical data—we must have experience of them before we can appreciate them. Progress in geography, as in other sciences, depends largely on performing systematic and similar experiments, and precisely recording the results, and interpreting them. Rough measurements should be replaced by exact ones. Let the children first of all measure the playground by paces, later by means of rods or tapes; let them first estimate time and direction by the position of the heavenly bodies, later let them tell the former by clock or watch, the latter by the compass, when they are able to understand and use these instruments.

The same remarks apply to atmospheric phenomena. First they should note if the day is clear or cloudy, calm or windy, hot or cold. But this is not enough. They should learn to know the chief form of clouds and to estimate cloudiness; to tell the direction and judge the force of the wind; to measure temperature, rainfall, etc. With the seasons they should measure and note for themselves the changing length of day and night, the variations of temperature, etc. Such things are simple to do, and can be carried out in every school, many of them without any special apparatus at all.

This study of weather is perhaps rather a part of general home lore than the more special home geography; but it is an essential experience for every geographer, and an admirable training for every child. It is necessary to be able to interpret the data of climatology in order to appreciate the different kinds of climate. To realise the geographical conditions of other lands, or of different parts of our own land, we must first of all have experience and understanding of the geographical conditions of our immediate neighbourhood. Every new stage of geographical instruction ought to be begun in the home region.

By the seashore the children naturally have a picture of the ocean, and see for themselves the capes and bays, and often the peninsulas and islands, isthmuses and straits. A lake or a pond in far inland places serves as a model of the mighty ocean, and the children's ideas may be made clearer by showing photographs, and, above all, good coloured pictures. Even a pool in a river can be made a model replacing the lake. But as Professor Kirchhoff points out, it is very difficult for a child of the plain to have a true conception of a mountain. In Scotland, however, this difficulty, which is so great in the North German Plain, does not arise. Everywhere in our land part of the course of a river can be studied.

In the large cities teachers often find it difficult to illustrate their lessons. It would be a great boon to them if the managers of public gardens tried to make a portion of their grounds of geographical interest—large models illustrative of various land surfaces. It would enhance rather than depreciate their beauty, if skilfully done. The botanical gardens in our great cities should be the natural and delightful scene of many a geography lesson. From the rock garden to the tropical hot-houses the teacher has many object-lessons for his pupils. But how many teachers in Edinburgh and Glasgow systematically make use of such opportunities of teaching geography? In cities, too, teachers have museums, where models, on a much smaller scale than could be adopted in gardens, can usually be seen, and art galleries, where landscape pictures are often invaluable. At all stages of geographical teaching the appropriate specimens in a good museum should be inspected. Perhaps few keepers of museums have thought of the needs of geographical students and teachers, and a remembrance of them would make these institutions of still greater value than they are at present. The city teacher himself is partly to blame, as he does not use gardens, museums, and galleries enough—indeed until recently the authorities over him did not permit his doing so in school hours as they now do.

But even in a town the judicious utilisation of such natural helps to the teacher as a rainy day will do much to aid the children to clear conceptions of the interaction of land and water, and of the atmosphere with both. In the country, of course, it is much easier to study the contours of the land, the river courses, and the atmospheric effects. From a study of these on the small scene of our own region, we learn to conceive grandeur on the whole world's stage.

The rocks of the country, the building stones of the city take us back into the past, and parts of the wonderful story of the shaping of our earth can be told with these as texts.

The world around us is full of living things, which attract the child. Ferns and flowers, beetles and butterflies, fishes and birds have their peculiar distributions. These, of course, need not be studied in detail, but general characteristics should be observed. Indeed, most children are easily made enthusiastic naturalists, and soon learn the peculiar habitats of different plants and animals. The seasonal changes of plant and animal should all be carefully observed, noted, and compared with those of light and temperature. Our diversified country makes such studies especially easy in Scotland, and shows many landscapes which may be considered typical of vast areas in other parts of the world.

Even at an early age children begin to understand some of the relationships between the various geographical conditions. The limiting of fish to water, and even of certain plants to river or marsh is soon appreciated. Many boys pride themselves on knowing the trout peculiar to different streams in their neighbourhood, and just where to find this or that bird's nest. There is no reason why their enthusiasm for such knowledge should not be fostered and directed.

The geography of mankind is the most complicated and the most difficult to explain in a rational way. But its elements must be learned in the actual world around us. It is important, indeed necessary, to do so. Many of its facts can be appreciated by young children, and the number of these increases as the boys and girls grow older. The water is the home of fish, and many men who live near it are fishers. The woods and fields are full of animal life, which the hunter seeks and destroys. The grasses of pasture-lands feed flocks and herds tended by shepherds. The forester looks after the trees, the farmer cultivates the fields, and the gardener the garden. The natural occupations are not difficult to classify in the country.

The settlements of the people may be studied,—the usually lonely hut of gamekeeper and forester and shepherd, the group of farm buildings with the hinds' cottages and farmer's house close at hand, the small village with its blacksmith to do the farmer's repairs, the tailor and bootmaker, the general store, usually that of grocer, draper, ironmonger, etc. combined—a universal provider on the small scale necessary in a country village—the station, the post-office, the inn, the surgery, the school and the church, all have a natural and necessary existence. Even the distribution and position of villages can be interpreted in many districts. There is no lack of geographical work for every class in a school in its own immediate surroundings. In the more advanced classes the greater environment of the school may be studied, the places in close touch with the home district.

The town is richer in human associations, but these are much more complex and difficult for a young pupil to understand than those in the country. The great seasonal drama, with its consecutive acts of leafing, flowering, fruiting, of sowing, growing, mowing, is little in evidence, save in the changing length of the day and its temperature, and most obviously in the fruiterer's window. These, however, may all be utilised by the teacher.

The mainstay of a great port or mining centre is easily understood, and the position of the town usually readily explained. Many manufacturing towns are placed where they would be expected, but a large number are not.

There are few towns whose geographical features are so well marked as Edinburgh; but in most towns certain districts can be distinguished and explained. The harbour, the river, the railway often determine industrial areas; the lie of the land, and the prevalent direction of the wind, sometimes explain the distribution of residential quarters.

Quite a profitable study for older pupils is the distribution of shops, of factories, and of public buildings of all kinds, civic and ecclesiastical. Again, for older pupils, the wider region in immediate touch with the home ought to be studied most carefully—the chief centre of its staple trade, the regions supplying its needs of food, dress, habitation, etc.

The artificial political boundaries come late in the study of home geography, and should not be used to limit it at any time.

In the country the work in higher classes might be the study of the morphology of the region in an elementary but systematic way, as is suggested and described in Professor Davis's pamphlets.[6] The comparison of various distributions of soil and plants, for instance, might be attempted by pupils interested in natural history. This is not so hard as some may imagine, for we must not forget that the youth attending natural history courses in our universities do not use their eyes or their imaginations nearly so much as boys of ten or twelve.

The more advanced work, of course, comes late in the school career. While the earliest geographical studies are around the home, the latest ones should also centre round it. Ever and anon during the geographical course a systematic survey of the local region should be made, each a little more complex and profound than the previous one. Home geography is the foundation course; it also forms the binding courses and the coping-stones of the geographical edifice.

2. GEOGRAPHICAL EXCURSIONS.

It is obvious that to put this in practice teachers of geography should have a much more careful training than they can procure at present anywhere in this country. They must also have much more liberty to make use of opportunities that arise, and cease to be confined by four walls. They should be encouraged to make local excursions, not in their spare time, but in school hours. Such work should be recognised as among the most serious efforts of the school, and not be stigmatised as a mere frivolous amusement, as is too often the case. Of course it may not be profitable in incompetent hands, but neither is any other kind of instruction.

Occasional excursions to the more interesting places in the larger environment should be encouraged, and looked on as the reward of the pupils who show themselves best fitted to profit by them by their progress through local excursions. It might be advantageous, in some cases, to ask some geographical authority in the districts visited to act as teacher on such an occasion. Our own Society might undertake to prepare a list of people interested enough in such matters to offer their services, should they be asked to show and explain their own district to children from a distance.

A beginning of such work was made in Scotland last year, through the generosity of Henry Beveridge, Esq., of Pitreavie, and the co-operation of the teachers and School Boards of Dunfermline Burgh and Parish. A series of excursions were arranged and carried out by Mr. T. R. Marr and the writer. They were limited to children in and above the sixth standard. Each party was not to exceed thirty, and at least two teachers accompanied the children to preserve order and to gain experience in conducting such excursions. Teachers were asked to specially prepare the children for these excursions, and each child had to write an account of it, which was examined by the conductor of the party.

The excursions were not purely geographical; indeed it was found that the majority of teachers preferred that they should be historical, while a few prepared the children for natural-history rambles.

In all excursions, however, considerable attention was paid to geography. The excursions were to Edinburgh, Stirling, Perth, the shores of the Forth, Inchcolm, etc., on Saturdays.

The results were encouraging, although it was obvious that the children were not accustomed to think in the presence of things, and that their memories were better trained than their intelligence. The most disappointing experience of the writer was to discover how few children understood a map, and that of maps they had seen only political ones—the most complex and uninstructive of all for boys and girls. The essays sent in showed that the children had profited by their excursion.

Mr. Beveridge again this year gave a sum of money sufficient to carry out a somewhat similar series of excursions, planned by Mr. Marr, and carried out by him, Mr. Goodchild, and the writer. The details of these excursions have been slightly modified since last year, and the teachers recommended to pay less attention to dates and trivialities when preparing their pupils, to their obvious advantage. The teachers themselves are taking part in this work this year, and are organising expeditions in the immediate neighbourhood of Dunfermline. Excluding the excursions carried out by the teachers, about ten have been made each year to places at some distance from Dunfermline.

It is to be hoped that other liberal-minded gentlemen will follow Mr. Beveridge's admirable example.

3. SUMMARY.

It may be useful to teachers of geography to quote the summary of the views of German authorities on Home Geography as given by Professor Goodison in the U.S. Education Report already mentioned :[5]—

"1. In the terrestrial and celestial phenomena of the home region types and analogies are to be sought for all the concepts that will be met with in advanced geographical study.

"2. Hence the limits of home geography are determined by these respective requirements of future work. As these include the relations of the earth to the sun, etc. (mathematical geography, so called), form, relief, hydrography, and other features of physical structure, organic life, man, etc., all these must find their beginnings and fundamental concepts in home geography.

" 3. The study of the geographical aspects of the home is obviously primarily a study of its physical features, hence to proceed by political divisions, township, county, etc., is to violate the fundamental idea of home geography.

" 4. A leading purpose of home geography is to train the pupil to map-reading by establishing a relation between geographical elements and their representation.

" 5. Each grade of geographical teaching (primary, grammar, and high school) should have its preparatory course in home geography.

" 6. Home geography once completed, the next step is to the globe as a representation of the earth. When the limit of the pupil's field of observation is reached the work of home geography is completed.

" 7. The order of geographical teaching may be stated as follows :— *First*, The pupil is led to observe the geographical fact. *Second*, He is led to describe it. *Third*, He is taught to represent it. This order is true of all geographical work, and is not restricted to home geography."

There is a difference of opinion about paragraph 6. Some teachers prefer to teach the geography of the native land after that of the home region.

We must take exception to the last sentence of the same paragraph, if it is unqualified. The systematic study of home geography may be temporarily suspended, but the geography of the home region must constantly be used to make more real that of other regions. As section 5 rightly points out, at all grades of geographical teaching the home geography should form the subject of a preliminary course. Even in the university this must not be forgotten; and geographical excursions, which are as necessary as geological, botanical, or zoological ones, will mainly be in the home region and its immediate surroundings.

We necessarily begin our study of geography with that of our own locality, and it is just as important to continue the study of it as it is to pass to the geography of other regions and of the earth as a whole.

The last paragraph quoted is not complete. Observation, description, and representation are not all that we should teach our children. As Professor Davis rightly remarks,[6] "Even in the earliest stage of obser-vational work, suggestion should be made of the physiographic processes by which the existing geographical forms are explained. At first, these suggestions should be very simple; but they may be carried to a good degree of advance during the grammar-school course " (10-14 years). Geography should be not merely descriptive, it should be scientific—and the more our pupils advance, the more scientific should their geographical knowledge become.

Papers referred to.

1. *On the Scope and Methods of Geography.* By Halford J. Mackinder, M.A.
   Proceedings of the Royal Geographical Society, vol. ix. N.S., p. 141.

2. *Geographical Methods: A Chapter of Suggestions.* By Arthur Montefiore, F.R.G.S.
   London, *Educational Review* Office, 1895. *Price 6d.*

3. *Didaktik und Methodik des Geographie-Unterrichts (mathematische und allgemeine Geographie)*, von Dr. S. Günther und Dr. A. Kirchhoff. (Sonderausgabe aus Dr. A. Baumeister's *Handbuch der Erziehungs- und Unterrichtslehre für höhere Schulen*).

München, Oskar Beck, 1895.   *Price 3 M.*

Only the latter part is referred to in the present paper.

4. *Die methodischen Fragen in der Geographie*, von Professor Dr. Ludwig Neumann.

Geographische Zeitschrift, 1896, vol. ii. p. 35.

5. *Recent Developments in the Teaching of Geography in Central Europe.*

This contains several articles. Those referred to in the present paper are—
"Recent Developments of Geographical Teaching in Europe," by Professor John Goodison of Ypsilanti Normal School (Mich.), pp. 289-313.
"European Schools," pp. 313-318.
There is a useful bibliography of German works at the end of the article, including works on Methods of Teaching Geography and on Home Geography.

Report of the Commissioner on Education for the Year 1892-93.
Vol. i., chapter vii.   Washington, 1895.

6. *The State Map of Connecticut, as an Aid to the Study of Geography in Grammar and High Schools.* By William Morris Davis, Professor of Physical Geography in Harvard University.

Connecticut School Document, No. 6, 1896.

Professor Davis has prepared similar pamphlets for other States which have such topographical maps.

(*To be continued.*)

GEOGRAPHICAL NOTES.

By The Acting Editor.

EUROPE.

Geography at the Universities.—The reports of the Lecturers on Geography at Oxford, Cambridge, and Owens College, Manchester, are published in the *Geographical Journal* for July. Mr. Mackinder lectured as usual at Oxford twice a week during the three terms. The attendance in the three terms was 56, 74, and 21 undergraduates respectively, besides a few ladies. It is satisfactory to hear that the University has provided for the continuance of the readership after the present arrangement with the Royal Geographical Society expires. Mr. Gunther, the geographical student for 1895, will shortly send in a report of the work he has accomplished in the Phlegræan Fields of Naples.

Mr. Mackinder also delivered a course of lectures at Gresham College, London, which was attended by nearly 300 students, mostly teachers and pupil-teachers. At the conclusion of the course 174 of these presented themselves for examination, and Mr. Chisholm, the examiner, reported that 16 per cent. merited "distinction," and that 48 per cent. had obtained half marks or more.

At Cambridge Mr. Yule Oldham lectured on Physical Geography, and the first part of the course, which appealed to candidates for the Geological Tripos, was attended by twenty men and some ladies. He also delivered forty lectures outside the University in connection with the Extension movement, which were attended by large audiences. A year ago some useful modifications were made in the geography papers of the Local Examinations by Mr. Oldham's advice, and these have been very successful. An important step has been taken by the syndicate which regulates the examinations, in placing geography as a special subject on the same footing as languages, history, and natural science ; and at least one college will include physical geography among the subjects for future scholarship examinations.

Mr. Herbertson's classes at Owens College were attended almost exclusively by first-year students of the Day Training College, and the Education Department's regulation, excusing the best geographical scholars, again showed its effect in keeping down the numbers. Evening classes were held in the rooms of the Manchester Geographical Society. Mr. Herbertson also delivered two courses of University Extension lectures. Considerable progress has been made towards forming a geographical collection at Owens College. At present, however, the work of the geographical lecturer is very discouraging. The Education Department's regulations leave him only the poorer material to work upon, students who feel that their geographical work is a penalty for not doing better in the subject at the Queen's Scholarship examination, and who don't care to devote to it more time than the bare minimum necessary.

The Bay of Biscay.—In the *Annales de Géographie* of July 15th Professor M. J. Thoulet summarises the results of a voyage in the *Caudan* during the latter part of August 1895. Zoological studies were the chief object, and therefore oceanography was only a secondary consideration. The latter, indeed, should lead the way, and serve as a guide to the natural sciences, for in every problem one should proceed from the more simple to the more complex.

It might be expected that under the water, where the fluid medium tends to obliterate all differences, topographical, chemical, and physical, the transitions would be very gradual. On the contrary, however, the diversities of the fauna brought up at closely adjacent points were very striking. These variations are evidently due to currents which modify considerably the temperature, salinity and density of the water, and the quantities of gases contained in it. Dr. Natterer of the *Pola* expedition noticed that in the Mediterranean, where the fauna is on the whole scanty, it was abundant over limited areas, and this circumstance he ascribes to the same cause.

The soundings of the *Caudan* prove the existence of a regular cliff bounding the continental plateau. Off the French coast the bottom gently slopes to a depth of 500 to 600 mètres (270 to 330 fathoms), and then falls suddenly to 2000 mètres (about 1100 fathoms), and thus the continental plateau here extends much beyond the 200 mètres (110 fathoms) usually assigned as its limit. The submarine cliff that skirts the shores of Languedoc and Provence in the Mediterranean commences at 110 to 160 fathoms. Professor Thoulet gives a sketch-map on which are given the course of the *Caudan*, isobathic lines, etc.

The surface temperature was always, except at one station a little south of lat. 47° and some three degrees west of the Loire, higher in the evening than in the morning. Of the temperatures below the surface three diagrams are given. The one shows the temperatures along the general course of the *Caudan*, sensibly parallel to the coast and to the submarine cliff, while the other two are short sections approximately at right angles to the former. Below the isotherm of 12°, or at most of 11° C. (52° F.), the variations are almost nil, and hence the superjacent stratum of forty-five to fifty-five fathoms is alone subject to seasonal or other perturbations. There was nowhere observed a *Sprungschicht*, or level where the temperature suddenly changes, so frequently present in lakes. The observations of Professor Thoulet on the temperature and density of the water confirm the conclusions of M. Hautreux. A branch of the Gulf Stream abuts with a feeble velocity on the continent of Europe on the latitude of the entrance to the British Channel. Running from north-west to south-east it passes over the surface of the Bay of Biscay, and, being of small depth, is easily affected by the winds which blow in different directions according to the seasons. The waters discharged by the Loire and the Gironde turn aside to the south-west, and, at least in summer, those carried by the tide over the shallows and sandy shores of La Vendée and Gascony form a sheet of warmer and salter water which likewise is diverted to the south-west, but below the superficial layer. The heat propagated through a thin layer of water must be considerable, and this, combined with radiation, etc., has probably considerable effect on the currents of the coast. M. Thoulet suggests that in future work in these waters a series of profiles of the continental slope should be obtained ; that soundings should be made below sixty-five fathoms in order to definitely establish the tranquillity of the water below the isotherm of 52° F. ; that specimens of water from great depths should be tested, and the isotherms compared with the lines of equal density ; and that the direction and velocity of the currents should be ascertained. .

ASIA.

The Mekong.—The progress made by the French in opening up the navigation of the Mekong has been reported from time to time in these pages. In the *Comptes Rendus* of the Paris Geog. Soc., Nos. 10-12, is an account given by Lieut. G. Simon of his voyage up the river to Kiang Khong with the gunboats *La Grandière* and *Massie*. These boats, sent out to Cochin China in sections, were set afloat at Saigon, whence they ascended to the first rapids at Khone. Here the river, running through a numerous group of islands, lowers its level nearly sixty feet. Khone is the largest of these islands, and across it the boats were carried by a railway. Above the falls the river contracts a little, but still encloses a number of islands, among which the current flows very swiftly in the rainy season. One of these islands, three hours' steam above Khone, is Khong, called also Sitan-Dong, of considerable area, abounding in rice fields, and occupied by a dense population of 10,000 to 12,000 souls.

From Khong to Bassac the stream has a wider navigation channel, the mean breadth of the bed being still 1300 to 1600 yards. The village of Bassac lies along the right bank of the river for a distance of a mile and a half or more, and contains 4000 inhabitants. On the left bank there are scarcely any people as far as the Se-done river, which skirts the northern side of the plateau of the Bolovens. After a thorough survey of the rapids of Kemarat, the *Massie* was successfully navigated up them to the town. These rapids extend for a distance of seventy-five miles to the village of Hona-Don-Tane, a little above Kemarat. The river, 650 yards broad above, narrows so as to leave a free channel of not more than twenty-eight, and the waters dash furiously over the rocks at the side. For six days the *Massie* had to struggle up a succession of such narrows, each fresh obstacle demanding a special examination. At Kemarat the boat had to wait three months till the water rose high enough to afford a passage over the Neng-Sa rapid and the bars that follow one another as far as Don-Tane. Beyond this place the way is open to Vien-Tiane, a distance of 370 miles.

La Grandière received some injuries at the foot of the Khone rapids. It was repaired and followed the other gunboat, but was stopped at one of the Kemarat rapids, not being able to make way against the current when the water was high. It was a month before the boat could rejoin the *Massie* at La-Khone and proceed in its company to Vien-Tiane, the old capital of Laos, and now the chief town of an important province. The town suffered severely from the attack of the Siamese sixty years ago. It formerly extended along the left bank of the Mekong for a distance of twelve to fifteen miles, but now most of it is in ruins.

The year is divided in Laos into two well-marked seasons—the rainy season from April 15th to October 15th, and the dry season during the other half of the year. The rain is generally abundant; 79½ inches were measured in 1894, while in 1895, an exceptionally dry season, the amount was only 46¼. The temperature is trying to Europeans only in March, April, May, and October; in June, July, August, and September the air is freshened by the rains; and from November to February the north-east wind so cools the air that the thermometer has been known to descend to 44° or 46° F.

On the way up to Vien-Tiane the gunboats steamed for three days up the Ngum, an important affluent of the Mekong. Large and numerous villages are situated on its banks, and the country abounds in game. Another tributary on the left bank, the Nam San, is navigable for a distance of thirty miles for about four or five months in the year. A short distance above Vien-Tiane the Mekong is again encumbered with obstructions, forming a series of rapids. These are only navigable during high water, and *La Grandière* had to wait till August, when it accomplished the passage with some very narrow escapes, and reached Luang Prabang, having navigated the Mekong for a distance of more than 1200 miles from the sea. An attempt made in September to steam up to Kiang Khong was unsuccessful. In October, when the water was lower, Lieut. Simon tried again, and this time, after a constant struggle, succeeded in five days in passing the forty-seven rapids between the two towns, having sometimes to partially unload the vessel. From Kiang Khong to Kiang Sen, forty miles farther, the navigation is for the most part troublesome, the river being encumbered with rocks. Finally, on October 25th, 1895, *La Grandière* completed its journey at Tang-ho, 1500 miles from the sea and a day's journey below Kiang Lap, which may be considered the limit of navigation.

Lieut. Simon concludes that the river must be navigated in reaches. Between Khong and the Se-mun the river is navigable from May to January by vessels drawing a mètre (3¼ feet) of water. Between Khone and Khong it can only be navigated with safety when the water is at least 1½ mètres deep, that is, in general, from June to the end of November. The rapids of Kemarat can be passed by powerful vessels from November to January and in May and June, when the water is at a moderate height. The stretch of 370 miles from Don-Tane to Sampana will be navigable throughout the year when some slight work has been done in removing rocks. From Sampana to Luang Prabang navigation is impossible when the water is low, and boats cannot ply frequently and regularly beyond the Nam Neun. If certain modifications be effected in the bed of the river, a steamer may travel without very great difficulty as far as Vien-Tiane, and very probably Luang Prabang, during three months in the year.

Nearchus in Sumatra.—After the voyage of Nearchus and Onesicritus from the Indus to the mouth of the Tigris, the two captains are not heard of again till shortly before the death of Alexander, when Nearchus is said by Plutarch and Quintus Curtius to have given his master an account of his voyage upon the ocean. Now, characters of pure Phœnician type found at Rejang in Sumatra are described by Marsden in his history of the island, and may date from a visit of ships of Alexander to these parts, of which written traditions also exist. The inscriptions

run from left to right, whereas Arabic is written in Sumatra in the reverse direc-
tion. Moreover, two of the characters are Cypriote forms, and, according to
Plutarch, there were Cypriotes among the crew of Nearchus.—*The Geogr. Journal,*
June.

The Formation of Petroleum.—Ochsenius' theory is that in some way or other a
bay is cut off from the sea by a bar ; then through evaporation it becomes very
salt, gypsum and salt are precipitated, and the lake becomes an accumulation of
strong lye. If by some catastrophe the bar is broken through, the lye is poured
into the neighbouring sea, killing all the living organisms within its reach, and
thus forming the material for a petroleum deposit. M. N. Andrussof has observed
a phenomenon differing in some essential particulars. Through the strait connecting
the Caspian sea with the Aji Daria or Karabugaz bay, the Caspian water, with a
salinity of 1·4 to 1·5 per cent., mixes with the highly saline water (16 to 17 per
cent.) of the Aji Daria, and all living organisms that come in contact with it—
Plankton, fragments of seaweed, and fishes—immediately die, and are either
mingled with the sediment at the bottom, or cast up on the shore, where they are
preserved from rapid decay by the saturated solution of salt. In this case there is
no sudden catastrophe as in Ochsenius' theory, and as the process is constantly
going on there is time for the accumulation of large quantities of organic matter.—
Globus, Bd. lxix. No. 11.

AFRICA.

The Lakes of Timbuktu.—The existence of lakes to the west of Timbuktu was
briefly noticed in vol. xi. p. 81. A description of the chief of these, Faguibine and
Tele, was given by Lieutenant Bluzet in the *Bull. de la Soc. de Géographie,* Trim. 3,
1895, and further information has been lately published in the *Comptes Rendus,*
Nos. 10-12. These lakes extend on the left bank of the river in a long succession,
beginning on the west with the Tenda lake, separated from the Kabara by a plateau
impregnated with iron. Then follow the lakes of Sompi, Takaji, Gaouati, Horo,
and Fati, and lastly the Gonndam creek, communicating with the Niger. North
of Gonndam is the Tele lake, and the great Faguibine, sixty-eight miles long, with
depths of a hundred feet. Lieut. Hourst was caught in a regular tempest on the lake,
from which he found shelter at the island of Taguilam in a harbour he named Poit
Aube. Important mountain masses look down on the lakes Tele and Faguibine at the
north and east. Rich cultivated fields skirt the shores, and the Tuareg Iguellad
have numerous slave villages on the southern coást of Faguibine ; on the north are
their camps and pastures. On this side, the Bonkor lake communicates with the
Faguibine at high water by a creek seven miles long ; its length is ten miles. At
its eastern extremity rises the mountain Tahakim, beyond which are two small
swamps and beyond these again the mountain Miziran. To the south of Faguibine
lie other lakes. The Daouna lies ten miles to the south and stretches parallel to
the western end of Faguibine for a distance of twelve or thirteen miles, and a much
smaller permanent sheet of water lies rather more than a mile to the north-east.
Still farther south is the Bankare, five and a half miles long and one broad, which
is filled only at high water. The mountain Tassermint or Sorma, north of the
eastern extremity of Daouna, really consists of two elevations ; the one nearer to
the lake is named Sarma-bibi, and the other known as the Sarma-kore, of lesser
elevation, lies two miles farther north.

The Origin of Lake Tanganika.—Mr. R. T. Günther, in describing the anatomy
of a jelly-fish found in this lake, has suggested changes in the physical geography

of the country as the explanation of the presence of these marine forms. Where these are found not far from the sea, in a lagoon in Trinidad, for instance, their presence may be explained by a former connection with the sea, and an excess of evaporation over rainfall after the connection was severed. If the marine forms reached Tanganika from the Atlantic, they must have found their way in when the lake region was 2700 feet lower than at present, and the Atlantic extended over the Congo basin. Dr. J. Cornet, in the *Mouvement Géogr.*, Nos. 25 and 26, combats this theory of the origin of the lake, which has at different times been maintained by various authorities. He doubts the marine origin of the molluscs now found in the lake. According to many naturalists who have examined specimens brought to Europe, and whose opinion is worthy of respect—P. Pelseneer among others—all the molluscs of Tanganika are representatives of species living in fresh water. Many of the shells are, indeed, in external appearance like those of marine molluscs, being abundantly provided with projections, spines, and protuberances ; but Tanganika, with its length of 420 miles, breadth of 50 miles, and a depth in some places exceeding 330 fathoms, is an inland sea with storms in which thin shells would be dashed in pieces. Accordingly, the individuals with the strongest shells would survive, and the present characteristics gradually arise from natural selection. As for the *Medusæ*, Dr. Cornet considers that their presence is no sufficient proof of the marine origin of the lake, seeing that they are found in Lake Urmiah and in the Upper Niger at Bammako. Again, the few fish found in the lake are of species peculiar to it, and representatives of fresh water genera common to the Niger, Congo, and other waters of Tropical Africa.

Geology alone can, in Dr. Cornet's opinion, solve the question. Had the sea at one time extended from its present shores into the part of Central Africa occupied by Tanganika, some traces of marine sediments would be still left in spite of the erosion which may have carried most of them away during the ages that have elapsed since the formation of the lake. But, with the exception of a narrow strip of Jurassic and Cretaceous deposits along the eastern coast, Tropical Africa is an enormous mass of archæan rock, overlaid by considerable tracks of horizontally stratified sandstone of continental origin. Nowhere, except in the north, has the country been submerged since the foldings which contorted the Palæozoic rocks in the Carboniferous age. Nor has the continent, except in the Atlas region, been subjected to the powerful folding of Tertiary times, which raised the Alps and the Himalayas. But, though tangential tension has been absent, radial fractures of great importance have occurred, the latest of which have produced a series of gigantic clefts nearly vertical, which make the structure of Africa, and especially its eastern portion, particularly interesting. These great vertical faults run in general from north to south, and the intervening ground has in some cases subsided, forming a great ditch, with steep walls. Tanganika is a sheet of water that has collected in a very low part of one of these ditches, which is prolonged northwards past the basins of the lakes Kivu, Albert Edward, and Edward, and along the course of the Nile. At the summit of Mount Rumbi the post-primary continental deposits may be seen at an elevation of 2985 feet above the lake, and opposite the mountain Giraud found the water 2123 feet deep. The subsidence to which Tanganika owes its origin must, then, have caused a vertical displacement of at least 5000 feet.

The subsidence of the Tanganika basin is of old date relatively to those which formed the lakes Nyassa, Manyara, Rudolf, etc. It is more recent than the red felspathic sandstones of the Congo basin, but older than the white sandstones. And to its ancient origin and various local circumstances it no doubt owes its specialised fauna, while Rudolf, Stephanie, and other lakes, which have been separated from the basin of the Nile at a later date, still retain the fauna of river.

AMERICA.

The Magdalena.—In a report of Mr. E. MacGregor, Vice-Consul at Barranquilla, some account is given of the navigation of this river. Between Arrancaplumas, a short distance above Honda, which is situated 603 miles above Barranquilla, and Yeguas, fifteen miles below Honda, the fall of the river is so great as to give rise to a series of impassable rapids, and therefore the traffic between these points is carried on by a railway belonging to a British company. It is proposed to continue the railway twelve miles farther, to Conejo, but the scheme seems to be in abeyance, for as long ago as 1884 it was expected that the line would be completed in the following year to La Dorada, seven and a half miles below Conejo. Above Honda, the navigation extends to Girardot, 718 miles from Barranquilla, and sometimes in high water to Purificacion, 754 miles, or even to Neiva, 830 miles. From Girardot, a railway runs a short distance north-eastwards to Juntas de Apulo, which is to be extended to Facatativa, in the north-west of the plain of Bogota. Forty stern-wheel steamers ply on the Magdalena, of tonnage varying from 30 to 300. They take seven or eight days to steam up the lower river, as it is not safe to proceed in the night. A safe passage between Barranquilla and Savanilla by the Boca de Cenizas, the chief mouth of the Magdalena, is greatly needed.—*Geographical Journal*, July.

The Shrinking of Lake Titicaca.—The plateau, on which the lake stands at a height of 12,645 feet above sea-level, falls towards the south like the hollow of a shell between two chains of the Andes. At its northern shore the water has most unmistakably receded, leaving productive lands, the possession of which has been disputed by the neighbouring land-owners by legal process and even by violence. Twenty-eight years ago the lake extended to the suburbs of Puno, while now it is about 550 yards distant, and the intervening land is laid out in fields. The cultivated lands in the bay of Guarico, in the district of Taraco, and in the pampas of Acora and Ilave are also much more extensive than in former years. Tradition affirms that the lagoon Umayo, five leagues from Titicaca and fifty feet higher, once formed part of the lake, so that the intervening pampas must then have been submerged. More reliable evidence is found in the fact that the rocks of soft stone and conglomerate standing at many points on the edge of the basin show signs of erosion by waves at heights far above the present level of the water. Also, at a distance of six leagues and more, in Ayavacas, for instance, similar marks are found as well as numerous fossilised fresh-water shells, limnites, planorbites, and ammonites, similar to species still living in the lake.

The streams that feed Titicaca descend from the terraced mountains, which consist for the most part of soil, gravel, and conglomerate, offering little resistance to the violent summer rains. Enormous quantities of detritus are carried down both by the perennial streams and by the numerous brooklets which appear only in the rainy season. At this time the lake is increased by the large volume of water poured into it, but three or four months later it sinks lower than ever, and the bottom is covered by a fresh deposit of mud. In course of time, the lake will probably be changed into small basins, and after that become simply a river, forming the upper course of the Desaguadero. The shoaling of the water may increase the relative evaporation. Certainly there is no increase in the volume of the Desaguadero, for, though a sluggish stream, it does not overflow its banks. The water which flows through it to the Poopo lake is finally evaporated in the swamps of Coipasa. The Desaguadero has been partially regulated within the last few years Since 1893 the Peruvian Corporation has used it for the transport of copper from the mines of Corocoro, and between Nazacara and the lake it has been enlarged to a breadth of forty-five feet and a depth of seven. The navigation could, it is affirmed, be extended to Oruro, were there sufficient freight to make it profitable.— *Globus*, Bd. lxix. No. 24.

Juan Fernandez.—As mentioned in vol. x. p. 658, Dr. Ludwig Plate has investigated the geology and other natural characteristics of this island. The rocks of the island are exclusively of volcanic origin, consisting of black basaltic lava, in which at some places whitish or reddish tuffs are imbedded. The lava cliffs have a singular resemblance to sedimentary rocks, clearly exhibiting stratification. On all sides the island descends to the sea in precipitous walls 300 to 1000 feet in height, except at the harbour of the settlement (Cumberland Bay), the Puerto Ingles, the Puerto Frances, and the Bahia de la Vaqueria, where valleys debouch on the coast, giving easy access to the interior. The stratification of the lava is distinctly seen from the sea, and, moreover, at intervals of sixty feet or more are dark vertical lines visible at a distance of more than half a mile. The stratification is no doubt due to successive eruptions; lava streams issued repeatedly from a submarine crater, and in the course of time these deposits were partly raised above the sea. The vertical lines are not so easily accounted for ; probably a lava flow left on cooling numerous fissures or clefts, which were filled up by the lava of the next eruption. In earlier geological times the island was undoubtedly much larger. An examination of the sea-bottom between Mas-a-tierra (Juan Fernandez) and Mas-a-fuera would probably prove that the two islands, which are now ninety-two nautical miles apart, were originally one, or are, at any rate, the highest points of the same submarine plateau.

Though only eleven and a half miles long by four and a half broad, Mas-a-tierra consists of two regions, the eastern mountainous and the western lower and flatter. These differ widely in climate, vegetation, and scenery. Their common boundary is marked by the crest of a range of mountains which runs from the most northern point south-eastwards to the Yunque, the highest summit of the island, 3040 feet. East of this line, the country is occupied by numerous ranges rising to heights of 1300 to 3000 feet, the directions of which have not been fully laid down on maps. Their slopes and the intervening valleys are for the most part densely wooded. The summits of the mountains rise into the moister upper air and cause abundant precipitation throughout the year, so that every valley has its brook, and in many parts the vegetation is of tropical luxuriance. The western half of the island, on the other hand, contains only one narrow range, skirting the north coast and culminating at a height of about 1300 feet. The greater part of this district lies only 300 to 600 feet above sea-level and, owing to the slight changes of temperature, there is often no rain for months together. Consequently there is no forest, and the broad undulating plains are either bare or covered with the yellow monotonous *Teatina* grass (*Avena hirsuta*).

As the land fauna of oceanic islands is always poor, Dr. Plate turned his attention chiefly to the sea organisms. In the tidal waters he found very few varieties, probably because tender organisms are liable to be crushed between the loose stones. The fauna below three fathoms was much more varied, containing examples of all classes of invertebrates, and differing in character from the shallow water fauna of the coast of the continent, only 360 nautical miles distant. It is distinct, not only from that of the Bay of Valparaiso in the same latitude, but also from the marine life of the whole coast from Arica to Cape Horn, which retains a remarkable uniformity through thirty-six degrees of latitude. In those cases where a relation to the continental fauna is observable, it seems to be due to immigration

from North Chili or Peru. These facts Dr. Plate illustrates by a number of
examples. Fish are remarkably abundant, but, as on the continental coast, the
species are few. Dr. Plate collected twenty-five varieties, half of which occur in
enormous numbers. To what extent the character of the fauna of Juan Fernandez
is due to immigration from the north is a question that may be studied in relation
to the ocean currents. The Humboldt stream is supposed by some authorities to
sweep past the island, while by others it is limited to a distance of one hundred
nautical miles from the continent. Dr. Plate took the temperature of the water on
the voyage from Juan Fernandez to Valparaiso, and found that it fell from 67° F.,
fifteen miles east of the island, to $56\frac{1}{4}°$ in the harbour of Valparaiso. Between 130
and twenty miles from the coast, there was a sudden fall of 4°, which evidently marked
the entrance of the vessel into the Humboldt current. The western boundary Dr.
Plate was unable to determine exactly, as it was crossed in the night, but it must
lie within 130 miles from the coast. This stream must give rise to counter currents
flowing south-west and south, as stated by Krümmel, and of their existence Dr.
Plate obtained decided evidence in his boat excursions round the coast ; also the
vessel in which he sailed to Valparaiso was carried southwards, and had to sail
northwards again to reach the harbour. This counter current is very feeble, and
clearly perceptible only when the wind is slight, and therefore it has not been
much observed. Probably it starts where the Humboldt stream leaves the shores
of the continent, between latitudes 5° and 10° S.

Dr. Plate concludes from the shallow water fauna that Juan Fernandez is an
oceanic island which has never been connected with the continent. Its volcanic
formation does not of itself prove this, for it is conceivable that, originally a part
of the continent, it might have been converted into an island by sinking. But in
that case its coast fauna would have exhibited a continental character.—*Verh. der
Gesell. für Erdkunde zu Berlin,* Bd. xxiii. Nos. 4 and 5.

MISCELLANEOUS.

A contract has been signed for the extension of the railway from Durban to
Verulam, nineteen miles long, to the river Tugela, fifty miles farther.

The Busumakwe lake, lying two days' journey to the south-east of Kumassi,
was visited in spring by Major Donovan. It is about eight miles long by six
broad. Around it stand a number of villages, the inhabitants of which catch fish
in the lake.

The greatest depth hitherto discovered in the northern part of the Red Sea,
1197 fathoms, has been sounded by the Austrian vessel *Pola,* in 38° E. long. and
22° 7′ N. lat. In the Gulf of Akabah a maximum depth of 704 fathoms was
sounded in 28° 29·2′ N. lat., and 34° 42·8′ E. long.—*Globus,* Bd. lxx. No. 2.

An astronomical observatory has been built by Herr Bischoffsheim on Mont
Mounier, a summit of the Maritime Alps, over 8800 feet high. It is provided with
meteorological instruments, and the records will be communicated daily by telephone
to Beuil, the nearest village, and thence by telegraph to the observatory at Nice.
—*Globus,* Bd. lxx. No. 2.

The Emperor of China has granted permission to construct a railway from
Tientsin to Pekin. The distance is seventy-eight miles, and the cost is estimated
at about £550,000. The country is flat, and the only difficulty will be to lay a
foundation firm enough to withstand the floods of the Pei-ho. Permission has
also been obtained for a line, sixty-five miles long, from Shanghai to Su-chau.

According to *Globus*, Bd. lxx. No. 2, the Russians intend to construct a **railway** from Merv up the Murghab and Kushk valleys to Kushk, a distance of 208 miles. Kushk lies only six miles from the Afghan frontier post Kara-tepe, and less than ninety-five miles from Herat. A second line is to run from Charjui along the left bank of the Oxus to Kerki. This place is 140 miles from Charjui, and about the same distance from Balkh, the centre of Afghan Turkestan.

The new works for the utilisation of the **Niagara** Falls are now completed. The water is received a mile and a quarter above the Falls into a canal 82 yards broad, and 11½ feet deep, and it works ten turbines, each of 5000 horse-power, placed in a shaft 150 feet deep. From the shaft the water is conducted into the river again by a tunnel rather more than 1¼ miles long. With the machinery already at work the total power now amounts to 60,000 horse-power.—*Geogr. Zeitschrift*, Jahrg. xi. Heft 5.

The Assistant Government Geologist of Victoria, Mr. James Stirling, has erected a high-level meteorological observatory on Mount St. Bernard, 5000 feet above sea-level. This is the highest station in Australia. Mr. Stirling has also made a series of observations on underground temperature in the deepest mines, of which one, Lansell's 180 at Bendigo, goes down to a depth of 2000 feet below sea-level. These may be usefully compared with the recent researches of Professor Agassiz in America.—*Symons's Monthly Meteor. Mag.*, June.

A new route has been opened between **Bogota**, capital of Colombia, and the outer world. The Rio Meta, a tributary of the Orinoco, is nearer to Bogota than the Magdalena, and a Frenchman, Joseph Bonnet, has since 1880 been considering the practicability of bringing the traffic of Bogota eastwards by this route. He has had a stern-wheel steamer built 130 feet long, 26 feet broad, and 4¼ deep, and made the voyage from Port of Spain in Trinidad to Cabugaro in Colombia between October 6th and November 4th. Cabugaro will be connected with Bogota by a short railway. The Government has made a contract with M. Bonnet and pays him a subvention, in return for which he carries officials and missionaries free. The undertaking will probably soon pay a profit, and regular communication will be of great advantage to the districts along the route.

The project of a canal from the Bay of Biscay to the Mediterranean has been for a long time discussed in France, and persons qualified to judge have expressed the opinion that the scheme is impracticable owing to the considerable changes of level, and the want of water sufficient to supply the locks. In 1894 a commission was appointed by the French Government to report on the feasibility of the scheme, estimate the traffic, etc. The results the commission has arrived at are published in the *Mouvement Géographique*, No. 23. It is estimated that the cost of construction per mile would be more than double that of the Baltic canal, and considerably in excess of that of the Suez canal, while the receipts, four shillings per ton being estimated as the maximum rate that could be charged, would not provide even for the expenses of maintenance and management, much less pay interest on the capital.

NEW BOOKS.

Through Jungle and Desert: Travels in Eastern Africa.   By WILLIAM ASTOR
CHANLER.   With Illustrations from Photographs taken by the Author, and
Maps.   London: Macmillan and Co., 1896.   Pp. 519.   *Price* 21s.

British East Africa is gradually being completely explored and mapped, and
this volume forms a not unimportant contribution to our knowledge of that region.
Mr. W. A. Chanler, a young American, accompanied by Lieutenant Ludwig von
Höhnel, has filled up an important blank between the explorations of Count Teleki
and Dr. Donaldson Smith near Lake Rudolph and Mt. Kenia.   From the geographical point of view, Lieutenant von Höhnel's maps are the most important
contribution in the volume.   He has, by his painstaking observation, greatly added
to his previous reputation, and the production of the maps by the Imperial and
Royal Military Geographical Institute in Vienna leaves nothing to be desired.
During the journey attention was paid to Natural History, and a valuable collection
made.   A new antelope was discovered (the *Cervicapra Chanleri*), a number of
new species of reptiles, and two new *Lepidoptera*.   Also four new genera, and
thirty-four new species of beetles, etc.

The author has given a sprightly and interesting account of his varied experiences, and his descriptions of the country, the natives, and his dealings with them
are clear and graphic.   He had considerable difficulty with his porters, and in the
end we are sorry to notice that he had great unpleasantness with the officials at
Zanzibar, who it appears would not punish the deserters from his caravan, and
seem to have treated him in a very cavalier fashion.   We regret that he left Zanzibar without the matter having been satisfactorily settled.   He writes: "As I found
all my efforts to obtain justice at the hands of the authorities unavailing, I decided
to return home."   If memory serves us rightly, this is the only instance of such an
unfortunate occurrence that has come to our notice, and we should like to hear
what General Sir Lloyd Matthews and Mr. Cracknel have to say on the subject.

Mr. Chanler believes that in some districts through which he journeyed Europeans could live with comfort.   For instance, he writes: "Here among the Embe
we were encamped about 5000 feet above sea-level, and the air was very bracing.
Both Lieutenant Höhnel and I found our appetites increased and our spirits rose.
I am sure that at this height above sea-level Europeans could live with comfort.
My negroes, however, appeared to suffer from the cold.   They imagine that no
country is healthy where they are not continually bathed in perspiration, although
at ease."   The volume contains a considerable amount of very interesting information respecting the habits and customs of the various tribes with whom the
travellers come in contact.   An extract or two may be given to illustrate the care
with which points of detail are dealt with.

"The inhabitants of this mountain range (the Jombeni), like all East Africans
who are not as yet converted to Christianity or Mohammedanism, had no clear idea
of the Deity.   They supposed there was a Supreme Being of some sort, and that it was
their duty to propitiate this Being before starting upon any enterprise.   He was
supposed to be a stern God, and, as far as I can gather, not altogether just; but in
no small measure open to the benign influence of bribery.   Before essaying any
enterprise, sacrifices were offered—not burnt offerings, but gourds filled with grain,
or some other small tribute—of value, however, to the giver.   These were hung
upon the trees or suspended from forked sticks in the neighbourhood of the offerer's
dwelling.   Connected with this supreme Being, and so closely as scarcely to be
disjoined, were the local deities—such as the genii of hills, holes, and rivers.   These

likewise were to be propitiated in some way, either by the inhabitants of the district where the hills, holes, or rivers were located, or by the person whose enterprise led him to the neighbourhood of such places."

The future was foretold by the wise men in two ways : either by means of a gourd, filled with different-coloured pebbles, or by means of two bits of leather shaped like sandals, and covered on one side with ornamental designs. The gourd and pebbles were used as follows. The old man, placing the pebbles in the gourd, shook it up, meanwhile uttering some magic word, and then allowed the stones to drop out one by one into his left hand. The message of Fate was conveyed by the manner in which they emerged from the hole in the gourd, and thus was determined the outcome of the enterprise under consideration. The strips of leather were used after the following manner. One strip was held in each hand, and the seer, closing his eyes, danced back and forth for a time, repeating words of supposed magic import, meanwhile beating the strips together. After a time a stick was thrown over each shoulder, and the position in which they fell upon the ground determined whether the omen was good or evil. Prophesying by this latter means is not confined to the old men, for the old women likewise claim skill in thus foretelling the course of events.

Lieutenant von Höhnel was unfortunately charged and wounded by a rhinoceros, so badly, indeed, that he had to be carried to the coast.

The volume is profusely illustrated from photographs taken by the author. They give a good idea of the country and the natives, but their value varies considerably.

The volume is further embellished by portraits of the travellers.

A String of Chinese Peach-Stones. By W. ARTHUR CORNABY. London : Charles H. Kelly. Pp. xv + 479. *Price* 10s. 6d.

This book might be best characterised by a string of epithets—charming, instructive, enthralling, romantic, picturesque, tragic, kaleidoscopic in its shifting scenes and adventures. In his introduction, the author states that his "object has not been to attempt anything like a novel, but by means of a series of character sketches, in which the details are drawn from the life, to picture the normal village life of Central China, to describe some leading incidents in the earlier Taiping Rebellion, and to indicate how Chinese character may be modified under the changes which come, and must come, even in the changeless East." Shortly described, the work is a romance, and as such will appeal to all lovers of a good story. But it is at the same time the work of a scholar and student of Chinese life and literature, and presents a lively and eminently readable picture of Chinese society, with its mysterious codes, customs, and conduct. The hero of the tale is the son of a farmer, Li Seng-teh by name, who is captured and forced to serve, as a mere lad, under the Taiping Captain Li (the Crouching Tiger). Seng-teh's father meets his death during a Taiping raid ; and soon after the Crouching Tiger is slain in fight. Young Li, now free, finds his way to where his widowed mother has taken refuge with Nieh, the Confucian schoolmaster, but arrives to find himself a lonely orphan. The teacher adopts him as a son, betrothing him to his little daughter of three, and then sends him forth to serve in the Imperial army. Seng-teh proves a brave soldier, passes through some thrilling adventures, and in his daring deed of scaling a fortress wall has two fingers of his left hand chopped off. He then gives up soldiering and proceeds to develop his artistic powers as a designer in a pottery. At a critical time, when suffering from fever, he encounters a medical missionary, who cures him completely, and in whose company he finally falls in with his old teacher and *fiancée*, now a girl well up in her teens. Nieh, having meanwhile come under the evil influence of opium, is almost a wreck both physically and morally. The foreign doctor puts him on his feet again, and restores to the Confucian B.A. something of his old self-respect. There is, of course, the fitting consummation of the early betrothal. Such is the skeleton of the tale round which the accomplished author develops his oriental lore.

In the earlier chapters Nieh, the distinguished graduate, gives informal discourses on the classics, and teaches a broad faith in Heaven's decrees. He strives to give natural explanations to the superstitious and magical tales which form so large an element in the Chinese mind. In both the Taiping and Imperial camps, the soldiers amuse themselves in telling stories, some of fairyland, but some very human in their interest. A favourite type is the story of how the man of wit, ofttimes most unscrupulously, gains the advantage over his duller brother. We meet also with varieties of Whittington's Cat, Aladdin's Lamp, and other familiar gems of folklore. Heroic and filial deeds are duly recounted. All are told brightly, and each presents to the thoughtful reader an important aspect of the Chinese mind. The book is, in short, a perfect mine of ethnographic information, and has a high scientific value. The String of Peach-Stones, it should be mentioned, is a charm bought from a Tao-ist pedlar by Seng-teh's father, and worn round the neck by the hero all through his chequered career. There may be little in a name, but there is no doubt that the charm is in the book. Numerous illustrations, bearing in many cases on the tales that are being told, beautify its pages ; and a quaintly-coloured picture of the village school forms an appropriate frontispiece.

Die Donau als Völkerweg, Schiffahrtsstrasse und Reiseroute. Von AMAND FREIHERR V. SCHWEIGER-LERCHENFELD. Mit 467 Abbildungen und Karten. Wien, Pest, Leipzig : A. Hartleben's Verlag, 1896. Pp. viii + 949.

None of the large European rivers deserves an exhaustive monograph more than the Danube, and we are glad to say that an excellent one has just been supplied by Amand Freiherr v. Schweiger-Lerchenfeld. His splendid volume of upwards of 900 pages should satisfy every one, be he geographer, historian, engineer, or traveller. The author sets out with a full account of the physical features of the river and the adjacent country, to which are added chapters on the Flora and Fauna. We notice that a steady decrease in the amount of water has been proved in the case of the Danube and its tributaries. With regard to the "Puszta," it may not be generally known that it is no longer a grassy plain or steppe, but that cultivated fields now predominate everywhere. Bears and wolves are still to be found in the Hungarian lowlands, and even jackals have recently been shot there. The second part deals with the inhabitants from prehistoric times down to the present day. The mythical expedition of the Argo, the battles of Roman, Goth, and Hun, the foundation of the Hungarian kingdom, the rise and growth of the Hapsburg monarchy, and the recent struggle of Russian and Turk are vividly told. The author considers next the Danube as a highway of commerce, giving a very valuable history—the first ever written, we believe—of the Navigation of the Danube. It is interesting to note that steam navigation was introduced in 1828 by two British engineers, John Andrews and Joseph Prichard. The demands of navigation, however, have necessitated very extensive engineering works, and the author's detailed account of the difficulties that had to be overcome is by no means the least interesting part of the book. The fourth and last section is purely descriptive. We are supposed to travel along the river from its source to its mouth, and under the author's able guidance no more enjoyable journey could be imagined.

The book is rendered more attractive by an abundance of beautifully executed illustrations, maps, and plans. We have no hesitation in saying that it will take rank as the standard work on the Danube.

Une Expédition avec le Négous Ménélik. Vingt mois en Abyssinie. Par J. G. VANDERHEYM, 1896. Paris: Hachette et Cie. Pp. 203.

In November 1893 the author sailed from Marseilles to Obock, a coaling station on the Red Sea, purchased by France in 1862. He went as agent of the "Compagnie commerciale Franco-Africaine," with instructions to visit its Abyssinian agencies at Djibouti, Harrar, and Addis-Ababa. Eventually he pushed on to Lake Abbai in the Pulamo country. Addis-Ababa is the residence of Menelik II., "Emperor of Abyssinia, King of Kings of Ethiopia, Elect of the Lord, victorious Lion of the tribe of Judah," as he styles himself. Menelik is fond of relating that he is the descendant of Solomon and the Queen of Sheba, although actually he is the natural son of the King of Shoa, Haélou, and a beggar woman taken into his palace. Menelik was born about 1845, and was elected Emperor of Ethiopia in 1889. Since 1892 he has resided at Addis-Ababa, a new and growing capital upwards of 7000 feet above the sea-level, and in the centre of the possessions and tributary states of the Negus. Menelik objects tó strangers coming to him empty-handed, and gave M. Vanderheym a charming reception, as he brought him several pieces of silk stuffs. His intelligent physiognomy pleases at the first glance. His grizzled beard covers his very black, pock-marked face. His clothes consisted of a coloured silk shirt, white cotton trousers, a very fine white cotton *chamma*, and a gold-braided black satin burnouse. An enormous broad-brimmed black felt hat and a white muslin headband concealed his baldness. He has enormous hands and feet, the latter naked and encased in unlaced Molière shoes.

Menelik recognises that if he desires to remain independent he must not resemble the petty Negro kings; and, being possessed of extraordinary intelligence and powers of assimilation, he does everything to promote his country's independence, and gain the respect of civilised nations. Thus, he does his best to prevent the mutilation of prisoners by his soldiers. He also endeavours to broaden the minds of the Abyssinian clergy, who resist all innovations. Judging by the photograph of her which the Empress invited M. Vanderheym to take, she is a large stout lady, with strongly marked features. Her complexion is light, and her maids of honour are always selected from the blackest coloured ladies in the Empire.

The author was present at a terrible massacre of a tribe of Oulamos, which had revolted from the Negus. The Emperor and his suite took part in the carnage, which, says the author, "resembled some infernal battue where human beings occupied the place of game."

M. Vanderheym has written an interesting book on a country which has lately attracted European attention. Sixty-eight illustrations, from photographs by the author, embellish the volume. There is also a sketch-map, which, considering the interest now taken in Abyssinia, might have been more elaborate. M. Jules Claretie furnishes a rhetorical and unnecessary preface.

Lettres du Brésil. By MAX LECLERC. Paris: Armand Colin et Cie., 1895. Pp. 268.

A collection of letters sent from Brazil to the *Journal des Débats* some weeks after the fall of the empire and the proclamation of the republic, this volume depicts Brazil during December 1889 and January 1890. The author devotes a chapter to "Brazil and France : Economical Questions," in which, as usual, England comes in for some tart criticism. Yet M. Leclerc shows that in 1889 the numbers and nationalities of sea-going ships sailing from the port of Rio de Janeiro were 526 British, 164 Norwegian, 156 German, 150 French, 103 North American, and 57 Italian. Further, whilst one-third of the exports of Brazil goes to the United States and one-third to Britain, only one-tenth goes to France and one-fourteenth to Germany. Forty-five per cent. of the imports are from Britain, and only seventeen per cent. from France. The chief export from Brazil is coffee, of which 245,000 bags were sent in 1889 to London, 158,000 to Hamburg, and 104,000 to Marseilles. Whilst Britain pays Brazil 128 millions of francs for her coffee, France pays only 82 millions. Moreover, Britain sells Brazil 176 millions worth of goods, whilst France sells only 96 millions. Two-and-a-half millions of francs represent the British capital invested in Brazilian enterprises, and M. Leclerc maintains that the Brazilians have grown weary of the financial yoke imposed upon them by England—a sentiment which is not uncommon among people who have borrowed freely. Brazil would perhaps like to banish her creditors as well as her Emperor.

Future Trade in the Far East. By C. C. WAKEFIELD, F.R.G.S., etc. London : Whittaker and Co., 1896. Pp. xii + 184. With Maps, Illustrations, etc. *Price* 7s. 6d.

This volume consists of a series of short and interesting sketches of the ports and places open to international trade in the Far East, along with which the author includes some of the ports in India and Ceylon.

Although the book bears the title of Future Trade, the sketches deal generally with the present position of trade in the places described, and are, the author tells us, based upon recent personal observation. He repudiates the intention of indulging in concrete prophecy ; and, as appears to us, wisely so, for, when he steps beyond the province of description and enters on that of prophecy, his manner changes and the voice becomes uncertain. As he says in his concluding chapter, for example, he detects a tendency, to which he makes specific allusion in his notes on Yokohama, to eliminate the intermediary trader, and to "deal directly with the producer," and thinks that this tendency is especially present in the minds of native and foreign manufacturers whose interests lie in China and Japan. Yet he finds that the Japanese will, in the future, welcome foreign capital, and will afford every opportunity for its investment in mercantile enterprise, and that the improved facilities for trade will continue to attract foreign capital thither, although at a lower percentage of return. Again, in treating of the Trans-Siberian Railway, he says :—"From the hour the new railway closes up the many sections on which work is now proceeding simultaneously, and connects them into one unbroken chain . . . it will be possible to deliver the wares of Great Britain in the territory of the Mikado within a fortnight, and in China two days earlier." "But," in the next sentence he finds that "although this consummation is possible, and might in some degree mitigate the troubles of the Western manufacturer who is now suffering from Asiatic competition," yet, when the increased cost of land transport through foreign countries is taken into account, this facility as to speed may do nothing for the British manufacturer after all.

His advice, in view of recent developments, that the Western merchant, whether in Japan, China, or India, can only hold his own by watchfulness, study, and enterprise, by "acclimatising himself," as he terms it, to the wants of the nation, is sound.

The small maps and plans of the countries and places described are well adapted to assist in following the descriptions in the text.

Annals of Colinsburgh, with Notes of Church Life in Kilconquhar Parish. By Rev. ROBERT DICK, Colinsburgh. Edinburgh : Andrew Elliot, 1896. Pp. 192.

This is a good specimen of a parochial history. The author gives an accurate account of the Burgh of Barony in the parish of Kilconquhar in Fife, founded by and named after Colin, third Earl of Balcarres (1662-1721) in 1707. The book also contains interesting notes of church life, the parish and dissenting ministers, various trades and manners and customs. Colinsburgh is a typical example of the rise and progress of a small industrious and recent Fife village, due (*first*) to the feuing system which gave every householder what was practically a freehold of his house and garden for a moderate annual payment ; (*second*) to the weaving trade commonly carried on here as elsewhere in Fife ; and (*third*) to the thrifty habits of the natives, for whom, it is told, the United Presbyterian Session " acted as bankers in a small way, and lent money to members in temporary need as the Kilconquhar Session had always done." There is scarcely anything in the volume which can be called geographical, except in so far as the history of the changes in the surface of any portion of the earth has its place, though it may be a very minute, almost in this case an infinitesimal, one in *Earth Lore,* the convenient translation of *Erdkunde,* for which we have to thank Professor Geikie. Here will be found everything that local knowledge can supply as to the growth of the agricultural hamlet of Rires, which consisted originally of the houses and lands (tofts and crofts) of small farmers, a smiddy, and, no doubt, later a public-house, into the present flourishing little village of Colinsburgh, the " shopping centre " of a small district, on or near the road between Anstruther and Leven, about three miles north from Elie. The author may be referred to Mr. Liddall's ingenious conjecture, in his book on the place-names of Fife, as to the origin of the name (Rerays in the oldest spelling) as " an enclosure for growing rye," and to the fact that it was, according to Mr. Liddall, one of the ancient small shires of Fife, of which a tentative list is given in the recent History of Fife and Kinross by Sheriff Mackay. A word of praise is due to the completeness of the index and the excellence of the printing by the local press of Mr. Russell of Anstruther.

Stanford's Compendium of Geography and Travel (new issue). *Asia: Vol. I., Northern and Eastern Asia.* By A. H. KEANE, F.R.G.S. London : Edward Stanford, 1896. Pp. xxiv + 527. *Price* 15s.

The new editions of these handbooks are greatly extended, besides being brought up to date by reference to all the latest books of travel and exploration. This is eminently the case in the present volume, which is compiled by an experienced and reliable author. He has searched for his material both in the narratives of older travellers and in those of the most recent times—British, Russian, French, etc.

The parts of Asia here described are the Russian territories, China, and Japan. The hydrography and orography are sketched in their general outlines, and the various races and tribes are noticed, with tables of the numbers of the populations and the adherents of the religious sects. A few remarks are made on the flora, fauna, and their distribution ; commerce and industries, roads, railways, and telegraphs, minerals, and other natural resources, all find a place.

One great difficulty in such a compilation is to make a proper selection of facts and to avoid the extension of the book beyond reasonable limits by dwelling too long on less important details. Mr. Keane has, we think, kept the balance well, and has incorporated into his book all, or almost all, that should be expected in a work of this description. Its accuracy can only be properly tested by long use. We have not detected any errors of importance, and of many parts of Central Asia our knowledge is still so fragmentary and uncertain that criticism has no sure ground to stand on.

The volume is illustrated and furnished with several maps.

Moderne Völkerkunde, deren Entwicklung und Aufgaben.   Nach dem heutigen Stande der Wissenschaft gemeinverständlich dargestellt.   Von TH. ACHELIS. Stuttgart : Verlag von Ferdinand Enke, 1896. Pp. viii + 487.

Ethnology as a science is of very modern origin, but it has already its history, and a history full of interest. The book under review by the well-known German ethnologist records this history for the general public. The author divides his work into three parts. In the first he reviews the writers in historical succession, grouping them under different heads according as they treat the subject from the points of view of Ethnography, Sociology, or Ethnology proper. This section of the book consists mainly of extracts from the various authors, and so enables the reader to judge for himself. The second part deals critically with the different branches of ethnological study, including law, religion, family, etc. Here our author follows the same method of making the writers speak for themselves. This is perhaps the fairest way as far as the writers are concerned, but it is apt to become wearisome to the reader. The third part attempts a scientific definition of "Völkerkunde," and an accurate delimitation of its province as compared with allied sciences. Geography and Ethnography, according to Dr. Achelis, describe the habitats of the different human races and the individual peoples, while "Völkerkunde" deals with mankind as an organic whole, traces its evolution, and establishes general social laws.

The book is a thoroughly conscientious piece of workmanship, rich in valuable bibliographic references. We notice that in a footnote (p. 123) there is mention of the French scientist, Romanes, and his book *L'Intelligence des Animaux*. Those who are able to master "Moderne Völkerkunde" will get a good idea of the development and the present state of ethnological knowledge.

Meyers Reisebücher.   Rom und die Campagna.   Von Dr. Th. Gsell Fels.   Vierte Auflage.   Mit 5 Karten, Plänen und Grundrissen, 63 Ansichten.   Leipzig and Wien : Bibliographisches Institut, 1895.   Pp. xii + 616 (or 1232 columns or half pages).   Price 13 M.

This is a fourth and thoroughly revised edition of what is probably the best guide-book to Rome. Baedeker's guide is good and trustworthy (as we know by experience), but this by Gsell Fels is fuller than Baedeker's, while it contains nothing that is not of distinct value to the man of scholarly tastes or interested in art. There is nothing superfluous, or which such a traveller will not be glad to have. The work, while containing all the information that any traveller will require, is not intended for the hurried tourist who is content with a rapid and cursory inspection of what is to be seen. The author has, as he says, made it his aim to provide such an outline of the historical development of Rome in all its aspects as will make a visit to it profitable and satisfying, and it is no exaggeration to say that he has been largely successful in his undertaking, and that his work has all the excellence which made so good a judge as Mr. Freeman give the highest praise to the same author's guide to Sicily. We have tested it in many places and found it always explicit and luminous. Middleton's *Remains of Ancient Rome* will still be indispensable. to the classical scholar and the student of architecture, but as a guide to Rome in all departments this is the one which will be found most helpful by every visitor who has time and inclination to make a thorough study of the city which beyond all others invites prolonged examination and abundantly rewards it. The maps and plans are numerous and good. The illustrations are most useful, but in some instances might now with advantage be replaced by photographs like those in Burn's *Ancient Rome*, which was reviewed in this *Magazine* for April 1895. The bulk of the book has been unnecessarily increased by the inclusion of sixty pages of advertisements, but as it is bound so as to be easily broken up into five separate sections, each complete in itself and with its own list of contents, this is a matter of less moment.

Brittany for Britons, with the newest Practical Information about the Towns frequented by the English on the Gulf of St. Malo. By DOUGLAS SLADEN. London : Adam and Charles Black, 1896. Pp. 173.

The Brittany coast was long a favourite place of residence for English families in reduced circumstances, who went there nominally to give the children a little French ! It has now become a fashionable resort for well-to-do Britons, who, with that energy which distinguishes them, have established golf links at La Guimorais (five miles from St. Malo), which "are the most sporting that the most ardent lover of golf novelties could desire," and at St. Briac, a very picturesque place near Dinard.

Mr. Sladen has written an admirable handbook for Britons visiting Brittany, the only jarring note in it being the chapter beginning "Mont St. Michel is not exactly a fraud." No doubt Westminster Abbey, when "overcrowded with Britons and worse," is spoilt, but the term "fraud" would under no circumstances be used in describing it, and Mont St. Michel is quite as impressive. Mont St. Michel on Michaelmas Day, when pilgrims reverently resort to it from all parts of France, is an extraordinary spectacle, and should not be missed by the visitor to Brittany, even although he only goes there to play golf.

Black's Guide to the English Lakes. Edited by A. R. HOPE MONCRIEFF. Twenty-second Edition. London : A. and C. Black, 1896. Pp. x+224.

This is a new and improved edition of a deservedly well-known guide-book— one of the best of its kind in existence, as the fact of its having reached its twenty-second edition clearly shows. It is provided with excellent maps, those relating to the various sections being very clearly printed in colours to show elevations. The introduction to the book contains some valuable hints to tourists in general. If there is a weak point in the book it is to be found in the glossary. Only those not intimately acquainted with the district would repeat the common mistake of translating "fell" as elevated land, instead of land *uninclosed*, whether elevated or not. It is nearly synonymous with the word "forest" in its original sense of land *outside* of the limits of cultivation. The author has unfortunately given his countenance to the spelling "ghyll" for a ravine. The original Scandinavian word "gill" for ravine, whether containing a stream or not, was altered into the modern abomination in the early part of the present century by Miss Martineau. A "borran" is a heap of stones cleared from the rough land, and a "thwaite" is a spot cleared of trees. Perhaps the editor may note these little matters for correction in future editions.

NEW MAPS.

ATLASES.

THE CHURCH MISSIONARY ATLAS, containing an account of the various countries
in which the Church Missionary Society labours and of its Missionary Operations. New edition (the eighth).

Church Missionary Society, London, 1896.

This volume is rather a book than an atlas. It contains 239 pages of letterpress
in which are general descriptions of the various countries and many useful statistics
concerning them. The history of each country is also sketched, especially in
relation to the work of the Society, and naturally the languages are enumerated.
The work therefore contains much useful information, while it is specially prepared
for the use of those interested in missionary work. Scattered through it are
thirty-two maps, whereon the stations of the Society are marked.

FORMÆ URBIS ROMÆ ANTIQUÆ. Delineaverunt H. Kiepert et Ch. Hvelsen.
Accedit Nomenclator Typographicus a Ch. Hvelsen compositus. Berolini
apud D. Reimer (E. Vohsen) MDCCCLXXXXVI. Pp. xii + 110. *Price 12 M.*

This work consists of three maps representing (1) Rome in the time of the
Republic ; (2) Rome from the age of Augustus ; both of these are on the scale of
1 : 10,000 ; and (3) the central part of the city as it was in the times of the Emperors,
on the much larger scale of 1 : 2500. On all three maps the principal streets of
the modern city are indicated in faint colour, which does not in the least interfere
with the clearness of the maps, while greatly helping the identification of the exact
position of ancient buildings and localities. It is well known that a magnificent
map of Ancient Rome on the scale of 1 : 1000 (of which some parts have already
been issued) is in progress under the superintendence of Signor Lanciani, but this
will take some time to complete, and is too large and too costly for many to acquire,
while its plan does not provide for the representing of many features of Republican
Rome which are indispensable to the student of pre - Augustan times. The
maps under review supply a distinct want, and will be invaluable to classical
scholars and students of Roman history, showing as they do for the first time on so
large a scale the results attained through the excavations and researches of the last
thirty years, which have thrown great light on many questions of Roman topography.
Even now not a few identifications are only probable, but those here preferred and
indicated represent on the whole the consensus of the best recent authorities. The
maps are beautiful specimens of the engraver's art — a pleasure to the eye, and
admirably clear. Great attention has evidently been given to correctness of
typography, but we notice one or two errors that seem strangely to have escaped
detection, and must prove rather vexatious to the authors. In map 2, Regio iv.
is by mistake given when it should be Regio vi., and Statium is a mistake for
Stadium. Accompanying the maps is a list, filling 102 pages, of all places, buildings, monuments, etc., of any note which are incidentally mentioned by ancient
authors or in inscriptions, with indication of their position on the maps (if shown
there), and with a reference to the document where they are mentioned, and to the
works of recent commentators who have discussed their site. The preparation of
this list must have cost the compiler much labour, and it cannot fail to prove of
great service to scholars.

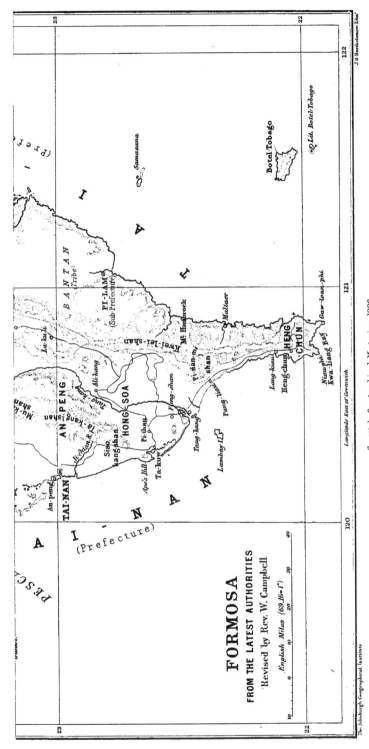

FORMOSA

FROM THE LATEST AUTHORITIES

Revised by Rev. W. Campbell

English Miles (69.16=1°)

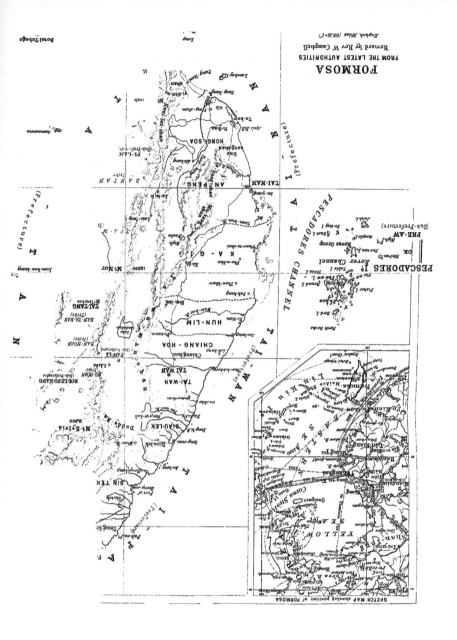

FORMOSA
FROM THE LATEST AUTHORITIES
Revised by Rev W Campbell

English Miles (69.16=1°)

SKETCH MAP showing position of FORMOSA

THE SCOTTISH

GEOGRAPHICAL

MAGAZINE.

NOTES ON ASHANTI.

By Major C. Barter,

Second-in-Command, Special Service Corps, Ashanti Expedition, 1896.

(Read before the Society in Edinburgh on May 7th.)

WHEN the Royal Scottish Geographical Society did me the honour to ask me to deliver a lecture on my experiences in Ashanti during the recent expedition, I warned their esteemed Secretary that the most I could offer, outside the military features of the campaign, would be a record of general impressions and of the local accounts and traditions which my memory had retained.

If with these disconnected notes of a personal observer I should have the good fortune to interest you at all, I must in fairness attribute a large share of the merit to a charming companion of our march, Canon Taylor-Smith, of Sierra Leone, who was attached to the Special Service Corps as chaplain, and who followed all its fortunes. Canon Taylor-Smith, in the course of his good work throughout several years in these regions, had acquired a vast fund of information on all points connected with them. When during the long hours of early march, or in camp under our hastily raised shelter of palm or plantain leaves, I listened to his interesting descriptions of the country and its people, I little thought that I should shortly be serving as his unworthy mouthpiece before a distinguished audience in Edinburgh. Having made this just acknowledgment, I will now pass to the subject-matter of my discourse; but as there appears to have existed a good deal of misapprehension in the mind of the public generally regarding the real causes that led to the despatch of the recent expedition, allow me, very briefly, to retrace the historical events which have now culminated in the complete subjugation to British power of the once mighty and dreaded Ashanti kingdom.

Since the commencement of the present century, the tribes on the coast line, of which the principal is the Fanti race, have been subjected, sometimes with only brief intervals of quiet, to the savage and desolating incursions of their warlike northern neighbours, the Ashantis.

As the Fantis and other tribes of the coast, in whose territories we had formed settlements, were our allies, we did what we could, sometimes by force of arms with inadequate means, and sometimes by remonstrances or by diplomatic action, to assist the nations who were friendly to us. Our efforts were not always successful, as the well-known disaster to Sir Charles Macarthy's force, early in the century, proves, nor was the action of the British authorities in the colony always judicious or well-timed. Be that as it may, the Ashantis on more than one occasion carried fire and sword to the very walls of the British forts on the coasts.

Our Gold Coast Protectorate was finally established in 1852, and extended to the river Prah, and for two decades peace was restored. But in 1873 fresh trouble arose. The Ashantis came down with large forces as far as the coast. Trade with the interior was entirely stopped, owing to the cruel and rapacious behaviour of the Ashantis, and the friendly tribes of the coast were decimated and terrorised, large numbers being carried off to slavery or sacrifice. So menacing was the attitude of the Ashantis, that it became necessary to decide whether the British settlements on the Gold Coast should be entirely abandoned, or the invaders driven back by a large British force, to be despatched from England.

As a result of the selection of the latter course, Sir Garnet Wolseley, in the beginning of 1874, after some severe fighting, occupied and destroyed Kumassi, which had been evacuated by the Ashantis. In the end peace was concluded by the Treaty of Fomanah, the principal conditions imposed on King Kofi of Ashanti being—

An indemnity of 50,000 ounces of gold.

Freedom of trade in Ashanti.

The abolition of human sacrifice.

The maintenance of a good road from Kumassi to the Protectorate.

From the date of this treaty, taught, no doubt, by costly experience, the Ashantis almost totally abstained from molesting our Fanti allies; but the non-observance by the Ashantis of the other conditions of the compact, and especially their continued and increasing interference with the important carrying trade through Kumassi, led to frequent representations being made to them by the British authorities on the Gold Coast—representations which were treated by the Ashanti rulers with ever-increasing contumely as they saw that the official admonitions were never sustained by force. Finally, last year, a mission which went to Kumassi under Captain Donald Stewart, with an escort of Houssa police, was treated with scant courtesy, and the escort was attacked on the way back, one of the Houssa police being killed. As the King of Kumassi continued to offer nothing but the same vague and ineffective promises, without attempting to put an end to the abuses complained of, it was finally decided to send out an expedition, not necessarily with the object of waging war upon the Ashanti people, but for the purpose of enforcing suitable conditions for the permanent establishment of a just and humane government in the country, and for the protection of the legitimate trading operations of Gold Coast Colony—operations which were practically impossible so long as the Ashantis maintained their attitude of sullen hostility and menace.

It is unnecessary for me to recall events of very recent date, or to enter into a detailed description of the military features of the expedition. Suffice it to say that it was excellently organised and equipped in England, the health and comfort of the force being minutely considered. Large and commodious transports were provided for the conveyance of the troops from England to the Coast, and the hospital ship stationed off Cape Coast Castle contained every arrangement and every appliance which could be devised for the wellbeing of the sick and wounded. On the Gold Coast itself, before the arrival of the force, Governor, now Sir William, Maxwell had by untiring efforts succeeded in collecting from every available source, extending as far even as Sierra Leone, a veritable army of native carriers, numbering some 15,000; and to this fact, as well as to the able conduct of the expedition, must be attributed the rapid and highly successful issue of the campaign.

I regret that I can only give a very brief and general geographical description of Ashanti itself; but if time allows of my doing so, I propose to offer, later in the lecture, a few notes which I have collected on the immediate Hinterland of Ashanti.

Ashanti, which, as you will see on the map, is situated a few degrees only to the north of the Equator, is essentially a forest kingdom, and lies entirely within the belt of primæval forest which extends for over a thousand miles along this part of the west coast of Africa. The forest, a part only of which is occupied by Ashanti, has a breadth of about 300 miles. The surface of the country, except in isolated hilly districts of small area, is undulating, the water-courses being generally broad and of a swampy character. Save where human hands have made a clearing, the forest, dense and tangled, everywhere covers the face of the land. The southern boundary of the Ashanti kingdom is the river Prah, which separates it from the sphere of Gold Coast Colony. To the north, the limit of Ashanti power has been the forest edge, some six marches north of Kumassi. Here begin rich prairie plains with a healthy climate and an abundance of big game; but these districts are under the dominion of the great chief Samory, who rose from the position of slave to be sultan, and now rules practically over the vast district which extends from the Coast Kingdoms to Timbuctoo. He it was who lately defeated the French expedition under Monteil, but who refused, it is said, probably with truth, to join the Ashantis against the British.

The kingdom of Ashanti is, or rather has, up till recently, been composed of a realm of seven provinces, each ruled over by a king or great chief. These provinces in the order of their importance are Kumassi, Kokofu, Juabin, Mampon, Beckwai, Insuta, and Adansi. The King of Kumassi has, as far as is known, always been the paramount chief; but the presence of the other six chiefs at his "stooling," or installation as King of Ashanti, has always been legally indispensable. In the case of Prempeh the formality had never been gone through, so that he was really only King of Kumassi.   He possessed, however, a controlling power undoubtedly greater than that of any of the other ruling kings.

It is impossible to offer a correct estimate of the population of the collective Ashanti kingdoms, but it lies probably between 150,000 and 200,000.

From a military point of view the outcome of the march to Kumassi was, of course, disappointing, as the Ashantis offered no resistance to the advance of the British force and the occupation of Kumassi ; but considering the question from a broader and more humane standpoint, it must be a matter of congratulation that the desired end was attained without the unnecessary loss of British lives, and without the inevitable slaughter of ignorant savages, a slaughter which would have been great in proportion to the bravery of our adversaries.

Nor can it well be denied that the ulterior objects of peaceful administration and development are best furthered by the firm display of force, not accompanied by the stern necessities of war, but by the tempering influences of a Christian-like civilisation.

I have seen it more than once stated that the expedition was unnecessarily strong in its composition, and that a heavy hammer was employed to crack a nut ; but I think it hardly doubtful that if a weak force had been sent, bloodshed would have resulted, for the Ashantis are a warlike and courageous race, and fought with great bravery against us in the last campaign, though the force then employed was considerably stronger than that which marched into Kumassi in January last.

In Ashanti, however, we found, in the result, that the nation's power had greatly suffered both in actual strength and in prestige by its defeats in the campaign of 1874.   There were plenty of fighting men, no doubt, and there existed in favour of war a very strong party, which included the fetich priests and some of the greater chiefs ; but many of the older people still lived to remember the disastrous past and offer wiser counsels of peace.   Two or three Ashantis had, further, visited England, and the tales of their travels must have convinced the king and his advisers that resistance would in the end be hopeless.   Besides this factor in favour of peace, there was, I am sure, another and very powerful one.   In olden days before 1874, the wars waged, invariably successfully, by the Ashantis against neighbouring nations had provided them with a never-failing supply from outside of slaves and victims for sacrifice.   After their defeats by the British force, fear of retribution kept them from openly raiding those tribes surrounding them who had come more or less under our protection.   But victims to fetich rites had still to be found, for the cruel custom of human sacrifice flourished as before, and there followed a constant drain from amongst the Ashantis themselves.   The chiefs were called upon to furnish contingents of victims.   They selected their enemies and the least influential natives.   One poor old woman, whom we questioned in a village not far from Kumassi, informed us through the interpreter that she was alone in the world.   She had had three sons ; they had all been taken for fetich sacrifice.   I believe the people were sick of the custom.   Around them they beheld tribes who by treaty with civilised nations were prohibited from offering human lives to fetich, yet the wrath of the evil spirit had not been visited on these. They flourished without the practice of decimating human oblations. I am convinced that a large proportion of the Ashanti people silently welcomed our approach, dreading only the vengeance of the native allies who accompanied us, and who for centuries had suffered from their pitiless aggression.

The combatant portion of the British force landed on the 28th December, and at once began its march inland. The first day's march, owing to the necessity of disembarking by daylight in surf boats, was in the afternoon, and very trying it was in the hot, damp atmosphere, though every effort was made to make the march easy for the men. For twenty miles inland the road lies through a sandy, undulating country, covered with scrub and low bush, affording no protection from the rays of a tropical sun. I think one of our youngest officers here aptly expressed his discomfort when he said that it was hot enough to make a sponge-cake perspire. After the first day, however, a start was always made at night or early in the morning, and the troops arrived in camp before the sun had risen much above the horizon. Ten miles or so from the coast we came upon the first trees, standing like sentinels of the great forest, watching our approach. Thicker and thicker they became, until, twenty miles from the coast, we reached the vast primæval forest itself. Never shall I forget our first plunge into its depths. We started at midnight, as the bells in England were ringing in the New Year, and more than one mind must have instituted a striking comparison between the merry sounds and the brightness at home, and the deadly gloom of the dark forest in front of us. Entering it suddenly at night from the open moonlit space which held the village where we had camped, the arched opening above the path seemed to form the entrance of a vast dungeon, and as I looked up I half expected to read overhead the inscription imagined by Dante over the gates of the infernal regions—"All hope abandon, ye who enter here." And so it must have often seemed to the poor captive who was being led back to Kumassi to slavery or execution.

It is really hard to accurately describe the first sensations of a march in the night through this dark forest. A thick vapour, which not a breath moves, covers the forest to the apex of the highest trees which, when they can be seen, tower over 200 feet overhead, and so heavy is the dew that the leaves are dripping as after a heavy shower. There is no silence in the forest. A thousand noisés strange to the white man's ear assail it on every side from myriads of members of unseen animate life, and above all dominates the ceaseless hissing trill of the crickets, whilst at intervals up in the trees is heard the complaint of the sloth, which, beginning with a pitiful sound of pain, traverses the whole gamut until its wail forms the most weird and discordant screech that it is possible to conceive. All are subject to the depression caused by the heavy, ill-smelling atmosphere, charged with damp and the odour of decaying vegetable matter, and the feeling is accentuated when the night is moonless and the narrow and treacherous path dives down into one of the swampy valleys where the bamboo clumps grow thick, excluding all light from above. Suddenly morning breaks. The light comes with a rush, and all is changed. The vapour cloud rises from the saturated litter which forms the forest bed, and seems to lift with it the oppression which weighs down the human mind. The unseen birds far up in the topmost branches break forth in song, and Tommy Atkins, a veritable child of nature, echoes the birds' *réveillé*. There is music in the air, and the pace becomes brisker, as though it were a return from a military funeral.

At this hour the forest is grand and beautiful in the extreme. The leaves are covered with dew, and every atom of wet shines in the morning light. It is a vast conservatory, filled to profusion with the richest tropical plants. The undergrowth is extremely dense, the intertwining of vines and the stems of plants offering a very wall to any advance. In many places a passage through can only be attained by arduous cutting. The luxuriance of vegetable growth, under the existing conditions of soil and climate, cannot well be imagined. As an instance, I may state that I have seen places which had been cleared absolutely to the ground covered three weeks later by growths higher than a man's head. This very luxuriance is the cause of the long-drawn height of all trees and plants. Each is stretching upward to the light in its response to the natural law of the survival of the fittest. It is very human in its aspect, this struggle of the plants for light. The strongest tribes push their way upward, regardless of the way in which they prejudice their weaker brethren, till they reach the light of day, far up, and parasite creepers that have profited by the rise of the giants send down their offshoots to feed on the earth below.

Some of these vines are remarkable for their perfect symmetry of form, and give a very accurate representation of carefully twisted strands. Some of them hang down, straight as a plumb-line, a stretch of over one hundred feet, and look like the cable hawsers of ironclads at anchor.

Perhaps the most striking forest effect is that presented in the bamboo groves, down in the low ground bordering the stagnant pools and oozing streams. The bamboos grow to the thickness of a man's leg, in immense solid clumps set wide apart. Overhead, some fifty feet, they form beautiful Gothic arches, sometimes of remarkably regular vaulted appearance. At night, when the grove is lit up by fires, the architectural effect is heightened, and the picture afforded by the red glow, which makes the surrounding objects stand out against the inky darkness behind, is extremely beautiful.

Every two or three miles the path leads to a clearing in the forest, where grow the trees and plants on which the natives' existence depends—the plantain and banana, the papaw, some cocoa-nut trees, and patches of yams. In the centre of the clearing is the village, the houses of which are all made of red or yellow mud with roofs of closely knitted branches or vines. Every habitation is built on the same design, only the extent of the habitation varying with the importance of the owner. Each has an entrance at one of the angles of the building, leading to a square courtyard round which are four rooms, all without walls on the inside, and therefore open to the court. The floor of the rooms is raised two or three feet above the soil, and is reached by one or two steps. In the houses of the chiefs there are two or more courtyards opening at one of the angles into each other, and the houses are larger, but all have a regular entrance to the front, and another, a bolt-hole really, in rear.

In the courtyard are mud cones forming the fireplace, and either in the centre, or along one of the walls, is generally to be found a stake, on the top of which is placed a bowl containing an offering to fetich—palm oil, decayed eggs, or a decomposed chicken. These fetich emblems are held in great respect; and some of the officers' native servants, even amongst those brought up in the comparative civilisation of the coast, could not be persuaded to remove these unsavoury objects of superstitious veneration.

Each village has at least one tree which is held sacred, partly, no doubt originally, owing to the grateful shade which it affords. Under its branches lie roughly hewn trunks of trees on which the natives squat and hold council, which they appear to do all day long when they are not engaged in eating or sleeping.

Travelling on through the forest along a path which, as far as the Prah, is of fair width in most places, and along which generally two and sometimes four men can walk abreast, the river Prah is reached, about seventy-four miles from Cape Coast Castle, and just half-way to Kumassi. The Prah is the southern boundary of Ashanti, and is looked upon as a sacred river by the natives. It has always been crossed with great reluctance by the Fantis and other southern tribes, who all lived in mortal fear of the Ashantis beyond; and we witnessed the incantations of several groups of natives, who would not move on from the farther bank until they had fired guns and sung prayers to propitiate the spirit of the river. Having finished the ceremony, they dashed forward in the direction of Kumassi with every appearance of zeal and reckless courage, but their active gait soon showed sign of flagging, and it is more than doubtful whether the appearance of one armed Ashanti would not have sent the whole party flying back.

From the Prah to Kumassi the path becomes very narrow, and often rough, and the men were often obliged to march in Indian file. The features of the forest are here still the same, save when the Adansi hills are reached, forty miles from Kumassi. The path crosses the crest of these hills at a height of 1100 feet, and at last, after many days, we breathed fresh, pure air again, but not for long, as the path descends again abruptly on the northern slope. This group of hills, the only one of any importance in this part of the country, will, probably, as the development of the country progresses, become a sanatorium, or "hill station," as it is called in India, for the whole district, or it may be made into a military station for some Houssa companies. A little farther, at Essian Kwanta, we witnessed an interesting ceremony. The King of Beckwai met the commander of the force, Sir Francis Scott, in solemn palaver to conclude a treaty, which placed his kingdom under British protection. A great show this potentate made as he passed through our camp, carried on a brass chair, under an immense umbrella of many colours, preceded by his band of tomtoms and ivory horns, and attended by minor chiefs; and when in the open place in the village he sat on his throne, surrounded by his principal chiefs in rich cloths, and his "doctors" and executioners in weird attire, the group formed was very interesting and picturesque. On his right sat his "chief linguist," or principal adviser, a cunning and sly-looking old savage, with a dignity of manner and calmness of speech which many a politician at home would be glad to possess. The negotiations did not last long. The King had crossed the Rubicon, and now had to stand or fall with us. But the terms were fair and just and readily agreed to—only on one point did the king and his minister make an obstinate stand. It was on the question of human sacrifice. By every subterfuge did they endeavour to obtain the insertion of a clause which might have served as an excuse for human sacrifice to fetich. But the strict rule which laid down that capital punishment could be inflicted only by sentence of a proper British tribunal was immutable, even for the crime of "lèse-majesté" against the persons of the King or his wives.

The submission of the King of Beckwai, one of the principal chiefs of Ashanti, left little doubt that the expedition would meet with no resistance, and we pushed on rapidly to Kumassi, which we entered on the 17th January, after crossing the sacred river Adra. As we neared the capital we found our path strewn with emblems of fetich, rough human figures carved in wood, goat-skins, dead chickens, and other symbols, probably, which we did not notice. These signs all had a meaning, though we did not understand them. Some of the interpreters declared that they were peace-offerings, and so they must have been, judging by the result. The Ashantis had decided not to fight, but King Prempeh had determined to offer us a brilliant reception; and as we marched up the great market-place in the centre of the town, we found him sitting in great state, surrounded by a great crowd of chiefs and courtiers in brilliant array, but all unarmed. Along the sides of the square were the lesser kings and chiefs, all seated under coloured umbrellas, on chairs of size. and quality suited to their respective ranks. Thus a small chief had a very small chair with a minimum of brass ornament and a small black umbrella, generally in very poor condition and with some of its ribs sticking out, while the kings were covered by huge umbrellas of velvet and silk, and sat on larger chairs covered with brass plates and nails. From every group rose an incessant din from savage musical instruments, the tomtoms especially sustaining the character of the orchestra with emphasis. With regard to these drums we were informed that there exists a perfect system of signals which allows of orders being rapidly transmitted in the forest, so that in war messages can be passed on with great ease and rapidity. We were told that every one of the apparently aimless beats which we heard formed part of a code which the natives could perfectly understand.

Of Kumassi itself there is not very much to tell, for its ancient glory was much shorn by the events of the war in 1874. It stands for the greater part on a plateau 400 to 500 feet above the level of the sea. It is a town in size, but it bears the character of a village throughout.

It stands in a large open space about one mile in diameter, and is surrounded by the dark forest. The so-called palace is a very poor affair, and consists of a succession of courtyards surrounded by high mud walls, which contain the rooms I have already described for the ordinary houses. The King's wives were lodged on the other side of a lane which the palace forms with a long narrow building where the ladies lived. Each wife had a room upstairs, each room being separate, and reached only by a narrow staircase from below. To each room was a small window opening on the lane. The ladies were, I was informed, very strictly guarded, and I may here state that when the King's wives took their walks abroad they were preceded by a court crier, who gave notice of their approach. As they passed, the people knelt with averted faces in the dust, for to look on them was death for a man.

By far the most commanding and interesting feature of Kumassi, a name which in Ashanti means the place of sacrifice, is the sacred grove, the Golgotha of the place, where the bodies of the victims to fetich were thrown to feed the vultures. This grove, especially when seen from some little distance, offers a magnificent specimen of Nature's gardening. It stands on the highest ground in the town, and is composed of a great variety of beautiful trees. The cotton-trees, with a girth at base of 40 to 50 feet, and as much as 250 feet high, tower above the others. Some graceful and tall palm-trees with plumed summits give lightness to the group; while below, close to the ground, the dark depths beneath the boughs suggest that Nature is trying to cover the crimes of humanity. Entering the grove, the soil is at once seen to be littered with the remains of thousands of human bodies. The character of the soil in places is changed by the dust of the crumbling bones, and skulls of every age lie about under the trees in hundreds. Busy in this gruesome spot might be seen our native allies, extracting the front teeth from the skulls to wear as charms, which are supposed to bring much good luck.

Before we left Kumassi the engineers had ruined the beautiful aspect of the grove to a large extent by blowing down the larger trees with gun-cotton, but this proceeding was necessary in view of its effect on the superstitious mind of the natives.

To the north of Kumassi, and about a mile from it, were situated the temples of Bantama, the home of Ashanti fetich, and the resting-place of the kings. The word temples must be taken here to describe the object of the buildings, and not their outward character, for they consisted merely of ordinary courtyards and houses, in some of which were stored the remains of the Ashanti rulers. I may mention that when a king dies he is first buried, or rather stored, in a special house in Kumassi. At the end of one year his bones are cleaned, bound together with gold, and then taken to Bantama. Great chiefs when they die are thrown into a sacred stream which runs not far from the town. It had been rumoured that the temples held great treasures. If so, the priests must have removed them before our arrival, for not a thing of value was found. The priests fled on the approach of the troops, and the temples were burned and their walls thrown down.

An interesting feature on the road in front of the temples was the sacred fetich tree of Kumassi, a fine banyan, affording tempting shade to the unwary traveller, who invariably paid the penalty of sitting under its branches with death. Still lying under the tree was found the large copper basin which received the blood of the victims who were decapitated at the frequent sacrifices which attended the visits of the king to the tombs of his forefathers. Beneath the mound on which the bowl rested the soil was black with the blood of years.

Each quarter, it is said, the King visited Bantama, and from twenty to thirty victims were put to death. As a rule, torture was not resorted to ; but when the necessary licence was given to the executioners, they vied with each other in refinements of savage cruelty.

Apart from the victims procured by raids or by contributions from dependent chiefs, others were furnished through the commission of specified crimes and misdemeanours. It was said, for instance, that any native finding a nugget of gold, and not at once forwarding it to the King, was condemned to death. In the same manner, every man who found any article of value on the market square, and did not deliver it up at the King's palace, forfeited his life. There appear, however, to have been certain saving clauses. A man arrested could save his life by rapidly reciting a formula containing the King's name, or if by running he could reach a certain spot before being caught, he went free. But the offender or victim really had little chance of escape. He was generally pounced upon, his hands instantly bound behind his back, and a knife stuck from cheek to cheek through his mouth to prevent him from uttering the redeeming formula.

At sundry festivals, notably that of the Yam, human sacrifices were indispensable, and at the death or coronation of a king a thousand victims were often offered in sacrifice. The death of even a minor chief always involved the execution of a number of slaves.

The stay of the British portion of the force at Kumassi was short— only three days—but events marched rapidly, and with the great palaver 'and the submission of the King and the Queen-Mother the object of the expedition had been attained. The great palaver, an account of which appeared in all the papers, is of very recent date, and I shall not therefore describe it, but a few individual descriptions of the chief native personages may be of interest. King Prempeh is a young man of about twenty-eight years of age, of not very intelligent aspect, but with a not disagreeable face. He does not seem to have been of a cruel disposition, but merely dropped in with the hereditary customs of his race as a matter of course. Like all other natives, he was very superstitious, and there is no doubt he was very fond of gin. The priests are said to have exercised great influence over him, and to have counselled war on the occasion of our march up country, a course from which he was only restrained, as I have already stated, by some of his principal chiefs and by the evident disinclination of the people to leave their homes to fight a stronger enemy.

Prempeh is of medium height, naturally well built, but effete from the sedentary nature of his office. He in this respect offers a strong contrast to his brother, who is a tall, wiry, savage-looking fellow, with dignified bearing and defiant aspect.

On state occasions the king used to wear a crown which in shape was something like Punch's hat. It was covered with ornaments worked out of gold plate, and behind was a tail with a gold knob, which the people struggled round to touch as he passed through the crowd carried on a chair; but whether they strove to touch the knob through loyalty or through superstition, I cannot say.

In the first preliminary palaver which King Prempeh had with Sir Francis Scott, before the arrival of the Governor, he appeared with a nut in his mouth, so strapped down that he could not speak. This appears to have been in accordance with the native custom, and the rule seems to have a logical reason, inasmuch as it prevents the highest authority from uttering decisive conclusions without due consideration. In consequence, the Chief Linguist, or principal adviser of the King, does the talking, the King approving or otherwise with a vague nod. In native negotiations the King delivers no decisive answer on the spot, the official result of his deliberations being rendered later, after due deliberation with his responsible advisers.

The right-hand man of the King, the Chief Linguist, as he was called, was a chief by name of Kokofuku, and a very fine fellow he was, six feet three inches in height, all muscle and bone, and with a fine deep voice. He had visited England, and was one of those who had recognised the uselessness of resistance. One of our officers having asked him in a quiet moment why the Ashantis had not fought, he candidly replied that our arms were better than theirs, or they would not have allowed us to enter Kumassi. I think Kokofuku must be a sportsman at heart.

It is said that the real ruler of Ashanti was the Queen-Mother. Judging from the attitude she assumed at the palavers, and from the evident understanding which existed between herself and the Chief Linguist, whom I have just mentioned, it is probable that the assumption is a correct one. A very disagreeable-looking object was this old woman, with a thin bony body and a cruel face, which lips unusually thin for a native and one long prominent tooth made very repulsive. And indeed her reputation bore out her looks, for rumour speaks of her with an evil tongue. She is credited on good authority with having possessed several dozen husbands, all of whom she had executed, with the sole exception of Prempeh's father. We were informed from a local source that when displeased she would wipe out her whole suite, and start with a completely fresh retinue; and it was said that changes were frequent, and that the formality of a month's warning was generally dispensed with. In her official capacity as Queen-Mother she was summoned to attend the great palaver before the Governor; but as she, like the King, failed to appear at the appointed hour, both had to be brought up under escort.

I think one of the most comical sights imaginable was that of the Queen-Mother entering the parade ground, shorn of all her state except one umbrella-bearer. As it happened, the escort which flanked her on each side with fixed bayonets consisted of two Yorkshire lads, men of my own detachment. They advanced in strict military attitude, but with a smile on their faces which spoke for all they thought. They had never been so close to a queen before!

Pity would be wasted on this old hag, the Messalina of Ashanti, who in addition to her other enormities, is credited with having murdered the last king in order that her son Prempeh might come to the throne. She is said, with the aid of Kokofuku, the Chief Linguist whom I have just described, to have forced down the king's throat a piece of chew-stick which he had in his mouth, and to have thus killed him by suffocation. Surely she may be properly called in more senses than one the *bête noir* of Ashanti.

With the King of Kokofu sitting on a fence and watching the development of events from a distance, the most powerful of the ruling kings, or great chiefs, was the King of Mampon, whose capital is situated some fifty miles north of Kumassi. This chief was said to command 10,000 warriors; but in the tour which the governor subsequently made in the direction of Mampon, this force was not apparent. It is true that their king was at the time accompanying Prempeh, the Queen-Mother, the King's brother, and the Chief Linguist to a castle seven miles from Cape Coast at Elmina, in the dungeons of which in former days the Ashanti kings were wont to imprison their notable captives.

It is doubtful if the King would now be glad to return to his throne from exile, for death is the penalty which an Ashanti king pays for humiliation by the enemy. Shortly after our last war King Kofi was murdered as a consequence of his defeat.

The succession to the throne in Ashanti is regulated in a manner somewhat strange to our ideas. It is not the son of the king that is his heir. The throne is always inherited through a female, passing either to the king's brother or to a nephew (a sister's child), so that there shall be no doubt that the reigning king has royal blood in his veins.

Turning to the principal characteristics of the tribes through whose territories we passed, we found that not much difference was to be noticed between the appearance of the Fantis who live near the coast and the Ashantis inland. Both are physically fine races. The latter possess a less purely negro type than the inhabitants of the coast-line. Their features are somewhat more aquiline, and to our ideas they are handsomer. Both races are intelligent, but in respect of courage the Ashantis are much superior to all their neighbours. The Fantis appear to be a morally run-down race. It is to be feared that in contact with European civilisation their animal spirit has been tamed, whilst they have at the same time failed to gain any compensating moral advantage from the connection. The Ashantis, on the other hand, have always shown themselves to be an untamed and warrior race.

To give the Fanti his due, however, it must be conceded that he did his work as a carrier both honestly and actively, a circumstance for which we were fully grateful. The women make as good carriers as the men, and I fear that in private life the ungallant male part of the population takes only too full an advantage of this quality in the women.

The customs relating to marriage in Ashanti do not, I fear, coincide with European ideas. Polygamy is, of course, universal, but I think that on the whole the people are more moral than might be supposed. The laws of the country are very strict as regards lapses from conjugal duty, but the main idea pervading them is evidently one rather of property than of propriety. Thus, there are heavy penalties for the man who tempts a wife. Indeed, it appears only too certain that some of the chiefs make handsome incomes from the penalties extracted for misdemeanours of this nature. The marriage ceremony of the country is of the usual simple and commercial nature amongst savages. The husband pays about £4 to the parents, and the wife is supposed to be able to regain her liberty at any time by repayment of the purchase price. The husband on his side can make over his wife to another man; and so great is the value of women for work, that the husband has seldom any difficulty in getting a wife off his hands, provided she be young and strong.

This simple process of divorce does not involve the same difficulty in respect to children as in civilised countries, for children are bought and sold like any other article, and are early put to labour. Any soil that is cleared will produce at least four crops a year without further trouble than the first planting, so it will be understood how great the value of labour, and therefore of women and children, must be. The master of one or two wives and two or three slaves can be idle from one year's end to the other.

The religion is, of course, universally fetich, though there exists a small proportion of people of Moorish descent, whose religion is a mixture of Mohammedanism and fetichism. The chief business of these people is to make and sell charms which are highly prized. A good many of them, bearing Arabic inscriptions, were found in the possession of the pure Ashantis.

The fetich priests form practically a separate caste. They become priests by hereditary descent, and are of both sexes. They marry, and the children are brought up as fetich-men.

The sacrifices made to fetich, whether they be animal or human, have solely for object the propitiation of the Evil Spirit. I was told that there is no belief in a Good Spirit or in a future life. The missionaries have a wide field of good work here open to them, and soon no doubt Christian missions of various denominations will establish themselves in the country. Before the 1874 expedition a few noble men had braved the dangers of climate and savage men to preach the Gospel, and had obtained a footing, but the outbreak of hostilities compelled them to leave before their work had prospered far.

The climate of Ashanti bears a very bad character, which, I think, is well deserved, though probably it is healthier on the whole than the climate of the districts bordering on the coast. Still, there is the forest everywhere, with the unceasing decay of vegetable matter and offensive atmosphere, and there are the malarious swamps in the low grounds. The temperature throughout, even in the hot season, is not excessive— generally 75° to 90° in the daytime. This fact is principally due to the Harmattan, a steady cool breeze which blows from a northerly direction during the winter months, and which is considered so healthy that it is often known as the Doctor. The nights are fairly cool as a rule, but the air is never free from the steaming dampness which weakens the

European system, and thereby predisposes it to the effects of fever, a malady which in one form or another is bound sooner or later to attack even the strongest European.

The evil effects of the climate in the case of a short stay in the country do not as a rule show themselves until the subject has left the country, and this fact was proved by the small death-roll during the actual stay of the expedition in the country. I need do no more than recall to your mind, in this connection, the sad loss which we sustained in the person of a Prince whom every one admired, and whom every one who met him liked for the genial and simple comradeship which marked his relations with all. Another very sad event, though appealing to the sympathy of a narrower circle, was the death, also through fever, of Major Ferguson of the Life Guards, a distinguished young officer, best known in connection with a mission some years ago to Lobengula, the King of the Matabele.

The water everywhere in the forest is dangerous to drink. Generally it is thick, sometimes stagnant, and as a rule of unpleasant odour. Even in the rare instances when we met with clear water, and this was chiefly in the hilly Moinsi district, it was impossible to be certain that the water was free from evil germs, and the wise precaution of boiling the water was invariably practised. The bad water was filtered, as best it might be, through native calabashes and wood, or precipitated with alum, and the men had ordinary small carbon filters which helped to clear the water; but I should recommend every traveller in these parts to take with him a couple of the best pocket filters. Some of us had the Pasteur filter, which can be procured in all sizes and which has been proved by analysis to almost totally exclude all minute organisms. I used one throughout, and perhaps owe my fortunate immunity from sickness in a large measure to drinking no water which had not passed through this pocket filter.

What the effect of this climate is on vermin insect life I do not know, but we were agreeably surprised to notice an almost total absence of mosquitoes, and the more personal parasites were nowhere visible. Whether the climate is too much even for them I cannot say, but it certainly was strange to notice the conspicuous absence of mosquitoes near the swamps in the hot damp weather presumably so suited to their needs. We saw few snakes, though there are plenty no doubt, and huge forest spiders, centipedes, and scorpions and barracouta rats, though they were seen now and then, kept in their proper places, and appeared, like the Ashantis, to be determined not to molest us.

The ants are a great plague, however, and penetrate everywhere, and at all times. It is no wonder that the numerous and enormous lizards which run about the walls and courtyards of the houses in the most tame and confident manner should remain unmolested by the natives, for they are the ants' worst enemies, and may be seen darting along the ground all day snapping up victims. They certainly keep the houses almost entirely clear of these small but persistent pests. Of bigger animals we saw nothing, as of course 'the unusual human activity along the paths drove them into the far recesses of the forest. The natives trap the leopards and kill the monkeys for their skins, and there are said to exist huge pythons, but they too kept at a discreet distance.

I would now, in conclusion, endeavour to briefly expose the results gained by the success of the recent expedition.

There is first the sentimental but important question of the abolition of human sacrifice and cruelty in this dark corner of the dark continent. The principal advantage gained has, however, been the opening of the trade route to the interior of Africa, and a reference to the map will show that the road through Ashanti is the most direct route from the Gold Coast to Timbuctoo. When a good road shall have been laid through Kumassi and Mampon to the fertile plains which lie to the north of Ashanti, a new and vast field will be opened to the enterprise of British commerce, and Gold Coast Colony, which paid the piper for the recent campaign, should soon earn a rich return for its expenditure. It may be expected that a flourishing trade will be opened with these regions, and that in exchange for our goods we shall draw to the coast, amongst other objects of trade, indiarubber, ivory, skins, kola-nuts and palm-oil, and gold.

Ashanti itself, once such a remunerative field for British commerce, will now be opened up again to our trade.

As regards the question of gold, it is impossible so far to say what resources of the precious metal the country may possess; but that alluvial gold is to be found in many places is certain, and many Europeans have come across the deserted holes three feet or so deep which the natives have worked with success to this—to them—convenient depth. In the hills, as well as on some of the higher ground, promising-looking quartz, including the white variety, was reported by qualified persons. However that may be, the question of the dreadful climate must always seriously prejudice the sustained working of mines by Europeans. Still, where the god Mammon offers rewards, votaries will always be found to worship him, at whatever risk, and possibly extensive clearings and drainage, and precautions as regards water and other sources of disease, may make certain districts more habitable. There are no serious engineering difficulties to be surmounted from the coast to Kumassi in view of the construction of roads or light railways; but it must not be forgotten that, owing to the extent and character of the great forest and the phenomenal growth of all vegetable life, it can never be cleared to any appreciable degree. The luxuriant tropical growth must ever remain as an enemy to health and an obstacle to any extensive agricultural development.

In view of the importance to British trade of the countries forming the Hinterland of Ashanti, I beg to present to you a few notes on them drawn from information procured from the most recent reports by English travellers in this region.

The principal territories directly inland of the colony are Gonja, Dagomba, Gruma, Gaman, Gurnshi, and Mosi. With these Lobi, Bona, Male or Chakosi, Pampamba, Bimbla, Borgu, Busiansi, and Massina have trade and political intercourse. With the exception of Massina, they lie more or less in the area watered by the Volta and its tributaries, between the Gold Coast Colony on the south and the Middle Niger on the north.   A range of hills stretching NNE. and SSW., called Boosu, is the eastern boundary of the area of the Volta.

The resources of the countries just mentioned are said to consist of gold in the western districts of Gaman, Lobi, and in branches of the Kong Mountains; ivory, horses, cattle, and slaves, brought down from Mosi and the region of the Middle Niger; shea-butter fruit, the pro-ducing tree growing abundantly in the plains; gum arabic, produced in enormous quantities in the plains, but which at present does not form an article of export to any extent.   The plains afford pasture to cattle, sheep, and horses.   Rice, guinea-corn, millet, and other cereals, as well as ground-nut, sweet potato, and cassava, grow luxuriantly in the rich plains. The people cultivate indigo, tobacco, and cotton.

A peculiar disease is said to afflict the cattle in epidemic form, and kills large numbers of them.

As regards articles of trade—silver, glass armlets, knives, powder, flints, Madras handkerchiefs, rum, gin, beads, grey baft, swords, and fish-hooks are accepted in exchange by the people for cowries, ivory, gold, cattle, and slaves.   A slave is worth between £3 and £7 in Salagha, and the following are said to be the usual prices of some of the articles sold there, in shillings and decimals of a shilling :—

				s.		s.
Brass rods, .	.	.	.	0·8	to	1·2
Copper rods,	.	.	.	1·0	„	1·5
Gum,	.	.	.	25·0	„	30·0
Powder, 20 lbs. keg,	.	.	18·0	„	30·0
Red yarn, .	.	.	.	1·0 per bundle.		
Kola-nuts, .	.	.	.	10·0 to 50·0 per 1000.		

The price of gold at Bona is £2 an ounce, as against £3, 12s. 6d. on the coast.

Cowries used as currency are the local medium of exchange, 1000 being reckoned as worth a shilling.

Cotton goods for native trade must be thick in texture, good in quality, and low in price to replace the native manufacture, though the present ordinary trade kind is largely used for native garments.   Traders must be ready to deal with the tolls exacted by the native authorities, as well as with the restrictions of the middlemen, on a satisfactory basis.

Ordinary travellers can cover on an average fourteen miles a day, but caravans travel more slowly, eight to ten miles a day being the usual distance.

The cost of transport through the forest kingdom of Ashanti to the plains of the Hinterland is £9 to £10 per ton for every sixty miles. The proper load for a carrier is about 60 lbs.

The ground-rock is principally sandstone.   Eruptive rocks are abundant, but not many schistose rocks are to be seen.   There are indications in some parts, however, of clay-slate, quartzite, etc.   Salt is made at Daboya from the alluvium on the right river bank of the Volta, and rock-salt is reported from the Pampamba country.   Nitre has been observed at Bole and Massina.   West of the main Volta gold is said to be abundant; and Lobi, rich in that mineral, is noted for the largeness of its nuggets.

To the east of the main Volta clay ironstone has frequently been observed, but the deposit is reported as not persistent over a wide area. The ore is, however, smelted in parts of Dagomba, Pampamba, Buem Kotokori, Gruma, and Mosi. The smelting is carried on with great secrecy; charcoal is obtained by burning a particular kind of wood, and blast is afforded by means of two leather bags made of goat's skin with a common orifice. The iron produced (generally soft and white, though a superior variety is occasionally obtained) supplies the material for forging arrowheads, spears, knives, and other small iron-work.

The geological formation on the west and north is composed chiefly of crystalline rocks, while Palæozoic sandstones and shales occupy the south and east. The middle portion has a plateau 1200 feet high, trending roughly north-west and south-east.

From the inquiries of travellers in the more interior countries of Gando and Mosi, there appears to be an analogy and probable synchronism of the geological formation of the Gold Coast Colony with the more inland territories.

The physical aspect of these districts is reported as generally undulating plain, with tall, coarse, reedy grass in the countries south of 8° 30′, spreading southwards into and terminating in the forest region of the Colony. Northwards the grass is more or less turf-like in character, with trees scattered singly and in clumps, treeless tracts being occasionally met with. The vegetation of the river banks is, however, denser, and the whole region affords room and food for the elephant, hippopotamus, deer, buffalo, and other animals. Lions exist in considerable numbers.

The plains afford facilities for transport by animals (generally horses and donkeys, but rarely camels), of goods brought down by caravans from the Mosi and Houssa countries, which travel from November to February. The whole region is said to be healthy.

The rains in these countries are said to be very heavy in September and October, the rivers reaching their highest water-level about the beginning of November. The natives possess notions about rain-making and rain-stopping corresponding to those ascertained by Livingstone to exist among other tribes of the interior of Africa.

The King of Dagomba is the most influential potentate in this district. In the days of extended Ashanti power, he furnished an annual tribute of one thousand slaves to the Ashanti king, and he was called by the Ashantis the Rich King. Between Dagomba and Mosi there is a belt of inhospitable barbarous tribes, through which caravans often have to force their way. The Mosi are a warlike tribe, and are for the greater part vassals of the King of Wagadugu. The old long Dane guns carried into Mosi, and those countries which use large numbers of horses, are shortened by cutting off a portion of the barrel and stock to adapt them for cavalry use.

In conclusion, I regret that I am precluded from dealing, even in a general way, with the political aspect and future of the rich countries which form the Hinterland of Ashanti. I may, however, state that, in the complicated negotiations which are at present being conducted by the three great colonising Powers of Europe regarding the partition of this portion of Africa, the interests of Great Britain are being carefully watched.

THE IGNEOUS ROCKS OF THE CAUCASUS.

By V. DINGELSTEDT, Corr. Member.

BEFORE treating of my special subject, indicated above, I may say a few words about the geological study of the Caucasus. The Caucasian mountains have not as yet attracted the attention of geologists in a degree corresponding to their real importance in the history of the formation of the earth's crust. There is but little record to be found in English geological literature concerning this extensive mountainous country, and whilst in every English text-book of geology important references are made to the geology of almost every other part of the earth's surface, there is comparatively little said about the Caucasus. The reason of this is certainly not to be found in lack of interest in the subject, but rather in the inaccessibility of the Russian language, and perhaps also in defective research and the scarcity of the positive results as yet obtained in the country, to which I would now draw the attention of the reader. I have already had occasion, in a description of Svanetia, to speak of the unfavourable conditions of such a country as the Caucasus for scientific research, and I have only to add now that geological studies, on account of their special character, requiring as they do great physical exertion, much courage, and varied knowledge, can only be pursued here under enormous difficulties. I do not imply, however, that this kind of study does not make some advance. Thanks to the efforts of the Government, a Board of Mining Engineers has been established for the purpose of prosecuting geological investigations independently of their proper task of controlling and supervising the exploitation of metallic ores, and an eminent geologist and palæontologist, M. Abich, has been engaged, also at the cost of the Government, who has worked indefatigably and with considerable success for many years. The Russian and German geological literature (M. Abich having written in German) is therefore not inconsiderable, and it is really to be regretted that, with the exception of a geological map of the Caucasus, recently published in *The Exploration of the Caucasus*, by D. W. Freshfield, and accompanied by a sketch of the geology of the country by Professor Bonney, there is little information on the subject printed in English.

I do not propose in the present paper to pass in review the whole geology of the country as at present known, discussing the ages of the strata forming the vast mountainous country between the Black Sea and the Caspian Sea, their stratigraphical or tectonical disposition, or their lithological character, but only to give an account of what is actually known concerning the igneous rocks in the Caucasus. Having supplied material for the sedimentary or clastic rocks of all ages, they still form the nucleus of a considerable part of the main chain of the Caucasus, appearing on its loftiest ridges; they lie at the base of Meskhi range, which forms a connecting link between the main chain and the Armenian highlands, of the Arsiani mountains in the Minor Caucasus, on the left bank of the Chorokh, and of a considerable part of south-western Transcaucasia. There is as yet considerable difficulty in the proper classification of the igneous rocks met with in the Caucasus, and, indeed, a rational classification will not be possible before a thorough microscopic examination of all the existing types has been completed. Mr. Abich distinguishes three principal kinds of igneous rocks, viz., granitic, porphyritic, and volcanic. We should prefer the division adopted by many English writers, viz., holocrystalline, hemicrystalline, and glassy.

In the granitic group M. Abich includes (1) granite proper, (2) pegmatite, in which mica is very scarce and often absent, and which might be better called aplite (old binary granite), and (3) protogine. It was long held that in the last mica is replaced by talc; but, according to M. Michel Lévy, protogine is only a chloritic variety of the granulite of French authors, that is, of granite with white mica or muscovite. The chlorite mistaken for talc is, according to the French authority mentioned, the result of the decomposition of mica.

The porphyritic group of M. Abich embraces the igneous rocks distinguished generally under the names of granite, syenite, and eurite. In the same group he has included also a large and varied series of igneous rocks without quartz, imperceptibly passing, through numerous changes, into volcanic rocks. He thus puts together in one group some holocrystalline rocks, as syenite and eurite, and some hemicrystalline, as quartz-felsite, felsite, porphyrite, etc. Under the general name of volcanic rocks Abich understands rocks principally distinguished by their dark colour, and composed of plagioclase, pyroxene, amphibole, magnesian mica, chlorite, and serpentine. To this group of rocks belong, according to Abich, some porphyries (as oligoclase-porphyry, labradorite-porphyry), some diorites, diabase, schalstein, various amygdaloidal rocks, amphibolites, hypersthenites, and ophites.

Rocks are thus grouped together differing considerably in structure and texture. They are often in contact with metamorphic rocks, and indeed some of them, as for instance amphibolites, are probably of metamorphic origin. But there is as yet much obscurity about this question of metamorphism.

In the category of volcanic rocks proper, M. Abich includes trachytes, basalts, rhyolites, various lavas which are of glassy structure, as also some dolerites, which are holocrystalline. These volcanic rocks often assume a porphyritic structure, and are thus scarcely if at all distinguishable from the porphyritic group of Abich defined above.

We give Abich's classification of Caucasian rocks because it has served, and continues to serve, as a basis for the classification of rocks adopted by the Caucasian Board of Mining Engineers and all official inquirers into the geology of the Caucasus; but, as already observed, we prefer for our purpose to divide the rocks according to their texture and structure into holocrystalline, hemicrystalline, and glassy.

According to this classification we may distinguish in each of these three principal divisions three groups of rocks—Acid or Orthoclase rocks, Basic or Plagioclase rocks, and an Intermediate group.

To the holocrystalline igneous rocks appertain granite, with all its varieties, granitite, granitic-porphyry, protogine, etc., eurite, of similar composition to granite and including such rocks as quartz-felsite or microgranulite, felsite, felsite-porphyry, keratophyre (porphyry), etc., as well as syenite with its varieties, then diorite, gabbro, and dolerite.

To the hemicrystalline class of rocks we shall reckon trachytes, porphyrites, andesites, and basalts.

To the glassy rocks belong certain basalts, tachylyte, obsidian, rhyolite, and certain andesites, etc.

It is scarcely necessary to observe that a rock may be wholly or only partly glassy, and that some rocks, as for instance quartz-felsite, quartz-porphyry, and others, are included by some geologists among the holocrystalline and by others among the hemicrystalline or hypocrystalline groups.

We shall now examine what kinds of rock are met with in those parts of the Caucasus which have been investigated by M. Abich, the Russian official mining engineers, and Mr. Ernest Favre, a Swiss geologist, who visited the Caucasus, and whose work, under the title *Recherches Géologiques dans la partie centrale de la Chaine du Caucase*, written in 1875, I have also made use of. Granitic rocks constitute the nucleus of the highest part of the main chain, viz., between Elbrus and Kasbek, for a distance of about 130 to 170 miles, between 42° and 44° E. long., as well as the Meskhi or Kartlo-Imeritian range; they appear in different parts of Transcaucasia—in the upper valley of the river Khram, an affluent of the Kura, on both banks of the Kwirila and its affluent the Katskhur, in western Imeritia, along the course of the mountain rivers Dsirula and Macharula in eastern Imeritia, in the spurs of the mountains that enclose the Cheremela and Marilissi valleys (in Sharopan), on the Bzyb river in the Black Sea province of Sukhum, etc.

Granites, supposed to be of comparatively less remote origin, build up also the mountain ridge of Pamb in the south of Kartalinia, and the range of Daralagos, between the towns of Ordubat and Migri on the Russo-Persian frontier.

We will now enter into some particulars about these granites, which under considerable varieties occupy so vast a tract of country. The granites of the main chain are the most extensive, but the least known in consequence of their elevation, difficulty of access, and considerable distance from inhabited places. The difficulties of research in these high regions are of such a serious nature that it was only in our time, and thanks to the efforts of M. Abich, that it was discovered that granite existed there, the base-rock being usually supposed to be trachyte. The granitic rocks constitute a mass of unequal breadth stretching from north-west to south-east for a distance of about 240 English statute miles. Beginning in the high regions, where the Kuban has its sources, they form on the north of Svanetia, as we had already occasion to observe in a former paper, a high range where no pass is lower than

9840 feet, rising into summits such as Tsalmag, Uzhba, Gwalda, Tetnult, Adish, and Namkwam, and continue south-eastwards towards the valley of the Terek, constituting the summits of Edemis-mta, Gurzievtsek, Burjula, and Adai-khokh; past the valley of the Terek they disappear, plunging deep under the mighty formation of crystalline limestones of metamorphic origin. The breadth of this zone, where granite largely predominates, varies from about fifty-five to fourteen miles, and even down to two and three-quarters of a mile, being usually broadest where it is lowest, and diminishing in breadth from north-west to south-east. These rocks terminate abruptly on the southern side, and have here but few ramifications, whilst they have a far more gradual inclination on the north side, and, thrusting themselves for a considerable distance under the almost horizontal Jurassic beds, produce a series of lateral ramifications separated by profound valleys. The granite of the Caucasus presents a considerable number of varieties, as we can better judge from the more thorough examination that has been made of this rock in more easily accessible parts of Imeritia and Kartalinia. Coarse-grained, medium-grained, and finely-crystalline varieties abound, and graphic granite, different kinds of protogine, resembling that of Mont Blanc, and granite passing into syenite and gneiss, also occur. The usual constituents of the granite of the main chain are yellow or pink orthoclase, gray quartz, and micas of the ferro-magnesian group. There are, however, other kinds of micas—black and greenish—which lend a greenish tint to the rock, usually light pink. We should also observe that the term granite is not employed by Caucasian geologists in a very strict sense.

We will notice all these differences more particularly in examining the granites that have been found in Imeritia and other regions in Transcaucasia, and especially in the Meskhi or Kartlo-Imeritian range of mountains, which constitute the watershed between the basins of the Black and Caspian Seas, and play such an important *rôle* in determining the climates of the eastern and western parts of this country.

In western Imeritia granites have been observed in the lower part of the Dsussa and Buji rivers, on the Kwirila and Cherimela, Macharula and Ghedsrula, on the road from the town of Kwirila to the village Katskhi, in the defile of Delicauris-ghele, on the lower course of the Katskhur. There are gray, dark-gray, and red varieties. The gray granites consist of gray quartz, light-gray orthoclase, and small quantities of alumino-alkaline micas, together with a little chlorite, which lends to the rock a somewhat greenish hue. The dark-gray granites, along the course of the Katskhur and Cherul, right affluents of the Kwirila, are of a microgranular structure, and consist of gray orthoclase, greenish oligoclase, gray quartz, and dark-green mica (biotite), and also of some hornblende and almandine. They are called granite by Russian official geologists, though, considering their complex composition, they should rather be regarded as the variety known under the name of granitite.

Red granite extends for a considerable distance along the course of the Kwirila, Macharula, and Ghedsrula, and is of a coarse crystalline structure, the largest crystals being those of red orthoclase; it also contains gray quartz, white mica, and some chlorite and almandine.

There are also some microgranitic varieties. In some places the red granite turns into keratophyre with albite, having a porphyritic structure, such as is well known in Ireland. It is noteworthy that the mass of red granite, keratophyres, syenites, and quartz-diorites has in western Imeritia a distinctly pseudo-laminated structure, with a south-eastern strike and a south-western dip of 35°. It is but seldom that granites here form elevations with beds of sedimentary rocks tilted up on their flanks, as is usually the case; they lie at the base of the sedimentary rocks, which are disposed more or less evenly upon them, being either horizontal, or but slightly undulating.

In the same defile of the Macharula, at the junction of the Ghedsrula and Dsirula rivers, are found, besides the red granite, large masses of gray granite with white orthoclase and oligoclase, a large admixture of quartz and black mica, as well as a light-gray variety, containing a small proportion of magnesian mica. These granites are easily disintegrated and crumble into quartz or mica sands. The granite has partially a porphyritic structure owing to the development of quartz crystals. At the confluence of the Marilissi and Cheremela the gray granites, extending to Molita, are traversed by small veins of amphibole-porphyry, in which pink and greenish crystals of oligoclase and hornblende are particularly conspicuous. The veins are only from seven to twenty-one inches thick. In eastern Imeritia, between the villages Martotubani and Boslebi, the mountain consists of a granite rock composed of medium-sized crystals of yellow orthoclase, light-gray oligoclase, gray quartz, and ferro-magnesian mica; by Russian official geologists it is called granulite. In the neighbourhood of Boslebi this granulite passes into a red, coarse-grained granite, which in its turn passes, a little farther up along the course of Kwirila, into gneiss of variable structure and consisting of light-blue, almost transparent, quartz, black ferro-magnesian mica, and a smaller proportion of gray orthoclase. Under atmospheric influence the gneiss partially loses its felspar and thus graduates into mica-schist with an admixture of iron-pyrites and, in rare cases, garnet. Farther up along the course of Kwirila, from Boslebi up to the village Tiri, granite of light-yellow and light-gray orthoclase, gray quartz, and white and black micas (muscovite and biotite) exhibit a porphyritic appearance from the presence of large felspar crystals.

. In the Meskhi and Satseretlo mountain range, as also on the right bank of the Cheremela, the granites mostly consist of medium-sized whitish orthoclase, quartz, and biotite; they sometimes contain amphibole, and are of gray colour and somewhat porphyritic in structure. They are very easily disintegrated by atmospheric agencies. They are here and there traversed by dykes or veins, from seven to thirty-two feet thick, of darker kinds of granite, containing biotite and magnetic hæmatite, and closely resemble melaphyre in external appearance. Besides the dark varieties of granite, reddish and red, gray and light-gray granites occur here, as also fine-grained syenite, or rather hornblende-granite. This last rock, into which the reddish granite easily passes, contains much magnetite. The red granite appears usually in thin sheets from two to six inches; it contains biotite, and in its fissures have been found grains of greenish epidote. The light-gray granites of quartz, pink orthoclase, and biotite are often rendered porphyritic by large crystals of the last-mentioned mineral.

The granites in this part of the Caucasus are mostly overlaid by limestones, considered to belong to the Lower Cretaceous system.

On the banks of the Cheremela are found two varieties of granite— one light-yellowish gray and the other dark. Besides its ordinary constituents, which are rather variable in quantity, either quartz or orthoclase predominating, the yellowish-gray granites contain some greenish felspar and iron-pyrites. They crop out between the Molita bridge and a place above the mouth of Vakhan river, where the Cheremela cuts its bed through a massive dyke of andesite. The dark variety is fine-grained, and its colour is due to mica; it is particularly abundant above and below Marilissi.

The granites forming the kernel of the Meskhi range and its western prolongation, called Satseretlo, are of vast extent; they appear on the western slope of this range, where they have formed all the summits in the basins of the Cheremela and Dsirula, and crop out frequently on the eastern side, rising in the elevations of the drainage area of the Cheratkhevi and Abanos-tskhali. These here form an elongated, slightly convex undulation, with its longer axis from south-west to north-east, in the direction of the strike of the sedimentary (Cretaceous or old Tertiary) beds, extending along both slopes of the range. The granite thus lies under a great anticline, now partially disrupted.

We have already spoken of the granites on the western, or Imeritian, side of the Meskhi range, so now we will turn our attention to those on its eastern, or Kartalinian, side. Here also they are very variable in colour, and in the relative quantities of the minerals entering into their composition, and even their structure is such that in some cases the propriety of retaining them in the class of granites is very doubtful. The principal mass is gray and of medium-sized grain, and consists of quartz, white orthoclase, and dark ferro-magnesian mica, with here and there an addition of oligoclase felspar, amphibole, and magnetite, as well as iron-pyrites. Crossing this in different directions are veins of several varieties of granite. The veins, mostly vertical, protrude through the principal mass, and are reddish, pink, gray, dark-gray, and even almost black. The reddish variety is due to the presence of red orthoclase felspar. There has also been found in this variety greenish oligoclase felspar, dark ferro-magnesian mica, quartz, and hornblende, which last mineral greatly increases in quantity, in some places to the exclusion of mica. There is also, here and there, a considerable admixture of magnetic iron-ore. The dark or black variety of granite owes its shade to the abundance of biotite. Other differences in colour come from the casual abundance of quartz as, for instance, in the upper part of the Eastern Kartalinian range, between the Likhi and Mansunary villages. In those granites iron-pyrites is often found in abundance.

Red and gray granites are seen at the sources of Cherat-khevi and along the valley formed by this stream up to the Ali village. The red consists of fairly large crystals of red orthoclase, transparent greenish oligoclase, colourless quartz, dark-greenish mica, and sometimes hornblende. These granites, which become more coarsely granulated down the course of the stream, present well-marked pseudo-stratification, and are fissured in many directions.

Gray granites met with in the narrow and deep defile of the Cheratkhevi are composed of whitish orthoclase, quartz, and black biotite, with a local addition of oligoclase and hornblende, and, in some places, contain iron-pyrites. Some of them have a porphyritic structure. They are traversed by large veins of fine-grained, pink granite, often passing into the so-called felsite, which consists of quartz and pink orthoclase only.

We must mention further that in the same region, in the valley of Abanos-tskhali, at the summit of Kajar, and along the course of the Lopnis, are found red granites, fine and coarse-grained, which are traversed by thick veins of melaphyre, containing a large proportion of magnetite.

The granites found in Sukhum district, near the Black Sea, underlie formations some of which are ancient and others apparently of more recent origin, the latter containing, according to the statement of the Russian géologist, M. Sorokin, rounded fragments of gray sandstone, which are thought to be of Jurassic age. The granites occur along an anticlinal fold, and build up some summits in the main range. They are traversed by other granitic veins, as also by dykes of lava. The fundamental granite is partly light-gray, as, for instance, in the mountains of Akhef, Alakhesh-ghia-akhai and Pridgal, partly reddish, as on the mountains Abjikho and Abge-dsykh, to the south of Lachta.

As yet we have spoken only of normal granites or such rocks as very nearly approach to them, as granitite, under which name Rosenbusch understands all biotite granite; but as was mentioned, we have besides in the Caucasus varieties of granite known under the names of protogine and graphic granite, and granite-porphyry. The first of these varieties has been met with on the main range, on the mountain Bach near Sukhum, in eastern Imeritia near the village Sagwine in the Cheremela valley, between the Chorokh and Tmer-khevi, in the mountains of Kenia, Chirkham and Kari-Khala and other places. Graphic granite (pegmatite of Haüy) occurs, apparently, in the main chain, but we are not sure of the variety of rock that has been found there, and as to granite-porphyry, or porphyritic granite, as it is called by Russian writers, it is found at the junction of the Kwirila and Cheremela valleys and in eastern Imeritia, between Boslebi and Tiri. Its constituents are light-yellow or gray orthoclase, gray quartz, biotite, and silvery, ferro-magnesian mica, as also, but not always, hornblende. This rock is sometimes intersected by a rock of the same nature but darker in colour.

After granite, by far the most important of the igneous rocks in the Caucasus, comes, among holocrystalline rocks, syenite, in which orthoclase predominates and free quartz is almost absent. The syenite group seems, however, to be not very abundant in the Caucasus, though it has been observed on the main range, in Svanetia and in western Imeritia, in the valley of the Kwirila, and on the banks of the Macharula and Ghedsrula rivers. In Svanetia the syenite is micro-granular, is composed of yellowish orthoclase, black and sometimes green micas, and amphibole, and has sometimes a gneissic structure, or passes into a true gneiss. The syenite of Imeritia is composed of white orthoclase and dark-green hornblende, and contains very little mica, whilst quartz is almost absent. Hornblende (or amphibole) is by far the most important constituent, whence the term hornblende-granite usually applied to it. This rock has a very distinct pseudo-bedding, with a constant south-eastern strike and a dip of 35°; its appearance is exactly that of a sedimentary rock.

After granite proper and its varieties we now come to the eurite-granite group, such as felsite, quartz-felsite, quartz-porphyry. Some of them are of hemicrystalline structure.

Felsite, having a porphyritic structure (felsitophyre), is to be met with in western Imeritia, along the lower course of Dsussa and Buji rivers, along the Katskhur and Cherul, in the Kwirila valley itself, and on the Cheremela. Its constituents are those of granite, but quartz and ortho-clase largely predominate. Felsite and quartz-porphyry are powerfully developed in the Cheremela valley, appearing either in the form of dykes or in beds or layers conformably disposed amidst Kimeridgian and other sedimentary rocks. The conformable disposition is especially striking between the Dsirula railway station and the Kwirila village.

The hemicrystalline igneous rocks of the Caucasus may be classified into two principal groups—the Andesite and Olivine-Basalt groups. To the former belong porphyrite, diabase-porphyrite, and porphyritic green-stone, as well as andesites: to the second appertain basalts, anamesite, diabase, dolerite, and melaphyre.

Porphyrites, included by some in the diorite and aphanite group of holocrystalline rocks, have played a large part in the building up of the lofty Svanetian mountains, and form the waterparting between the Nenskra and Larakvakva basins. They overlie Lower Cretaceous limestones. The ground-mass of these rocks, consisting of pyroxene and oligoclase is interspersed with large crystals of dark-green and even black pyroxene. The rocks are traversed by fine veins of zeolite and calcite.

The same kind of rock is to be met with on the left bank of Dsirula river, in eastern Imeritia, about five hundred yards to the north of Khunevi village; it contains iron-pyrites.

The rock recognised by Caucasian geologists as diabase-porphyrite is met with in Svanetia. It is composed of green plagioclase, pyroxene (augite), chlorite, and magnetite. It is interstratified with trap-tuffs, passing into diabase conglomerates.

Porphyritic greenstones, or, more exactly, diorites, are found in the basin of the Kwaga and Lagram-sagher rivers in Svanetia and on the upper course of the Ingur; they build up the range parting the waters of the Larakvakva and Sunturi, and are particularly developed in the lower course of the Nenskra (Khube) and Khaisht rivers. Through the greenish ground-mass of Svanetian diorites are scattered large crystals of yellow orthoclase. Stilbite and limonite occur in cavities. The Ingur rock, containing large and magnificent crystals of pyroxene, is called by Russian geologists pyroxenic porphyry. This rock consists of an admixture of oligoclase and pyroxene, amongst which very large crystals of the same dark-green or even black pyroxene are conspicuous. Fine veins of zeolite and calc-spar traverse the rock.

As to andesites proper, this kind of rock has been noticed in the Akhalzykh-Imeritian range, in the valley of the Cheremela, and the high plateau of Akhalkalaki. The andesites of Akhalzykh appertain either to hornblende-andesites (trachytic type), or to augite-andesites (basaltic type), with or without quartz. Hornblende-andesites are found north of Kutais, along the banks of the Rion. They have been forced through argillaceous slates, considered the most ancient of the sedimentary rocks of the Caucasus, and also through rocks of more recent origin (Oxfordian, Lias); they are fine-grained, and somewhat greenish-gray. The predominance of large hornblende crystals gives them, moreover, an almost porphyritic structure.

Augite-andesite is found near Gumati, in the Kutais Government, and on both banks of the Cheremela above the mouth of the Biolis-khevi, near the fortress Tskheris-tsikhe. It shows light-gray spots on a foundation generally black, and consists of light-gray plagioclase, black augite, and magnetite. As accessories it contains olivine, mica, and iron pyrites. It assumes a spheroidal aspect, the spherical masses being composed of concentric layers. It is subject to very rapid decomposition. In the Tertiary sandstones of Akhalzych appear spherical nodules (half an inch to eight and ten inches in diameter) consisting of felspar, biotite, and quartz. The andesites, originating from volcanic action, have not, according to the testimony of Russian geologists, even deranged the stratification of the overlying sedimentary beds. The whole plateau of Akhalkalaki also, about six thousand feet in altitude, is covered with a mighty deposit of volcanic rocks, mostly andesites, interspersed with extinct craters, such as Kanchicar, Tauk-viteli, Karagos, Samsar, Godorebi, the Greater and Lesser Abul, etc., attaining heights of some ten thousand feet.

We have now to deal with the second group of hemicrystalline igneous rocks, in which we include basalts, anamesite, dolerite, and melaphyre. Basalts have been found to the south-east of Kutais, where they have made their way through deposits partly of Neocomian and partly of Upper Cretaceous age. A little westward, in the vicinity of the villages Dedalauri and Gushtibi, basaltic rocks appear cutting through Gault, and at the Gordi village, on the right bank of the Tskhenis-tskhali river, through various Cretaceous beds consisting of marls and limestones. These basalt-rocks, however, are of no great importance. For the most part they appear as true basalts, but also in the form of more porous anamesites (Kutais and Gordi), and dolerites (Kutais), and amygdaloids. Basalts have been also observed in the vicinity of the following villages, on the western side of Meskhi mountains—Nagorevi, Chalastavi, Banoja, Gumra, Dedalauri, Didguabuna, Goghelauri, and Eto—and they form the range of Kachaber on the right bank of the Chorokh, south of Batum, and are also found on the same bank of the river from Kapandidi up to the mouth of the Ajara.

Anamesite is extensively developed in western Imeritia, between the villages Sagwine and Sanakhshire; it has a vertical columnar structure, divided into blocks concave at the top and convex at the base; the diameter of the columns varies from one to two feet. On a fresh fracture the rock has a greenish hue, but its weathered parts are green, yellowish, reddish, and even brownish-red. It has such a fine-grained structure, that it is impossible to distinguish its elements with the naked eye. It is very brittle and breaks with sharp edges and sometimes with a conchoidal fracture. It contains some olivine and secondary constituents, as zeolites.

The igneous rocks of volcanic origin occurring in western Imeritia, on the right bank of the Kwirila, between the villages Kwatsikhe and Tskhilati, are considered by Russian geologists to be diorites. They contain quartz, fine-grained crystals of white and greenish oligoclase, and prismatic and sometimes needle-like crystals of dark-green hornblende. Occasionally, the abundance of white oligoclase gives the rock a porphyritic structure.

A similar rock extensively developed in the valley of the Ingur in Svanetia, is, according to the same authority, diabase. It contains greenish plagioclase, fine-grained prismatic crystals of augite, as well as a little chlorite and fine grains of magnetic iron-ore. Its structure is usually porphyritic from the preponderance of plagioclase.

The diorites often alternate with trap-tuffs. Russian geologists, in distinguishing diorites from diabase, appropriate the first name to those rocks of igneous origin that contain quartz, plagioclase, hornblende, and alkaline micas, reserving the latter for the rock containing plagioclase, hornblende, and quartz. It is a rather misleading nomenclature, as in true diorites quartz is practically absent, and the term diabase, introduced by Hausmann, is applied to rocks of any grain containing pyroxene (augite), labradorite, and chlorite. It is allied to diorite, gabbro, aphanite, and dolerite.

Whatever term be applied to them, the rocks we are dealing with, designated in the Caucasus as diorite and diabase, certainly appertain to the diorite and aphanite group, since plagioclase and amphibole or hornblende are their principal constituents. They have upheaved and torn asunder the Jurassic and Cretaceous formations in the upper valley of Ingur river, between the Lebiovahashi and Larakvakva defiles, have built up a range running parallel to the main chain between the Ingur and Tskhenis-tskhali valleys, and have formed three groups of hills—the Urulash, Omiash, and Sakeri.

The northern and southern limits of these igneous rocks have not been exactly determined, but they have been found in the valleys of Giogiori, and Dsirula, in the mountain range of Tarichon, along the upper courses of the Lagianuri, Aski-tschali, and Ritse-uli rivers, and in parts of the valleys of the Tskhenis-tskhali, Tskhal-tsiteli (Krassnoi), Giuruchuli, Chikauri, Kwirila, and Cheremela.

The same rock appears in the form of a dyke, with a granitoid texture, on the eastern slope of the Meskhi range, to the north of the village Balta. It is supposed here, as also in the Cheremela valley, where it is extensively developed in the form of dykes and sheets, to have burst through sedimentary rocks belonging to the Upper Jurassic system (Kimeridgian), while in Sukhum, in the Bsyb mountain range, this rock appears among somewhat older Jurassic strata (Oxfordian). Again, igneous rocks held to be diabase crop out in many other parts of the Sukhum province, either in the form of thick sheets, covering older clay-schists and Oxfordian beds, or in isolated peaks and summits, rising up through granitic rocks, or in veins and dykes piercing through granite, schists, and Oxfordian beds, as, for instance, on the Achav-chara pass.

Diabase, or rather dolerite, is also met with on the Rion river and between the Ajaris-tskhali and Machakhelis-tskhali. It is here composed of plagioclase, augite, and chlorite, the last mineral giving the whole a greenish tinge. The rock contains besides fine-grained magnetic iron-ore, iron-pyrites, and chlorophaeite, sometimes also epidote. It has mostly a porphyritic structure. Its extension in the Kutais province is generally concomitant with Upper Jurassic sedimentary rocks.

Before leaving this group of hemicrystalline igneous rocks, we must mention melaphyre, a rock included by some authors among basaltic andesites. Melaphyre is usually understood to be a black compact basaltic rock without olivine; but the rock called by this name by Caucasian geologists contains olivine with plagioclase, augite, and magnetic iron-ore, and must, therefore, be included in the Olivine-Basalt group. It is met with in eastern Imeritia, at the mouth of the Dsirula river, on both slopes of the Dsirula valley between the villages Sakasria and Nadbeuri, in the Cheremela valley between the Sagandzili and Mari-lissi villages, between the Ingur and Tskhenis rivers on the upper course of the Kwirila, on the middle course of the Giorgiori and Dsirula rivers, as also on the eastern slope of Meskhi mountains, in the Kajar range, and elsewhere. It is a compact, dark-green, almost black rock, in which black basaltic augite and transparent felspar (plagioclase) are easily discern-able. It is fine-grained, and is often interstratified with, or overlaid by, Upper Cretaceous deposits (Gault or Albian). In some parts this igneous rock cuts through andesite. Elsewhere, it is covered by beds of dark-green sandstones. On the eastern side of Meskhi mountains, melaphyre, or what is considered such, presents different varieties imperceptibly passing from one into the other. Thus, at the entrance of the Prona defile it has a dark-gray colour, and shows, dispersed through its ground-mass, crystals of white plagioclase, dark augite, and magnetic iron-ore. Elsewhere, to these constituents are added olivine and hornblende, and the rock becomes compact; it is much fissured, and is very apt to slide down the slopes into the valley. Farther up this defile melaphyre assumes a light-green colour, is compact and contains crystals of gray plagioclase, dark-green augite, olivine, and hornblende. It is sometimes of amygdaloidal structure, the cavities being oval or round and contain-ing concretions of calcite and rock crystal. At considerable altitudes, higher up in the same defile, are met with black varieties of melaphyre, composed of black augite, transparent plagioclase, and brilliant crystals of magnetic iron.

We have now, in the last place, to deal briefly with those igneous rocks that are almost wholly glassy, and those which are mixed up with sedimentary rocks and imbedded in a powdery or earthy matrix. We mean obsidian and trachyte, and also agglomerates and tuffs.

Rhyolite-glass, called obsidian, is met with on the Zanga river on the road from Delijane to Erivan; it is greenish and glassy. Another glassy rock of a similar nature, called teschenite, has been found on the Rion river, in the vicinity of the village, Opurch-kheti. It was first discovered by M. Tchermak, in 1872. There are two distinct varieties, the one white and spotted with dark-green, and the other almost black. The white variety consists of plagioclase, apatite, and analcime; the dark variety is made up of at least five or six minerals, viz., augite, felspar, nepheline, magnetite, iron-pyrites, and chlorophaeite. Both varieties have a pseudo-laminated structure. It crops out among the Liassic beds. This rock has been found only in Australia, Silesia, and Moravia, and was quite recently discovered by Macpherson in Portugal, where it cuts through Cretaceous deposits.

Trachyte-glass is not always distinguished by Caucasian geologists from the andesite we have dealt with before; it is found in the defile of Kudar, on the Giorgiora river and near Gurshe$_y$i village, on the Glora-tskhali, one of the left affluents of the Upper Rion.

Trachytic lava in the valleys of Kuban, Malki, Baksan, Aragva, and Liakhva overlies beds of conglomerate and boulder-stones, and is some four or five hundred feet thick. Lava of more recent origin has profoundly modified the former orography of the Minor Caucasus, forming the large plains of Alexandropol, Akhalkalaki, Lori, the high tablelands of Karabach and Agmangan, and the mountain Alagos. Most of the summits of the Minor Caucasus are really built up of trachytic masses overlaid with more recent lava. The last weathers easily, and forms a good soil, whilst the permeable trachytic rocks give origin to a number of springs. We may mention here, that the most powerful manifestation of eruptive action took place on the northern slope of the main chain, viz., from Kasbek, Elbrus, and Sheltrac, between Baksan and Cheghem. On the southern slope the volcanic activity has especially manifested itself in the country of Kelli or Goudori, compared by M. Abich to Auvergne. In the Minor Caucasus and Armenia the extinct volcanoes are scattered almost everywhere; but they are especially concentrated on intersection of the principal axis of upheaval, where we find Ararat, Alagos, Agmangan, and other mountains. The era of eruptive action was presumably at the beginning of the Pleistocene period, but the volcanic forces are not yet completely at rest, as is testified by numerous hotsprings, naphtha-springs, solfataras, and not infrequent earthquakes.

The hot-water and petroleum springs are found on lines running west and east and north-west and south-east, in the same directions as the principal ranges. Of the former set are the hot springs of Chechnia (between the Terek and Sunzha rivers), Tiflis, Abano, Borjom, and Abbas-tuman. The latter, include the springs of the Apsheron penin-

sula, of Kunakh-kent, Akhty, Ghenaldon (north of Kazbek), the sulphur
springs between Derbent and Chil-yurt, the carbonic acid springs between
Kasbek and Elbrus, and the sources that gush up in the valleys of the
Arpachai in Transcaucasia. It is remarkable, however, that the most
abundant and best known mineral springs of the Caucasus, those
in the neighbourhood of Besh-Tan, are ranged along quite a different
line, running from south-west to north-east. Petroleum and brine
springs are found at both extremities of the main chain, as well as
between the Kura and the Alazan, between the Sunzha and Terek, and
along the Kuban.

All kinds of tuffs and agglomerates must be very abundant in
the Caucasus. Remarkable are the diabase-tuffs, containing felspars,
augite, magnetic iron-ore, and sometimes imbedded epidote. Augite
appears in the form of distinct prismatic crystals, about 0·15 inch
in size. Epidote is found in considerable quantities in the form of
yellowish-green, transparent granules. Carbonate of lime is usually
present in these tuffs, in veins or nodules; and also some hydrated double
silicates, such as zeolites, form occasionally aggregates of distinct, beautiful
crystals. Apophyllite is usually white or pink; natrolite takes the form
of long, needle-like prisms, either colourless or brown-yellow. Zeolites
are often met with in brecciated and conglomerated diabasic rocks,
filling up the fissures and cavities or bounding the margins of the diabase
fragments.

Tuffs form often compact beds, which are broken by fissures, giving
them the appearance of a columnar structure; they have also sometimes
a loose texture and always a course, uneven, earthy fracture.

We have just spoken of brecciated diabase; like other fragmental
volcanic deposits, it is very abundant in the volcanic regions of
Transcaucasia, but must be considered as sedimentary, as indeed many
tuffs and ashes may also. We have come thus to the end of our very
brief and incomplete account of the igneous rocks in the Caucasus.
We are well aware of its shortcomings, but we hope it may nevertheless be of some interest to students of the Earth's crust.

GEOGRAPHICAL NOTES.

By The Acting Editor.

EUROPE.

The Second Italian Geographical Congress.—The first of these meetings was held
in 1892 at Genoa to commemorate the discovery of America by Columbus, and it
was then decided that a Congress should be held every three years. Accordingly
the second took place in Rome from September 22nd to 27th, 1895. The Proceedings, forming a bulky volume of more than 850 pages, have now been issued.
Several papers were read at the general meetings, while the rest were read in the
sections, of which there were four—Scientific, Commercial, Educational, and Historical.

At the general meetings reports were given of the work of the Istituto Geografico Militare and of the Italian Hydrographic Office, and also of the progress made in the compilation of a geological map of the kingdom. As these reports are accompanied by lists of the sheets issued, and of other publications of the departments concerned, students of the topography, hydrography, or geology of Italy will find them useful for reference.

The future of the colony of Eritrea formed the subject of a paper by Baron Franchetti, which has already been noticed in these pages ; and Signor G. Boggiani gave an account of the Caduvei, a tribe on the Upper Paraguay. This isolated tribe has many interesting customs which have been described by Signor Boggiani at greater length in a large volume. They display great ingenuity in design, and paint their bodies with elegant patterns, using as a pigment the juice of the genipap (*Genipa oblongifolia*), which has the property of sinking into the epidermis so deeply that the ornamentation is not obliterated by washing for six or seven days. The juice of the arnotto (*Bixa orellana*) is also used, producing a bright red tint, but it is not so lasting. The practice here mentioned is of particular interest because, as Signor Boggiani showed in another paper read in the Scientific section, it was in vogue among the ancient Peruvians. Many of the Peruvian mummies, preserved in the Prehistoric and Ethnographic Museum at Rome, exhibit designs executed after death, some in bluish black, and others of the reddish hue of cinnabar. Several authorities have held that the designs were made by tattooing, but Signor Boggiani argues that they were painted with the juices of the plants now used by many South American tribes. The Caduvei have raised the art to a perfection probably never attained even by the ancient Peruvians.

The paper in the Scientific section of, perhaps, most general interest was one read by Dr. Olinto Marinelli on the distribution of the population of Sicily in relation to distance from the sea. Following the method first introduced by Dr. Rohrbach (see vol. vii. p. 213), he drew lines of equal distance from the coast (*isochoric* curves) at intervals of 5, 10, 20 km., etc., measured the areas enclosed by them, and ascertained the number of inhabitants in each zone. Hence he distinguished in Sicily three principal zones, namely, a coast zone within 5 km. (about three miles) of the shore, containing a population of 702 persons to the sq. mile ; secondly, an intermediate zone between 5 and 10 km., containing 262 to the sq. mile ; and an inland region with 199 to the sq. mile. Professor Marinelli also calculated the populations of the zones for the southern, northern, and eastern slopes respectively, and also the mean distance of the population of each slope from the coast. On the northern and eastern slopes the influence of the sea is very marked, about half the population being concentrated in the coastal zone, while in the intermediate zone the densities are sensibly the means for the whole slopes. On the southern slope, on the other hand, the intermediate zone has a low density of population, much less than in either of the other two divisions, and the inland region has a small population. Many and complex influences may continue to modify the distribution of population. Elevation, malaria, history, water, soil, etc., are no doubt all factors in the result at present exhibited.

One more paper may, perhaps, be noticed here. It is a sketch of the geology and mineralogy of Brazil, according to the most recent studies, by Professor Vincenti Grossi. The base of the great Brazilian plateau is formed of metamorphic rocks belonging to two great series. The more ancient, consisting of crystalline rocks, such as granite, syenite, gneiss, and mica-schists, has been referred to the Laurentian system, while the second, less crystalline in structure, is composed of quartzites, schists, and limestones,'which are probably Huronian. The Laurentian system is especially developed in the Serras do Mar and da Mantiqueira, forming the principal peaks. It is not particularly rich in minerals of economic value, and is in this respect far inferior to the Huronian. It contains extensive deposits of iron ore, and a little gold is found in its upper strata. In the east of Minas Geraes it yields precious stones in abundance, and rich deposits of graphite are known to exist in the same region. The Huronian system is especially characteristic of the Serras do Espinhaço, da Canastra and da Matta da Corda, and the mountains of Goyaz, and appears also in conjunction with the Laurentian in the tableland of the south of Minas Geraes, in the southern parts of the Serras do Mar and da Mantiqueira, and in the hilly parts of the valley of the Upper Paraguay. The predominating rocks are the chloritic and micaceous schists and quartzites known as itacolumites, and their ferruginous variety called itabirite. These rocks contain the great mineral wealth of Brazil; the abundance of iron of the first quality is extraordinary. In others of the series are found gold in quartz veins, arsenic, and, less frequently, copper, bismuth, lead and antimony.

The Silurian system is also represented in Brazil, and Devonian and Carboniferous rocks are found from the Amazons to Rio Grande do Sul, and probably in the west of Minas Geraes and in Matto Grosso. Mesozoic formations are known to exist in the northern states—Amazonas, Piauhy, Ceará, Rio Grande do Norte, Pernambuco, Sergipe, Alagôas, Bahia, and, perhaps, in the west of Minas Geraes, where the Cretaceous strata are often rich in fossils. The Tertiary formation is represented at various points on the plateau by small basins of fresh-water deposits which sometimes contain lignite, while the Quaternary occurs in fluvial and lacustrine deposits, and in the limestone caves famous for the remains of extinct mammals.

Many of the other communications were made to the Congress, which, though deserving of attention, cannot be referred to within the limits of a note.

ASIA.

Alagos.—M. Pastukhof, whose meteorological observations on Mount Ararat have been noticed in the *Magazine*, made an ascent of the neighbouring Alagos in 1893, of which he read a report to the Caucasian Section of the Russian Geographical Society in March last. He made for the western peak, but on the first day was prevented from reaching it by a thunder storm, and had to spend the night on a grassy flat 11,620 feet above sea-level. Next day M. Pastukhof and his attendants continued the ascent, reaching, 300 feet above their camping ground, a wide, steep slope plentifully covered with hailstones, which became smaller as the party mounted to the summit of the western peak, 13,167 feet above sea-level.

The name Alagos is probably derived neither from the Tartar *Allagos* (God's eye) nor Alagos (blue eye), but is a corruption of the old Armenian name Arakad (Arakas?). The mountain has four summits ranged along an arc of 270 degrees. The highest has a height of 13,436 feet; it is the most northerly, is very sharp and rocky, and lies on the intersection of the parallel of 40° 31' 29·9" N. lat. with the meridian of 61° 51' 49" E. long. The south-eastern summit is 12,732 feet high, and the south-western 12,810. There is also a smaller peak on the western flank of the highest summit. All these peaks with the connecting saddles form a crater with its longer axis lying from north-west to south-east, and having a length of 2300 yards, a breadth of 1900, and a depth below the lowest saddle of 1050 feet. On the south-eastern side of the crater is a huge cleft forming the head of a valley through which flows the Dadaly-chai, draining the snowfields within the crater.

At a time now very remote the Alagos was one of the mightiest volcanoes of the world, from which lava flowed in enormous streams, reaching to the river Araxes, Kars, and the mountain Karniyarykh, and forming the plain of Abaran, 7000 feet high. The area covered by the lava is about 3400 square miles. After the volcanic activity had passed away, huge glaciers made their appearance on the crater bottom. At the present time the snowfields cover an area of only $2\frac{1}{4}$ square miles. Even this extent, with an average depth of $24\frac{1}{2}$ feet, gives a volume of over 1,500,000,000 cubic feet. But this is nothing to the ancient glaciers, one of which has left in the Gysal-dara valley a lateral moraine with an average height of 420 feet. It may be estimated that this glacier had a volume thirty-seven times that of all the permanent ice and snow now on the mountain. Six large glaciers then flowed down from the Alagos, besides several of less striking dimensions. The miserable remains of glacier and snowfield now end at a height of 11,000, whereas the glacier above mentioned descended as low as 8000 feet.

It has been stated in some geological text-books that there is a lake on the summit of Alagos, but of such a feature M. Pastukhof saw no sign. On the flanks of the mountain, however, are scattered forty-two lakes of various sizes, the greater number lying on the north and west slopes. The largest and deepest is the Kara-gol on the south, about $2\frac{1}{2}$ miles from the south-west peak and at an elevation of 10,521 feet. Its periphery measures nearly $1\frac{1}{2}$ miles. To the west of it lies the Boku-gol, fifty feet lower, and to the north-west of this the Tanysh-gol at an elevation of 10,060 feet. Again, to the west-south-west of Tanysh-gol, by the bridle-path from Echmiadzin to Alexandropol, lies an insignificant pool fed by warm springs. In July the temperature of the water was $64\frac{1}{2}°$ F., while that of the air was 52°.

In spite of the great altitude of the western summit, several varieties of plants grow there, among them a very odorous forget-me-not, and winged insects fly through the air and fall victims to small birds. The fulgurites and shattered stones, of which there is a great number, are a curious sight; they are all highly magnetic, attracting the northern pole of the needle.—*Globus*, Bd. lxx. No. 6.

Western China.—M. C. E. Bonin has travelled across the eastern boundary of Tibet, from Tali to Ta-tsien-lu, exploring a great deal of country hitherto unknown to Europeans. Leaving Tali on November 8th, he marched directly to the Yang-tse-kiang, passing by Likiang. This large town, now the residence of a Chinese prefect, was formerly the capital of the Mossos, the least known and most interesting of the peoples who form the advance-guard of the Tibetan invasion. One day's journey north of Likiang, and seven from Tali, flows the Yang-tse-kiang, spanned by a large iron bridge, which affords a passage to merchants and mandarins travelling between Tali and Sze-chuen. M. Bonin, wishing to enter Tibet, crossed the river higher up at the ford of Ashi, where it is forced to turn from east to north by a lofty snow-clad peak rising to the north of Likiang. At Ashi the Blue river flows at an elevation of about 5900 feet, with a breadth of 220 yards and a current of four knots an hour. For several days M. Bonin followed the left bank of the stream upwards in search of a passage through the wall of mountains, and on December 1st crossed by a pass 14,400 feet high to the plateau of Tsong-tien in Tibet. Four days later he reached the town of Tsong-tien. It was destroyed twenty years ago by the Chinese expedition which started from Tali to take the field against the Tibetans, and therefore the present town is new. Between Tsong-tien and Yunning-fu the path, not more than eight inches wide, crosses a succession of precipitous mountains and narrow gorges, sometimes winding about at an enormous height above the abyss. From this path the course of the Yang-tse-kiang was clearly visible. From the Ashi ford the river flows northwards, passing round a huge mountain mass terminated at the south by the Likiang mountain and on the north by the Kuatyn. The summits of this ridge, covered with eternal snow, rise to heights of 16,000 to 20,000 feet. Having passed the Kuatyn, the river flows southwards again through the Yong-pe country, and then eastwards, after receiving the waters of the Pe-shui-kiang. This tributary, which descends from the Likiang heights, has been mistaken for the main river.

Yunning-fu is situated in a valley formed by an amphitheatre of mountains, and is inhabited by Chinese of Yunnan, Kutsongs (Tibetans of Tsong-tien), Mossos, and Si-fan. A day's journey to the north of it lies Kien-su on the direct trade route to Tibet. Here M. Bonin had to wait until he had obtained permission of the king of Meli to pass through his territory. Whereas at Lhassa and other chief towns of Tibet there are always Chinese officials, the king of Meli, who is also a lama, has until recently refused to allow even Chinese traders to pass through his dominions. The situation of his palace is striking. A snow-covered mountain about 13,000 feet high descends to the river of Meli, here about 7200 feet above sea-level; about half-way up its slope, at about 10,000 feet, stands the lamasery, formed of houses of three stories with white walls pierced by windows, balconies, verandahs, and roofs with ridges of gold. At the time of M. Bonin's visit the king was gone on an expedition against the king of Litang, from whom he wished to wrest a part of his territory. His subjects are Si-fan. Their industry and commerce are entirely devoted to the maintenance of eighteen large monasteries in which the king resides in turn.

Crossing a chain of mountains, the Tsi-tseng-hu, by a pass 17,000 feet high, M. Bonin came to the Yalong-kiang. On the way he passed several dwellings occupied by herdsmen guarding troops of yaks. The highest of these was at an altitude of 16,560 feet, and is probably the highest inhabited spot in the world. The Yalong-kiang, in Tibetan Nag-chu (black water), is the principal affluent of the Yang-tse-kiang on the left bank. Four days' journey beyond it M. Bonin entered the territory of the king of Kiala, whose capital is Ta-tsien-lu. His subjects are chiefly Man-tze, a branch of the Si-fan race, but among them are some Lolos, and here M. Bonin noticed among the women the fashion of collecting their hair into a horn, mentioned by Mr. Colquhoun but never observed by him. Several chains of mountains, higher than any met with before, were crossed between the Yalong-kiang and Ta-tsien-lu. The highest saddles were one of 17,000 feet near the river and two of 17,400 feet near Ta-tsien-lu. The route from Ta-tsien-lu to Cheng-tu is well known.—*Comptes Rendus* of the Paris Geogr. Soc., Nos. 13 and 14, 1896.

The Recent Great Wave on the Japanese Coast.—Professor John Milne has sent to the *Geographical Journal* a note on the wave that recently killed 27,000 persons in Japan. The wave clearly originated at a short distance from the east coast, about half-way between Tokio and Yezo, being probably due to some disturbance at the southern edge of the Tuscarora Deep. Impinging on the coast with a height of eighty feet at some points, it stretched for about 300 miles from south-west to north-east. To the north of the famous Matushima Archipelago the coast is fiord-like in character, and in the bays and estuaries there are several important towns and fishing hamlets, almost all of which have been destroyed. In the town of Kumaishi, which had 6000 inhabitants, very few survive. Earthquakes and sea-waves have in past times devastated the east coast of Japan. To the south of the region lately visited by the great wave lies Kamakura, once the capital of the empire, with a million inhabitants, which was frequently a victim to such disturbances, 30,000 persons having thus lost their lives in 1293, and the history of

Shikoku and the cities of Tosa and Osaka is full of records of similar disasters. All along the slope of the eastern coast, which sinks suddenly to the depths of the Pacific, earthquakes are frequent, and probably this mighty monocline is intersected by the submarine continuation of the anticline which forms the backbone of Yezo. Observations of the times at which the disturbances were felt at different places enable their sources to be determined approximately, and lead to the conclusion that the Yezo anticline is yet in process of construction, and that a submarine range of mountains may be slowly growing up.

It is curious that the seismograph at Shide, Isle of Wight, indicated a disturbance on June 15th, but when the 17th was given in the newspapers as the date of the Japanese disaster, it was supposed that it was some other movement which influenced the seismograph, and that the wave sent no message to Europe. The newspapers, however, were at fault, and not the instrument.

From *Science*, July 17th, we learn that tidal waves appeared on June 15th on the Pacific coast of the United States, the height of the waves at the mouth of Rogue river being three to six feet between two and three o'clock in the afternoon. On the west coast of Hawaii the disturbance was more marked, the height of the waves at Keauhou being thirty-five feet.

The Chin-huan of Formosa.—The Japanese have lost no time in making friends with the wild aborigines of Formosa and sending parties into the unknown eastern part of the island. Two or three expeditions started from Shin-ten-gai, seven and a half miles from Tai-pei-fu, the last of which left the town in February last and came to the village of Kushaku, situated on a plain of a little more than a square mile beside the Tai-keika river, the boundary of the Chin-huan territory. Then the party crossed five steep mountains, where bears, stags, and monkeys abounded, and came to the village of Togasha, ten miles from Kushaku. This is one of the eight *sha*, or villages, of the Chin-huan known by the name of Marai. In these villages there are two hundred houses with eight hundred inhabitants. The people of the Maraisha hold no communication with other Chin-huan villages, but when they meet other natives they treat them in a friendly manner. The Chin-huan are very similar to the Malays and the natives of the Philippines, and have much in common with the Japanese. Unmarried women are not tattooed, and when they put on Japanese costume they are hard to distinguish from Japanese women. All the sounds used in the Chin-huan language can be represented by the Japanese alphabet.

The descriptions given by the expedition of the habits and customs of the Chin-huan agree with those in Mr. Dodd's paper (vol. xi. p. 559), but a few new details are added. Tattooing is a mark of distinction. A man who has never killed an enemy may not be tattooed, and if not tattooed he cannot marry. Those who have a large number of skulls in their houses are much esteemed. Youths live in their parents' house until they are sixteen years of age, when they move into a bachelors' house, which is to be found in every village. The parents of a bride are not informed until the marriage is consummated, when the bridegroom sends them a present of cows, stags' antlers and skins, and other articles prized by the Chin-huan. Then a great feast is held, which among the wealthier families continues for several days. Two other feasts are held—in November, when a Chin-huan exchanges his summer for his winter dress, and when he puts on a newly-made garment. The Chin-huan has no doctor and no medicine; when he falls ill, he calmly awaits his death. Few people above forty years of age are seen. When any one dies, he is wrapped up in his garment and buried with his weapons and other personal belongings in or beside his house, which is then deserted. No one remembers a dead man or troubles himself about his resting-place. The Chin-huan has no calendar; he knows that a year has passed when a particular flower blooms again, and that another month is completed when the moon is full. Most of the natives do not know their own age. A Japanese named Hiyama, formerly a member of parliament, has married a Chin-huan woman, and consequently can visit the Chin-huan villages and learn the customs of the people.—*Globus*, Bd. lxx. No. 6.

AFRICA.

Changes in the Volume of the Nile.—Several authors have remarked that in former times the volume of the river must have been much greater than at present. Professor Edward Hull, during a recent visit to Egypt, sought for evidence bearing on the question, and, at the suggestion of Sir Joseph Prest-wich, paid particular attention to the terraces. He remarks, in the *Quarterly Journal of the Geological Society*, vol. lii. part 2, that only by personal examination can one realise the extent of the erosion by river action in the Libyan region after it was raised out of the sea at the close of the Eocene period. The Miocene period was the special epoch of elevation, disturbance of strata, and denudation over all the Egyptian and Syrian region, though there was a second period of great fluviatile inundation beginning in later Pliocene times, and lasting through the Pleistocene period. At the present time erosion has almost ceased, and the cultivated terraces bordering the river, as well as the plain of Lower Egypt, are rising higher above the sea-level owing to the deposition of sediment, showing that the bed of the river is also rising.

From Aswân to Cairo the fall, six inches per mile, is just sufficient to keep the fertilising sediment in suspension when the river is not at its lowest. The Nile valley, ten miles broad on an average, cut down through a tableland 800 to 1000 feet high capped with Eocene limestone, and extending above Cairo for a distance of 300 miles, shows the enormous duration of the Miocene erosion.

The river flows across the escarpments of the Eocene limestone, of the Cretaceous limestone farther south, then of the Nubian sandstone, and lastly of the granitic and schistose rocks of Aswân. Zittel has suggested that it once entered the Red sea, but this was possible only in early Pliocene times, when Lower Egypt was submerged to a depth of about 220 feet, and the entire region below this level, from the Red sea to the Mediterranean, was covered with water. The physical features of the region are quite consistent with such an hypothesis.

Professor Hull gives details of the results of his examination of faults, the levels of the ancient Nile, old river channels, terraces, etc. He also refers again to the second period of inundation mentioned above, which he calls the "Pluvial Period." In 1883-84 he came to the conclusion that the fine valleys of the Sinaitic peninsula and Arabia Petræa, now dry and covered with alluvial deposits, once constituted the channels of an extensive river system draining into the Red Sea, and the dry river valleys which open into the Nile likewise point to a period of heavy rainfall extending from the Pliocene down to recent geological times.

The decreased volume of the Nile can only be due to a diminution in the rainfall. Through 1200 miles of its course the river now runs through a region wellnigh waterless, where it is subject to a constant drain by evaporation, and the Abyssinian floods alone enable it to reach the sea during the hotter months. A slight climatic change would cause an enormous expansion or contraction of the river, seeing that its basin is estimated at 1,100,000 square miles. During the Miocene period, when the infant river was beginning to form its bed, and the land was higher relatively to the surface of the Mediterranean, the climatic conditions may have been altogether different from those of the present day. At any rate, there are good grounds for believing that in the subsequent Pliocene period, when Lower Egypt was submerged to a depth of 200 feet, there must have been an increased humidity and a lower temperature, and these changes must have been accentuated in Pleistocene times. The precise effect of the Glacial Period of Europe on the mean temperature of subtropical Africa cannot be estimated, but Professor Hull believes that the climate was then similar to that of Europe of to-day as regards temperature and rainfall. Hence the vastly greater volume of the Nile waters is easily accounted for.

The Chakwati Lake.—Inland from Kifmangao, on the coast between Dar es Salaam and the mouth of the Rufiji, lies the Chakwati Lake. It was first seen, probably, by Herr Preuss, who erected the East African telegraph line. In October last Dr. Oscar Baumann visited Kifmangao, a miserable village on a mangrove creek, which, nevertheless, is of some importance as an emporium for trade in india-rubber and copal, and, after crossing low swampy land covered with light bush, gradually ascended the Kibunpuni hill, one of a range running parallel to the coast. Beyond lies the hollow in which the Chakwati Lake is situated, a yellowish-brown sheet of water running from north to south. It is not more than fourteen to sixteen feet deep, and has no visible tributary or outlet, but its water is drinkable, with a scarcely perceptible saline flavour. It contains shad and eels and other fish in abundance, which are caught by the natives. Hippopotami also live in it, but no crocodiles. The banks are sandy and covered on the south and east by fields, on the other sides with bush. No rocks are visible. A slight rise of the ground on the west separates it from the smaller Kiputi lake. The natives in the Kiputi village and in the settlements on the Chakwati lake are Wadengereko, a tribe which speaks a language different from that of the Wasaramo or the Rufiji people. In the absence of rocks it is difficult to discover the origin of the lake; perhaps it is the residue of an old river or estuary, or an old mouth of the Rufiji. From Dr. Baumann's sketch-map in *Petermann's Mitt.*, Bd. xlii. No. 6, it seems to be about two and a half miles long by three-quarters of a mile broad.

AMERICA.

The Climate of Hudson Strait.—Mr. F. F. Payne read a paper on this subject at the Canadian Institute in March. He was sent up to the Strait in 1884 for the purpose of inquiring into the feasibility of establishing a line of steamers between Hudson Bay and Europe, and continued his observations to the autumn of 1886.

The first sign of spring is probably the change in the colour of the lichens, which begin to grow brighter in March. The first snowbird appeared at the beginning of April, and in early May caterpillars were seen crawling over the rocks, though the temperature was ten degrees below freezing-point. About May 20 there was a marked rise in the temperature, and plants of thirteen species put forth their leaf-buds, though in 1886 snow continued to fall up to June 17th. In the first half of June the vegetation made great progress, and the snow was nearly gone, while the temperature reached as high as 45°, and averaged 37°. Between the 15th and 30th the average was 39°, and 60° was once registered. The ice in the Strait was much broken in the month of May, and was driven backwards and forwards by the winds and tides, sometimes leaving a belt of open water. Throughout June it was still more or less compact.

July 1st may be considered the first day of summer, and on the 18th the maximum temperature for the year, 68°, was registered. The prevailing winds in this month were the north-east at the western entrance of the Strait, and east elsewhere, and no doubt much ice was drifted into Hudson Bay and there melted. By July 22nd, 1886, the ice was generally open, and fifteen days later there was little to be seen at Cape Prince of Wales. By July 15th several birds were fledged, the fox, hare, and lemming had assumed their summer garb, and seals and white whales spent much of their time basking in the sun. On July 30th the ground was frozen to a depth of 48 inches at Stupart's Bay, and to 27 inches at the western entrance of the Strait. Fogs are frequent in July, 147 hours being recorded in 1886. The rainfall for the month was only ·39 inch. The average temperature for August was 43°, and the rainfall was 2·5 inches. On several days the thermometer rose to 60° or 65°, and the Eskimo were thoroughly exhausted by the heat. On August 20th the approach of autumn was announced by the drooping of the plants, and by September 12th nearly every plant had ceased to show signs of life. The average temperature for September was 33°. Early in the month the wild geese began to fly southwards, and on the 26th winter set in, all the lakes being frozen over. In 1885, when, according to the Eskimo, the season was normal, the ice was compact at Nottingham Island on October 3rd, at Cape Prince of Wales on November 14th, and at Cape Chudleigh on November 23rd. The ice was quite open occasionally until about the end of December. On the Strait, owing to the presence of the water, the temperature does not fall so low as in the interior ; the minimum during the winter 1885-6 was − 45°, while − 65° was registered at Fort Chimo, Ungava Bay. The means for December to March were − 66°, − 24·4°, − 26·2°, and − 11° respectively. By April 17th the ice on the lakes at Cape Prince of Wales had reached a thickness of 72½ inches.

The Kansas River.—The Kansas river is formed by the junction of the Smoky Hill and Republican forks at Fort Riley, about 140 miles from its mouth in the Missouri. Its drainage area lies entirely in a region of plains, and its basin extends from eastern Colorado to the Missouri, with a length of 485 miles and a maximum breadth of 200. Its area measures 61,440 square miles, of which 34,526 are in Kansas, 17,454 in Nebraska, and 9459 in Colorado. The altitude of the basin falls from 5000 feet in Colorado to 750 at Kansas City, the average being about 2500 feet. Gauge readings have been made for several years at Lawrence, but not in sufficient numbers to determine the mean annual flow. The minimum discharge is probably a little over 500 second-feet. The mean annual rainfall of the basin varies from 10 inches at its western border to nearly 40 at the Missouri, averaging, perhaps, 20 inches. Thus it differs from the typical western stream, which rises in the mountains where the precipitation is great, and carries water down to arid plains. Rising in the most arid portion of the basin, the feeders of the Kansas river are insignificant until they reach a region where irrigation is not required. If the rainfall conditions of the basin were reversed, the usefulness of the river for irrigation would be many times increased.

The three principal rivers that flow into the Kansas are the Blue from the north ; the Republican from the north-west ; and the Smoky Hill from the west. The Blue has a drainage basin of 9490 square miles, and has the greatest volume of any of the tributaries of the Kansas. The discharge at Rocky Ford, five miles from its mouth, has been gauged, and the minimum found to be about 300 cubic feet per second. Next comes the Republican, with a drainage area of 25,837 square miles, and a minimum flow, at Junction City, of about 200 feet. The difference between the drainage areas and discharges of these streams is very striking ; the Republican rises at the extreme west of the Kansas basin and runs for hundreds of miles through arid sandhills, whereas the Blue drains the rainiest parts of the basin. The Smoky Hill, rising in Colorado, drains an area of 20,428 square miles, and has two tributaries, the Saline and the Solomon, draining 3311 and 6882 square miles respectively. Its low-water flow is about 140 feet per second. About 6600 horse-power are at present obtained from the Kansas and its tributaries, the most important mills being on the Solomon and Blue rivers. But only a small proportion of the available sites are utilised, and there is much room for improvement in this direction.—*The National Geographic Magazine*, May.

The Minerals of British Guiana.—Gold-mining was commenced in 1863 at Wariri on the Cuyuni river by the British Guiana Gold Company, but, owing to unskilful management and the situation of the mine on the disputed territory, the undertaking was abandoned in a few months, and nothing was done for twenty years, when placer mining was commenced. In 1884 the yield of gold was only 250 ounces ; it increased year by year till it reached a maximum of 137,629 ounces in the financial year 1893-94. In 1894-95 there was a decrease owing to the transfer of capital from placer to subsoil mining, which is not yet sufficiently developed to affect the production. The most productive rivers in order are the Essequibo, Barima, Cuyuni, and Potaro. The output from the Essequibo has fallen off, want of easy transport being the cause, while that of the Barima and Cuyuni has increased. Labour is not easily obtained. As to other minerals, Mr. H. J. Perkins, Crown Surveyor and Acting Commissioner of Mines, says that iron is the most abundant. Silver is often associated with the gold in the quartz rock, copper pyrites has been found with galena in Demerara, and diamonds and sapphires occur in many of the gold placers.—*Imperial Institute Journal*, August.

The Xingu.—In a letter to M. Reclus, which he has kindly allowed us to read, M. Henri Condreau writes from the mouth of the Ambe that he has passed the great bend of the Lower Xingu, ninety miles long, travelling by land along the chord of the loop, which is only about thirty miles long. He was organising an expedition on the Upper Xingu to the region of the great falls, fully 600 miles up the river, the route for half the distance running through country occupied by india-rubber gatherers, civilised but with a very bad reputation, and for the other half through a desert where dwell peaceful Yurunas and hostile Carajas. The voyage up the river will probably be accomplished in sixty days, if all goes well. But the Indians have to be reckoned with. Not long ago the Assurinis, a tribe never heard of before, murdered a peaceful fisherman and wounded another. M. Condreau expects to obtain much new information ; Dr. von den Steinen's map needs to be completed, especially as regards the main affluents, of which nothing is known. Then there are the Indian tribes—the Yurunas and Penas, of which little is known, and the Assurinis, Ashipayes, Araras and Carajas, hardly known at all.

The Rapids of the Madeira : Construction of a Road.—The navigation of the river is open up to San Antonio, but above this town is interrupted for a distance of fully 230 miles by twenty-six cataracts, three or four of which the Indian canoe-men, skilful and courageous as they are, cannot pass. In 1871 Church obtained the concession necessary for a railway, and in 1884 the Bolivians confided the direction of a survey to Sr. Julio Pinkas, which also led to no practical results. Since then the economical conditions have been greatly modified. Manaos, then a small place, is now a town of 40,000 inhabitants, San Antonio increases daily, and Para is become one of the calling-stations of the great lines of steamers connecting

Europe with South America. M. E. G. de Blaymont, therefore, proposed to the Government of Matto Grosso to open a road for wheeled traffic through the forest from San Antonio to the fall of Guajara-merim, the highest of the series. This might later on be replaced by a railway. M. de Blaymont commenced the work at Guajara-merim in January 1895, and by March had cut a way through the forest for a distance of forty miles. Being then incapacitated by illness, he was obliged to have recourse to other contractors, and a company was formed under the title of Mercado Ballivian and Co.

This road must have a most important effect on the trade of the country. India-rubber is the chief product of the Beni, Madre de Dios, Orton, and the affluents of the Madeira. The collection has increased rapidly during the past five years, and 10,000 to 12,000 tons pass the rapids annually. Again, on the Mamoré, on the runs of Trinidad, Los Reyes, etc., are immense herds of cattle, while Manaos has to obtain its meat from Argentina. The cattle arrive in bad condition after a voyage of twenty to twenty-five days, and fetch an enormous price. In the plains of Majos an animal costs eight shillings, while in Manaos it is worth £10 to £12. Considering, then, that the consumption of Manaos is about 2000 beasts a month, large profits might be obtained on this article alone, were there an easy means of transport. Besides the above, ipecacuanha, coffee, and cocoa might be exported. The coffee and cocoa plants cover very large areas.—*Bull. de la Soc. de Géogr. Comm. de Paris*, Tom. xviii. Fasc. 5.

The Guaira Falls of the Parana.—These falls have never been thoroughly explored, and have not been duly appreciated (see vol. vii. p. 621). Count Pietro Antonelli visited them in the autumn of 1895, travelling overland from Rosario, and found that they far exceeded in grandeur the descriptions hitherto given of them. The waters of the Parana, divided by islands into numerous channels, arrives at the brink of the falls with a scarcely perceptible current. The edge forms an arc six miles long, over which the water is precipitated from a pool nearly ten miles broad, forming at least twenty-two irregular falls 80 to 100 feet high, and collects at the bottom into a channel not more than sixty-five yards long. Count Antonelli has obtained some photographs of these beautiful falls, some of which are reproduced in the *Memorie della Soc. Geogr. Italiana*, vol. vi. part 1. One visitor calculated that about 530,000 cubic feet of water pass over these falls per minute, but they have yet to be fully investigated.

Dr. Steffen on the Rio Manso.—A letter of Dr. Steffen to the editors of the *Geographische Zeitschrift* (Jahrg. ii. Heft 6) describes briefly a visit to the Rio Manso, a tributary of the Rio Puelo lying at about 41½° S. lat. This river, like the Palena and Puelo, breaks through the chains of the cordillera by winding gorges where navigation is impossible. Its upper course lies in broad open valleys between the watershed and the central elevation to the west. The ground is excellently suited for grazing, and colonists have made their way from the Patagonian side into the Upper Manso valley. In this valley were seen cattle that, shut up by the rushing stream on the east and the high cordillera on the west, had become wild. Probably they are the remains of the herds of the Indians who formerly dwelt in this country. The ranges that confine the valley of the Rio Manso can be surmounted without difficulty. The greatest height to which the traveller had to ascend on his way to the upper valley of the river was about 5350 feet. The cordillera that forms the watershed rises in some places in steep dentated crests 6500 feet or more in height, but it has many breaks where it can easily be crossed to the Patagonian side.

The Magellan Territory of Chile.—Councillor Cruz has been sent by the Chilian Government to examine the condition of the settlements and lands in .Patagonia and Tierra del Fuego. The Magellan Territory (*Territorio Magellano*) extends from 47° S. lat. (Cape Tres Montes) to Cape Horn in lat. 56° S., and includes the western part of Patagonia as far as the watershed and the adjacent islands, the lands lying on the Strait of Magellan, and the western part of Tierra del Fuego, having a total area of nearly 75,300 square miles, of which some ten million acres are suitable for sheep-grazing, twenty-five or thirty millions for cattle, while the rest is rock and desert. The population consists of about 8000 civilised men and 3000 to 4000 savages, who are divided into four tribes differing in physique, language, and customs, but all leading a nomad life. The nomads of the mainland are the Patagonians and Tehuelches, who have at present made no use of the reservations assigned to them. The Yaghans wander in boats along the canals to the south of the Strait, subsisting on fish and shell-fish. The Alakalufs, closely resembling the Yaghans in person and habits, inhabit the channels to the west of the strait. The Onas of the north and east coast of Tierra del Fuego are as robust as the Patagonians, and therefore it has been suggested that they emigrated from the mainland or were thrust out of it. In language and customs, however, they differ widely from the Tehuelches. They seem to have little capacity for civilisation.

Chile has devoted her energies to the development of this territory, and grazing, commenced first on the mainland, yielded encouraging results. Consequently, the industry was extended to Tierra del Fuego, the Government assisting the movement by granting concessions of large tracts of land for long periods. Sheep-grazing has been remarkably successful in the territory of the Onas, where it has led, however, to some difficulties. The chief food of this people is a small rodent, the *Cururo*, which is exceedingly numerous in the great treeless plains,' and multiplies almost as fast as the rabbit. Where sheep have taken possession of the ground, the rodent disappears, and the Onas, deprived of their usual food, have found that sheep furnish even more palatable meat, and have learned the value of their skins. The graziers and the Onas have therefore become deadly foes, and the Government will have to devise means of reconciling their opposing interests.—*Globus*, Bd. lxx. No. 3.

ARCTIC REGIONS.

Dr. Nansen's Advance towards the North Pole.—The *Fram* left Vardö on July 21st, 1893, and on August 13th Dr. Nansen arrived at Vardö in the steamer *Windward*, of the Jackson-Harmsworth expedition, which made the voyage from Franz Josef Land in the short time of fifty-five hours. Though he did not actually reach the North Pole, he advanced much farther northward than any previous explorer. Leaving the Yugor Strait on August 4th, the *Fram* arrived off the mouth of the Olenek on September 15th, but was prevented from putting in for dogs by rocky scars previously unknown. The New Siberia Islands were passed on September 18th, and on the 22nd the vessel was made fast to a floe in lat. 78° 50′ N. and long. 133° 37′ E., and allowed to be enclosed in the ice. Thence it drifted gradually, on the whole, north-westwards. During the winter the temperature was uniformly low, with little variation; the minimum recorded was −62° (probably C. = −80° F.). At about nineteen degrees from the Pole the depth of the sea sank to 1600 to 1900 fathoms, and hence the theory of a shallow Polar basin is proved to be erroneous. On October 21st, 1894, the 82nd degree was passed, and on Christmas Eve the 83rd. On March 3rd the *Fram* reached its highest latitude, viz., 84° 4′; and eleven days later Dr. Nansen and Lieut. Johansen left the ship in lat. 83° 59′ N., and 102° 27′ E. (position marked on map), with dogs, sledges, and kayaks, to travel as far as possible towards the Pole. For more than three weeks they marched across the ice, until, on April 7th, in lat. 86° 14′, it became so bad that further progress seemed impossible. Dr. Nansen made an excursion on *skier* up to 86° 25′ to examine the ice. It was, however, everywhere in the same condition, drifting freely and rapidly in a northerly direction. No land was visible in any direction. The following day the explorers shaped their course for Franz Josef Land, and on August 6th reached some ice-capped islands in 81° 38′ N. lat., and

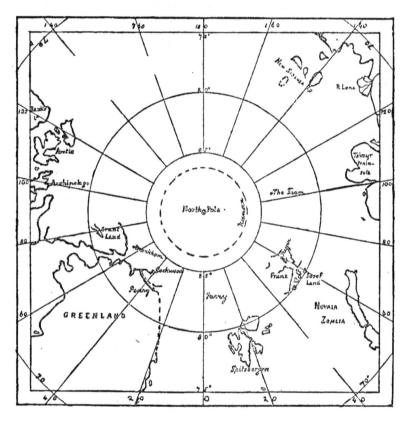

about 63° E. long. Through these they made their way in kayaks, and on August 26th came to land in 81° 13′ N. lat., and 56° E. long., where they determined to encamp for the winter. They built a hut of stones, earth, and moss, and subsisted for ten months on bear meat and blubber. On May 19th of this year they started again, and having travelled through the islands of Franz Josef Land southwards, were making for Spitzbergen, when they fell in with the Jackson expedition to the south-south-east of Cape Flora. Owing to the discrepancies of Payer's map, and to the fact that their watches had stopped, the travellers were then very uncertain of their whereabouts.

Dr. Nansen's voyage, though not extended to the very Pole itself, has been most successful. The *Fram* exceeded his expectations, resisting the pressure even when the ice, as on January 4th and 5th, 1895, was piled up high above her bulwarks. Nor was the traveller disappointed in finding a northward drift, caused by the discharge of the Siberian rivers, across the Polar regions. Professor Mohn, who happened to be at Vardö when Nansen arrived, pronounces the scientific results of the voyage to be magnificent.

The accompanying map shows the farthest points attained by the chief Arctic explorers during the present century. In 1827 Parry reached in boats and sledges a point north of Spitzbergen in lat. 82° 45′ N. This record was not beaten for nearly half a century, until Commander (now Admiral) A. H. Markham and Lieutenant Parr, of Sir George Nares' expedition, reached in 1876 the latitude of 83° 20′ 26″ N., to the north of Grant Land. Six years later, in 1882, Lieutenant Lockwood of the Greely expedition made another small advance, up to Lockwood Island off the north-west Greenland coast, in latitude 83° 24′. Now Nansen has reduced the distance yet to be covered towards the Pole by over three degrees, or 210 miles, leaving only some 240 miles to be traversed on a future occasion. The broken circle on the map, with Dr. Nansen's distance from the Pole as radius, permits a comparison to be easily made with other incursions into the Polar regions.

On August 19th the *Fram* arrived at Skjærvö in Norway, Captain Sverdrup and his crew being all in excellent health. The vessel had drifted to the north of Franz Josef Land as far as 85° 57′, or nearly as far north as Nansen, and to a far higher latitude than any other ship. Open water was reached at lat. 81° 32″ long., 11° 40′ E. The greatest depth sounded was 2185 fathoms.

GENERAL.

Temperatures and Currents of the Pacific Equatorial Zone.—Professor Krümmel describes in *Petermann's Mitt.*, Bd. xlii. No. 6, the results obtained by Dr. C. Puls from an examination of the log-books in the Deutsche Seewarte. The parts where special difficulties have hitherto been encountered are the neighbourhood of the Bismarck Archipelago, where strong easterly streams are reported ; the Equatorial counter-current, which by some authorities is extended across the whole ocean ; and the corner between Lower California, the Galapagos, the Gulf of Guayaquil, and Panama. The southern Equatorial current has always its greatest strength to the north of the Equator. Its lowest temperature occurs as a rule just to the west of the Galapagos, along a narrow strip lying along the Equator between 90° and 105° W. long. Between June and December the temperature is below 72° F., and between September and November even less than 68° (minimum 62°). Between December and March this cold area moves westwards till it lies between 140° and 150° W. long., where it has a relatively high temperature, never, however, reaching 77°, and in April returns to the Galapagos with a falling temperature. This cold strip has nothing to do with the Peruvian current, but is due to a local welling-up from below, which is in all probability a compensating movement caused by the flowing off of the surface water north-westwards and south-westwards.

Dr. Puls has detected a remarkable disturbance of the south Equatorial current in the month of March. To the west of the Galapagos, just at the time when the cold water area is displaced far to the east, this current entirely fails in many years, as also the trade-wind, which gives place to calms permitting easterly currents to develop. The same phenomenon is observed in the Atlantic on the Equator, though less frequently.

The disturbances of the Equatorial current in Melanesian waters are also clearly defined by Dr. Puls. In the summer of the southern hemisphere easterly currents are strongly marked, flowing from December to February along the north coast of

New Guinea, the Bismarck Archipelago, and the Solomon Islands, and to the east of the latter. From May to November easterly currents have been observed only close to the Solomon Islands, between 163° and 170° E. long., and then only feeble ; they disappear entirely in July and August.

The Equatorial counter-current is present in all months of the year, though it is sometimes very narrow and thin during the winter of the northern hemisphere. It has not been doubted that in summer and autumn it flows with a powerful current over 5 degrees of latitude and 155 of longitude. In all respects it much surpasses the Guinea stream.

Dr. Puls' greatest merit lies in his exposition of the movements of the water between Central America and the Galapagos Islands. Berghaus has laid down monsoon currents, flowing north-westwards in summer along the coast of Central America and Mexico under the influence of the south-west monsoon, and south-westwards during the winter months. But the log-books show such movements only in Mexican waters ; in our winter a branch of the cold Californian stream (the head of the north Equatorial current) flows south-eastwards past Cape S. Lucas as far as the Gulf of Tehuantepec, and then turns round to the west, while in summer very warm water flows north-westwards from the Gulf of Panama, causing sudden changes of temperature at Cape S. Lucas, where it meets the Californian stream, $3\frac{1}{2}°$ colder even at that season. South of 10° N. lat. the currents are very remarkable. Dr. Puls asserts that there is a great difference here between the arrangement of the currents from January to May and from June to December. In the latter period not only the water of the Equatorial current entering the Gulf of Panama is driven by the south-west monsoon to the north-west, but water is also drawn up from the south between the Galapagos and the mainland. From January to May, on the other hand, a part of the warm water from the Gulf of Panama flows southwards and about the Equator turns westwards, accompanies the cold Peruvian current north-westwards past the Galapagos, and then is carried northwards and eastwards by the Equatorial counter-current. Hence an elliptical whirlpool is formed along the parallel of 3° or 4° N. lat. and between 82° and 90° W. long. In summer such a southerly movement of the water along the mainland is prevented, according to Dr. Puls, by the prevailing south-west monsoon. Professor Krümmel differs from Dr. Puls on this question. The isotherms run close together north-westwards in all the months of the year, and show a difference of 11° to 13° in a distance of 250, up to the northern side of the Equator. The current can hardly run at right angles to the isotherms, for it is impossible that the colder water can be so much raised in temperature within so short a distance. From the isotherms alone, then, it may be gathered that no water flows northward into the Gulf of Panama, but that the currents are the same during the south-west monsoon as in winter. Professor Krümmel believes also that Dr. Puls has been misled with regard to the directions of the currents along the parallel of 1° N. lat. and between 82° and 83° W. long. by the directions of the winds, which are marked on the map, and by their number render it indistinct. The local cooling of the surface by the showers of the monsoon must also be taken into consideration, and the eddies of warm water which set northward along the shore may also have influenced Dr. Puls.

Very interesting is the behaviour of the water which runs north from the Gulf of Panama and in summer is found beyond 105° W. long. The wind then blows over the Central American coast from the north, or even north-west, on the northern side of the small cyclonic area which on its southern side produces the rainy winds of the south-west monsoon. Yet the current makes its way in the teeth of this wind. In winter and spring the current is driven from the pe⋅insula of Nicoya westwards along the 10th parallel of N. lat. by the *Papagatlos* or *Papagayos*, strong north or east winds which prevail at those seasons, and with such velocity that the cold deep water is drawn up to the surface, where the temperature, instead of the normal $78\frac{1}{2}°$, sinks to 77° or occasionally to 73°. So also the similar *Tehuantepec* wind drives the south-easterly current at Cape S. Lucas southwards towards the *Papagayo* drift, and the colder water welling up brings the temperature of the surface down to 75°. Dr. Puls confines the *Tehuantepec* cold area to December and January, whereas the *Papagayo* continues to March and even, with diminished strength, to April. In the Gulf of Panama the cold water may be found in March close to land, observations having been made of temperatures ranging from 68° to $89\frac{1}{2}°$.

A Proposal for the Organisation of an International System of Seismic Stations.— A number of scientific men of nearly all European countries have combined to promote a scheme for the systematic observation of earth-tremors, and have issued a small pamphlet explaining their aims. For many years slight movements imperceptible to the senses have been noticed at several stations in Europe. They often last for several hours, extending for thousands of miles without sensibly losing their intensity, and they take place in countries not usually subject to ordinary earthquakes, and when no perceptible disturbances have occurred in Europe. Consequently those interested in these tremors have sought for an explanation in seismic catastrophes that have occurred in distant countries, and have at length succeeded in placing beyond doubt the existence of a connection between the two sets of phenomena.

Many of the more important earthquakes occur in Japan, and as seismical observations are regularly recorded in that country, the exact time may be determined that the movement takes to arrive in Europe. Thus, a violent shock which occurred at Tokio in April 1889 was perceived at Potsdam, 5600 miles distant, after an interval of thirteen minutes. Again, the great earthquake which shook the western districts of the Argentine Republic on October 27th, 1894, was noticed at Rome, 7100 miles from the centre of disturbance, seventeen minutes after it had been perceived at the Observatory of Santiago in Chili ; and a little later at Tokio, 10,800 distant.

These and other examples show that the part of the movement which causes the first perturbations in Europe travels with a velocity of about ten kilomètres (6·2 miles) per second (v_1). The later disturbances, in the case of very distant earthquakes, do not make their appearance until thirty or forty minutes later. They take the form of long, low waves, which have been found, in a few cases favourable for observation, to be twenty-five to thirty miles long, and only an inch or two high. These waves are propagated with a mean velocity of 1·7 miles per second (v_2).

The above values for v_1 and v_2 refer to movements the origin of which is at a distance of about a quarter of the Earth's circumference. In general, they are less when the distances from the seismic centres are shorter. For a distance of 3000 miles v_1 does not exceed three miles, and 2·2 miles for 930 miles. The law of decrease for v_2 is not easily determined for long distances ; at 900 miles it does not exceed 1·5 miles, and decreases rapidly towards the epicentre. It seems, therefore, beyond doubt that the movement travelling with the velocity v_1 passes through the Earth, and its great rapidity is probably due to the fact that vibrations are propagated more quickly in the depths of the Earth than at its surface. As for the long waves of velocity v_2, it would seem from observations at long distances, that they pass chiefly over the surface, though observations at short distances show that they are also partly transmitted through the globe with a velocity increasing with the depth.

An important instrument for registering these perturbations is the horizontal pendulum. It is movable round an axis almost vertical, and, if displaced when the ground is stationary, returns rapidly to its position of equilibrium, but it is put into violent agitation by the least tremor of the ground or alteration of level. By means of a small mirror attached to its extremity, its movements may be photographed on to a revolving slip of paper. In Italy very long and heavy vertical pendulums are used, which register the movements by means of pencils attached to the bob. They produce very detailed diagrams of the first phase of the disturbances, but require more than thirty feet of paper to indicate the whole movement. The second phase is shown better in the diagrams of this instrument than in those of the horizontal pendulum. The English bifilar pendulum is an instrument of great sensitiveness. It consists of a mirror attached by two strings of unequal lengths to two points, one of which is vertically above the other. A movement not in the plane of the mirror causes it to revolve. As it has not yet been furnished with a photographic recorder, it is impossible to say exactly what its capabilities are, and whether it can register the initial movements, which are of so much importance.

The results hitherto obtained with inefficient means encourage the promoters of the present scheme to propose, in the first place, the installation of a network of seismographic stations so situated as to record systematically the propagation of terrestrial movements from the great seismic centres, and, as a commencement, they would select about ten stations arranged more or less symmetrically with regard to a semicircular arc starting from Japan, which is not only one of the most important seismic countries, but also possesses the best organisation for the observation of earthquakes. Tokio, situated in 35° N. lat. and 140° E. long., has for its antipodes a point off the coast of South America, a little to the east of Buenos Ayres. Accordingly the following stations are suggested :—Shanghai, 16° from Tokio ; Hong-Kong, 26° ; Calcutta, 47° ; Sydney, 69° ; Rome, 89° ; Tacubaya, in Mexico, 102° ; Natal, 121° ; Cape of Good Hope, 136° ; Santiago de Chile, 154° ; and Rio de Janeiro, 167°. At all these stations the time can be ascertained with precision. The cost of the instruments and their installation and maintenance for a year would probably amount to £50 for each station. When once properly adjusted they may be intrusted to any one, even if he have no scientific knowledge.

A necessary adjunct to the system of observation is a centre for the collection and publication of the reports of earthquakes throughout the world. It is therefore intended to publish these reports from time to time in the form of gratis supplements to Professor Gerland's *Beiträge zur Geophysik.* Information of all kinds regarding earthquakes perceptible at a considerable distance will be made known through this medium, and the value of these indications will be greater when accompanied by exact details about the position of the epicentre and numerous time observations made in the neighbourhood. Secondly, observations will be included of shocks, perceptible or imperceptible, registered by sensitive instruments—horizontal, vertical, and bifilar pendulums, seismographs and seismoscopes of different forms, etc.—whether they arise from a known origin or not. Under this head the reports of the proposed stations will be the most important.

The great value of observations of this nature need not be insisted on. As it is almost certain that the vibratory movement caused by an earthquake is propagated through the Earth with a velocity depending on the density and elasticity of the strata at different depths, the observations will throw light on the condition of the interior of the Earth, where direct observation will probably never be possible, and at the same time they will assist the seismologist by furnishing information about this inaccessible region. The promoters of the scheme will be glad of assistance from friends of science in all countries, and will receive reports and observations of seismic phenomena. Those who are desirous to help in any way towards the realisation of the scheme are requested to communicate with Professor Gerland of the Strasbourg University.

Sun- Spots and Volcanic Eruptions.—Signor Luigi de Marchi has discovered a remarkable correspondence in the occurrences of these two phenomena. The question was discussed by E. Kluge in 1863, and his figures show the periods for the years 1818 to 1840 much more clearly than De Marchi's, but the great variations during the years 1860-1875, shown in the latter's figures, are a remarkable confirmation of the law. The tables of Kluge and De Marchi are given in *Himmel und Erde* for August, as well as curves showing the frequency of spots and cruptions. In these there certainly is apparent a conspicuous parallelism. Indeed, the eruptions during the years 1840-1875 were twice as numerous at the times of spot minima as at the spot maxima. Against this evidence the result of Poey's investigation into eruptions in general, and earthquakes in the Antilles and Mexico, namely, that there is at most a very small increase in these phenomena during periods of spot minima, is of less weight.

MISCELLANEOUS.

Mr. Goodchild will deliver a course of twenty lectures on the Rivers, Mountains, and Sea-Coast of Scotland, at the Rooms of the Photographic Society, Castle Street, Edinburgh, on Fridays, at 8 P.M., commencing on October 16.

Mr. Albert F. Calvert has fitted out an expedition, under the command of Mr. L. A. Wells, surveyor of the Elder expedition, to continue the exploration of West Australia. The party left Cue, in the Murchison District (lat. 27° 25' S. and long. 117° 52' E.), in May last.

We have received the *Annals of Scottish Natural History* for the quarter ending in July. It contains reports on the occurrence of rare birds and fishes in Scotland, and on *Hymenoptera* and *Hemiptera* collected on Ben Nevis by Mr. W. S. Bruce, with other interesting articles, besides the usual zoological and botanical notes, etc.

Dr. Max Schoeller, who has travelled with Dr. Schweinfurth in Abyssinia, is about to undertake a journey into Central Africa. Having marched by Kilimanjaro and Kavirondo to the north shore of Victoria Nyanza, he will explore the countries to the north of Mount Elgon, the Baringo Lake, and Kenia, and possibly he may try to prolong his journey thence to Abyssinia, being provided with the necessary means.—*Mitth. der k.k. Geogr. Gesell. in Wien*, Bd. xxxix. Nos. 4 and 5.

When a new viceroy arrives in Canton, a list of the inhabitants is handed to him. From such a list, prepared in June 1895, it appears that the town and suburbs then contained 499,298 inhabitants—336,754 males and 162,544 females. These figures do not include the inhabitants of the Ho-nam quarter on the opposite side of the river, who may be estimated at 30,000, nor the 20,000 who dwell in boats on the river.—*Bull. de la Soc. de Géogr. Comm. de Paris*, Tome xviii. Fasc. 7 and 8.

In *Petermanns Mitt.*, Bd. xlii. No. 6, Dr. J. Rein identifies Flatey, the island whence Erik the Red sailed to Greenland, and which gave its name to the cele-

brated *Flateyjarbók.* It is not, as he has learned from certain books and from Dr.
Thoroddsen, the island shown in maps and charts off the north coast of Iceland,
but one of a numerous group in the Breidhfjördhr, on the west coast. They are
frequently visited, being a good fishing-station and a breeding-ground of the eider
duck, and some of them are inhabited. Flatey is about a mile in circumference,
and has a population of 150.

In the *Annales de l'Université de Grenoble,* M. Collet remarks that, if it be
established that there is always a diminution of the force of gravity in the neighbourhood of lofty ranges, this circumstance may be employed to determine the site
of ancient chains now planed down by erosion. Thus the deficiency of gravity on
the parallel of Bordeaux and on the plains of Southern Russia may indicate
the sites of offshoots from the main range to which M. Marcel has given the
name Hercynian. Erosion has long ago removed the relief, without, however,
obliterating the details of structure and composition of the base.—*Ciel et Terre,*
July 16th.

Mr. Peary started in July for Greenland. It was his intention to extend his
survey, already completed from Cape York to Cape Alexander, as far north as
Littleton Island. He will also endeavour to enter Jones Sound, which has not
been entered to any distance since Inglefield's voyage in 1852. On his way south
he will probably visit Cumberland Sound, and spend a week in Hudson Strait.
He is accompanied by two parties of scientists, of which one, headed by Mr. Ralph
S. Tarr, will make geological collections and glacial observations in the neighbourhood of Devil's Thumb, Melville Bay, while the other, under Prof. A. E. Burton,
will make pendulum observations and study the glaciers at Umanak Fiord.—*Bull.
of the Am. Geogr. Soc.,* vol. xxviii. No. 2.

NEW BOOKS.

From North Pole to Equator: Studies of Wild Life and Scenes in Many Lands.
By the naturalist-traveller, ALFRED EDMUND BREHM. Translated from the
German by MARGARET R. THOMSON. Edited by J. ARTHUR THOMSON, M.A.
London : Blackie and Son, Limited, 1896. Pages xxxi + 592. *Price 21s.*

Brehm was a naturalist thoroughly alive in all his senses. A true biologist, he
studied the nature and habits of plants and animals in their natural habitats. The
result was a series of remarkable volumes on animal life invaluable to the naturalist
and of interest to all who read them. But he also used to give brilliant lectures
on the many types of life he had experienced, and these popular descriptions of
scenes in many lands were well worth collecting and translating.

Brehm had seen much of the world, from the frozen deserts of the north to the
arid regions of Africa, and even beyond, for he had glimpses of the skirts of the
dense primæval forests of equatorial lands. Wherever he went, he saw, heard, and
felt acutely, and further had the gift of translating his impressions into forcible
and picturesque phrases, which convey very vivid pictures to the reader.

It is very difficult to notice features of special interest in a work so full of
interesting matter. The general reader, perhaps, will be most fascinated by the
chapter on love and courtship among birds ; the geographer will not find a single
lecture that does not bear on his favourite subject and make him more enthusiastic
about it than ever.

Even the frozen wastes become interesting when looked at through the eyes of one who sees them not devoid of all beauty, not a lifeless wilderness, but periodically stirring with life in wondrous fulness. "The year is well advanced before the tundra begins to be visibly peopled. . . . Hesitatingly the summer visitors make their appearance. The wolf follows the reindeer, the army of summer birds follows the drifting ice blocks on the streams. Some of the birds remain still undecided in the regions to the south, behave as if they would breed there, then suddenly disappear from their resting-place by the way, fly hastily to the tundra, begin to build directly on their arrival, lay their eggs, and brood eagerly, as though they wished to make up for the time gained by their relatives in the south. Their summer life is compressed into a few weeks." . . .

Here is Brehm's description of these barren lands :—"The tundra is neither heath nor moor, neither marsh nor fen, neither highlands nor sand-dunes, neither moss nor morass, though in many places it may resemble one or other of these. . . . In my opinion the tundra most resembles one of these moors which we find—and avoid—on the broad saddles of our lofty mountains ; but it differs in many and important respects even from these boggy plateaus ; indeed its character is in every respect unique."

We might make similar quotations from the pictures of the steppes of Asia and of inner Africa, of the primæval forests of Siberia and Central Africa, of the Sahara and its fringes, and the life of the inhabitants of desert, steppe, and forest. There is no more graphic description of the seasonal vicissitudes of continental lands near the tropics than that which begins the chapter on the steppes of inner Africa, a description too long to quote here.

Against one thing we must protest, and that is the abuse of the word winter. "As the sun travels to the north, the winter sets in rapidly," we are told of the upper Nile region . . . "the long and terrible winter of the African interior—a winter which brings about by heat the same dire effects as are wrought in the north by cold." Winter is used to express the resting-time of plants and animals, and not the astronomical and climatic season, for which it should be properly reserved.

Brehm has certainly not lost anything in the hands of his translator, Mrs. Thomson, who has done the work with great skill ; and given us a contribution to English literature as well as to the literature of travel written in English. Mr. Thomson in his introductory essay classifies the various naturalist-travellers into romantic, encyclopædic, general, specialist, and biological types, and rightly puts Brehm in the last group. He also gives a very brief outline of his life.

Mr. Thomson has done us a further service by compiling and printing a list of other works in English by naturalist-travellers, most of which should be in every public and school library, where Brehm undoubtedly should have a prominent place.

At the end of the book Mr. Thomson has inserted many notes serving to explain, emphasise, or slightly modify the text. In one on the origin of the steppes he quotes Seebohm as saying, " No one, so far as I know, has suggested a climatic explanation of the circumstances," while, as a matter of fact, climatic explanations are very vigorously supported, especially by German geographers against the views of many Russians, who maintain that the nature of the soil is the prime factor.

This is a book the teacher will turn to again and again, for in few others will he find the graphic descriptions that help us so much to realise what we have not seen for ourselves, and to give a clear account of them to others.

There are many excellent illustrations in the text, but no map.

Climbs in the New Zealand Alps. By E. A. FitzGerald, F.R.G.S. London :
T. Fisher Unwin, 1896. Pp. 363. *Price* 31s. 6d. *net.*

In the summer of 1894, when scaling the European Alps with Sir Martin
Conway, Mr. FitzGerald planned the expedition to New Zealand detailed in this
magnificent volume. Sailing in October 1894 from Brindisi, he eventually reached
New Zealand in December. His party left Christchurch by rail for Fairlie Creek,
the nearest station to the Southern Alps of New Zealand, of which an admirable
map by Stanford, from the latest Government surveys, with additions by Mr. Fitz-
Gerald, is appended to the volume, throughout which are interspersed most
excellent illustrations from original drawings and from photographs. The author
gives a graphic account of mountaineering amid the New Zealand Alps, including
his ascent of Mount Tasman (11,475 ft.). His companion, Zurbriggen, ascended
Mount Cook (12,349 ft.). As to Mr. FitzGerald's literary style, the following
paragraph, describing his farewell glimpse of the ranges of the Southern Alps, may
be selected as an example :—" The whole chain of great snowy peaks stood forth
clearly against the crimson west, and was again mirrored in the placid lake at our
feet. So peaceful was this evening scene, that, as I gazed, I could scarcely realise
that these white, glittering peaks had been the theatre of so much hardship, so
much privation, and so much peril during the last few months. I felt amply
rewarded, however, for the long marches and hazardous climbs, for the cold, the
wet, and the general discomfort of my sojourn, by the closer knowledge that I had
obtained of these majestic heights, these seemingly impregnable fastnesses of ice
and rock, by the sights which had been unfurled before my eyes of the wonderful
contrasting zones of glacier and vegetation on the west coast, unparalleled by all
that I have yet seen or heard described in the extraordinary proximity in which
one climatic region is brought to another."

*The Portuguese in South Africa, with a Description of the Native Races between the
River Zambezi and the Cape of Good Hope during the Sixteenth Century.* By
George M'Call Theal, LL.D. of the Queen's University, Kingston, Canada,
etc., etc. With Maps. London : T. Fisher Unwin, 1896. Pp. xvi + 324.
Price 6s.

Dr. Theal tells us in his preface that the publication of this volume has been
rendered necessary by recent events. When, a few years ago, he wrote his *History
of South Africa* in five volumes, and even when, two years ago, he wrote *South
Africa* in " The Story of the Nations " series, the expression " South Africa " meant
Africa south of the river Limpopo, but now, mainly through the enterprise of Mr.
Cecil Rhodes and the Chartered Company, the term has a much wider signification,
the boundary having been removed from the Limpopo to the Zambezi, or between
400 and 500 miles farther north. The starting-point of the history has thus been
moved back from the arrival of the Dutch Van Riebak in Table Valley, in 1652, to
the arrival of the Portuguese Da Nhaya in Sofala, in 1505. The present volume
has therefore been written in order to fill the gap between these events and
dates.

But the author's task has involved much more than that of filling the gap of a
century and a half between the dates mentioned. He has gone back to prehistoric
times, and has found interesting material for comment and speculation in the proofs
that have been found of an early " stone age " in South Africa. This is followed
by an account of the early inhabitants—the Bantu, the Hottentot, and the Bushman—and of their characteristics, their superstitions, and their social customs.

The foundations having been thus laid, the superstructure, which forms the proper subject of the book, begins with the exploration of the west coast of Africa in the fifteenth century, and specially with the discovery of the Cape of Good Hope by Bartholomeu Dias in 1486. Thenceforth the narrative proceeds on historic lines, the prominent landmarks being the discovery of Table Bay, the occupation of Sofala by the Portuguese, the death of Lourenço d'Almeida, the exploration of the lower Limpopo by Lourenço Marques, and the disastrous expedition of Francisco Barreto in the Zambezi Valley, which ended with his death at Sena. Thereafter the French, the English, and the Dutch appeared as rivals of the Portuguese on the southern coast of Africa. This brings us down to the beginning of the seventeenth century. From that point the narrative runs rapidly over a period of nearly two centuries, including civil wars, conflicts with native tribes, and an account of the beginning of the slave trade with Brazil.

This brings us to the middle of the nineteenth century—a point which marks the revival of Portuguese activity in South-eastern Africa. The last chapter of the book, which treats of the history of South Africa since 1850, will seem to many, in view of recent events, the most interesting section of the volume. It refers to the new value which Delagoa Bay and the coast north of it acquired after the settlement of the Boers in the Transvaal. Similarly, the northern part of the Portuguese Coast has more recently obtained new importance from the opening up of the interior by the Chartered Company. In both cases the revived activity of the Portuguese has been signalised by the construction of a railway from the coast to the interior—in the south, from Lourenço Marques, in the other from Beira. The volume closes with a prediction of increased prosperity for Portuguese South Africa; but the events of the last two or three years do not come within the author's scope, and consequently no notice is taken of the recent and present disturbances in that quarter.

Dr. Theal writes calmly and dispassionately, without partiality, and also without any glow of enthusiasm; but his book is evidently the result of patient labour and research, and it may be accepted as a trustworthy guide both on matters of fact and on matters of opinion. The maps, which are disappointingly meagre, are also poorly executed.

In the Kingdom of the Shah. By E. TREACHER COLLINS, F.R.C.S. London : T. Fisher Unwin, 1896. Pp. 300.

The writer of this book paid a two months' visit, in the spring of 1894, to Ispahan, for the purpose of treating H.R.H. the Zil-es-Sultan for an affection of the eyes, from which he had been suffering for some time. Mr. Collins, accompanied by his wife, made the outward journey by way of Bushire and Shiraz, returning home by Teheran, Resht, the Caspian, and the Caucasus. The book, which is pleasantly written, gives on the whole a pretty accurate account of what the writer saw during his journey over a well-known track, and his short sojourn in the former capital of Persia. Fifty years ago it would no doubt have been a welcome addition to the few works then extant on the modern conditions of life in that country. As it is, it contains little or nothing that is not to be found in the recent works of writers like Mr. Curzon and Dr. Wills, who had far greater opportunities than Mr. Collins, not only of seeing the country and its people, but of correcting early impressions in the light of later experience.

In common with many other travellers who have published books on countries with whose language they were not conversant, Mr. Collins has fallen into some rather meaningless mistakes in native words and proper names—such as *Satib* for *Sartip, Mak-Sud-Beggi* for *Maksud Beggi, Faith Ali* for *Fath Ali, Zille Sultan* for *Zil-es-Sultan,* etc. But a mistake of a totally different kind that he appears to have made during his journey is one against which we feel bound to protest—the too ready recourse to the argument of the riding-whip in disputes with natives not in a position to think of retaliating. At page 285, for instance, after narrating a dispute with a Chaparji (owner of post-horses) about delay in furnishing animals, he goes on to say : " It being evident he was lying, and having a long and difficult stage before us, I was obliged to resort to physical force and use my whip on him. He then at once produced five horses, and as we left had the cool impertinence to ask for a present. The led horse, a poor beast with only two sound legs, delayed us considerably. Fortunately after going about a farsakh ($3\frac{1}{2}$ miles), we met some others returning, and were able to change it for the best of them." Mr. Collins was probably unaware that five is an unusually large number of fresh horses to expect to find at every post-house, and evidently quite misunderstood the postmaster's efforts to induce him to remain until a sounder animal than one of those he insisted on starting with had returned from the stage in front. The nature of the impression left upon the mind of the unfortunate Chaparji by the incident thus complacently related by Mr. Collins may easily be guessed.

Description of a Journey in Western China. By G. E. GRUM-GRZHIMAILO, assisted by M. E. GRUM-GRZHIMAILO. *Vol. I., Along the Eastern Tian-Shan.* (In Russian.) St. Petersburg : C. M. Nicolaief, 1896. Pp. xii+547.

It is now some five years since the two Russian brothers returned from their journey in Central Asia, and during the interval their journey has often been referred to, and the conclusions they deduced from their observations discussed. Other travellers also have visited the Tian-Shan and added greatly to the discoveries of the Messrs. Grum-Grzhimailo. This work, therefore, does not possess the charm of novelty in the same degree as if it had been published soon after the journey was made. In a geographical library it is, nevertheless, a valuable addition, for the journey described was most fruitful in scientific results.

In the present volume, the travellers describe their exploration of the Tian-Shan range, Turfan, and the Chol-tau, which was briefly noticed in vol. vii. of the *Magazine.* They were struck with the majestic appearance of the Dos-megen-ora, and especially of the Bogdo-ola. Of the latter they say : " In all Central Asia there is no spot more picturesque and at the same time more mysterious and grand. The gigantic mountain 'holding up the clouds and obscuring the sun and moon,' as the Chinese say, visible from five towns, but best from Central Jungaria, whence it seems indeed like a ' throne' or a truncated cone rising to a wondrous height from beyond the mass of snow-clad mountains, was now before us in its entirety, unconcealed by foothills. Its base was washed by the turquoise-blue waters of a lake, the banks of which were wild cliffs clothed with forest, and above them, on our side, emerald pastures and spruce groves, while opposite were deposits of particoloured stone. And all this scene was framed in a circle of mountain summits, interrupted only at the north by the wild, narrow gorge of the Khaidajan river. What a marvellous and mysterious place ! And this is in the heart of the Gobi desert, which embraces with two broad arms this Parnassus of the Turkish and Mongolian tribes, hitherto unseen by Europeans."

The brothers obtained a couple of specimens of the wild horse of Jungaria, *Equus Przhevalskii.* They think that Przhevalski mistook the wild ass for this animal, from which it can hardly be distinguished at a distance, owing to the similarity of colour and size. The illustration shows how much domestication has improved the breed.

The book is well printed, and many of the illustrations are excellent.

New Ground in Norway. Ringerike—Telemarken—Sætersdalen. By E. J. GOODMAN, Author of *The Best Tour in Norway,* etc. London: George Newnes, 1896. *Price* 10s. 6d.

The contents of this work hardly justify the title. The districts described can hardly be termed *new ground* to British tourists, though they are not so much visited as the western parts of Norway. Nevertheless, the descriptions will be useful to those tourists who intend visiting these interesting parts of the country, and are very pleasant reading to those who have seen them. The Vestfjorddal, the Telemarken Canal, Dalen, and the Sætersdal are delightful, especially in June, when the days are long and tourists comparatively few ; and Mr. Goodman is a very competent guide. He travelled leisurely, picking up information whenever he had an opportunity, and this he retails in an easy style. We can hardly credit his statement that the Rjukanfos is eight hundred feet high; at the little Krokan hotel, we were told that the estimates usually given were far beyond the mark. Nor can we quite agree with Mr. Goodman as to the cleanliness of the Sætersdal people. There is no doubt much improvement of late years, especially where tourists put up.

The volume is adorned with fifty-six excellent illustrations from photographs by Paul Lange. They do great credit both to the photographers and the reproducers. The publishers also have done their work well.

From Cairo to the Soudan Frontier. By H. D. TRAILL. London : John Lane, 1896. Pp. 256. *Price* 5s.

This book is a reprint of articles which appeared in the columns of the *Daily Telegraph,* and records the author's impressions derived from a couple of brief tours in Egypt during the winters of 1894-5, 1895-6.

The sixteen chapters it contains are exceptionally well written and interesting, although the book is not geographical in any sense of the word. The author evidently believes in the extension of the Egyptian frontier. He remarks on the improvement which has taken place in the native troops under British supervision, and refers to the Dervish raids and their results. At the present time it will be doubtless read with interest.

The Wild North Land ; being the Story of a Winter Journey, with Dogs, across Northern North America. By Colonel W. F. BUTLER, C.B., F.R.G.S., Author of *The Great Lone Land,* etc. Tenth Edition. London : Sampson Low, Marston, and Co., 1896. Pp. x+358. *Price* 2s. 6d.

The journey here described was accomplished in 1872. Colonel Butler travelled from the Red River to Athabasca, and through the cañon of the Peace river into British Columbia. The present edition is clearly printed, though in a cheap form, and is similar to the edition of *The Great Lone Land* issued a year or two ago. It is illustrated and provided with a map.

Tantallon Castle. The Story of the Castle and the Ship, told by ELIZABETH ROBINS PENNELL. With illustrations by W. L. WYLLIE, A.R.A. ; W. HATHERILL, R.I. ; JOSEPH PENNELL ; A. S. HARTICK, N.E.A.C., and D. B. NIVEN, A.R.I.B.A. Edinburgh : Printed for the Castle Mail Packets Company, by T. and A. Constable, 1895.

The history of Tantallon Castle is here shortly sketched; and its namesake the vessel is described. The chief feature is the illustrations, in which not only the castle and ship are portrayed, but also one or two scenes in Africa.

Die Liparischen Inseln. VI^tos. Heft.—*Alicuri.* Prag : Heinr. Mercy, 1896.

Alicuri is the smallest of the group, having an area of little more than two square miles, and is also the least important. But its insignificance has not prevented the author from describing it with the same thoroughness as the larger islands. Alicuri consists of a round cone rising to a maximum height of 2215 feet above sea-level. Like Filicuri, it is bare and rocky.

Der Vesuv und seine Geschichte von 79 *n. Chr.*—1894. Von Dr. J. SCHNEER und
    VON STEIN-NORDHEIM. Carlsruhe : G. Braun. Pp. 70.

This publication is intended as an outline of the history of Vesuvius from the date of the Plinian eruption to the present day, and has been compiled chiefly for the use of tourists and others who wish to obtain a general idea of the chief features of interest in connection with the celebrated Neapolitan mountain. It aims at covering much the same ground as the late Professor John Phillips' book on the same subject, but can hardly be said to take the place either of that or of the almost equally good account of Vesuvius given in Murray's *Handbook to Southern Italy and Naples.* It appears to have been one of the objects of the authors to reproduce various illustrations relating to Vesuvius that have been published since the early "sixties." This has been done without giving either the dates or the sources whence the figures have been copied. The copies, moreover have been executed in a coarse style of drawing and engraving.

Minerals, and how to study them. By E. S. DANA. Second edition, revised.
    New York : Wiley and Sons. London : Chapman and Hall, 1896. 8vo,
    pp. 372. *Price* $1.50.

This is a book for beginners in Mineralogy by one who is an acknowledged authority in that science. As mineralogy is generally presented in text-books it is not apt to captivate any one who has not some special reason for attempting its study. And yet, if properly taught, the science should have many charms for beginners. We quite agree with the author that no subject is better fitted to cultivate the powers of observation, and at the same time to excite an active interest. In this little work we welcome the best introductory text-book of mineralogy with which we are acquainted. It is simply and clearly written, and its perusal we feel sure will induce many to begin the study of minerals, from which more ambitious works are apt to repel.

*Die atmosphärische Luft. Eine allgemeine Darstellung ihres Wesens, ihrer
    Eigenschaften und ihrer Bedeutung.* Von Dr. ADOLF MARCUSE. Berlin :
    Friedländer und Sohn, 1896. Pp. 77. *Price* 2 *M.*

Dr. Marcuse was one of the competitors for the Hodgkin Prize for the best popular outline of Meteorology, and his work was awarded an honourable mention by the committee and recommended for publication by the adjudicating committee.

The author divides the book into an introduction and three chapters. The first chapter is called "Statical Atmospherology," and deals with pressure, temperature, humidity, and the optical, electric, and acoustic phenomena of the atmosphere. The next chapter on "Dynamic Atmospherology" discusses variations of temperature, pressure and humidity, clouds, rainfall, thunderstorms and atmospherical electricity, the movements of the atmosphere, winds and the wind-laws. The last chapter applies the information thus gained to questions of climate and weather, and especially in their relations to marine, agricultural, aeronautical and medical problems.

The book is very well arranged, and clearly and accurately, if not very brilliantly, written.

Report of the Sixth International Congress held in London, 1895. Edited by the Secretaries. London: John Murray, 1896. Pp. xxxvi + 806 + 190.

The editorial work has been done by Dr. Mill, who was assisted in the revision of the proofs by Mr. B. V. Darbishire, M. Gustave Korts, and Dr. A. Markoff. Miss Cust gave help in the preparation of the index.

The papers, the chief of which were referred to in the *Magazine* last year, are printed in the original languages—French, German, Italian, and, of course, English —and are accompanied by their diagrams and maps. A Diary of the Congress, and a Catalogue of the Exhibition are added. The whole forms a valuable record of the proceedings, and a digest of information relating to all branches of geography.

The Statistical Year-Book of Canada for 1895. Eleventh year of issue. Ottawa : Government Printing Bureau, 1896. Pp. 965. Map, Tables, and Index.

As usual, the Year-Book published by the Canadian department of agriculture is well up to date. The statistics are carefully compiled, and the book shows the progress made during the year. Once more we note with pleasure that Canada makes rapid progress. The "Record" contains, in addition to the usual chapters, the first of a series of biographical notes of advocates of the principles of Confederation, the third of a series on "Countries with which Canada deals," and an extended analysis of the industrial, mechanical, and manufacturing returns of the census. A digest of the treaties Canada has made with her Indian tribes is appended to the synopsis of the treaties made by the mother-country in which Canada is specially interested. We can again commend Mr. George Johnson, the editor, on his work. The map shows the proposed route of the British submarine cable from Canada to New Zealand and Australia. We trust that it will soon be laid.

The "Contour" Road Book of Scotland. By HARRY R. G. INGLIS. Edinburgh : Gall and Inglis, 1896. Pp. x + 281. *Price 2s.*

It is no easy matter to give travellers by road a thorough notion of the character of the route. For cyclists especially, the important point is not the change of level along any particular stretch, but the way in which this change takes place. Contoured maps, unless on a very large scale, cannot show the short sharp rises and falls that are a great hindrance to the progress of the cyclist. In this book the gradients are given, and, knowing that one of 1 in 20 is a stiff hill, and that one greater than 1 in 15 is dangerous to descend on a bicycle, the traveller can at once perceive the difficulties before him. A general description of the road, distances, and a brief note of the objects of interest, etc., are added, as well as a sectional road map of Scotland. The roads are also shown in profile, but as the dimensions are small and the vertical scale is necessarily much exaggerated, only a very general notion of the changes of level can be acquired by comparison.

All this information is contained in a very small volume easily carried in the pocket.

NEW MAPS.

EUROPE.

SCOTLAND, New "Three Miles to Inch" Map of ——, for Tourists, Cyclists, and General Reference.   Sheets 4, 5, 7.   *Price* 1s. 6d. *each, on cloth.*
W. and A. K. Johnston, Edinburgh and London.

—— Reduced Ordnance Survey of ——.   Sheets 12 and 13.
John Bartholomew and Co., Edinburgh.

IRELAND, Tourist's Map of ——.   Ten miles to an inch.   *Price* 1s. ; *cloth* 2s.
John Bartholomew and Co., Edinburgh.]

FORMA URBIS ROMÆ.   Consilio et Auctoritate Regiæ Academiæ Lynceorum Formam dimensus est et ad Modulum 1 : 1000 delineavit Rudolphus Lanciani, Romanus.   Fasciculus Quartus.   *Mediolani, apud Ulricum Hoepli*, 1896.
Presented by Hugh Rose, Esq.

AFRICA.

SOUTH AFRICA, Wall Map of ——.   50 inches by 42.
W. and A. K. Johnston, Edinburgh.

A clear and well-executed map of South Africa up to Matabeleland, accompanied by a small *Handbook* in which a general sketch of the physical and political geography of the country is given.   We are not aware that Pietersburg has yet any railway connection, while, on the other hand, considerable progress has been made with the line from Komati Poort to Leydsdorp.

AMERICA.

NÖRDLICHES VENEZUELA.   Von Professor Dr. W. Sievers.   1 : 3,000,000.   I. Übersicht der geologischen und tektonischen Verhältnisse.   Nebenkarten : 8 geologische Profile durch Mittel- und Ost-Venezuela.   Verhältniss der Länge zur Höhe wie 1 : 7·4.
Petermanns Mitteilungen, Bd. xlii. Tafel 10.

LAGO TITICACA.   Plano formado sobre los Trabajos de Pentland, Raimondi, Agassiz, etc.   Por Rafael E. Baluarte, Cartógrafo de la Sociedad Geografica de Lima, 1893.   Escala de 1 : 500,000.
Presented by the Geographical Society of Lima.

ATLASES.

MOLL'S ATLAS OF SCOTLAND.   A set of thirty-six new and correct maps of Scotland, divided into its Shires, etc.   By Herman Moll, Geographer.
Reprinted from the first edition of 1725 *by R. S. Shearer and Son, Stirling*, 1896.

Moll's *Atlas of Scotland* may be said to have been the first popular work of its kind.   The work of Gordon of Straloch and Timothy Pont, as published in Blaeu's *Atlas*, is here reduced to a more compact form and amplified from the surveys of John Adair, besides being revised to date (1725) by "the generous informations of some curious Noblemen and Gentlemen."   To the antiquarian and student of Scottish topography the work will always be of great interest.   For the present edition of this interesting work we are indebted to the enterprise of Messrs. Shearer of Stirling, who have reproduced it by photo-lithography.

THE SCOTTISH

GEOGRAPHICAL

MAGAZINE.

ADDRESS TO THE GEOGRAPHICAL SECTION OF THE BRITISH ASSOCIATION.

(*Liverpool*, 1896.)

By MAJOR DARWIN, Sec. R.G.S., President of the Section.

IN reviewing the record of geographical work during the past year, all other performances pale in comparison with the feat accomplished by Nansen. It is not merely that he has gone considerably nearer the North Pole than any other explorer, it is not only that he has made one of the most courageous expeditions ever recorded, but he has established the truth of his theory of Polar currents, and has brought back a mass of valuable scientific information. When Nansen comes to England, I am certain that we shall give him a reception which will prove how much we admire the heroism of this brave Norwegian.

Besides the news of this most remarkable achievement, the results of a considerable amount of useful exploratory work have been published since the British Association met last at Ipswich. With regard to other Arctic Expeditions, we have had the account of Lieutenant Peary's third season in Northern Greenland, from which place he came back in September last, and to which he has again returned, though without the intention of passing another winter there. In October the *Windward* brought home more ample information as to the progress of the Jackson-Harmsworth Expedition than that communicated by telegram to the Association at Ipswich, and on her return from her remarkably rapid voyage this summer she brought back the record of another year. As to geographical work in Asia, Mr. and Mrs. Littledale have returned safely from their explorations of the little-known parts of Tibet; the Pamir Boundary Commission, under Colonel Holdich, has collected a great deal of accurate topographical information in the course of its labours; Dr. Sven Hedin continues his important researches in Turkestan ; and the Royal Geographical Society was glad to welcome Prince Henry of Orleans when he came to tell us about his journey near the sources of the Irrawaddy. As to Africa, the most important additions to our knowledge of that continent are due to the French surveyors, who have accurately mapped the recently discovered series of lakes in the neighbourhood of Timbuktu, Lake Faguibine, the largest, being found to be sixty-eight miles in length ; Dr. Donaldson Smith has filled up some large blanks in the map of Somaliland ; and Mr. and Mrs. Theodore Bent have investigated some interesting remains of ancient gold workings inland from the Red Sea. In other parts of the world less has been done, because there is less to do. Mr. Fitzgerald has proved for the first time the practicable character of a pass across the Southern Alps, thus supplementing the excellent work of Mr. Harper and other pioneers of the New Zealand Alpine Club ; and Sir W. M. Conway has commenced a systematic exploration of the interior of Spitzbergen, a region to which the attention of several other geographers is also directed.

It is impossible in such a brief sketch to enumerate even the leading events of the geographical year, but what I have said is enough to remind us of the great amount of valuable and useful work which is being done in many quarters of the world. It is true that if we compare this record with the record of years gone by, we find a marked difference. Then, there was always some great geographical problem to be attacked ; the sources of the Nile had to be discovered ; the course of the Niger had to be traced ; and the great white patches on our maps stimulated the imagination of explorers with the thought of all sorts of possibilities. Now, though there is much to be learned, yet, with the exception of the Poles, the work will consist in filling in the details of the picture, the general outlines being all drawn for us already. Personally I cannot help feeling a completely unreasoning regret that we have almost passed out of the heroic period of geography. Whatever the future may have in store for us, it can never give us another Columbus, another Magellan, or another Livingstone.` The geographical discoverers of the future will win their fame in a more prosaic fashion, though their work may in reality be of even greater service to mankind. There are now few places in the world where the outline of the main topographical features is unknown ; but, on the other hand, there are vast districts not yet thoroughly examined. And, in examining these more or less known localities, geographers must take a far wider view than heretofore of their methods of study, in order to accommodate themselves to modern conditions.

But even if we confine our attention to the older and more narrow field of geography, it will be seen that there is still an immense amount of work to be done. We have been filling in the map of Africa during recent years with extraordinary rapidity, but yet that map is likely to remain in a very unsatisfactory condition for a long time to come. Englishmen and other Europeans have always shown themselves to be ready to risk their lives in exploring unknown regions, but we have yet to see how readily they will undertake the plodding work of recording topographical details when little renown is to be won by their efforts. It should be one of the objects of geographical societies to educate the public to recognise the importance of this work, and General Chapman deserves great credit for bringing the matter before the International Congress last year in such a prominent manner. He confined himself to four main recommendations. (1) The extension of accurate topographical surveys in regions likely to be settled by Europeans. (2) The encouragement of travellers to sketch areas rather than routes. (3) The study of astronomical observations already taken in the unsurveyed parts of Africa in a systematic manner, and the publication of the results. (4) The accurate determination of the latitude and longitude of many important places in unsurveyed Africa. I am certain that all geographers are in hearty accord with General Chapman in his views, and it is, perhaps, by continually bringing the matter before the public, that we shall best help this movement forward.

Not only do we want a more accurate filling in of the picture, but we have yet to learn to read its lessons aright. The past cannot be understood, and still less can the future be predicted, without a wider conception of geographical facts. Look, for example, at the European Colonies on the West Coast of Africa. Here we find that there have been Portuguese settlements on the Gold Coast since the year 1471, the French possibly having been established there at an even earlier date; whilst we English, who pride ourselves on our go-ahead character, have had trading factories on the Coast only since 1667. I have here a map showing the state of our geographical knowledge in 1815. Why was it that Europeans had never, broadly speaking, pushed into the interior from their base on the coast, which they had occupied for so many centuries? That they had not done so, at least to any purpose, is proved by this map. Why had four centuries of contact with Europeans done so little even for geographical knowledge at that time? The answer to this question may be said to be mainly historical; but the history of our African Colonies can never be understood without a study of the distribution of the dense belt of unhealthy forest along the shore; of the distribution of the different types of native inhabitants; and of the courses of the navigable rivers—all strictly geographical considerations.

Geography is the study of distribution, and early in that study we must be struck with the correlation of these different distributions. If we take a map of Africa, and mark on it all the areas within the Tropics covered with dense forest or scrub, we shall find we have drawn a map showing accurately the distribution of the worst types of malarial fever; and that we have also indicated with some approach to accuracy—with, however, notable exceptions—the habitat of the lowest types of mankind. These are the facts which give the key to understanding why the progress of European colonisation on the West Coast has been so slow.

Along the coast of the Gulf of Guinea we find settlements of Europeans at more or less distant intervals. All along, or nearly all along this same coast, we find a wide belt of fever-stricken forest, fairly thickly inhabited by uncivilised Negro and Bantu tribes. Inside this belt of forest the country rises in altitude, and becomes more open, whilst at the same time there is a distinct improvement in the type of native; and the more we proceed inland the more marked does this improvement become. There appear, in fact, to have been a number of waves of advancing civilisation, each pressing the one in front of it towards these inhospitable forest belts. Near the coast the lowest type of Negro is, generally speaking, to be found; then, as the more open country is reached, higher types of Negroes are encountered. For example, the Mandingoes of the Senegal region are distinctly higher than the Jolas inhabiting the mouths of the Gambia : and the Hausas of the Sokoto Empire are vastly superior to the cannibals of the Oil Rivers. In both these cases the higher types are probably not pure Negroes, but have Fulah, Berber, or Arab blood in their veins, for we see in the case of the Fulahs how they become absorbed into the race they are conquering. Near the Senegal River they are comparatively light in colour, but in Adamawa they are hardly to be distinguished by their features from the Negroes they despise. Thus the process appears to have been a double one; the higher race driving some of the lower aboriginal tribes before them out of the better lands, and, at the same time, raising other tribes by means of an admixture of better blood. These waves of advancing civilisation seem to have advanced from the north and east, for the more we penetrate in these directions the higher is the type of inhabitant met with, until at last we reach the pure Berbers and the pure Arabs. Thus there are two civilising influences visible in this part of Africa; one coming from the north and east—a Mohammedan advance—which keeps beating up against this forest belt and occasionally breaking into it; the other, a Christian move-ment, which, until the middle of this century, was brought to a dead halt by this same obstacle. The map of Africa, showing the state of geographical knowledge in 1815, makes it clear that, except in a few cases where rivers helped travellers through these malarial regions, nothing was known about the interior. No doubt much has been done since those days, but this barrier still remains the great impediment to progress from the West Coast; and those who desire our influence to spread more effectively into the interior must wish to see some means of overcoming this obstacle. On the East Coast of Africa the conditions are somewhat different, as there is comparatively little dense forest there; but the districts near that coast are also usually unhealthy, and how to cross those malarial regions quickly into the healthy, or less unhealthy, interior is the most important problem connected with the development of tropical Africa.

Other influences have been at work, no doubt, in checking our pro-gress from the West Coast. In old days the European possessions in these districts were mere depôts for the export of slaves. As the white residents could not hope to compete with the natives in the actual work of catching these unfortunate creatures, and as the lower the type the more easily were they caught, as a rule, there was no reason whatever for attempting to penetrate into the interior, where the higher types are met with. But though this export trade in human beings is now no longer an impediment to progress, the slave-trade in the interior still helps to bar the way. When the forest belt is passed, we now come, generally speaking, to the line of demarcation between the Mohammedan and the Pagan tribes, and here slave-catching is generally rife; when it is so, the constant raids of the Mohammedan chiefs keep these border districts in a state of unrest, which in every way tends to impede progress. Thus, a mere advance to the higher inland regions will not by any means solve all our difficulties, but it will greatly lessen them; and it is universally admitted that, the more communication with the interior is facilitated, the more easy will it be to suppress this terrible traffic in human beings. By the General Act of the Brussels Anti-Slavery Conference of 1890-91, it was agreed by the assembled delegates that the construction of roads, and, in particular, of railways, connecting the advanced stations with the coast, and permitting easy access to the inland waters and to the upper courses of rivers, was one of the most effective means of counteracting the slave-trade in the interior. Here, then, we have the most formal admission which could be given of the necessity of opening up main trunk lines of communication into the interior.

But not only does geographical knowledge help to demonstrate the necessity of improving the means of communication between the coast and the interior, but it helps us to decide where it is wise to make our first efforts in this direction. In the first place it is essential to note that, if the continent of Africa is compared with other continents, its general poverty is clearly seen. Mr. Keltie, in his excellent work on the Partition of Africa, tells us that 'at present (1895) it is estimated that the total exports of the whole of Central Africa by the east and west coasts do not amount to more than £20,000,000 sterling annually.' For the purposes of comparison it may be mentioned that the export trade of India is between sixty and seventy millions sterling annually, and that India is only about one-seventh or one-eighth of the area of the whole of Africa. On the other hand, the trade of India has been increasing by leaps and bounds largely in consequence of the country being opened out by railways, and there is every reason to hope that somewhat similar results would occur in Africa under similar circumstances, though the lower civilisation of the people would prevent the harvest being so quickly reaped. But, however it may be as to the future, the present poverty of Africa is enough to demonstrate the necessity of pushing ahead cautiously and steadily, and of doing so in the most economical manner possible.

M. Dècle, in an interesting paper read before the International Geographical Congress in London last year, strongly advocated the construction of cheap roads for use by the natives, taking precautions to prevent any traffic in slaves along them. His suggestions are well worthy of consideration; but the cost of transport along any road would, I should have thought, soon have eaten up any profits on the import or export trade to or from Africa. What must be done in the first instance is to utilise to the utmost all the natural lines of communication which require little or no expenditure to render them serviceable; in fact, to turn our attention at first to the rivers and to the lakes. I have already pointed out that the early maps of Africa prove that the rivers have almost invariably been the first means of communication with the interior, and, until this continent is rich enough to support an extensive railway system, we must rely largely on the waterways as means of transport.

It may be as well here to remark that geographical knowledge is often required in order to control the imagination. I do not know why it is, but almost every one will admit that, if he sees a lake of considerable size depicted on a map, he immediately feels a desire to visit or possess that locality in preference to others. A lake may be of far less commercial value than an equal length of thoroughly navigable river, and yet it will always appear more attractive. Look at the way in which the English, the French, and the Germans are all pressing forward to Lake Chad; and yet Lake Chad is in reality not much more than a huge swamp, and in all probability it is excessively unhealthy. Again, it is probable that the Albert Nyanza will prove to be of comparatively small value, because the mountains come down so close to its shores. Of course, the great lakes form an immensely important feature in African geography, but we must judge their commercial value rationally, and without the bias of imagination.

To develop the traffic along the rivers and on the lakes is the first stage in the commercial evolution of a continent like Africa. But it cannot carry us very far. Africa is badly supplied with navigable rivers, chiefly as a natural result of the general formation of the land. The continent consists, broadly speaking, of a huge plateau, and the rivers flowing off this plateau are obstructed by cataracts in exactly the places where we most want to use them—that is, when approaching the coasts. The second stage in the commercial evolution will therefore be the construction of railways, with the view of supplementing this river traffic. Finally, no doubt, a further stage will be reached, when railways will cut out the rivers altogether, for few of the navigable rivers are really well suited to serve as lines of communication. This last stage is, however, so far off that we may neglect it for the present, though it must be noted that there are some parts of Africa where there are no navigable rivers, and where, if anything is to be done, it must be entirely by means of railways.

Thus, as far as the immediate future is concerned, the points to which our attention should be mainly directed are (1) the courses of the navigable parts of the rivers, and (2) the routes most suitable for the construction of railways in order to connect the navigable rivers and lakes with the coast. As to the navigable rivers, little more remains to be discovered with regard to them, and we can indicate the state of our geographical knowledge on this point with sufficient accuracy for our purposes by means of a map. Of course, the commercial value of a waterway depends greatly on the kind of boats which can be used, and that point cannot well be indicated cartographically.

As to the railways, we must study the physical features of the country through which the proposed lines of communication would pass. All the obstacles on rival routes should be most carefully surveyed when considering the construction of railways in an economical manner. Great mountain chains are seldom met with in Africa, and from that point of view the continent is as a whole remarkably free from difficulties. But drifting sand is often a serious trouble, and that is met with commonly enough in many parts. Wide tracts of rocky country also form serious impediments, both because of the cost of construction, and also because the supply of water for the engines becomes a problem not to be neglected. Such arid and sandy districts are of course thinly inhabited, and we may therefore generally conclude that where the population is scanty, there railway engineers will have special difficulties to face. On the other hand, dense forests are also very unsuitable. We have not much experience to guide us, but it would appear probable that the initial expense of clearing the forest, and the cost of maintenance, in perpetually battling against the tropical vegetable growth, will be very heavy, for it will not do to allow the line to be in constant danger of being blocked. The dampness of the forest, which will cause all woodwork and wooden sleepers to rot, will be no small source of trouble; and the virulent malarial fevers, always met with where the vegetation is very rank, will add immensely to the difficulty both of construction and of maintenance. The health of the European *employés* will be a most serious question in considering the construction of railways in all parts of tropical Africa, for the turning up of the soil is the most certain of all methods of causing an outbreak of malarial fever, and the evil results would be most severely felt in constructing ordinary railways in dense forests. In making the short Senegal railway, where the climate is healthier than in many of the districts farther south, the mortality was very great. Perhaps we shall have to modify our usual methods of construction so as to mitigate this danger, and in connection with this subject I may perhaps mention that the Lartigue system seems to be specially worthy of consideration—a system by which the train is carried on a single elevated rail. This is, perhaps, travelling rather wide of the mark of ordinary geographical studies, but it illustrates the necessity of a thorough examination of the environment before we try to transplant our own methods to other climes.

We may, however, safely conclude that we must as far as possible avoid both dense forests and sandy and rocky wastes in the construction of our first railways.

Then as to the lines of communication, considered as a whole, rail and river combined, we must obviously, if any capital is to be expended, make them in the directions most likely to secure a profitable traffic. In considering this part of the question, it will be seen that there are several different problems to be discussed : (1) trade with the existing population in their present condition ; (2) trade with the native inhabitants when their countries have been further developed with the aid of European supervision ; and (3) trade with actual colonies of European settlers. To many minds the last of these problems will appear to be the most important, and in the end it may prove to be so. But the time at my disposal compels me to limit myself to the consideration of trade with the existing native races within the Tropics, with only an occasional reference to the influence of white residents. We must, no doubt, carefully consider which are the localities most likely to attract those Europeans who go to Africa with the view of establishing commercial intercourse and commercial methods in the interior; and there can be no doubt that considerations of health will play a prominent part in deciding this point. Moreover, as the lowest types of natives have few wants, the more primitive the inhabitants of the districts opened up, the less will be the probability of a profitable trade being established. For both these reasons the coast districts are not likely in the end to be as good a field for commercial enterprise as the higher lands in the interior, for the more we recede from the coast the less unhealthy the country becomes, and the more often do we find traces of native civilisation. To put it simply, we must consider both the density of the population and the class of inhabitant in the districts proposed to be opened up. Of course, the exact nature of the products likely to be exported, and the probability of demands for European goods arising amongst the natives of different districts, are vitally important considerations in estimating the profits of any proposed line of railway; but to discuss such problems in commercial geography at length would open up too wide a field on an occasion like this.

If the importance of considering the density of the population in the different districts in such a preliminary survey is admitted, we may then simplify our inquiry by declining to discuss any lines of communication intended to open up regions where the population falls below some fixed minimum—whatever we may like to decide on. Of course, the question of the greater or less probability of a locality attracting temporary white residents is very important, but unless there is a native population ready to work on, there will be little done for many years to come. Politically it may or may not be right to open up new districts by railways for the sake of finding outlets for our home or our Indian population, but here I am considering the best lines for the development of commerce, taking things as they are. What, then, shall be this minimum of population? The population of Bengal is 470 per square mile; of India, as a whole, about 180; and of the United States, about 21 or 22. If it is remembered that the inhabitants of the United States are, per head, vastly more trade-producing than the natives of Africa, it will be admitted that we may for the present exclude from our survey all districts in which the population does not reach a minimum of eight per square mile; it might be right to put the minimum much higher than this. On the map now before you, the uncoloured parts show where the density of population does not come up to this minimum, and we can see at a glance how enormously this reduces the area to be considered. The light pink indicates a population of from eight to thirty-two per square mile, and the darker pink a denser population than that. Of course, such a map, in the very imperfect state of our knowledge, must be very inaccurate, as I am sure the compiler would be the first to admit. On the same map are marked the navigable parts of rivers. I should like to have shown the dense forests also, but the difficulty of giving them with any approach to correctness is at present insuperable.

Here, then, is the kind of map we want in order to consider the broad outline of the questions connected with the advisability of attempting to push lines of communication into the interior. The problem is how to connect the inland parts of Africa, which are coloured pink on this map, with the coast by practicable lines of communication at the least cost, with the least amount of dense forest to be traversed, and, in the case of railways, whilst avoiding as far as possible all thinly populated districts.

It is, of course, quite impossible here to discuss all the great routes into the interior, and I should like to devote the remaining time at my disposal to the consideration of this problem as far as a few of the most important districts are concerned, confining myself, as I have said, to trade with existing native races within the Tropics. Taking the East Coast first, and beginning at the north, the first region sufficiently populous to attract our attention is the valley of the Nile, and parts of the Central Sudan. Wadai, Darfur, and Kordofan are but scantily inhabited, according to our map, and this is probably the case now that the Khalifa has so devastated these districts ; but without doubt much of this country could support a teeming population, and is capable of great commercial development. The Bahr-el-Ghazal districts are especially attractive, being fertile and better watered than the somewhat arid regions farther north. These remarks remind me how difficult it is at this moment to touch on this subject without trenching on politics. Few will deny that the sooner this region is connected with the civilised world the better, and it is only as to the method of opening it up, and as to who is to undertake the work, that burning political questions will arise. The geographical problems connected with the lines of communication to the interior can be considered whilst leaving these two points quite on one side.

A glance at the map reminds us of the well-known fact that, below Berber, the Nile is interrupted by cataracts for several hundred miles, whilst above that town there is a navigable water-way at high Nile until the Fola rapids are reached, a distance of about 1400 miles, not to mention the 400 to 600 miles of the Blue Nile and the Bahr-el-Ghazal, which are also navigable. The importance of a railway from Suakin to Berber is thus at once evident, and there is perhaps only one other place in Africa where an equal expenditure would open up such a large tract of country to European trade. This route, however, is not free from difficulties. Suakin is hot and unhealthy. Then the railway, about 260 miles in length, passes over uninhabited or thinly inhabited districts the whole way. Though the hills over which it would pass are of no great height, the highest part of the track being under 3000 feet above the sea, it is often said that the desert to be traversed would add greatly to the difficulty of construction. According to Lieut.-Colonel Watson, R.E., however, these difficulties have been greatly exaggerated, for the water-supply would give no great trouble. The sixth cataract, between Metemma and Khartum, would make navigation for commercial purposes impossible when the waters are low ; it is probable that this impediment could be overcome by erecting locks, but it is impossible to estimate the cost of such works. Then, again, the Nile above Khartum is much obstructed by floating grass or *sudd*, making navigation at times almost impossible ; but it was Gordon's opinion that a line of steamers on the river, even if running at rare intervals, would keep the course of the stream clear ; this, however, remains to be proved.

If the canalisation of the sixth cataract should prove to be too costly an undertaking, then it would be most advisable to carry the railway beyond that obstacle.   This might be done by prolonging the line along the banks of the Nile, or by adopting an entirely different route from Suakin through Kassala.   I hope we shall hear something from Sir Charles Wilson as to the relative merits of these proposals during the course of our proceedings.   Proposals have also been made for connecting the Nile with other ports on the Red Sea, and all of these suggestions should be carefully examined before a decision is made as to the exact route to be adopted.   But in any case, considering the matter merely from a geographical standpoint, and putting politics on one side—a very large omission in the case of the Sudan—it would appear that one or other of these routes should be one of the very first to be constructed in all Africa.

Passing farther south, it is obvious from the configuration of the shore, and from the distribution of the population, that the lines of communication next to be considered are those leading to the Victoria Nyanza, and on to the regions lying north and west of the lake.

Two routes for railways from the coast to the Victoria Nyanza have been proposed, one running through the British and the other through the German sphere of influence.   Looking at the matter from a strictly geographical point of view, there is perhaps hardly sufficient information to enable us to judge of the relative merits of the two proposals.   Both run through an unhealthy coast zone, and both traverse thinly inhabited districts until the lake is reached.   The German route, as originally proposed, would be the shorter of the two ; but there is some reason to think that the British line will open up more country east of the lake, which will be suitable for prolonged residence by white men.   Sir John Kirk, in discussing the question of the possible colonisation of tropical Africa by Europeans, said : " These uplands vary from 5000 to 7000 feet in height, the climate is cool, and, as far as known, very healthy for Europeans.   This distiict is separated from the coast by the usual unhealthy zone, which, however, is narrower than elsewhere on the African littoral.   Between the coast zone and the highlands stretches a barren belt of country, which attains a maximum width of nearly 200 miles.   The rise is gradual, and throughout the whole area to be crossed the climate is drier and the malarial diseases are certainly much less frequent and less severe than in the regions farther south."   These very advantages, however, may have to be paid for by the greater difficulty of railway construction.   Putting aside future prospects, the maps show that the populous region to the west of the lake makes either of these proposed lines well worthy of consideration, though it would perhaps be rash to predict how soon the commerce along them would pay for the interest on the capital expended.   What will be the fate of the German project I do not know, but we may prophesy with some confidence that the British line, the construction of which has been commenced, will be completed sooner or later.

The two lines of communication we have discussed—the Suakin and the Victoria Nyanza routes—are intended to supply the wants of widely separated districts; but, looking to a more distant future, they must sooner or later come into competition one with the other, in attracting trade from the Central Sudan. Before this can occur, communication by steamboat and by railway must be opened up between the coast and the navigable Nile by both routes. This will necessitate a railway being constructed, not only to the Victoria Nyanza, but also from that lake, or round it, to the Albert Nyanza; and, as the Nile is rendered unnavigable by cataracts about Dufile, and as the navigation is difficult between Dufile and Lado, here also a railway would be necessary in order to complete the chain of steam communication with the coast. If goods were brought across the Victoria Nyanza by steamer, and taken down the Nile in the same manner from the Albert Nyanza to Dufile, this route would necessitate bulk being broken six times before the merchandise was under way on the Nile; by the Suakin route, on the other hand, bulk would only have to be broken twice, provided the sixth cataract were rendered navigable. Thus, if this latter difficulty can be overcome, and if the *sudd* on the Nile is not found to impede navigation very much, this Nyanza route will certainly not compete with the Suakin route for any trade on the banks of the navigable Nile until a railway is made from the coast to Lado, a distance of over 800 miles as the crow flies, and certainly over 1000 miles by rail. It must be remembered also that the Nyanza route passes over mountains 8700 feet above the sea; that the train will have to mount, in all, nearly 13,000 feet in the course of its journey from the coast; and that a difficult gorge has to be crossed to the eastward of the Victoria Nyanza. From these facts we may conclude that it will be a very long time before the Nyanza route will draw any trade from the Central Sudan.

The line through the British sphere of influence runs to the northern end of Victoria Nyanza, but from Mr. Vandeleur's recent expedition into these regions we learn that a shorter route, striking the eastern shore of the lake, is under consideration. To lessen the expense of construction would be a great boon, but if we look to the more ambitious schemes for the future, something may be said in favour of the original proposal as being better adapted to form part of a line of railway reaching the navigable Nile.

With regard to the comparison between the German and British routes to the Victoria Nyanza, the latest accounts seem to imply that the Germans have practically decided on a line from the coast to Ujiji, with a branch from Tabora to the Victoria Nyanza. This would be a most valuable line of communication; but it seems a pity that capital should be expended in competitive routes when there are so many other directions in which it is desirable to open up the continent. If the Germans wish to launch out on great railway projects in Africa, let them make a line from the south end of Lake Tanganyika to the northern end of Lake Nyasa, and thence on the coast; they would thus open up a vast extent of territory, and Baron von Schele tells us that a particularly easy route can be found from Kilva to the lake. Such a line of communication, especially if eventually connected with the Victoria Nyanza to the north, would be more valuable than any other line in Africa in putting an end to the slave-trade, as it would make it possible to erect a great barrier, as it were, running north and south across the roads traversed by the slave traders.

A line through German territory connecting Lake Nyasa with the sea would, no doubt, come into competition with the route connecting the southern end of that lake with the Zambesi, and thus with the coast. The mouths of the Zambesi, though they are passable, will always present some impediment to commerce. But, after entering the river, navigation is not obstructed until the Murchison Rapids on the Shirè River are reached. Here there are at present sixty miles of portage to be traversed, and this transit must be facilitated by the construction of a railway if the route is to be properly developed; Mr. Scott Elliot tells us that 120 miles of railway, from Chiromo to Matope, would be necessary for this purpose. Beyond this latter point there is a good waterway to Lake Nyasa. Thus a comparatively short line of railway would open up the lake to European commerce, and this route is likely to be developed at a much earlier stage of the commercial evolution of Africa than the one through German territory above suggested. It will be seen that these routes connect fairly populous districts with the coast, and it must also be recollected that the high plateau between Lake Nyasa and the Kafue River is one of the very few regions in tropical Africa likely to attract white men as more or less permanent residents.

Farther south we come to the Zambesi River, which should, of course, be utilised as far as possible. But this line of communication to the interior has many faults. The difficulties to be met with at the mouths of the Zambesi have already been alluded to. Then the whole valley is unhealthy, and white travellers would prefer any route which would bring them on to high land more quickly. Moreover, the Kebrabasa rapids cause a serious break in the waterway, and, as the river above that point is only navigable for canoes, it is doubtful if it would ever be worth making a railway for the sole purpose of connecting these two portions of the river.

As the population of the Upper Zambesi valley is considerable, and as the country farther from its banks is said to be likely to be attractive to white men, there can be no doubt of the advisability of connecting it with the coast. This naturally leads us to consider the Beira route as a possible competitor with the Zambesi. A sixty centimètre railway is now open from Fontesvilla to Chimoio (190 kilometres), and it is probable that, during the course of the next two years, the construction of the railway will be completed from the port of Beira itself as far as the territory of the Chartered Company. This will form the first step in the construction of a much better line of communication to the Upper Zambesi regions than that afforded by the river itself. It is true that the gauge is very narrow, and that the first part of the line passes through very unhealthy districts; but this line will nevertheless be a most valuable addition to the existing means of penetrating into the interior of the continent. It is needless to say that the object of the railway is to open up communication with Mashonaland, not for the purposes now suggested.

South of the Zambesi the map shows us that there are no regions in tropical Africa where the density of the native population reaches the minimum of eight per square mile. Here, however, we come to the gold-fields, where there is attractive force enough to draw white men in great numbers within the Tropics, and where, no doubt, some of the most important problems connected with railway communications will have to be solved in the immediate future. But, for reasons of time and space, I have limited myself to the discussion of districts within the Tropics, where trade with the existing native races is the object in view. The Beira railway does not in reality come within the limits I have imposed on myself, except as to its future development. Had time permitted, I should like to have discussed the route leading directly from the Cape to Mashonaland, its relative merits in comparison with the Beira railway, and to have pointed out where the two will come into competition the one with the other. But I must pass on at once to consider the main trunk-routes from the West Coast leading into the interior of Africa.

Passing over those regions on the West Coast where railways would only be commenced because of the probable settlement, temporary or permanent, of white men—passing over, that is, the whole of the German sphere of influence—we first come to more dense native populations near the coast towns of Benguela and St. Paul de Loanda. The latter locality is the more hopeful of the two, according to our map, and here we find that the Portuguese have already constructed a railway leading inland for 191 miles to close to Ambaca. The intention of connecting this railway with Delagoa Bay was originally announced, and I am not aware to what extent this vast project has now been cut down, so as to bring it within the region of practical proposals. A further length of 35 miles is, at all events, being constructed, and 87 more miles have been surveyed. The Portuguese appear to be very active at present in this district, as there are several other railways already under consideration ; one from Benguela to Bihe, of which 16 miles is in operation, another from Mossamedes to the Huilla plateau, and a third from the Congo to the Zambesi. It is difficult to foretell what will be the outcome of these schemes, but our population map is not very encouraging.

Next we come to the Congo, and here there is a grand opportunity of opening up the interior of the continent. In going up this great stream from the coast we first traverse about 150 miles of navigable waterway, and afterwards we come to some 200 miles of cataracts, through which steamers cannot pass. Round this impediment a railway is now being pushed, 189 kilometres of rails (117 miles) being already laid. Then we enter Stanley Pool, and from this point we have open before us—if Belgian estimates are to be accepted—7000 miles of navigable waterway. If this fact is correct, and if the population is accurately marked on our map, then there is no place in all Africa where 200 miles of railway may be expected to produce such marked results. The districts traversed are unhealthy, and the natives are, generally speaking, of a low type; but in spite of these drawbacks, which, no doubt, will delay progress considerably, we may confidently predict a grand future for this great natural route into the interior.

To the north of the Congo, the next great navigable river met with is the Niger. Again, granting the correctness of the population map, it can be seen at a glance that there is in all Africa no area of equal size so densely inhabited, and no district where trade with the existing native population appears to offer greater inducement to open up a commercial route into the interior. Luckily little has to be done in this respect, for the Niger is navigable for light-draught steamers in the full season as far as Rabba, about 550 miles from the sea; here the navigation soon becomes obstructed by rocks, and at Wuru, about 70 miles farther up the river, the rapids are so unnavigable that even the light native canoes have to be emptied before attempting a passage, and there are frequent upsets. From Wuru the rapids extend to Wara, after which a stretch of clear and slow-running river is met with. Above this, again, the Altona rapids extend for a distance of 15 miles; then come 15 miles of navigable waterway, and then 20 miles more of rapids are encountered. Yelo, the capital of Yauri, is situated on these latter cataracts, above which the Middle Niger is navigable for a considerable length. The Binue is also navigable in the floods for many miles, the limits being at present unknown; part of the year, however, it is quite impassable except for canoes. The trade with the Western Sudan, which has been made possible by the opening up of this river, is still only in its infancy, and to get the full benefit of the waterway a line of railway ought to be carried from Lokoja to Kano, the great commercial centre of Hausaland; Mr. Robinson's recent journeys over this country, which we hope to hear about at a later period of our proceedings, have served to confirm the impression that no great physical difficulties would be encountered. The political condition of the country may, however, make the construction of this railway quite impossible for the present; for here we are on the borderland between Mohammedanism and Paganism, where the slave-trade always puts great impediments in the path of progress, but where the same circumstances make it so eminently desirable to introduce a higher condition of civilisation. The only drawback to the Niger as a line of communication to the Western Sudan is the terribly unhealthy nature of the coast districts which have to be traversed. Any man who finds a means of combating the deadly diseases here met with will be the greatest benefactor that Africa has ever had; but of such a discovery there are but few signs at present.

It is perhaps too soon to speculate as to the best means of opening a trade route to Wadai and the more central parts of the Western Sudan; for we may be sure that little will be done in this direction for years to come. Several competing routes are possible. From the British sphere, we may try to extend our communications eastward from the navigable parts of the Binue. The French, on the other hand, may push north- wards from the Ubangi; whilst, in a later stage of commercial evolution, the best route of all may be found through German territory, by pushing a railway from the shore in a direct line towards Bagirmi and Wadai. To compare the relative merits of these trunk-lines is perhaps looking too far into the future, and traversing too much unknown country, to make the discussion at all profitable.

Proceeding northwards, or rather westwards, along the coast, we find ourselves skirting the belt of dense forest already described as being the great obstacle to advance in this part of Africa. It is to be hoped that this barrier will be pierced in several places before long. Naturally we turn our attention to the different spheres of British influence, and here we are glad to learn that there are several railways being constructed or being considered, with a view to opening up the interior.

At Lagos a careful survey of a railway running in the direction of Rabba has been made, and the first section is to be commenced at once. To connect the Niger with the coast in this way would require 240 miles of railway, but the immediate objectives are the towns of Abeokuta and Ibadan, which are said to contain more than a third of a million inhabitants between them. No doubt the populous coast region makes such a line most desirable: but whether it would be wise to push on at all quickly to the Niger, and thus to come into competition with the steamboat traffic on that river, is a very different question.

Surveys have also been made for a railway to connect either Kormantan or Apam on the Gold Coast with Insuaim, a town situated on a branch of the Prah. It is believed that the local traffic will be sufficiently remunerative to justify the construction of this line. But, looking to the further prolongation of the railway into the interior, it appears possible that those who selected the route were too much influenced by the desire to reach Kumasi, which is a political rather than a commercial centre. According to the views I have been advocating to-day, the main object of a railway in this quarter should be the crossing of the forest belt, and if, as there is some reason to believe, that belt is exceptionally wide and dense in this locality, the choice of Kumasi as a main point on the route will have been an unfortunate selection. A little farther south, nearer the banks of the Volta, it is probable that more open land would be met with, and moreover that river itself, which is navigable for steam launches from Adda to Akuse, would be of use as a preliminary means of transport. It is to be hoped that the merits of a line from Accra through Odumasi will be considered before it is too late.

I am now approaching the end of my brief survey of tropical Africa, for the best method of opening communication between the Upper Niger and the coast is the last subject I shall touch on. With this object in view, the French have constructed a railway from Kayes, the head of steam navigation during high water on the Senegal, to Bafulabé, with the intention of ultimately continuing the line to Bamaku on the Niger. Unexpected difficulties have been met with in the construction of this railway, and, as the Senegal River between Kayes and St. Louis is only navigable for about a quarter of the year, it would hardly appear as if the selection of the route had been based on sound geographical information. No doubt the French will find some other practicable way of connecting the Upper Niger with the coast, and surveys are already in progress with that object in view. It may be worth mentioning that the Gambia is navigable as far as Yarbutenda, and that it affords on the whole a better waterway than the Senegal; it is possible, therefore, that a railway from Yarbutenda to Bamaku might form a better means of connecting the Niger with the coast than the route the French have selected.

At Sierra Leone a railway is now being constructed in a south-easterly direction with a view of tapping the country at the back of Liberia. But here, as in the case of the Gambia route, political considerations are of paramount importance; for no doubt the best commercial route, geographically speaking, would have been a line run in a north-easterly direction to some convenient point on the navigable part of the Upper Niger. If such a railway were ever constructed, it would connect the longest stretch of navigable waterway in this region with the best harbour on the coast. But the fact that it would cross the Anglo-French boundary is a complete bar to this project at present.

Proposals for connecting Algeria with the Upper Niger by rail have often been discussed in the French press, the idea being to unite the somewhat divided parts of the French sphere of influence by this means. If the views here sketched forth as to the necessity of selecting more or less populous districts for the first opening up of lines of communication into the interior are at all correct, these projects are simple madness. For many a year to come Algeria and the Niger will be connected by sea far more efficiently than by any overland route, and I feel sure that, when the details of these plans are properly worked out, we shall not find the French wasting their money on such purely sentimental schemes.

I must now conclude, and must give place to the other geographers who have kindly undertaken to read papers to us on many interesting subjects. All I have attempted to do is briefly to sketch out some of the main geographical problems connected with the opening of Central Africa in the immediate future. Such a review is necessarily imperfect, but its very imperfections illustrate the need of more accurate geographical information as to many of the districts in question. Many blunders may have been made by me in consequence of our inaccurate knowledge, and, from the same cause, many blunders will certainly be made in future by those who have to lay out these routes into the interior. In fact my desire has been to prove that, notwithstanding the vast strides that geography has made in past years in Africa, there is yet an immense amount of valuable work ready for any one who will undertake it.

Possibly, in considering this subject, I have been tempted to deviate from the strictly geographical aspect of the case. Where geography begins and where it ends is a question which has been the subject of much dispute. Whether geography should be classed as a separate science or not has been much debated. No doubt it is right to classify scientific work as far as possible; but it is a fatal mistake to attach too much importance to any such classification. Geography is now going through a somewhat critical period in its development, in consequence of the solution of nearly all the great geographical problems that used to stir the imagination of nations; and for this reason such discussions are now specially to the fore. My own humble advice to geographers would be to spend less time in considering what geography is, and what it is not; to attack every useful and interesting problem that presents itself for solution; to take every help we can get from every quarter in arriving at our conclusions; and to let the name that our work goes by take care of itself.

SUTHERLAND PLACE-NAMES.

By JOHN MACKAY, As. Inst. C.E.

LOCAL names are usually descriptive, historic, or personal. They are not to be regarded as arbitrary signs. They have a meaning and a history, though, owing to the absence of early documents, the history may be lost and the meaning may not be accurately ascertained.

Physical features, such as the names of mountains and rivers, are frequently the memorials of extinct or ancient races. Not only is this the case in Sutherland, but it is the same all over Europe, where local names are met with bearing witness of races departed or absorbed.

In Sutherland, place-names are chiefly descriptive; a few there are, historic and personal, recording battlefields and the names of commanders that were slain and interred where they fell. The stone and the cairn were raised to commemorate the event and the men. We also find that a religious spirit has actuated the inhabitants in causing places to bear the name of a local hermit or a saint, as we find in other parts of Scotland, Ireland, Wales, Cornwall, and France. Even in India a large number of towns take their names from the temple of some deity.

The names of places, like those of the streets of towns, are endowed with extraordinary vitality, frequently surviving the race or the nation that imposed them, and often defying the accidents of conquest and even of time, while furnishing information of an unexpected character. In Sutherland they disclose the existence, extinction, or absorption of four different races and four different languages,—Iberian or Basque, Caledonian Pictish, Norwegian, and Gaelic.

Of the first, few footprints remain, only two river names—the Iligh and the Naver. Of the second, there are several records in river and place names, proving that the Brythonic language had been spoken in that region previous to the commencement of the Christian era. Of the third, there are many all around the west, north, and south coasts, and even for some miles inland, along some of the valley rivers, and on the frontiers adjoining Caithness on the east; while in the interior Gaelic holds undivided dominion over mountain, lake, river and place names, and the preponderance all round.

Sutherland itself may be termed a name of position, given it by the Norsemen: "Sudrland," the land to the South. From the eighth to the twelfth century that portion of Scotland north of the Dornoch Firth and the Oykell river formed only one province under the name "Catenes." On the expulsion and subjection of the Norsemen by William the Lion, the province of Cateness was divided into three districts. Caithness, east of the Ord and the ridge of mountains extending from it to the Northern Ocean, was left to the Norse Earls of Caithness, while the northern portion, from the river Halladale to Kylescow on the west, was called Strathnavernia, and became "the territory" of the Mackays, styled in the native language "Duthaich Mhic-Aoidh," granted to their first chief for services rendered to William in subduing the Norsemen, 1196-98. The southern portion, or the Norse Sudrland, was conferred by the same monarch upon Hugh Freskyn, a Morayshire proprietor, who also assisted William in the same expedition. This territorial name was adopted by Freskyn's successors as the family title and surname.

To keep this Paper within reasonable limits, only names supposed to be of Iberian origin, Pictish, and Norwegian can be adverted to.

The river names,—Abona (the Dornoch Firth or the estuary of the Oykell), Il-a (the Iligh), Nabar-os (Naver)—of Ptolemy's map are the most ancient records we have of geographical names in Sutherland which have come down to us from remote times, and which for eighteen centuries or more have as nearly as possible preserved their primitive forms and pronunciations to this day. The Abona, being so evidently of Pictish origin, will be alluded to hereafter in that class.

The Il-a, or Iligh, has been referred to an Iberian origin, from the Basque word *Ill*, water, owing to its resemblance to the first syllable in Il-a. It may be right to show that the term given by Ptolemy has as great similitude to a word in the language then spoken by Northern Caledonians, now generally admitted to have been a dialect of the Brythonic, of which we have still the living remains in Welsh, Cornish, and Armoric; Welsh, *lli*, *llif*, stream, flood; Cornish, *lif*, flood; Armoric, *liv*, flood. The Welsh pronounce *lli*, *llif*, as if written *thli*, *thliv*. The Gaelic equivalent is *lighe*, flood or fulness of water, and pronounced nearly similarly to the Welsh. This river is subject to floods from its extensive watershed, and the numerous lakes round about its sources form reservoirs which give it abundance of water and regulate its flow.

The Nabar-os or Naver is also presumed to be Iberian from the Basque words *nava*, *naba*, high table-land, and *erri*, country—high table-land country; cognate is Navarre, a province of Spain, a Celto-Iberian territory. It may apply to this river as rising in and flowing through high lands.

PICTISH PLACE-NAMES.

Abona—this name is Pictish or Brythonic. Its form in Welsh would be Y-bwn-aw. *Y*, the; *bwn*, head or end; and *aw*, flowing water; the end or head of the flowing water or tide. We have this name still preserved in modern Bon-ar. At the lower end of Loch Ness is Bun-aw, or Bon-aw, with the same signification. Awe, a lake in Assynt; Welsh, *aw*, flowing water; Gaelic, *ath*, ford; a place where water is shallow and flows fast.

Esc—stream, running water; Welsh, *wysg*, stream; Cornish and Armoric, *isge*, stream; Gaelic, *uisge*, stream, water. There are several Esks in Scotland. The same word appears as river-names in England, as Esk, Exe, Axe, Ax, and Usk. Several hamlets and townships in Sutherland, situated on small streams, are named from their situation, Aber-esc-ag, Over-esc-ag, Shiber-esc-ag; the terminal syllable is a Gaelic diminutive.

Ochtow—a place-name in Strath Oykell, situated on high ground above the river, means high side; Welsh, *Uch*, high, and *tu*, side, Uchtu; Gaelic, *Uachdar*, upper part; Welsh and Armoric, *Uchder*, upper part; Cornish, *Uch*, above. Here we have in the far north a remnant of the Caledonian language similar to place-names in Pictavia from the Forth to the Spey, and similar to Welsh, Cornish, and Armoric. The Ochil Hills, Auchterarder, etc., are cases in point.

Pit.—This is a prefix met with throughout Pictland, meaning a small portion of cultivated land, a hollow, frequently a sheltered hollow. In Assynt, where arable land is very scarce, the natives enclose and cultivate patches of land in dry hollows; they call such an enclosure *poot*; Welsh, *pwt*, a short part or portion, also *peth*, a small part; Cornish, *pyth*, a small part.

Pitfure—*pit*, a small cultivated plot of land; *for, fur*; Welsh, *ffor, fur*; Cornish, *for*, a way, a road. Pitfur would therefore mean a plot of cultivated land near a roadway.

Pittentrail.—Welsh and Cornish, *pit-yn-trai*; *pit*, enclosed land; *yn*, at or near; and *trai*, ebb tide; the enclosed land or habitation at or near the ebb of the tide. W. *trai*; Corn. *traith*; Arm. *traez*; Manx, *traih*; Irish, *traigh*; Gael. *traigh*; shore at ebb tide.

Pitgrudie—*pit* and *grut*; W. *grit*; the gritty or stony enclosure.

Oykell, river, the Eccial of the Norsemen, probably the same as Ochil; Welsh, *Uchel*; Corn. *uchel*; Arm. *uchel*; high, lofty. This river is flanked on both sides by lofty mountains, and possibly it derived its name, like the Naver, from the high lands in which it has its source and through which all its course runs.

Proncy.—This place is situated on a protuberance of a long acclivity facing the Dornoch Firth; near it on a higher site are the ruins of a Pictish tower. In Wales, Cornwall, and Brittany a great number of place-names begin with *bron*, signifying breast, pap, protuberance: *b* and *p* being labials frequently interchange; *b* becomes *p* and *p* becomes *b*. Here we have *p* adopted for *b*, and the *bron* becomes *pron* for the first syllable; the *cy* is the *sedd* of the Welsh, seat; the *sedh* of the Cornish, seat, habitation. Proncy may be of the same origin, signifying the habitation on the elevated ground. On the same acclivity are two other Proncies —Proncy-cruaidh (G. adj. hard or sterile) and Proncy-mban (Lower Proncy). Old forms: 1360 Proncey, 1448 Pronnsy, 1525 Pronnse, 1536 Spronnse, 1562 Pronsie, 1563 Pronnse, 1566 Prompse, 1616 Pronsie.

Rogart—old form Roth-gorthe—purely a Brythonic name, from *rhoth*, circular hollow, and *gorthe* or *gorthir*, high lands, signifying the circular hollow flanked by high lands, perfectly descriptive of the township of Rogart. The old church is situated high up on one of the flanks of these slopes looking down on the circular basin below, once a marsh frequented by herons, now drained and called the "Lon-mor" (big meadow).

Rhi—slope, declivity, is a common prefix in Sutherland local names; W. *rhiw*, pro. *rioo*, slope, declivity. Gaelic, *ruigh*, run, also slope. The affinity of the Welsh word with the Gaelic, the near similarity of pronunciation, the position of the *i* before the *w* in the Welsh, and after the *u* in the Gaelic word marks the difference in the pronunciation, the one being phonetically *rioo*, the other *rooi*. The Sutherland form, *rhi*, pro. *ree*, would indicate its Brythonic or Pictish descent. On the Ordnance maps of Sutherland this prefix is given as *reidh*, Gaelic, plain, level. This is a mistake for *rhi*, or *ruigh*, slope, declivity. The old form should be adopted, as *Rhi-linn*, slope to the pool; *Rhi-lochan*, slope to the little lake; *Rhi-an-daggie*=*Rhi-an-dach-aidh*, slope to the homestead, etc.

Uidh.—This is a word commonly applied in W. and N. Sutherland to a stretch of a river where the flow is scarcely perceptible, and to that part of a stream which leaves a lake before breaking into a current. Welsh, *gwy*, water gently moving; Cornish, *gy*, water, river: hence the Wye in Hereford and Derby, Surrey, Sussex, and Dorset, all of them rivers smoothly flowing. May we not infer that this too is a remnant of the British or Brythonic language in the far north? It is also the terminal syllable, as *ie* or *y*, in several place-names, as Strath-y (Strath-uidh).

NORSE PLACE-NAMES.

Parish of Assynt.—Old forms of this parish name, 1445 Assend, 1509 Assint, 1600 Assyin, 1640 Assynt, modern form. Various definitions of this word have been hazarded. The *Statistical Account*, 1795, states that two brothers in remote times, named Unt and As-unt, contended for the mastery of the district. Unt was slain, and As-unt, proving victorious, obtained the mastery, and thenceforth gave the district his name. Another is from two Gaelic words, *as* (out), and *innte* (in), "out and in," in reference to its sea-coast line being very much out and in. The words certainly represent the native pronunciation of this parish name, but a more probable origin of the term may be assigned to the Norse word, *Asynte*, "seen from afar," as the Norse rovers would see and regard it from the Northern Ocean or the Minch when sailing past the coast, or viewing it from Lewis and Skye. Its mountains towering aloft into the skies would be very conspicuous objects on the horizon, especially the Canisp, Suilven, Glasven, Quinag, and Stack.

Canisp, from *kenna*, well-known, and *ups*, house-roof shape; the well-known roof-shaped mountain; *isp* is simply a change in the position of *s* to ease pronunciation.

Kirkaig—church, from Kirkja; here was a church founded by the Culdee monk Maolrubha, a disciple of Columba, destroyed by the Norse pirates in the eighth century.

Traligill—the devil's ravine; from *träll*, a fiend, and *gil*, ravine. This ravine is near Inch-na-damph. The depth and narrowness of it excludes the light; the stream running through disappears in the limestone caverns, and issues out in some places to be lost to view again in others. The Norsemen, although very brave, were very superstitious, believing in evil spirits, goddesses and gods. The name they gave this ravine represents their opinion of it.

Cleit.—G. rugged height; N. *klettr*, rock; Arm. *clet*, rock. This rugged, rocky island stands out in the sea from Lochinver, and is 120 feet high.

In studying Icelandic it is interesting to find many words very similar to the Gaelic of N.W. Sutherland, leading to the inference that they were introduced into Icelandic literature from that region. From the dissertation of the *Corpus Poeticum Boreale* it might be inferred that, at least, parts of Iceland had been colonised by the N.W. inhabitants of Scotland when the Druids were persecuted and probably expelled after the diffusion of Christianity, or that Norsemen, who had lived in Assynt and districts round about it, had returned to Iceland and Norway, carrying with them tales, legends, and terms peculiar to the N.W. of Scotland. The Norse pirates were given to capturing women and carrying them home.

Soyea.—An island in the bay of Lochinver, forming a breakwater to it, 100 feet high; from *sae*, sea, and *ey*, isle; the sea isle.

Crona.—The three isles, from *threnn*, triple, and *ey*, isle; threnn-ey, throna, eventually pronounced Crona.

Oldany.—The old island, from *aldinn*, old, and *ey*, isle, from being the largest in size on the coast. In the English Channel we have Alderney, old island; Guernsey, rugged island; Jersey, grassy island; Sark (Sercque), temple isle; Herm, serpent isle.

Rafn.—A place where sea-weed accumulates.

PARISH OF EDDRACHILIS.

Kylestrome—compound word; G. Kyle, *caolas*, strait; strome, N. from *stromr*, stream, in reference to the rapidity of the water rushing in and out through the strait at the flow and ebb tides.

Beinn Strome—takes its name from its proximity to the strait.

Laxford—Salmon Firth; from *lax*, salmon; and *fiord*, firth; as celebrated now for salmon as the Norsemen found it.

Sandwood—sandy lake, from *sandr*, sand; and *vatn*, fresh water. The lake gives the place adjoining its name.

Gisgil—roaring ravine; from *gjosa*, gushing, noisy; and *gil*, ravine.

Handa—the sandstone island; from the oblique case of *sandr*, sandstone; and *ey*, island.

Scourie—a place were birds resort to; from *skorrie*, bird; and *ey*, island; as Chels-ea, Cherts-ey.

PARISH OF DURNESS.

Durness, dyr-nes—the deer promontory; from *dyr*, deer; *nes*, promontory.

Keol-dale—the hinds-dale; from *kollu*, hind, or hornless deer; and *dalr*, dale. The genius of the Norse or Scandinavian language puts the substantive word last, the adjectival word first, while the Gaelic puts the adjectival word last and the substantival first, as in Dal-more, Dal-beag.

Parph—a turning away, in reference to a promontory; the land receding in rounding it, from Hvarf, Cape Wrath. Cape Farewell in Greenland was named by the Norse navigators *Hvarfs-gnipa*, meaning the peak of the receding land.

Hope, N. *Hop*—a lake of fresh water into which the sea comes in at high tides and goes out at the ebb. This lake is only a few feet above sea-level. The lake gives its name to the adjoining noble mountain.

Erribol—an arm of the sea running eleven miles inland; from *eyrri*, gravelly, shingly banks; and *ból*, habitation, hamlet, township; the township on the shingly banks; *ból* has the same signification as the Gaelic *baile, bal*.

Arn-bol—the eagle township; from *arnar*, belonging to an eagle, and *ból*.

A-Chleit—the rugged rocky height, in Bal-na-cill bay, G. and N.

I-Ghoil—*I, ey*, isle; *ghoil*, oblique case of *goil* (G.), boiling; the isle of the boiling, in reference to the boiling of the waves meeting around it from different directions.

Hoan—the plant island; from *hvoan*, the plant angelica, esteemed by the Norsemen for flavouring their ale.

Choarie—corruption of the Norse word Kviar-ey, the fold island; cattle were wont to be swum to this isle for keep and protection. It grows excellent grass, being entirely composed of limestone. It is only a score yards from shore. It is sometimes called in modern maps, An-coir-eilean; this should be An-curra-eilean, the heron isle.

Far-out Head—corruption of the N. word *forad*, a dangerous place or precipice.

Heilim, oblique of *holmr*—an islet or peninsula in a river, lake, or bay; here it is a very small peninsula in Loch Erribol.

Sango-mor, Sango-beag—sandy bay, from *sandr*, sand; and *gjá, geo*, creek; *mor* and *beag* are Gaelic adjectives.

Musal—mossy land, from *mosi*, moss; and *fial* or *fell*, rough land; mosi-fell, moss-fell. Musal is in Strathmore, or anciently Strath-Urradal, so named from a Norse commander, said to be one of Haco's captains, who was there slain with several of his men when on a plundering expedition on the return of the fleet from Largs in the autumn of 1263.

The Parish of Tongue—from *tunga*, a spit of land jutting out into the bay in the shape of a tongue. The natives to this day pronounce this word, like the Norsemen, phonetically—*toong-a*.

Borgie—a fort; from *borg*, a fort.

Lamigo—the lamb's creek; from *lamb*, lamb; and *gjá*, creek.

Blandy—place of meeting; from *blanda*, meeting.

Coldbacky—cold ridge; from *kaldr*, cold; *bakki*, ridge.

Conasaid—the lady's residence; from *kvenna* and *sida*.

Falside—the fell or hill, residence; from *fial*, fell; and *sida*.

Hysbackie—the house-back; from *husa-bakki*.

Kirkiboll—church township, *kirkju-ból*.

Melness—benty grass promontory; from *melr*, benty grass, and *nes*, promontory; so it is to this day, no change, no alteration, for seven centuries.

Modsary—muddy moorland; from *moda-seyra*.

Ribigill—spelled in ancient charters Riga-bol, Rege-bol; from *rygjar-bol*, the lady's house and farm. In its modern form the letters *b* and *g* have been transposed, not an unusual circumstance, to favour pronunciation.

Scrabster—outlying homestead; from *skara-böl-stadr*.

Skerray—isolated rock in the sea, covered at high tides; *skerja*.

Skinnid—bleached, withered; in reference to the grass grown on it; from *skinni*. In charters this name is written Sgianaid, phonetically.

Scullomie—a court or place or hall in which rents, fines, rates, taxes, or dues were wont to be paid; from *skulda-mot*, *skulda-domr*.

Slettel—flat, smooth land; from *slett lendi*, flat land.

Talmine—toll free; from *tollr*, toll, and *minnr*, less or free. Contracted to Talminn, Talmine; the terminal *r* changed to *e* for Gaelic euphony.

Torrisdale—Thor's dale, the *dal* of Thor.

PARISH OF FARR.

Boursay—the wall isle; from *bjarr*, wall, and *ey*, isle, in reference to its perpendicular sides. It is quite close to the mainland. There is a Birsay in Orkney.

Baligil—the township near the ravine; *böligil*.

Dal-langal—long meadow. Here is a confusion. *Dal* is G. meadow; *langal* is a G. contraction of N. *langa-dalr*, long dale. From both languages we have tautology in Dal-langdal. The Norsemen called the dale Langdale; the natives, in retaking possession on the expulsion of the Norsemen, named it after their own way from the Norse, Dal-langdal. The same thing occurs in Strath-Halla-dale.

Dal-harald—This place-name is G., and was so named by the natives to commemorate the battle fought upon this Dal in 1196, when Harald Maddadson, Earl of Caithness and Sutherland, was severely defeated by Reginald of the Isles, son of the redoubtable Somerled, ancestor of the MacDonalds. The field of battle is on the east or right bank of the Naver, two miles below Lochnaver. It was a fair field for the contest of heroes. Judging from the numerous tumuli and cairns still to be seen on this field of strife, the contest was one of great severity and slaughter. A pillar of stone called "Clach-an-righ" commemorates the victory and Harald's position in the battle, and several smaller ones, reared, no doubt, where commanders fought, fell, and were interred.

Kirtomy—place of thin scrub or brushwood; from *kjor*, copse-wood, and *tomr*, valueless.

Leac-biurn—G. and N. compound word; *leac*, G. flagstone; *Biurn*, the name of a Norse noble buried where he fell after the battle of Dalharald. His name being Björn, this place-name should be Leac-Björn.

Langdale—long dale; from *langa*, *langi*, long, and *dalr*.

Mudale—moorland dale; from *mosa*, moorland, and *dalr*.

Skail—a hall, or a better than ordinary dwelling, from Skali, hall; a pretty place in Strathnaver. Here the "Sagairt Ruadh" (red priest) officiated in his day. He was regarded as a prophet.

Swordley—place of mowing grass; from *svordr*, sward, and *lja*, mown grass.

Syre—sour or swampy land; from *saur*, swampy land.

Armadale—the bay dale; from *armor*, an arm of the sea or bay, and *dalr*, dale.

PARISH OF REAY.

Bighouse—a barley or bigg store in Norse times; *bigg hus*. In Strath-Halla-dale there are two places of this name, Lower and Upper Bighouse.

Forsinard—N. and G. *fors*, N. torrent, *an airde*, G., in the height; in contra-distinction to

Forsinain—*forsamhan*, the torrent below, or lower torrent.

Melvich—benty grass bay; from *melr*, benty grass, and *vik*, bay.

Port Skerra—port of the hidden rocks; from *port*, opening, and *skerja*, isolated rocks in the sea, hidden at high tides.

PARISH OF KILDONAN.

Rimistal—from Raumsdalr in Norway, a dale and district in that country.

Borroboll—barley township; from *barr*, barley, and *bòl*.

Dalial—small meadow; from *daela*, small dale.

Duible—bog township; from *dy*, bog, and *bòl*. Old forms, Daypoll, Doweboll, Dweboll.

Eldrable—the beacon township; from *eldr*, beacon, and *bòl*. The adjoining beacon mountain gives its name to the township.

Gearnsary—pasture upon which cattle are grazed at so much per head.

Grodsary—pasture round ruined buildings.

Helmsdale—terrace at the mouth of the dale; from *hjalli*, terrace at a mountain foot, *munn*, mouth, and *dalr*, dale. N. form, Hjalmundal.

Learable—muddy township; from *leir*, muddy, and *bòl*.

Lerwick—muddy bay.

Marril—farm or township near the sea. N. form, Marbaeli, contracted in pronunciation to Marril.

Suisgill—the noisy ravine; from *suis*, roaring like the sea, and *gil*, ravine.

Ulbster—Ull-bol-stadr, the homestead of Ulli.

PARISH OF LOTH.

This parish name is of great antiquity. The tribe that inhabited the district is named "Logi" by Ptolemy. The inhabitants are called by their neighbours "Lo-aich," as those dwelling in Helmsdale are called "Il-aich." The most remarkable feature in the parish is its principal river, which falls 1500 feet in its short course of six miles. When in flood it flows with amazing rapidity; hence it would deserve the Gaelic appellation of *luath*, swift.

Slet-dale—small dale or glen; from *slet*, small, and *dalr*.

Rimisdale—Raumsdalr; province and dale in Norway, signifying "dale of giants."

PARISH OF CLYNE—has no place-name of Norse origin.

PARISH OF GOLSPIE.

Backies—N. *bakki*, eminence, ridge, bank. This township is situated on a terrace 500 feet above sea-level, overlooking the Golspie river.

Golspie—1570, old form Gospye, 1581 Golspe, 1682 Golspie, its present form; native pronunciation, Goysbie, Goilsbie. The name is evidently of Norse origin, from *geil*, *gil*, narrow valley, and *bu* or *by*, village—village at the narrow glen.

Uppat—high part, pronounced Oopait; from *upp*, high, and *att*, quarter or part.

Unes—*u-nes*, not a promontory. It refers to the land between Golspie and Littleferry. The form of this land is more like a peninsula than a promontory; from *ú*, a negative prefix, and *nes*, promontory.

ROGART PARISH.

Fleet river—from *fljot*, estuary; has the Latin signification of *ostium*. Compare the Fleets in Kent and Middlesex.

[LAIRG PARISH—has no Norse place-names.

CRIECH PARISH.

Migdale—*myg-dalr*, moist dale.

Ospisdale—Ospis, name of a Norse commander slain and interred here. An obelisk by the roadside records the event.

Spinningdale—*spenja-dalr*, attractive dale.

PARISH OF DORNOCH.

Cyderhall—corruption of *sudrha'*, the south hall, 1222-45. *Siwardhoch* (*Sigurd-hangr*)—Siward's grave. Here the redoubtable Sigurd Eystein-son, who subdued Caithness, Sutherland, and parts of Ross (875), was interred, and a great cairn raised over his grave. Native pronunciation, Shee-er-a, 1557, Sytheva.

Embo—native pronunciation, Erribol; shingly banks.

Skelbo—native pronunciation, Skerribol, 1222-45 Scelleboll, 1529 Skailbo; from *skerja*, rock in the sea, and *böl*, the township near the sea rock. Here was a Norse castle or fortress, situated quite close to south shore of the Fleet estuary, near the present Littleferry, which, after the expulsion of the Norsemen (1196-8), became the property of Hugh Freskyn and his successors, and of the Duffus family, a junior branch. The Norsemen were great in castle-building wherever they went, in Normandy, Sicily, England, Wales, Ireland, and Scotland, to protect themselves from the native population, as we are now doing in Matabeleland.

Skibo—1222-45 Sceitheboll, 1275 Schytheboll, 1548 Skebo; from *skipa*, ship, and *böl*. From the Dornoch Firth runs a narrow bay right up to Skibo Castle. Into this arm of the sea the Norse rovers came in their long galleys, and at Skibo built a fort, which gradually extended to a castle with a deep moat all round it. Hugh Freskyn granted this castle, with several davachs of land, to Bishop Gilbert for his residence, and it continued to be the residence in Sutherland of the bishops of the diocese for several centuries.

GEOGRAPHICAL EDUCATION.

By A. J. Herbertson, F.R.S.E., F.R.G.S.

(Continued from p. 422.)

II. The Use of Maps in Education.

It is not enough to teach a child to see what is around him. He should also be trained to express the results of his observations, so that others may understand them. Of course this can be done by verbal accounts of them and records of such descriptions. But geographers have long ceased to be contented with mere verbal descriptions, indispensable as these are. There are many symbols much more comprehensive and direct in meaning than our alphabetical ones. Even the ideographs we commonly use, such as the Arabic numerals, are not so readily intelligible as diagrams whose proportions are relative to the values of these numerals. For instance, it is much more graphic to compare different distances by drawing lines and making the length of each proportional to a distance, than to express this difference in arithmetical symbols. The proportion between the length of the line and the actual distance is called the *scale* of the drawing. This method of drawing to scale gives exceptionally valuable results when applied to geographical description. By taking note of directions and making various combinations of lines drawn on the same scale, the relative positions of places are fixed, and what we call a plan is produced. A map may be regarded as an elaborated plan.

These diagrams or maps are the most evolved form of geographical record, and are of all degrees of complexity. They cannot be read without any instruction, as many people seem to imagine. It is just as absurd to expect a child, without any training, to decipher the hieroglyphs on some old Egyptian monument, or a page of print, as it is to expect him to understand an ordinary political map at first sight. The child must begin with the simplest plans and gradually proceed to the use of maps of increasing complexity.

Professor de Lapparent has recently delivered a most instructive address on the "Art of Map Reading," which every teacher should

study.[7] He points how very few legible maps exist. By this he does
not mean that the lettering is indistinct, for lettering, although convenient in some cases, is not necessary on a map. He means that the features
of a country are not clearly and accurately shown. Given legible maps,
however, wonderful stories can be read by those knowing the map language.
Some of these the French Professor outlines in his interesting paper.

1. Maps of the Home Region.

The best way to learn to understand a map is to make one. This is
not achieved by copying accurately the maps in an ordinary atlas, nor even
by copying that of the home district. It is best done by the children
actually making one from measurements they have taken themselves.
The teacher should not do these things for them and then let the children
copy his work, but supervise their observations, advise them how to record
them, and then revise their plans. The fault in many schools is that the
children only copy the teacher's work, thus missing the great point—the
relation between the concrete fact and the graphic representation of it.

Professor Kirchhoff quotes with approval Dr. Max Ebeling's advice
that the children should measure first of all the schoolroom and draw it
on a scale of 1 : 100, and later, the school and playground, and draw them
on a scale of 1 : 500.* After the children have made these maps the
teacher should draw on the blackboard a map of the schoolroom on a
scale of 1 : 10, and of the school and grounds on a scale of 1 : 50. In this
way children would realise the value of different scales and gradually
become accustomed to their use.

On every plan the cardinal points should be clearly marked.

But such work should not be confined to schoolroom and grounds.
It is just as important to know how maps of the physical features of a
country are constructed, by trying to make them, as it is to study these
features themselves before reading descriptions of them. Once the preliminaries of planning are mastered in the school, Dr. Kirchhoff insists
on work out of doors. The pupils should take the measurements for
themselves and afterwards make maps from them. Distances can be
measured by counting paces, and directions must be carefully noted. At
first the work would be very simple, but more and more should be
attempted as the pupils gain experience. After the regular outlines of
the schoolroom and buildings have been measured and mapped, the probably more irregular one of the playground should be attempted, then
those of a straight piece of roadway, cross-roads, a curved road, a river's
banks, and so on. At last the older boys might make rough surveys,
using plane-table, compass, and ruler.

Such a course need not be continuous. During good weather

---

[7] " L'Art de lire les Cartes Géographiques." Par M. A. de Lapparent. *Revue scientifique*,
28 mars. 1896.

* The difficulty of adopting simple scales, like $\frac{1}{100}$, with our system of units hinders our
boys and girls from doing this until they have mastered the reduction of values in one unit
to its equivalent in another. It would be a great saving of time, if from the first the metrical
system of units were adopted in teaching map-drawing.

occasional excursions for this purpose would be made.   As far as possible
every new kind of map should be constructed or at least studied first in
the home district by the pupils.   Of course once the pupils are thoroughly
familiar with topographical maps, the standard topographical maps of the
district should be accepted and worked upon.   The teacher's aim should
be to make his pupils thoroughly conversant with maps—able to read
them, and to know the meaning of every symbol.   He should teach them
to use maps as they would use books—to be able to read and to interpret
them.   In the case of maps, much more than in that of books, many
people can read them superficially, but few really understand them.

Once more we find that geographical teaching must begin with the
home region.   The first maps should be made from nature and not copied.
As far as possible the need for a new symbol should be realised by the
pupils before they use it themselves, or are expected to understand its
meaning in the works of others.   All the different kinds of maps should
be studied first in local sheets, whereon the pupils can see the representation of the reality they can observe for themselves.   Only when this is
done can the children understand maps of other regions.

The map must be prominent in all excursions.   It must be read in
the open air with the actual region in view.   In the excursions of Dunfermline children we were surprised to find how puzzled the children
were by the ordinary political map, which was the only kind any of
them had ever seen.   They did not know how to put the map on the
ground in the proper position.   Most of the children who had not
already been shown Dunfermline from Edinburgh Castle were quite at a
loss to point it out from it when the position of both was shown them
on the oriented map.   These Dunfermline children and their teachers
were assuredly no worse than their neighbours.   It had never occurred
to the teachers to give the children such exercises.*

Teachers themselves need teaching in these matters.   Many have
never been taught the principles of construction of different kinds of
maps, nor been trained to use them.   How then can we expect that the
children will be instructed in these matters, as they should be, if their
teachers themselves are not?

The teacher often cannot read maps.   He devotedly sticks to the
politically coloured one which he saw at the normal school, and in one or
other of the cheap and nasty atlases which unfortunately are still the
favourites, in some training colleges at any rate.   It cannot be too
emphatically repeated that the political colouring is confusing, and that
such maps should be kept for senior pupils.   The simple uncoloured
topographical map of the district is the first map to use, once its elements
have been understood.   It would be a great help to teachers if a series
of simplified regional maps on a large scale could be produced.   If the

---

* Even if they had thought of it, perhaps they would not have been encouraged to try it
in school.   A teacher, not in Dunfermline, told me the other day that one of Her Majesty's
Inspectors of Schools and his assistant both complained of the time spent in teaching
geography, which they thought wasted, as the children surely could learn as much geography
as was necessary incidentally.   This happened in Scotland in the year of grace 1896 !

teacher be enthusiastic he may, however, make such a map for himself from the Ordnance Survey sheets.

A map showing the condition of the surface of the land round the school is easily made, and children can make it for themselves. On uncoloured copies of the topographical map of the district or on tracings from it, the children might tint the ponds and rivers blue, and the marshes, say, a light blue, the roads dark brown, the barren land light brown, cultivated land light yellow, pasture land light green, woods dark green, and buildings vermilion.

The problem of representing elevation is not so easy. Children are best introduced to the third dimension by making sections of table, bench, room, school buildings, etc., where the outlines are regular. Then they will more readily appreciate the difficulties of showing heights and hollows on a map. The ordinary local map is drawn as if the land were quite flat. Marks may be chosen to represent mountains, but that is not sufficient. If a map of the district, coloured or hachured, darker for the steeper slopes, lighter for the slighter gradients, be taken out and studied in the field, it will probably be the best way to continue this study, after it has been begun in school by outlining and colouring or shading a ball, an egg, and other natural objects. The meaning of the various symbols used for gradients in hachuring should be explained as the pupils advance. They should be able to distinguish between a map that actually gives some idea of the mountains and one where the shading is almost meaningless, which few people in this country seem able to do, perhaps because so many of our maps belong to the latter category.

Although contour maps, as ordinarily coloured, do not appeal very forcibly to the boy or girl, if the tinting is graded between two contour lines they become much more graphic. Contour maps are invaluable, however, for the pupils can draw sections of the country by their aid. In such sections the vertical and horizontal scale should be the same and not exaggerated, as is quite possible in the case of a small area.

A good relief model of the school district should be made if possible. But it too should not have the heights magnified relatively to the horizontal distances. A large scaled photograph of such a model might be distributed to the schools in a city, where it should be the duty of the School Board to see that at least one large model of the city and the region around it exists within their bounds.

Every school should possess plans and sections of the school-room, and buildings and grounds on various scales, large-scale maps of the immediate neighbourhood—the 1 : 2,500 Ordnance map if possible, and certainly the 1 : 10,000 and the 1 : 63,360—with the school in the centre. Various copies of the last two, or tracings of them, variously coloured, should also be at hand.

Teachers who desire to use maps of the home region in schools, and to use maps freely in studying the geography of other countries, should read Professor William Morris Davis's book, *Governmental Maps for Use in Schools* (New York : H. Holt and Co., 1894. Thirty-five cents), or his pamphlet on the "State Map of Connecticut," already mentioned.

The concluding paragraph of Professor Davis's pamphlet may be quoted :—" Lack of experience in the use of the state map in field-teaching is, for the present, a necessary result of the recent completion of the map. Three, four, or five years hence, a similar lack of experience will be interpreted, by those who understand the advances in the teaching of geography now in progress, as the result of the neglect of the opportunities for self-improvement that every teacher is in duty bound to use to the utmost. Fifteen or twenty years hence, it will be a reproach to the school in which the younger teachers of that time began their education, if they do not bring from it an acquaintance with home geography and the geography of the state, such as the proper use of the state map in the grammar schools will surely develop."

How long our Ordnance Survey maps have been available for such purposes is perhaps best left unsaid.

2. MAPS OF WIDER AREAS.

When once a pupil has learned to read and understand a map of, and in, his own district, he may be shown similar maps of other regions, and be taught to read and interpret them. The map should have a more important *rôle*, and the text-book a minor one in geographical teaching than at present. Text-books are guides for the teachers rather than for the taught. The teacher should make the children tell him about the geography of a country from maps, and trace the inter-relationship of various elements by comparing the different maps which illustrate each of them.

At present it is very uncommon to find any wall-maps hung up in a school except those coloured to show political boundaries. The tints used are often so crude and inharmonious that any inspector of schools with a sane colour-sense would order many such maps to be burned, to prevent them doing permanent injury to the children's powers of seeing. Apart from this, these maps are not of very great importance. A wall map should be a graphic diagram, and not a comprehensive collection of every geographical hieroglyph. One map for one thing is infinitely preferable to one map for everything.

Before using the usually complicated wall-map of any country the teacher should first draw its outline on the blackboard or use a prepared map in simple outline. Mountains, rivers, and other features should be marked in different colours, and the map gradually built up; every outline being simplified as much as possible. Only after this has been done should the wall-map be exhibited. The present plan of using the politically coloured maps preserves the delusion that to be a geographer is to know topography. The teacher too often is contented with pointing out, or asking the children to point out, the position of this or that place on the map. In teaching the geography of different regions, as well as in teaching home geography, the political map must be kept in the background. Professor Goodson says : " Maps giving political divisions only, with a few indications of mountain chains . . . are absolutely worthless for any intelligent work in geography."

Wall-maps, as well as those in our atlases, should show us in a condensed and generalised diagram the various surface features of a country, its highlands and lowlands, its climatic elements, its vegetable covering, the distribution of animals and of man, and all his activities. The construction and use of such maps must be learned in the home region; and then, and only then, can the child proceed to study the geography of other regions from such maps.

Last winter, when visiting a large Board school in Manchester, the writer listened to an admirably arranged lesson on the climates of Australia, their causes and effects. It was illustrated by an ordinary political map and a silhouette map. On the latter the coast line, mountains, and rivers were first drawn as the children answered the questions put to them. It would have been better had the teacher not hung up the other map at first, but waited until he had filled in the particulars on the silhouette map. Indeed the political map should never have appeared at this lesson. The children were questioned about winds and rain and plants, without seeing any relevant maps either prepared before or during the lesson. Indeed, the teacher did not know such maps could be procured. He was delighted to see atlases like Bartholomew's *Australasia* (Nelson) and *India* (London: Constable), and the *Atlas général Vidal-Lablache* (Paris: Colin), which is so full of suggestive sketch-maps invaluable to the teacher. It was an eloquent commentary on the geographical equipment and training of his college that he had never seen any good geographical maps. He informed the writer that it was only when he began to teach, and had to give lessons in geography, that he became interested in the subject by reading and using Professor Meiklejohn's well-known textbook.

An objection raised by teachers is that maps other than political ones cannot be bought on a large scale for class purposes; while mapmakers say that it would never pay them to print such maps as wall maps. Some teachers would gladly draw and colour such maps themselves if the outlines of coasts, rivers, mountains (also contours if possible) were printed on uncoloured maps of sufficient size to be used as wall-maps. The cautious, conservative map-maker, however, seldom sells such maps, and cannot be induced to keep them in stock, although many people would make use of them if they could be had at a reasonable price.

Many teachers, however, have scant time allowed for the geography lesson, and still less for its preparation, and these teachers declare that they have no opportunity of making maps even if they are wanted.

In most schools children are taught map-drawing, and many make very neat and accurate maps. These are usually political maps, which are far more difficult to draw than the simpler lines of maps illustrating climatic condition, or the distribution of plants, animals, and man. Let it be the reward of the good map-drawer that he is chosen to make one which will be added to the school collection of maps. In this way, with an expenditure of very few shillings, ten or a dozen maps illustrating the geography of each continent and smaller regions could be acquired by the school.

With such maps the teacher would no longer confine his attention to the pointing out of places. Such maps, supplemented by photographs and pictures as much as possible, rather than a text-book, should be the basis of the geographical lessons. As already said, the pupil should read the maps and deduce the relationships between them, under the teacher's guidance. He should not merely repeat the ideas picked up from the teacher, whose proper function is to lead his pupils, by judicious questions, to discover them from the maps for themselves. At first only simple relations would be understood, but gradually more and more complex ones could be discovered as the pupils progressed, and learned to read more accurately the information shown in the maps.

3. MAP-DRAWING IN SCHOOLS.

There is much discussion as to the value of map-drawing in schools. As has already been pointed out, mere mechanical copying is more or less wasted labour. The pupils must learn to make maps before they copy them, to feel the need for, and understand, each symbol before they use it. The objections to map-drawing by children disappear when the maps drawn are not laborious copies of the topographical map, but diagrams of distribution.

The shapes of countries alter with the projection used in representing a spherical on a plane surface. Hence the absurdity of expecting children to know the exact shape of every country. The maps the children draw should be simplified as much as possible, and political boundaries should never appear on their maps until an advanced stage. They should construct their own network of parallels and meridians of straight or broken lines, and the coast, river, mountain, and other lines copied should be simple curves. The edge of a plateau, for instance, might be indicated by a single brown curve convex to the shore, and a mountain range by a double line of curves with the concavities inwards. Simplified maps for copying are published (*e.g.* by Longmans and Co.), and should be used as samples more freely than at present; only the copy must not be a tracing, the maps serving merely as guides to the children in their drawing of the country.

Children should not be confined to drawing maps showing topographical features. Climate and distribution maps of vegetation, animals, man, occupations, etc. should be drawn. Such maps are much easier to make than the political ones commonly copied by children at present, and in addition have some educative effect. A boy who can sketch from memory, in rough outline, a dozen maps showing physical features, the rainy and dry regions, the hot and cold regions, the desert, pasture, and forest lands, the sparsely and densely peopled regions, etc., knows far more about its geography than the boy who can reproduce the sinuosities of coast and river, the intricate zigzags of political boundaries. The true geographer is he who has a clear comprehension of all the parts and aspects of a country, and their relationships, not he who has merely a minute knowledge of the names of its natural features and artificial boundaries.

(To be continued.)

BRITISH ASSOCIATION, 1896.

THE sixty-sixth meeting of the Association was held at Liverpool from the 16th to the 23rd September, under the presidency of Sir Joseph Lister.

The Geographical Section, which was presided over by Major Leonard Darwin, Hon. Sec. R.G.S., held its first meeting on Thursday, the 17th, in St. George's Hall, when the President delivered an address (published as the first article in this number of the *Magazine*), which, after a brief review of the geographical work of the year, and some remarks on the principal objects still to be aimed at, was devoted to the development of Africa.

A vote of thanks to Major Darwin was proposed by Count Pfeil and seconded by Sir Charles Wilson.

Then followed a paper from Mr. H. S. Cowper, who made a journey in Tripoli last March for the purpose of examining the megalithic ruins.

The route taken was by the Wadi Terr'qurt, a fine valley running parallel to the Wadi Doga, by which he entered the hills in 1895. He then proceeded to the districts of Ghirrah and Mamurah, south of Ferjana, through which runs a great wadi, the Tergilat. This reaches the sea at Kam, twelve miles south-east of the ruins of Leptis Magna, and is undoubtedly the Cinyps of Herodotus. On reaching the coast a week was spent at the ruins of Leptis and the Kam district, and the return journey was made to Tripoli by sea.

A large number of interesting views were shown on the screen.

The Rev. J. C. Robinson then read a paper on "The Land of the Hausa"; after which Mr. John Coles, Map Curator R.G.S., gave a concise history of the application of perspective drawings and photographs to surveying. He also described two new surveying cameras, and referred to the method of photographic surveying now extensively employed in Canada.

Mr. H. N. Dickson gave an account of the progress of marine research in the North Atlantic, which called forth some interesting remarks from Professor Hjort of Christiania; and Dr. H. R. Mill, Librarian R.G.S., then explained his "Proposed Geographical Description of the British Isles," which was fully described in the *Geographical Journal* for April. Dr. Mill's scheme is to provide for each sheet of the Ordnance Survey (one inch) a memoir containing a succinct account of the physical geography of the district, and of the influence exercised by geographical conditions on the utilisation of natural resources, sites of towns, and movements of population.

The meeting of Friday, the 18th, opened with a paper by the Rev. W. K. R. Bedford on "Old Tapestry Maps of England"; after which Dr. Tempest Anderson gave an account of the Altels avalanche, which in September 1895 descended on to the valley of Spitalmatte, on the Gemmi route, covering a large area, and causing the death of six persons. Dr. Anderson showed a series of beautiful lantern views to illustrate his address.

Lieut. G. Vandeleur, of the Scots Guards, then read a paper on
Uganda; and was followed by Dr. F. P. Gulliver, who, in treating of the
"Coast Forms of Romney Marsh," explained how Dungeness might have
been formed by successive stages, and showed that the history of the
district closely accorded with his theory.   Mr. Montefiore Brice gave an
interesting account of the last year's work of the Jackson-Harmsworth
Expedition, illustrated by numerous lantern views; after which Mr. F. G.
Scott-Elliot read a paper on the "Influence of Climate and Vegetation
on African Civilisation."

For this purpose Mr. Elliot divided the continent into four main divisions :—

I. *The wet jungle*, which is marked roughly by the presence of the oil or cocoa-
nut palm, numerous creepers—especially the *Landolphia* (rubber vines)—and such
forms as sesamum, *Cajanus indicus*, and manihot as cultivated plants. This region
is characterised by great heat and continuous humidity, without a season sufficiently
dry to leave a mark on the vegetation.   Limited by the direction of the prevalent
winds, and other meteorological considerations, and by the elevation, it extends to
three thousand feet, but often ceases below this level.   Everywhere it is inhabited
by small tribes of a weak enfeebled character, and on the lowest stage of civilisation, which have been subdued by Arabs and Europeans without difficulty.

II. *The deserts.*—Characterised by xerophytic adaptations, by *Zilla, Mesem-
bryanthemum, Capparis sodada,* etc.   The climate is severe, and is distinguished
by possessing no proper rainy season whatever, while the tribes dwelling in this
division are exceedingly healthy and vigorous.

III. *The acacia and dry grass region.*—Characterised by acacias, tree euphorbias, giant grasses, or frequently grassy plains in which each tuft of grass is isolated.
The climate is marked off from all the remaining regions by distinct dry and wet
seasons ; the dry season occupies from five to nine months, and leaves a distinct
mark on the vegetation.   This region occupies practically all Africa between three
thousand feet and five thousand feet, and also extends below three thousand feet
wherever the above climatic conditions prevail.   It is everywhere rather densely
inhabited, but there has not been a swarming centre here, and no emigration in
large numbers has taken place from this acacia region.   The nations inhabiting it
have fallen under Arab and European domination with scarcely a struggle.

IV. *The temperate grass and forest area.*—This region is distinguished by having
at no season of the year such drought as leaves a permanent mark on the vegetation, by a moderate rainfall, by moderate heat, etc.   The grass resembles the turf
of temperate countries, and the forest shows the same sorts of adaptation as occur
in temperate countries.   This region is found between four thousand six hundred
feet and seven thousand feet.   In this zone are situated the only places in Africa
that have acted as swarming centres of population.   The character of the native
races inhabiting it is vigorous and turbulent, and raiding is often carried on.   The
differences in climate, vegetation, and abundance of wild and domestic animals
explain why it is that these races only have, except in one instance, resisted both
Arab and European.

The work of the day concluded with a paper by Mr. Vaughan
Cornish on Sand Dunes.   Mr. Cornish said that

in the sorting of materials by wind the coarser gravel is left on stony deserts or
sea-beaches, the sand is heaped up in dune tracts, and the dust (consisting largely

of friable materials which have been reduced to powder in the dune district itself) forms widely scattered deposits beyond the limits of the dune district. Three principal factors operate in dune tracts, viz. (1) the wind ; (2) the eddy in the lee of each obstacle ; (3) gravity. The wind drifts the fine and the coarse sand. The upward motion of the eddy lifts the fine sand and, co-operating with the wind, sends it flying from the crest of the dune. The backward motion of the eddy arrests the forward drift of the coarser sand, and thus co-operates with the wind to build the permanent structure of the dune. Gravity reduces to the angle of rest any slopes which have been forced to a steeper pitch either by wind or eddy ; hence in a group of dunes the amplitude cannot be greater than (about) one-third of the wave-length. This limit is most nearly approached when the wind blows alternately from opposite quarters. Gravity also acts upon the sand which flies from the crests, causing it to fall across the stream-lines of the air. To the varying density of the sand-shower is due the varying angle of the windward slope of dunes. When there is no sand-shower the windward becomes as steep as the leeward slope. When the dune tract is all deep sand the lower part of the eddy gouges out the trough, and, when the sand-shower fails, the wind by drifting, and the eddy by gouging, form isolated hills upon a hard bed. In a district of deep sand, negative dunes ("Suljes") may be formed. The encroachment of a dune tract being due not only to the march of the dunes (by drifting) but also to the formation of new dunes to leeward from material supplied by the sand-shower, it follows that there is both a "group velocity" and a "wave velocity" of dunes. Since the wave velocity decreases as the amplitude increases, a sufficiently large dune is a stationary hill, even though composed of loose sand throughout.

Where material is accumulated by the action of tidal currents, forms homologous with the ground plan of dunes are shown upon the charts. The vertical contours and the movements of subaqueous sand dunes are conditioned by the different tactics of sand-shower and sand-drift.

A short meeting was held on Saturday, the 19th, when Mr. A. J. Herbertson read his paper on "World Maps of Mean Monthly Rainfall."

For practical purposes it is almost as important to know how rainfall is distributed throughout the year, as to know the total annual precipitation. The best way to show this is to make maps of mean monthly rainfall that will be comparable. Each month must be considered one-twelfth of a year and the average monthly rainfall reduced accordingly. This is being done at present by Dr. Buchan, Secretary of the Scottish Meteorological Society, and Mr Herbertson, and, so far as they are aware, this is the first attempt to do so for the whole world.

Mr. J. W. Moir, who was to have read a paper on the "Climate of Nyasaland," was not present ; but abstracts of his observations were laid on the table. The Report of the Committee on African Climate was presented, and some remarks on the subject were made by Mr. C. G. Ravenstein. In the absence of Mr. J. Howard Reed, his paper on "Practical Geography in Manchester" was read by Mr. B. V. Darbishire :

The author believes the Manchester Geographical Society has demonstrated that geography is popular among the people. Mr. Eli Sowerbutts, secretary of the Manchester Society, commenced giving popular geographical lectures some years ago. The demands for work of this kind grew to such proportions that a body of prominent members of the Society, including the chairman, took up the lecturing work, which has increased year by year ever since. The lecturers now form an organised body of expert geographers and practised speakers, who freely volunteer their services for the purpose of spreading reliable geographical information. The lectures are all given in a popular manner, and are mostly illustrated by lantern views. During the past five years over three hundred lectures have been delivered in Manchester and the surrounding districts, and over ninety thousand hearers have been reached. The audiences are principally of the working class, but also include the members of many well-known literary and scientific clubs, and students of continuation schools. The lectures given include such titles as : "Shaping of the Earth's Surface by Water-action," "Map Projection," "India," "China, Corea, and Japan," "Polar Exploration," "Across the Rocky Mountains," "Canada," "Across Africa with Stanley," "Uganda," etc. Applications for lectures are made to an hon. secretary, who conducts all correspondence and makes arrangements with the local societies and clubs and the lecturers. The engagement of halls, printing, and similar matters are carried out on the spot by the local people. This system has proved so satisfactory, and the enthusiasm of the voluntary workers has been so well maintained, that no hitch has ever occurred. The terms on which the lectures are given are very simple. Any member of the Manchester Geographical Society or any affiliated society is entitled to apply for lectures. Lantern apparatus and volunteer operator are supplied when required. A nominal fee is charged for each lecture, travelling and lantern expenses being added when incurred. Any balance in hand at the end of each season is applied to the upkeep of lantern plant and the making and purchase of new slides. Another important branch of voluntary work consists in the analysis of some two hundred British and foreign scientific journals. This is most useful for scholars and students. It enables them to follow up, with ease, the literature on any special subject. It has received the commendation of several high authorities. The Manchester geographers intend to follow up the work they are doing and hope to more fully occupy the field. They are conscious that there is ample room for development. The author feels sure they would be glad to hear of similar organised effort in other parts of the country.

After the reading of the paper Mr. Eli Sowerbutts further explained the part taken by the Manchester Geographical Society in the teaching of geography; and the day's proceedings were brought to a close by an address on "Canada and its Gold Discoveries" from Sir James Grant, who, as a member of the Canadian Parliament, introduced the Billfor the construction of the Canadian Pacific Railway.

On Monday, the 18th, Mr. W. A. L. Fletcher gave an account of his journey towards Lhasa with Mr. and Mrs. Littledale, and showed some interesting photographic slides. He was followed by Mr. F. W. Howell on his explorations in northern Iceland, and Dr. K. Grossmann on the less-known interior of the island. Both of these gentlemen exhibited excellent lantern views. Mr. George G. Chisholm then read a paper on the "Relativity of Geographical Advantages," which will appear in the Magazine.

Mr. Chisholm was followed by Mr. Ralph Richardson, on the various boundary lines between British Guiana and Venezuela attributed to Sir Robert H. Schomburgk.

After adjournment for luncheon, Sir Martin Conway gave, to a crowded meeting, an account of his recent journey in Spitzbergen; and this was followed by Mr. H. W. Cave's paper on "The Present Condition of the Ruined Cities of Ceylon." In Mr. Cave's absence the paper was read by Mr. Ferguson, and was profusely illustrated by a series of beautiful photographic slides.

The last address on this day was delivered by Professor J. Milne, F.R.S., on "Earthquakes and Sea Waves." The Professor explained and illustrated the delicate instruments by which even the smallest movements of the Earth's crust are automatically recorded, and predicted that we should shortly hear of some considerable disturbances in the Far East, which they had indicated.

The last day's proceedings were opened by Mr. A. C. Fitzgerald's paper on the "Southern Alps of New Zealand," where he did some remarkable climbing, and nearly lost his life through an accident. Major-General Sir Charles Wilson then read a most interesting paper on the Egyptian Sudan, in which he advocated the line from the neighbourhood of Berber to the Red Sea as that by which the trade of the Sudan should be opened up. Mr. W. A. Andrews followed with "The Teaching of Geography in Relation to History," pointing out the influence of geographical conditions on the development of a nation, and the social life of its people, and how necessary a knowledge of these conditions is to the thorough comprehension of any historical epoch.

Mr. E. Odlum then dealt with the boundary between British Columbia and Alaska.

After luncheon, Mr. Scott Keltie, who had just returned from Christiania, gave a graphic account of Dr. Nansen's reception in the Norwegian capital, and read a paper by Professor Mohn on the scientific results to be anticipated from Nansen's observations. Mr. A. J. Herbertson exhibited an ingenious apparatus constructed to illustrate map-projections :—

Every teacher of geography experiences a difficulty in trying to give his pupils a vivid idea of the various map projections. This is in part overcome by using a candle and a skeleton hemisphere formed of a wire network of meridians and parallels, and, if possible, with an outline of the continents, such as the author has recently devised, and Messrs. Philip make. By altering the position of the lighted candle, different projections of the network can be thrown on a flat screen, and the pupils can see the different distortions of the network that result for themselves. By using half a cylinder or half a cone, various cylindrical and conical projections can be illustrated in the same graphic way.

Mr. B. V. Darbishire exhibited and explained a New Population Map of South Wales, which shows how the land features have influenced the distribution of population; and the proceedings of the meeting were brought to a close by Mr. Herbertson reading the "Report of the Committee on Geographical Teaching."

The meeting of the Geographical Section was a successful one, and the hall allotted to it was truly palatial both in dimensions and decoration.

GEOGRAPHICAL NOTES.

By The Acting Editor.

EUROPE.

Determination of the Force of Gravity in Germany.—Measurements have been taken at twenty-two stations on the line Kolberg—Schneekoppe by the Royal Geodetic Institute in Potsdam. Along this line, which passes through Arnswalde, Gröditzberg, Grunau, and Giersdorf, considerable variations of gravity and deviations of the plumb-line were observed. Measuring them by the thickness of the mass which would produce them, Dr. Helmert finds that in the Pomeranian lake area the excess of the force of gravity corresponds to a stratum 690 feet thick. Between the Kleistberg and the valley of the Netze there is a deficiency of 330 feet. Under the valley of the Oder, from Bomst as far as twelve miles north of Gröditzberg, there is a subterranean mass 1000 feet thick, though the surface is sandy. The heavy basalts of the Gröditzberg and the equally heavy Silurian clay-slates of Ludwigsdorf, on the other hand, do not prevent this excess of gravity from gradually vanishing. Still further appears a deficiency of mass, averaging 650 feet in thickness, which extends under the Schneekoppe, and here, too, nothing in the nature of the ground on the surface would indicate the change. From this fact, and the sudden variations sometimes occurring in the force of gravity, Professor Helmert concludes that the disturbing masses do not lie near the surface, but not deeper than twelve or eighteen miles. As, however, the results may be explained by many different assumptions with regard to the disturbing masses, it is not safe to dogmatise on the question. The deviation of the plumb-line is very marked in the neighbourhood of the mountains. The greatest deviation was observed at the station Alter Bruch on the flank of the Schneekoppe, at an elevation of 3008 feet.—*Geogr. Zeitschrift*, June.

Sunshine in Germany.—A table of the sunshine at twenty-nine towns is published by Herr Helmuth König in the *Annalen der Hydrographie*, Heft vii. The sunshine has in all cases been registered by the Campbell-Stokes apparatus, and the records of most of the stations extend over six to eight years, while a few have been kept for as long as ten or twelve years. The monthly amounts have been reduced to thirty days, while the yearly totals refer to the full year of 365 days.

In discussing the results, Herr König points out that the duration of sunshine increases unmistakably though irregularly from west to east. The stations close to the coast and their hours of sunshine are as follows : Emden, 1770 hours ; Meldorf, 1706 ; Kiel 1510 ; Rostock, 1693 ; Kolberg, 1773 ; Dirschau, 1801. An uninterrupted increase is therefore evident on the Baltic coast in a direction from west to east. On the North Sea coast, on the other hand, there is a diminution in the same direction. Jutland and the Danish isles seem to form a boundary as regards insolation between the North and Baltic Seas, for the yearly mean at Copenhagen is only 1204 hours, or less than at Hamburg, which has 1333 hours ; while Emden, Heligoland, Meldorf, and Bremen have more sunshine than the western stations on the Baltic. So, too, the stations beyond the German boundary, in Brussels and on the south coast of England, have a still longer duration of sunshine. In general, places inland receive less sunshine, as shown by the figures for Ellewick, 1698 ; Cassel, 1614 ; Magdeburg, 1603 ; Breslau, 1691. There are, in this case also, some notable exceptions. There is, further, an increase of sunshine towards the south, independent of the law of decrease with distance from the sea. For Vienna, Padua, and Rome the figures are 1816, 2057, and 2431, respectively, and if to these be added the maximum in Europe, namely, 2908 hours at Madrid, it will be seen that the insolation increases more regularly and more rapidly towards the south than along a parallel of latitude.

The topographical situation must have an influence on the duration of sunshine, as clouds are formed to a great extent on mountains. On the other hand, valleys lose sunshine owing to the overshadowing heights and their cloud-caps, as well as through the mist that often rises in valleys. Thus, the Inselsberg, 2972 feet above sea-level, has 1522 hours, while Erfurt, 2300 feet lower, has 1622. The difference is chiefly made up during summer and autumn. Again, the Sonnblick and Obir have 1531 and 1642 hours respectively, as against 1816 at Vienna, the proportions of the possible duration being 34, 37, and 41 per cent.

In general, the yearly period exhibits great regularity. The hours of sunshine gradually increase from the minimum to the maximum, and as gradually decrease. The position of the minimum is remarkably constant. Except at Ellewiek and Poppelsdorf, where it occurs in January, the minimum always appears in December—that is, in the shortest days, when, the sun being low, the rays have to penetrate a thicker stratum of air. The absolute amount varies in general from thirty to forty hours, or 13 to 17 per cent. It is still less at Marggrabowa, Kiel, and Eberswalde (10 per cent.), Hamburg (9 per cent.), and Stuttgart (3 per cent.), while it is above these limits at Leobschütz, Erfurt, and Emden (20), Chemnitz (21), Jena (22), Ellewiek, and Poppelsdorf (22 to 23). The maximum has a less uniform character. It does not fall in the period of the longest days, for almost everywhere, at least in Northern Europe, there is a falling off in the duration of sunshine in June and July. Over the whole of North-Western Europe there is a pronounced maximum in May, generally followed by a secondary maximum in August. As the latitude becomes lower and the longitude increases, the chief maximum moves more and more towards the summer, until at last the positions of the two maxima are reversed. South France, Switzerland, the Adriatic, and the whole of Southern Europe have the chief maximum in July or August. At the German stations the August maximum averages something under 50 per cent., and in few cases reaches 52. In the south much higher values are attained: Vienna, 54; Zürich 57; Trieste, 66; Montpellier, 67; Lugano, 67; Rome, 75; and Madrid, 84. On the mountains the duration of sunshine is more equally distributed throughout the year. On the Sonnblick the monthly minimum is 115 hours, while the maximum is 151; and on the Obir and Säntis the range is much the same. Again, the winter months are very sunny, the procentual maxima occurring at this season.

The daily period is also very simple in its variation. On the whole, the duration of sunshine follows the course of the sun, rising in the morning to a maximum in the middle of the day and then declining. Little sunshine is registered in the early morning and late in the evening, owing to the low altitude of the sun, but as a rule twice as much is registered towards sunset as near sunrise. The increase is much more gradual than the decline, which is very marked from four o'clock. At high stations the variations follow in reversed order. Sunshine is most frequent and of longest duration between one and two o'clock at most stations, but at Rome and Madrid the maximum is reached between ten and eleven o'clock, and at high stations an hour earlier. At the beginning of the year the daily maximum is usually at mid-day, and is reached at an earlier hour as the summer approaches, returning to its former position at the end of the year. At the German coast stations, however, the maximum has a tendency to move to the hours of the afternoon. One peculiarity in the daily variations is, that in the warmer seasons there is often a depression in the curve of sunshine at mid-day, so that the maximum is separated into two parts by an intervening minimum. The cause probably lies in the formation of clouds which, as shown by hourly observations at Görlitz and Potsdam, is greater between eleven and two o'clock than earlier or later.

The Arend Lake.—In the north-west corner of the Altmark, in lat. 52° 54′ N. and long. 11° 30′ E., near the town of Arendsee, lies the lake of the same name. Its peculiarities have now been thoroughly and scientifically investigated by Dr. W. Halbfass. Its absolute depth, and relative depth compared to its area (max. 162·4 feet), is remarkable, exceeding that of almost all the lakes of North Germany. The great unevenness of its bottom is in striking contrast to the flatness of the surrounding country, and its form is a regular oval. These peculiarities are probably due to the landslips of 822 and 1685, to the former of which the lake is generally supposed to owe its existence. It lies at a height of seventy-seven feet above sea-level, on a sandy heath distinctly different in character from the region of the Baltic lakes. Dr. Halbfass believes that the Arend Lake lies on the southern or south-western edge of a river-bed of the second Glacial Period, and may have been originally a river lake. A comparison with soundings made in 1786 shows that no perceptible change in the level of the lake has taken place during the last hundred years. In general the depth diminishes from south-west to north-east. The area of the lake is 1369 acres, and its volume 5721 million cubic feet, whence the average depth is about ninety-six feet. Of the lakes of North Germany only six surpass the Arend Lake in maximum depth, and not one in average depth. As the average depth amounts to more than the half of the maximum, the basin belongs to the caldron-shaped cavities of Prof. Penck's classification. Its mean slope is 5° 5′ (Lake of Constance only 3°), and the depth is 3·73 per cent. of the radius of the surface (Lake of Constance 1·92). If the old notion that a landslip brought it into existence be erroneous, it probably owes its origin to subsidence. A full description of the lake, with a map and sections, is given in *Petermann's Mitt.*, Bd. 42, No. 8.

Huns in Switzerland.—The Eifisch Valley (Val d'Anniviers) is one of the southern lateral branches of the Rhone valley. It is a narrow, deep hollow, 22 miles long, excavated by erosion, and drained by a rapid torrent, the Navisence, which issues through a rocky portal that closes the valley to the outer world. In this valley are situated several villages, of which Chandolin, at an elevation 6463 feet, is perhaps the highest parish in Europe. On the origin of the inhabitants many different theories have been propounded. The geologist Eduard Desor holds them to be of Arab race, while Fröbel and J. von Tschudi regard them as Celts. Herr A. K. Fischer, who has lately published a book on this people, maintains that they are Huns, following the local tradition. He shows that during a scouting or foraging expedition, a small band, becoming separated from the main body, established themselves in the Val Tournanche, or some other valley on the southern flank of the Pennine Alps, whence on the invasion of the Lombards in the second half of the sixth century they fled into the uninhabited Eifisch valley. Here they lived for centuries in complete isolation, and though on the introduction of Christianity among them, probably in the eleventh century, they came into contact with their neighbours to some extent, they have preserved in great measure up to the present day the language and customs of their race.

Herr Fischer points out similarities between their dialect and that of the Hungarians, who, he shows, were a sister nation of the Huns. As in Hungary also, the family life of the Eifisch people is based on a patriarchal foundation, and the whole community is knit together by strong patriotic feeling. Marriage and . burial ceremonies, ornaments, etc., all point to the same origin.—*Aus Allen Weltteilen,* July-August.

The Pitch Wells of Zante and the Origin of Petroleum.—On page 426 the case of Karabugaz Bay was cited as showing how petroleum may possibly be formed from decaying organic matter. The other theory, that petroleum is of intratelluric origin, is expounded by Prof. K. Mitzopulos of Athens, in *Petermann's Mitt.,* Bd. 43, No. 7, in connection with the pitch springs of Keri in Zante. These springs, described by Herodotus, are an attraction to all visitors to the island. The hollow of Keri lies in the southern extremity of the island, and here at the edge of a swampy field are two springs of good water, at a temperature of 57° F. From the bottom, fifteen to eighteen inches deep, rise bubbles which, bursting at the surface, spread over the water an iridescent film of petroleum and leave pitch on the bottom. All around the springs the ground is covered with dirty solid pitch.

In former times eruptions of pitch have taken place ; and on Jan. 25, 1895, another eruption threw out stones which caused Prof. Mitzopulos to doubt the organic origin of pitch, and the cause usually given of the formation of the hollow of Keri, namely, the contraction of a coal-bed lying at an unknown depth. The ejected stones were of recent volcanic origin, being pebbles of pumice, a substance which had never before been found on any of the Ionian Islands or on the west coast of the Peloponnesus or Central Greece. From the information he received concerning this eruption, Prof. Mitzopulos is convinced that these recent volcanic products overlying the pitch belong to beds situated at some unknown distance below the surface and overlaid by the Tertiary strata, and they are certainly of sub-aërial formation. Submarine eruptions also occurred on Feb. 6th, and it is probable that the whole bay from Keri to Geraka, 5 miles long and 3 to $3\frac{1}{2}$ broad, was an active crater in the Tertiary period. The melting of the loose and porous pumice, and the enormous pressure of the superincumbent Tertiary strata, may have produced a gradual subsidence.

It cannot be disputed that hydrocarbons may be evolved by distillation from animal and vegetable matter, but so many millions of buried bodies and trees would be necessary to produce the enormous quantity of petroleum and hydrocarbon gas at Baku and in the United States, that many scientists have doubted the organic origin of this fuel. Various hydrocarbons may be produced by the direct action of water on ferrous carbonate, and lately the combination of hydrogen with calcium has been effected in a simple manner by the Wilson Aluminium Company in the United States. An electric current is made to pass through a molten mass of lime and anthracite, whereby calcium carbide is formed and carbonic oxide given off $(CaO + 3 \ C = CaC_2 + Co)$. If water is allowed to act on the combination the result is acetylene and lime. Although we do not know for certain the chemical composition of the heated interior of the Earth, it is very probable that hydrogen is present in combination with various minerals, especially calcium, and from these combinations acetylene and carbonic oxide may be formed by the addition of water, which either filters down from the upper strata or is given off from the magma. Hence carbonic acid, which streams out in many volcanic regions, may be produced by the action of the heated carbonic oxide on various oxides and salts, and the acetylene through the addition of atoms of hydrogen be converted into the hydrocarbons constituting naphtha, petroleum, and asphalt. Sulphuretted hydrogen may also occur as a secondary volcanic product, if silicum disulphide be present.

AFRICA.

Uhehe.—Colonel Baron v. Schele describes this district of German East Africa in the *Mitth. aus den Deutschen Schutzgebieten*, Bd. ix. Heft 2. Its eastern and southern boundaries are clearly defined by the mountains, the passes of which lie about 6500 feet above sea-level. On the other sides the limits are not known exactly, especially towards Ubena and Ussangu, countries which are politically dependent on Uhehe and are probably similar in physical features. From the Gombo lake, south-east of Mpapwa, to the Ruaha stretches, in a southerly direction, a range of mountains called Rubeho. Its slopes, towards the Gombo lake and the valley of the Mukondokwa on the north, and to the valley of the Myombo and Yovi, are extraordinarily steep and render the approach to Uhehe very difficult. Equally steep is the section between the Ruaha and the Ulanga, where the range turns south-westwards to join the Livingstone mountains. The range consists of several parallel chains, all likewise steep. After it is passed the plateau of Uhehe is reached, which has an average elevation of 3900 to 4900 feet, and appears as an undulating country traversed by separate ranges rising 1000 to 1600 feet above it.

While the north-eastern part of the boundary range is drained into the Mukondokwa, the entire south-eastern part of the plateau belongs to the drainage area of the Rufiji. The hydrography of the southern part of the plateau is not thoroughly known. The Livingstone range is certainly the watershed between Nyassa and the Ulanga, but where this river has its chief source has not been ascertained. The greater part of the water from the plateau probably flows into the Ruaha.

The steep slopes of the mountains are not suitable for fields and plantations, but the valleys, which have heights reaching up to 2600 feet, have an excellent alluvial soil, in which coffee, tea, cocoa, and perhaps tobacco, might be cultivated. The crests and slopes are bare or moderately wooded, and offer a profitable field for the exertions of the forester. In German East Africa the afforesting of the mountains is a pressing necessity, for one of the greatest drawbacks to cultivation is the unregulated drainage. The rainfall is sufficient, but unequally distributed, and forests would check the too rapid drainage in the wet season. The mountains are composed chiefly of gneiss and quartz; the cultivable soil is red laterite, and in the valleys there is some black vegetable earth. The plateau is better adapted for field cultivation and grazing. In the valleys rice and cotton could be grown in large quantities, and sesamum, ground-nuts, etc. on the more elevated lands.

To what race the Wahehe belong is still uncertain—whether they are akin to the Mahenge and Wabena, or whether they are Zulu immigrants. They are not the original owners of the country, for of these the residue is still found in the more remote parts of the mountains. The Wahehe are a remarkably fine people, gracefully built and with finely cut features. They are brave and warlike, but as lazy as Negroes. Only the crops actually necessary to support life are cultivated, and these are chiefly tended by slaves obtained by raids into the lands of the neighbouring tribes. The pure Wahehe have no domestic industries except the weaving of mats and the manufacture of their simple wooden utensils and weapons. Their chief wealth consists in cattle, sheep, and goats. In the mountains the people live in separate families, but on the plateau they are collected in large *tembes*. These are capable of defence, and are sometimes several hundred yards in length, with hundreds of inmates. Kwiringa, which was taken a few years ago by the Germans, consisted of a large number of such *tembes* surrounded by a wall with bastions, and contained 4000 to 5000 inhabitants.

The Seychelles.—In the *Verhandlungen der Gesell. für Erdkunde zu Berlin*, Bd. xxiii. No. 6, Dr. A. Brauer gives a sketch of the islands from observations made during a visit the chief object of which was to determine the zoogeographical relations of the group, and to carry out special zoological investigations for a detailed study of coral reefs. The Seychelles are said to have been discovered in 1528, or, according to other reports, in 1505, by Pedro de Mascarenhas, who named them the Mascarene Archipelago. For more than two hundred years they remained unnoticed by Europeans, until French settlers in Madagascar, under the leadership of Picault, rediscovered them in 1742, and named them Îles de la Bourdonnais, after the Governor of Mauritius, Mahé de la Bourdonnais. The first settlers are said to have been French creoles from Réunion and Mauritius, who arrived in 1793, and again changed the name of the group, calling them the Seychelles after a French officer. They were occupied by Great Britain in 1794, and formally ceded by the French in 1814.

The group, lying between 3° 33′ and 5° 35′ S. lat., and between 55° 16′ and 56° 10′ E. long., consists of thirty islands, with an aggregate area of 102 square miles; only eighteen are inhabited. It stands on a submarine bank ten to forty fathoms deep, which slopes rapidly down to a depth of 1500 to 2000 fathoms towards the Amirante Islands. The conditions of the islands are so similar that they may be described together.

Whereas Mauritius and Réunion are of volcanic origin, and the Chagos archipelago to the north, and the Amirante, Aldabra, and other small islands between the Seychelles and Madagascar, are of coral formation, the Seychelles are entirely built up of granite, which gives a definite character to the islands, and reveals its presence in their outward appearance. The summits of the mountains are generally rounded, or consist of masses of blocks which often assume the form of ruins. The flanks rise in terraces, and steep precipices fissured into the most fantastic forms meet the eye of the traveller in all parts of the mountains. The greater part of the surface of the ground is covered with blocks of all sizes and shapes, which, though they render a great part of the land useless for cultivation, preserve the fruitful soil from being swept away by the heavy rains. Laterite, produced by the weathering of the granite, is everywhere present; a thick layer of humus is found only where the old forest still remains. In Silhouette the mountains rise steeply from the sea to heights of 2600 to 2800 feet, and the Morne Seychellois of Mahé to nearly 3200 feet. On the rest of Mahé, and on the other islands, they are lower, and are generally separated from the sea by broad plains with a soil of coral sand. Almost all the islands are begirt with coral reefs, which make them difficult of access. Between the northern point of Mahé and St. Anne there is an entrance through the reef into a roomy basin deep enough for large vessels. In opposition to the hitherto prevailing opinion that the Seychelles have sunk and are still sinking, Dr. Brauer states that raised coral reefs on almost all the islands he visited show that an upward movement has taken place and still continues. An upheaval of 80 feet can be proved, but probably this is much below the actual rise. Several of the smaller islands have recently been connected with the larger ones, and at many places more land could be reclaimed from the sea without great difficulty.

The climate is excellent, the heat being generally quite endurable in spite of the proximity of the Equator. The mean yearly temperature is 81° to 84°, the yearly range 18° to 21½°, and the daily 11° to 12½°. The lowest temperature is 73½° in the lowlands, and 68° in the mountains, and the highest 84° to 88°, rising only in the months of April, May, and November, the seasons of calms, sometimes to 97°. The summer is the rainy season, especially from December to April, while in winter scarcely any rain falls except in the mountains, where it must be fairly abundant, for the streams never run dry. In 1895 the rainfall was 98 inches. The long and heavy rains sometimes cause great landslips, one of which in 1862 buried several houses and their inhabitants under heaps of blocks and mud. With a good climate, the islands are in general healthy, but leprosy seems to be very prevalent, 10 per cent. of the inhabitants, according to competent authorities, suffering from this terrible disease. The sanitary condition is also favoured by the good water flowing down from the granite, which descends in Mahé in no fewer than 133 streams. In the southern part of Mahé and on some of the other islands the water is not so good, and causes dysentery and mild fevers.

The vegetation thrives luxuriantly in the rich, well-watered soil. The coastal zone is nearly everywhere occupied by coco-palms. Up to 1000 or 1300 feet the forests have nearly disappeared, having given place to grass and bush, or, especially on Mahé, to cultivated fields. The most extensive remains of the forests which once covered the islands are found in the central part of Mahé, on Silhouette and Praslin, and are now carefully protected by the Government. They contain several varieties of good building-timber, palms, tree-ferns, etc., and are bright with orchids and other flowering plants. The fauna is scanty, but contains a large number of peculiar forms, among them the huge land-tortoise (*Testudo elephantina*), the proper habitat of which is the Aldabra islands.

Since the islands came into the possession of Great Britain, planting has become the chief occupation of the inhabitants. At first coco-palms and sugar were mostly cultivated ; but the latter, owing to competition with beet sugar, has given place to cocoa, vanilla, cloves, and (more recently) coffee. The prosperity of the islands suffers, however, through the degeneracy of the creoles, who, with few exceptions, are boastful and talkative, listless, depraved and lazy. Palm oil and vanilla are therefore the chief exports, for plants which require some little atten-tion soon deteriorate. Consequently the creoles live in great poverty, and most of them have fallen into the hands of usurers.

Though the Seychelles have now been under British rule for a century, the sympathies of the inhabitants are still French, and the language is also French, or the corrupted patois spoken by the creoles, the British Govern-ment having guaranteed the maintenance of the French religion, laws, and language.

A Lake in Fernando Po.—In December last P. Joaquin Juanola, starting from Con-cepcion, marched to the village Balacha (Boloco), situated at an elevation of about 1600 feet. The following day he attained a height of 4430 feet above sea-level, and there found a lake lying in a hollow which appeared to be the crater of a huge extinct volcano. A steep descent of 650 feet led down to the shore of the lake. A large stream, forming a picturesque cascade, pours its waters into the lake, but no outlet was visible. The length of the sheet of water is about 1300 yards, and its maximum breadth 850 to 900. It lies in a hollow among the mountains, which rise 1000 feet above it. Monkeys and ducks are numerous along its banks ; and the Kroomen who accompanied P. Juanola saw some bulky form in the water which they declared to be a hippopotamus, and he himself saw in the dis-tance disturbances which might be caused by the gambols of some large animal. —*Bol. de la Soc. Geogr. de Madrid,* Tomo xxxviii. Nos. 1-3.

AMERICA.

New Districts in Canada.—On p. 103 the division of the northern territory of Canada into provisional districts was briefly recorded. The boundaries of these districts are given in the *Statistical Year-Book of Canada* as follows :—(1) The District of Ungava is bounded on the north by Hudson Strait, on the south by the Province of Quebec, on the east by the dependency of Newfoundland on the Labrador coast, and on the west by Hudson Bay and Ontario, all islands within three miles of the coast being included. (2) The District of Franklin embraces all the Arctic islands from Baffin Bay, Smith Sound, and Robeson Channel on the east, to Banks Land and Prince Patrick Island on the west, its southern boundary running from Cape Best through Hudson Strait, Fox Channel, the Gulf of Boothia, Franklin, Ross, and Simpson Straits, Deane Channel, and Dolphin and Union Strait. (3) Yukon is bounded on the west by Alaska, on the south by British Columbia, and on the east by a line drawn from the western mouth of the Mac-kenzie river, along the 136th meridian, to the range of mountains north of the Pelly river, and then along the range to the Liard river on the British Columbian boundary. (4) Mackenzie extends from Yukon eastwards to the 110th meridian, and southwards to British Columbia and Athabasca. To the latter it is proposed to add the territory north of Saskatchewan as far east as the 100th meridian. No mention is made of the 470,000 square miles said to have been added to Keewatin.

The Population of the Argentine Republic.—The second census of this republic was taken on May 10, 1895, and the Commission has made public some pre-liminary details, which are given in the *Bol. del Instituto Geogr. Argentino*, Tomo xvii. Nos. 4-6. The fourteen provinces, including the capital, have an aggregate area of 610,936 square miles, with a population of 3,850,011. The density is, therefore, 6·3 persons to a square mile. The territories of Misiones, Formosa, Chaco, Pampa, and others extending down to Cape Horn, have an area of 500,143 square miles with a population of 102,979. These figures give a density of ·2 per square mile ; but the more populous Misiones, with nearly 2·9 to the square mile, raises the ratio for the territories, the remainder having barely ·13 inhabitants to the square mile. The total area of the republic is, then, 1,111,079 square miles, and the inhabitants, including 90,000 whose domicile is not specified, and 50,000 in foreign countries, number 4,092,990. There results a density of 3·7. The largest and most populous state is Buenos Ayres (the capital being excluded), with 921,225 inhabitants, but the density, 7·85, is less than that of Entre Rios, 10·15. The population of the city of Buenos Ayres is 663,854. At the time of the first census, in 1869, the population of the country was only 1,830,214. During the twenty-six years the increase has accordingly been 123 per cent., or 4·7 per cent. annually.

AUSTRALASIA.

Artesian Water in Queensland.—In noticing Mr. Jack's investigation, in vol. xi. p. 486, a possible leakage of the water into the sea was mentioned. On this question, one of great importance to the colony, Mr. J. P. Thomson read a paper before the Queensland Branch of the Royal Geographical Society of Australasia in January. Mr. Jack remarks that leakage may arise, either from the strata at the base of the Lower Cretaceous formation dipping seawards and then rising to the ocean bottom, or converging to it at a lower angle ; or, secondly, by the beds crop-ping out at gradually lower levels till they reach the sea. Between Warrnambool and the mouth of the Murray, fresh water certainly does well up in the sea, but this proceeds from the Lower Tertiary rocks and has no connection with the Lower Cretaceous strata. Whether the latter crop out at the sea bottom at a greater distance from the shore it is impossible to say, but there is no evidence of leakage in this manner. Mr. Thomson also affirms that his experience leads him to believe that there is no leakage of true artesian water into the sea.

Those who maintain the contrary ask what becomes of the enormous quantities of water absorbed by the bibulous beds, a small fraction of which is drawn off by the artesian wells. The rapid disappearance of flood waters is pointed to as evidence of the truth of the theory. But it is no uncommon occurrence for water to disappear in some sections of a river channel to reappear in other places. And, again, the rivers have not been accurately gauged, so that it is impossible to say what proportion of the rainfall is carried off by them. Mr. C. H. Russell has stated that only one and a half per cent. of the rainfall of the Darling basin passes Bourke along the river channel. But, according to the Chief Engineer for Water Conservation, New South Wales, the drainage area of the Darling above Bourke is 74,760 square miles, the mean annual rainfall twenty inches, and the discharge at Bourke 6557 cubic feet per second, or about six per cent. of the rainfall. There is also the evaporation to be considered, which, in the Bourke district, one of the hottest and driest in the country, must account for a large proportion of the rainfall. At the Enoggera reservoir, near Brisbane, the water evaporates at the rate of $\frac{1}{4}$ inch per day, and in Central Australia the evaporation has been estimated from reliable experiments at one inch a day. The so-called lakes of Central Australia are salt, and this fact bears testimony to the enormous quantity of water drawn from the ground by the dry atmosphere.

The physical structure of the country is such as to prevent any extensive leakage into the sea. Except at the head of the Australian Bight where limestone crops out, and on the southern shores of the Gulf of Carpentaria, the whole rim of the great central basin consists of the older Palæozoic rocks. The basin of the Darling is bounded on the north-east and south by the Palæozoic rocks of the Great Dividing range, which constitute an impregnable barrier against the sea, while on the west are the Flinders and Gray ranges.

Lastly, Mr. Thomson, though not inclined to deny that the water-bearing beds may crop out at the sea bottom in the Australian Bight and in Torres Strait, believes that the pressure of the denser waters of the ocean would prevent any extensive leakage at considerable depths.

Dampier Island.—Herr Georg Kunze, a missionary, visited this island, which lies off the coast of German New Guinea, in March 1893. It is named by the inhabitants Krakar, and contains about 2000 Papuans distributed among sixty villages and settlements. Separated by the central mountains, the natives on the north and south speak different languages, and can only communicate with one another for the purposes of trade by the mediation of a few individuals who understand both dialects. Their huts are poor compared to those on the mainland, as there is no good building material on the island. Leaves of the pandanus are used for thatching, as the better kinds, those of the nipa and sago palms, are not procurable. The natives manufacture shell bangles, articles of tortoiseshell, bags of fibres, basket-work shields, etc., for trade with the mainland. The northern half of the island maintains a lively traffic with the coast of New Guinea from Cape Croisilles to Franklin Bay, while the other half trades with the coast from Cape Croisilles to Friedrich Wilhelm-Hafen.

Seen from the south, Dampier Island appears as a huge truncated cone wooded to the summit, but its outline changes on every side. There are no proper rivers ; the natives obtain a scanty supply of fresh water from torrents. Birds of paradise, parrots and cassowaries are not to be found, and the larger quadrupeds are pigs and dogs. The island is almost entirely surrounded by reefs. Since June 1895 an old crater, the exact position of which is unknown, has broken out afresh. The natives state that it was active many generations back, and caused great destruction by pouring out ashes.—*Petermann's Mitt.*, Bd. 42, No. 8.

GENERAL.

Atmospheric Dust.—In April Mr. E. D. Fridlander discussed before the Royal Meteorological Society (see *Quarterly Journal*, July) observations taken by him in many parts of the world with a form of Mr. Aitken's pocket dust counter. He first dealt with the pollution of the atmosphere by artificial causes. Whereas the average number of particles in the air in the Pacific Ocean from October 30th to November 6th inclusive was 540 per cubic centimètre, on November 7th, at about 350 miles from Auckland, it rose to 1229, and on November 8th, about fifteen miles from Great Barrier Island, it was 1972. Again, the average from all the observations in the Pacific was 613, while that of the North Island of New Zealand and the polluted area around was 1336. High mountains, also, unless rising from practically uninhabited regions, are liable to pollution by air carried up from below. On La Paraz, the highest point of the ridge forming the northern wall of the Vallée des Ormonts, Switzerland, the average dustiness at a height of 8360 feet was 2062, whereas 4400 feet lower, at a point out of the direct stream of polluted air from the village of Ormonts Dessus, it was 1958.

As regards the effect of elevation, the local influences are so marked, and the records at present obtained so scanty, that it can only be said that in general in a calm atmosphere the number of particles per cubic centimètre diminishes with the height. The most satisfactory observations in this respect were made on the Bieshorn, which forms part of the chain containing the Rothhorn and Weisshorn. Here the number of particles fell fairly uniformly from 950 at a height of 6700 feet to 157 at 13,600 feet. An abrupt change between 10,665 and 11,000 feet may be ascribed to imperfect mixing of the upper and lower strata of air due to the form of the mountain at this part.

Over the Pacific and Indian Oceans the air is remarkably pure, especially over the latter, where the average number of dust particles for seven out of nine days was less than 500, and for five of these less than 400. Much lower values were obtained by Mr. Aitken at Kingairloch, but almost invariably in sunset and passing showers. The purifying effect of fog and rain was illustrated by several observations of Mr. Fridlander. The lowest value in the Indian Ocean, 210, was recorded after much rain ; and in the Pacific, on October 24th, the average dustiness was 529, whereas on the 25th, which was a showery day, it fell to 303.

Of the precise nature of the dust particles over the ocean little more can be said than that they are to a large extent salt particles produced by the evaporation of the water from fine spray. While on the Indian Ocean the average dustiness was 512, it was 613 in the Pacific, where the weather was not quite so calm, and in the Atlantic with still rougher weather it rose to 2053. Some dust is probably contributed also by volcanoes, for 9470 particles were recorded on the shore of Lake Taupo, New Zealand, in the direct line of the smoky air blowing from the active Ngauruhoe.

Mr. Fridlander had also much to say on the absorption of light by the dust particles, and concluded that the absence of the ultra-violet rays from the direct rays of the sun may in no small measure be due to their action.

MISCELLANEOUS.

The first census of the population of the Russian Empire will be taken in 1897.

The Government of the Argentine Republic proposes to take possession of the South Shetland Islands, and for this purpose is organising an expedition which will set out in December.

Of the 59,000 acres gained by the draining of Lake Copaïs (see vol. x., p. 203) 17,300 acres are now cultivated by the peasants of the neighbouring villages, 6200 are temporarily occupied, and 990 are reserved for agricultural experiments. The remainder consists of beds of reeds, or is in part utilised as pasturage.—*Revue Scientifique*, August 29th.

An observatory is to be established on the Schlagendorferspitze in the Carpathians, 8048 feet high. The project was started by the Carpathian Club, and is supported by the Hungarian Academy and the Royal Society of Natural Science. This will be the most eastern high-level observatory in Europe.—*Aus Allen Weltteilen*, July-August.

Dr. Étienne Ritter has calculated the area, depth, etc., of Lake Maggiore from the *Carta Hydrografica del Verbano*, and has published his figures in *Le Globe* (*Mémoires*, Tome vii.). The area is 81·7 square miles, the volume 8·9 cubic miles, and the average depth 576 feet. These results differ slightly from those of Professor Marinelli given in vol. xi., p. 138.

The population of Greenland was decimated by smallpox in 1733, and was estimated at only 5100 in 1789. In 1886 it was 9983, according to M. Rabot, and the census of 1890 gave 10,207 inhabitants, 4821 males and 5386 females. The Europeans numbered only 309. Between 1881 and 1890 there were 3671 births and 3222 deaths.—*Revue Française*, September.

A tunnel, which will be the largest in the world, is about to be constructed under Pike's Peak in Colorado. Starting from Colorado City, it will emerge near the placers of Cripple Creek. The principal tunnel will be more than twenty-two miles in length ; there will also be tunnels on either side having a combined length of twenty-five miles. The extremities of the main tunnel will be at an elevation 7480 feet above sea-level.—*Mouvement Géogr.*, September 6th.

The temperature of the Earth's Crust has been tested in New South Wales by Mr. Everett. In a boring at Cremorne, which was carried down to a depth of 2926 feet, it was found that the temperature increased at the rate of one degree Fahrenheit for every 79¼ feet. As this is a rate somewhat lower than observed elsewhere, it was supposed that the waters of the adjacent sea cooled the ground, but it was proved that the temperature of the water was higher than that of the subsoil at depths of forty-six and sixty-two feet.—*Mouvement Géogr.*, September 6th.

Mr. Wragge's Report on the Meteorology of Tasmania has been sent to us by Mr. Albert J. Allom, by whom it has been printed and circulated. It is accompanied by a circular in which Mr. Allom exposes the unfortunate position of the Tasmanian Meteorological Service, which, established in 1882 by agreement with the other Australian Colonies, was maintained up to the end of 1894 at a cost of £300 a year. For the last two years the Government grant has been reduced to £90, and in 1895 the Department was struck off the estimates. The

Tasmanian stations are provided with excellent instruments and are efficiently supervised by Mr. H. C. Kingsmill, who receives no renumeration for his services, except free quarters, light and fuel. It is unnecessary to insist on the importance of this southern outpost of meteorological science, and especially of the high-level observatory on Mount Wellington. It will be much to be regretted if the Government of Tasmania cannot see its way to maintain the service at least in its present state. The £600 a year estimated by Mr. Wragge as sufficient to provide for an efficient service does not seem a large sum to devote to such a purpose.

In *Himmel und Erde* for August a new instrument for the measurement of very small and sudden changes of atmospheric pressure is described. It has been invented by the electrician v. Hafner-Alteneck, and is called by him a variometer. It consists of a bottle closed by an india-rubber stopper through which pass two glass tubes. One of these has its upper end bent nearly horizontal, and in the lowest part of this section lies a drop of coloured petroleum. The extremity beyond is left open. The upper end of the other tube is drawn out to a very fine point. When a sudden change of pressure takes place, the air cannot flow in and out through the small opening with sufficient rapidity to restore equilibrium within the bottle, and hence the drop of petroleum in the other tube is slightly displaced in one direction or the other. The bottle is enclosed in a jacket so that changes of temperature may produce no sudden changes in the volume of air in the bottle. So delicate is this instrument, that the decrease of pressure with elevation may be demonstrated by placing the bottle on the floor and then lifting it as high as the arm will reach. Small changes may often be detected, especially on cloudy days, which do not affect the barometer.

NEW BOOKS.

The Cruise of the "Antarctic" to the South Polar Regions. By H. J. BULL. London : Edward Arnold, 1896. Pp. 243. *Price 15s.*

After many years' enthusiasm had brought Mr. H. J. Bull no nearer his desire of furthering exploration in Antarctic seas, suddenly he found the first great difficulties of finding a ship and men vanish when he met the late Mr. Svend Foyn, an enthusiastic and wealthy Norwegian whaler, who offered to finance an expedition which was to search for the right whale in the regions where Ross had recorded its presence in great numbers. The whaling did not prove very profitable, but on the voyage to Australia many seals were killed and brought in a considerable sum.

No scientific work was done on the outward voyage ; and unfortunately the invitation to Mr. W. S. Bruce, who went to the Antarctic with the Dundee whalers, to join the expedition, reached Scotland too late for him to catch the *Antarctic* before it sailed from Melbourne.

Just before leaving Australia Mr. Bull was approached by Mr. Borchgrevink. Through the exertions of the former a place in the ship was obtained for the latter. Mr. Bull explains how he acted as "guide, philosopher, and friend' to the young Norwegian, and that he was very much hurt when his pupil rushed off to Europe and claimed all the credit at the Geographical Congress and elsewhere

The chief results of the expedition from a scientific point of view were the finding of lichens where no vegetable life had previously been observed, the landing on the Antarctic continent, and the finding and bringing back of some of its rocks.

The pack was estimated at 500 miles broad, and took thirty-eight days to pass through. This delayed the ship so much that it was not prudent to go nearly as far south as was possible.

Mr. Bull publishes the meteorological observations of the voyage, which show that the temperature was warmer than during Ross's visit.

One thing seems plain, that good scientific work will not be done on these whalers with the ordinary captain, but that for success either the scientific head must have considerable power in the direction of the expedition, or a captain with moderately scientific interests should be in command of the vessel ; best of all, that both conditions should exist. A purely scientific expedition under Government or private officers would accomplish most.

The Land of Gold: the Narrative of a Journey through the West Australian Gold-fields in the Autumn of 1895. By JULIUS M. PRICE. London : Sampson Low, Marston, and Co., 1896. *Price* 7s. 6d.

The contents of this little book have already appeared in a series of letters in the *Illustrated London News,* of which the author is the special artist-correspondent. Mr. Price paid a short visit to Western Australia in 1895, and the information here given is of the superficial character that pertains to the writings of the average globe-trotter. Western Australia has of late increased in population and importance, in consequence of the development of the gold-fields of the Coolgardie, Murchison, and Pilbarra districts, at such a rapid rate that statistics and numerical details become antiquated in a few months. This book does not therefore profess to deal with statistics, but describes in a sketchy way the author's experiences on his arrival in Perth, the capital, and in his travels through the dreary tracts of bush and desert that surround the gold-fields beyond. The writer visited the rich mines of Hannan's district, and while in that locality met Herr Schmeisser, the celebrated German mining expert, whose report was at that time looked forward to with much interest. The cautious Teuton was however not to be " drawn " by the author, in the interviewer's sense, at all events, and preferred to keep his opinion of Western Australia to himself. The book may be of interest to any one who thinks of visiting "the land of sin, sand, sorrow, sore eyes, and Sir John Forrest," as they say out there, and some solid information may be obtained from the appendix and the map at the end of the volume.

The Scenery of Switzerland and the Causes to which it is due. By Sir JOHN LUBBOCK, Bart. London : Macmillan and Co., 1896. 8vo, pp. 480.

In this volume the author endeavours to sketch the origin of the surface-features of Switzerland for the benefit of those who are not specially versed in geology. Like everything that comes from Sir John Lubbock's pen the book is clearly and brightly written. The subject he takes in hand is one of no ordinary difficulty, yet he succeeds in giving a very readable account of the principal geological phenomena of the " play-ground of Europe," tracing out the several parts played by rock-structure and erosion in determining the outlines of mountains and valleys. The book will be welcome to those who have already made themselves familiar with the scenery of the Alpine lands, and will prove an admirable travelling companion to every tourist who would fain know something about the secrets of the hills. Numerous geological diagrams and sections and a general topographical map illustrate the work, but we miss an index—an omission which ought to be supplied in the next edition.

Die Insel Mafia, und ihre Kleineren Nachbarinseln. Von Dr. OSKAR BAUMANN.
Leipzig : Verlag von Duncker und Humblot, 1896. Pp. 38 and Map.

Aided by a grant from the Leipzig Geographical Society, Dr. Oskar Baumann
has undertaken a detailed survey of the Zanzibar Islands. This monograph is the
first fruits of his labours.

As might be expected from so excellent an explorer, the work has been extremely well done. It is expected that by the end of 1896 all the islands will have
been explored.

Mafia has belonged to Germany since 1890. It is a triangular island having
an area of 4349 kilomètres. It is flat, its highest ground being only 50 mètres
above sea-level. It is a coral island, having a climate rather cooler than the
neighbouring mainland, but wetter. It is well watered, and has some seventeen
small lakes. The island is covered with "bush." Hippopotami are very numerous,
which is curious, as they are unknown on Zanzibar and Pemba. The inhabitants
are Wambwera and Shatiri. They till the soil, herd cattle and are very peaceful ;
even the boys do not quarrel. There are about 6000 inhabitants. The exports from
Mafia amount to 150,000 R., the imports 130,000 R. The map is drawn with care.

Publications de l'Etat Indépendant du Congo. No. 3.—*Le District d'Upoto et la
Fondation du Camp de l'Aruwimi.* Par LE LIEUTENANT D'HANIS. No. 5.
—*Le District de l'Arouwimi et Ouellé.* Par L. ROGET. No. 10.—*Le
Mayombe.* Par M. FUCHS, Inspecteur d'État. Bruxelles : Société Générale
d'Imprimerie et Publicité, N.D.

The authors of these pamphlets give a short sketch of the surface features and
the hydrography of the districts, their climate, fauna and flora, the native inhabitants and their customs, and the useful products, cultivated or wild. Lieutenant
D'hanis also narrates the circumstances attending the formation of the camp on
the Aruwimi, after the station at Stanley Falls was taken by the Arabs in 1886.
This entrenched camp was founded at Basoko near the mouth of the river, to hold
the Arabs in check and serve as a base for exploring expeditions. In all the
pamphlets, the ethnographical details occupy a considerable proportion of the
space, while the geographical part is chiefly confined to the rivers and their
navigable capacities. Much useful information is contained in these pages,
especially with regard to products.

The Life and Travels of Mungo Park in Africa. Edinburgh : W. and R.
Chambers, 1896. Pp. 304. With Illustrations.

A new edition of Mungo Park's travels is a thing to be heartily welcomed at
the present day. Although the fame of this great pioneer of African exploration
can never be forgotten, there is always the risk that newer and perhaps less
illustrious names may partly overshadow his ; simply because of that "one touch
of nature" which, as Shakespeare (in an often misapplied passage) wisely says,
"makes the whole world kin,"—namely, "that all, with one consent, praise new-
born gawds . . . and give to dust that is a little gilt more laud than gilt o'er-
dusted." This observation is not made here with any idea of detracting from the
honour justly paid to more modern explorers such as Livingstone, Stanley,
Thomson, and others, but rather as a reminder of what we and they owe to their
brave *devanciers*, of whom none was greater than Mungo Park.

The present edition of his *Life and Travels* is a tastefully got-up volume, with
such exceedingly good illustrations that one regrets they are so few in number.

The "Map of Park's Route" (p. 18) might also with advantage have been drawn upon a much larger scale. It is a good one, but almost too minute to be of use. Prefixed to the well-known "Travels" by Park himself is a brief sketch of his life, which has the merit of adding such recent information as the names of his descendants living in 1895, and the whereabouts of various articles associated with him. Chapter XXIX., on the "Present position of the Niger Territory," also constitutes an interesting epilogue to the story ; since it shows how we as a nation are indirectly reaping the fruits of the toil, privation, and danger undergone by Park, at the final cost of his life.

There is every reason to hope that this bright and modern-looking volume will prove at least as attractive as less authentic stories of adventure, in the eyes of the rising generation, to whom nothing ought to be more interesting than this narrative of one of the greatest of our British explorers.

The Pioneers of the Empire, being a vindication of the principle and a short sketch of the history of Chartered Companies, with special reference to the British South Africa Company. By AN IMPERIALIST. London : Methuen and Co., 1896. Pp. 139 and Sketch-map. *Price* 1s.

An Imperialist divides his brochure into three chapters, The Principle of Chartered Companies, a Sketch of English Chartered Companies, and the British South Africa Company. It is a vigorous counterblast to those "Little Englanders" who hate the name of Empire, and a defence of Chartered Companies. "In a country governed by talkers," says the author, "it is possible that their (the Little Englanders') invective may secure a wider hearing than the position of the critics deserves." He goes on, "It is well to remind those short-sighted people that nature has confined the ancestral estates of the English race within the narrow boundaries of our sea-girt isles ; that our population is developing at a rate which, but for the outlets caused by the existence of the Empire, would be appalling ; that we cannot grow on British and Irish soil sufficient of the necessaries of life to sustain our population ; that we depend less and less for our wealth on the products of our soil, and more and more upon our manufactories ; that foreign countries close the doors of their markets more rigorously every day against the importations of British-made goods, by hostile tariffs and bounties and all the artifices of protection."

Our author believes in the eventual supremacy of the Anglo-Saxon race, and thinks that it is incumbent on us to see that we have sufficient extra room in which to work out our destinies.

From this it will be seen that the brochure is of a political nature, and that we can only call attention to it without expressing any opinion as to the justness or otherwise of the author's views. It may, however, be said, that within the limits he has set himself he gives an accurate if somewhat brief account. Matters of fact are correctly stated, and the author is evidently well up in his subject.

All politicians will do well to read the little work. Do people realise that the British South Africa Company has added 750,000 square miles to the Empire?

Holiday Fortnights at Home and Abroad. By GEORGE ERNEST PHILIP, Glasgow. Paisley : J. and R. Parlane, N.D. Pp. 199.

Rambles among the English Lakes, in the Channel Islands, Scotland, Ireland, Wales, the Ardennes, the Riviera, and North Italy, and Norway, are all described in this small volume. No wonder, then, that the descriptions are sketchy. From the preface it appears that these chapters first appeared in the columns of a newspaper, and this fact may account for the want of detail. About a page is devoted to Florence, its picture galleries, churches, and all ; and barely two are allowed to

Venice. The book contains thirty-eight illustrations, not of a very high order, and is clearly printed on good paper.

Gazetteer of the Sialkot District. By Captain J. R. DUNLOP SMITH, Settlement Collector. 1894-95. Compiled and published under the authority of the Punjab Government. Revised Edition. Lahore : *Civil and Military Gazette* Press, 1895. Pp. 195+li. *Price Rs.* 3-10-0.

The first edition of the *Gazetteer* was declared in its preface to have been compiled from inadequate materials, and a more complete edition was contemplated when the district should come under resettlement. The *Gazetteer* has now been entirely rewritten by Captain Dunlop Smith, who conducted the recent "Settlement." He has taken a wide range of subjects, far beyond the ordinary range of settlement inquiries, and has furnished a very valuable compendium of historical, statistical, and economic particulars, a few of which will be briefly referred to.

The importance of the district of Sialkot is indicated by its being first in number of inhabitants among the thirty-one districts of the Punjab, and second in the amount of its land revenue, though only ninth in the extent of cultivation. In general aspect it is a treeless plain, sloping down from Kashmir and confined between the two important rivers, the Chenab and the Ravi. The town of Sialkot is believed to date from about the Christian era, but, being of only moderate importance, its continuous history is little known.

The Mohammedans, who constitute 61 per cent. of the population, are said to be the descendants of the original conquerors and of converts from Hinduism. Hindus, form 34 per cent. of the population, and Sikhs the remaining 5 per cent. More Christian missionaries are at work here than in any other part of the Punjab, and their converts are said to be developing a spirit of independence, demanding that wages shall be paid under more definite agreements.

Among the tribes and castes the Jats, who in religion are Mohammedans, Hindus, and Sikhs, are foremost as landowners and cultivators, and make the best soldiers. The Rajputs are numerous, but inferior to the Jats in many respects. The leading families are Sikhs, and many of their members have distinguished themselves in military service under the Indian Government.

Sialkot is one of the most densely populated districts of the province, the density, according to the census of 1891, being 569 to the square mile. The landowners, as a whole, live in comfort, and the working classes are not badly off, though their condition is becoming worse under the pressure of an increasing population. Small holdings are the rule, and the rent is paid in kind, being usually a fixed share of the produce. Wheat is the chief crop, occupying about two-fifths of the area under cultivation. The paper manufacture, for which Sialkot was famous under the Mughal Emperors, has declined, but shows signs of improvement. Articles of damascened iron are sent in considerable quantities to Europe and America, and cloth is largely manufactured from fine wool brought from Kashmir and Tibet.

As Settlement Officer, Captain Dunlop Smith is naturally interested in the systems of revenue assessment of former rulers. The Mughals introduced an assessment based on land measurement, while their successors, the Rajputs and Sikhs, levied shares of the produce in different proportions, and on a variety of systems. On the British occupation a summary money settlement was introduced, based on the average grain collections of the Sikhs. This was soon followed by a regular survey, by which the imposts were considerably reduced, and a second was made ten years later. The last settlement, completed after inquiries extending over seven years, has resulted in a large increase of revenue. Other cesses amounting to about 25 per cent. have been imposed in addition to the land revenue.

The British Colonies, Dependencies, and Protectorates: Physical, Commercial, Descriptive, Political, Social, Historical. By GEORGE GILL, F.R.G.S. London: George Gill and Sons, N.D. Pp. 471. *Price 2s. 6d.*

Mr. Gill has made another attempt to produce a geography combining the features of a gazetteer and a statistical year-book. It is a creditable performance in its way, and if geographical books are to be judged by the number of accurate tit-bits they contain, then this one will take a high place among them. But a student would make a mistake in trying to learn geography from it, although he will find it a convenient and serviceable book of reference.

Despite much care in securing accuracy there are a few slips in the book. A note about Mounts St. Elias and Fairweather, on page 232, states that both peaks are in Alaska. The former is in Canada, and if Mr. Gill wants the height as ascertained by triangulation, it is just over 18,000 feet; while Mount Logan, also in Canada, and 19,514 feet above the sea-level, is the highest point in North America.

The many sketch-maps, whose source is not usually acknowledged, are useful, but sometimes are rather carelessly edited. For instance, the map of Africa (p. 129) is unduly, or at least prematurely, reddened: not only Egypt, but Tripoli and all lands east of Fezzan and Lake Chad are coloured British crimson; while, perhaps to make up for this, the Ashanti hinterland is French violet; and the maps on pages 75 and 109 do not agree in all particulars.

Latitude and Longitude, and How to find Them. By W. J. MILLAR, C.E. London: Charles Griffin and Co., 1896. Pp. 53. *Price 2s.*

The publishers are issuing a series of small volumes intended to assist young officers in the mercantile service to obtain a more scientific knowledge of the sciences on which the art of navigation depends. The treatise on latitude and longitude explains how to find these co-ordinates, as well as the azimuth, time, etc. The introductory chapter might have been omitted, and the student referred to another volume of the series on *Practical Algebra and Trigonometry,* for it is hardly possible to teach in a few pages all the mathematics required.

On the whole, the explanations are clear enough, but in one or two cases we should advise an alteration ; the azimuth, for instance, might be better defined. Again, it should be shown how the expression for the hour angle may be converted into a form adapted for the use of logarithmic sines, cosines, etc. Throughout the book the processes of calculation are carefully explained, but here the student is given only a rule.

The book will probably be acceptable to young officers, containing, as it does, all they require to know on the subject, and no more. Captain Weir's ingenious geometrical solution of the azimuth problem might have been referred to.

Year-Book of the Scientific and Learned Societies of Great Britain and Ireland. London: Charles Griffin and Co., 1896. Pp. 262. *Price 7s. 6d.*

It is not the first time we have noticed this publication, which is now in its thirteenth year. The volume is a little larger, as may be expected, but in form and arrangement it remains the same. It is a very useful book of reference for those interested in science.

Black's Shilling Guide to the English Lakes. Edited by M. J. B. BADDELEY. Twentieth Edition. Pp. 116. Maps. London : A. and C. Black, 1896.

The fact that this book has reached its twentieth edition is a sufficient proof that it satisfies the requirements of a large number of the persons for whom it was originally intended. In this case the judgment of the many has not been at fault.

Ireland : its Health-Resorts and Watering-Places. By D. EDGAR FLINN, F.R.C.S., D.P.H., etc. Second Edition. Dublin : Fannin and Co., 1895. Pp. x + 180.

The visitor to Ireland, whether for health or pleasure, will obtain much assistance from this little book in making choice of a suitable halting-place. Watering-places, inland health-resorts, and mineral springs are described with details of temperature, humidity, analyses of waters, etc. Those who wish a change from Buxton or Harrogate or the crowded watering-places of Great Britain can have a choice of many health-resorts in Ireland, where of late years much has been done to render its beauties more accessible to visitors.

The Guide to South Africa for the Use of Tourists, Sportsmen, Invalids, and Settlers. Edited annually by A. SAMLER BROWN and G. GORDON BROWN. London : Sampson Low, Marston, and Co. ; Cape Town : J. C. Juta and Co., 1896. Pp. xlv + 418. *Price 2s. 6d.*

The edition for 1896-97 is larger than the former (*rev.* vol. xi. p. 550) by more than sixty pages. There is no great innovation, but the additional space is taken up by more detailed information with regard to means of travelling, hotels, etc., and a description of some of the more remote parts of the Kalahari desert. The work can be thoroughly recommended.

The Rhine from Rotterdam to Constance. Thirteenth Revised Edition. Pp. xxiv + 422.

Paris and Environs, with Routes from London to Paris. Twelfth Revised Edition. Pp. xlvi + 417. Leipzic : Karl Baedeker, 1896.

Not only are the *Guides* of Herr Baedeker brought up to date, but also in almost every edition further additions and improvements are made. More information is given of those retired spots into which British tourists are, year by year, penetrating in greater numbers ; and the maps, always good, are more numerous and more carefully executed. These remarks apply to the present volumes. In the *Rhine Guide* there are several small additions to the text, while the maps and plans of towns have been considerably increased in number ; and the *Guide to Paris* is equally improved.

Jamaica in 1896. A Handbook of Information for intending Settlers and Others. Kingston : Institute of Jamaica, 1896. Pp. 85.

This pamphlet, now issued for the second year, is mainly an abstract of the *Handbook of Jamaica,* which we have noticed more than once. The information selected is chiefly such as is wanted by new settlers in the island, relating to cultivation, cost of articles and labour, etc. With a view to assisting young men who wish to learn plantation work, the Institute of Jamaica publishes here a list of proprietors who are willing to receive articled pupils, with particulars. But the Institute does not make itself in any degree responsible for the *bonâ fide* nature of these offers.

NEW ATLAS.

ATLAS DE GÉOGRAPHIE HISTORIQUE. Par une Réunion de Professeurs et de Savants sous la direction géographique de F. Schrader. Contenant en 55 feuilles doubles 167 cartes en couleurs, accompagnées d'un texte historique au dos de 115 cartes, figures, et plans en noir dans le texte. *Librairie Hachette et C^(ie)., Paris*, 1896.

The extended title of this work will help to convey its scope. It is a work of general excellence. The epochs (on the choice of which the merit of such an undertaking must largely depend) are fixed with judgment, and the main and the subsidiary maps are carefully planned and are usually explicit. Thus, the coloured maps dealing with the Roman world towards the end of the Republic indicate (a) the possessions of Rome in 146 and in 30 B.C.; (b) the Roman and the non-Roman world at the time of the invasion of the Cimbri; and (c) the East of the age of Mithridates; while in the text are uncoloured maps to show Africa in the time of Jugurtha and of Julius Cæsar, Pontus in the time of Mithridates, and the sites of the battles of Aix (Aquæ Sextiæ), Chæronea and Pharsalus. Or, again, Europe after the Peace of Utrecht is represented by four coloured maps : (a) Western Europe (with the several acquisitions marked); (b) Northern and Eastern Europe ; (c) Italy from 1714 to 1748; (d) the Low Countries and Lorraine from 1714 to 1748. The maps accompanying the text are determined on the twofold principle of elucidating the text and supplying links for the coloured maps. There are several plans of battles and of cities. As a rule, the colouring is distinct, and the maps are not overcrowded. There are a few exceptions (for which intricate periods are partly to blame), such as map 32—"Germany in 1648."

The text is a well-written universal history in little ; the editor has been ably assisted by his twenty collaborateurs. We have noticed a few errors. Thus, it is not strictly accurate to say (map 19) that Athelstane "finally brought about the unity of England." The date of the Treaty of Troyes (map 28) is misprinted *1410*. To students who are not British it must be misleading to represent Wales as a principality (conspicuously marked) in 1494 (map 29). The language of the text is almost invariably terse. Occasionally the judgment is striking, but apt to be overstrained, as when it is said of the loss of England's French possessions, that it helped "to turn Great Britain"—"England" one should rather say—"towards its natural destiny and its true sphere of expansion—the sea." Sometimes, but rarely, the space at the writer's disposal might have been better utilised ; for instance, the character-sketch of Frederick II. of Prussia might have been omitted in the interests of proportion. The liberal manner in which the writers have viewed their task is favourably shown from M. A. Debidour's account of France in 1789—a compact piece of writing touching on *l'ancien régime* in general, the position of the clergy, the *noblesse* and the commons, the state of the provinces, the central power, finance, justice, and other topics. The editor has a very suggestive review of the world of to-day, especially from the economic and moral standpoint. We do not know of any historical atlas likely to interest both the student and the general reading public more than this now before us.

CHARTS.

MONTHLY CURRENT CHARTS FOR THE INDIAN OCEAN. From information collated and prepared in the Meteorological Office.

Presented by the Meteorological Society.

THE SCOTTISH

GEOGRAPHICAL

MAGAZINE.

NOTES ON THE YUKON COUNTRY.

By Alexander Begg, C.C.

Occupying the north-westerly portion of the continent of North America, west of the 141st meridian west of Greenwich, is the territory of Alaska, which up to 1867 was known as Russian America. In 1867 the territory was purchased by the United States from Russia for $7,200,000. On the eastern side of the 141st meridian, the country belongs to the Dominion of Canada, and from the 60th degree of north latitude to the Arctic Ocean forms a part of the Canadian North-west Territories.

The 60th parallel of north latitude is the northern boundary of British Columbia, and is referred to here, because by a treaty made between Great Britain and Russia in 1825, the eastern boundary of Russian America was fixed at the 141st meridian, and was continued south-easterly along the Pacific coast as far as the 56th parallel of latitude, including Prince of Wales Island and the other islands northerly and westerly of the channel mentioned in the treaty, and ceded at that time to Russia.

Alaska naturally falls into two divisions: the western lying to the west of the 141st meridian, and extending thence to the Pacific, to the Bering Sea, and to the Arctic Ocean; the eastern extending south-easterly from the 141st meridian, along the Pacific coast, as already mentioned, to the 56th degree of latitude. This vast territory contains about six hundred thousand square miles; being nearly eight hundred miles long from north to south, and about seven hundred east and west. It has a sea-coast longer than the combined seaboard of the rest of the United States on the Atlantic and the Pacific. Neither does this include the line of the Aleutian Islands, which extend for hundreds of miles from the Alaska peninsula towards the Asiatic coast.

The interior of Alaska is principally drained by the Yukon river and its tributaries, and is known as the 'Yukon country.' The main river, the largest in North America flowing westward, is navigable for eighteen hundred and fifty miles, from its mouth to the junction of the Pelly and Lewis rivers, by 400-ton stern-wheel steamers drawing four feet of water. The Porcupine river, which joins the Yukon near the Arctic Circle, at the old Fort Yukon of the Hudson Bay Company, is navigable for one hundred miles. It flows, first northerly, then westerly, through British territory, to the 141st meridian; thence south-westerly until it joins the Yukon, near the 145th meridian. One of the tributaries of the Porcupine takes its rise near Peel river, a tributary of the Mackenzie, which reaches the Arctic Ocean between the 134th and 136th meridians.

The main river, at the old Fort Yukon, is about eight miles wide; but it gradually narrows, up stream, to about three-quarters of a mile, at the junction of the Pelly and Lewis rivers, as far as which it bears the name Yukon. Navigation is continued by the Pelly for fifty miles farther. Along the Lewis and Five Fingers, a 150-ton steamer, with powerful machinery, would be able to pass on three hundred miles farther, by Hootalinqua river to the head of Lake Teslin, in British Columbia. By Taku river and inlet the distance from Lake Teslin to the Pacific is given at twenty-six miles by the trail.

Passengers or freight for the Upper Yukon are obliged to land at St. Michael's, which is about sixty miles to the north of the usual entrance to the river. This has been the principal trading-post of the Alaska Commercial Company, and the outfitting-post for their stations on the river, for the past twenty years.

It may be asked why a location for a town has not been made nearer the mouth of the river. In explanation it may be said that the Yukon flows into Bering Sea through several mouths, that farthest north being nearly one hundred miles distant from the most southern channel. The various terminal branches turn from a westerly to a northerly direction, and reach Bering Sea near latitude 62° north, a short distance south of Bering Strait; river steamers are therefore obliged to steam out into the open waters of Bering Sea to receive freight or passengers. So far as is known, there is not a suitable location for a town where the high water on the breaking up of the ice in the river does not overflow.

The Yukon is shallow at its mouths, the greatest depth found being only eight feet. The ice passes out, and leaves the river free for navigation, about the middle of June, but it does not leave an approach open to St. Michael's till several days later. If a station could be located with easier access to the river some advantage might be gained, but until the ice had run out, no progress could be made up the river.

St. Michael's is, strictly speaking, a native town. Apart from the buildings and store of the Alaska Commercial Company and the residences of its employees, a church and the residence of its pastor, the houses are those of the natives. Enormous supplies of goods are shipped there every year for the trading-posts and missions on the river. During. the months at the opening and closing of the season of navigation, St. Michael's presents an air of bustle and business activity not found at any other of the frontier Alaskan towns.

A new company, the North American Transportation and Trading Company, is making arrangements to build warehouses and a trading-post about a mile south of the former town. This company has established trading quarters at Fort Cudahy, near where the international boundary crosses the Yukon. As might be expected, the life artery of this western division is the river from which it takes its name, which has served as the highway of nations and tribes for many centuries, long before the white man, with his improved means of transport, accomplished the feat, marvellous in their eyes, of traversing in one brief season the distance from its deltoid mouth to the Hudson Bay Fort at the junction of the Yukon and Porcupine rivers.

The natives inhabiting the banks of the great river belong to two tribes. The Indians of the interior, of the Athabascan stock, occupy the banks of the Yukon and its tributaries eastward of the Anvik river. These are called Ingalits. The hardy Eskimo, or Innuit, on the other hand, live along the coast of Norton Sound and on the lower Yukon and the Kuskokwim delta. At an early period, the Eskimo advanced across the divide between the great river and the sea, and followed its course up nearly to Nulato, settling along the banks of the Chageluk river; but they were not allowed to hold peaceable possession, for the Ingalits rallied from all directions and drove the intruders back, far down the river. From time to time the Eskimo advanced again, and many are the traditional tales of bloody battles and years of war between the tribes. The result is that no Eskimo will ascend beyond the mouth of the Anvik, at the present time, nor will an Ingalit descend beyond that point unless accompanied by white men.

When passing up or down the river during the busy season—that is, the brief summer—the traveller would form an incorrect estimate of the population were he to base it on the number of those living on the banks of the river, for he would find, were he to make a short excursion into the almost impenetrable forests and over the hills and mountains, that along the river only exist the conditions necessary for life throughout the year. The small rivulets of the interior, and the vast swampy plains, covered with snow for seven or eight months of the year, are only visited by the trapper, when the skins of the marten, mink, and musk-rat are in their prime. Along the upper reaches of the Yukon and the Tanana the inhabitants are less dependent upon the river, and fish and game are more abundant.

A recent writer says:—"For hundreds of miles from the sea the Yukon flows through low, level tundras, or mossy morasses, resting upon a foundation of clay. The shifting current of the river eats away the shores on either side with astonishing rapidity; the dull thud of caving banks is constantly heard by the traveller. Stepping upon the shore the explorer must jump from hummock to hummock, or wade around from knee to waist deep. In many places the ice never disappears within a few inches of the surface, being protected from the rays of the sun by a non-conductive carpet of sphagnum. Wherever there is a slight elevation of ground in all this watery waste the wretched natives have located their villages, the dwellings consisting of excavations in the ground roofed over with mounds of sods. Here they fish during the summer, and hunt the mink and the moose in the winter."

The following is the latest account of the proceedings of Mr. Wm. Ogilvie, Chief Dominion Surveyor, who is engaged in delimiting the international boundary between Alaska and the North-west Territories of Canada. The particulars were sent by Mr. Albert M'Kay, one of Mr. Ogilvie's staff, to a friend in Calgary. In his letter he explains to his friend that at Fort Cudahy, where he was writing, June 24th, 1896, only three mails are received *per annum.* His letter only reached Calgary in August. It says :—

" Mr. Ogilvie's party left Victoria, B.C., 9th July 1895. They had along with them three cedar, lightly-built Peterboro' canoes ; these they packed, along with their baggage, from Taiya (the head of navigation) *viâ* Chilcoot Pass to the headwaters of the Yukon at Lake Lindeman, a distance of about twenty-five miles, during which they reached an altitude of 3375 feet from sea-level. Thence they proceeded from Lake Lindeman along several lakes and Lewis river. Leaving Lewis river at Fort Selkirk, they went along the Yukon to Forty Mile Creek, the distance together being 590 miles, in the canoes, with the exception of three portages, where everything had to be carried. The journey was made in twenty days.

" To find a suitable place for winter quarters Mr. Ogilvie's party went down the Yukon thirty-seven miles farther, where they built and fitted up a camp and erected an observatory. Here a number of astronomical observations were made to determine the geographical position. This Mr. Ogilvie found to be 140° 54' 8" west longitude, and 64° 41' 6·4" north latitude—a little less than two degrees south of the Arctic Circle. The line of demarcation (141° W.) between the United States and Canada was found to be a few hundred yards farther down the Yukon. Where it crosses, the great river narrows to about 1300 feet, discharging a volume of water of 100,000 cubic feet per second."

Mr. M'Kay further writes that the long winter nights and short days passed by comfortably. The party had abundance of fresh meat, as a herd of nearly 1000 cariboo had made a *corral* for themselves at no great distance from the camp. Eighteen carcasses were speedily secured and kept frozen until required. The lowest reading of the thermometer for December 1895 was, on the 4th at dawn, 54·1 degrees below zero, and at 1.30 P.M. 46·2 degrees. On January 4th, 1896, at dawn 62·7 degrees, and at 1.30 P.M. 64·2 degrees. There were three periods of three to six days each, with the mercury in the neighbourhood of 60 degrees below zero (Fahr.).

On February 19th Mr. Ogilvie broke up camp, to carry out his instructions from Ottawa, and establish a portion of the international line between Alaska and the North-west Territories. Starting from where the line, on the 141st meridian, crosses the Yukon, he proceeded due

south, cutting out the line through the forest over mountain and valley·
for a distance of sixty miles, food and luggage being hauled on tobog-
gans, and Mr. Ogilvie also drawing his astronomical and scientific instruments.

As winter travel in the Yukon country is done by the toboggan, it
may be interesting to describe the sledge or carriage.   It should be about
seven feet four inches long, seven inches high, and sixteen inches wide, of
strong but light timber, and with runners shod with either brass or steel,
the former being preferable, because the sledge will glide over the snow
more smoothly in intensely cold weather, while steel is inclined to grind
and lug very much, as if it were being hauled over sand.   When the
weather is cold, if water is taken into the mouth and held a moment and
then blown over the runner, a coating of ice will immediately form ; and
if this process is repeated when necessary it is surprising how much
easier and more smoothly the sledge will draw.   It is preferable to adopt
the Eskimo mode of making sledges for Yukon travelling.   They use no
nails or bolts, binding the joints together with strong cords or thongs.

The survey party, as they proceeded, intersected Forty Mile Creek,
twenty-three miles from its mouth, where it joins the Yukon.   This gives
Canada twenty-three miles of the creek, which, with its banks and surrounding country, are highly auriferous.   The work of demarcation, so
far completed in that trip, occupied nearly two months, when the party
returned to Fort Cudahy.

The work which has to be performed by the survey parties is by no
means child's-play or as agreeable as a picnic.   Mr. Ogilvie, in describing some of the work performed on the south-eastern coast of Alaska,
says :—"The labour in connection with the survey is extremely hard ;
there are many giddy precipices, yawning chasms, and frowning cliffs to
be surmounted.   A man must be of iron frame and constitution to stand
it."   As an instance, Mr. Ogilvie and the members of his party on one
occasion started at two o'clock in the morning ; walked at least fifteen
miles through snow often waist deep ; forded a river running with ice water
thigh deep one hundred yards in width ; crossed several smaller streams
by fording ; traversed the slope of the Mendenhall glacier, meandering
here and there over its surface to avoid the deep crevasses, to the summit
of a mountain at its head, 6100 feet above sea-level ; spent two hours on
a giddy cone on its top, from which there was a sheer descent on one side
of 2000 feet ; and returned by the way they went, tired, hungry, wet, and
cold, arriving back at camp at twelve o'clock the succeeding night.
Thus every member of his party on that occasion endured twenty-two
consecutive hours of the most severe physical labour, apart from the cold
and suffering.   Twenty-five ascents were made by his party during the
summer (1894) ; several of them were to heights of 6000 feet and
upwards, many up to 5000, the majority over 4000 feet.

The method adopted in carrying out this mountainous survey work—
photo-topography—has lately excited much public interest.   The system
pursued can hardly be understood by ordinary readers without mathematical training, but the following facts stated by Mr. Ogilvie will be
interesting.   The idea is older than photography itself, for the attempt

was at one time made to compile a chart from sketches of country drawn in the dark camera. Photo-topographic surveys of a kind were made in France, Germany, and Italy a great many years ago, and in the United States at least one text-book was written on the subject; but its development to its present state of perfection, as adopted by the Alaskan Boundary Commission, is due to the Surveyor-General of the Dominion of Canada, Mr. E. Deville. He took up the subject some years ago, and during the years between 1884 and 1893 Messrs. J. J. M'Arthur and Mr. W. E. Drewry, acting under his instructions, tested his ideas and devices in actual work along the line of railway in the Rockies and westward.

To sum up the method and express it in a popular manner, it is the solution of the sides of a triangle. Almost every one knows that with a given line of any finite length, and the direction of any given point from each of its ends, that point can be located. Now, suppose a line and any point or any number of points which it is desired to locate with reference to its ends. Then suppose photographs taken from each end of this line, each of which shows all those points in its field. It is easy to see that after the observer has oriented those views— that is, located the direction of the centre of the picture from the point of sight—and knowing the focal length of the lens used in the camera, he can lay off or project on the picture-plane the direction of any point or points in each picture. Lines drawn in those directions from the ends of the base will intersect each other in the position of any desired point, so long as it can be distinguished in the view. Its elevation with reference to the station occupied is determined on the same principle that elevations are found with the transit or theodolite—that is, from angles of depression or elevation.

Juneau, situated on the mainland coast, twelve miles west from Taku Inlet, and opposite Douglas Island, is the largest town in Alaska, and the commercial metropolis. It is built at the base of a mountain which rises almost perpendicularly for nearly three thousand feet, forming a most picturesque background. The town contains about 2000 inhabitants. Nearly all the ground available for building purposes is occupied. Sitka, on Baranof Island, continues to be the capital of Alaska. It was formerly the capital of Russian America, and was founded under the administration of Alexander Baranof, who held office as governor of the Russian American colonies from July 27th, 1791, until January 11th, 1818; his predecessor, the first governor, having held office from August 3rd, 1784, until July 27th, 1791. At that time, the seat of Government was on Kadiak Island, five hundred miles west of Sitka. Governor Banarof built a large castle in 1813. The city boasts of its Greek Church, with its dome painted blue and chime of bells. The Industrial School and the Jackson Museum are visited with much interest by tourists. It contains a large collection of Indian curios secured by Dr. Sheldon Jackson, who is a zealous worker in the cause of education and in missionary work among the natives. The population of Sitka is computed at 500.

In the interior of the Yukon country, the towns are chiefly mining camps and trading establishments, namely, Forty Mile City, Fort Cudahy, and Circle City. The North American T. and T. Company have a large establishment at Cudahy, which is three-quarters of a mile across the creek from Forty Mile City. It was established in 1892, and transacts a very large business. The company own a steamer which plies on the river. Another United States company,—the Alaska Commercial Company, whose extensive establishment is at Forty Mile City—have two steamers on the Yukon, and each company is building another steamer ready for next year's trade. They do an enormous trade with the miners. "So far," says Mr. M'Kay in his letter, "they are unable to supply the demands. The Alaska Commercial Company does, by far, the bulk of the trade. They sold out all their supplies last fall, before the winter fairly set in, and had to buy $30,000 worth of merchandise from the North American T. and T. Company. The last-named company made, outside of that, a cash sale of $3500 last autumn.   The two companies must have sold (between Circle City and Forty Mile) to miners and others, not less than $680,000 worth of merchandise. The currency is principally gold-dust, taken at $17 per ounce. 1400 miners are supposed to have wintered here, and some six hundred more are coming in." Mr. M'Kay instances two of his former acquaintances from near Calgary, one of whom cleared $16,000 last season, and the other is computed to be worth $25,000 ; but he remarks that, like other places, only a few energetic persons do make money.

Since Mr. Ogilvie has run the boundary line, which undoubtedly is correct, it is clear that Forty Mile City and Cudahy are on British territory. This fact, taken in connection with Dr. Dawson's report, is very encouraging.   He said :—"I do not maintain that the country is suitable for immediate occupation by a large, self-supporting agricultural community, but hold that agriculture may before many years be successfully prosecuted, in conjunction with the natural development of this great country, of which by far the most valuable part lies east of the Alaskan boundary."

The Dominion Government has stationed a detachment of the Northwest Mounted Police at Fort Cudahy, built and occupied in 1895. The force is well mounted, and, in all, numbers twenty officers and men. Their presence is found to be exceedingly useful in restraining evil-disposed persons. A recent writer, describing Forty Mile City, says :— "The price of restaurant board is $12 per week, and whisky costs $9 per gallon, or fifty cents per glass over the bar. This town on the Yukon is an ideal '49 mining camp ; its saloons, gambling houses, concert halls, etc., give it an air of bustling activity, from which, however, the element of lawlessness is almost entirely eliminated. Miners' law prevails, and justice is fairly and impartially administered. The entire Yukon valley bears an enviable reputation for peace and morality. Simple, but effective, self-adopted rules of government are found amply sufficient to ensure order, and they are universally respected."

A collector of Her Majesty's Customs has been sent to Fort Cudahy, so that British law and order may be established and maintained throughout this remote and valuable country. In the House of Commons, Ottawa (September 1896), a sum was voted to open up a route from British Columbia to the headwaters of the Yukon.

THE PORTUGUESE IN ANGOLA.

IT is but seldom that information is obtained of the Portuguese settlements on the west coast of Africa, and therefore a long article by M. Ch. Delannoy in the *Bull. de la Soc. R. Belge de Géographie,* of which the following is an abstract, probably contains information that is new to most readers in this country.

GEOGRAPHY.

The name Angola, in its more restricted application, denotes only the small territory lying between the rivers Dande and Cuanza, but it is now generally accepted as the name of the whole province extending from the Congo to the Cunene, with the seat of administration at Loanda. Its boundaries towards the Congo State and German South-west Africa are laid down exactly in the most recent maps. The eastern frontier is defined by M. Delannoy, as far as it is determined by treaties between the Congo State and Great Britain, and as laid down on maps. Along the Zambesi, however, the frontier is not yet delimited above the Katima rapids, though it was stipulated in the treaty of 1891 that the Barotse country should belong to Great Britain.

The length of Angola, from north to south, is about 750 miles, and its greatest breadth, from the ocean to the Kasai, 500 miles. Its area is 395,000 square miles, or about $3\frac{1}{4}$ times that of the British Isles. The surface is exceedingly uneven. A chain of mountains runs through the colony from north to south, filling with its plateaus, secondary chains, and long slopes, almost the whole space between the ocean and the rivers Kwango and Cunene. Between the Congo and the Cuanza the chain expands into a vast plateau, lying fully 3300 feet above sea-level, which occupies the interval between the coast and the Kwango. Towards the river it terminates in a steep slope, while on the west it is connected with the coast by a series of terraces. To the south of the Cuanza and its tributary, the Lucalla, the range recedes from the coast and increases in height up to the colossal group of the Bihé, which is the centre of the orographical system of Angola; it contains summits rising to heights of more than 6500 feet, such as the Lovili and Elonga. Most of the crests are extensive, spreading out into plateaus. Behind the Bihé mountains, the Tolla Mangongo, a chain of less elevation, passes round the sources of the Cuanza and extends to the plateau of the Congo. To the south the chains follow one another without interruption, but with diminished height, as far as the Chella or Serra da Neiva, which has plateaus fully 3300 feet above sea-level. The southern part of this chain has been deeply eroded by the affluents of the Cunene, which flow in fertile but often unhealthy valleys.

The transition from the littoral to the great elevations of the interior is effected, both to the south and north of the Cuanza, by parallel chains, gradually increasing in height. Dr. Walwitsch distinguishes three regions —the coast zone, up to a height of 1060 feet; the region of primeval forest, rising to 2600 feet, and the jungles, which extend up to 3300 or 3600 feet.

Accordingly, the arrangement of the mountains is not favourable to the formation of large navigable streams, and, except the Cuanza and Cunene, all the rivers are short and torrential, often dry in summer. The Cuanza, the most important as regards colonisation, rises in the same region as the Zambesi, behind the lofty mountains which form the backbone of Angola. It flows at first northwards until it meets a break in the mountains through which it finds a passage to the ocean. As far as Cambambe it is encumbered with rapids, but below this point its current becomes gentle, and it is navigable by steamboats down to the coast, a distance of 120 miles. Its mouth is obstructed by a bank which renders navigation difficult.

The Cunene, rising in the same group of mountains and flowing south to the extreme bounds of the colony, has a much longer course than the Cuanza, but is of much less importance, as it is not navigable. Its length is estimated at 750 miles, and its drainage area at 105,000 square miles. Near its confluence with the Caculovar it traverses a lacustrine region, 3500 to 3655 feet above sea-level, where it spreads out its waters in great inundations during the rainy season, and consequently carries little water down to its mouth. Sometimes it is unable in summer to cross the bank of sand at its mouth, and coasting vessels search in vain for an entrance.

Among the tribes most capable of following industrial pursuits are, first of all, the Ba-Fyots, called usually by the Portuguese Cabindas, from the name of their chief village. They are good sailors, laborious and apt at all kinds of employment. The Ma-Voumbous also are very intelligent, energetic traders, and skilful artisans. In type and in certain of their customs they bear some resemblance to the Jewish race, so that some authors regard them as half-caste Jews and Negroes formerly expelled from the Portuguese possessions on the Gulf of Guinea. South of the Congo, as far as the Dande, dwell Ba-Fyots who formerly constituted the famous kingdom of the Congo, but they are much inferior to the Cabindas and have hitherto entirely escaped European influence. All the country down to Mossamedes is inhabited by the Bunda race, divided into numerous fractions, often hostile to one another. They are in general of good physique and intelligent, and would make excellent workmen if they were willing to undertake regular work. They have a decided taste for commerce, and readily enter into relations with European traders. Behind the chain of Tolla Mangango live the Bangalas, often called Cassanjes by the Portuguese, from the name of the chief place in their region, a robust and hardy people who have frequently given trouble to the authorities. It is they, nevertheless, who bring down the products of the interior to Ambaca or Punto Ndongo.

Though the whole of the territory of Angola is divided into concelhos, or cantons, the authority of Portugal is by no means upheld in all the regions included in the administrative system. Ten or fifteen years ago little was known of Angola but the coast and a few small districts in the interior, and even in 1880 Serpa Pinto and his companions had great difficulty in collecting porters for their expedition in Benguella. It was then that Portugal began to make serious efforts to extend her dominion over this vast territory, and, thanks to Serpa Pinto, Major Carvalho, who led an expedition in 1884-88 from Loanda to Lunda, Chavanne, Wissmann, and other foreigners, Angola is no longer the unknown country it formerly was.

But the colonisation zone is far from being conterminous with the sphere of exploration. The Congo district is very much neglected, the natives being absolutely independent of the Portuguese authorities, while the Ambriz district is of no greater value than it was fifteen years ago. Loanda, on the other hand, has made more rapid progress, for the construction of the railway from Loanda to Ambaca has brought into it a large number of Europeans, and efforts have also been made towards an effective occupation of Cassanje, which extends to the Kasai. The regions south of the Cuanza are still beyond European influence, and the steamers which ply on that river have often to be defended against the attacks of the natives. A very different state of affairs is found in the Bihé mountains, and care has been taken to keep open a route thence to Benguella. Great progress towards the establishment of colonies has been made in Mossamedes, and doubtless all the suitable lands will in time be occupied, though the cultivated patches are at present insignificant compared to the virgin lands around them.

The territory effectively occupied by the Portuguese may be estimated at about 3800 square miles, an area sufficient to satisfy the demands of colonists for many years to come. As for the population, it has been estimated at widely different figures, from 200,000 up to a million, and therefore it may be safely said that nothing is really known of their number.

HISTORY.

In the year 1484 [1] Diogo Cão, who was sent out by John II. of Portugal to continue the exploration of the west coast of Africa beyond Cape St. Catherine, discovered the estuary of a great river. At the promontory on the southern side he erected, according to Portuguese usage, a pillar bearing the arms of Portugal, the date and the name of the discoverer, and he called the river Padrão, this being the name by which these pillars were known. The river has changed its name to Zaïre or Congo, but the Portuguese discovery is still commemorated by the name of Cape Padrão.

In 1490 missionaries were sent to the Congo, who were well received by the powerful king then ruling over the whole country between Loanda, the Congo, and Cabinda. Of what race this king was who showed such capacity for state organisation and civilisation history has failed to discover. Soon after their arrival at the capital, Ambassi, the missionaries began to build a church, and the name of the village was changed to San Salvador. Not long after, the king and queen were baptized, receiving the names of João and Leonora, which were those of the Portuguese

---

[1] In a memoir on Diogo Cão, published by the Lisbon Geogr. Soc., Sr. L. Cordeiro produces evidence to prove that the voyage took place in 1482-3.

sovereigns. For a century the town of San Salvador had almost a European appearance; European workmen were imported to teach the people how to build houses, schools, and churches, and the Negro nobility assumed the titles used in Portugal. But early in the seventeenth century there was a marked decline in evangelical enthusiasm, the bishops ceased to visit San Salvador, and in 1814 Don Garcia v. complained that missionaries no longer came to his capital. Certainly the kingdom of the Congo was during this period gradually declining until it became no more than a Portuguese military post, while to-day it is a *concelho*.

One cause of the failure of the Portuguese to civilise this country no doubt was that the missionaries, being deprived of any civil or military support, were unable to protect themselves or their flocks in the midst of the frequent and sanguinary wars that devastated the country. A greater check was given to their work by the introduction of the slave-trade. As early as 1548 it was found that there were more than ten European houses engaged in this trade on the Congo, and that four hundred to seven hundred slaves were shipped annually. Again, the Portuguese colonial empire expanded so rapidly in the sixteenth century that the resources of the country, with its 1,100,000 souls, were insufficient to maintain it properly, and the demand for missionaries being greater than the religious ardour of the country could satisfy, men crept into their ranks whose characters seriously injured the cause which they were enlisted to serve. Lastly, the missionaries seem in many cases to have misunderstood the real needs of the Negro population. They were content to exercise their religious functions, and made no attempts to set before the natives the advantages of European civilisation, to teach them useful trades, and thus ameliorate their condition.

Such is the history of the northern part of the Portuguese territory. It was not till eighty years after the discovery of Diogo Cão, in 1559, that a certain Paulo Dias de Novaes received orders to enter into relations with the king of Angola. The objects that the Portuguese had in view are not precisely known, but it is very probable that they wanted slaves for Brazil and their other colonies. In 1575 Paulo Dias founded a colony on the small island in the bay of Loanda, which was shortly after moved to the site of the existing town, and, aided by reinforcements from Portugal, he advanced to the mountains of Cambambe, supposed to contain silver, and subdued all the native chiefs of Quissama and Illamba. Peace was maintained with the natives by the succeeding governors until Portugal passed, in 1580, under Spanish rule, when the governors were constantly urged to push into the interior of the country in search of gold. At the beginning of the seventeenth century the Portuguese authority was respected as far east as Cabembe, on the Rio Cuanza, 100 miles from Loanda, and as far south as São Felippe de Benguella, recently founded in honour of the King of Spain.

This southern extension was made for the purpose of searching for gold in Benguella, where it was believed to exist, and when none or very little of the precious metal could be found, the Government ceased to take any interest in Angola.

In 1641, the Dutch, then masters of Brazil, seized Angola also. But the Portuguese, retiring to the mountains, defended them against the invaders until 1648, when the Brazilians, who had expelled the Dutch, came to deliver the colony which supplied them with the greater part of their slaves.

Before these events occurred, in 1640, Portugal had thrown off the Spanish yoke, and in consequence some order was introduced into the administration of Angola. During the most brilliant period of this colony, 1660 to 1760, a governor-general resided at Loango, surrounded by a staff of functionaries and magistrates. At Benguella resided a governor subordinate to the one at Loanda, the whole territory more or less under Portuguese control was divided into districts, and small posts were distributed throughout the country, commanded by officers who also possessed administrative and judicial powers. This system might have performed its functions admirably, had it not been for the corruption of the officials.

During this period there was absolutely no commerce except in slaves, and the sole aim of the Portuguese was to penetrate farther into the interior where their raids would be more productive. Meanwhile, the Jesuits did something towards the amelioration of the condition of the natives, but they were expelled in 1760, and with them the civilising movement died out.

The state of the colony underwent little change for the following eighty years, the lamentable condition of the finances of Portugal, the Napoleonic wars, and the civil war that followed soon after, crippling the action of the mother-country; and the partial suppression of the slave-trade in 1842 gave a crushing blow to Angola, which had hardly any other resources. In 1844 the commerce with Portugal was estimated at only £73,312, the chief exports being wax, ivory, orchella, and copal. The value of the coffee exported was less than £1000. At length the occupation of the port of Ambriz by the British, to prevent the exportation of slaves, roused the pride of the Portuguese, who in 1854 decreed the abolition of the trade, and even opened the country to scientific exploration, Dr. Walwitsch being authorised to investigate its flora. But this renewal of activity did not last long; the Portuguese rapidly lost ground and were unable to maintain their authority in the interior, and it was not till 1869 or 1870 that they began to work diligently for the development of their unfortunate colony. Angola, though still behind other European colonies, is now really progressing, and may expect a brilliant future if not hampered by the policy of the mother-country.

We shall not follow M. Delannoy through the long and detailed account of the present legislative and administrative system, but after a very short sketch pass on to the effects of the system on the progress and well-being of the country. Portugal has recognised the right of the colonies to be represented in Parliament, and of the twelve colonial deputies has allotted two to Angola. The electorate is very large, the franchise being extended to all heads of families, black or white, who know how to read and write, for which qualification an assessment of about four shillings is considered an equivalent. But, though the legislative system is thus assimilated to that of the mother-country, the colonies are not governed exactly in the same way. No law applies to the colonies unless a special provision is inserted, and, as a matter of fact, almost all measures for the guidance of the colonies take the form of decrees, which the executive was empowered to issue by an act of 1852. Angola has nominally the right to manage certain of its local affairs, to be consulted with regard to new laws, and to be represented on the administrative councils; practically, however, the Home Government is still absolute.

At the head of the colonial administration is the Governor-General, residing at Loanda. He has military and civil powers, except in judicial matters, the law not expressly giving him the right to preside over the tribunal. He appoints all officials whose salary is below 300 milreis (about £70), and all the higher officials, subject to confirmation by the Crown. Another important personage is the Colonial Secretary, who is likewise appointed by the King, and is head of the Government departments. In the various branches of administration the Governor is assisted by four councils, which represent the Crown rather than the colony, and are entirely under the control of the Governor.

The territory of Angola is divided into five districts—Loanda, Benguella, Mossamedes, Ambriz, and Congo—each of which is in charge of a military governor, appointed by the King, who also discharges civil functions. Each district is again divided into *concelhos*, a division intermediate between those known in Europe as cantons and communes. The *concelho* is presided over by an administrator appointed by the Governor, and is ruled by an elected council of five to seven members. As, however, in many of the *concelhos*, the inhabitants, exclusively natives, are not capable of forming a council, the administration is in such cases intrusted to an official who exercises civil and military powers.

A court of appeal is established at Loanda, the jurisdiction of which extends to the Cape Verd islands, San Thome, and Principe, and there are six tribunals of first instance, frequently composed of unpaid magistrates recruited from the neighbouring planters. In the *concelhos* the chiefs act as magistrates.

The army is composed of European troops, the ranks of which are filled chiefly by military convicts, and which actually number only 1462 men (besides the battalion of the Congo)—a number far too small to maintain Portuguese authority over such an extensive territory—and a few battalions of black militia.

Portugal has not yet learned that the colonial civil servants ought to be selected with even greater care than those at home, because they cannot be so closely supervised and directed. The Portuguese colonial officials are not obliged to produce any evidence of capacity; there are no colleges to train them, and no examinations to test their powers. Those in the lower ranks are often sent to Africa as a punishment. Then, the pay of the officials, though slightly raised within recent years, is wretchedly small. The Governor-General receives only £1334 annually, on which he has to keep up the dignity of the representative of Portugal; and the other salaries are proportionally lower. Out of these salaries the officials sent into the interior have to provide means of transport for themselves and their baggage, and support themselves in countries where European goods fetch enormous prices. Even in the towns many articles which are necessaries of life to Europeans in tropical countries are very dear. Consequently, the whole staff of officials, civil and military, is sunk in poverty.

It is in the *concelhos* that the administration is most defective, precisely the divisions where it should be maintained at a high standard, because here it comes into contact with the natives. A *concelho* may be very rich, like that of Melanje, and cover several thousand square miles, and yet the *chéfe* may be only a simple sergeant of the Portuguese army, who has been raised to the rank of officer on passing into the colonial army. Nevertheless, he is administrator, judge, military commander, postmaster, civil officer, treasurer, and generally doctor and chemist. To assist him in discharging these functions he has a clerk, one or two messengers, a corporal with nine soldiers and some trumpeters. With these resources at his disposal he has to maintain order in the *concelho*, keep open the trade routes, and carry into practice the measures adopted by the Government for the civilisation of the Negroes.

But these chiefs introduce hatred of European authority, instead of civilisation, into the countries submitted to their control. They supplement the insufficient pay allotted to them with contributions from the natives extorted by a thousand illegal devices. Being both military officers and judges, they hold the native completely in their power. The superior judge, to whom the native has the power of appealing, is simply a trader without legal knowledge and not inclined to quarrel with an officer. Loanda is often too distant, and there the resistance must be overcome of protectors whom the *chéfes* procure by pecuniary means. Owing to the iniquitous proceedings of the *chéfes*, the neighbourhoods of the Portuguese posts have been abandoned by their inhabitants. Ambaca, Duque de Bragança, and Caconda, formerly populous centres, are now deserted.

On the other hand, it would be a mistake to suppose that the sums voted for the administration are inadequate, and that the colonies are stripped of officials. There are more than one thousand direct agents of the Government in Angola, but the central administration absorbs an undue proportion, while the country districts are undermanned. It would also be advisable to abandon the system of government by army officers, to which Portugal is so much attached. These men are rarely fitted by their training to conduct the regular administration of old colonies. This reform will be difficult to carry out, as Portugal has a surplus of officers at its disposal and therefore seeks to find them occupation abroad.

There is still another hindrance to the progress of the colony, namely, the transportation of criminals. Brazil formerly served as a penal settlement, and now Angola bears the greater part of this disagreeable burden. When the colony was only a slave-market the presence of convicts was of little consequence, but now that a better state of things has been introduced, the transportation gives the colony a bad reputation. Moreover, the convicts are seldom employed by the Government, but are placed at the disposal of private individuals, who are answerable for their conduct. Thus, transportation is little more than banishment under the surveillance of the police, and colonists show a disinclination to settle in Angola.

THE NATURAL WEALTH OF THE COUNTRY.

The nature of the productions and the state of the natives, who are too little advanced in civilisation to possess or demand articles of trade, indicate Angola as a colony for plantation. To derive a profit from the resources of a country it is necessary to organise the utilisation of the soil and collect in a methodical manner the products that the land yields of itself. Fortunately for Portugal, Angola possesses great natural wealth, and, on the whole, its soil may be considered one of the most fertile in the world. Up to the present time only two plants have been properly cultivated—coffee and sugarcane; all other products having a commercial value have simply been gathered in as nature produces them.

The chief natural products exported are palm oil, indiarubber, coffee, palm-nuts, ground-nuts, and ivory. Cotton grows wild, and hitherto has not been cultivated. Indiarubber is collected in large quantities, but the stores will soon be exhausted unless the Government takes measure to regulate the exploitation. The oil palm, *Elaïs Guincensis*, is widely distributed in Angola. Its fruit yields an excellent oil, used in the preparation of foods and in other industries, and a nut not less prized. The tree has been supposed to need no cultivation, but this has been proved to be a mistake, and it has been found that the oil is more abundant when the fruits are gathered with care. The ground-nut also yields an oil, but inferior both in quantity and quality to that of the palm.

Coffee grows wild in the northern part of the territory, especially in the basin of the Kwango. As, however, the bushes are widely scattered, and the collection of the berries is somewhat difficult, certain tribes of the Encoje district, encouraged by the price this article fetches, have commenced to cultivate the plant regularly.

Until Angola is better known, it is impossible to define exactly the geographical distribution of these products, and to determine the quantities supplied by the various regions of the province. Some rough indications are afforded by the returns from the ports where the goods are shipped.

There are four chief outlets for the natural products of Angola. The Portuguese Congo, of which the chief localities are San Salvador and Cacongo, with the detached district of Cabinda on the northern side of the Congo, sends its exports to Landana, Cabinda, Ambrisette, and the river ports. The port of Ambriz drains the valley of the Loge. To Loanda is sent the produce of the valleys of the Cuanza and Dande, while Benguella is the port of embarkation for the vast and fertile regions of the interior which encompass the Bihé mountains.

The exportation from the Congo is very important, especially considering that, the climate being very unhealthy, the European element is very small, and therefore the commercial activity of the district is due to the natives. In the year 1888-89, the exports—palm oil and nuts, coffee, indiarubber, ground-nuts and ivory—amounted to 6795 tons, and in the following year rose to 7864 tons. In the former year the value of the oil exported was £158,840; of the indiarubber, £68,750; of the coffee, £25,036; while in 1889-90 the value of the indiarubber rose to £101,486 and of the coffee to £51,700. The increase in the value of the exports is the more remarkable because, since 1885, the Congo State has become a formidable competitor with the Portuguese colony, and the exports pay duties from which they were formerly exempt.

The port of Ambriz owes its importance entirely to the coffee harvested by the natives of the Encoje district. The value of the coffee exported has fluctuated considerably, but rose to £111,520 in 1890, falling to £67,300 in the following year in consequence of drought and an epidemic of smallpox among the natives. The total exportation from Ambriz, including copal, wax, and indiarubber, rose to £176,263 in 1890, the first year in which the value of 1880, when duties were imposed, was surpassed.

Loanda, the capital of the Portuguese possessions, and the best port on the coast, owes its importance rather to the facilities of penetrating into the interior offered by the river Cuanza. It is by this waterway that the coffee of the Cazengo plantations reaches the coast, and it is the proximity of this river that makes Loanda of more importance in the commerce of natural products than Ambriz, in spite of the fiscal advantages granted up to 1892 to the latter port.

Except as regards ivory and gum, the production of which has been steadily diminishing for years, the exploitation of the natural products shows signs of improving, though at present the cultivation of coffee only has received attention, and troubles in the Upper Cuanza country have hindered progress. In consequence, the exportation of oil fell off, but is again on the increase. Palm nuts are steady, but indiarubber promises to turn out a most important article of commerce. The export increased year by year, until in 1892 it attained a value of £95,826.

From Benguella principally indiarubber and wax are exported. Since a military expedition in 1889-90 has established Portuguese authority in the Bihé country, the export of the former has made great strides, and the value attained in 1891 the sum of £188,122, being ten per cent. greater than in 1888-89. Wax and orchella, on the other hand, have fallen off rapidly. But ivory has to a large extent compensated for the decline in the export of these articles, the quantity exported having risen from a little more than twelve tons in 1888 to over nineteen tons in 1891, worth about £16,000. During 1890, the last year for which complete returns are available, the total value of the exports was £836,080, orchella, guano, ivory, timber, and hides being excluded, as the quantities vary and are comparatively small.

The results of native labour as shown by the above figures are truly astonishing, considering the slight efforts made by Portugal to develop its colony. No roads have been made to supply the want of navigable waterways, and all goods must be transported along narrow paths almost impracticable. The military posts in the interior are insufficient to maintain order, and therefore trade is frequently interrupted by disturbances among the natives. A scientific expedition to Muene Puto (Kasongo) in 1884 had the effect of considerably increasing the quantity of merchandise transported from the districts east of Malanje to Loanda, by rendering the routes safer.

Besides the natural products of its land, Angola possesses a source of revenue in the fisheries on the coast, which at present are confined to Mossamedes. Three or four thousand tons of salt fish are annually exported from Mossamedes to other parts of the province, and are sold at such a low price that on the Congo fifty Negroes can be fed on fish and rice for the sum of 3s. 1d. per day.

The soil also is rich in mineral deposits. Iron, copper, and gold are found in all parts, but at present little is known of these deposits, and it is uncertain whether they can be worked profitably. The natives of Cazengo have worked iron from time immemorial. Gold mines on the Lombize, an affluent of the Cuanza, and copper mines at Ambriz and Bembe are in the hands of English companies, which, however, are not showing much activity, and there are limestone quarries on the Upper Cuanza. The concessions are granted by the Portuguese on terms calculated to crush an infant industry, the price of the minerals extracted being fixed at Loanda.

AGRICULTURE AND MANUFACTURES.

Agriculture in Angola has not yet attained the same importance as the collecting of the spontaneous produce of the soil. While the colony was simply a slave-market, agriculture was completely neglected, and for years after Portugal was not in a position to assist its unfortunate colony. Now, indeed, a better day seems to be dawning, and the feeble and badly directed efforts made for the better utilisation of the land have had most encouraging results.

The principal crop, and the only one that furnishes exports, is coffee; next in importance is sugar, mostly used for the distillation of rum. Food-stuffs are also grown and cattle are grazed, these being the industries engaged in by the white colonists established in Mossamedes. The natives also own numerous herds in the elevated country south of the Cuanza. Between this river and the Congo there are no horned animals, and attempts to introduce them have failed.

Coffee is cultivated in the districts of Cazengo and Golungo Alto, lying 200 miles from the coast in volcanic regions 2600 to 2900 feet above sea-level. The British Consul, Mr. Newton, estimated the yield of the chief plantations of Cazengo in 1888 at 514 tons, and of Golungo Alto at 136; but these figures probably refer only to the plantations belonging to Europeans, for four times the quantity was shipped at Loanda.

The great hindrance to the development of plantations is the cost of transport. The coffee is carried by porters to Dondo on the Cuanza, eighty-seven miles from Cazengo, and thence in the boats of the Cuanza Company to Loanda, the freightage of a ton of coffee from Cazengo to Loanda varying from £7, 10s. to £10, which amounts to thirteen to nineteen per cent. of the value of coffee. When the railway from Loanda to Ambaca has been extended to the coffee district, the transport will not cost more than 48s. to £3 a ton, and this reduction will undoubtedly have a very favourable effect on the prosperity of the coffee industry. In 1890 and 1891 the value of the coffee shipped at Loanda exceeded £240,000, in 1892 there was a large increase, and 1893 was expected to be a still better year. Angola coffee is inferior to that of Java or Brazil, but the profits derived from it are considerable, though the planter realises only 2d. to 2½d. per lb.

At present only a very small part of the country where the plant grows wild is occupied by plantations, and that part is the worst as regards climatic conditions. Coffee is found in a wild state as far as Cassanje, near the Kwango, and the vast plateau of Malanje, between Cazengo and Cassanje, at an elevation 3600 to 3900 feet, contains lands as fertile as those now under cultivation, and perfectly healthy. Cazengo has been chosen for plantations, because it is comparatively near to the Cuanza, but if the railway is prolonged beyond Ambaca, there will be no obstacle to the laying out of plantations in districts where the climate is suitable to Europeans.

Sugarcane does not play the same part in the economy of Angola as coffee. As already mentioned, its products—sugar and rum—are consumed locally. The cane requires more heat, and is grown on the damp low-lying lands. Hence Cazengo, where the region favourable for the coffee plant commences, is the extreme limit for the sugarcane. Most of the plantations are on the right bank of the Cuanza below Dondo. There are others near Benguella, and a few small ones near Mossamedes. The industry is protected by heavy import duties on foreign spirits and sugar, but the cultivation of the sugarcane is hardly advanced enough to supply the distilleries with sugar for the manufacture of rum, and a considerable quantity of sugar is imported at Loanda in spite of the protective duties.

The situation in Angola is not favourable for the spread of agriculture. The defects of the administration, the unsatisfactory state of the finances, the want of roads, harbour works, and other public works, the bad organisation of the medical service, which discourages European immigration —all these causes tend to check enterprise. Nor is the land system such as to palliate the influence of these causes to any extent. Under pretext of attracting colonists lands have been granted too freely, without inquiry into the means of the colonist to bring the land under cultivation, and thus large tracts lie fallow which the Government has alienated without advantage to the colony. Land is sold at very low rates, and exemption of duties is granted for five years on machinery, tools, and building material, and also of direct taxes for a period of years. Moreover, a decree of 1891 appears to authorise concessions of lands without any equivalent from the purchaser. As a matter of fact, however, the land is generally granted in large areas to companies.

Another drawback is the scarcity of labour. When slavery was abolished in 1869 the slaves were bound to work for their old masters till 1878. The planters then sought to recruit native workmen, and the system now in vogue is practically slavery. The tribes living on Portuguese territory still possess large numbers of slaves, who do not dream of claiming the protection of the Government. These slaves are bought by the planters and are made to sign a contract, and as most of them are indebted to their masters they never acquire their freedom. Still the planters complain of want of hands, and no wonder, for the Negro is not naturally inclined to work, and only does so to obtain the necessaries of life. The hoe and the mattock are still the only implements used for turning up the soil, and therefore the work is very exhausting. Left to himself the Negro is quite able to cultivate coffee, as is proved by the exportation at Ambriz of coffee from Encoje, and therefore the parcelling out of estates among Negroes would probably be more successful than the cultivation of extensive lands under the direct management of the proprietor. At present the plantations are not in a flourishing condition, owing to defective methods of agriculture, the number of Europeans thoroughly acquainted with colonial cultivation being very few.

Manufacturing industry in Angola is in a very elementary stage. At the ports there are a few brick and tile works and limekilns, and in Loanda there are one or two tobacco factories, and a small spinning and weaving factory at Mossamedes. A few years ago machinery was introduced for the manufacture of cotton, but the new duties on the raw material will inevitably ruin this industry. Certainly Angola will never become a manufacturing country, but certain industries must be established on a small scale to satisfy the needs of the colony itself. The most insignificant undertaking is expensive, whether it be to obtain a supply of water, construct a railway, or set up plant, simply because the artisans must be brought from elsewhere.

Transport and Commerce.

Owing to the arrangement of the mountain ranges, Angola possesses only one navigable river, the Cuanza. Small boats ply along it between Loanda and Dondo, a distance of 125 miles. But this means of communication is far from perfect, the level of the water being subject to great fluctuations, so that the voyage to Dondo is sometimes prolonged to ten days. Besides, in accordance with Portuguese practice, the Company of the Cuanza enjoys a monopoly of navigation on the river, and avails itself of its position to levy heavy contributions from the commerce ; its boats are wretched and its tariff very high.

In the absence of waterways nothing remains but to make roads or, still better, railways. The Portuguese, recognising the necessity, granted in 1886 a concession to a company for a railway from Loanda to Ambaca, and in 1890 decided to make surveys for railways from Benguella to the Bihé country, and from Mossamedes to the European settlements on the high plateaus. But railway construction is dear in tropical climates, and in a country so diversified as Angola ; consequently only the line from Loanda to Ambaca has actually been commenced. The existing concession is for a distance of 225 miles, but the survey has been continued to Malanje, 93 miles from Ambaca. This latter section will be more easily constructed, for the line will traverse level country, whereas between Loanda and Ambaca it encounters numerous changes in the relief of the ground, and rises to a height of 2697 feet above sea-level ;

but the multitude of small streams which intersect the plateau of Malanje will necessitate a number of bridges, etc.   The gauge is small, being only about forty-one inches; the locomotives, supplied by Belgium, are of two types, the larger ones intended to run on the first section, from Loanda to Cassoalalla, 142 miles, at the rate of twelve and a half miles an hour, while the others will attain a speed of only nine miles an hour.   The undertaking, owing to financial losses, has cost the company £9304 per mile.   The line was opened to Ambaca at the beginning of the present year, but as yet it is impossible to estimate the traffic.   Large profits are expected on the completion of the line to Malanje, for it is hoped that the commerce which now takes the route of the Kasai through the Congo State will then be attracted to Loanda.   The sections already opened have earned rapidly increasing sums for freightage, and the natives have shown appreciation of a rapid means of locomotion.

The port of Loanda, already the most important in the colony, has acquired additional value as the terminus of the railway.   The harbour is the finest on the west coast of Africa; it is sheltered from the ocean swell by a small island, so that the shipping and unshipping of goods is never interrupted.   All it wants is quays of sufficient length.   Of the other ports, Ambriz and Benguella are rather open roadsteads, which would be worthless were not tempests very rare.   Mossamedes has a pier connected with a small line of railway, and Ambriz had one formerly, but it decayed and had become unserviceable in 1877.   Angola is in communication with Europe by means of the Portuguese steamers of the Empreza Nacional the steamers of the British and African Steam Navigation Company, of the Woermann Line (Hamburg), and of Les Chargeurs Réunis (Havre).

The most usual form of exchange is barter, though as far as Ambaca the natives are well acquainted with the value of money.   Part of the trade with them passes through the hands of the planters who exchange their rum for natural produce, and with this obtain the articles of European origin which they stand in need of.   The trade with the natives farther up the country is in the hands of a few Portuguese and German firms established in the ports or at Dondo, who send agents into the interior.   The chief articles dealt in are woollen and cotton goods from England, guns from Belgium, and powder from Germany.   The guns and powder pay high duties, but the trade in them does not seem to be affected thereby.

Trade meets with no troublesome fiscal barriers within the country. No rates are levied on goods, and there are no road tolls to pay.   So much the greater are the custom duties levied on the external trade.   The Congo district, being in the zone of free trade established by the Con- férence de Berlin, has to conform in all respects to the Congo State. Ambriz, up to 1892, was almost a free port, but in that year higher protective duties were imposed at all the ports, and Ambriz was placed on much the same footing as the others.   Goods sent from or to Portu- guese ports pay considerably less than others.   Only coal, mineral waters, vessels of over 200 tons, railway carriages, and agricultural implements are now exempt, whatever country they come from.   The coasting trade, as in times past, is reserved for Portuguese boats.

After 1874, when slavery still existed, the external commerce of Angola decreased until 1889, when it began to expand, in consequence, doubtless, of the public works then originated, and the attention that from that time the Portuguese Government paid to its colony; and in 1891 the value of the imports was £1,144,400, and of the exports £862,692, being an increase of about thirteen per cent. on the trade of 1890. The movement of navigation shows a corresponding advance. All the statistics received at present refer, however, to years before the new tariff was put in force, and the most competent judges expect that the effect will be disastrous for Angola, while the Congo State will reap the benefit. The tariff of 1881 was burdensome enough to interfere considerably with the trade, and many goods would not bear the slightest additional duty. The planters were obliged to set aside no inconsiderable part of their income for the purchase of European goods, and many vessels ceased to call for supplies at Angola ports in consequence of the high prices, resorting instead to Walvisch Bay or St. Helena.

The protection of Portuguese industries aimed at by the framers of the new tariff could only be of advantage to an industrial country, or one likely to become such, whereas Portugal exports only wine and cork. The only consequence to Portugal of the differential duties in its favour is, that foreign merchants cause their goods to be transhipped in a Portuguese port in order to take advantage of the lower rates. Thus the small Portuguese mercantile marine obtains a certain profit, but it is at the expense of the colony, for the freightage from Lisbon to Loanda is higher than from Liverpool to Loanda, though the distance is much less.

M. Delannoy has a chapter on the finances of Angola, a subject somewhat complicated owing to the system on which the colonial budgets are kept. He concludes that revenue and expenses are now in equilibrium, but whether this equilibrium is stable is a question more difficult to decide. It depends on whether the commerce is maintained in its present state. The estimate of the customs for 1894-95 is £288,600, or £68,000 more than in 1889, from which it may be inferred that the new tariff has not had the bad effect that many prophesied—that, though it has checked the upward movement of trade, it has not made it go back. But another danger lies in the low rate of exchange, which Portugal is unable to remedy until it can improve its own financial position.

THE CLIMATE AND THE COLONIES OF THE SOUTH.

As the capital, São Paulo de Loanda, is the only town where there is an observatory, it is impossible to give more than a general sketch of the meteorology of the colony from the narratives of travellers. The variations of temperature from one season to another, and also the daily variations, increase with the distances from the Equator and the sea. The rainfall, and with it the luxuriance of the vegetation, diminishes in intensity from north to south. Down to Ambriz the forests extend to the shore, then they gradually recede, never passing the line of greatest elevation, and cease almost completely in the district of Mossamedes. The Congo district, very warm, damp, and covered with dense vegetation, and the coastal zone to a few degrees south of Benguella, are considered unhealthy. At San Salvador, the only place in the Congo district where records have been regularly kept, the mean temperature from January to April is 75° F., with a maximum of 81° and a minimum of 61°. It then falls gradually to 66° in August, rising again to 79° and 77° in November and December. The mean annual temperature is 76°. The humidity is not very high, if the mean, 77°, be taken as a test. But the variations are very great, and in summer 99° is reached. The sky is rarely clear, though the rainfall is only 35 inches, or little more than at Oporto.

In the districts of Malanje and Cassanje, at altitudes of 3786 and 3100 feet respectively, the temperature is more variable. The mean temperature at Cassanje is 77° for the months October to December and 75° in January and February. Between October and January the hygrometer has indicated from 45 to 91 degrees of humidity. Travellers have extolled the salubrity of this region, but observations have not yet been made throughout the year. Much the same has been said of the Bihé country, but MM. Capello and Ivens noticed great ranges of temperature; not infrequently it freezes at night, when during the day the temperature rises to 82° or 86°. Mossamedes is the only district of which it can be said with certainty that it possesses a climate suitable for Europeans, and here only have white families permanently established themselves.

Loanda consists of two quarters—the native town situated on the shore, and the European built on the nearest elevation, 200 feet above sea-level. The surrounding country is arid, and there are no marshes near the town, but the banks of the Peixe and Bungo, which have been receptacles for refuse from time immemorial, exhale pestilential vapours when the water is low. Being situated on the coast, Loanda lies in a zone of low pressure. The mean for seven years was 29·76 inches, the maximum being 29·93, and the minimum 29·65. The variations of temperature are much greater, at least the diurnal varia-tions, for the monthly means are fairly constant. The mean maximum is 84°, the mean minimum 32·8°, and the annual mean 77°. The absolute maximum and minimum observed were 92° and 56°. The nights are often fresh, and during six months in the year it is pleasant to sleep under a blanket. The mean humidity varies from 82 to 92, with a maximum of 95·2 and a minimum of 75·7. The sky is generally cloudy, being absolutely clear on only ten days or so in the year, but yet the rainfall is small. The rainy days are only about twenty-five in the year, and the quantity of water that falls does not exceed 5·65 inches. The heavy rains from January to March render the air moist, but August is no drier, though it owes its humidity not to rain but to the *cacimbo*, a mist which rises in the night and is not dispersed till long after sunrise. In general the season of the *cacimbo* is most favourable to immigrants fresh from Europe, while, on the contrary, acclimatised colonists then feel the cold at night.

It is difficult to draw any very definite conclusions from the mortality as to the influence of the climate of Loanda on its inhabitants, for, strange to say, their number is not exactly known. The town cannot be described as decidedly unhealthy, and indeed its salubrity has been much increased since 1888, when, in connection with the erection of railway stations and waterworks, the sand was removed from the streets and the shore cleansed of its impurities.

The only coast town where Europeans can at present live with safety and bring up their children is Mossamedes. Its outward aspect is agreeable; its streets are regular, well built, and adorned with trees. And yet it owes none of its beauties to Nature, for the country is sandy and dry. The success of agriculture in the *concelho* of Mossamedes is dependent on the rains, which unfortunately are very irregular. Out of a total area of 7530 square miles only 2200 acres, near the town, are cleared or occupied by buildings. But Mossamedes has other resources in its fisheries and its rich deposits of guano and nitrate of soda, while it will acquire yet greater importance as an outlet for the trade of the European colonies in the interior.

These colonies are scattered among the affluents of the Cunene on the western edge of the great plateaus. All are situated in the Mossamedes district except Caconda, which is a little nearer to Benguella, but its trade will probably take the route to Mossamedes, which traverses healthy plateaus, whereas the road to Benguella is somewhat unhealthy. The other colonies, Humpata, Huilla, Sa da Bandeira, and S. Pedro em Chibia, lie more to the south on plateaus not lower than 5300 feet, among the affluents of the Caculovar and of the Rio Chimpumpunhune. Huilla, almost at the centre of the group, stands 5670 feet above sea-level, and is distant from Mossamedes about seventy-eight miles.

The plateau is as a whole very fertile. Its soil, a siliceous clay, is adapted for crops of both tropical and temperate climes. The rivers are subject to regular floods, which cover the banks with rich mud and form magnificent meadowland. Huilla was founded as a military colony in 1857. It did not succeed, and in 1881 was handed over, with 5000 acres on the Rio Mucha, to French missionaries. About ninety-five acres are now under cultivation, producing most of the vegetables and cereals of Europe. The mission makes wine and liqueurs, and intends to take up the cultivation and preparation of tobacco.

In the same year that Huilla was founded, the Government authorised some Boers from the Transvaal to settle at Humpata, where they founded the colony of S. Januario, fifty miles north-west of Huilla. The number of these colonists was 270, forming fifty-seven families. They brought with them 840 draught oxen, 120 horses, 3000 sheep, and 2160 cattle. Unfortunately Humpata is situated on a very high plateau, where there is little pasturage, so that a large proportion of the cattle died, and the Boers, averse to agriculture, fell back on hunting. In 1884 the colony was reinforced by colonists from Portugal and Madeira, and is now apparently prospering. About 250 acres of land are cultivated, the crops being cereals, sweet potatoes, and yams.

Sa da Bandeira lies about fifteen miles from the other colonies on a small plateau of 4200 acres at an average altitude of 5150 feet. The colonists numbered 940 in 1891. The soil is a compact clay and ought to be manured with lime, but the expense is too great; 995 acres are under cultivation. The colonists also possess 500 cattle and 300 pigs, besides goats, horses, asses, etc.

S. Pedro em Chibia was founded as late as 1885, by forty-two colonists from Madeira. It is situated on the Chimpumpunhune, and has the advantages of a better climate and soil than any of the other colonies. The necessaries of life are produced in abundance and rum is distilled, the only article that finds a sale at that distance from the coast.

Caconda possesses a very similar climate and much more good soil, the colonisable lands extending over about 7400 acres. Only six or eight plantations have been laid out, and therefore the colony can hardly be said to be formed.

But few meteorological observations have been made on the plateaus. Caconda apparently has a mean temperature of 69°. The highest recorded was 88°, and the lowest 42°. Huilla seems to be cooler, for the means during three years ranged from $59\frac{1}{2}$° to 61°. At both places the rain is abundant, and distributed over a fairly large number of days. The fall at Caconda in 1889 was $68\frac{3}{4}$ inches, and there were 135 rainy days.

The plateaus are undoubtedly suitable for European colonisation, but there are two sources of danger to the future of the colonies. The first is the hostility of the natives, who should be restrained by a strong military force. Had it not been for the Boers, Mossamedes would have fallen long ago into the hands of the natives. The other difficulty is transport. Owing to the cost of carriage, combined with heavy duties, European goods can hardly be purchased at all by the colonists. The Boers sometimes make an excursion to Walvisch Bay, where the prices are more reasonable. It is now hoped that a railway will be constructed through the aid of the Mossamedes Company, which has obtained a grant of fifty-seven million acres in the south of Angola, where it proposes to trade, work mines, and encourage colonisation.

GEOGRAPHICAL EDUCATION.

By ANDREW J. HERBERTSON, F.R.S.E., F.R.G.S.

(*Continued from p.* 528.)

III. GEOGRAPHY AS A SCHOOL SUBJECT.

THE question, Should geography be taught in schools? can have but one answer for readers of the *Geographical Magazine.* Nevertheless there are people who dispute the propriety of teaching a subject which can best "be picked up, mainly by general reading, after the pressure of competitive examination is over," and therefore would waste time in school. For such people it has no educational value whatever, and the headmaster of an English public-school recently wrote: "As to history, geography, and science, where pupils *do* nothing but listen to what they are told and reproduce it, they are useless for mental discipline, however valuable and interesting they may be as a collection of facts."

It is, unfortunately, only too true that in many schools pupils do *do* no geography. They read a few pages of a dull catalogue known as a geography book, commit to memory as much as possible, and reproduce what they can at school the next day. Probably they have an atlas at home, published perhaps a quarter of a century ago, or a cheap modern one. In the former case the printing is good but the political boundaries are wrong, and many important features and places are unmarked. In the latter, the printing is almost indecipherable, for many cheap atlases are printed from the same plates as the antiquated atlas, now almost worn-out, and bought cheap from a firm that would not use them any longer for its own work, and then patched up. From such atlases the pupils have occasionally to copy a map at home or in school, and that constitutes their practical geography!

Such, no doubt, is the sort of geographical teaching which the headmaster just quoted had in his mind when he condemned geography as useless for mental discipline. And until such theory and practice of geographical education have been replaced by sounder conceptions and methods we cannot expect a very different verdict. This, teachers of geography freely admit.

A most successful teacher of geography in a secondary school, Mr. Hewlett, in an address to the Manchester Geographical Society,[8]* recently said, "If Geography has had a bad name amongst schoolmasters as a subject of no educational value, the reason is that it has been judged by the bad methods followed, and in ignorance of the capabilities of the study when right methods are adopted."

But geography in competent hands is of the greatest possible educational value.

Sir Archibald Geikie says in the opening chapter of his book on *The Teaching of Geography* [9] :—"Instead of being a mere exercise of the memory, as it has so often been treated, geography steps at once into a foremost place among school subjects as an instrument for training various mental qualities that are hardly reached at all by the other branches of an ordinary curriculum. In the first place, and above all, it calls out into active exercise the observing faculty, which is otherwise left well-nigh dormant in the ordinary tasks of the school. It stimulates the reasoning powers, by teaching the value of classification and co-ordination of facts and the methods of scientific induction. It affords ample exercise of the memory, but not in the mere mechanical way implied in the learning by rote of tables of figures and pages of statistics. . . ."

So much, then, for its claim on the ground of educational discipline.

Of the practical value of geography, it is hardly necessary to speak, for nobody denies it. In business affairs it is an essential knowledge no man can lack with impunity. The Rev. F. R. Burrows says of the clergyman [10] : "I do not think he is fit to talk of missions until he knows where missions are ; and if he has grasped his geography aright, he will learn and teach his people the causes of the slow growth of Christianity

---

* The numbers refer to the publications noted at the end of this article.

in regions where it comes in contact with religions—bone of the bone and flesh of the flesh of the people. He will not expect in Peking the converts of Uganda, or the success of Travancore in the capital of Persia. He who knows geography knows missions." For every citizen it is necessary before he can understand the vast empire of which he is a member, or the community of interests that unites all men together.

But many who recognise the practical, and even in a measure the educational, value of geography, would deny it a place on the school programme. When its claims for recognition as a school subject are pressed, they object that time-tables are already too full, that some subject must be sacrificed, and that geography can be most easily dispensed with. They have no higher conception of geography than as one more routine task to be undertaken by masters and pupils; one more added to the many isolated subjects that have to be studied.

Many complain of the uncoördinated nature of our present school programmes. This is very just in the main; but the introduction of adequate geographical teaching will not intensify this condition. It will tend to remedy it. In the hands of a capable teacher, geography has the advantage of giving a connection to some of the otherwise uncoördinated subjects on the school syllabus.

Its advantages as a unifier of knowledge have been pointed out again and again by many writers, not a few of them being experienced teachers as well. The most recent article, by the Rev. F. R. Burrows, has just appeared in the *Parents' Review*,[10] and eloquently pleads for geography as the "element of cohesion" in our educational system.

In the preface to *A Short Geography of the British Isles,* the late John Richard Green admirably expresses the same idea, and points out the important position geography must occupy in any rational system of education. Rightly studied, "it is the natural starting-point for all the subjects of later training. History strikes its roots in geography; for without a clear and vivid realisation of the physical structure of a country, the incidents of the life which men have lived in it can have no interest or meaning. Through history again politics strike their root in geography, and many a rash generalisation would have been avoided had political thinkers been trained in a knowledge of the earth they live in, and of the influence which its varying structure must needs exert on the varying political tendencies and institutions of the people who part its empires between them. Nor are history and politics the only studies which start naturally from such a ground-work. Physical science will claim every day a larger share in our teaching; and science finds its natural starting-point in that acquaintance with primary physics which enables a child to know how earth and the forms of earth came to be what they are. Even language, hindrance as its premature and unintelligent study has been till now to the progress of education, will form the natural consummation of instruction when it falls into its proper place as the pursuit of riper years, and is studied in its historical and geographical relations."

Assuming, then, that geography is entitled to an important place in the school curriculum, (1) as a valuable means of educational discipline,

(2) as a subject of the highest practical utility, and (3) as a cohesive force binding together previously uncoördinated branches of study, how is it to be taught?

It is important for us to distinguish between the claim for an extension of geographical teaching in schools, which we have just been considering, and the desire for better methods of teaching geography, which is equally important. Both are urgently demanded. Changed regulations can extend the time given to geography in schools. Better teaching can be obtained by giving teachers, both embryonic and full-fledged, an acquaintance with better methods.

In geography, as in other subjects, we want to change from the teaching by words to the teaching by the things the words are supposed to describe. Geographical teaching, to be worth anything, must be real and practical. It is not enough to appeal to the ear alone. The eye must be trained as well, first of all by seeing the actual land surface from which the first lessons in geography must be taken.

The need for a thorough training in home lore in general, and home geography in particular, in early school days has already been discussed, and the necessity for frequent geographical excursions emphasised. An interesting paper on "Geography from Nature,"[11] by Mr. Dodge, of the Teachers' College, New York, describes how such excursions can be made profitable round that city. Teachers will derive many useful hints from this article, which has just been published.

The next question is, How are we to pass from the geography of the visible region around into that beyond? Some go on to the geography of the native land, others prefer to consider the world as a whole and the principal divisions of land and water.

I believe that from an early age children should be made familiar with the globe and the continents and oceans. The globe should always be in the schoolroom during a lesson in geography. But if possible a school should possess more than one globe, and the globe a child should first study should have its surface very simply marked, and not, as in the ordinary one, covered with every sort of sign and crowded with names. A slate globe with latitudes and longitudes sunk in the surface, and the outline of the continents in white paint, is an essential part of school apparatus, and can be made to serve many purposes.

The globe, too, should be used much more than at present by *every* pupil for solving many problems, which used to be commonly given in schools long ago, and might well be revived.

There is a real difficulty for most people in passing from the globe to the map, in understanding how the curved surface is expressed on the flat. After trying many methods, I have devised an apparatus for illustrating projections graphically by throwing the shadow of a skeleton hemisphere on a screen. By modifying the positions of the light and the hemisphere, and by bending the screen into a half cylindrical or conical shape, and then unbending again after tracing the shadows, various projections can be shown. The making of the network of a map on the flat surface in this way is very effective, and also gives an idea of the different distortions made in the earth's surface by various projections.

If the globe should be introduced at an early stage of geographical teaching, the systematic geography of each continent in turn need not be begun at the same time. The details of the different divisions of the earth's surface should be filled in gradually.

For some time home geography should be the systematic geography of the pupil. The good teacher will constantly use what the child sees with his own eyes to tell him about something similar in another part of the world. A waterfall, or even the rapids in one of our rivers, will lead the teacher to mention other waterfalls and rapids. Photographs and drawings of the waterfall familiar to the child should be shown and compared with the reality. Then photographs and drawings of other water-falls may be exhibited and their characteristics compared with those of the representations of the neighbouring falls. The study of a little stream may be made the basis of the study of a great river.

As children grow older this should be carried out more systematic-ally. Maps as well as views of the local river should be examined by its banks, and from the heights that give a view of the surrounding land. Then, in school, maps and views of the great river basin should be studied, and compared with the maps of the familiar stream.* So unknown mountain systems can be compared with the known hills, and other features may be studied in the same way.

The home region should be used to teach the first notions of geography, and then these should be applied to unknown regions, first likenesses and then differences being noticed. Gradually general laws should be pointed out, and at first as far as they are illustrated in the locality of the school.

The systematic teaching of the geography of other lands will in this way become more interesting, more intelligible and more educative, when it is begun ; for the good teacher will not commence systematic *Länder-kunde* too early. The pupil will be prepared to understand it, and will have an interest in it, because he will have already correlated some of the geography of the new country to what he is familiar with at home, and he will be incited to discover resemblances and differences to what he already knows in what he is now learning.

To make this more impressive, the teacher should reverse the process he has hitherto adopted. The maps and pictures of the unknown land should be studied systematically, and the pupil be led to discover for himself the salient features. He should constantly be brought to com-pare the strange land he is now discovering with the familiar ; to try to find within his own ken in miniature or in larger scale something of the nature of this new region, or in marked contrast with it. On the next class excursion the suggestions made in the class should be examined in the field, and confirmed or modified.

A word of warning should be added against the tendency in many school programmes to parcel the world into different areas, to begin with England or Scotland, British Isles, Europe, British Possessions, and so

---

* The importance of having maps drawn to scales related in some simple proportion is again obvious.

on, the advance being in progressing from countries near, and of special interest to us, to more remote lands. But this is not enough. It is necessary to treat geography differently in higher classes, to make it more intensive as well as more extensive. In a properly arranged curriculum the earth's surface will be studied at least twice, the second time more in detail, and what is even more important, with a profounder study of the general principles.

The French programmes permit this,[12] and the instructions to teachers emphasise it. Formerly the world was studied three times in French secondary schools—once in the preparatory classes, once in the grammar-school division, and once in the senior division of the schools. This is still practically retained with two express stipulations: "que l'enseignement aille vraiment en *s'élargissant* et en s'élevant, et qu'à chaque période l'élève apprenne non seulement *plus*, mais *autrement*"; and further, "qu'on ne sacrifie pas tout à la symétrie de cette distribution, et que . . . on se préoccupe avant tout de l'importance *actuelle* des sujets." The greatest modification recently made in the French programmes is a wise one. In the highest class of the French Lycées, the boy, now almost a man, once more takes up the geography of his own country and studies it in a more thorough way in the new light thrown on it by much he has learned since his last lessons. He goes to college or business with a thorough knowledge of the geography of every province of his country.

For all of us, as has been said in a previous article, not merely the geography of our native land, but even more the geography of our immediate home region should be the last as well as the first division of geography to be studied.*

The furnishing of schools with adequate apparatus for teaching geography is an equally important matter.

Of globes and maps account has already been taken ; and incidental mention has been made of photographs and other illustrations. The supply of cheap photographic reproductions is now large ; but it is very unfortunate that so few are of much use to teachers of geography. The ordinary books are full of pictures of buildings, with no arrangement of any kind. One page may show the Opéra in Paris, the next the Bourse in Berlin, and the next the Union Depôt in New York. Unfortunately the reproductions are printed on both sides of the paper and cannot be classified. A little thought and a little arrangement would have made these books really useful, and increased their sale enormously. The only systematic series of pictures advertised has been arranged, not in this

---

* A valuable report on geographical education has recently been published by the National Education Association of the United States of America, and may be commended to the notice of our educational authorities as well as to that of individual teachers. But in dealing with advanced work in schools, the Committee curiously neglect all branches of geography but "physiography," or geomorphogeny and meteorology, both of which have been so capably developed by Americans in recent years. Recent progress in other branches of science besides those mentioned fully warrants the inclusion of their geographical aspects into advanced geographical teaching.

country, which has more colonies than any other, but by the energetic professor of colonial geography in the Sorbonne, Dr. Marcel Dubois.[14]

To show a series of pictures to a class, there is no doubt that the optical lantern is the best means to employ. The supply of lantern slides is now abundant, but again there is a lack of systematic arrangement. At a moderate estimate, about fifty per cent. of the views in any set of slides are of little value. Exception must be made and attention called to the slides prepared by members of the Geographical Association, more particularly by Mr. Dickinson, one of the secretaries. If every secondary school subscribed to the Association, and geographical masters joined it, and so acquired the right to borrow these slides, which teachers of geography will find of great value, its resources would be largely increased, and its usefulness extended.* Mr. Dickinson has recently published two lectures showing how effectively the lantern can be used in making geography really a subject of educational value in our schools.[15]

Of models, specimens, and other appliances for illustrating lessons in geography, and for hints as to practical work, I may refer to Sir Archibald Geikie on *The Teaching of Geography*,[9] a book which should be read through by every teacher of geography before beginning each school year. It is full of wise and admirable suggestions, and if only they were adopted in our schools, geography would become one of the best-taught and most effective of all educational subjects.

The text-book should occupy a minor position, although its pictures and sketch-maps should be often referred to. The teacher should himself be the text-book. Geography has suffered acutely from that painful practice of the incompetent teacher, to prescribe the reading at home of a few paragraphs of a text-book, and ask questions on them the following day. The teacher must do much more than this. He must be trained himself to do the things his pupils should learn to do.

Standard works on various aspects of geography and selected books of travel should be found in every school library, and should be referred to as often as the text-book. The teachers should constantly read or have read descriptions by famous travellers and statements by leading authorities.

In conclusion, it must be pointed out that satisfactory improvement in the teaching of geography in our schools can be brought about only by improvement in the training of teachers in this subject. Little effort is made to do this at present, and the teacher enthusiastic about geography has to find out everything for himself. Competent and skilled teachers are needed for geography as much as for other, subjects. The delusion that this is not the case is akin to that other, that anybody can read a map without teaching. Imagine the outcry there would be if it were the custom to appoint the distinguished science graduate a master of

---

* The Secretaries of the Geographical Association are Mr. B. Bentham Dickinson, M.A., Bloxam House, Rugby; and Mr. J. S. Masterman, M.A., 55 Campden House Road, Kensington.

ancient languages although he had not opened a Greek or Latin book since he left school. Yet few seem impressed with the ludicrousness and ineptitude of intrusting the teaching of geography to one who has never concerned himself with the subject since he learned a little topography at school.

It is in the hands of the educational authorities to remedy these matters for the coming generation of teachers. For those who have lacked the opportunities in the past, Saturday courses and Summer courses might well be subsidised by School Boards and County Councils. In the schools themselves much might be done, and is being done in a few cases, to help the teacher and the teaching of geography, by giving to the upper classes University extension and other courses of lectures by geographical specialists.

BIBLIOGRAPHY.

8. *The Position of Geography as a School Subject.* By E. G. W. Hewlett, M.A.
   Manchester Geographical Society's Journal, xi. (1895), pp. 255-263.

9. *The Teaching of Geography.* By Sir Archibald Geikie, LL.D., F.R.S.
   London, Macmillan & Co.

10. *On Some Methods of Teaching Geography.* By the Rev. F. R. Burrows, M.A.
    Reprinted from the *Parents' Review*, 1896. London, G. Philip & Son.

11. *Geography from Nature.* By Richard E. Dodge.
    Bulletin American Geographical Society, xxviii. (1896), pp. 146-156.

12. *Enseignement de la Géographie*, pp. 89 à 104. *Programmes*, pp. 147 à 151.
    Instructions, Programmes, et Règlements. Ministère de l'Instruction publique et des Beaux-Arts. Enseignement secondaire.
    Paris, Imprimerie Nationale. 1890.

    See also the account of Geographical Teaching in France, *La Géographie dans les Écoles et à la Université.* Par E. Levasseur.
    Report of the Sixth International Geographical Congress, pp. 27-71. London, 1895.

13. *Report of the Committee of Ten on Secondary School Studies ; with Reports of the Conferences arranged by the Committee.*
    American Book Company. New York, 1894, pp. vi+249.

    An account of the work of the Geographical Committee will be found as under :
    Reports of a Conference on Geography. By Israel C. Russell.
    Bulletin of the American Geographical Society, xxvii., 1895, p. 30.

14. *Album Géographique.* Par Marcel Dubois et Camille Guy.
    Tome 1er.—*Aspects généraux de la nature.* Paris, Colin.

15. *Geography as a School Subject (an attempt to show that Geography can be taught as a training of the mind.)* By B. Bentham Dickinson, M.A., F.R.G.S.
    Printed for the Geographical Association by A. J. Lawrence, Rugby, 1896.

CRETE.

THE position of this island, which has of late drawn the attention of Europe, not for the first time, is too well known to require description. Bounded north and south by deep seas, it is connected on the west with Cerigotto, and on the east with Kasos, by channels only 430 fathoms deep, and thus forms a link in the chain of islands bounding the Ægean sea on the south, and connecting the south of Greece with Asia Minor. In form it is very elongated, being 160 miles long by six to forty miles in breadth.

This form is due to the mountains which run through the whole length of the island, and expand into three considerable masses—in the centre Mount Ida, now called Psilorites (ὑψηλόν ὄρος), the White Mountains in the west, and Dicte or Silia at the eastern extremity. The White Mountains (Ἄσπρα Βουνά) rise, like Mount Ida, to between 7000 and 8000 feet above sea-level; in average height they are superior to the latter, though their culminating points are probably lower. They owe their name either to the snow which clothes their summits, or to their white limestone cliffs. Their slopes are very abrupt, and access to the villages is often only to be gained by following the beds of torrents, and when these are full of water the villages are quite cut off from communication with the country below. Mount Ida, situated in the centre of the island, where it is broadest, presents a lofty, isolated summit 8060 feet high, almost always capped with snow, and, with its massive buttresses, makes an imposing appearance.

Almost the whole surface of the island is occupied by the mountains and their spurs, between which the towns and villages on the coast are enclosed, and thus isolated from one another. There is only one plain of any extent—that of Messara, to the south of Mount Ida. This is the granary of the island, and it is watered by the Hieropotamo, which always contains water even in summer. Small level tracts lie up among the mountains, and verdant valleys nestle in the slopes of Mount Ida.

In ancient times nearly the whole surface of the island was covered with dense forests. Of these nothing is left, and almost the only trees are those cultivated for the sake of their fruits and other products. But all these are found in a wild state in one part or other of the island. Chestnuts are to be seen in the extreme west; evergreen oaks and cypress are confined to the high valleys of the White Mountains; the oak that yields valonia grows in the province of Retimo, west of Mount Ida; the Dicte mountains contain the pinaster and carob; and a grove of dates exists in the south-east. In many parts the ground is brilliant with flowers of varied colours.

No bears or wolves are found in Crete, and their absence is accounted for among the Cretans by a tradition that all large animals of prey were banished by St. Paul. The Cretan ibex still exists in diminished numbers; it is a peculiar species only found elsewhere in Antimelos and in Skopelos and Jura, near Mount Pelion. Small game is also scarce.

Before the dawn of history, Crete was inhabited, and, moreover, by a people considerably advanced in the arts of civilisation. Situated between Europe, Asia, and Africa, it was a commercial emporium in very ancient times, as it was in the Middle Ages and still later centuries, and through Crete the learning and arts of the Egyptians were communicated to Greece and Etruria. Of whatever race the earliest inhabitants may have been—Greeks or other Pelasgian tribes—it is certain that Doric colonies were founded in Crete at a very early period. Herodotus says that Barbarians possessed the island before Minos took possession of it. At any rate, Homer mentions that Crete, ἑκατόμπολις as he styles it, sent eight vessels, under Idomeneus, to the Trojan war, and even at the present day some traces of the old Doric may be distinguished in the Cretan dialect. Indeed, until the insurrection of 1866 the Sphakiotes, a remnant of the old race, had preserved their independence and purity of blood in the White Mountains, which are often named after them.

The Cretans first submitted to a foreign yoke in B.C. 67, when their ports being a resort of Isaurian pirates, they were subdued by Quintus Metellus, who received the title of Creticus. For about twelve and a half centuries Crete formed part of the Roman Empire, except from 823 to 961 A.D., when it was in the hands of the Arabs. Retaken in the latter year by Nicephorus Phocas, it was finally lost to the Byzantine Emperors in 1204, when Constantinople was taken by the Crusaders, and in the division of territory which followed, Crete fell to the share of Venice. It again changed masters in 1668; after a long and obstinate defence, Venice was obliged to cede it to the Turks. Since then it has always remained in Turkish hands, though the Cretans have made many attempts to shake off their oppressive yoke. After the Greek war of independence Crete was assigned by the Allied Powers to Mehemet Ali, Viceroy of Egypt, but was restored to Turkey in 1841. A rebellion broke out in 1866, after which a constitution was granted, and another in 1878, during the Russo-Turkish war, in consequence of which it was decreed that the governor should always be a Christian.

What renders the dissensions in Crete more deplorable is that they are due not to differences of race but only of religion. The Mohammedans of the island are not Turks but Greeks, whose forefathers adopted the religion of their rulers to avoid persecution or obtain material advantages. And perhaps partly in consequence of the bitterness that exists between the adherents of the rival creeds, parliamentary government, however restricted, proved impracticable. Another insurrection broke out in 1889, and from that time until 1894 Crete was governed despotically by Mohammedan pashas. In that year, under pressure from the European Powers, the Turkish Government again appointed a Christian governor, whose entrance on office was the signal for fresh dissensions, which culminated in the late insurrection.

All the nations who successively held possession of Crete have left traces of their rule in some form or other, without, however, modifying perceptibly the type of the original Greek inhabitants. For the Romans held the island as a naval port to maintain the tranquillity of the eastern Mediterranean, as a halting-place between Rome and the East, or between Constantinople and Africa, while the Venetians valued it only as a commercial emporium, and therefore were little concerned about the interior of the island and its inhabitants. On the southern side are the ruins of Gortyna; not far from Khanea is the site of another ancient Greek city, Cydonia, while Megalocastron was the port of Cnossus. The name Candia, by which Megalocastron as well as the whole island was till recently known in this country, was introduced by the Venetians, though it is not of Italian origin.   During the Arab occupation the town was fortified and surrounded by a moat (*khandak*), and hence the Venetian name.   Both this town and Khanea are still surrounded by thick walls built by the Venetians, and the Lion of St. Mark still surmounts the sea gate of Khanea.

The frequent disturbances which have occurred under Turkish government have naturally had a disastrous effect on the prosperity of the people and on their industries.   Crete is no longer the populous place it appears to have been in ancient times.   It contains a population of only some 295,000, of whom about 100,000 are Mohammedans, over an area of 3000 square miles.   The density of nearly 100 per square mile is, however, greater than in most other provinces of the Turkish Empire.   But with its fertile soil the island could support a much larger population.   In the insurrections not only have a considerable number of lives been lost, but many families have been reduced to beggary, and fields, orchards, and vineyards have been devastated.   Consequently the production and trade of Crete are small.   Cotton, tobacco, and other crops are much neglected by the peasants.   The vines, which in Venetian times yielded the famous Malmsey, still produce good wine, and the oranges are in great repute, and are exported in large quantities.   The land is often owned by the cultivators, but when a property is large, the *métayer* system generally prevails, the tenant paying half the produce.

GEOGRAPHICAL NOTES.

By The Acting Editor.

EUROPE.

The Climate of Edinburgh in 1895.—In March Mr. R. C. Mossman read before the Royal Physical Society a report of his observations during last year, which appears, with tables, in the Society's *Proceedings*.   The monthly values were compared with averages for periods ranging from 25 to 132 years.   When the accumulated mass of meteorological data have been reduced, means for over 100 years will be available for most of the elements, but the work cannot be completed for some time.

Considerable divergences from the normal occurred in the year under consideration.   The great frost beginning in January attained its greatest severity in February, the mean temperature of the first three weeks being 28·2°.   The absolute minimum was 11·9° on the 8th, the lowest reading in February since 1823, while the average temperature of the month was 31·2°, being 7·8° below the fifty years' mean.   Since 1764 only two colder Februarys have been experienced, that of 1838 with a mean of 29·8°, and of 1855 with 30·6°.   From December 29th, 1894, to March 13th, 1895, the maximum temperature was only 46·3°.   The latter month was the dullest for seven years, and owing to the large amount of cloud the temperature by day was 1·0° above the normal, while the nights were 1·8° warmer than usual.   Influenza was very prevalent during the month, and many children fell victims to measles.   July was marked by some days of very low temperature and heavy rains, 1·63 inches falling on the 26th, and an average of an inch per diem on the three days ending with the 27th. September, on the other hand, was unusually fine and warm, the mean temperature being 58·6°, which since 1764 was surpassed only by the September of 1846, when the mean was 59·3 ; and the maximum, noted on the 25th, 78·3°, was the highest recorded so late in the season during the past fifty-five years. November also was mild, but wet and stormy ; the thermometer never fell below 33·3°, the highest minimum for the last fifty-five Novembers.

Weather and Mortality in Edinburgh.—Mr. R. C. Mossman has compiled tables of the meteorological records and the mortality during the seventeen years 1878-1894. Some diseases have a much closer connection with the weather than others, and climatic changes take some little time to affect the constitution. As, moreover, no information as to the duration of the illness is obtainable, only the most general results of the inquiry can be discussed. The mortality from all causes is high from about the middle of November to the end of January, and is low from the beginning of July to the end of September. During the former period temperature falls rapidly, with a considerable increase in rainfall and humidity, while the variability of temperature is at a maximum. The period of low mortality is marked by high temperature, low humidity, and small variability of temperature. During the four weeks ending in the middle of February the death-rate falls rapidly, but further improvement is retarded until the end of May, which confirms the general impression of the prejudicial effect of easterly winds.

Infant mortality has its primary maximum from November to January and a secondary maximum, due to diarrhœa, in August ; the absolute minimum is reached about the end of June. The maximum and minimum of deaths of persons sixty years old and upwards occur at the same seasons as those of the total mortality. The winter maximum seems to be due to respiratory complaints.

The various diseases are also discussed separately or in groups, and a table of curves is appended to the paper, the parts of the curves above the mean line being coloured red and those below blue.—*Proceedings of the Royal Society of Edinburgh*, vol. xxi. No. 2.

Hydrographic Work in Switzerland.—Professor Paul Chaix writes :—" Observations have for the last twenty years been made in the fall, depth, width, discharge, and fluctuations of the rivers of Switzerland by means of a network of stations spread over the country and connected by a very careful series of levellings.

" The records of the measurements taken have been published by the Federal Hydrographic Office in half-yearly tables. The reduction of these records, for the purpose of exhibiting the variations in the volume, etc., of the rivers, will be a long and difficult task. A special report [1] has just been issued on the part of the Rhine basin between the sources of the river and its confluence with the Tamina, near Ragaz. This is one of the fourteen hydrographic regions into which our country has been divided, and it is provided with fifteen stations, the altitudes of which were measured by the late General Dufour, the standard level being a horizontal line, marked N.F.R.P.N., on a colossal granitic boulder rising in the harbour of Geneva, and called the Pierre du Niton. The height of this line above the level of the sea has been given as 376·86 mètres (1236·40 feet), but it will probably be corrected to 373·54 (1225·51 feet), and even this figure can only be provisionally accepted as long as the neighbouring countries have not completed the levelling of their territories and come to an agreement on the mean sea-level.

---

[1] A copy of this report has, through the kindness of Professor Chaix, been presented to the Society's Library.

"The depth of the water in the Rhine has been carefully measured in cross sections at every mètre of its width, and also the banks to a sufficient height to allow of the changes in the area of the cross sections being calculated for any given rising or falling of the level of the stream.

"The engineers have also measured the extent of the whole basin of the river, including the drainage areas of the most insignificant tributaries, for this purpose drawing on maps contours at intervals of 300 mètres (984 feet). The whole surface reduced to standard level (the inclination being, therefore, neglected) of the part of the basin in question has been found to be 1720 square miles, of which 393, being more than 2400 mètres above sea-level (7874 feet), may be considered to lie within the region of eternal snow. Of the whole basin 64 square miles are actually occupied by glaciers and the snow-fields that feed them, 401 square miles are covered with rocks and shingle, 302 with forests and plantations, and 4 are occupied by diminutive lakelets, leaving only 951 square miles of poor soil for agriculture.

"By the patient labour of the engineers the surfaces, elevations, and watersheds of 94 tributaries of the Vorder Rhein have been ascertained, of 100 streams feeding the Hinter Rhein, and of 79 affluents of the united river between Reichenau and Ragaz—in all 273 over an area of 1720 square miles. Of these only the Medelser Rhein, the Glenner, Plessur, Landquart, and Tamina are of any importance. The basin of the Vorder Rhein covers 585 square miles, of which 49 belong to the Medelser Rhein and 147 to the Glenner. To the south lies the Hinter Rhein, with a drainage area of 653½ square miles, while the Plessur, the Landquart, and the Tamina drain 102, 238, and 57 square miles respectively.

"The measurements of the two main branches of the river show that they have a breadth varying from 115 to 140 feet, and maximum depths of 22 to 39 inches only. Below their confluence at Reichenau the breadth of the Rhine is from 138 to 236 feet, and its maximum depth 23 inches to 6 feet 7 inches.

"While freely acknowledging the conspicuous services that the engineers MM. Epper and E. Rod have rendered to science, I am much grieved that their immense labour should be connected with an obnoxious scheme for converting the streams with which Nature has endowed the Swiss Cantons into a socialistic financial monopoly, whereby the free Swiss will be compelled to part with his birthright, and his sublime Alps will be lowered to the level of the Ural, the gate of a prison.

"In the Geographical Department of the National Exhibition at Geneva, which will shortly be closed, is a series of tables compiled by the Meteorological Service of Switzerland. The central office at Zürich is kept in touch with the meteorological departments of other European States, and the records received are checked by M. Billwiler, and after comparison with the observations of Geneva, Bâle, Zürich, Einsiedeln, Lugano, and Sion, are published daily in the Journal de Genève.

"The energetic director of the office at Zürich has contributed to the exhibition at Geneva a series of comprehensive tables bearing on all the elements of meteorology. The mean monthly temperatures are here shown for thirty years, 1864 to 1893, at specially selected stations, of which the chief are Lugano, Geneva, Zürich, the Chaumont above Neuchâtel, Sils-Maria between Val Bregaglia and the Engadine, and Mt. Sentis. The highest mean, 50·5° F., has been recorded at Lugano, 902 feet above sea-level, and the lowest, 36° on Mt. Sentis (8202 feet). Geneva's mean, 49°, surpasses that of Zürich, 47°.

"Another table shows the rainfall at Sion, Geneva, Bâle, Einsiedeln, and Lugano. The heaviest monthly rainfalls are 4·4 inches at Geneva, and 8·1 inches at Lugano. June, July, and August are the wettest months in Einsiedeln and Zürich, while Geneva receives most rain in October. The dry season occurs at about the same time over the whole country, the minima being recorded in January. Geneva, the Engadine, Sion, and Sierre, with 2·4 inches, are the driest parts. The Breithorn, the Jungfrau, the Finsteraarhorn, and the Titlis have a monthly fall of 8 inches or more.

"The daily insolation, and the direction, frequency, and velocity of winds and storms are also exhibited on the tables."

ASIA.

The Red Sea.—Professor Luksch has published a further report on the work of the *Pola*. Three main lines of soundings, along the courses of the submarine cables, have appeared on the British Admiralty charts. 103 new soundings were taken by the *Pola*, 57 in the Red Sea itself, 7 in the Gulf of Suez, and 39 in the Gulf of Akabah, where none had been made before. With the material obtained a chart has been constructed with isobathic lines of 100, 270, 550, 820, and 1100 fathoms. Between Ras Mohammed, the extremity of the Sinai peninsula, and the latitude of Jedda there are two depressions of over 550 fathoms, separated in about lat. 25° 30′ by a ridge rising to 320 fathoms. The northern depression, about 160 miles and 20 to 40 broad, has a maximum depth of 635 fathoms, in lat. 26° 8′ N. and long. 35° 27′ E., and slopes steeply up to 40 fathoms at the entrance of the Gulf of Suez and 70 fathoms in the Straits of Tiran, leading into the Gulf of Akabah. The Gulf of Suez slopes gradually down, but its depth nowhere exceeds 45 fathoms, whereas the Gulf of Akabah contains an area below the 500 fathom line measuring 50 miles long by 10 broad. The other large depression extends south of Jedda and beyond the region visited by the *Pola*. It varies in width from 20 to 60 miles; the deepest sounding made in it was 1200 fathoms. Depths of 100 to 300 fathoms were found everywhere just outside the coral reefs. Professor Suess holds that the soundings of the *Pola* prove that the Gulf of Akabah is a continuation of the Jordan fissure, helping to connect it with the "East African Channel" south of the Red Sea. The temperature and salinity observations of the *Pola* supplement those published by the Meteorological Office. The transparency of the water was on the whole low, the discs ceasing to be visible at depths of 130 to 170 feet.—*Geogr. Journal*, Oct.

The Marine Survey of India.—In April 1895 the *Investigator* and *Nancowry*, having completed the survey of Palk Strait, sailed for Bombay, where they were laid up for the summer and repaired. In October they started for Karachi, and the survey of the coast from that port to Cape Monze was taken in hand, the last survey having been executed as long ago as 1848-49. Later on work was commenced at the mouths of the Indus, from Hajamro to the neighbourhood of Goria Creek—a distance of ninety miles. Many changes were found to have occurred since last year in the direction of the deepest water over the bar of the Kediwari mouth. It seems that this mouth was opened only eight or nine years ago; previously the chief mouth was about fifteen miles farther northward. Enormous changes were also met with near the Kakaiwari mouth, which in 1877 was one of the principal outlets of the waters of the Indus. At present the principal discharge takes place ten miles east of the former Kakaiwari mouth, and the spits opposite the latter were found to have disappeared, while farther east a deposit of mud, sand, and clay had been laid down over a large bay fifteen miles long by six broad at the widest part, the depths over it varying from two to eight fathoms.

On leaving the Indus delta Commander Oldham proceeded to the Swatch, a singular deep gorge off the mouths of the Indus. This depression, running from south-south-west to north-north-east, is connected with the deep water beyond the 100-fathom line, and gradually shoals inwards from a depth of 600 fathoms. On the east it is bounded by a ridge of hard sand thirteen to fifteen fathoms below the surface, and on the west by sand and stiff mud at a depth of forty to forty-five fathoms. The bottom is covered with fine soft mud.

Having returned in February to Bombay, the *Investigator* was employed in the re-survey of the harbour in the neighbourhood of Karanja, while the *Nancowry* proceeded to Thana Creek and Bassein Creek.

The surgeon-naturalist, Mr. A. R. S. Anderson, gives the following details of the trawlings :—

Stations.	Position.	Depth in Fathoms.	Deposits, Bottom and Surface Temperatures.		
199	8° 40′ N., 81° 27′ 35″ E.	800-637	Green Mud.	41°—42·5° F.;	83·5° F.
201	8° 29′ 5″ N., 81° 31′ 35″ E.	320-296	do.	49·0°;	84·5°
202	7° 4′ 40″ N., 82° 2′ 45″ E.	695	do.	42·0°;	85·0°
203	5° 50′ 30″ N., 80° 25′ 30″ E.	364	do.	48·0°;	85·5°
204	6° 50′ 20″ N., 79° 36′ 20″ E.	186-817	Broken Coral.	53·0°;	84·8°
211	23° 0′ 0″ N., 65° 6′ 8″ E. (?)	609-620	Grey Mud.	46·5°;	75·0°
212	21° 43′ 45″ N., 68° 0′ 0″ E.	111	Pteropod Ooze.	63·5°;	74·0°

At station 199 *Eucopia sculpticauda*, new to the Indian fauna, having been obtained previously only from deep water in the eastern Pacific Ocean, was collected, and at station 204, *Trichopeltarion ovale*—a new species—and *Oxypleurodon stimpsoni*, new to the Indian fauna, were brought up by the trawl, as well as a beautiful strawberry pink Alcyonarian of a species hitherto undescribed. An *Acanthephyra* caught at station 203 showed no signs of life, and is probably a true bottom dweller.

The use of glycerine instead of alcohol for the preservation of marine specimens was tried, and it is expected that the colours of deep-sea organisms will be better preserved in this antiseptic.

At Poona charts were drawn of Palk Strait and of the Indus delta coast, and tides and high-water were tabulated for Kankasanturai. A tracing was also made showing the connection of the *Investigator* triangulation in the neighbourhood of Palk Strait with that of the Ceylon Government.

The Fauna of Borneo.—At the International Zoological Congress held at Leiden in September 1895, Herr J. Büttikofer stated that during his expedition in the Kapuas basin he had obtained hardly any new forms ; the fauna was really identical with that of Sarawak and other parts of Borneo. Very striking is the preponderance of tree-dwellers among the mammals of Borneo, including not only *Quadrumana*, flying animals and squirrels, but also the numerous varieties of tree-shrews and several beasts of prey. Of sixty-six species observed by Herr Büttikofer, fifty-two are tree-dwellers. This large proportion of tree-dwelling mammals, which does not occur in the same latitude either in the neighbouring Celebes, or in Africa or America, cannot be ascribed to animals of prey living on the ground, for there is a remarkable scarcity of such, even the tiger and leopard being absent. Herr Büttikofer believes that it is due to the circumstances that almost all the island is covered with forest, and that the great plains are almost yearly inundated by the rivers.—*Globus*, Bd. lxx. No. 14.

AFRICA.

Futa Jallon.—This country, now placed under the direct control of the French authorities, lies at the back of the territory known as the Rivières du Sud. Sometimes the mountains of Futa Jallon are opposed to the plains of the Rivières du Sud. But this is an incorrect representation of the country, for the mountains, at many points at any rate, extend to the coast. The mountains of Sierra Leone attain a height of 2300 feet, while the peak of the Kakoulima, near Konakry, rises to 2600 feet above sea-level. These elevations extend as far as Cape Verga, beyond which the great plains of Senegambia stretch northward to the desert. The boundary between the region of plains and the uplands of Futa Jallon runs from Cape Verga in a direction N. 30° E., and strikes the Senegal a little above the confluence of the Faleme. It should be remarked that the mountainous district of the coast follows a line running N. 30° W., for a number of physical features follow one or the other of these directions, being connected with conjugate systems of fractures.

The Futa Jallon elevations run in parallel lines from south-east to north-west, separated by hollows which are often of considerable breadth, and are interrupted transversely by narrower fissures in the direction south-east to north-west. Naturally erosion has modified this primitive plan, but it is still easily perceptible, even if only a small area be examined. Hence the courses of the rivers are very capricious, and those of the Senegal and Niger in particular, which describe similar curves in contrary directions. These two rivers may be said to be akin, for the highlands of Futa Jallon are prolonged to the hinterland of Sierra Leone in the south-east, embracing the mountains whence the Niger flows. Here the Daro mountain marks the southern limit of the mass, while its northern is the Linsan in Tamgué. The line joining these mountains lies exactly in the direction N. 30° W., but the watershed lies rather more to the west, at any rate in the central part of the highlands.

The waters of the western slope are collected to form the three large rivers, Niger, Senegal, and Gambia, while on the eastern slope the streams flow directly to the adjacent sea without uniting to form any important waterways. Behind Kakoulima lie plateaus in which the Konkouré, the chief river of the French territory, has excavated a deep channel. Up to Bouramaya, twenty-five miles from the sea, the river flows through flat, marshy country and is navigable by boats drawing ten feet of water; above this point it is obstructed by rapids, and is enclosed by bare vertical walls of rock. These narrow profound channels are characteristic of the other rivers of the country also. The upper course of the Konkouré traverses, in a deep winding valley, the plateau of Labaya, which has a mean elevation of 1500 feet, and culminating points of 2600. On the south-west the plateau descends to the bank of the Badi, an affluent of the Konkouré, sixty-five yards broad and sixteen to twenty feet deep even in the dry season, but much encumbered by rocky bars.

The highlands are composed of granitic and schistose rocks; sandstone occurs only in strips along the edge. On the high parts hard furrowed laterite, impenetrable to the roots of plants, covers the ground, and therefore the plateau is bare. In the valley of the Badi, however, the vegetation is as luxuriant as on the coast. A few patches of fine forest are also met with in the valley of the Konkouré on the Labaya plateau.

Beyond this plateau sandstone appears again, forming in the great plain of the Tené and Bafing tabular mountains with precipitous flanks. Towards the north-east the plateau of Labaya is connected with those of Bentaya and Kahel, which

have an altitude of fully 3300 feet. Except the difference of altitude, these
plateaus have all the characteristics of the Labaya. The Kakrima, like the
Konkouré, has cut out a deep trench. This river, rising in the neighbourhood of
the sources of the Gambia and Rio Grande, is certainly the most important artery
of the Konkouré system. The main stream lies more to the south-east, having for
its principal feeders the streams descending from the heights to the north of Teliko,
where rise also the Senegal and an important affluent of the Little Skarcies. The
altitude is moderate—barely higher than the Labaya plateau. Here is the border
of the comparatively low belt separating the high terraces of Futa Jallon from the
plateaus of Falaba and Limba, which affords an easy passage to the Niger. A route
from Konakry across this country would be the shortest line to reach the great
river of the Sudan from the coast.

These highlands of Futa Jallon are comparatively healthy, and, being near the sea
and at no great distance from Europe, are well adapted for colonisation.—*Tour du
Monde*, September 19th.

AUSTRALASIA.

Geography in Australia.—On July 22, the Queensland Branch of the Geographical
Society of Australasia held its anniversary meeting, when Mr. G. P. Thomson, the
President, delivered an address on Geography in Australasia. The first of the four
branches of the Geographical Society of Australasia was founded at Sydney in the
year 1883 chiefly through the exertions of Mr. E. Marin La Meslée, one of the warmest
advocates of the exploration of British New Guinea. About the middle of the
same year the Melbourne branch was established, the promoters being Baron Sir
Ferd. von Mueller and Mr. A. C. Macdonald, who have also directed the subsequent
operations of the Society. In 1885 an expedition, organised by the Sydney branch
and equipped at the cost of New South Wales, Victoria, and Queensland, was sent
to British New Guinea under the command of Captain H. C. Everill, and in the
same year the branches of the Society in Adelaide and Brisbane were founded, the
latter through the exertions of Mr. Thomson, who was for many years its Hon.
Secretary. The branch at Adelaide has done some good work, especially in connection with the Elder Expedition, while the Brisbane branch justly prides itself on being
the most active of all. Though receiving no assistance from the Government, as is
the case with the other branches, it has held meetings regularly during the eleven
years of its existence, published *Proceedings*, and promoted geographical investigations and education. It has been fortunate in receiving encouragement from the
Governors of the colony, Sir Henry Norman having shown a lively interest in its
work, while Lord Lamington, like the President a Fellow of our own Society, has
already shown his desire to promote its usefulness.

As Mr. Thomson pointed out in his address, there is still much work to be done
in Australia. Expeditions have of late years been sent out to the little-known
districts of South and West Australia. But it is not so much information concerning the topography of the country that is wanted, especially in Queensland, as
investigations into the available areas for agricultural and pastoral industry and a
more thorough examination of the mineral resources. The geographical societies in
Australia will therefore find useful work in collating and disseminating information
of this nature, and in promoting geographical education in schools and colleges.
The Queensland branch has endeavoured to encourage the study of geography in
the schools of the colony, and has urged on the other branches the necessity of an
intercolonial conference to consider what steps should be taken with this object,
but unfortunately has not been supported except by South Australia.

Mr. Fitzgerald in New Zealand.—At the meeting of the British Association in Liverpool Mr. FitzGerald read a paper on his experiences in the Southern Alps. Of the general conditions of mountain climbing in that country he remarked :—"In New Zealand the mountains themselves are not very high, and the snow-line is, on the other hand, much lower than in the Himalayas and Andes, so that the condition of things is in a way similar to Switzerland. The glaciers are larger than those in Europe, and, owing to the excessive rainfall, always in a very active and dangerous condition, while the rocks are all of a very crumbly structure. More than this, the prime requisite of every climber, effective and trustworthy porterage, is simply unavailable. A mountain is ascended by a gradual advance, and therefore a supply of food and fuel, from which to start or to which to return, to recuperate one's strength if successful, or, if defeated, for refreshment and a new start, is absolutely essential. In our Antipodean Colony the porter is simply not to be had. Those who are prepared to risk their necks climb for their own pleasure ; those who are not prepared to do so refuse to climb for anybody else's pleasure. I am not, I need scarcely remark, of herculean build, and I find it quite enough to carry my ten stone of humanity into the rarified regions you pay me the compliment of desiring to hear about to-day. I am comforted, however, by the reflection that my guide and friend Zurbriggen, who is the strongest man I know, also shares my disability to combine the duty of pack-horse and polar bear. In New Zealand, therefore, and it is a matter which future climbers should bear in mind, one must either carry one's own food or go without. I chose the latter alternative, and Zurbriggen and I left our camp near the Hermitage for the ascent of Mount Sefton (our most difficult climb in New Zealand) with exactly one box of sardines, a few biscuits, and half a pound of tallow candles."

Mr. FitzGerald then described his ascent of Mount Sefton :—"Our starting point," he said, "was a bivouac about five thousand feet above the sea-level. We left soon after midnight on February 14th, resolute but scarcely hopeful after our five unsuccessful attempts. The bright moonlight revealed a somewhat appalling prospect. Huge masses of glacier, broken into tottering *séracs* and precipitous ice-falls towered above us in wild and seemingly impossible confusion. We advanced rapidly, however, choosing our route among numerous obstacles, and reaching by daybreak the base of the great rock *arête* which must necessarily form the final ascent from the west side. Here we halted for a moment, and braced ourselves for the struggle by consuming our sardines and biscuits, and thus strengthened and refreshed, proceeded to cope with as rotten and crumbling a wall of rock as ever I expect to come across. From this point until we stood on the summit never once were we secure.

"Here the ledge upon which we stood would crumble beneath our feet : there the very rock we grasped in our ascent would give way under our hand : at one moment the whole *arête* would quake ominously under our advancing tread ; or again a stone, whizzing by us from the dizzy height above, would force us to dodge from its path and cower under some friendly ridge till the peril was passed."

It was during this ascent that the accident happened with which all readers of Mr. FitzGerald's book must be familiar. A boulder, giving way under Mr. FitzGerald's hand, dashed him from his position on the face of a precipice, and but for Zurbriggen's strength and presence of mind both climbers would have fallen down some six thousand feet. At about 10.30 A.M. they stood on the summit of the mountain, where they discovered a route across the

Alps. On this subject he remarked :—"In ascending this peak from the Hermitage I was leaving behind me all that arid tableland known as the Mackenzie country, which the great ranges then stretched out beneath my feet divide from the west coast, a country of astonishing fertility and luxuriance, but almost wholly cut off from communication by these mountains and an unapproachable coast. Standing on the summit of Mount Sefton that day I descried a practicable passage in this barrier, the pass to which the Government of New Zealand has done me the honour of giving my name, and which I trust may in the future facilitate the opening of these new and rich western valleys to travellers. I have not time to-day to describe how Zurbriggen and I, ten days later, crossed the ranges by the very route we now traced out for ourselves, or how we spent two and a half days without food in the Copeland valley, struggling with the thick bush and creepers."

The following extract explaining Mr. FitzGerald's plans for the future is interesting :—"Impelled by the scarcity of unclimbed European peaks, I leave in three weeks for South America to try to climb the Mountain Aconcagua, which rises to a height of nearly 23,000 feet, and is the highest mountain in America. In fact, outside the Himalayan range,[1] it is the highest mountain in the world. Our plan is to proceed by boat to Buenos Ayres, and thence to Mendoza. From there we shall strike up towards Cordilleras de los Andes. Our party will consist of my friends Mr. Barrow, who was with me for part of the time in New Zealand, Mr. de Trafford, Mr. Stuart Vines, and Mr. Philip Gosse. My old guide Zurbriggen will accompany us, together with four porters from Switzerland. We are taking our provisions and equipment ready packed in our panniers, so that when we arrive we shall only have to buy mules, and as there are ten of us, we hope to be independent of local help. My idea is to cover as much of the country as possible, to ascend several peaks, and to bring back as many geological, botanical, and zoological specimens as possible. I hope to make a rough triangulated survey of the country I go through, and take many photographs. It is my intention to ascend Aconcagua by gradually moving slowly upwards and establishing several camps.

"We have spent this summer in Switzerland, practising heliographic and flag signalling, and therefore, by leaving one of the party at each camp, we expect to keep up communication and to facilitate the supply of provisions ; while by the same means we hope to report our ascent to London immediately on reaching the summit, should we be successful."

GENERAL.

Anthropology at the British Association.—Anthropology, as far as it deals with the spread of nations over the world and the gradual formation and changes of trade routes, is a subject closely related to geography. On the whole, the papers read at the late meeting of the British Association in Liverpool were rather of an archæological than a geographical character, and, therefore, few of them can be quoted from at length. The President of the Section, Mr. Arthur J. Evans, in his opening address, dealt with "The Eastern Question in Anthropology." Till recent years, he remarked, the glamour of the Orient pervaded all inquiries into the origin of European civilisation. Recent investigations have brought about a reaction. Stricter application of phonetic laws and the increased detection of loan-words has

---

[1] And the Kuen Lun, where Mr. Littledale recently measured summits 25,000 feet in height.—ED.

reduced the Aryan stock of culture to very narrow limits, and entirely stripped this linguistic family of any trace of a common Pantheon. If the Aryan existed as a common ancestor of the Indo-European races, it must have been far back in the mists of the Stone Age. The Phœnician has met with the same fate ; it is no longer believed that he landed on the coast of Cornwall, and built the dolmens of the north and west. Primitive trade took the form of inter-tribal barter, rather than of commerce between remote localities. It has been shown that in all parts of Europe centres of primitive metallurgy existed, and that the implements and utensils of the Bronze Age were not imported wholesale by Semites or " Etruscans."

Mr. Evans then turned to the early Ægean culture and especially to Crete, where he has collected evidence of an early and intimate connection with Egypt, going back to the Twelfth Dynasty and to the earlier half of the third millennium before our era. In these early times a Chaldæan influence cannot certainly be excluded from Ægean art, but whatever was borrowed was developed into new forms. Before the days of Phœnician contact, a system of writing had been worked out in Crete, which the Semite only carried one step further.

The discussion which followed on the second day of the meeting was chiefly directed to the identification of the people who produced the Mycenæan works of art and to the determination of dates. Dr. Montelius mentioned that Mycenæan pottery was known in Italy as far back as the fifteenth century B.C., as proved by the presence of scarabs with the name of Amenhotep III., and Prof. Flinders Petrie referred the beginning of the use of copper tools in Egypt to 3500-3000 B.C., while iron, he said, was worked in Egypt and the Assyrian highlands as far back as 1000-600 B.C. The source of the Iron Age in Europe was stated by Prof. W. Ridgway to be Hallstatt in Austria, where alone articles of iron are found gradually replacing those of bronze. Close by was one of the most famous iron mines of antiquity ; Strabo mentions the ironworks of Noreia.

Mr. J. L. Myres read a paper on " Sergi's Theory of a Mediterranean Race," supposed to have given rise to four chief stems—the Iberians of the Peninsula, the Ligurians of Italy, the Pelasgi of Greece and Asia Minor, and the Libyans of North Africa. The centre from which this race expanded over Egypt and Europe is supposed to have been situated in the upper valley of the Nile, among the Abyssinians, Somalis, and Gallas. Another paper read by Mr. Myres was of more geographical interest than most that were read in the Section. Its title was " Cyprus and the Trade Routes of South-Eastern Europe." The Copper Age seems to have begun in Cyprus before the end of the Stone Age elsewhere ; the types were very simple, and were associated with pottery imitated from gourd bottles and ornamented with incised lines filled with white earth. Both the types of instruments and the pottery have been found at Hissarlik and in Central Europe. The route seems to have been overland, and hence the importance of Hissarlik as commanding the ferry of the Dardanelles.

Mr. G. L. Gomme and Professor E. Odlum dealt with more modern subjects, the former discussing the value of folklore in ethnology, as shown in particular by survivals of fire-worship in the British Isles. He maintained that when any given custom and belief, having undergone the double process of analysis of component elements and classification of individual examples, reveals a distinct parallel between its radical elements and the elements of a custom or belief occupying a place in the cultus of a barbaric or savage people, its right to a genealogy dating back to a prehistoric cultus may then be discussed. Mr. Gomme then compared the rites and ceremonies practised in Britain at Christmas, New Year's Eve, and Easter with the system of fire-worship belonging to early Aryan tribes elsewhere.

Professor Odlum described the customs of the "Coast Indians of British Columbia and Alaska," and mentioned the Tsimphians in the vicinity of Port Simpson as the most wonderful linguists he had ever met.

A report was read of the committee appointed to direct an Ethnological Survey of Great Britain and Ireland, in which great progress has been made. Mr. Read, keeper of the Ethnological Department of the British Museum, proposed the establishment of an Imperial Bureau of Ethnology, referring to the excellent work done by the American Bureau. This proposal, as well as one by Professor Flinders Petrie for an ethnological museum, was received with approval, but both may for the present be considered rather as counsels of perfection.

The Height of the Atmosphere and the Zodiacal Light.—Of the fall of temperature in the higher regions of the atmosphere nothing is known from observations. As far as our knowledge reaches, the temperature falls about one degree Fahr. for every 273 feet. Now the upper surface of the atmospheric envelope must be absolutely cold, that is, the air must have lost 460 degrees of heat in cooling from the zero of our scale. Air, therefore, which leaves the earth at a temperature of 80 degrees loses 540 degrees as it ascends to the upper surface of the atmosphere, and, taking one degree as the loss for every 273 feet, we obtain about twenty-eight miles as the height of the atmosphere. This figure may be incorrect for various reasons. In the upper regions, where clouds are absent, the decrease of temperature must follow the law of a mixture of oxygen and nitrogen only, and, further, the composition of the atmosphere may be different in those regions. In pure hydrogen, temperature diminishes about fifteen times slower than in our atmosphere.

Again, the surface of the atmospheric envelope may be disturbed by undulating movements as the hydrosphere is. Herr Gruson has lately suggested, in a book of travel, that such tides may be caused by the attraction of the moon, precisely as tides are caused in the waters of the ocean, and he considers that the zodiacal light is a prolongation of twilight due to the greater height of the atmosphere when the tidal wave is passing the meridian. The sun may also produce waves, but they must pass the meridian at midday and midnight, whereas at new and full moon the ebb of the lunar tide, and at the quarters the crest of the tidal wave, is found in those parts of the Earth where there is twilight. Herr Gruson recorded, in Cairo, fifty-five minutes twilight at full and new moon, and ninety-five minutes at the quarters. These figures give approximately for the heights of the atmosphere twenty-nine miles and eighty-eight miles respectively. The latter value is surprisingly large, while the former does not differ much from the height calculated above.

In discussing the theory in *Globus*, Bd. lxx. No. 14, Professor M. Möller first considers the point that these waves have not been noticed within the strata of air accessible to observation, and shows that the movement of the tidal wave at the Equator, travelling nearly 1500 feet per second, would extend only to about $26\frac{1}{2}$ miles from the surface, and therefore would not descend to the Earth. Some modification would be caused by the elasticity of the air, but even then the undulation at the Earth's surface would be too weak to be detected.

The development of large waves in the seas and oceans is restricted by various causes. The depth is in some seas too shallow to allow large waves to be developed, or they are stopped by the coasts before they have had time to attain their full dimensions. But the conditions of the atmosphere are very different. Here there is depth enough for the tidal wave to be perfectly developed, and no shores to check its progress. Yet it may be that these movements sometimes descend near to the surface of the Earth, for the tidal wave, caused by the sun, ought at the Equator to reach a little lower than that due to the moon, and the combined action of the two bodies may, to some extent, disturb the atmosphere to a depth of more than twenty-seven miles. Possibly, also, a secondary wave may be produced, however imperfectly, in the lower strata.

In this way the theory of wave formation can explain many of the circumstances of these movements, but it has not data on which to base calculations of the height of the waves on the surface of the atmospheric envelope, the velocity of the vibrations of particles in the upper regions of the air being entirely unknown. Certainly the fifty-nine miles given by Herr Gruson as the height of the crest of the wave above its trough seem excessive. Only careful observation can decide the point, and it is desirable that Herr Gruson's theory of the connection between the appearance of the zodiacal light and the position of the moon should be tested by a long series of observations, and that attention should be paid to the incandescence of meteorites, which, in accordance with the theory, ought to take place at varying heights.

The Colour of the Water of Lakes and Seas.—Nearly fifty years ago Bunsen stated that pure water, when sufficiently deep, has a blue colour. But when Tyndall proved that the blue of the sky was not necessarily the colour of the gases of the atmosphere, but might be caused by the reflection of the light of the sun by minute transparent particles, and Soret and Hagenbach showed that the blue light emanating from certain lakes was polarized, like the light from the sky, it was supposed that the colour in both cases might be due to the same cause. Professor W. Spring considers this opinion erroneous, believing that water itself is blue. If the colour were due to impurities only, physically pure water would reflect no light, and a lake or sea sufficiently deep to absorb all luminous rays would appear black and not blue; but the blue of the Mediterranean is purest and most intense in its deepest parts, and the same holds good of the Lake of Geneva. To experiment on the transparency of water and its illumination Professor Spring had two tubes made, each 26 mètres (85¼ feet) long, which, if required, could be placed in line so as to form a column of water 170½ feet long. He found that distilled water does not contain sufficient particles to alter its transparency in a stratum of 85¼ feet, and, having admitted the light from a gas jet into the tube at a certain point, he ascertained that the water was illuminated, but only to a distance of about 6½ feet from the source of light. He concluded then that the illumination is not entirely due to the presence of material particles in suspension. Professor Spring then arranged experiments to show the influence of convection currents on the transparency of the water, and the results convinced him that a lake of pure water will appear luminous and of a blue colour as soon as convection currents are set in motion, and that the colour will be more brilliant as the currents increase in intensity, while particles tend to turn the blue to green because they more readily absorb the most refrangible rays. This conclusion is in harmony with the observation of Professor Forel, that fresh water lakes are more transparent in winter than in summer, for in summer the difference of the temperatures at the surface and bottom is greater, and the agitation due to various causes prevents the layers of different densities from lying uniformly one above another, so that the convection currents, following no definite lines, diffuse the light in all directions. Professor Forel explains the phenomenon in accordance with the accepted theory, supposing that the water in summer holds a greater number of particles in suspension than in winter. To this Professor Spring objects that the denser water in winter would be more apt to hold particles in suspension. He does not, however, reject Professor Forel's explanation altogether, but considers that his own experiments show that other causes may be at work, and that the phenomena observed in lakes are not as simple as generally supposed.—*Ciel et Terre,* August 1st.

MISCELLANEOUS.

The production of amber in Germany during the year 1894 was 220 tons, being about fifty tons more than in the preceding year. Most of it came from the mines of Palmnicken and Kraxtepellen, six tons only having been dredged up along the coast.

Mr. George Gould, son of the deceased millionaire, intends to send out next year an expedition to the Arctic regions with the object of reaching the North Pole. His plan is to divide the distance into stages, and to continue the work year after year until the goal is reached.

The *Bulletin of the United States Geological Survey*, No. 131, contains Mr. Newell's report on the work of the Division of Hydrography during the years 1893 and 1894. Tables are given of the height and discharge of a number of rivers, and particulars concerning wells are added, besides information on a variety of subjects connected with irrigation.

An expedition under the command of General Glukhovskoi has left St. Petersburg for the purpose of examining the country between the Amu Daria and the Caspian. If the commission comes to the conclusion that the river can be led to the Caspian along its ancient bed at a less expense than would be incurred by the construction of a railway between Orenburg and Tashkent, the latter scheme will be abandoned.

The Committee of the British Association for the Climatology of Africa believes that numerous observations still remain unpublished, and therefore invites all residents in African countries to communicate original registers of meteorological observations, or copies published in newspapers or abstracts, which may come under their notice. "Instructions for Observers" and blank registers are supplied free of cost on application to the Secretary at Burlington House.

Professor Marinelli, in a note presented to the Instituto Veneto, remarks that in most text-books of geography the authors give Monte Croce, a well-known pass in the Carnic Alps, as the most northern point of Italy, without distinguishing between the two passes of the same name. One of them, Monte Croce di Comelico, lies in lat. 46° 39′ 3″ N., and the other, Monte Croce di Carnia, in lat. 46° 36′ 3″ N. It now appears from a map of Comelico Superiore surveyed in 1889, but not published till some years later, that the Cima di Vanscuro, 8789 feet high, lies in lat. 46° 40′ 8″ N., and is therefore the most northern point of the kingdom. —*Boll. della Soc. Geogr. Italiana*, vol. ix. Fasc. 9.

Dr. Sven Hedin has discovered ruins of large towns buried in the sand to the east of Khotan, between the Keria and Shahiar rivers. He calculates that they must have been overwhelmed about one thousand years ago by the encroaching sands. One of these towns extended for a distance of two and a half miles, and contained a large number of houses. The framework is of wood, and is filled in with reeds daubed over with clay. The walls are covered with white mortar, on which are painted human figures, horses, dogs, and flowers, with a skill indicating considerable culture among the inhabitants. Small images of Buddha, and remains of poplars, apricot, and plum trees were also found.—*Globus*, Bd. lxx. No. 13.

Work has been commenced on the line from Vossevangen to the Taugevand, one of the sections of the railway which is ultimately to connect Bergen with Christiania. Great difficulties are presented in this part of the country, the railway having to ascend from 180 feet above sea-level at Voss to some 4260 at the

Taugevand, or 4080 in a distance of 46¼ miles. The most formidable task will be to carry the line from the Rundal to the Moldaudal, for which purpose a tunnel, 5¼ miles long, will be excavated at an elevation of 2820 feet. The tunnel will, it is expected, be finished in the October of 1903. When this part of the line is completed, there will be no further difficulties on the route to Christiania.

According to *L'Universo* the Prince of Monaco has discovered a bank fifty-five miles to the south of the Azores, where depths of 1000 to 1600 fathoms were expected. On letting down the sounding apparatus only 132 fathoms were found. This bank, which has been named after the *Princess Alice*, has a perimeter of about thirty miles, and two culminating points at depths of 42 and 104 fathoms. Fish was found in abundance, and, as soon as the discovery was made known, four vessels were made ready in the Azores to exploit this source of wealth, which may prove of great value to the inhabitants of the archipelago, as the islands, being simply volcanic peaks, are surrounded by deep waters. Unfortunately the distance is such that only strongly built vessels can fish on the bank.

Up to 1892 the railway system of **San Salvador**, Central America, consisted only of a line some thirty miles long, from the port of Acajutla to the unimportant village of Ateos, about equidistant from the capital and from Santa Ana, the chief commercial centre. In that year the Government became convinced of the necessity of extending the line to Santa Ana, and in July last this section was thrown open to traffic. The branch to San Salvador will probably be completed by next June. The population of the Republic is dense for Spanish America, being 108 to the square mile, or fifteen times the density in Guatemala. The country yields tropical products in abundance, coffee and indigo included. The foreign merchants are steadily growing in numbers, and their energy and capital will promote the future prosperity of the country.— *South American Journal*, Sept. 26th.

NEW BOOKS.

The Water of Leith from Source to Sea. By JOHN GEDDIE. Illustrated by JOSEPH BROWN. Edinburgh: W. H. White and Co., Ltd., 1896. Pp. 220.

Not many even of the great rivers of the earth have had their beauties celebrated in so sumptuous a volume as that in which Mr. Geddie and Mr. Joseph Brown have, with pen and pencil, delineated the charms of the Water of Leith. Yet the Water of Leith is in itself an insignificant, and was till lately a most unsavoury, stream—so much so, that the most famous of its admirers calls it "that dirty Water of Leith." Mr. Geddie says truly that it can scarcely pretend to the name of river, seeing that its sources are barely twenty miles distant from the sea. It is, in fact, a mere burn or streamlet, and as a burn it is surpassed in picturesqueness and beauty by scores of streams both in the Highlands and in the Lowlands of Scotland.

To what, then, does the Water of Leith owe its exceptional fame? Chiefly, it must be said, and that without disparagement to its intrinsic merits, to its proximity to Edinburgh, and to the historical and personal associations thence arising. On its banks there have sprung up thriving villages—Balerno, Currie, Juniper Green, Colinton, Slateford, and Stockbridge, the last now a part of Edinburgh ; and along its course there are many mansions and country houses—Cockburn, Glenbrook, Malleny, Dalmahoy, Ravelrig, Curriehill, Hailes, Redhall, and others, which have been the retreats of eminent lawyers, literary men, and merchants ; while the villages and the mansions have been interspersed with notable centres of industry, such as paper-mills, corn-mills, and snuff-mills. The number of famous characters thus associated with the historic stream is quite phenomenal. The list includes Queen Mary and Bothwell, Lord Brougham and Sir Walter Scott, Lord Cockburn and Lord Jeffrey, De Quincey and Christopher North, Leonard Horner and Henry Mackenzie, David Hume and Sir Archibald Alison, David Roberts and the brothers Lauder, Robert Louis Stevenson, whose country it was in a special manner, and last, though not least, David Masson, who in his retirement has found a congenial retreat in a quaint old house at Juniper Green.

To this extraordinary wealth of interest full justice has been done by Mr. Geddie and Mr. Joseph Brown. It must in truth be said, however, that Mr. Geddie's narrative is much more satisfactory than Mr. Brown's illustrations. A few of the pictures are charming, such as " Harper Rig Farm," and " Haughhead Ford " ; but most of them are disappointing ; while some, such as "The Upper Valley from the Old Lanark Road," are positively bad. To Mr. Geddie's poetical prose nothing but praise can be given. His style is not only well informed and rich in literary allusion, but it is also crisp and sententious, reminding us strongly of the prose-poetry of Alexander Smith, though it lacks the melody and the indescribable charm of Louis Stevenson.

Mr. Geddie seems to us to take liberties, which are scarcely warrantable, with the physical geography of the region with which he deals. For example, he says, parodying the well-known couplet about the common source of the Tweed, the Clyde, and the Annan, that

> " Clyde and Almond, Tweed and Leith
> A' rise in ae snaw wreath."

Now about the common source of the Almond and the Leith there is no question ; but between the Leith and the Medwin, which is the tributary of the Clyde referred to, and between the Leith and the Lyne, which is relied on as the link with the Tweed, nature has interposed the not inconsiderable water-parting of the Pentland Hills. It must also be noted that Bonaly and Redford have no connection with the Water of Leith, but are on the Braid Burn, while Hatton House stands near to the Gogar Burn. Some allowance, however, may be made for poetical licence.

The conspicuous defect of the volume is that it has no map of the region with which it deals. This is the more regrettable because many places mentioned in the text are not noted even in Bartholomew's great *Atlas of Scotland.*

Wild Life in Scotland. By J. H. CRAWFORD, F.L.S. Illustrations by JOHN WILLIAMSON. London : John Macqueen, 1896. Pp. 280. *Price 8s. 6d.*

Mr. Crawford writes pleasantly about wild life in Scotland, with a competent knowledge of natural history, not too obtrusive, and with a very charming poetic feeling. His chapters form a series of prose idylls on "Spring Bird Life," "Early Burn-fishing," "Gulls and Divers," "Marine Mammals and Predatory Fishes," "Loch Leven and Loch Tay," "The Stag," "Grouse and Partridge," and similar subjects. Mr. Crawford's paragraphs are somewhat disjointed, and he indulges in some eccentricities of construction and punctuation ; but on the whole he writes well. For his impeachment of Scottish patriotism on the ground that the Tweed "is not even under the Scots Fishery Board, and is classed as an English river," there is not sufficient warrant. It is quite true that the Tweed "is not even under the Scots Fishery Board," and that for the sufficient reason that the Tweed is not exclusively a Scots river. But it is not the case that it "is classed as an English river." The truth is that the Tweed, being partly a Scottish and partly an English river, is regulated, as regards the salmon-fishing, by special Acts of Parliament, known as the Tweed Acts, under which the Tweed Commissioners form a special board consisting of the riparian proprietors on both sides of the border. As regards legal title, or the right to fish, the Scots law applies to the Scottish bank, and the English law to the English bank. The standard authority on this question is Mr. Charles Stewart's treatise on "The Law of Scotland relating to Rights of Fishing"; and there we read that "The fisheries of the river Tweed, though in the first instance subject to the general Acts of England and Scotland, have been in almost every instance specially exempted from their operation ; and the whole course of the river and its tributaries, in those parts where it forms the boundary between England and Scotland, as well as in those parts where it is wholly within either country, has been placed under special legislation." As a matter of fact, the general Scottish Acts apply to the Tweed, unless that river is specially exempted ; but that is a very different thing from classing the Tweed "as an English river." It must be added that the book is beautifully printed, and that Mr. Williamson's illustrations are admirable.

Annals of Garelochside: Being an Account, Historical and Topographical, of the Parishes of Row, Rosneath, and Cardross. By WILLIAM CHARLES MAUGHAN, author of *Rosneath Past and Present*, etc., etc. With Illustrations by ALEX-ANDER M'GIBBON, Esq., Architect. Paisley : Alexander Gardner, 1897. Pp. xii + 331.

Encouraged by the success of his *Rosneath Past and Present*, Mr. Maughan has enlarged his scheme so as to include the neighbouring parishes, and the result is a large and handsome volume with the general title of *Annals of Garelochside*. The book gives proof of extensive and careful reading, and of some original research. Its production has evidently been to the author a labour of love, and that is certainly the best spirit that can animate the writer of a local history. It has at the same time its dangers, as it is apt to lead the author to go too minutely into details, and to dwell at undue length on matters of secondary importance. Such details, however, will delight local readers, though they may seem tedious to those at a distance. The hand of the amateur is revealed in the loose structure of the style, and also in the compliments to personal friends. A practised writer would scarcely have made Dr. Story a relative of William the Third, or have subjected Sir George Mackenzie to the torture he designed for Principal Carstares (p. 259). Nevertheless, the book is a valuable one, and belongs to a class which we should gladly see extended.

The United States of America, 1765-1865. ("Cambridge Historical Series.") By EDWARD CHANNING, Ph.D., Assistant Professor of History in Harvard University. Cambridge : University Press, 1896. Pp. 352. With three maps.

Dr. Channing is master of the facts of the period of which he writes. Indeed, the only real fault we have to urge against his work is due to the fulness of material presented by him. Interpretative power might be more bounteously displayed. Dr. Channing is too much the accurate chronicler and narrator, too little the living historian ; and his readers will get the impression that he has been even over-diligent in collecting, and has deemed it a duty to present everything—even a "story" disbelieved, yet commented on at disproportionate length (p. 89). As a result, the significant is sometimes merged in the insignificant.

But the book reveals a thorough grasp of the course of events, and it is conspicuously impartial. The balance could not easily be held more fairly between this country and her American colonies at the time of the schism and in subsequent disputes. With regard, however, to the war of 1812, Dr. Channing allows less than he should for party cries and interests in America; and, further, the Atlantic cable would in all probability have made war impossible, as a distinguished fellow-countryman of the author has expressed it. The doctrine of "natural equality" and other doctrines underlying the United States Constitution were commonplaces with certain sections of Puritan politicians, from whom they were inherited more naturally than from either Hooker or Locke, in spite of the special and undoubted influence of the latter (pp. 87-88). The system of lettering adopted in the maps does not commend itself to us as facilitating reference and understanding. The date of Chatham's death is given as 1788 (p. 341) instead of 1778. There are a few grammatical inaccuracies, *e.g.*, in the use of the participle. The Bibliographical Note (pp. 336-341) is good, but the absence of a list of works dealing with industrial and economic growth is suggestive of a want in Mr. Channing's own work. There is no other book published in this country which fills the place ably filled by this volume of the "Cambridge Historical Series."

Bohemia. By C. EDMUND MAURICE. ("Story of the Nations Series.")
London: T. Fisher Unwin, 1896. Pp. 533. *Price 5s.*

"Few countries," says the author, "have been more strangely misunderstood by the average Englishman than Bohemia has been. The mischievous blunder of some fifteenth century Frenchman, who confused the Gipsies, who had just arrived in France, with the nation which was just then startling Europe by its resistance to the forces of the Empire, has left a deeper mark on the imagination of most of our countrymen than the martyrdom of Hus or even the sufferings of our own Princess Elizabeth. The word 'Bohemian' has passed into newspaper slang; and it has been so often quoted in the slang sense by people who ought to be more careful in their language, that it has really hindered the study of the real country which it misrepresents."

As the birthplace of Jan Hus, whom we call John Huss, Bohemia will ever be a centre of interest to lovers of religious freedom. He was born at Husinec in 1369, and the tiny cottage where he first saw the light (and of which a photograph is given in this volume) still remains. In 1403 he was elected Rector of the University of Prague, and during the same year the national reform movement began to connect itself generally with questions of ritual and doctrine. "The exact point in history," says Mr. Maurice, "at which the doctrines of Wyclif gained influence in Bohemia is very difficult to fix. The marriage of Anna, the sister of Wenceslaus[1] in 1381, to Richard II. of England, undoubtedly produced close contact between the two countries." Whilst Huss had studied Wyclif's works and expressed his belief that much good was to be learned from them, it was not till 1408 that the clergy of Prague denounced him as a Wyclifite. From then onwards his enemies, ecclesiastical and personal, pursued him until they burned him at the stake in Constance in 1415.

Mr. Maurice brings his story of Bohemia down to the "fall of National Independence in 1620." At the same time, as Austria acquired Bohemia by marriage, its independence may be said to have been preserved; and the vigorous, if not actually rebellious, steps taken to this day by the Czech party show that they have no idea of losing their distinctive nationality. A society for the cultivation of Bohemian

---

[1] The first of this name, St. Wenceslaus, is immortalised in the English Christmas carol.

literature has existed since 1831, and all attempts to Germanise the country have been stoutly, and to a large extent successfully, resisted. Mr. Maurice states that the Bohemians are equally determined not to be absorbed by Russia.

The Czechs were a Slavonic tribe which acquired such an ascendency in Bohemia that in the ninth century the name "Czech" was commonly applied to the whole Slavic population of Bohemia. At this day, owing to the *double entendre* involved in the name "Bohemian," the term "Czech" is often used for and by the Bohemians. Mr. Maurice is strangely silent on this subject, and does not make any reference to the Czechs. He complains of the injury done to the Bohemians by calling them "Bohemians." The injury is easily remedied by calling them "Czechs," an ancient and honourable appellation.

Madagascar before the Conquest—the Island, the Country, and the People. With Chapters on Travel and Topography, Folk-lore, Strange Customs and Superstitions, the Animal Life of the Island, and Mission Work and Progress among the Inhabitants. By the Rev. JAMES SIBREE, F.R.G.S. With Maps and Illustrations from Photographs. London : Fisher Unwin, 1896. Pp. 382 and Map. No Index. *Price* 16s.

This is the Rev. James Sibree's third book on Madagascar, and is the outcome of thirty-two years' work amongst its people. It is evident, therefore, that he writes with an adequate knowledge of his subject, and does not give us mere travellers' tales. It is a considerable time since a volume of this character came before us for review. It may be taken to be a monograph on Madagascar, and is of great value, as it deals with the physical geography, geology, fauna and flora, as well as the ethnology, of the island. Every page is of interest, and shows that we have not a mere sketch, but a well-considered, well-weighed opinion of a qualified and conscientious observer. In reading the book one is of necessity, as it were, compelled to think of how much real influence mission work has had upon the country. France has, in part at any rate, conquered Madagascar, and we have seen how a State, brought up to a certain pitch of civilisation, can behave when brought into conflict with an European Power ; it is, therefore, of the greatest interest to compare "Madagascar before the Conquest" with what we have heard of the French invasion and its results. Have the self-sacrificing endeavours of the missionaries been of real advantage to the natives ? have they really penetrated more than skin-deep? These questions must arise, and we think that a careful perusal of this book may give, to some great extent, an answer to the question. What that answer is the curious reader will probably find out.

The folk-lore of Madagascar is of great interest, and Mr. Sibree gives in some detail much information on the subject, which may well be compared with the folk-lore in the great African Continent. He shows where the traditions are analogous. Many of the superstitions we recognise as like those met with on the mainland ; some, however, seem to have another origin. The methods of divination used by the people are of exceptional interest ; the old geomantic figures are used, although the plan of using them differs somewhat from that employed in the Middle Ages. With regard to *vintana*, it seems, like the fate of the Greeks and Romans, an invisible power that makes itself felt always and everywhere. The destiny of a man (his *vintana*) depends on what day he was born (partly also on what time of day), or, rather, on what constellation of the Zodiac governed the day of his birth. They have also lucky and unlucky days, and the different months are thought to have their peculiar characters according to the constellations they are named from. Probably *vintana* is really the key to the whole system of idolatry in Madagascar and to everything connected with it.

The interest in the subject is kept up from the first page to the last, and we are sure that no one taking up the volume will lay it down until it is finished. The illustrations are very good and add much to the value of the volume.

Blandt Nordpolen's Naboer. Af EIVIND ASTRUP. Kristiania: H. Aschehoug and Co.'s Forlag, 1896.

The journey made by Peary and Astrup across the inland ice of Greenland is one of the most remarkable in the history of Arctic exploration. In the year 1891 Astrup, being in Philadelphia, read in the newspaper a notice of the intended expedition to Greenland, and immediately offered his services to Lieutenant Peary. At that time he had not completed his twentieth year, and therefore was still quite a young man when, after so much experience amidst snow and ice in high latitudes, he met with such a sad and lonely death in his native country.

In this work he has not attempted to write an exhaustive account of the two expeditions in which he took part, having left the scientific results to be discussed by those who are in possession of the material. Consequently these pages possess attractions for a far wider circle of readers, who may obtain much instruction as well as amusement from them. The Eskimo of Smith's Sound, their habits, dress, methods of hunting and travelling are dealt with, as well as the experiences of the expedition. The most interesting chapters are naturally those containing the narrative of the march to Independence Bay and back, when the two explorers covered more than 1200 miles in ninety days, and ascended to 8000 feet above sea-level. The exploration of Melville Bay, noticed in vol. xi. p. 314, is also described in this book.

A number of illustrations, from photographs and drawings by the author, are a great assistance to the text in representing the scenery, etc., though some of them are rather rough. Especially striking is a coloured picture of the travellers' camp on the inland ice, which, tinged with pink by the rays of the sun, stretches unbroken to the horizon.

The Work of the Tibetan Expedition, 1889-90, *under the Command of M. V. Pievtsof.* Part III. (In Russian.) St. Petersburg: Tipografia M. Stasiulevich, 1896. Pp. 126.

This part, recently published by the Imperial Russian Geographical Society, contains descriptions of several interesting excursions by MM. Roborovski and Kozlof into districts lying away from the main route of the expedition. Here we have details of the former traveller's explorations on the northern boundary of the Tibetan plateau, along the upper course of the Cherchen river, from Nia between the Russian range and the Uzu-tagh to the upper Keria Daria, with an attempt to reach Tibet across the latter chain, and a journey from Mandalyk in the Cherchen valley to the tableland between the Moscow and Kolymba ranges on the north, and the Przhevalski range, known in its western part as the Akka-tagh, on the south. On this occasion M. Roborovski seems to have left the Cherchen valley by the same route as Mr. Littledale, as far as can be judged from the latter's map in the *Geographical Journal.* He followed the Muzlyk-su, descending from the Tokus Dawan, but turned eastwards to the Ulug-su, whereas Mr. Littledale appears to have kept more to the west.

M. Roborovski's narrative is most interesting, the country being thoroughly described, as well as its flora and fauna. In most parts yaks, antelopes, and other large game were abundant. The extremes of temperature to which travellers in these regions are exposed will be realised when it is stated that, while the thermometer marked 94° F. in April between Cherchen and Nia, it fell below 8° on the Uzu-tagh in May.

M. Kozlof's journeys were to the Konche Daria and the Bagrash-kul, a part of Central Asia he also has visited more recently (see vol. x. p. 540, and vol. xi. p. 311).

The maps are very useful, and one good feature in them is that points astronomically determined are marked. They would be improved by the addition of a few meridians and parallels.

Grundzüge der Pflanzenverbreitung auf der iberischen Halbinsel. Von MORITZ WILLKOMM. Leipzig: Engelmann, 1896. Pp. xiv + 395. *Price 12 M.; bound 13.50 M.*

This work will take a high rank amongst the numerous botanical monographs which have been directly or indirectly inspired by Professor Engler and the large body of capable botanists which he has collected at Berlin. It is the first of a new series of plant-distribution text-books under the joint editorship of Professors Engler and Drude. The author, Herr M. Willkomm, has continually published botanical papers on Spain and Portugal from the year 1844 until 1893, and the present work is, as one would expect, a thorough monograph of the subject. The title is perhaps not well chosen, for it is to be feared that many English readers would not expect a book of some 400 pages, dealing with the distribution, agricultural and commercial plants of Spain and Portugal, from the above "Grundzüge, etc." The first chapter is devoted to the geography (mountain and river-systems), geology, and climate of the peninsula. A very instructive rainfall map gives reasons for the division of the country into six areas, which are as follows :—1, Pyrenees ; 2, North Atlantic ; 3, West Atlantic ; 4, South Atlantic ; 5, Mediterranean ; 6, the Central Tablelands of Spain. The second chapter is perhaps the most valuable, as it gives full details of the characteristic vegetations or formations which are found in the flora. The idea of such "formations" is not generally understood in Great Britain, but their nature may be gathered from the headings—Cistus Hedges, Tamarisk Clumps, Inland Sands (plains and river-shingles), Moors, etc. A map gives the limits of the beech, orange, olive, oleander, esparto grass, *Pinus pinaster*, date-palm, fig and almond, *Abies alba*, and other plants : and this map alone ought to be of the greatest value to any one interested in the commercial products of the country. The remaining chapters are devoted to the special climatic areas mentioned above, and contain long lists of species. An appendix gives an account of the changes which have been introduced by cultivation, and includes a long list of the more useful plants in cultivation. There are two plates (heliogravures), and twenty-one woodcuts. It will thus be obvious that the book is indispensable to the botanical tourist, and is also of great commercial importance. It might have been more judicious to introduce woodcuts of the characteristic scenery or "formations" instead of figures of the more interesting and rare plants : but there certainly should have been a third map giving the provinces and general geography of the country, with the climatic areas marked upon it. Many of the details of plant-distribution are also unfamiliar to most English readers, but this is perhaps inevitable.

Alpine Notes and the Climbing Foot. By GEORGE WHERRY, M.A., M.C. Cantab., F.R.C.S., etc. Cambridge: Macmillan and Bowes, 1896. Pp. xi + 174. *Price 5s. net.*

In the first part of this book ascents in Switzerland, Savoy, and Dauphiné are described. The chapter on the climbing foot is an inquiry into the position of

the foot, when a guide is climbing a steep slope.   It is possible that there should
be some structural difference in the foot of a man who has climbed from childhood,
which would enable him to place the whole sole on a steep rock.   Dr. Wherry's investigations, however, tended to show that there is little difference between the
angle made by the foot with the shank in an amateur climber and a guide.   It
seems, then, to be a matter of mere habit.

*Dolomite Strongholds—the last untrodden Alpine Peaks. An Account of Ascents
of the Croda da Lago, the Little and Great Zinnen, the Cinque Torri, the
Fünffinger Spitze, and the Langkofel.* By the Rev. J. SANGER DAVIES, M.A.
Second Edition.   London : George Bell and Sons, 1896.   Pp. 176.   *Price*
3s. 6d.

"While some of the Dolomites afford the easiest ascents of any mountains of
equal height in the whole Alpine chain, others are so steep and smooth-sided that
they offer the best rock-climbing in Europe, and for some years presented the last
untrodden peaks of the Alps."   In this volume the triumph of ascending the last
remaining virgin peaks is described.   Certainly the singular illustrations exhibiting the method of climbing these last unconquered Dolomites prove that those
who achieved success not merely deserved it, but possessed acrobatic powers which
approached the miraculous.   We venture to think that the feat displayed in the
picture representing the passage of the "Chimney-Breast on the Little Zinne" is
probably unequalled in the annals of the Alpine Club.

Bibliographie de l'Année 1895 (*Annales de Géographie, No.* 23).   Paris : Armand
Colin et Cie., 1896.   Pp. 288.

Every annual volume of the *Annales de Géographie* concludes with a bibliography of the preceding year.   Books, articles on geographical subjects, and maps,
are here recorded, with short notices of the more important works.   Every one who
has occasion to search for information on any country or subject of general
geography, geology, or climatology, etc., knows how useful such compilations are.
We miss a few works which were published in 1895.   It is impossible, of course, to
mention every work on geography and its allied sciences, but we cannot understand
why Keane's *Africa* (Stanford's *Compendium*), and Keltie's *Partition of Africa*
are omitted, while Siever's *Europa* is inserted, or why Hesse-Wartegg's *Andalusien*
finds a place in the *Bibliographie*, but not his *Korea*.   Girard's *Géographie
Littorale*, Haas' *Quellenkunde*, and Ratzel's *Zur Gebirgskunde* (*Anthropogeogra-
phische Beiträge*) are also books one would expect to find.

On the whole, however, the work merits great praise, and is a useful book of
reference.

Nouveau Dictionnaire de Géographie Universelle.   Ouvrage commencé par M.
VIVIEN DE SAINT-MARTIN et continué par LOUIS ROUSSELET.   Supplément,
Fasc. 1-4—Aa-Berlin.   Paris : Librairie Hachette et Cie., 1895-6.

As the first volume of this gazetteer appeared in 1879, while the last was
issued only last year, the early volumes are of course out of date, and hence this
supplement is much needed.   New articles have been added on many parts of
Africa, Alaska, etc., and many names appear for the first time in the Supplement.
The original work gave full information about well-known places, but was rather
defective as regards small and remote spots.   How far the Supplement supplies
the deficiencies of the original, we shall find out by frequent reference, and as the
parts appear : so far it promises well.

NEW MAPS.

EUROPE.

LONDON ENVIRONS, Map of ——. 50-mile radius.

COUNTY OF SURREY.

BRIGHTON AND SUSSEX COAST.

KENT WATERING-PLACES. *Price, cloth,* 2*s. each.*

EASTBOURNE, Plan of ——.

HASTINGS AND ST. LEONARDS, Plan of ——. *Price* 1*s. each.*
Reduced Ordnance Maps for Tourists. By J. G. Bartholomew, F.R.G.S.
W. H. Smith and Sons, London.

EDINBURGH, Bartholomew's Pocket Plan of ——, and Suburbs.
Geographical Institute, Edinburgh.

SCOTLAND, The North Part of Great Britain called ——. With Considerable
Improvements and many Remarks not Extant in any Map. According to the
Newest and Exact Observations. By Herman Moll, Geographer, 1714.
Reprinted 1896. *R. S. Shearer and Son, Stirling.*

Moll's Map is here reproduced in facsimile, with the views of towns and all the
remarks on the fisheries, etc. The printing is sharp and clear. The copy from
which the map is taken is in the Signet Library, and it should also have been stated
that this is a late copy, lest the ignorant should suppose that General Wade's
roads were already made in 1714. On the copy in the Society's library no roads
are marked in any part of the country.

AFRICA.

LE TRANSNIGÉRIEN. Le Bandama et le Bagoé. Carte levée et dressée de 1892 à
1895 par le Capitaine Marchand. Deux Feuilles. Échelle de 1 : 500,000.

CÔTE D'IVOIRE. Levée et dressée par H. Pobéguin, Administrateur Colonial,
1895-96. 8 Feuilles. Échelle de 1 : 500,000.
Service Géographique des Colonies, Paris.

AMERICA.

UNITED STATES, Geological Survey of the ——. Charles D. Walcott, Director.

GENERAL MAPS. Scale 1 : 62,500.

CALIFORNIA : San Francisco.

COLORADO : Aspen.

FLORIDA : Citra, Ocala, Panasoffkee, Tsala Apopka, Williston.

LOUISIANA : Bodreau.

MICHIGAN : Ned Lake, Perch Lake.

MINNESOTA : Duluth.

NEW HAMPSHIRE : Crawford Notch.

NEW YORK : Amsterdam, Chittenango, Oneida, Oriskany.

NEW YORK—VERMONT : Fort Ann, Whitehall, Willsboro.

Scale 1 : 125,000.

MARYLAND—VIRGINIA : Nomini.

NORTH CAROLINA : Hickory.

OKLAHOMA : Kingfisher.

SOUTH DAKOTA : Aberdeen, Byron.

TENNESSEE : Wartburg.

TEXAS : Alpine, Chispa, Marfa, Sherwood, Sierra Blanca.

VIRGINIA—WEST VIRGINIA : Tazewell.

WYOMING : Crandall Creek, Dayton, Ishawooa, Laramie.

SPECIAL MAP. Scale 1 : 25,000.

COLORADO : Cripple Creek.

VENEZOLANISCHEN GEBIRGLANDES, Karte des ——, zwischen Coro und Trinidad.
Nach eigenen Aufnahmen in den Jahren 1884-85 und 1892-92 in Massstabe
von 1 : 1,000,000 bearbeitet u. gezeichnet von Prof. Dr. W. Sievers.

Mittheilungen der Geographischen Gesellschaft in Hamburg, Bd. xii.

ATLASES.

HANDY REFERENCE ATLAS OF THE WORLD.   Edited by J. G. Bartholomew,
F.R.S.E., F.R.G.S., etc.   Fifth Edition.

John Walker and Co., London, 1896.

Besides the maps contained in the former editions, which have of course been
revised, a large number have been added of special localities, plans of towns, etc.
As these are printed on the backs of the original maps they do not add to the size
of the volume.   Some additions have been made to the "Geographical Statistics."

SCHWEIZ, Topographischer Atlas der ——, in Massstab der Original-Aufnahmen
nach dem Bundesgesetze vom 18 Dezember 1868 durch das eidg.   Topo.
Bureau gemäss den Direktionen von Oberst Siegfried veröffentlicht.   Massstab
1 : 25,000, Lieferung, XLVI.

Nr.	51 bis Hagnau.		Nr. 447	Versoix.
„	64 bis Vor der Argen.		„ 448	Meiner.
„	300 Mont la Ville.		„ 449	Dardagny.
„	365 Jaun.		„ 449 bis	Chancy.
„	417 Samnaun.		„ 469	L'Etivaz.
„	446 bis Hermance.		„ 502	Vergeletto.

Presented by Professor Paul Chaix.

CHARTS.

CLEVELAND HARBOR AND CUYAHOGA RIVER, Lake Erie.

PELÉE PASSAGE, Lake Erie.

CLEVELAND HARBOR and Approaches.

DETROIT RIVER, from Detroit River Lighthouse to Mamajuda Lighthouse.

SOUTH CHICAGO.

LOWER CALIFORNIA, West Coast of ——. Abreojos Point to Cape San Lazaro, 1896.

MERSINA ROADSTEAD, Asia Minor.

The U.S. Hydrographic Office, Washington.

THE SCOTTISH
GEOGRAPHICAL
MAGAZINE.

PLACE-NAMES OF BALQUHIDDER.

Communicated by Mrs. CARNEGIE of Stronvar.

IT has been suggested that it would be of interest to readers of this Magazine to see a list of place-names in Balquhidder parish, many of which refer to small spots, eminences, corries, etc., which are not marked in any map, and are known only to the inhabitants of the locality. No small proportion also have fallen into disuse, and in the next generation will be wholly forgotten. It occurred to me, therefore, some little time ago, that it would be well to make a collection of them before they were lost for ever; and I applied to Duncan Lamont, gamekeeper, Stronvar, whose personal knowledge of every spot in the parish was thoroughly to to be relied upon. He also consulted some old people in the glen as to some names which now linger only in the memories of the oldest inhabitants. The Rev. David Cameron, minister of Balquhidder, was kind enough to bring his Gaelic scholarship to bear in translating these names, and I think it will be found that they have been analysed, to quote the words of the late W. F. Skene, "upon sound philological principles and a comprehensive observation of facts."

AUCHLESKINE FARM.

Auchleskine—Field or land belonging to Scone; or more probably—the field with the cleft or rent made by the cataract.

Tom-na-Croiche—The knoll of the gallows.

Tom Banach—The whitish or grey knoll.

BALLIMORE FARM.

Ballimore—The large hamlet.

Allt-a-Ghlinne Dhuibh—Burn of the Black Glen.

Bealach Stacach—The rugged pass.
Baille Meanach—The mid village.
Clach Ghlas—The grey stone.
Craig-an-Fhiodha—The rock of the piled timber, or of the timber.
Creag Mhor—The great rock.
Creag-an-Tuill—The rock of the hole.
Coire Creagach—The rocky dell.
Caol Bhealach—The narrow pass.
Creag Chaorunnach—The rock of the mountain ash.
Dal-an-lagain—The field of the hollow.
Dalchopagach—The field of the dock weeds.
Tom-an-Eisg—The knoll of the fish.
Carn Dubh—The black cairn.
Coire Seiceach—The dell of hides or skins.
Glen Fathan—The little glen, or, coltsfoot glen.

BLARCRICHE FARM.

Blarcriche—March field.
Allt-na-Sgitheiche—Burn of the hawthorn
Coire Chaillich—The old woman's dell.
Coire-nan-Saighead—The dell of arrows.
Eas Cheataig—
Garbh—The rough or rugged portion.
Gleann Sgithiche—The glen of the hawthorn.
Stob breac—The spotted peak.
Taobh-na-Coille—The side of the wood.

GARTNAFUARAN FARM

Gartnafuaran—The cultivated land with many springs.
Beinn Luidh—Hill of herbs.
Creig-an-Fhithich—The raven's rock.
Craggan Chaorun—The rock of the mountain ash.

KIRKTON FARM.

Allt-a-choire bhain—Burn of the wan-coloured whitish dell.
Carn Chailean—Colin's cairn.   Cailein Uaine, "Green Colin," son of
Sir Duncan Campbell of Glenorchy, killed here by the Buchanans about
A.D. 1514.
Coire Ban—The white corrie.
Coire-na-Moine—The mossy corrie, or corrie of peat.
Creag-a-Bhuic—The rock of the buck.
Creag Mhullaich—The rock of the summit.   *Mùlach* = top.   The
real name is *Creag Mhalaich*—The rock of the brow of the mountain.
Gleann Eirionnaich ⎫
Loch an an Eirinnaich ⎪ Eirinach, a wedder goat.
Leum-an-Eirinnaich ⎬ The leap of the wedder goat, the name of
Eirionnaich ⎭   the rock at top of Kirkton Glen.
Lurg-nan-Coisichean—The ridge of the traveller.

Meall-an t-Seallaidh—The eminence of views, or watch hill.
Meall-an-Fhiodhain—The eminence of the cheese vat.
Sitheag—A little fairy knoll.
Tom na Buaile—A fold for cattle.

LEDCRIECH FARM.

Ledcrich—The land sloping to the march.
Allt Chrue—The burn of the hazel nuts (*chnoth*), or cows (*chrudhe*).
Coire Meann—The kids' dell.
Creag nan Saighead—The rock of the arrows.
Creag an Iasgair—The fisher's rock.
Creag Bhreac—The spotted rock.
Creagan Bhreac—The little spotted rocks.
Glen Chruie, as above—The glen of the hazel nuts (*chnoth* or *chnuth*).
Leth Immer—The half of a ridge of land.

LIANACH FARM.

Lianach—Many meadows.
Bealach-a-Chonnaidh—The pass of the firewood or fuel; or whins, sometimes called *connadh*.
Clach Mhor—The great stone.
Creag-a-Chonnaidh—The rock of the firewood or fuel.
Cuil—A nook.
Glen Shoinie (Gleann Shithean)—The glen of the fairy knolls.
Luachrach—Abounding in rushes.
Sitheag Riabhach—The grizzled knoll.
Mullach-an t-Samhraidh—The summer height.

MONACHYLE TUARACH.

Monachyle—The wooded hill.
Monachyle Tuarach—The hill of the wood having a northern exposure; or simply, Southern Monachyle.
Allt-a-choire Odhair—Burn of the dun-coloured dell.
An Caisteal—The Castle.
Ceann-na-Ban-Tighearna—The lady's head.
Coire Odhar—The dun-coloured dell.
Coire Beith—The birch-wood dell.
Na Slochdan—The dens or pits.

MUIRLAGGAN FARM.

Muirlaggan—Many hollows.
Allt Corrach—The precipitous burn.
Allt Craoibh-na-spuinge—Burn of the spongy tree, or, of the booty.
Bealach Driseaig—Pass of the little bramble bush.
Tom-na-cloiche—The knoll of the stone.

SRON-SLANY FARM.

Sron-slany—The promontory of Leny, or the marshy ground (*leana*).
Ath-a-Bhriogaise—The ford of the breeches.

Ceann-na-Coille—The head of the wood.
Coille-na-Sroine—The wood of the promontory.

STRONVAR.

Stronvar (Sronbhar)—The promontory of the pointed hill, or, the pointed promontory.
Baille-a-chnoic—The hamlet on the knoll.
Baille-an-luig—The hamlet of the hollow.
Clach Coimheadaidh—Watch stone.
Cnoc-an-Daraich—The little hill of oak wood.
Croit Ghobhain—The smith's croft.
Culearn (Cul Fhearn)—The nook of the alder-trees.
Dunbeag—The little round hill.
Dun More—The great round hill.
Eilean Dubh—The Black Island.
Mulan Fraoich—The heathery knoll.
Roinn Uinnseann—The ash point.
Tom-nan-Deur—Knoll of the "drops," so called from a house on the spot where whisky was sold.
Tom-na-Faidhreach—The market knoll.
Tom-na-Drochaid—The bridge knoll.
Tom-an-Dathadair—The dyer's knoll.
Tom-a-mhoid—The knoll of the Court of Justice.
Tom adhairc—The horny knoll.
Tom-an t-Sabhail—The barn knoll.
Tom-nan-Carnan—The knoll of sledges—so called, because on account of its steepness sledges only could be used here.   It might also be knoll of small cairns.
Tigh-na-Croiteag—The house of the croft.

TULLICH FARM.

Tulloch (Tulaich)—A hillock or eminence.
Bealeach-nan cabrach—Pass of the antlered deer.
Creag-an-taxman or Craigallda—The taxmen's rock, the fierce rock.
Coire-na-Tulaich—The corrie of the hillock.
Tom Tulaich—The knoll of the hillock.
Tom-na h-Ath—The knoll of the kiln.
Leacann Sgridain—A steep, shelvy, rugged side of a hill—a stony slope.

IMMEROIN FARM.

Immeroin—Hugh or Ewan's ridge.
Allt-an-Spuit Dhuibh—Burn of the black waterfall.
Beinn-an t-Sithean—The hill of the fairies' knoll.
Dalreach—The dark grey (grizzled) plain or field.
Druim-na-Ceardaich—The ridge of the forge.
Lon-ant-Sithean—The meadow of the fairies' knoll.
Sput Dubh—The black waterfall.

INVERLOCHLARIG FARM.

Inverlochlarig—The confluence of the stream from top of pass,—near Loch Larig.

Stob-a-choin—The hound's peak.

Amar stob-a-choin—The trough or channel of the hound's peak.

Bealach coire nan-saighead—Pass of the dell of arrows.

Drumlich—The ridge of the flaggy declivity or slope.

Bealach coire-an-Laoigh—The pass of the calf's dell.

Stob-an-duibhe—The peak of the blackness, or alders :—dwarf-alders, used for dyeing, and called *dubhaidh*.

Airidh Gaoitach—Windy shealing.

Coire nan Eilid—The corrie of hinds.

Sithean-a-chatha—The fairy knoll of the battles.

Bealach-nan-corp—Pass of the corpses (through which the Macgregors passed to bury their dead in Iniscailleach).

Parlan Hill—Macfarlane's hill.

Beinn-a-chabhair—The hill of help.

Allt-a-chuilinn—Burn of the holly.

Beinn Chroin—The hill of the parted hoof (*chrodhain*).

Allt Earb—The burn of roes.

Inshag Earb—The little island, or detached field of roes.

Beinn Tulaichean—Hill abounding in little heaps or knolls.

Bealach Glas—The grey pass.

Clach-na-Guiseach—The stone of the long slender grass.

Coire Earb—The corrie of roes.

Cruach ardrain—The high heap-like hill.

Craig-nan-sputan—The rock of the spouts, or small cascades.

Sgairneadh-an-airgid—Heap of loose silvery stones on hillside.

Allt-a-chearnaig—Burn of the small corner.

Creag artair (ard-tir)—The rock of the summit.

Stob Inver-Chearnaig—Peak of the small corner.

Meall-an-Lochan—The eminence of the little loch.

Na Staidhrichen—The steps.

Creagan Liath—The grey rocks.

MONACHYLE MOR FARM.

Coire-a cheothich (cheathaich)—The corrie of mist.

Creag-na h-Iolaire—The eagle's rock.

Immer Riabhach—The brown or grizzled field.

Lag-an-Fhuarain—The hollow of the spring.

Meall-na-Dige—The eminence of the fence.

Meall Monachyle—The eminence of the wooded hill.

Meall Reamhar—The broad or thick eminence.

Creag-na-speireag—The rock of the sparrow-hawk. ⎫
Meall-gainnmheach—The sandy eminence.   ⎬ On Monachyle
Stob Monachyle Beag—The peak of the little   ⎪   Beg Farm.
  wooded hill.   ⎭

Coire-na-ceardaich—The corrie of the smithy or forge.

Allt-na-ceardaich—The burn of the smithy.

Meall-nan-aighreag—Eminence of cloud berries.
Coire Uain—The green corrie.
Coire Luainie—Restless or changeful corrie (?).
Creagan-nam Putan—Rocky place of the young moorfowl.

CRAIGRUIE FARM.

Craigruie—The king's rock ; or the rock of the outstretched mountain base.
Ard-Latha for Ard-Lotha—The height of the fillies.
Bealach-an t-sneachdaidh—The pass of snow.
Faradh Dubh—The black ladder.
Tom-a-Bhuachaille—The shepherd's knoll.
Tom-na h-Iolaire—The eagle's knoll.
Uamh-an-Righ—The king's cave.

LAGGAN FARM.

Laggan—Hollow.
Stronyre—The promontory of Strathyre.
Creag Liath—The grey rock.
Allt Stronyre—Burn of Stronyre.
Creag-na-coimh Sheilg—The rock of the meeting of the hunters, or joint hunt.
Coire Buidhe—The yellow corrie.
Ard-nan Damh—The height of the stags.

KEIP FARM.

Keip—A heap or the heap.
Lochan Buidhe—The yellow tarn.
Beul-an-ath—The margin of the ford.

CREAGAN (CREGGAN) FARM.

Creegan (Cridheagan, or Cridhe-an-lagain)—The heart of the hollow, a sheltered spot in the curve of a hill.
Allt-a-Chreegan—The burn of Cregan.

KING'S HOUSE FARM.

Tigh-an-Righ—King's house.
Allt Stad Fhaochail (Allt Stairc a phuill dhuibh)—Burn of the stepping-stones of the black bog.
Ciste Bhraidhean—The coffer of the querns.

IMMERVOULIN FARM.

Immervoulin—The field of the mill.
Allt Tigh an Eas—Burn of the house of the ravine, or waterfall.
Tom-a-Challtain—The hazel knoll.
Tom-nan-Cearc—The knoll of the hens.
Tom-na-Moine—The knoll of the peat-moss.
Tigh Bhruce—Bruce's house, a shieling frequented by Bruce, the Abyssinian traveller, when passing the summer, as he usually did, at Ardchullerie.
Sean thalamh—The old land.

EDENAMPLE FARM.

Edenample, from Eadain-a-phuill—The brow or face of the marshy ground; or, Eudann-an-teampuill—the place of worship on the face' of the hill.

Coire-na-Cloiche—The corrie of the stone.

Creag-a-Mhadaidh—The hound's rock.

Coire Chorish—(?)

Coire Fhuadaraich—The lively corrie, or exiles' corrie (or to where sheep and cattle were driven).

Creagan-nan--Gabhar—The rocky place of the goats.

Choire Mheobhith—The corrie of lively life.

Coille-a-Mhaoir—The bailiff's wood.

Coille-na-Criche—The march wood.

Allt-a-Choire-Fhuadaraich—The burn of the lively corrie, or exiles' corrie (?).

Firach (am-Fireach)—The top or acclivity.

GLEN BEITH.

Glen Beith—Glen of the birch.

Carn Liath—The grey cairn.

Allt Iubhar—The burn of the yew-trees.

Tigh-na-Dalloch—The house of the field.

Dalveich (Dal-Bheathaich)—The field of or by the birch wood.

Ard Bheathaig (Bheithaig)—The height of the little birch wood.

AUCHRAW.

Auchraw, from *traigh*—The field by the shore or beach; or from *rath*, a fort.

Tom Ranaich—The knoll of ferns.

Leckine (Leachdain)—The steep shelving ground.

Allt-na-Gallanaich—Burn running among many branches of trees.

Meall-a-Mhadaidh—The hounds' eminence (said to be where the last of the bloodhounds used in hunting the proscribed Macgregors was killed by a party of them who turned on their pursuers).

Laggan Ruidleach, Laggan-ruadh-tulaich—The hollow of the reddish-coloured hillock.

EDENCHIP.

Edenchip—The face of the heaped or bulky hill; literally, the face of hummocks or small knolls (cf. *ceapach*—block or shoe last).

Lairg Mhuiltibh—The wedder pass.

Drochaid Ceann Droma—The bridge over Kendrum water.

Allt-Ceann-Droma—Burn of the head or end of the ridge.

Creag Mac Ranaich—Rock of the bellower, or bellowing rock; from *ranaich*, to howl or roar.

MISCELLANEOUS NAMES.

Creagan Laiste—The shimmering, or illuminated, rocky place.

Cam Chreag—The crooked rock.

Ardoch (Ardach)—The high field.

Auchtow (more properly Achtubha)—Field or land producing thatch, *i.e.* sprits or rushes.

Auchtowmore—As above, with *mor*, great.

Beannach Aonghais—Blessing Angus. St. Angus is said to have introduced Christianity into the district. The tradition regarding the name of this place is that the worshippers from the north-east on arriving here, where they had the first distinct view of the place of worship, were wont to implore a blessing upon Angus. The phrase employed to invoke the blessing was, according to tradition, "Beannaich Aonghes san Aorainn." The site is still called the Aorainn by the old people.

Balquhidder—The town or district of the back-lying country (?).

Balbhaig—The still or silent-flowing.

Beinn Vorlich (Beinn mhoir luig)—The hill of the great hollow.

Calair (Culair)—The stream coming in behind.

Glen Buckie—The glen of the bucks.

Glen Ogle (from Ogluidh)—The dismal or terrible glen, or high glen.

Loch Earn. Recent students of place-names give the derivations Loch-Ear-Abh—The loch of the easterly river, and of the Ernai (?).

Loch Lubnaig—The bent loch.

Loch Voil (Loch-a-Bheothuil)—The loch of the quick-running flood.

Loch Doine (Loch na Dubh Amhain)—The loch of the Black River.

Rinacraig (Rudha-na-Craige)—The promontory of the rock.

Ruskachan (Rusgachan or Riasgachan) from *riasgach*—The marshy land where coarse grass grows.

Strathire—The pronunciation of this word in Gaelic, which is strath-theo-thir, makes its meaning more likely to be the "strath of the warm country," than what would seem the more obvious meaning (strath-fheoir) the "grassy strath."

Stuic-a-Chroin—*Stuic*, a lesser hill jutting out, as it were, from a greater; Stuc-a-chrodhain—hill of the cloven hoof.

Tom-nan Ainil (Aingeil)—The knoll of fires.

BIRD MIGRATION IN THE BRITISH ISLES:
ITS GEOGRAPHICAL AND METEOROLOGICAL ASPECTS.[1]

By W. EAGLE CLARKE, F.L.S.
(Natural History Department, Museum of Science and Art, Edinburgh.)

GEOGRAPHICAL.

General.—In passing from their summer to their winter haunts, birds proceed from a northern to a southern clime, and *vice versâ* in the spring. It does not at all follow, however, that these seasonal haunts are reached

---

[1] From the Digest of Observations on the Migration of Birds at Lighthouses and Light-vessels, 1880-1887, presented to the British Association (Liverpool Meeting, 1896), by Wm. Eagle Clarke.

by a simple movement from north to south, or the reverse. Each species or individual of migratory bird has its particular summer and winter resorts, and these do not necessarily lie in the same·meridian—indeed this is often far from being the case. To attain these particular seasonal habitats, many of the voyagers must depart more or less considerably from a direct course. This is especially the case in Western Europe, where, owing to the south-western extension of the land-masses, and the consequent irregularity of the coastline, various more or less devious routes must be, and are, followed. The interposition of the British Islands between the north-western portion of the Continental Area on the one hand, and Iceland and Greenland on the other, is an important additional factor in this deviation.

The geographical distribution of birds during migration on the British and Irish coasts, and the routes traversed, naturally depend upon the nature of the particular movement.

The chief and most interesting movements, from the geographical standpoint, are the intermigrations between our Islands and Europe. There are, however, a number of movements between the various sections of the British and Irish areas, which are of considerable importance.

Intermigration between Britain and Northern Continental Europe.—Between Britain and Continental Europe travel a host of migrants, which are either birds of passage on, or winter visitors to, our shores. The former visit our eastern coastline in spring when journeying to their northern summer haunts lying to the north-east of Britain, and again in autumn when returning to their winter quarters to the south of our Islands. The winter visitors are chiefly individuals from the ranks of certain species of the birds of passage which winter in the British area and emigrate to the north-east in the spring.

In the autumn these numerous migrants cross the North Sea and arrive on the east shores of Britain at points between the Shetland Isles and the Humber, or the northern seaboard of Norfolk. All the movements do not necessarily cover this extensive stretch of coastline, but such is not infrequently the case. Indeed, as a rule, they are recorded from the greater part of the region indicated. It is possible to define the southern limit on the coast at which these birds strike Britain, with a considerable degree of precision. No section of the British coast is so well equipped with light-stations as that which lies between the north coast of Norfolk and Dungeness. In addition to an average number of lighthouses, there is a fleet of lightships off the coast, which are most favourably situated for recording the movements of birds crossing the North Sea to the English coast. These lightships have furnished the Committee with some of the most carefully kept records to be found among the returns, and it is a very significant fact that these great autumn immigratory movements are not observed at these south-eastern lighthouses and lightships. Evidence of a particularly important nature, in this connection, is also afforded by the records kept at the Outer Dowsing lightship, the most isolated of the stations in the North Sea, situated about thirty-eight miles E.S.E. of the mouth of the Humber. At this station these important movements are not observed—another significant fact, indicating unmistakably that these migrants pass to the northward or westward of this lightship.

The conclusion at which I have arrived, after a long and careful study of the records, is that these immigrants and emigrants from and to Northern Europe pass and repass between this portion of the Continent and Britain by crossing the North Sea in autumn in a south-westerly direction, and in spring in a north-easterly one,[1] and that, while the limit to their flight in the north is the Shetland Islands, that on the south extends to the coast of Norfolk.[2]   During these movements the more southern portion of the east coast of England is reached *after* the arrival of the immigrants on the more northern portions.

It is to be remarked also, as bearing upon this important point, that *all* the species occur on migration in the Orkney and Shetland Islands, but not in the Færoes.[3]   And, further, *all* the British birds of passage to Northern Europe are either summer visitors to Scandinavia or are regular migrants along the western shores of that peninsula.

After arriving on our eastern shores, these immigrants from the north—some of them after resting for a while—move either down the east coast, *en route* for more southern winter quarters, or, if winter visitors, to their accustomed haunts in Britain and Ireland.   A few occur as birds of passage on the west coast and in Ireland, which they reach by over-land routes across Britain, and then pass southwards to their winter quarters.   The west coasts, however, do not receive *directly* any immi-grants from Continental Europe.

Intermigration between the South-east Coast of England and the Coast of Western Europe—"East and West Route."—This is one of the discoveries of the inquiry.   It has been already shown that the more southern section of the east coast of England does not receive immigrants *direct* from Northern Europe. .   There is, however, a considerable amount of migration of a particular description, and on the part of certain species, observed at the lightships and lighthouses between the Kentish coast and the Wash. During the autumn, day after day, a stream of migrants, often of great volume, is observed off the coast, flowing chiefly from the south-east to the north-west at the more northerly stations, and from east to west at the southerly ones, across the southernmost waters of the North Sea. This will be hereafter mentioned as the "East and West Route."   From the stations off the mouth of the Thames as a centre, the birds either sweep up the east coast, sometimes to and beyond the Tees (many proceed-ing inland as they go), or pass to the west along the southern shores of England.   These important immigrations set in during the latter days of September, reach their maximum in October, and continue at intervals

---

[1] The direction varies.   It is probably more westerly (in autumn) or easterly (in spring) at the most northern British stations, and south-south-westerly (in autumn) or north-north-easterly (in spring) at the stations on the east coast of England.

[2] The formation adopted by the migrants during passage would seem to be an extended line—perhaps a series of lines—whose right wing extends to the Northern Islands, and its left wing to the coast of Norfolk.

[3] A few species occur in the Færoes on migration, but these are also summer visitors to those islands and to Iceland.

until November. They are chronicled with wonderful precision and regularity in the returns from the stations on the south-east coast of England. They are renewed during winter on occasions of exceptionally severe cold, but the birds then pass to the westward along our southern shores.

There are some remarkable features associated with these movements : (1) They are frequently observed for several or many consecutive days; (2) they often occur when there is an almost entire absence of bird migration on other parts of our shores; (3) the movements appear to be entirely confined to the daytime, and are usually timed as from soon after daylight to 1 P.M., sometimes until 3 P.M.—this being probably due to, and indicative of, the shortness of the passage; (4) the autumn migratory flocks are chiefly composed of Larks in vast numbers; 'Black Crows' (Rooks) very many; Grey Crows, many ; also numerous Redbreasts, Gold-crests, Chaffinches, Greenfinches, Tree-Sparrows, Swallows, Starlings, and occasionally Woodcocks ; and during the winter Larks, various Thrushes, and Lapwings ;[1] (5) and lastly, on certain occasions these immigrants, while passing northward along the English eastern seaboard, actually *cross* the movements of " coasting " emigrants proceeding southwards.

Whether this east to west stream is a branch of one that passes down the coast of Continental Europe, or whether it has its source in Central Europe, is a matter of conjecture.[2]

The conclusions relating to these continental migration-routes have been chiefly based upon the autumn data, because the information for that season is much more voluminous and complete. When, however, we come to examine the information relating to the spring movements, with a view to ascertaining how far they corroborate the conclusions so clearly indicated by the autumn chronicles, it is satisfactory to find decided evidence that the birds retrace their flight to the north and east along precisely the'same lines as those along which the autumnal southerly and westerly journeys were performed. Thus, in the spring, these birds depart from the same sections of our eastern seaboard as witnessed their arrival in the autumn.

Intermigration between Heligoland and Britain.—Much prominence has been given in some of the Annual Reports issued by the Committee, and in Herr Gätke's book, *Die Vogelwarte Helgoland*, to an intermigration between Heligoland and the east coast of England by a direct east-to-west autumn, and it is to be presumed west-to-east spring, movement. Herr Gätke most obligingly communicated the details of the bird-movements observed on Heligoland for four of the years (1883-1886) during which the inquiry was being prosecuted over the British area. These two sets of data have been carefully examined and compared, and it has been found that

---

[1] It is probable that such species as the Golden Oriole, Hoopoe, etc., which occur annually during spring and autumn migration in southern and south-eastern England, and the Black Redstart as a winter visitor, are birds that proceed along this route to and from our Islands.

[2] There are no essentially northern species recorded for this route, and the occurrence of the Rook so frequently and in such numbers is suggestive of a Central (Western) European source.

the dates of the chief movements of the species common to Heligoland and Eastern Britain seldom if ever correspond, and do not bear out this theory; that particular species which are irregular as migrants in Britain, such as the Ortolan Bunting, and others, occur regularly, often indeed in "rushes," at the more favoured isle off the mouth of the Elbe; that other species, which are very rare on our British shores, occur in Heligoland as regular migrants and in considerable numbers, as *Motacilla flava, Anthus Richardi,* etc.; while species common to both islands occur in "flights like clouds," in "hundreds of thousands," "thousands upon thousands," in "marvellous numbers," "astonishing flights," and so on, at Heligoland, at periods when there is not a single observation for the same species on the English shores. A study of the phenomena of migration at the stations on the east and west sides of the North Sea compels the investigator to come to the conclusion that Heligoland and Britain draw their migratory hosts from different sources. The ordinary movements of any common migratory bird occur in each month of its seasonal flight-periods, and the mere coincidence of the species being observed simultaneously in *ordinary numbers* on both sides of the North Sea has no significance whatever. It is not impossible or improbable that birds may *occasionally* cross the German Ocean by an east-to-west flight in the latitude of Heligoland, but our data lead us to believe that such cases are the *rare exception* and not the rule.

Intermigration between Britain and Færoes, Iceland, and Greenland.— The Færoes, Iceland, and Greenland are the summer home of several Palæarctic species which occur as birds of passage on the British coasts. The majority of these visit Iceland, and Greenland claims only two or three of them (Wheatear, White Wagtail, and Whimbrel). It is natural that these birds being strictly Old World species, our Islands should lie in the course of their migrations. It is quite possible that these migrants may pass along both the eastern and western coasts of Britain and the coasts of Ireland. Here, at any rate, we have evidence that these birds are observed on passage on our western shores. It may be that some of the birds proceed also along our eastern seaboard, but this is a point difficult to determine. There is good evidence, however, that important movements of Redwings, Wheatears, and Whimbrels are observed on the western coast of Great Britain and the Irish coasts (both east and west as regards the passage of the Whimbrel), which are not observed elsewhere. Such a fact points to the independent nature of these west coast flights, and indicates that, in some instances at least, the western route alone is followed.

It is thus evident that, so far as concerns the movements of the birds of passage to and from their northern breeding haunts, the British east and west coast migratory movements are very distinct in their characters. The west coast does not receive immigrants direct from Europe, nor do these continental breeding species depart from its shores in the spring. Indeed, it is quite remarkable how rare, or comparatively rare, certain well-known east-coast species are on the western portion of our shores.

With the movements of the British migratory birds next to be considered it is quite different, for, with the exception of a few species whose summer haunts are much circumscribed in our Islands, the movements are not only common to both coasts, but the great emigratory flights are usually simultaneously observed on the east, west, and south coasts, and also on those of Ireland.

The west coast of Great Britain and the Irish coasts are thus only under much migration during the great autumn departure movements from our shores, and to a less extent during the return movements in spring.

Intermigration between Great Britain and Ireland and the South, etc.—Having shortly described the migratory movements between the British Islands and Northern and Western Europe, undertaken by birds of passage and winter visitors to our Islands, the routes on our coasts along which the summer visitors[1] travel to and from their breeding quarters in Great Britain and Ireland now demand attention in their geographical aspect. It will be convenient also to refer to the routes between the different portions of the British area under this division.

The autumn or emigratory movements will be described—but it is necessary to remark that the data clearly indicate that the spring migratory movements along our western shores are simply return movements, on the part of the same species, along the same lines of flight as those laid down for the autumn.

The movements of these groups of migrants will be treated of under the various sections of our coasts. The first movement on the part of all emigrants among British birds is to the coast, which is reached in some cases, no doubt, by particular inland routes.

East Coast of Great Britain.—The emigratory movements on the east coast are very simple in their geographical aspect. When the coast is reached, the emigrants follow the coastline southward, gathering strength as they go, and finally quit our shores at various points on the south coast of England.

It is during such autumnal movements that the more southern coastline of Eastern England, and its off-shore fleet of lightships, record night migration. The ranks of the British emigrants are, as we have said, recruited as they fly onward, and if a great movement should be in progress, the causing-influence will affect also many birds of passage which may be sojourning on our shores. Two wings of the migratory army thus combine, and a great "rush" to the south is the result.

West Coast of Great Britain.—The emigratory movements which pass down the west coast are far from being so simple in their geographical details as those observed on the east.

That such should be the case is not surprising. Here we have Ireland, the Isle of Man, the Hebrides, and an extremely irregular coastline exercising their varied influences. In addition, there are intermigrations between these off-lying isles and the mainland, and often movements of an independent nature in some portion of the western area.

The general route followed by these departing birds has its northwestern source in the Outer Hebrides, and after leaving Barra Head it

---

[1] Those birds which are only partially migratory are included in this category.

joins an important stream from the Inner Hebrides at Skerryvore.   The course then followed is *viâ* Dhuheartach, Islay, the Wigtonshire coast, the Isle of Man, Anglesey, and the South Bishop (off Pembrokeshire). Finally, the south-western coast of England is reached (possibly in part by an overland route across Devonshire and Cornwall) between the Scilly Islands and Start Point.

In its course southward considerable tributaries, so to speak, are received at Cantire, Arran, the Ayrshire and Wigtonshire coasts, and the Solway, of birds passing down the west coast of Scotland.   At the Bristol Channel emigrants are received from western England and Wales, and often also important contributions are added from the south-eastern coast of Ireland.

In connection with these movements there are several more or less important features to note.   (1) The English shores of the Irish Sea, —*i.e.* the coasts of Cumberland and Lancashire—lie off the main line of these movements; (2) the north coast of Ireland, which seems to lie right in the course of the birds, and which would naturally be expected to come in for a considerable share of such movements, appears to be only occasionally affected by them; (3) the Irish contributory movements, when they occur, are chiefly, nay almost entirely, observed on the southern, and especially the south-eastern coasts; (4) the south-western coast of England and Wales—*i.e.* from the mouth of the Bristol Channel to the Land's End and the Scilly Isles—appears to be especially affected when there are considerable movements on the southern and south-eastern coasts of Ireland, implying that there is much intermigration between these particular portions of the English and Irish coasts.   Sometimes, however, these emigrations from Ireland only affect the south-west coast of England from the Bishop's Rock (off Scilly) to Start Point.

Irish Coasts.—The Irish chronicles have been most excellently and carefully kept, and the returns of specimens killed against the lanterns at the stations have been larger and more valuable than those furnished from the coasts of Great Britain.

The coasts of Ireland do not constitute in themselves a main highway for birds, though they participate, along with the western shores of Great Britain, in certain movements to and from the far north on the part of the section of the birds of passage already alluded to.   Indeed, the majority of the migrants observed on the shores of the sister isle are probably the migratory members of her own avifauna.

The movements of departing birds during the autumn at the southern and south-eastern stations have already been mentioned, and when migration is going on at this part of the coast there is often recorded an emigratory movement along the western coast from Slyne Head southwards, which probably forms a contributory stream to the general movement to the south.   These Irish emigrations, as a rule, occur simultaneously with similar movements passing down the western coast of Great Britain, and the two streams meet and unite at points between the Bristol Channel and the Scilly Isles.   Some of the Irish autumnal flights, however, are quite independent of these general movements.

There is much evidence to show that not only do the autumnal emigrants depart from the south-east coast of Ireland *en route* for more southern winter-quarters, but also, strange to say, that many birds (*e.g.* Thrushes, Redwings, Blackbirds, Chaffinches, Greenfinches, Linnets, Starlings, Larks) almost simultaneously *enter* that country by this very same section of her shores, in order to winter within her limits. These immigrants are often observed arriving from the south-east in great numbers for several days in succession. The English west-coast observations also bear evidence that such movements proceed across St. George's Channel in a north-westerly direction. These cross-channel flights are usually observed during the daytime, but sometimes the arrival of certain of these birds on the Irish coast takes place during the night.

According to the records it is only occasionally, as already stated, that the southerly autumnal movements from Western Scotland are observed at the northern Irish stations. Now and then, however, there is evidence that a considerable number of birds do arrive on, or skirt, the north coast of Ireland during the more pronounced west-coast emigratory flights.

Independently of, and in addition to, these main Irish migratory movements, Thrushes, Larks, and Starlings occur in October and November on the northern coasts of Ireland from Tory Island to the Maidens as immigrants from Scotland. These are to be correlated with movements of the same species observed at the Rhinns of Islay and the Wigton coast. Larks, too, are often recorded for this route during the daytime.

There are also autumnal movements between Ireland and England and Wales by an east to west flight across the Irish Sea, on the part of Starlings, Chaffinches, Greenfinches, Larks, and sometimes of various species of Thrushes. Anglesey is the chief Welsh point, and Rockabill (off the north coast of Co. Dublin) the main Irish station at which these departures and arrivals are observed.

The migratory movements observed on the west coast of Ireland are neither many nor important, and consist almost entirely of movements on the part of emigratory Irish birds. There are, however, remarkable immigrations from home sources witnessed on the west coast and its offlying islets during great cold and snow.

South Coast of England.—It is much to be regretted that observations relating to the migrations of birds on the southern coast of England *as a whole* were not obtained by the Committee. The data bearing upon this important English coastline are from a few stations on the south-eastern and south-western portions only.

This information points to (1) a considerable amount of migration taking place between these portions of the coastline and South-western Europe; and (2) important movements passing *along* the entire coastline from east to west in autumn and probably *vice versâ* in spring.

The south coast is naturally the great scene of the arrival and departure of migratory birds of all descriptions, but the movements along shore are, perhaps, in some of their aspects, more interesting. Regarding these last, much remains to be ascertained concerning their precise nature and the destination of some of the birds travelling along this route.

In the autumn this coasting stream of birds has its source chiefly in the immigratory movements from the Continent across the southern waters of the North Sea by the East and West Route, of which it is but a continuation. It is possible, also, that British emigrants, after passing down the east coast of England, may turn to the westward and skirt the south coast, but this is not shown with certainty.

The continental immigrants strike the Kentish shore, and, as has been already stated, some pass to the north along the east coast of England, while others pursue a westerly course along our shores of the Channel. The stations on the south-western coast again record these migrants, and the probable destination of many, perhaps most of them, is Ireland, on whose south-eastern shores the birds are chronicled, almost simultaneously, as arriving in great numbers from the south-east.

It is possible, however, that some of these birds—the Skylark especially—may reach a much more remarkable destination, for one branch of the stream sweeps northwards, being observed at the mouth of the Bristol Channel, at Anglesey, and at the Isle of Man stations, proceeding to the west and north-west, probably to Northern Ireland; while on the Wigtonshire coast and at the rocks of Dhuheartach and Skerryvore these birds are noted as moving in the direction of the Outer Hebrides.

The great autumnal movements from east to west along the south coast of England are renewed in winter, when that season is characterised by periods of unusual cold. At such times it is possible that this western stream is composed in part of native emigrants which have passed down our eastern coasts, as well as of birds of continental origin.

Channel Islands.—Records from the Hanois lighthouse, situated some two miles off the west coast of Guernsey, were furnished for each of the years of the inquiry, and afford some useful information. These, when compared with the English and Irish chronicles, show that on nearly every occasion on which considerable migration was observed at this station in the autumn, there was also much emigration going on practically simultaneously on the south-west coast of England. It is necessary, however, to state that a number of important movements on the south-west coast of England do not appear in the records for Hanois, indicating, perhaps, that many movements to the south in autumn, and to the north in spring, pass to the westward of this station. In the spring, Swallows are observed passing to both the north-east and north-west in great numbers during April and May, and a number of other summer birds are recorded on passage.

METEOROLOGICAL.[1]

When studying bird migration in connection with meteorological conditions, it is only necessary to consider the weather peculiarities prevailing in the area where the movement has its origin. The weather prevailing upon the shores reached after an extended flight does not affect the movement. An extensive series of comparisons instituted

---

[1] The following is only a brief abstract of the section in Mr. Eagle Clarke's report.

between the records of migration and the *Daily Weather Reports* of the Meteorological Office, shows an intimate connection between weather and bird migrations.

During the spells of genial weather that occur in both the spring and autumn periods of migration, the movements are of an even and continuous nature. If the weather at such times becomes slightly unsettled, it is a matter of indifference to the migrants; but if it becomes more markedly variable, their movements are somewhat quickened.

Certain weather conditions, however, have a very decided influence in either hindering or promoting migration. Unfavourable conditions of a pronounced nature may render migration impossible, while, on the other hand, favourable weather following unfavourable periods may tempt birds to start on their way. A decided fall of temperature warns them to move southwards, and such cold spells are characteristic of anticyclonic periods when the weather is calm and highly favourable for a prolonged flight. The most important factor is temperature, and it regulates migration, other meteorological conditions being favourable.

From the end of September to early in November the autumn migrations into our islands from the north-east are sufficiently pronounced to be associated with weather changes. In ordinary seasons the period named is characterised by a series of great immigratory movements simultaneously performed not only by many species, but also by a vast number of individuals. It has been ascertained that *all* these great movements are due to the prevalence of weather conditions favourable for emigration in north-western Europe. These conditions are the result of the following type of pressure distribution—namely, the presence of a large and well-defined anti-cyclone over the Scandinavian Peninsula, with gentle gradients extending in a south-westerly direction over the North Sea. On the other hand, cyclonic conditions prevail to the westward of the British area, with a low pressure centre off the west coast of Ireland, or, occasionally, farther south. Consequently the weather is clear and cold, with light variable winds, over Norway and Sweden; while in Britain the sky is overcast, and moderate to strong easterly winds are experienced. Previously to this period cyclonic conditions usually prevail in Scandinavia, which check migration and at the same time warn the birds that it is time to depart. The fall of temperature on the formation of the anticyclone is another incentive, and it is not surprising that there is a rush southwards as soon as the weather becomes favourable.

The gentle pressure gradients do not always extend entirely across the North Sea, and the emigrants may fall in with bad weather before they reach our shores. If the western cyclonic system is too close to Britain, or if the depression is exceptionally deep, strong winds prevail on the eastern coasts, and the birds perform the latter part of their journey under trying conditions. The autumnal immigration by the "East and West Route" are subject to much the same influences. Of course, in addition to the favourable and unfavourable conditions described above, there are intermediate phases, whose influences ar easily determined by a study of the two sets of phenomena.

The great spring emigratory flights, and most of the lesser ones, are undertaken under precisely the same type of pressure distribution as is so favourable to the autumn immigrations, namely, high pressure to the north-east of our islands over Norway and Sweden, with gentle gradients to the south-west. As in the autumn, the favourable periods usually follow spells of weather decidedly unpropitious for bird migration.

The importance of winds in connection with bird migration has been much over-estimated. The direction of the wind seems to be a matter of indifference; but its force may arrest migration or carry birds out of their course. Birds do not emigrate when the wind is exceptionally high, but they pay no attention to the direction from which it blows. It is true that particular winds (easterly) almost invariably prevail during the great movements, and hitherto they have been considered as direct incentives to migration. Such is not the case, and it may be at once stated that these supposed favourable breezes are simply another direct result of the pressure distribution favourable to the movements. As far as direction is concerned, westerly winds would be equally favourable to the emigration, but they are produced by cyclonic disturbances to the north or east of the British Isles, that is, over the areas from which our autumn immigrants come.

Gales have already been alluded to as arresting migration or rendering it impossible. They also sometimes sweep certain marine species out of their course, and hence these birds sometimes appear on our coasts in exceptional numbers. In foggy weather, characteristic of high pressure systems, birds are frequently killed in great numbers against the lanterns of the light-stations, and often lose their way.

THE COUNTRY OF THE YAKUTS.

IT is singular that a people inhabiting a country where the climate is so severe as in the north of Siberia, and even the necessaries of life are in some seasons difficult to obtain, should preserve their racial characteristics and habits in the face of invaders of far superior culture. This is, however, the case with the Yakuts; their nationality is of such vitality that, far from showing any sign of disappearing before the advance of Russian civilisation, it impresses its stamp on the invader, and the Russian immigrants become "Yakutised." The Yakuts have, therefore, naturally attracted the attention of Russian ethnologists, and much has been written on them and their country in the Russian language. One of the most complete accounts of this people has lately appeared in a large work [1]—or rather the first volume—written by M. V. L. Seroshevski, who frequently visited the country between 1880 and 1893. Before discussing the Yakuts he describes the country they inhabit, both from his own observations and from those of former travellers.

---

[1] *Yakuty. Opyt etnograficheskavo Izsliedovaniya.* V. L. Seroshevskavo. Tom. I. St. Petersburg, 1896.

If with the mouth of the most westerly arm of the Lena, in lat. 72° 30′ N. and long. 123° 50′ E. as centre, a circular arc of 1060 miles radius be drawn across the continent, it will fairly accurately limit the territory of the Yakuts, though it will include certain districts never inhabited by them, and leave out some of their most advanced posts in the south and on the shores of the sea of Okhotsk. To the west stretches the low marshy plain west of the Yenisei, to the east a narrow, rocky strip of littoral, and, farther north, the spurs of the ranges occupying the Chukche territory and the Kamchatka peninsula. Lastly, to the south lies a broad mountainous zone, with narrow, deep valleys, sharp crests, and ranges and ridges resembling congealed sea breakers. The line defined above passes continuously through such mountainous country, across which the Yakuts, wherever they came from, could only with difficulty have made their way, and could not have lingered long owing to the absence of pasturage. Their branches joining on to one another and interlacing, the mountain ranges encircle the Yakut country like a gigantic crown, 200 to 300 miles in breadth, with its broken ends dipping into the Arctic Ocean. Innumerable collateral branches and offshoots, running for the most part from north to south, are crowded together within the bounds of the territory in question; while others, on the outer side, stretch southwards towards the heart of the continent and coalesce with other systems, or expand into high plateaus. Great variety of relief may be expected in a territory which covers about 1,300,000 square miles. The bounding elevations have also very different altitudes; the highest part apparently lies in the south between long. 112° and 128° and between 57° and 60° N. lat. As the ranges approach the Arctic Ocean on the north, they decrease in height, the eastern section being loftier and more diversified than the western. While the highest southern part of this zone has an average elevation of 4000 feet, the eastern continuation rises to 3000, with northern spurs along the river Kolyma of 2000 feet. The western wing has a height of 2000 feet, with northern spurs rising to 200 or 300 feet on the right bank of the Yenisei, north of the Dudina.

The offshoots which traverse the territory within this bounding line of elevations also in general sink towards the north and west. They follow a meridional direction, as may be seen from the courses of the rivers, sloping on the left of the Lena to the west and on the right to the east. Within the semicircular bulwark the country quickly loses its mountainous character, until north of 60° N. lat. and between 116° and 137° E. long. it assumes the form of an elevated plateau surmounted by chains of low hills and isolated eminences. From a height above sea-level in its southern part of 1500 feet it descends northwards to 1400 feet or even 1000 feet. All this country is slightly undulating, and is bestrewn with lakes and intersected by deep river-valleys. North of 64° it again rises and assumes a mountainous character.

This second mountainous zone extends from west to east through the whole breadth of the territory, and its extremities intermingle with the spurs of the circumscribing mountain zone. Known in their western part by the general name of Syverma, these ranges rise to a mean height of 2000 feet, or the same as that of the neighbouring ranges on the

lower course of the Lower Tunguska. Eastwards the Syverma passes
into a plateau containing a whole system of rather large alpine lakes,
among them the south-western lake Siurungna. The height of this
district is, perhaps, 3000 feet above sea-level, though the only summit
accurately measured—the mountain at the southern end of Siurungua—
rises to 2792 feet. East of the lake region stretches the watershed of
the Olenek and Viliui, a deeply undulated country connected by its south-
western spurs with the mountains on the middle course of the Lower
Tunguska. Summits and ridges fill up the interval between the Olenek
and Viliui and reach nearly up to the Lena valley, beginning, however,
sixty miles from the lower course of the Viliui and breaking off at the same
distance from the Aldan, a tributary of the Lena.

The south-eastern part of the Verkhoiansk range continues the
easterly direction. The mean height is about 4000 feet, but its highest
points, which are apparently in the east, where it adjoins the Yablonoi
mountains, are much loftier, Cherski mentioning heights up to 7794 feet
above sea-level.

The northern offshoots of the central mountain zone run in meridional
directions, forming the watersheds between the rivers flowing into the
Arctic Ocean. In their southern parts they take the form of well-
defined ranges with sharp summits more or less broken and eroded,
while near the sea the mountains are more rounded. Sometimes they
disappear, giving place to plateaus 1000 feet high, studded as on the
southern side with isolated undulations and numerous lakes, as, for
instance, between the Alazeia and Kolyma. The southern spurs, on the
other hand, are short and are broken off sharp at the valleys of the
Viliui and Aldan.

Thus the central range lies between similar plateaus of about the
same height on the north and south. The plateau on the northern side
is open to the sea, and its parts are clearly defined by offshoots from the
mountain ranges. Therefore the basin of each of the principal rivers
has its own mountainous region, its plateau region, and a low stretch
near the sea, and does not encroach on its neighbour's ground. The
southern plateau is more homogeneous; the watersheds are ill-defined
and the whole country forms a slightly undulating plain stretching east
and west, with an area of some 220,000 square miles, surrounded on all
sides by mountains rising to a height of 2000 to 4000 feet above the
sea-level and 1000 to 3000 feet above the plateau.

The rivers of the northern plateau flow to the ocean; those of the
southern plateau to the Lena. Only the last traverses both mountain
zones and both plateaus, and through the arms of its delta pours more
than 350,000 cubic feet of water per second into the Arctic Ocean, which
it collects from its enormous basin of 880,000 square miles. The valley
of the river is 4300 miles long, beginning outside the Yakut territory in
mountains 3000 to 4000 feet high, near Lake Baikal. As far as
Kirensk it flows through a rather open valley deeply excavated in a high
plateau.

The head-waters of the Lower Tunguska flow at first in the same
plateau near the Lena. The valleys of the two rivers form, as it were,

the same hollow for some distance. They are separated by a wooded ridge not more than thirteen miles broad, which rises 1627 feet above sea-level, 811 feet above the Lena, and 562 above the Tunguska. On either side the rivers the country is gently undulating. Below Kirensk a mountainous wedge is interposed between the rivers, gradually rising northward and spreading out, so as to force the Lower Tunguska to incline towards the west and the Lena towards the east. The former, near the Yenisei, forces its way through a country of block mountains; while the Lena, breaking through a rocky bulwark that bars its passage between Kirensk and Vitimsk, again emerges on to a plateau country very similar to that it has passed through in the south. Thirty miles above Yakutsk the Lena turns straight to the north and preserves this direction as far as the conference of the Aldan, where, probably following the direction of the western offshoots of the Verkhoiansk mountains, it inclines towards the west up to the mouth of the Viliui. After this its course is on the whole northwards, with slight deviations to either side.

Except some not very dangerous rapids where the river is contracted, there are no other impediments to navigation on the Lena, no troublesome shallows or bars, for a distance of fully 2700 miles. At Kachugskoie the river is 70 yards broad, but after receiving a number of small tributaries and uniting with the Kirenga it widens out to 700 yards, while beyond the mouth of the Vitim it is seldom less than two-thirds of a mile, and below the Olekma a mile. Near Yakutsk it spreads out to two or two and a half miles in one or two places, and where there are islands the extreme breadth is seven to ten miles across—at Siktiakh, 265 miles from its mouth, as much as twenty miles—below which it contracts again till it breaks up into the arms of the delta. From the Viliui downwards the depth of the navigation channel is probably nowhere less than 35 feet, while in many places it sinks to 140 feet and more. The fall of the river thirty miles below Kachugskoie is 1·16 feet per mile, 660 miles from that place it is 0·59 feet, at Yakutsk, 1230 miles below Kachug-skoie, only 0·14 feet, and still less farther down.

The tributaries of the Lena, large and small, number about 1000. All that are of any importance enter the main stream above the Aldan on the right bank and the Viliui on the left. The Molodu, on the lower course, deserves mention because in old times Cosacks and traders used to ascend it to cross over to the Olenek. The Vitim and the Olekma rise far to the south in the north-eastern part of Transbaikalia, in the same latitude as the Lena, 54° N. Their valleys are there more open than in their lower and middle courses, where they pass through the rocky mountainous barrier forming the southern boundary of the Yakut territory. In this part the stream is swift and the rapids numerous, so that the rivers are navigable only for two or three hundred miles from their mouths. Pasture lands are rare, and the inhabitants, consequently, few. Strictly speaking, they cannot be called Yakut rivers like the Aldan and Viliui. The latter are also the largest of the affluents of the Lena, and are geographically interesting as belonging to the central basin enclosed between the outer and inner mountain zones of the country.

The Aldan has a length of quite 1300 miles, of which 935 are navigable. Its breadth, where it flows in a single channel, near its mouth, does not exceed two-thirds of a mile. In swiftness it is the second river of Eastern Siberia, the Angara being the first. There are no rapids in its bed, but numbers of sunken rocks and stony banks; and when the water is high, covering these obstacles, it fairly hisses with the rapidity of its current. Rising in 56° 15′ N. lat. and 123° 23′ E. long., not far from the valley of the Olekma, at a height of 3000 feet, the river flows east-north-east parallel to the Lena and 260 to 330 miles to the east of it. At about 52° 30′ N. lat. it reaches the outskirts of the inner mountain zone, and, turning sharply westward, falls into the Lena at 63° 15′ N. lat. Except in its upper course, it flows constantly at the border of the mountain and the plateau. Besides the Amga, which flows from the plateau between the Lena and Aldan, the latter receives several mountain streams, one of which, the Maia, on the left bank, is navigable at high water for a distance of 330 miles.

The Viliui rises in 65° 45′ N. lat. and 102° 53′ E. long., in the same elevations whence flow the Olenek, Anabara, and Khatanga. Its upper course is swift and full of rapids. At high water the river is navigable up to Suntar, a distance of 748 miles; but in autumn only to Viliuisk, or half the above distance. The upper tributaries of the Viliui are unimportant, but the brook Kampedziai, entering the bend of the river near Suntar on the right side, deserves mention, because on its head-waters are salt springs, lakes, and pans capable, if properly worked, of supplying the whole Yakut territory with cheap salt. The Viliui, from its affluent the Chona, flows east-north-east parallel to the Lena at a distance of 260 to 330 miles from it, and its chief bends correspond in a remarkable manner to those of the latter. It skirts the northern edge of the Lena plateau on the west as the Aldan does in part on the east. The mouths of these two rivers are less than 130 miles apart, and their valleys are rightly considered by Reclus to be virtually one; where it intersects the valley of the Lena is the lowest point of the Yakut plateau. The largest tributaries are on the left bank below Suntar, and after receiving the Viliuiska the river widens out to 660 yards, and encloses islands. Its total length is 2000 miles.

The rivers of the northern plateau—the Piacina, Khatanga, Anabara, Indigirka, Yana, Kolyma, and others—have all one common trait; after emerging from the mountains they flow for some distance over a high, fairly level plateau, occupying broad open valleys, and then, near the sea, again enter narrow rocky ravines. In their lower course they break through some slight undulations running from west to east. Most of the valleys have a twofold form, which is most pronounced in their middle section. They consist of a spacious hollow, five miles across, or up to several times that breadth, eroded to a depth of two hundred to three hundred feet in the plateau. At the bottom of this flows the river, in a channel some hundreds or thousands of feet broad, and sixty to a hundred feet deep, or more. In the middle course of the river the bed is divided by islands into several arms. The islets are seldom as high as the banks of the river, and are formed exclusively of alluvial deposits of recent origin. Both the banks of the river channel and the flanks of the outer valley are steep, and in many cases exhibit traces of old water levels. Below them, on the bottom of valley, are scattered pools, most of which are elongated in the direction of the river. Their narrow, riband-like form, and the heaps of alluvium on their bottoms, prove that they were once channels of running water. They generally lie over one hundred feet above the river, and are seldom reached by floods. The present rivers, therefore, wind their way through their old deposits, and, possibly, gradually moving their bends from one side of the valley to the other, have several times furrowed and turned over the bottom.

The lakes of the territory are for the most part scattered over the plateaus, though there are alpine lakes, and even whole series, surrounded by mountains, and lying at considerable heights above sea-level. The lakes of the plateaus lie more thickly in their lower parts than on the mountain slopes. On the southern plateau there are lakes everywhere, but their number increases towards the north, and they are particularly abundant in the part adjoining the Lower Viliui. They are numerous also in the northern part of the Amga-Lena plateau.

But the country *par excellence* for lakes in all this section of Siberia is the lowland district between the Lower Kolyma and Indigirka, which also lies at a lower elevation above sea-level than any other tract. The lakes have a diameter of only a few leagues, but their dimensions are not so striking as their number, the natives saying that they are like the stars in the sky. Consequently fishermen can perform long journeys by dragging their boats from one lake to another. Sometimes the lakes are separated only by low, narrow isthmuses, so that the water can flow from one to the other, if the levels be different. This usually happens when the ground has been cracked by frost in winter. In the spring the water passes along the cracks into the neighbouring lake, and the latter over-flows into the next. The uncovered land, especially in the earlier years, forms excellent meadows and hayfields. Such lakes, therefore, are much frequented by the Yakuts, especially those from which the water has been only partially drained off.

The plateau lakes lie high above the rivers, but seldom feed them. Many lakes are connected by streams, but these do not always reach the main river, frequently losing themselves in extensive swamps. Should a connection with the main stream be formed, the days of the chain of lakes are numbered; the brook soon erodes a deep channel, and the water is drained off, leaving a hollow of meadow land, which soon attracts the Yakuts.

Of dry, even steppe-land there is little. What there is, is mostly found in the valleys of the large rivers and on the beds of drained lakes.

To sum up, the southern part of the territory consists of a huge depression extending east and west, and covering an area of fully 220,000 square miles. Its surface, slightly undulating and sloping somewhat up towards the north, lies one thousand feet or more above sea-level. It is surrounded on all sides by a belt of mountains two hundred to three hundred miles broad. The depression is covered with sand, clay, and löss, and its surface is diversified by a confusion of small open hollows and flat hills. In each of the rounded hollows lies one or more lakes, with banks more or less marshy, and in each little dell there flows a stream. The mountains bounding the plateau on the southern side sink insensibly to its level, while those on the northern side rise from it in short, steep slopes. The three largest rivers of the Yakut country—the Lena, Aldan, and Viliui—meet in the lowest part; after receiving the two latter, the Lena skirts the Verkhoiansk mountains on their western side, and makes its way through the highlands in front of it to the sea.

The northern plateau is divided into two sections, differing both in situation and character. West of the Lena it consists chiefly of massive elevations of uniform character intersected by deep hollows, and traversed by mountain chains cast about without any regular order. East of the Lena the plateau is bounded on the south by the mountains of the periphery forming an unbroken chain, the ends of which, dipping in the Ocean, reappear again in the New Siberia Islands. The eastern plateau has something in common with the southern depression as regards relief of the surface, but the isolation of its parts is greater, for the ranges running in a meridional direction, and separating the basins of the Yana, Indigirka, Kolyma, and other rivers, are parallel and more pronounced; they are steeper, higher, and more continuous. Something more like a general uninterrupted plateau is found only close to the sea in the form of a narrow, tundra-covered, gently undulating strip, as well as in the swampy lowlands of the lower Kolyma and Alazeia. Each river basin, then, forms a complete whole, its southern and even its middle section mountainous, and its lower flat, swampy and abounding in lakes.

Through the midst of the country flows the Lena, entering the sea by numerous arms like the roots of an immense tree. Near the sea it has scarcely any affluents and forms a smooth strong trunk, three to five miles in breadth. The crown of this huge tree, composed of such great boughs as the Aldan, Viliui, Vitim, and Olekma, fills with its thousand secondary branches all the vales, dells, and ravines of the southern Yakut plateau. On the northern plateau nine large complicated basins and one hundred and twelve of more simple outline discharge their waters immediately into the ocean. All this drainage of the Yakut territory, from fully two thousand rivers and streams and some one hundred thousand lakes runs in one direction—to the north.

The climate of this part of Siberia, notwithstanding its enormous extent and considerable differences of latitude, is fairly uniform. Winter begins in all parts nearly at the same time. In the early days of September the weather turns cold and frost sets in, and in the first half of October the country is entirely covered with snow, which does not disappear until spring. By the end of the month the rivers are congealed, the lakes covered with ice, snow has fallen to a depth of six inches, and winter rages in its full severity.

On the northern plateau the rising wave of cold is checked by the moderating influence of the Arctic Ocean, which is frozen over only in ᶜ middle of winter, and consequently the lowest temperature is not ᵃched until January or even February. From the ocean the line of maximum cold passes in a south-south-easterly direction between the meridians of 112° and 122°, about, and attains its culmination in the neighbourhood of Verkhoiansk, where the mean.winter temperature is −54° F. (see also *S.G.M.*, vol. xi. p. 77). This is the pole of cold in the northern hemisphere. It is singular that from this pole of cold the temperature rises more rapidly towards the north than towards the south; Sagastyr and Ust-Yansk are warmer than Markhinskoie and Olekminsk. Eastwards and westwards, also, of the line above mentioned, the cold gradually diminishes with the distance.

The winter in the Yakut land is certainly subject to little variation, and is the same over the whole country. The degree of cold and its duration may differ in various parts, and from one year to another, but everywhere and always the winter is the calmest, clearest, driest, and most constant of the seasons. Sudden changes of weather and surprising bounds of temperature, characteristic of the spring, summer, and autumn of this region, do not occur in winter. The day temperature differs little from that of the night. Strong winds are scarcely ever experienced; light local breezes, with interrupted currents and no well-defined direction, blow most frequently. Nowhere is the winter so dry as in this corner of Siberia, except in Mongolia. The amount of precipitation is at most very small, the country to the east of the valleys of the Aldan and Yana receiving even less than the western districts. From Wild's map it appears that in the east, during the months of December, January, and February, the amount of precipitation is equivalent to 0·4 inches of water, being about one-twentieth of the annual, while on the west, in the neighbourhood of Yakutsk and Olekminsk, and northwards to Turukhansk it is 0·8 to 1·6 inches, or one-tenth of the annual. The maximum, two inches, being one-eighth of the annual, occurs in the comparatively damp valley of the Viliui.

The winds of spring yield in strength only to those of summer, and in duration to those of autumn. In the north they produce the notorious snow-storms that render travelling in the *tundras* exceedingly difficult and even dangerous. Sometimes the traveller has to lie in a hut for days together, rolled up in deerskins and without fire or warm food. It occasionally happens that the sky is clear, but then the wind carries with it a cloud of icy particles swept from the surface of the snow, hiding even the sun. After such storms the even surface of the tundra or of the broad northern lakes exhibits a series of wrinkles and runnels like ripples on water. In the heart of the continent, within the forest limit, the winds blow more softly; but even there, especially in spring, they do not fail to modify the distribution of the snow, a matter not unimportant in agricultural and pastoral districts. If they set in before the snow begins to settle down and become compact owing to the warmth, they sweep it into the lower spots, clearing the tops of hillocks and the slopes exposed to the wind. The soil on the exposed parts quickly dries and therefore does not yield the usual quantity of corn or grass. A windy spring is on this account undesirable.

Violent winds blow in spring in the districts of Yakutsk, Viliuisk, and Olekminsk, especially from the north, or south-west; they are dry and cold.   The south winds are very warm and promote the thawing of the snow.   East and north winds are accompanied by bad weather and snow; in winter these winds are warm, and cold in spring and summer. The snow that falls in spring is nearly three times as much as in winter, except in Viliuisk where there is less in spring, and in Turukhansk, where the quantities are about the same.   The snow rarely reaches a depth of three or four feet, and is usually much less.   In autumn, winter, and spring, the fall of snow in the neighbourhood of Turukhansk is equivalent to 7·9 inches of water, whereas in the east, in the basins of the Yana, Indigirka, and Kolyma it is only 3·15 inches.   In other parts the quantity lies between these figures.

In the beginning of April, the shade temperature at noon is not in-frequently as high as 18° above freezing-point.   The snow begins to thaw and turn into slush.   Vegetation begins to sprout, birds appear, and life and movement revive after the winter sleep.   The spring advances with the rapidity usual in high latitudes.   On the southern parts of the plateau the mean day temperature in May seldom sinks below freezing-point, and that only in the first half of the month.   In the latter half of the month snow remains only in deep wooded hollows.   In the north, the advance of spring is retarded by two or even three weeks.   The small streams are relieved of their ice at the end of May, and the larger rivers in the early days of June.   Immediately after, strong westerly winds begin to blow almost daily, and continue fairly regularly for several days.   These cold winds are the last manifestations of winter; when they die away summer comes in.

The summer rains increase in quantity with the distance from the sea, in a south-westerly direction, so that the minimum fall (two to four inches) occurs at the most northern limit of the continent, the mean (six inches) in the valleys of the Aldan and Viliui, and around Yakutsk and Turukhansk, and the maximum (seven to eight inches) in the valleys of the Olekma and Vitim, and the adjacent parts of the basins of the Lena and Lower Tunguska.   As regards time the rainfall is very unevenly distributed.

The hottest month in the year is July, and the highest temperatures are found in places lying a little to the south of the pole of cold.   Here the annual range is not far from 180 degrees (Verkhoiansk 88·4°, — 88·8° ; Markinskoe 100·2°, — 78·7°).   Though the thermometer sometimes falls at night to 35° or 37°, it very rarely sinks below freezing-point.   The heat in the sun in June and July attains to truly African fierceness ; it is not uncommon to register 118° several feet above the ground..   The stones and sand become so hot that it is impossible to walk on them barefooted, and the heat penetrates even through the skin boots of the Yakuts.   Sunstrokes seem, nevertheless, to be quite unknown.

A continuous covering of permanent snow does not appear earlier than the beginning or even the middle of October.   At this time the precipitation is great everywhere.   At Turukhansk, during twenty-one days, as much as is equivalent to over 1·1 inches of water has been measured, or half the winter snow ; and in Sredne Kolymsk 0·34 inches, or four-fifths of the winter fall.   By the end of the month the winter has come in full force, and the thermometer never rises above freezing-point. Properly speaking, there is in this country neither spring nor autumn. Only a part of May has the characteristics of spring, and October is the only autumn month.

Intimately connected with the severity of the climate is the layer of perpetually frozen soil that lies at a certain depth all over the country. Near the pole of cold it extends down to fully 400 feet. Farther south it diminishes in thickness and, indeed, disappears in some places, as on the Vitim plateau, where it is only found in patches on the bottom of narrow deep valleys. The depth also changes considerably from one place to another.

Forest, chiefly coniferous, is the predominating surface feature. A comparatively narrow strip—thirteen to twenty miles broad—of muddy-green *tundra* of moss and lichen intervenes between the forest and the Arctic Ocean, and the crests of the higher mountains overtop the forests with their bare, black, gray, yellow, or reddish crags. Beside the brooks and lakes is seen a thin border of light-green meadow and marsh plants, and grayish green *Cytisus* and willow. On the whole the summer dress of the country is fairly variegated. Forest extends as far north as 68° or 69° on the eastern rivers, and 70° to 72° on the Lena and those farther west. On the Khatanga, under the shelter of the far-projecting Taimyr peninsula, it is found as far north as $72\frac{1}{2}°$, this being the most northerly wooded spot in the world. But in these high latitudes the forest is poor, consisting of stunted trees, fifteen or twenty feet high, covered with long-bearded mosses, which put forth very little verdure and afford no shade. Such is its character for nearly sixty miles from its edge, and some patches of similar distorted trunks are found as far inland as Yakutsk. The summits of the inner range are bare above 2000 or 3000 feet, while in the south the woods often ascend to 3000 or 4000 feet. Only on the southern side of the central range do trees grow freely, and here first appear larches, firs and picea, cedars and other trees which are not found on the northern plateau.

The birch, poplar, alder, and aspen, though they occur locally farther north, on the ranges of the central chain, form woods only in the south. After the larch, the pine is certainly the most widespread form of tree. It grows on the Yenisei from 55° to 65° N. lat., and on the spurs of the southern Aldan range to a height of 3500 feet. It was a valuable tree to the Yakuts before they became acquainted with cereals, for they made meal of the bast. The Siberian spruce and fir exceed in height all the other trees of the forest.

There is also a large variety of shrubs and plants suitable for fodder. The cultivation of cereals seems to have gradually spread over the country during the past 250 years.

The above extract gives only a very faint idea of the mass of information on the country and its flora contained in M. Seroshenski's work. Nor is it possible here to say anything of the fauna, which includes a large number of fur-bearing animals, many game birds, such as ptarmigan, capercailzie, etc., and several species of fish, the *Salmonidæ* being represented by many varieties.

As regards the Yakuts themselves only a few words can be said concerning their original home. As is well known, it has been much disputed whether they are an indigenous race or have migrated from more southern latitudes, the latter view being more generally accepted at the present day. According to their own traditions, they are descended from two famous marauders, who, having made themselves objectionable in their native country, were obliged to fly northwards, and, on reaching the river Lena, sailed down it to a point a little to the south of Yakutsk. M. Seroshevski sees a confirmation of the truth of this legend, in its main outlines, in the fact that the Yakuts have names for the lion, camel, serpent, tiger, and other animals which do not exist in the present home of the people, and that these names, as well as others, have a resemblance to those used among the Turkish and other tribes of the south. But he confesses that it is very difficult to discover their original home, and almost impossible to decide at what period they migrated into their present territory.

To the ethnologist the book is full of interest, as it describes minutely the mode of life of this people, and their occupations and industries, with illustrations of their implements, dress, ornamental designs, villages, etc.

PROCEEDINGS OF THE ROYAL SCOTTISH GEOGRAPHICAL SOCIETY.

MEETING OF COUNCIL.

AT a meeting held on November 17th, the following ladies and gentlemen were elected Members of the Society :—

Buchanan, Mrs.	Chalmers, Sir D. P.
Burn-Murdoch, Miss.	Chapman, General E. F., C.B.
Glendinning, Miss E. A.	Cowan, Colonel H. S.
Gordon, Mrs Glegg.	Cumming, John.
Hay, Miss Jane.	Davidson, Charles.
Macnair, Miss Jean.	Dewar, Thomas.
Meikle, Miss J. M.	Drysdale, J. W. W.
Melville, Miss Balfour.	Edmond, John.
Pirie, Miss E. C.	Finlay, Herbert B.
Robertson, Miss C. L.	Glasgow, The Earl of.
Sullivan, Miss.	Hamilton, The Rev. W. A. D.
Talbot, Miss Mary.	Henderson, A. Duff.
Thomson, Mrs. E. A.	Hilson, J. Lindsay.
Wyld, Miss.	Law, G. H.
Allan, Major-General W.	Logan, Alex.
Alexander, W. Lindsay.	M'Arthur, J. S.
Arthur, Alex. T.	Maitland, Keith R.
Braid, Robert.	Martin, T. S.
Brigham, David.	Milne, Alex.
Bryden, H. B.	Moir, J. W.
Carruthers, J. G.	Moray, The Earl of.

Paterson, James.	Smith, Joseph.
Robertson, The Rev. A. E.	Stewart, Charles.
Rose, Hugh G.	Stewart, J. Lindsay.
Sanderson, Kenneth.	Stibbe, Godfrey.
Scott, Robert A.	Trevor, Alfred C.
Scott, The Rev. W. D.	Wade, T. Callender.
Shepherd, W. H.	—In all fifty-five.

Ordinary Diplomas of Fellowship were conferred on Mr. J. A. T. Ross Cormack and Captain D. V. Pirie, M.P.

It was also decided to confer the Society's Medal on Dr. Nansen, and to make him an Honorary Member.

The following gentlemen were elected Corresponding Members for the present session:—

Anderson, R. Hay, Esq.,	*Mexico*	Mohn, Professor H.,	*Christiania*
Ballivian, Don M.,	*Bolivia*	Mulhall, Michael G., Esq.,	*Buenos Ayres*
Begg, Alex., Esq.,	*British Columbia*	Murray, Charles, Esq.,	*Assam*
Bell, Dr. Robert,	*Ottawa*	Penck, Professor,	*Vienna*
Bodio, Professor Luigi,	*Rome*	Pittier, Professor,	*Costa Rica*
Chaix, Professor Paul,	*Geneva*	Russell, Professor I. C.,	*Michigan*
Corstorphine, Professor,	*South Africa*	Seidlitz, Mons. N. de,	*Tiflis*
Dingelstedt, Mons V.,	*Geneva*	Steffen, Dr. Hans,	*Chile*
Gannett, Henry, Esq.,	*Washington*	Thomson, J. P., Esq.,	*Queensland*
Harrington, Mark W., Esq.,	*Washington State*	Wagner, Professor Dr. H.,	*Göttingen*
		Wells, Colonel H. L.,	*Persia*
Johnston-Lavis, Dr.,	*Monaco*	Wild, Dr. J. J.,	*Melbourne*
Macdonald, Dr. W. H. B.,	*East Africa*	Willoughby, Rev. W. C.,	*Bechuanaland*
Margerie, Mons. de,	*Paris*	Woodford, C. M., Esq.,	*Pacific Islands*
Mason, W. B., Esq.,	*Japan*	Wragge, Clement L., Esq.,	*Brisbane*
Mello, Professor Carlos de,	*Lisbon*	Yate, Captain A. C.,	*India*

THE ANNUAL BUSINESS MEETING

was held in the Society's Hall on November 18th, Professor Copeland presiding.

On the recommendation of the Council, the Marquis of Lothian was re-elected President, and the Vice-Presidents were re-elected, and the Earl of Stair was added to their number.

The following twelve Members of Council, who retire by rotation, were re-elected:—Professor Butcher, Mr. John Geddie, Principal Grant Ogilvie, Professor Knott, Mr. W. C. Smith, Dr. Scott Dalgleish, Mr. Coutts Trotter, Dr. G. A. Turner, Mr. R. S. Allan, Sir W. Renny Watson, Mr. L. F. U. Garriock, Sir David Stewart.

In room of the other eight members who retire by rotation, and to fill two other vacancies, the following gentlemen were elected:—Mr. Spencer C. Thomson, Colonel John M. Trotter, Mr. H. M. Cadell, Dr. John Kerr, Mr. J. R. Middleton, The Hon. J. Abercromby, Mr. John Harrison, Mr. Paul Rottenburg, The Rev. Dr. Ewen, Mr. I. J. Weinberg.

All the other office-bearers were re-elected. On the motion of the Chairman, seconded by Dr. W. G. Black, the report of Council for the past session was adopted.

Dr. Karl Grossmann lectured on "The Less-known Interior of Iceland," at Edinburgh and the Branches. On the 5th of the month he addressed the Dundee Branch, when Mr. John Robertson presided, and Mr. Garriock proposed a vote of thanks to the lecturer. On the following evening Dr. Grossmann appeared before the Aberdeen Branch, Professor Pirie taking the chair. The Glasgow Branch met to hear Dr. Grossmann on the 9th, when Mr. Paul Rottenburg presided. Lastly, on the 10th, the lecture was repeated in Edinburgh ; Dr. John Murray was in the Chair.

A meeting was held in Edinburgh on November 26th, when Captain Joseph Wiggins described his expeditions to the Yenesei river. Mr. Robert Cox, M.P., presided.

MEETING IN DECEMBER.

On December 10th Dr. Charles Sarolea will deliver an address on "French Colonial Expansion in Africa," in the Society's Hall, Edinburgh, at 4.30 P.M.

THE NANSEN MEETINGS.

Dr. Nansen will lecture before the Society in Edinburgh on February 12th ; at Glasgow on the 15th ; at Dundee on the 17th ; and at Aberdeen on the 18th. He has accepted an invitation to a banquet to be held in Edinburgh, on February 13th.

GEOGRAPHICAL NOTES.

By THE ACTING EDITOR.

EUROPE.

The Gypsies of Hungary.—As a first step towards remedying the evils arising from a wandering population, a census of the Gypsies, both nomad and settled, was taken in Hungary on January 31, 1893. Not only were they enumerated, but information of all kinds relating to their mode of life, occupations, capacity for various kinds of employment, etc., was collected. In the census they were grouped under three headings—settled Gypsies ; those who stay for a long period in one place ; and wandering Gypsies. Many are not easily classified, for there are Gypsies who regularly pass the winter in the commune to which they belong, often dwelling in tents, and in summer wander elsewhere to obtain work, and many of them to the same district every year. Again, it is hard to define how long a residence in one place qualifies a Gypsy for the second category. His occupation sometimes retains him for years in one locality, and yet he does not make it his fixed abode. Nor is the possession of a tent a criterion of the wandering Gypsy, for some stroll about without tents, while others occupy them though they remain in one place.

The total Gypsy population was found to be 274,940, which figure is three times that obtained in the census of 1890 ; but in the latter only those were reckoned Gypsies who acknowledged the Gypsy language as their mother tongue. Many, no doubt because of the aversion of the race to contact with officials, and from other causes, were omitted—in Budapest no return was made owing to a mis-understanding of the scope of the inquiry—and, therefore, the total number may be taken to be in round figures 280,000. 243,432 of the 274,940, or nearly nine-tenths, are settled, about 7½ per cent. sojourn for a long time in one place, while only 8938 are constantly on the move.

Transylvania is the classic land of this people. From here they advanced five centuries ago into the heart of Hungary, and thence spread over all Europe. They still form five per cent. of the inhabitants, being perhaps numerous here because the adjacent Roumania contains the largest proportion of Gypsies in any country in Europe. They dwell in large numbers in the communes on the Küküllö rivers, forming in the county of Nagy-Küküllö ten per cent. of the population. From Transylvania the country most thickly peopled by Gypsies extends in two directions—north-westwards between the mountains and the Alföld and south-westwards into the most southern corner of Hungary, lying between Roumania and Servia. In Western Hungary they are few, for neither the unfruitful mountains nor the low plains are favourable to their distribution. Where the people are poor there is little to pick up and little demand for smith's work and other Gypsy industries, while in the Alföld the primitive occupations of this people find no scope.

The settled Gypsies sometimes live apart from the other inhabitants of the commune, sometimes mingled with them ; in some communes those who follow a despised or unclean occupation live apart, the others being scattered among the inhabitants. The percentages of those in each of these classes are about fifty-two, forty, and eight. The chief condition of a settled mode of life is the possession of land. Only 3·7 per cent. of the Gypsies in Hungary cultivate land on their own account, the larger part of them being gardeners. Education is a most useful means of bringing about the settlement of this people, but naturally it is impossible to exact school attendance from the children of the wandering Gypsies, and among the others much opposition is shown to the school regulations. Consequently, of the 58,747 children who are of an age to attend school, 40,624 receive no education.

Much information has also been collected by the commission relating to the manner of life, customs, religion, and language of the Gypsies. As a rule they adopt the form of religion of the people among whom they live, without exhibiting any personal predilection. As regards language the same rule holds good, and consequently more than half the Gypsies are even ignorant of their national language. Their occupations are numerous, including a variety of small crafts, as in other countries. They are not capable of hard field work, and their want of education and of self-control prevents them from learning a trade under a master in a regular manner. Still, there are in the towns some Gypsies who are masters in their trades. From these details it is evident that Hungary has a somewhat difficult task before it in civilising the Gypsies and rendering them useful members of society, though the fact is encouraging that the number of those without any declared occupation is comparatively small.—*Mitth. der k.k. Geogr. Gesell. in Wien*, Bd. xxxix. Nos. 6 and 7.

The Rivers and Lakes of Sicily.—Professor D. Vinciguerra has visited Sicily for the purpose of investigating the condition of the waters in connection with fishing, and in his report has published some interesting notes on the inland waters. The rivers, though some of them drain basins of considerable extent, hold a scanty supply of water in summer, while they become destructive torrents in rainy winters. The water, traversing chalky lands impregnated with sulphur and salt, dissolves part of their constituents, and thus becomes rich in sulphuretted hydrogen, and especially chloride of sodium, to which is due the name Salso, applied to two of the Sicilian streams. The only exceptions are the rivers that descend from the more elevated summits of the Hyblæan group, which flow over the compact lime-stones forming the peninsula of Syracuse, and those which rise on the flanks of Mount Etna. In some streams the tide ascends to a distance of several miles from the sea, and therefore freshwater fish are not plentiful, while marine species are often noticed at long distances from the mouths. The lakes, generally called in Sicily *Biviere* or *Bevaio* (drinking pools), are not numerous ; the most important is the Lentini, in the province of Syracuse, created, or at any rate enlarged, by a dyke raised in the Norman period across the valley through which flows the Gallicci or Trigona. It has a perimeter of under nineteen miles, and an area of about four and a quarter square miles. The depth is small, being only six feet when the water is at its highest. The lake Pergusa, almost in the centre of the island, is fed only by rainwater and one or two subaqueous springs, and has no outlet. Formerly it had a perimeter of three and three-quarter miles, and a depth of fully thirty-two feet ; but after the great drought of 1862-64 its circumference contracted to a little over half a mile, leaving a large part of its bed dry, while the water over the remainder is nowhere deeper than twenty-three feet. The water at 62° F. has a density of 1007, and is very briny, so that neither men nor horses will drink it. Another lake near Terranova, formerly the Lacus Coccanicus, is fed by a canal from the river Durillo and from the cliffs of the neighbouring mountains. It is about five miles in circumference, and is divided into two basins, one of which has a depth of twenty-six feet, while the other becomes swampy at times. The water is brackish, and is drunk by cattle but not by horses. Besides the above, there are several other pools and marshes. Of all these sheets of water, those best adapted for sea-fish breeding are two, which lie between the villages Ganzirri and Faro, to the north of Messina. The larger one, near Ganzirri, communicates with the sea by a canal 220 yards long, and the other has also a canal of about the same length connecting it with the sea. In former times there were other outlets, now silted up. The Lago di Faro has a depth of nearly 115 feet, and the Ganzirri of not more than twenty-six feet. The Marinello lake, close by Capo Tindaro, is of recent origin ; fifteen or twenty years ago it was a small, natural harbour, protected from the sea by a dune, with one opening at the south, through which fishing-boats could enter and find shelter in bad weather. Now, owing to the accumulation of sand or the fall of detritus from the mountains, it has been in great part filled up. The present lake is 330 to 440 yards long, by 160 broad ; its maximum depth is about thirteen feet, and its density at 60° F. is 1024, while that of the sea is 1027 at 66°.—*Boll. della Soc. Geogr. Italiana*, vol. ix. Fasc. 10.

ASIA.

Meteorological Observations at Tiberias.—Mr. Glaisher gives, in the *Quarterly Statement* of the Palestine Exploration Fund, the record for 1893. The minimum atmospheric pressure was 30·226 inches, in August, and the maximum 31·22 inches, in November, and the range, therefore, 0·994 inches. The mean for the year was 30·632 inches. The temperature reached or exceeded 90° on 164 days in the year, and on eighty-one days attained to 100° or more, whereas in Jerusalem, the thermometer rose to 90° only on thirty days, and to 100° on one day. The maximum temperature at Tiberias was 108°, on July 18th ; at Jerusalem, 104·5°, on July 19th. The winter, on the other hand, was warmer than at Jerusalem, the thermometer falling as low or lower than 40° on six nights in the year, as against sixty-five nights at Jerusalem. The lowest record was 36°, on January 30th, while at Jerusalem it was 27·5°, on December 23rd. The yearly range was 72° at Tiberias, and 77° at Jerusalem. The lowest mean daily range was 17·1°, in February, and the highest 28·3°, in June. For Jerusalem the figures were 11·8° and 25·8°. The mean daily range for the year was 23·7° at Tiberias, and 19·7° at Jerusalem. At both places the mean temperature gradually rose month by month from February to July, and then as steadily declined. At Tiberias the annual mean was 72·4°, and at Jerusalem 61·7°. The mean humidity was 71·7°.

The rainfall for the year was 25·62 inches, as against 30·54 at Jerusalem. No rain fell from May 18th till December 10th, except two very slight showers of 0·05 inches each. If these showers be ignored, there were 205 days on which no rain fell. The heaviest monthly fall was 8·85 inches, in January.

The Kura and Aras.—Professor Gustave Gilson, of Louvain, writes to the *Mouvement Géographique*, October 18th, on a remarkable change which has recently taken place in the relation of these rivers. The Kura, in its lower course, enters a broad valley which gradually expands into a plain, the southern part of which bears the name of Mughan. On this plain the Kura was, until lately, joined by the Aras (Araxes), but, since the inundations of last spring, the Aras has made its way through the marshes and lagoons of the Mughan directly into the Caspian, and now falls into Kyzyl-agach bay. The soil of the Mughan is formed by rich alluvium carried down by the two rivers, and the Russian Government has long desired to bring it under cultivation. Floods in some parts and great drought in others were the difficulties to be overcome, and to drain, and at the same time irrigate, this tract a canal was projected from the Aras to the Caspian. This canal has been formed by the river itself, and only a few simple engineering works are needed to render it permanent.

The Ancient Course of the Oxus.—This question has frequently been discussed, and has occupied the attention of Russian officials since the conquest of Khiva. In 1872 General Stebnitzki discovered the Uzboï, which skirts the southern border of the Ust-yurt plateau in the direction of the Caspian, but he did not venture to affirm that it was the old bed of the Oxus. In 1875 Lupandin visited the country between the Uzboï and the Sary-Kamysh, and reported that the ground sloped regularly from the lakes to the Caspian ; and then it was considered almost certain that the Oxus had formerly flowed in this direction. The belief was, however, shaken in the following year, when the engineer Hellmann established the fact, by a series of exact levellings, that the Sary-Kamysh lakes are at a lower level than the Caspian. Again in 1879 a commission was sent out under General Glukhovski, the results of which have been published by M. Konshin in the *Zapiski* of the Tiflis branch of the Russian Geographical Society.

From the Khiva oasis two ancient arms of the Oxus run towards the salt lakes of Sary-Kamysh—the Kunia Daria and the Daudan—having a length of 150 to 180 miles. West of the lakes there is no trace of water, and southwards the ground rises about 260 feet in 125 miles, as far as the wells of Bala-Ishem. Beyond this the channel of the Uzboï is clearly defined, but, instead of comparatively rapid slopes of two or three feet to the mile, there are level stretches so that the change of level is only about twenty feet in one hundred miles. Instead of rising to the Sary-Kamysh lakes, the ground slopes down towards them, with a change of level of about 260 feet in 125 miles, and therefore the waters of the Oxus must have filled the Sary-Kamysh depression before they could reach the Uzboï.

The basin of the Sary-Kamysh, separated from that of the Caspian by the threshold of Bala-Ishem, has a subsoil of grayish friable clay, which is easily reduced to fine powder. The clay contains a large quantity of vegetable fibres. Below this layer, eight or ten inches deep, lies a kind of peat of reeds, which in turn covers a dry friable mud lying on a compact red clay, an Aralo-Caspian deposit or marl and hard limestones of Tertiary age. The surface is entirely covered with white or grayish sand, which, as well as the friable clay, contains shells of molluscs in abundance. The depression contains hills of hard, compact red clay, which bear great traces of the action of water, and were evidently islands. In one of the lowest parts of the depression is a row of salt pans, hidden under sand containing gypsum and salt, and covered by a layer of the valves of *Cardium edule*, a salt-water mollusc now living in the Sea of Aral. On the other hand, the *Cyrena fluminalis*, which is now found in the Amu-Daria, is absent, whence it may be concluded that the Sary-Kamysh basin was once connected with the Aral through the Aïbughyr, now dry, and that the water of this salt basin never became entirely fresh, in spite of the constant influx of the water of the Amu-Daria.

Thus the Oxus once poured into the eastern part of the basin by a delta with an area of 7700 square miles, and at a certain epoch the overflow made its way to the Caspian by the bed now known as the Uzboï. The water of the Uzboï was brackish and quite unfit for irrigation. The small size of the channel caused M. Bogdanovich to suppose that it was only a furrow excavated by rain. But it contains a number of shells, identical with those of molluscs still living in the Caspian Sea or the Sary-Kamysh. At no period can the Amu-Daria have flowed through a delta to the Caspian, either at the peninsula of Darja or round the Balkans. All this region has been formed recently, as shown by characteristic sedimentary deposits.

The Uzboï may be divided into two sections. From the Caspian to the wells of Igdy it was simply a gulf of the sea, now dried up; while from Igdy to Bala-Ishem it was the drainage channel of the Aral and Sary-Kamysh basin. Moreover, the ridge of Tertiary rocks separating the Caspian basin from the Sary-Kamysh is at a higher level than that between the latter and the Sea of Aral, so that the water that filled the Uzboï would be brackish, resulting from the mixture of the water of the Amu-Daria with that of the lakes. For, as the Uzboï has far smaller dimensions than the Amu-Daria, the latter could not pass through the Sea of Aral as the Rhone does through the Lake of Geneva. Along the course of the Uzboï, 250 miles long, no shells characteristic of the Amu-Daria have been found, nor traces of river sediment, nor any remains of irrigation works or cultivation in bygone times, and hence this channel cannot be the branch of the Amu-Daria which flowed into the Caspian, and, according to Arab historians from the tenth to the fourteenth century, was skirted by magnificent fields.

Vestiges of former extensive cultivation are only to be found in the neighbourhood of the Kunia-Daria, an ancient watercourse which branches off from the delta of the Amu-Daria and runs for 120 miles through the Khanates of Khiva, Kunia-Urgench, Deu-Kensken, etc. The network of irrigation channels extends even through the Sary-Kamysh hollow, terminating at the escarpment of the Ust-yurt plateau. It is, therefore, probable that when the Amu-Daria was in flood a part of the water, escaping by the Kunia, reached the Sary-Kamysh lakes and raised the level of the water twenty to twenty-five feet, so as to flood the adjoining fields.

Another theory is that the Amu-Daria communicated with the Caspian by the Charjui-Daria, a name applied by the Turkomans to a series of valleys dividing the Kara-Kum plateau into two and commencing near Charjui to end at the Uzboï, where the channel emerges near Bala-Ishem from the Sary-Kamysh basin. MM. Konshin and Lessar examined these valleys in 1885 and found no trace of fluvial deposits, no shells characteristic of the Amu-Daria, and no remains of irrigation. Between Bala-Ishem and Mirza-Chile, they noticed only isolated pools covered with a saline efflorescence and deposits of gypsum, indicative of the presence of sea-water. The Charjui-Daria may have been an old shore, but not a river bed.

Lastly, a third branch of the Oxus, the largest and most ancient, has been supposed to have commenced at Kelif on the upper course of the river and, pass-

Merv, to have extended to Askhabad, skirting the eastern slope of the Trans-caspian mountains, and to have finally entered the Uzboï near the Balkan mountains. If this were so, the Oxus must have entirely covered the desert of Kara-Kum, 600 miles long by 250 broad, and thus formed a delta of unprecedented extent. The only basis for this hypothesis is the presence of the names Shor and Chink (*chink* in Turkoman means an abrupt bank). If the Amu-Daria had flowed through the deserts of Turkestan, these would be covered by a layer of gray sediment in which would be found turf and river shells, whereas they are entirely, with the exception of the Sary-Kamysh, covered with dunes of red and black sand.

M. Konshin therefore rejects all the theories regarding the ancient branches of the Oxus, and condemns the project of a waterway between Moscow and Central Asia to be brought into existence by converting these "ancient arms" into natural waterways.—*Annales de Géographie*, No. 24.

Changes in the Tarim River.—In *Petermanns Mitt.*, Bd. xlii. No. 9, Dr. Sven Hedin describes the changes that have taken place in the Lob-nor and the Tarim river since the beginning of last century, as shown by the Wu-chang-fu map, a copy of which was published by Dr. Wegener in the *Zeitschrift der Gesellschaft für Erkunde zu Berlin* (see vol. x. p. 36 of the *Magazine*). The lake then lay more to the north, with its longer axis running east and west;, afterwards it was gradually rounded off, and the water expanded southwards so that at length the lake stretched from north to south. Subsequently the lake contracted, the water flowing through it to the present site, further to the south-east. The remains of the old lake were noticed by Dr. Hedin during his recent visit. The Konche-daria then flowed directly into the river much to the north of its present mouths. For the last nine years the Lob-nor has shown a tendency to return to its more northern position, for the southern lakes have been choked up with mud, sand, and decaying vegetation, so that their bottom lies at the same level as the old Lob-nor, and it is probable that the old lake will continue to increase at the expense of the more recent basin. It would seem, then, almost as if the two basins were alternating.

Were not the volume of the Yarkand river and its tributaries, the Aksu-daria, Cherchen-daria, Khotan-daria, etc., so large, the river would not be able to make its way through the most inland desert of the Earth, where the rainfall is very small, but would be lost in the sand, as is the case with the Keria river. The transport of mud still goes on, especially during late summer when the water is high, and now the differences of elevation over the Lob-nor district are very small. Between Kashgar (4035 feet above sea-level) and Lob-nor (2590 feet) the fall is only 1445 feet, while the interval is 13 degrees of longitude, and most of this fall occurs in the western part of the basin, near the mountains, so that in the east it is very slight.

Another agent is the wind, which blows usually in the Lob-nor district from the east, east-north-east, or north-east. The east-north-east is most prevalent, and blows with great strength, darkening the air with dust, and gradually driving the sand dunes westward.

The changes that have taken place in the Lob-nor region are seen at a glance from a series of small maps which accompanies Dr. Hedin's article.

The Rapids of the Yang-tse-kiang.—At present Ichang, 1090 miles from the sea, is considered the limit of steam navigation. In 1891, Chung-king, 370 miles further up the river, was opened to foreign commerce, but under restrictions which prevented any attempt to navigate the river by steam. All political obstacles have now been removed by the treaty between China and Japan, and only those of natural origin have still to be overcome. Between Kwei-chou-fu and Ichang the Yang-tse-kiang cuts its way through a mountainous country, forming at certain points rapids, which are very dangerous owing to the irregularity of the rocky ridges, bristling with sunken rocks, which bar the river. M. Eysséric has examined these rapids, the chief of which are at Shan-tan-pien, Sin-tan, and Yeh-tan, the last being the most violent of all. M. Eysséric measured the velocity of the current at mean water-level, and found that, while at the anchorage of Ichang the velocity was 2·5 knots an hour, at the Shan-tan-pien rapids it was 4·5, at the Sin-tan 6·5 to 7, and at the Yeh-tan 7 to 8·9. These measurements were taken as the junk was being towed along the bank; in the middle of the stream, where deeper water available for steamboats is found, the current must be still more violent. The Rhone and Rhine, which are rapid rivers, have a velocity of three knots an hour, and hence the velocity of the current will alone be a great obstacle to navigation by steamers, while a still greater is the danger from sunken rocks.

M. Eysséric thinks that it is impossible to improve the bed of the river by removing the rocks, as has been done on the Danube, and that a system of towage is likewise impracticable. The construction of a lateral canal he also believes to be impossible.—*Annales de Géographie*, No. 24.

AMERICA.

Ice-cliffs in Alaska.—Professor Russell, in the paper published in the *Magazine* two years ago, mentioned the ice found in Alaska covered by vegetation. Lieut. J. C. Cantwell has noticed a still more remarkable form of this phenomenon on the Kowak river in the north-western part of the peninsula. After a tortuous course of about 550 miles, most of it within the Arctic Circle, the Kowak flows into Hotham inlet, which opens into Kotzebue sound. About eighty miles from its mouth are cliffs of ice 80 to 150 feet high, and one of them measured as much as 185 feet. On the tops of all the cliffs is a layer of black silt-like soil, six to eight feet thick, on which grow mosses, grass, and characteristic Arctic shrubs consisting of willow, alder, and berry bushes, and a dense forest of spruce-trees, fifty to eighty feet high, and four to eight inches in diameter. The cliffs cannot have been formed by the stream, for the marks on those composed of soil and rock show that the water has never reached a level sufficiently high. The heat of the sun and the erosion of the river constantly cause masses of soil and tree-laden ice to fall into the stream. On one of these ice-cliffs, 500 yards from its face, was discovered a lake about a mile in diameter. The water was clear, but on being disturbed became exceedingly turbid, owing to the decayed vegetable matter deposited on the bottom. The country is mostly rolling *tundra*, with innumerable lakes and streams. There is no evidence of glacial action, and rocks *in situ* are not met with till one hundred miles further upstream. How the ice-cliffs were formed, Lieut. Cantwell is unable to explain.—*The National Geographic Magazine*, vol. vii. No. 70.

Dust Storms in North America.—Professor Udden has for some years been collecting data on sand and dust storms in the Western States. As these take place for the most part in thinly peopled districts, the information is somewhat scanty and not always very reliable. Still Professor Udden has succeeded in obtaining much that is new and interesting. The regions chiefly subject to the storms lie west of the Mississippi, where the arid and the semi-arid tracts favour their formation. Of thirty-eight storms which were observed in the years 1894 and 1895 only one was noticed on the east of the river. Of the rest, the greater number occurred in the Pacific States—California, Arizona, Oregon, and Washington —nine falling to the share of California. There is a marked increase towards the south-west, due evidently to the conditions of climate and soil. In Yuma every strong wind unaccompanied by rain raises great clouds of dust, and in 1893 six regular dust storms were reported from that town. In Ontario, California, the annual number is estimated at twelve to fourteen. For the plains to the east of the Rocky mountains the annual minimum may be taken to be two, in the Great Basin and on the western slope five, while the maxima are four and twenty respectively.

The data for calculating the extension of the storms is very meagre ; for the few that have been observed in several places, the distance traversed is found to be between 75 miles and 380 miles, the average being 205. Respecting the time during which a storm prevails at one place, direct observations give a mean of a little more than fourteen hours. Estimates in cases where there were no direct observations yield a much longer average, namely, thirty-two hours. Probably the latter period applies to unusually strong and prolonged storms, whereas the shorter average is for short tornado-like winds. The most accurate period may perhaps be obtained by taking the mean of the two, say, twenty-four hours. And this value agrees with the values obtained for the length of the course of the storm and the velocity of low pressure systems, which on the average is less than eleven miles an hour.

Observations on the transparency of the air were used to measure the amount of sand transported, by comparison with artificial storms produced by blowing a known quantity of dust into the air. Direct measurements were also made by holding a bottle with a mouth of known area against the wind for a certain length of time, and weighing the dust collected. Neither of these methods can be expected to yield very accurate results ; the direct measurements gave 6·49 grammes in a cubic metre. Nevertheless they are close enough to show that the effect of these sand storms must be enormous. Taking the minimum number of storms for the Western States, namely, two a year, it is found that with a velocity of thirty-one miles an hour in the lowest 6500 feet of the atmosphere, the wind blowing over a million square miles of open country would transport 854 million tons of material to a distance of about 1340 miles, and so the work done is about 1,144,000 million mile-tons. This is about a three hundred and thirtieth of the work done by the Mississippi as calculated by Messrs. Humphrey and Abbott, and therefore seems insignificant in comparison. But it should be remembered that in sandy regions a great part of the material transported is not carried through the air, but is rolled and swept along the ground. And, besides, on days when no storms are registered, considerable quantities of dust are transported by light winds. Lastly, the minimum number of storms has been taken for this calculation, and a higher velocity of wind would greatly increase the result, as the transporting power increases as the sixth power of the velocity.

The sand takes the polish off windows, and the paint off railway carriages, and planes off the softer wood from telegraph posts. It is usually fine enough to be termed dust, though in many cases grains of sand are mixed with it and occasionally gravel. Even pebbles were observed on two occasions.—*Globus*, Bd. lxx. No. 18.

The Southern Extremity of Lower California.—About a year ago we gave an account of M. Diguet's explorations in Lower California (vol. xi. p. 533), and now we have an opportunity of adding a few details on the extreme southern point gleaned by *Globus* (Bd. lxx. No. 17) from a report of Messrs. Frank, Vaslit, and Eisen in the *Proceedings* of the California Academy of Sciences.

The climate is moderate ; it never freezes on the flat land up to a height of about 800 feet. In the San Jose valley, the region exempt from frost extends northwards to La Palma, and only slight frosts occur in January and February at Cadueno and Miraflores, which are situated at the upper end of the valley at a height of 1000 feet. Along the coast, from Todos Santos on the Pacific to La Paz on the Gulf, frost has never been observed, and in the mountains it does not freeze from April to December, even at heights of 3900 to 5900 feet. It seldom rains during the cold months, and snow has never been seen even on the highest peaks, which rise to a height of, perhaps, 7800 feet. In consequence of the mild climate, tropical plants—such as coffee, oranges, etc.—thrive on the coast and in the San Jose valley as far as La Palma or beyond. The warm months are from June to September, and in the first half of the latter, the hottest time of the year, the thermometer seldom rises above 89½° F. Being surrounded on all sides by the sea,' the atmosphere of this region is remarkably clear, and in the sierras becomes, if possible, still more pure and transparent, so that certain prominent points, such as El Taste, would be excellent sites for observatories. The salubrity of the country is perfect ; malaria is unknown, and even yellow fever has never broken out in San Jose.

The rainfall is more abundant and more regular than in any other part of the peninsula. It rains in the summer months, but much less than on the opposite mainland ; and the rains, which come from the south, set in rather later. They begin in July or August, and last till October or November. In January rain comes down from the north, but it does not last long. In general, the amount of rain increases from north to south, and from the level country to the mountains. In La Paz, for example, the fall is barely two inches in the year, while only a few miles further south there is rain enough to produce good pasture.

The valley of San Jose del Cabo, forty miles long by two to three broad, is the best watered of the whole peninsula. Water is always found in every gorge of the sierra, even in the dry season, though there are singularly few springs. Eisen found only three—a hot sulphur spring at Agua Caliente, north of Miraflores ; a spring of beautifully clear and cool water at San Bartolo, which pours out five cubic feet of water per second ; and a third near La Palma. Besides the San Jose, the only other river that always reaches the sea is the Todos Santos, rising in the Sierra Laguna ; its water irrigates many thousand acres of land near the town of Todos Santos, the second of importance in the country. The only lake is one half a mile long and a furlong broad, near Santiago.

The most interesting part of the point of the peninsula is the sierra, which rises gently from the Cabo San Lucas to the Monte Troyer (5196 feet) and El Taste, and further north, in Monte Limantour, reaches a height of 6003 feet. The mountains, running in parallel chains with passes 3000 to 4000 feet high, render the country difficult to traverse ; the culminating points are 6000 to 7000 feet in height,'and Santa Genoveva probably attains to 7800 feet. The main sierra, from El Chinche to the Sierra Laguna and as far as Triunfo, consists of granite, and almost everywhere exhibits marks of glaciers. Moraines are conspicuous in many parts, particularly between Miraflores and San Bartolo. The coast mountains, from the San Jose river to La Paz, seem to consist chiefly of red volcanic rocks. Herr Eisen believes that the whole extremity of the peninsula has been rising since the end of the Glacial Period. At that time the sierra was probably a low island separated from the mainland by a broad arm of the sea.

Except in a few spots, the whole of this country is covered thickly with bush and low trees, interspersed in the mountains with cactuses of various kinds. Among useful plants is the *Lysiloma candida*, the bark of which is dried and exported as a dye-stuff. Blooming plants are remarkably numerous. A week after the rains have commenced insects swarm in great numbers, and lizards of several varieties come out of their subterranean nests. Snails also abound. The rattlesnake is not so common as generally supposed; a roasted piece of the Pitahaya cactus (*Cercus Thurberi*) laid on the wound is said to be a remedy against the bite of this reptile.

Chile and Argentina.—According to the new agreement signed on April 7th, the frontier between these States is to be drawn as far as 23° S. lat. along the course indicated by the treaties of 1881 and 1893. From 21° to 23° the boundary is defined by the truce of 1884 between Chile and Bolivia, and is so drawn on all good modern maps. From the Cerro (?) de Zapaleri (or Sapaleri) the boundary runs in a straight line to the volcano Licancaur. But Zapaleri lies nearly a degree of longitude east of the main cordillera of the Andes, which Argentina can claim, in accordance with the treaties, as the boundary as far as 23° S. Bolivia is to take part in the determination of the boundary from 23° to 27°, but Chile does not relinquish any part of the Puna de Atacama, or undertake to move the boundary stone from San Francisco to Tres Cruces. The frontier from 27° to 52° will be delimited in accordance with the treaties, and any points of dispute referred to her Majesty the Queen as arbitrator. Many small differences will no doubt arise, but the only question of importance is whether the watershed throughout its length, or while it lies within the Andes, corresponds most closely to the boundary as indicated by the treaties, or the central line of elevation. Neither by the new agreement, nor by the commercial treaty ratified on May 1st, does Bolivia obtain a port on the Pacific.—*Verh. der Gesell. für Erdkunde zu Berlin*, No. 7.

Tierra del Fuego.—Dr. Otto Nordenskiöld and Herr P. Dusen gave last summer an account of their explorations, shortly described on page 324, to the *Deutscher Wissenschaftlicher Verein* in Santiago. The main island, Dr. Nordenskiöld stated, is divided into three zones. The southern is a highland country, with its lower parts covered with dense forest; the middle portion is low and comparatively level, with forest only on the hills; and the northern zone is quite treeless. The cordillera consists of three different ranges, of which the two southern, separated by the valley of Admiralty Sound and the Tagnano lake, 120 miles long, is built up of crystalline schists. Each of these ranges is virtually a plateau carved into small ridges by the numerous streams which have sunk their valleys deeply into it. To the north of the main chains extends a row of hills, consisting of folded Tertiary strata.

Besides the vegetation, the absence of the tabular form in the elevations distinguishes the central region from the northern. In both, the bed-rock is of Tertiary origin, which is usually covered with Quaternary deposits. Among the latter a boulder clay, identical with that found in European ancient glacial regions, is very abundant, and the presence of other morainic detritus up to the highest parts of the level country, and in the valleys and bays, also indicates that the whole island was once covered with ice. Many problems on the glaciation, formation of valleys, etc., will perhaps be solved when the expedition continues its work this (southern) summer.

Herr Dusen dealt with the vegetation. The northern and eastern parts of the main island are comparatively dry, treeless, and windy, and possess a flora which is poor in species and uniform throughout the region. From the Rio Grande changes begin to appear, forests being characteristic of this country. They consist chiefly of beech (*Fagus pumilio*), and shelter a thriving vegetation but poor in species. In the valleys forms occur which are not met with to the north of the river. In the western, more rainy, region the vegetation is very different. Woods of beech (*Fagus betuloides*) and a *Magnoliacea* (*Drymis winteri*) grows on the coast and up the ravines to about 1000 feet above sea-level. Between 1000 and 2000 feet *Fagus antarctica* is found. At the snow limit, about 2300 feet, a few patches of liverwort are the only representatives of the vegetable world. Nowhere, not even in damp tropical countries, are mosses so luxuriant as in the more rainy parts of Tierra del Fuego.—*Globus*, Bd. lxx. No. 18.

AUSTRALASIA.

West Australia.—On November 18th, 1895, an expedition under Mr. S. G. Hübbe left Oonadatta, the terminus of the northern railway of South Australia, for the purpose of crossing the country to the Musgrave ranges, and thence proceeding to the Coolgardie goldfields, the object being to find out whether it is possible to drive cattle in this direction. As far as the Musgrave ranges there was great drought ; and the camels became so weak that they could not carry their loads. Water was first found in abundance at Glen Ferdinand. An excursion was then made southwards, through the Everard ranges, to the Mound Springs, which were found dry by Lindsay. A boring, fifty-four feet deep, brought up salt water, quite useless for man or beast. Returning to the Musgrave ranges, the party moved westwards to the Litannia Springs, which were also dry, and then north-westwards to Opperanna, a spot noted by the South Australian Survey party as likely to yield a permanent supply. Then the Mound ranges were passed, and the Tomkinson mountains reached. A camp was pitched at the Crowther or Teizi Springs, in lat. 26° 5′ N., and long. 129° 37′ S. The Harriet Springs, north-west of Mount Kintore, were dry, and only at Mount Aloysius was good water procured, but it did not seem to be permanent. The Barlee Springs yielded a small quantity, and at Alexander Springs (26° 2′ S. lat., 124° 46′ E. long.) was a creek with a row of holes containing enough to last a herd of cattle for six months. At Mount Worsnop, the pillar erected by John Forrest was set up again. Marching south-westwards from this point, the expedition entered new country, and rested two days by a creek with good water. After travelling ninety-three miles they came across the tracks of prospectors, and soon after reached Niagara, the most northern settlement in the Coolgardie goldfields.—*Petermanns Mitt.*, Bd. xlii. No. 9.

New Zealand.—The *Report of the Department of Lands and Survey*, 1895-96, contains, as usual, an appendix giving information on recent exploration. The first subject discussed is the route over the Southern Alps. Mr. Brodrick, the district surveyor, reports that the pass named after him has been crossed by Mr. D. Matheson with pack-horses. This pass, discovered by Mr. Brodrick in 1890, and described as the Huxley pass in vol. vii. p. 330, is only 5300 feet high, and eighteen miles from Lake Ohau. Mr. Brodrick mentions that the Southern Alps were crossed long ago by Whitcombe and Browning, and that there is a coach road over Arthur's pass. The object of the surveyors, therefore, is not to cross the Southern Alps for the first time, but to find a practicable route for tourists through the fine scenery of this section, and to open a convenient road for them from the west coast to the Hermitage. Mr. FitzGerald's pass, at the head of the Copland valley, is quite unsuitable ; it is 7180 feet high, and the route crosses a snow-field.

Mr. Wilson describes the Fox glacier, his report being accompanied by a map and diagrams. During twenty-eight days, from July 6th to August 3rd, he was able to measure the movement of the glacier. Its average rate of advance varied from about nine and a half to thirteen inches a day at different parts of its surface. Mr. C. Douglas also writes on the Fox valley, its vegetation, fauna, geology, etc., while Mr. T. Mackenzie describes a river falling into Supper Cove, Dusky Sound (Western Otago.) This river, named after Mr. Mackenzie, flows through Lake Ida, a mile and a half long, and forms several cascades and rapids. The Spey river was also followed to its mouth in Lake Manapouri.

German New Guinea.—The *Verhandlungen der Gesell. für Erdkunde zu Berlin*, No. 7, contains letters from Dr. Carl Lauterbach, the leader of the Kaiser Wilhelm-Land expedition. In May he made an excursion to the Oertzen mountains from the village of Erima, north-west of Stephansort, and Dr. Kersting succeeded in climbing the Tajomanna, the highest summit ; the height is 3500 feet. Towards the west no high ranges were visible, but only low ranges running one behind the other from north-west to south-east. The mountains are composed of bluish-green shales alternating with conglomerates, and have a strike from north-west to south-east, with a dip of 70° to 80°.

In June the party ascended the Gogol river, called further inland the Narua. Behind the Oertzen mountains the country is traversed by mountains and hills of moderate height, apparently diminishing towards the north. The rocks are sandstone and soft, dark-coloured shales, in some places alternating with conglomerate. The country is fairly thickly peopled, and provisions are procurable. Presently the Narua divided into two brooks of nearly equal volume. The one flowing from the south-west it was impossible to follow, and therefore the party ascended the more westerly branch, reaching the neighbourhood of its source at a height of 1600 feet above sea-level. The source lies on the northern flank of a mountain 2900 feet high, at a distance from the coast of twenty-five miles in a straight line, and double that distance by the river. From the summit of a mountain, called by the natives Sigaun Yanu, Dr. Lauterbach saw a great range, perhaps the Arthur Gordon, to the west, and ten parallel chains in the interval, with heights ranging from 400 up to 6000 feet, all densely wooded. On the northern horizon, also, rose a range some 10,000 feet high, which probably lies on the Augusta river. West-south-west by west, at a distance of six or seven miles, was observed a mighty elevation, 13,000 to 16,000 feet high, and apparently running east and west, which was probably the Bismarck mountains. These mountains Dr. Lauterbach saw, it may be remembered, in 1890 (see vol. viii. p. 100). Having discovered a river running south-west, the party started again on June 19th and followed it seven or eight miles. Here a station was formed, at an elevation of 2000 feet, and at a distance of sixty miles from the coast. Dr. Lauterbach intended to continue his march in a southerly direction, and, having ascended one of the highest mountains, search for a way to the Huron Gulf.

The Ellice Islands.—The Ellice islands, which lie on the line of the projected cable between Australia and America, were annexed by Great Britain in 1893, at the same time as the Gilbert, or Kingsmill, islands. The inhabitants of the latter are a restless, warlike people, always at feud among themselves, so that the Resident has enough to do to keep the peace in the sixteen islands. Consequently, news of the Gilbert islands is now and then heard, whereas the quiet Ellice islanders, who never fight, and own no weapons but a few flint-locks to shoot birds, are little known.

The most south-easterly of the group, Sophia island (Nurakita), lies 600 miles north-west of Samoa, and is visible from a distance of twenty miles, being the highest of the islands. A few years ago it was uninhabited, though the natives of the neighbouring Nukulailai affirm that it had a good many inhabitants in old times. About three and a half miles in circumference, and containing few coco-palms, it would have attracted no attention but for the existence of a rather valuable bed of guano. Nukulailai (Mitchell I.), eighty to ninety miles further north, is an atoll of thirteen low islands, with a continuous reef surrounding a lagoon five miles long by three broad. The narrow girdle of land, barely a mile across, is densely covered with coco-nut palms, and presents a charming aspect from the sea. Thirty years ago it was peopled by four hundred natives, when a barque and brig approached one day, and three hundred of the unsuspicious, peaceful inhabitants were carried off to toil in the guano-fields of the Chincha islands and of Peru, and not one of them lived to return. Now the population is 120, and is slowly increasing as on the other islands. A few hours' sail to the north-west lies Funafuti (Ellice), a long chain of thirty-five islands enclosing a large lagoon, with two good entrances at the south-west and north-west. Kotzebue sailed, with a strong wind, from one end of the lagoon to the other, and found that unless a ship sailed ten knots an hour it could not make way against the stream. Once in the lagoon, and out of the course of the stream, men-of-war find ample room to anchor. Six miles from the south-western entrance is situated the chief island of the chain, with the native village and good anchorage. Fifty years ago almost every island was in-habited ; now, all the natives, four hundred, live in Funafuti. They express great satisfaction at being British subjects, and, doubtless, in a few years, when they are fully assured of the good intentions of the authorities, will willingly lease their coco-nut woods to dealers and copra buyers. For forty years millions of coco-palms have borne fruits which have fallen unheeded to the ground, the natives collecting only enough to procure from the dealers the few articles of clothing, tobacco, etc., which they required. Only one trader was permitted to live in the village, and all attempts to bring the nuts into the market proved fruitless.

The archipelago, consisting of nine groups of islands, lies between 5° 35' and 11° 20' S. lat., and between 176° and 180° E. long. It seldom suffers from drought or hurricanes, though the drought which prevailed in 1892 on the Gilbert islands, and has not yet passed away, has to some extent injured the Ellice islands also. Great destruction was caused among the palm-trees by a violent storm in 1890.

Since last year the Gilbert islands have been divided into thirteen districts, with magistrates and police, under the British Resident at Tarawa ; and the Ellice group is divided into eight districts. The external commerce of the Gilbert islands, in 1894, was £14,900, and of the Ellice islands, £4,900. The expedition under Professor Sollas will shortly visit Funafuti to examine the coral reefs by boring, and thus test the Darwinian theory.—*Petermanns Mitt.*, Bd. xlii. No. 9.

GENERAL.

Cloud Formation.—Many scientists have studied the condensation of atmospheric moisture on particles of dust. How the atmosphere would part with its moisture if there were absolutely no nuclei to initiate condensation has not yet been ascer-tained. According to Von Bezold and Professor Cleveland Abbe, the editor of the *Monthly Weather Review* (May), there are in the ascending portions of every cloud regions supersaturated with moisture, and the strained molecular condition thus produced suddenly gives way, accompanied by the production of large drops of rain and electric phenomena. Mr. C. T. R. Wilson has also investigated the formation of cloud without dust, and made known the results in a paper read before the Philosophical Society of Cambridge (England), in 1895. He found that after a small number of expansions of the air, due to the removal of dust particles by condensation of water upon them, no further condensation takes place until the expansion exceeds a certain amount. When this limit is passed, condensation invariably takes place again. The ratio of the final to the initial volume, when the critical limit of expansion is reached, is 1·258 : 1, the initial temperature being 62·06° F. This corresponds to a fall of temperature of 46·8° F., and a vapour pressure of 4·5 times the saturation pressure for a plane surface of water.

Mr. Wilson has now added a new fact to his former results. In 1868, Professor Tyndall found that a dense cloud was formed when a powerful beam of light penetrated a tube full of pure vapour. The cloud, when freshly formed, was of a brilliant blue, but became white as the particles increased in size. Mr. Wilson has experimented with Röntgen rays, and finds that they have the effect of greatly increasing the number of drops when the expansion exceeds the critical value. Hence they appear to furnish nuclei of condensation similar in action to dust particles.

MISCELLANEOUS.

A *List of Marine Records of the late East India Company, and of subsequent date, preserved in the Record Department of the India Office* has lately been issued. The list of voyages will be useful to students, and the introduction contains many interesting details of the most important voyages of the company's ships.

M. E. A. Martel, the well-known subterranean explorer, has discovered in the grotto del Drach, Majorca, more than half a mile of new galleries, in the midst of which is a lake 200 yards long and 30 feet deep. The water is salt, as it oozes in from the sea. The temperature is 67° F., while the mean annual temperature of the place is only 60°.—*Mouvement Géographique*, No. 39.

On the supposed existence of Phœnician inscriptions at Rejang in Sumatra (see p. 425), *Globus*, Bd. lxx. No. 7, gives a statement of Herr J. D. E. Schmeltz, that the Rejang is a branch of the great Indian group of characters, and that the alphabet came into existence 860 A.D., whereas the Phœnician dates from about 500 B.C. Innumerable traces of Indian colonisation are found in Indonesia.

On January 1st a salt shower was experienced in parts of Utah and Wyoming, between Ogden and Evanston, which are 112 miles apart. At Almy the amount of salt that fell was calculated by Dr. C. T. Gamble to be 27¾ tons, being rather more than three tons on each of the nine acres on which the town stands. The shower lasted nearly two hours. An official of the Rio Grande Western Railroad states that saline showers are frequent when the wind, blowing from the west, sweeps over the Great Salt Lake, with its area of 2050 square miles. What is remarkable is, not that there was salt in the rain, but that it was carried to a distance of more than sixty miles.—*Ciel et Terre*, October 16th.

Dr. S. E. Dawson, whose *Voyages of the Cabots* was noticed in vol. xi. p. 266, has forwarded another pamphlet containing various additions to his former work. Among other questions he returns to the landfall of the Cabots, answering the arguments of Judge Prowse, who maintains that Cape Bonavista, Newfoundland, was the land first sighted. The Judge's arguments are that it is the unbroken tradition of the coast, and that on a map of 1616 by John Mason, a captain in the Royal Navy, are inscribed against Cape Bonavista the words, "First found by Cabot, A Cabote primum reperta." The objections are that this map appeared 119 years after Cabot's voyage, and that the name Bonavista is Portuguese and is not laid down on any map earlier than that of Gaspar Viegas in 1534.

NEW BOOKS.

Im Australischen Busch, und an den Küsten des Korallenmeeres; Reiseerlebnisse und Beobachtungen eines Naturforschers in Australien, Neu Guinea, und den Molukken. Von RICHARD SEMON, Professor in Jena. Mit 85 Abbildungen und 4 Karten. Leipzig: Engelmann, 1886. Pp. xvi+569. *Price M.* 15 ; *bound* 16·50.

The narrative before us is, we are told, intended as the popular version of a more strictly scientific work, which has already gone through six editions, and is expected to see twenty more. The appreciation of the intelligent "general" reader will not, we venture to think, lag behind that of the scientific. The author shows himself to be a man of singularly genial temperament, and withal a many-sided man, fully alive to the dangers of specialisation, to which travel is, he finds, an antidote, while also affording endless interest in the study of human nature, and the discovery of the good which underlies its various manifestations. Although the greater part of the book is taken up with his special researches—the purport and tenor of which are adequately explained with singular clearness and a minimum of technicalities—it abounds in interesting facts and ingenious theories ; and besides all this, there are the spice of personal adventure ; details of the rough bush life while encamped for months in the wilds of Queensland ; and clear descriptions of the physical characters of the country, both there and amid the more attractive scenery of some of the lesser-known Malay Islands, especially Ambon. The home life and farming operations of his squatter friends are full of interest for him. He was thrown in close contact with an Australian tribe, and discusses very carefully their mental characteristics in relation to other races.

The people of New Guinea he considers on the whole the most interesting and the most artistic race he has ever met. On the question of a distinct "Papuan" race, he admits that there is not yet to hand sufficient material for a decisive opinion. It may anyhow, we think, still be assumed that the term Papuan, as applied to the inhabitants, should have rather a geographical than an ethnological significance.

The author's main object in visiting Australia was the study, first, of the ornithorhynchus and more especially the echidna, and, by forming a very full embryonic series, to ascertain the history of their development and their relationship to the placental mammalia on the one hand, and the more distant reptilia and birds on the other ; and, secondly, of the habits and life-history of the lung fish (*Ceratodus Fosteri*). His adventures, his failures and successes, with which the reader will heartily sympathise, are capitally told. His explanation of the causes which have led to the Ceratodus, a type formerly of wide distribution, being now confined to two North Queensland rivers, the Burnett and the Mary, and specially flourishing there, is a good example of the writer's ingenious but cautious reasoning.

The disappearance of the larger marsupial carnivora from Australia, while they swarm in Tasmania, he attributes to the introduction into Australia, by man, at a very remote period, of the dingo.

He has a keen eye for natural beauty, and can see much that is pleasing in the —to a less discerning vision—monotonous Australian forest.

He mentions, as a curious instance of instinct, that whereas in Australia the mound-building birds (*Talegalla sp.*) deposit their eggs, as is well known, in a great heap of leaves, which by decomposing generate the heat necessary to hatch them, the *Megapodius tumulus* in New Guinea, where such artifical heat is not required, constructs its mound simply of sand. Many other interesting problems and wonders of natural history are discussed, as well as larger questions, such as the growth of coral reefs, and the great caves in coralline rocks, in the formation of which the author considers that running water as a solvent plays only a very subordinate part. On questions of geographical distribution, including the singularly isolated character of the Celebesian fauna, he agrees in the main with Mr. Wallace.

British anthropologists, while recognising all the importance and interest of Dr. Eugène Dubois' discovery in Java, will hardly admit that "die Mehrzahl der kompetenten Beurtheiler" consider that it has settled the question of the Missing Link.

We have hardly, perhaps, done justice either to the attractive personality of the writer, or to the charm, as well as the value, of his book. Its dedication to Professor Ernst Haeckel recalls, as a worthy subject of comparison, the delightful volume on Ceylon produced under somewhat similar circumstances by that master of natural science.

Monomotapa (Rhodesia). Its Monuments and its History from the Most Ancient Times to the Present Century. By the Hon. A. WILMOT, Member of the Legislative Council, Cape of Good Hope. With Preface by H. RIDER HAGGARD. Maps and Plates. London: T. Fisher Unwin, 1896. Pp. 260. *Price 6s.*

In his researches in the history of South Africa, the Hon. A. Wilmot is working backwards. About two years ago he published *The Story of the Expansion of South Africa,* which was a systematic history of the settlement of Cape Colony, and of the events which transferred it to the British Crown. The volume now issued deals with still earlier times, in which the history of Bechuanaland is gleaned from ancient monuments—from the remains of temples and ruined cities. Mr. Wilmot has availed himself of the pioneer work of Herr Mauch, who, in 1871, discovered the colossal ruins of Zimbabwe, as well as of those of Mr. Theodore Bent, who described these ruins in his admirable work on *The Ruined Cities of Mashonaland,* in 1891. But at the instance, and with the help, of Mr. Cecil Rhodes, Mr. Wilmot prosecuted independent researches in the Propaganda College and in the Vatican at Rome, and also in the archives at Lisbon. The results of these researches have scarcely realised the expectations with which they were undertaken. Indeed, no very important addition has been made to the discoveries of Herr Mauch and Mr. Bent, referred to above, if we except the account of the martyrdom of Father Silveira, of which a translation is given in the appendix to Mr. Wilmot's volume. That appendix also contains a list of twenty-six documents found in the archives of the Vatican concerning South Africa, as well as translations of all the documents found in the archives of the Propaganda which relate to Monomotapa and the adjacent countries. The Lisbon papers have already been utilised by the Portuguese historians of South Africa. Mr Wilmot's most important discovery was that of a map of Africa, of date 1623, in which Bechuanaland bears the name, *Monomotapæ Imperium,* whence the name "Monomotapa" has been given to the region and to the present volume. This map is reproduced by Mr. Wilmot, by permission of the Vatican authorities, and is extremely interesting. The region so described is well filled in, both with physical features and with settlements or "towns"; and if Mr. Wilmot had done nothing else, his discovery and reproduction of this map would have laid geographers under an important obligation to him. When he seeks to establish a similarity between the remains at Monomotapa and ancient monuments found in Sardinia, and ascribes to them a Phœnician origin, he enters on more questionable ground. If the Sardinian

Nuraghe and the temple at Zimbabwe are of Phœnician origin, the round towers of Iceland and the so-called Pictish tower of Mousa in Shetland must also be of Phœnician origin. That conclusion, however, will not be accepted by skilled antiquaries without considerable misgivings.

Mr. Wilmot's researches, both at Rome and in Lisbon, show conclusively that, in the early historic period, Monomotapa was famous, not only for its colossal buildings, but also for its wealth in gold, and in this respect the ancient record may be but a forecast of modern times. As Mr. Rider Haggard says in his interesting preface, "it is legitimate to hope—it seems probable even—that in centuries to come a town will once more nestle beneath these grey and ancient ruins [of Zimbabwe], trading in gold, as did that of the Phœnicians, but peopled by men of the Anglo-Saxon race."

Mr. Wilmot's style cannot be commended; it is involved, prolix, and laboured. But the conspicuous defect of the book is the want of an index.

In New South Africa: Travels in the Transvaal and Rhodesia. By H. LINCOLN TANGYE. London : Horace Cox, 1896. Pp. viii + 432.

In this volume Mr. Tangye has given us a very interesting, graphic, and lively account of his travels through the Transvaal and into Rhodesia, as far as to Salisbury and beyond it, with a detour in the direction of the remains of the famous Temple of Zimbabwe. It is perhaps his misfortune rather than his fault that he has been betrayed into the discussion of South African politics ; for while his descriptions of scenery and native life are excellent, his political comments are very questionable. He excuses, if he does not actually defend, Dr. Jameson's raid, and he makes himself the champion out and out of the Johannesburg Uitlanders, and the thick-and-thin defender of the Chartered Company in all its exploits and adventures. Into these speculations we do not propose to follow him, but we may be allowed to say that, in our opinion, Mr. Tangye writes much better—not only much more freely and spontaneously, but also much more correctly, from the point of view of the critic of style—when narrating his experiences as a traveller, and when describing the scenes through which he passed, than he does when he assumes the *rôle* of a political critic. His descriptions of the journey "Across Desert and Veldt" and of "Johannesburg the Golden" are very fresh and entertaining ; and the "Rambles in Rhodesia," which constitute a second part of his volume, are both lively and instructive. They give an excellent idea of the resources of the country, of its industries in gold-mining and agriculture, as well as of its scenery and its antiquities. The illustrations are beautiful; there is an excellent sketch-map of the author's route ; and the book is furnished with a comprehensive index.

Der Amazonas: Wanderbilder aus Peru, Bolivia, und Nordbrasilien. Von DÀMIAN FREIHERRN VON SCHÜTZ-HOLZHAUSEN. Freiburg im Breisgau : Herdersche Verlagshandlung, 1895. Pp. xix + 444. *Price 7 M.*

This is a second and enlarged edition of Count von Schütz's *Amazonas*, which was published in 1883, the year of his death, and which contained the result of his observations during nineteen years of travel and residence in South America. More than half of the book is devoted to Peru, but there are three chapters relating to the fascinating region of the Amazon and its great tributaries, into whose recesses the Count penetrated, and with regard to whose wild inhabitants he has much to say. Not the least interesting portion of the book is that which relates to the German colony of Pozuzo, visited by Von Schütz in 1864. The territory thus colonised is situated in the province or department of Huánuco, whose chief town, now possessing a population of 7500, was founded by Pizarro. From Lima it lies north-east, among the mountain valleys of the eastern slopes of the Peruvian Andes, watered by the upper courses of the streams which eventually reach the Rio Maranon and the Amazon. The first German settlers came to Pozuzo in 1857, one hundred of their number being Tyrolese and the remaining fifty hailing from the banks of the Rhine and the Moselle. In 1868 there was a fresh accession, also chiefly from the Tyrol. Of the Pozuzo colony, now numbering four or five hundred souls, an excellent account is given (pp. 221-284 and 410-425); and there is a life-like portrait (p. 268) of Pastor Joseph Egg, "the untiring minister and doctor of the colony," to whom, indeed, as "the true friend of Count von Schütz," this edition of *Der Amazonas* is dedicated.

One admirable feature of the book is the appended Bibliography (pp. 427-438), containing all the noteworthy works relating to Spanish America, from Oviedo's *Las Indias Occidentales* of 1525 down to the latest publications of 1895.

The written descriptions are embellished by numerous illustrations; and, in addition to a sketch-map of the neighbourhood of Pozuzo (p. 222), there is appended a map of Central South America, with insets of Southern Chile and the Galapagos Islands. As a frontispiece, there is what appears to be a very faithful likeness of Count von Schütz himself.

Streifzüge durch Grossbritannien. Von GUSTAF F. STEFFEN. Aus dem schwedischen von Dr. OSKAR REYHER. Stuttgart: Hobbing und Büchle, 1896. Pp. 387. *Price 7 Marks.*

This is not the first Swedish book on Great Britain which Herr Steffen has translated, for his *Aus dem modernen England* met with considerable success in Germany. In the attractive work before us many of the leading features of England, Scotland, Ireland, and the Channel Islands are described and illustrated, particular attention being bestowed upon English manufacturing and mining districts. As regards Scotland, Edinburgh, Rosslyn Chapel, the Forth Bridge, the Highlands, and Balmoral are sketched with originality and ability. Let us see what is said about Edinburgh by our Swedish critic. He extols its natural beauty, and declares that "it is one of the most delightful surprises of Edinburgh that those who have become disgusted with English civilisation may recover here from their feeling of weariness without abandoning the society of mankind. We can wander a whole day up and down Edinburgh and enjoy novel and splendid sights without feeling that the vicinity of over a quarter of a million of people and their dwellings and workshops does any violence to the natural beauty of the place."

Dr. Reyher is not, however, such an admirer of the character of the people. "The tone of Edinburgh society is, though vigorous and lively, somewhat dry. Even although we may be charmed with the panoramic beauty of city, country, and sea spread out before us as we stand on the Calton Hill, and though a dazzling summer sun may scatter the sea-fog and give us lovely peeps of the Highlands to the north of the Forth, we can never forget that from the people born and bred in this country we have received sour theologians, harsh moral philosophers, and hair-splitting metaphysicians, as well as poets and artists. Richer than the Englishman in intellectual gifts, the Scotsman in the realm of poetry takes up a middle position between the Englishman and the often highly æsthetically endowed Celts of Ireland and Wales. He is, it is true, a better musician than the Celt, but generally not a better poet or painter. His intellect—in itself much more profound than the Englishman's—is above all things analytic. Apart from this, the typical Scot appears to be a sensible, if somewhat heavy, man, who thoroughly understands how to control both himself and others, and who is certainly not inferior to the Englishman either in enterprise or wisdom."

The women of Midlothian were also noticed by our sharp-eyed critic as he drove to Rosslyn. "This type of woman," he remarks, "is stronger than is to be seen in Northumberland and other parts of Northern England, and is radically different from the Southern English type. Here are no lean bony creatures who resemble dressed-up lads rather than women. At the same time, the Scottish women, in comparison with the Southern English, have generally coarse and irregular features. Why the land of John Bull so often displays noble womanly countenances, perhaps the most distinguished in Europe, united to deteriorated bodies, is a question which we must refer to physiologists or anthropologists."

Generalisations such as these can scarcely be made reliably during mere flying excursions, but the following may be given, for what it is worth, as the author's opinion of English character :—" The English nation possesses great aptitude for trading, but less for thinking ; it has much self-control, but little depth of soul ; it has much cleverness, but less wisdom ; it is able to meet calamities, if they are not of long duration, but it sinks in the long-run into lamentable social conditions ; and it is endowed with a perseverance which, whilst displaying a certain confidence, exhibits likewise a foolish conservatism."

Autour du Tonkin. Par HENRI PH. D'ORLÉANS. 4me édition. Paris : Calmann Lévy, 1896. Pp. 535. *Price 3.50 Frs.*

In more than one arduous journey besides the one here described, Prince Henry of Orleans has made his mark as a distinguished and capable Eastern traveller. On the occasion before us, he travelled up from Haiphong through the north-west parts of Tonkin to Luang Prabang, that important political and commercial centre at the eastern angle of the Mekong ; thence down the river to Pak-la, and thence westward, through the forest region of Eastern Siam, to the river Meping and Bangkok. The interest as well as the value of the edition before us is materially lessened by the absence of a map, which is essential to the understanding of some of the most important points of which the author treats, viz., the political frontiers of the region ; the districts occupied by the various aboriginal or immigrant tribes, as the Khas, the Laos, the Meos, and others, of whom he gives some interesting details ; and lastly, the different trade routes, leading to certain eastern and western outlets, the respective merits of which he discusses, and which are of pressing importance to ourselves and others.

Of the three routes which now serve the region of which Luang Prabang is the centre, two run westward : the first, *via* Pak-la on the Mekong to Outaradit on the Menam ; the other, descending the Mekong further to Nongkai, branches off by Korat to Bangkok. The author admits that both of these routes, though somewhat longer than the road to Hanoi, possess greater advantages. He feels sure, however, that the Mekong can be connected directly with the Annamese coast by rail, when Tourane will become the Bombay of Annam. On the map, no doubt, it is the shortest route, and the engineering difficulties are said to be not very great. But the country served is by all accounts of far less commercial importance than that which would be served by the Korat line, which is promoted by the Siamese Government. "Let us, however," says the Prince, "oppose their schemes by fair fighting—railway against railway, steamer against steamer." This is all as it should be ; by all accounts, however, it is opposition of a much less legitimate kind which has hitherto prevented the construction of the Korat-Bangkok line.

The heavy charges he brings against the French administration of Tonkin, including peculation and gross jobbery ; French troops going baré-foot ; native discontent, caused by acts of spoliation and fiscal tyranny, are perhaps more interesting to the French reader. Of the Siamese he always expresses himself unfavourably, the head and front of their offending being their (surely pardonable) obtuseness in not at once recognising those French "rights" with which they have recently come into collision with such serious results. A pleasanter topic is the social life of the Laos, which is not without a certain poetic grace ; there are evidently considerable artistic and intellectual elements in the character, shown notably in their games and amusements. Not only is the position of the women good, but there is also, the writer declares, a chivalrous feeling towards them, which one associates only with mediæval Europe.

A History of Egypt during the Seventeenth and Eighteenth Dynasties. By W. M. FLINDERS PETRIE, D.C.L., LL.D. With 160 Illustrations and 4 Maps. Pp. 353. London : Methuen and Co., 1896. *Price 6s.*

This is the second of the promised seven volumes of the great work of the Edwards Professor of Egyptology in University College, London. It arranges, describes, and pictures the materials from mummy-pits, obelisks, statues and temples, from which the annals of the two most brilliant of the Dynasties of Egypt are pieced together. The special value of this volume lies in the account of the campaigns of Tahutmes III. in Syria, 1503-1449 B.C. The wonderfully detailed and graphic stories are illustrated by outline maps of the king's approach to Megiddo, where the great battle was fought, of sites in Southern Syria, of Northern Syria, and of Syria generally, under Amenhotep IV. There is a chapter on the geography of the Syrian campaigns. Mr. Flinders Petrie, improving on the work of Mariette, Brugsch, Max Müller, Conder, and even Maspéro, attempts to verify the list of Tahutmes III. of 119 places in the Upper Ruten country or Palestine. Three versions of this list exist on his monuments. The first circuit of places is followed northwards through Galilee to the Damascus road, thence from Damascus to Megiddo again. Other circuits radiate out from Megiddo, across Jordan, and include one up to the Jerusalem region. Migdal, on the road to Egypt, was the centre of the later operations of the mighty monarch, whose half-century's rule marks the culmination of the Egyptian power which rose against the Hyksos' oppression. Deposits of mummies, like the famous find still being worked at Deir el Bahri, may be expected to yield more data for a complete history of Egypt. Meanwhile, this work is invaluable to Biblical students as well as other scholars.

The History of Mankind. By Professor FRIEDRICH RATZEL. Translated from the second German edition by A. J. BUTLER, M.A. With Introduction by E. B. TYLOR, D.C.L., F.R.S. Vol. I. London : Macmillan and Co., 1896. Pp. xxiv+486.

The History of Mankind is scarcely the proper title for the English edition of Professor Ratzel's *Völkerkunde.* The book is in fact a survey of the stages of culture, reached by what Ratzel calls the *Naturvölker.* The Professor objects to the term " Savages " ; to denote races poor in culture, as opposed to cultured and civilised races properly so called, he employs the word *Naturvölker,* rendered by the English translator " Natural Races." But Professor Ratzel warns his readers against the evolutionist view of regarding his " natural races " as the oldest strata of mankind now existing. " We call them races deficient in civilisation, because internal and external conditions have hindered them from attaining to such permanent developments in the domain of culture as form the mark of the true, civilised races and are guarantees of progress." He even admits, in some cases, a relapse from a higher state of culture.   In agreement with his fundamental conception, he does not group the races according to anthropological and physiological characters.   In dealing with each group, he gives first a short sketch of the physical and mental qualities of the races comprised within it, and then treats in greater detail of their dress, weapons, implements, their social condition, and their religion. The reader need be in no fear of German metaphysics or wild evolutionist speculations.   The Professor strives to give only well-ascertained facts, and a conservative spirit pervades his book throughout.   So he says with regard to religion : "Ethnography knows no race devoid of religion, but only differences in the degree to which religious ideas are developed"; and speaking of the family he asserts : "All attempts to prove the existence of absolute promiscuity may be regarded as unsuccessful."

The English edition is not a literal translation ; the three original volumes are condensed into two.   But one would like to know on what principle their condensation is done ; sometimes one has a feeling as if important passages had been omitted. We can scarcely help thinking that the translation has not been done by the same hand throughout.   While generally no fault could be found with the language, now and then pages are met with which have a distinct foreign flavour about them, and some passages are rather confused.   Let us hope that the second volume will be free from such blemishes.   A special feature of Professor Ratzel's book are the illustrations, which are reproduced in the English edition.   They are beautifully executed, and the collection where the original objects are at present to be found is always indicated.   The first volume contains, besides several chapters on the principles of ethnography, the American Pacific group of races, comprising those of Oceania, the Australians, Malays, and Malagasies.   Dr. Tylor has contributed a very interesting introduction.

In Germany Professor Ratzel's book is considered the standard work on ethnography ; we venture to predict for the English edition the same distinction in this country.

Leitfaden zur Geschichte der Kartographie in tabellarischer Darstellung.   Von Dr. W. WOLKENHAUER, in Bremen.   Breslau : Ferdinand Hirt, 1895.   Pp. 93.

Notwithstanding the great advances in cartography during recent times, and the historical interest connected with its early conception and development, it is somewhat remarkable that no one has as yet attempted a complete and systematic history of the science.   The Columbus Jubilee of 1892 brought forth several valuable contributions relating to the early cartography of America, notably Kretschmer's great work, published by the Berlin Geographical Society. Nordenskiöld's *Facsimile Atlas* has also revealed to us the great wealth of material to be dealt with.   Mr. Collingridge's researches on the discovery of Australia disclose to us a fascinating field of speculation.   The Scottish Geographical Society, in its *Atlas*, traces the gradual progress of the mapping of Scotland ; while in its *Magazine* Mr. Bartholomew's papers on the mapping of the world show the point reached by cartography at the present day.   But a complete general work, tracing the evolution of the cartography of the world, yet remains to be done.   Looking at the atlases of Mercator and Ortelius, we should like to know why the interior of Africa is thus mapped in detail with numerous cities and towns, what authority these geographers had for their information, and why their work was ignored by subsequent cartographers ?

In the *Deutsche Geographische Blätter*, 1883, Dr. Wolkenhauer published his *Zeittafel zur Geschichte der Kartographie*. So far as we know, it is the first attempt at a general summary of the best-known maps and atlases from earliest times to the present day. That short sketch he has now amplified and enriched with critical notes. He divides it into seven sections : I. Period of the Earliest Cartography ; II. From the Discovery of the Mariner's Compass to the Time of Ptolemy ; III. From the Time of Ptolemy to the Reform of Cartography ; IV. The Reform of Cartography ; V. The Transition Period ; VI. Period of Triangular and Geodetic Surveys ; VII. Modern Cartography.

Under these headings he notes the leading works of each period. So far as it goes his work is interesting, but it does not go far enough. The incompleteness of his treatment can best be realised from the fact that the leading works of modern cartography only extend to one hundred and forty entries, and deal mostly with educational publications of technical rather than scientific merit ; while our leading English atlases are not even mentioned. Dr. Wolkenhauer's sketch is too incomplete for a reference work or for use in comparative study, but it is suggestive, and a more complete work on the same lines would be most valuable.

Complete Geography. By ALEX. EVERETT FRYE. Boston, U.S.A., and London : Ginn and Co., 1896. Pp. 208. *Price* $1.25.

The recognised size of a text-book of geography in America is a large quarto. This size, notwithstanding its many advantages for the treatment of the subject, has never found favour in this country. But consideration of the fact reveals that we have a good deal to learn from American methods of geographical teaching. In the American geography wealth of illustration, in the form of pictures, maps, and diagrams, is considered a necessary adjunct to the text. And not only so, but this scheme of illustration, both in respect of practical aid and artistic beauty, receives as much study and consideration as the writing of the text. Unfortunately this is not so in English text-books ; and, although many are well illustrated, it will be found impossible to rival the American books in this respect, unless a larger size of page is adopted.

Frye's *Geography* may be instanced as one of the best examples of the American type of text-book. It deals with geography in its widest sense. It conveys a comprehensive view of the Earth as the home of man, its forms of land and water, its climate, vegetation, animals, its peoples and their towns and states, tracing through all the scientific connection and sequence. The language of the text is simple, concise, and effective, while the subjects of illustration are well selected to present the most characteristic and representative forms. The book is well supplied with maps, which are in two sets, and very properly distinguished as maps for study and maps for reference—the one set being very simple and the other full of detail. In borrowing special features from the recent work of English cartographers, the author might well have acknowledged his indebtedness to them with the same courtesy as he acknowledges assistance from American sources.

Man and his Markets: A Course in Geography. By LIONEL W. LYDE, M.A. London : Macmillan and Co., 1896.

This book is of a distinctly good and welcome type. In it some of the leading topics of commercial geography are treated in an interesting manner, that provokes and demands thought on the part of the reader. There is no appeal to mere verbal memory. The illustrations are numerous, and most of them are good. A few, however, are but slightly connected with the text. The " view " of Gibraltar is very poor, and that of the Acropolis is too old for a present-day book. The text is not free from error. Thus it is said that the Iberian peninsula has been known in English history as "*The* Cape"; an extraordinary confusion is made as to the origin of the name "Archipelago"; the Belem Tower, Lisbon, is called "Betem"; and the earth's "revolution" is confused with its "rotation."

Chamonix and the Range of Mont Blanc. A Guide by EDWARD WHYMPER. London: John Murray, 1896. Pp. xiv + 189. *Price 3s. net.*

There is much information in this excellent little volume not usually found in guides, the object of the writer being "to give in a small compass information which some may desire to have at home, and that others will wish for on the spot." Accordingly, besides the usual details relating to guides and ascents, hotels and conveyances, etc., we have a sketch of the history of Chamonix and a condensed record of the various ascents of Mont Blanc and of the most noted accidents. Especially interesting is the account of the work of MM. Eiffel and Vallot and Dr. Janssen in erecting observatories. The illustrations are numerous and good, many being quite new, and the maps are very useful.

Western Australian Year-Book for 1894-95. By MALCOLM A. C. FRASER, Registrar-General of Western Australia, F.R.G.S., F.R.C.Inst. Perth: Richard Pether, Government Printer, 1896. Pp. 393.

Mr. Fraser has considerably extended his *Year-Book*, chiefly by fuller details under almost every heading. The goldfields, of course, demand more space, and the chapters on the useful timber, the fauna and flora, are amplified. Mr. Saville-Kent also contributes an interesting report on the fisheries, and the late Baron von Mueller's list of extra-tropical West Australian plants (*Vasculares*) is printed in full. The book is a very useful summary of all that concerns the products, agriculture, mining industry, etc., of the colony.

The Island of Capri: A Mediterranean Idyll. By FERDINAND GREGOROVIUS. Freely translated by M. DOUGLASS FAIRBAIRN. London: T. Fisher Unwin, 1896. Pp. 155.

Gregorovius, the historian of the city of Rome, has spent, at intervals, considerable time in Capri, evidently knows it thoroughly, and, moved by a great liking for it, has been inspired with the desire to combine his impressions in a picture which would do justice to all sides of his subject—the history, the physical aspect, and the people with their simple, natural life. The result is by no means so good as the high reputation of the author led us to expect. Whether our disappointment is due to the author or to the translator we cannot positively say, not having the original at hand for comparison; but we are inclined to think that both are in some measure responsible. Gregorovius has evidently not the power to form a clear, definite picture of natural scenery, and to describe it in concise, apt terms so as to make others realise it, nor the lightness of touch and graceful fancy which enable an artist to combine a variety of impressions into a speaking and attractive picture. The surprising thing is that the sketch of the people and their life is the most successful part of the book, while the physical and historical are the weakest and least satisfactory—the former being so vague and general as to convey no definite impression—the other rather confused and lacking in insight and convincingness. The translator, however, has not done him justice—evidently not being at home in ancient or modern history. There are not a few sentences which seem simple enough, yet, when carefully considered, it is hard to say what meaning they are intended to convey.

NEW MAPS.

WORLD.

SUBMARINE CABLES OF THE WORLD, with the principal connecting Land Lines ; also Coaling, Docking, and Repairing Stations.

TRACKS FOR FULL-POWERED STEAM VESSELS, with distances in nautical miles.

U.S. Hydrographic Office, Washington.

EUROPE.

SCOTLAND. Sheet 2. Dumfries and Solway.

NORTH DEVON.

KILLARNEY LAKE DISTRICT.

Reduced Ordnance Survey Maps. *Price, cloth, 2s. each.*

The Edinburgh Geographical Institute.

ATLASES.

EUROPE, Historical Atlas of Modern ——, from the Decline of the Roman Empire, comprising also Maps of Parts of Asia and of the New World connected with European History. Edited by Reginald Lane Poole, M.A., Ph.D., Lecturer in Diplomatic in the University of Oxford. Part I.

A good historical atlas on an extended scale has long been wanted in this country, the few that have been published being hardly detailed enough to meet the requirements of advanced students. The present work is modelled on the well-known atlas of Spruner and Menke, and is being compiled by many of the best authorities. It will be noticed fully when more of the parts have been issued.

INDIA, Statistical Atlas of ——. Second Edition, 1895. Printed by the Superintendent of Government Printing, Calcutta.

London: Edward Stanford.

A new edition of this atlas, first published in 1886, has been compiled from the data supplied by the last census. The maps show the physical configuration, geology, climatology, races, and languages of India, and also illustrate a variety of subjects connected with the natural resources of the country, the industries and social status of its inhabitants, etc. To these are added explanatory chapters written by officials having special knowledge of the subjects dealt with.

OESTERREICHISCHEN ALPENSEEN, Atlas der——. Mit Unterstützung des hohen K. K. Ministeriums für Cultus und Unterricht, herausgegeben von Dr. Albrecht Penck und Dr. Eduard Richter. II. Lieferung : Seen von Kärnthen, Krain, und Südtirol. 10 Karten und 32 Profile auf 9 Tafeln hauptsächlich nach eigenen Lothungen entworfen von Prof. Dr. Eduard Richter.

Stich und Druck von Ed. Hölzel's Geographischem Institut, Wien, 1896.

We hope shortly to receive the numbers of Professor Penck's *Geographische Abhandlungen* containing an account of the methods employed in the construction of these elaborate maps, and a description of the lakes, and shall then publish an extended notice of the work of the Austrian limnologists.

ROYAL SCOTTISH GEOGRAPHICAL SOCIETY.

REPORT OF COUNCIL.

TWELFTH SESSION, 1895-96.

The Council has the honour to submit the following Report :—

MEMBERSHIP.

The changes in the number of Members have been as follows :—

Number on 31st October 1895,	. . . .	1325
New Members added,	. . . . . .	180
		1505
Deduct by Deaths, ... 25		
„ Resignations, ... 55		
Struck off the roll by the Council (subscriptions in arrear), 11		
		91
Number remaining on 31st October 1896,	. . .	1414

This represents a net increase of 89 Members.

Of the present Members, 773 are on the Edinburgh list ; 283 on the Glasgow list ; and 95 and 78 on the Dundee and Aberdeen lists respectively ; 185 Members reside in England or Abroad. 211 are Life Members.

DIPLOMAS OF FELLOWSHIP.

The Council, acting on the recommendation of the Education Committee, has conferred the Ordinary Diploma of Fellowship on the following gentlemen :—Colonel P. Dods ; John Elliot Shearer, F.S.A.Scot. ; Lionel W. Lyde, M.A. ; and Rev. R. Stewart Wright.

FINANCE.

The Council submits the annexed Financial Statement.

From 31st October 1895 to 31st October 1896.

FUNDS AT CLOSE OF LAST ACCOUNT :—

£1000 Glasgow and South-Western Railway 4 per cent. Funded Debt, cost, 29th May 1885,		£1102	19	9
£533, 6s. 8d. North British Railway 3 per cent. Consolidated Lien Stock, cost, 30th December 1890,		496	2	0
£300 New Zealand Government 3½ per cent. Inscribed Stock, cost, 30th May 1894,		301	9	10
		£1900	11	7
Deduct—Balances due to Bank of Scotland on Accounts Current at Edinburgh, £149, 12s. 6d., and at Glasgow, £5, 12s.,	£155 4 6			
Less—in hands of Honorary Treasurer, £6, 10s. 5d., and of Secretary, £6, 10s. 5d.,	14 6 5	140	18	1
		£1759	13	6

ORDINARY SUBSCRIPTIONS :—

24 Life Members at £10, 10s.,		252	0	0
1 Annual Subscription at £1, 1s., for year 1890-91,	£1 1 0			
3 Do. Do. 1893-94,	3 3 0			
28 Do. Do. 1894-95,	29 8 0			
1136 Do. Do. 1895-96,	1192 16 0			
28 Do. Do. 1896-97,	29 8 0	1255	16	0
DIPLOMA Qualification Fees,		4	4	0
FEES received for Society's Diploma—4 at £1, 1s.,		4	4	0
LECTURE TICKETS sold,		8	13	4
MAGAZINES sold,		59	4	4
DIVIDENDS and INTERESTS, £64, 6s. 8d., *less* paid to Banks, 4s. 2d.,		64	1	6
FURNISHINGS sold,		1	16	4
		1649	19	2
		£3409	**12**	**8**

MAGAZINES, Printing, Paper, Lithographing, etc., from Nos. 9 to 12 of Vol. XI., and Nos. 1 to 8 of Vol. XII., all inclusive, and for Postages, Delivery, etc.,

		£619	13	4
Less—Received during the year for Advertising,		60	0	0
		£559	13	4
GENERAL PRINTING, £32, 2s. 9d., and List of Members and Laws, £18, 2s. 6d.,		50	5	3
BOOKS and other Furnishings for Library,		32	0	2
RENTS of HALLS, and other expenses of Lectures,		224	18	6
RENTS of COUNCIL ROOM, Taxes, Gas, Repairs, Insurance, etc.,		128	18	6
GRANT FOR ROOM IN GLASGOW, for use of Members,		31	0	0

SALARIES :—

To Secretary, year to 15th October 1896,	£200 0 0			
„ Librarian and Clerks,	272 0 0	472	0	0
MISCELLANEOUS, including Stationery, General Expenses, Advertising, Cleaning, Postages, and Office Furniture,		59	14	4
EXPENSES connected with attending Meeting of British Association, and applying to Government for Grant,		9	4	8
		£1567	14	9

FUNDS AT CLOSE OF THIS ACCOUNT:

£1000 Glasgow and South-Western Railway 4 per cent. Funded Debt, cost, 29th May 1885,		£1102	19	9
£533, 6s. 8d. North British Railway 3 per cent. Consolidated Lien Stock, cost, 30th Dec. 1890,		496	2	0
£300 New Zealand Government 3½ per cent. Inscribed Stock, cost, 30th May 1894,		301	9	10
		£1900	11	7
Deduct—Balances due to Bank of Scotland on Accounts Current at Edinburgh, £82 6s. 9d., at Glasgow,	£93 2 3			
£10, 16s. 6d., Balance due to Secretary,	0 8 0			
	£93 10 3			
Less—in hands of Honorary Treasurer,	34 16 7	58	13	8
		1841	17	11
		£3409	**12**	**8**

MEETINGS.

During the Session the Society held 33 meetings, of which 17 were in Edinburgh, 6 in Glasgow, 4 in Dundee, and 6 in Aberdeen. These meetings were addressed by Captain F. D. Lugard, C.B., D.S.O.; Captain F. E. Younghusband, C.I.E.; Mr. J. Theodore Bent; Rev. C. H. Robinson; Mr. C. E. Borchgrevink; Dr. A. Markoff; Rev. Walter Weston; Sir David P. Chalmers; Miss Kingsley; Prof. Clifford Allbutt, F.R.S.; Mr. James Troup; Mr. G. Seymour Fort; Mr. A. Montefiore; Mr. R. M. Routledge; Captain F. R. Maunsell, R.A.; M. Lionel Dècle; Major C. Barter; Rev. W. Campbell. Mr. Lionel W. Lyde, M.A., on the invitation of the Education Committee, delivered a special lecture on the "Teaching of Geography"; and Mr. J. C. Oliphant gave a Christmas lecture for young people.

THE SOCIETY'S MAGAZINE.

The *Magazine* has appeared as usual each month in the year, and the Council desires to record its obligations to the contributors of articles, and also to the following members who have rendered valuable assistance to the editors :—The Hon. J. Abercromby; J. G. Bartholomew; J. T. Bealby; G. Bickerton; W. B. Blaikie; Dr. James Burgess; H. M. Cadell; Dr. Scott Dalgleish; Dr. Th. Delius; Captain Dunlop-Smith; Dr. R. W. Felkin; J. G. Goodchild; D. P. Heatley; A. J. Herbertson; Prof. C. G. Knott; J. W. M'Crindle; Sheriff Æ. Mackay; D. Mac-Ritchie; Ralph Richardson; C. Robertson; Colonel Sconce; A. R. Scott; General Sir R. Murdoch Smith; W. C. Smith; J. Arthur Thomson; Coutts Trotter; James Troup; A Silva White.

LIBRARY AND MAP DEPARTMENT.

During the past session 305 books, 46 pamphlets, 24 atlases, and 263 map-sheets have been added to the Library and Map-room. Three new exchanges have been arranged. The number of volumes borrowed by members was 1700. The expenditure on the Library amounted to £32, 0s. 2d., which includes the cost of binding.

While the number of books borrowed from the Library has been rather less than in the preceding session, the number of inquirers who have visited the Library in search of information has shown no falling off.

The Council again records its thanks to Foreign and Colonial Governments for the official publications which they present to the Library, and to private donors, among whom are the following :—Miss G. Milne Home; Miss Graham; Mrs. R. Innes; Mr. M. J. B. Baddeley; Dr. Alex. Buchan; Mr. Robert Cox, M.P.; Mr. Andrew Mackintosh; Mr. C. S. Robertson; Mr. Hugh Rose; Dr. George Smith; and Mr. Coutts Trotter.

The Society's Rooms were closed for cleaning from 31st August to 19th September.

PLACE-NAMES COMMITTEE.

The regular meetings of this Committee have been discontinued, as the Director-General of the Ordnance Survey has decided that it is unnecessary to revise the whole of the names in each county. Only doubtful names will, therefore, be referred to the Committee in future. A few such references were made during the past session and were disposed of.

THE SOCIETY'S BRANCHES.

The Council has again the pleasure to acknowledge the services rendered by the Honorary Officials of the Branches at Glasgow, Dundee, and Aberdeen. A room has been leased as the headquarters of the Glasgow Branch.

INDEX: VOL. XII.

In the following Index the ALPHABETICAL ORDER *is adhered to throughout. The m important references and all sub-headings of Articles are indicated by small capita Names of Books and of Vessels are in Italics; titles of Papers in deeper ty Contraction,* rev. = *Review in the Magazine.*

MAPS AND ILLUSTRATIONS.

END OF VOLUME XII.

PRINTED BY T. AND A. CONSTABLE, PRINTERS TO HER MAJESTY
AT THE EDINBURGH UNIVERSITY PRESS